PASSIVE ELECTRONIC
COMPONENT HANDBOOK

Electronic Packaging and Interconnection Series
Charles M. Harper, Series Advisor

ALVINO • *Plastics for Electronics*

CLASSON • *Surface Mount Technology for Concurrent Engineering and Manufacturing*

G. DI GIACOMO • *Reliability of Electronic Packages and Semiconductor Devices*

GINSBERG and SCHNOOR • *Multichip Modules and Related Technologies*

HARPER • *Electronic Packaging and Interconnection Handbook, 2/e*

HARPER and MILLER • *Electronic Packaging, Microelectronics, and Interconnection Dictionary*

HARPER and SAMPSON • *Electronic Materials and Processes Handbook, 2/e*

HWANG • *Modern Solder Technology for Competitive Electronics Manufacturing*

LAU • *Ball Grid Array Technology*

LAU • *Flip Chip Technologies*

LICARI • *Multichip Module Design, Fabrication, and Testing*

Related Books of Interest

BOSWELL • *Subcontracting Electronics*

BOSWELL and WICKAM • *Surface Mount Guidelines for Process Control, Quality, and Reliability*

BYERS • *Printed Circuit Board Design with Microcomputers*

CAPILLO • *Surface Mount Technology*

CHEN • *Computer Engineering Handbook*

CHRISTIANSEN • *Electronics Engineers' Handbook, 4/e*

COOMBS • *Electronic Instrument Handbook, 2/e*

COOMBS • *Printed Circuits Handbook, 4/e*

J. DI GIACOMO • *Digital Bus Handbook*

J. DI GIACOMO • *VLSI Handbook*

GINSBERG • *Printed Circuits Design*

HARPER • *Active Electronics Handbook, 2/e*

JURAN and GRYNA • *Juran's Quality Control Handbook*

JURGEN • *Automotive Electronics Handbook*

MANKO • *Solders and Soldering, 3/e*

RAO • *Multilevel Interconnect Technology*

SZE • *VLSI Technology*

VAN ZANT • *Microchip Fabrication*

To order or receive additional information on these or any other McGraw-Hill titles, in the United States please call 1-800-722-4726. In other countries, contact your local McGraw-Hill representative.

PASSIVE ELECTRONIC COMPONENT HANDBOOK

SECOND EDITION

Charles A. Harper, Editor in Chief

McGraw-Hill

New York San Francisco Washington, D.C. Auckland Bogotá
Caracas Lisbon London Madrid Mexico City Milan
Montreal New Delhi San Juan Singapore
Sydney Tokyo Toronto

Library of Congress Cataloging-in-Publication Data

Harper, Charles A.
    Passive electronic component handbook / Charles A. Harper. — 2nd
ed.
        p.   cm.
    Rev. ed. of: Handbook of components for electronics. © 1977.
    Includes bibliographical references and index.
    ISBN 0-07-026698-0
    1. Electronic apparatus and appliances—Handbooks, manuals, etc.
I. Harper, Charles A. Handbook of component for electronics.
II. Title.
TK7870.H28424   1997
621.3815'4—dc21   97-10743
                                                                        CIP

McGraw-Hill

A Division of The McGraw·Hill Companies

1 2 3 4 5 6 7 8 9 0   DOC/DOC   9 0 2 1 0 9 8 7

ISBN 0-07-159029-3

The acquiring editor for this book was Steve Chapman, the editing supervisor was Bernie
Onken, and the production supervisor was Pamela Pelton. It was set in Times Roman by Jana
Fisher through the services of Barry E. Brown (Broker—Editing, Design and Production).

McGraw-Hill books are available at special quantity discounts to use as premiums and
sales promotions, or for use in corporate training programs. For more information, please
write to the Director of Special Sales, McGraw-Hill, 11 West 19th Street, New York, NY
10011. Or contact your local bookstore.

This book is printed on recycled, acid-free paper containing a minimum of 50%
total recycled fiber with 10% postconsumer de-inked fiber.

DEDICATION

This book is respectively dedicated to Leo E. Wilson, who contributed the chapter on Transformers and Inductive Devices to both this new second edition and the earlier first edition. Leo, a long term personal friend and associate passed away after completing his chapter for this edition.

Charles A. Harper, Editor

CONTENTS

PREFACE

This new and totally revised edition of the *Handbook of Components for Electronics* addresses the need for detailed information in this field. The number and types of component parts, as well as the advances in basic component types, have very greatly expanded since the publication of the earlier edition. In a field where a myriad of segmented data sheets provides the primary source of information for engineers who design electrical and electronic products, this single source reference will be invaluable. To better serve the user audience in this field, this new edition is divided into two conveniently organized volumes. This second *Passive Component Handbook* volume covers passive components, while the recently published first volume *The Active Component Handbook* covers active components. These volumes are the culmination of extensive efforts by leading experts in this field. The result is the most complete and up-to-date reference text to be found—a must for the desk of anyone involved in any aspect of electronic and electrical design or any aspect of component specification or utilization, and for every reference library.

This *Handbook of Components for Electronics* was prepared as a thorough and comprehensive sourcebook of practical data, guidelines, and information for all ranges of interests. It contains an extensive array of property and performance data for all the important component groups; these are presented as a function of the most important design and performance variables. Further, it presents comparison data and guidelines for the best component and selection trade-off design decisions; extensive test and reliability data; detailed listings of important specifications and standards and a wealth of data and information on dimensions, configuration, and mechanical and functional performance.

In addition to its thorough contents, the chapter organization and coverage of *The Passive Electronic Component Handbook* are equally well suited to user convenience. Subjects covered range from simple to complex passive components, to interconnection and cooling, to protection of components in stress environments. The first two chapters are devoted to the basic passive components, namely resistors and capacitors. This is followed by a series of chapters which cover transformers and inductive devices, relays and switches, batteries, fuses and protective components, and then filters and transient voltage protective devices. Next come chapters on connectors, connective devices and connector technology, and on cooling and thermal management devices. Finally, a chapter is devoted to the increasingly important subject of electrostatic discharge protection of component assemblies.

Length and coverage of a Handbook of this magnitude are necessarily measured compromises. Inevitably, varying degrees of shortages and excesses will exist, depending on the needs of the individual user. In spite of the tremendous effort that has been made to minimize such shortcomings, it is our greatest desire to improve each successive edition. Toward this end, any and all comments will be welcomed and appreciated.

Charles A. Harper
Technology Seminars, Inc.
P.O. Box 487
Lutherville, MD 21094

CONTRIBUTORS

Drexler, H. Bennett, Orlando Division, Lockheed-Martin Corporation (Chap. 2, Capacitors)

Driscoll, M.M., Electronic Sensors and Systems Division, Northrop-Grumman Corporation (Chap. 7, Filters)

Dunlop, D.D., Jr., Electronic Sensors and Systems Division, Northrop-Grumman Corporation (Chap. 1, Resistors)

Harris, L.H., Consultant (Chap. 10, Component Handling with ESD Control)

Kwon, D., Electronic Sensors and Systems Division, Northrop Grumman Corporation (Chap. 9, Electronic Device Cooling)

Lindquist, C.E., SAN-O-Industrial Corporation (Chap. 6, Overcurrent Protective Components)

Mroczkowski, R.S., AMP, Incorporated (Chap. 8, Connector and Interconnection Technology

Porter, R.F., Electronic Sensors and Systems Division, Northrop-Grumman Corporation (Chap. 9, Electronic Device Cooling)

Urquidi-Macdonald, M., Department of Engineering Science and Mechanics, The Pennsylvania State University (Chap. 5, Batteries)

Vale, C.R., Electronic Sensors and Systems Division, Northrop-Grumman Corporation (Chap. 7, Filters)

Wilson, L.E., Consultant (deceased) (Chap. 3, Transformers and Inductive Devices)

Wright, E.R., Leach Corporation (Chap. 4, Relays and Switches)

ABOUT THE EDITOR

Charles A. Harper is the President of Technology Seminars, Inc., a leading electronics industry seminar provider, and a widely known electronics educator. He has written and contributed to many technical books, including *The Electronic Packaging and Interconnection Handbook.*

CHAPTER 1
RESISTORS

Donald D. Dunlap, Jr.

With Credits to Edward L. Hierholzer, H. Bennett Drexler, and John H. Powers, chapter authors in the First Edition

Northrop-Grumman Corporation

1.1 INTRODUCTION

1.1.1 Scope

This chapter covers standardization, selection, application, and reliability issues of resistors used in electronic equipment design.

1.1.2 Sources of Information

Although this chapter updates the information presented earlier editions of this handbook, information provided by the manufacturers is also necessary to determine proper component selection.

The means of presenting this information has changed. Traditional sources of information, such as the following, remain available:

- *Electronic Engineers Master (EEM)*, Hearst Business Communications, Garden City, New York, annual.
- *Thomas Register of American Manufacturer and Thomas Register Catalog File*, Thomas Publishing Company, New York.

Distributors, such as Newark Electronics, Digi-Key, Allied-Avnet, and RESCO/ Baltimore publish condensed catalog information for many manufacturers. Electronic means (such as "ReCalZ" by Information Handling services) provide information (such as selection software and manufacturer's literature) that is available to the designer's personal computer or at a centralized workstation/library. Also, organizations, such IEEE, DESC (Defense Electronic Supply Center), and the manufacturers will or can provide information via Internet or computer bulletin board services. Modeling programs are also commercially available to assist the circuit designer.

1.1.3 Standardization Considerations

Need for standardization   First, almost rhetorically, a designer or a company must ask themselves why standardization is necessary. After all, this desire conflicts with the designer's ability to exactly fit an application requirement with the exact part required to do the job. Although not a trivial consideration, making a design fit the standard parts might allow for more economic, producible, and supportable designs. Often, resistor manufacturers (for that matter, most component manufacturers), can and do offer better pricing for parts bought in volume. Established processes and high-volume production generally allow manufacturers to produce better-quality parts than those from short production runs. System supportability on older production programs can also suffer when nonstandard parts are selected. As presented later in this chapter, standard values are available as defined by a decade table that defines resistor values and tolerances. Careful examination of the needs for special values, power ratings, matching, temperature coefficients, tolerances, etc., by a designer or company can result in lowering of part types and part numbers being stocked and procured.

Published standards   Industry and military standards and specification have served customers and manufacturers of components well as a starting point for standardization and component selection. These standards provide a means to define part types and performance for a particular type of component. When standards allow for a defined configuration and part requirements for a part type or part number, competitive procurement is then possible. Standards also provide guideline and requirements for quality assurance provisions (i.e., acceptance inspection, testing, sampling, rejection criteria), shipping and handling requirements, marking, and substitutability definitions.

Military standards and specifications have been historically required in the product assurance and parts control program section of many government contracts. These standards are available individually, by subscription from the stocking center designated by the Department of Defense, or from companies who package industry, military, and international standards in a manner compatible with the designer's personal computer or workstation.

Most military specifications have an associated Qualified Product List (QPL), which is revised as needed to list suppliers, addresses, Commercial and Government Entity (CAGE) address codes along with the types, values, tolerances, and "slash sheets" for which a supplier is qualified. Qualification requirements are defined within a particular specification. The aforementioned "slash sheets" (so called because of the slash bar separating the particular specification from the general governing specification) or "detail sheets" will define particular subsections of resistors or other component types defined by a particular specification. Usually, configuration, power rating, or other significant parameters are used to separate part types within a specification.

Military standards (such as MIL-STD-199, Resistors, Selection and Use, and MIL-HDBK-798) provide application and advisory information. Also, they provide a condensed comparison of part types useful for narrowing the selection of resistors. Inclusion within MIL-STD-199 has also served as means of standardizing such selection and informing users of the active or inactive status of part types and specifications.

Resistors specified by the military, like many other military parts, have experienced a reduction in sourcing in some styles. Some military specifications have been renamed performance specifications, and the testing in other specifications have been reduced to sampling small amounts, so long as few or no failures have been reported. DESC has also published drawings that were originally intended as interim documents, but are often used when no source wishes to maintain qualification to the closest military specification. Commercial item description drawings (CIDs) and the DESC drawings serve as a means of standardizing part numbers for military and government OEMs (Original Equipment Manufacturers), such as Northrop-Grumman.

In addition to military and government standards, industry associations have provided a similar function. Some of these organizations are the American National Standards Institute (ANSI), International Electrotechnical Commission (IEC), Electronic Industries Association (EIA), Aerospace Industries Association of America (AIA), and military agencies of the Department of Defense (DOD). National Aerospace Standards (NAS) specifications and standards are published by AIA. Publication references and designations of these organizations are used throughout this chapter.

Preferred values Fundamental standardization practices require the selection of preferred values within the ranges available. Values have been defined in Department of Defense publications, by industry standards, and in other trade publications, using a system proposed by Charles Renard in 1870 for use by the French army to reduce the proliferation of cordage sizes he found to be specified for balloon moorings.

Decade progression The system is based on preferred numbers generated by a geometric progression devised to repeat in succeeding decades. The general geometric progression is defined by:

$$N = ar^{n-1}$$

where N is the nth term, a is the first term, and r is a chosen common ratio. If r is chosen to be the kth root of 10 and the first term is set at unity, then:

$$N = 1 \times (\sqrt[k]{10})^{n-1}$$

where k can be selected to provide a desired scale graduation. If, for instance, three values per decade are desired, k is 3 and the common ratio becomes 2.154. The three rounded off values are 1.00, 2.15, and 4.64. Standard decades for resistors and capacitors have been chosen, having 3, 6, 12, 48, 96, and 192 terms, with common ratios being the appropriate roots of 10. The 192-value-per-decade system has use for high-precision capacitors and resistors, but the large number of values tends to defeat the standardization purpose. Table 1.1[1] shows the preferred number decade values with appropriate numbers of significant figures used to designate resistors, capacitors, and Zener diodes.

Service variability The service variability of a resistor or capacitor, frequently called *end-of-life tolerance*, is an overall value tolerance composed of factors caused by purchase tolerance, lapsed time, and stress, and short-term excursions from the local environment. Service variability constitutes a more useful and realistic factor in part selection for a given application than does the use of the purchase tolerance alone. Decade-value progressions used in the procurement of components should cover incremental commensurate with part variability.

Aggressive standardization practices base the decade common ratio on expected service-life variability, rather than on purchase tolerances, thereby decreasing the number of stock values. Thus, if a component purchased with a 5-percent tolerance is actually found to remain within a ±20 percent range in service, the service variability is 20 percent. Observing the values shown in the 10-percent decade, with 12 values per decade, you can see that steps are essentially 20 percent between values. A standard value for circuit design selection, therefore, should be an alternating value in the 5-percent decade, or actually, in the 10-percent decade. A resistor or capacitor of a particular marked value is considered to be allowed to assume a value anywhere within the expected service tolerance band.

Statistical distribution techniques can be applied to variation if more than one of the parts in question are significant to a particular circuit variation, but for each individual part the total variability must be considered possible. The use of a decade ratio graduated

TABLE 1.1 Standard Decade Values (Industry and Military Standards), Preferred Values for Resistors, Capacitors, Zener Diodes

*	±1%	±2%	*	±1%	±2%	*	±1%	±2%	*	±1%	±2%	*	±1%	±2%	*	±1%	±2%	±5%	±10%	±20%
1.00	1.00	1.00	1.47	1.47	1.47	2.15	2.15	2.15	3.16	3.16	3.16	4.64	4.64	4.64	6.81	6.81	6.81	1.0	1.0	1.0
1.01			1.49			2.18			3.20			4.70			6.90			1.1		
1.02	1.02		1.50	1.50		2.21	2.21		3.24	3.24		4.75	4.75		6.98	6.98		1.2	1.2	
1.04			1.52			2.23			3.28			4.81			7.06			1.3		
1.05	1.05	1.05	1.54	1.54	1.54	2.26	2.26	2.26	3.32	3.32	3.32	4.87	4.87	4.87	7.15	7.15	7.15	1.5	1.5	1.5
1.06			1.56			2.29			3.36			4.93			7.23			1.6		
1.07	1.07		1.58	1.58		2.32	2.32		3.40	3.40		4.99	4.99		7.32	7.32		1.8	1.8	
1.09			1.60			2.34			3.44			5.05			7.41			2.0		
1.10	1.10	1.10	1.62	1.62	1.62	2.37	2.37	2.37	3.48	3.48	3.48	5.11	5.11	5.11	7.50	7.50	7.50	2.2	2.2	2.2
1.11			1.64			2.40			3.52			5.17			7.59			2.4		
1.13	1.13		1.65	1.65		2.43	2.43		3.57	3.57		5.23	5.23		7.68	7.68		2.7	2.7	
1.14			1.67			2.46			3.61			5.30			7.77			3.0		
1.15	1.15	1.15	1.69	1.69	1.69	2.49	2.49	2.49	3.65	3.65	3.65	5.36	5.36	5.36	7.87	7.87	7.87	3.3	3.3	3.3
1.17			1.72			2.52			3.70			5.42			7.96			3.6		
1.18	1.18		1.74	1.74		2.55	2.55		3.74	3.74		5.49	5.49		8.06	8.06		3.9	3.9	
1.20			1.76			2.58			3.79			5.56			8.16			4.3		
1.21	1.21	1.21	1.78	1.78	1.78	2.61	2.61	2.61	3.83	3.83	3.83	5.62	5.62	5.62	8.25	8.25	8.25	4.7	4.7	4.7
1.23			1.80			2.64			3.88			5.69			8.35			5.1		
1.24	1.24		1.82	1.82		2.67	2.67		3.92	3.92		5.76	5.76		8.45	8.45		5.6	5.6	
1.26			1.84			2.71			3.97			5.83			8.56			6.2		
1.27	1.27	1.27	1.87	1.87	1.87	2.74	2.74	2.74	4.02	4.02	4.02	5.90	5.90	5.90	8.66	8.66	8.66	6.8	6.8	6.8
1.29			1.89			2.77			4.07			5.97			8.76			7.5		
1.30	1.30		1.91	1.91		2.80	2.80		4.12	4.12		6.04	6.04		8.87	8.87		8.2	8.2	
1.32			1.93			2.84			4.17			6.12			8.98			9.1		
1.33	1.33	1.33	1.96	1.96	1.96	2.87	2.87	2.87	4.22	4.22	4.22	6.19	6.19	6.19	9.09	9.09	9.09			
1.35			1.98			2.91			4.27			6.26			9.20					
1.37	1.37		2.00	2.00		2.94	2.94		4.32	4.32		6.34	6.34		9.31	9.31				
1.38			2.03			2.98			4.37			6.42			9.42					
1.40	1.40	1.40	2.05	2.05	2.05	3.01	3.01	3.01	4.42	4.42	4.42	6.49	6.49	6.49	9.53	9.53	9.53			
1.42			2.08			3.05			4.48			6.57			9.65					
1.43	1.43		2.10	2.10		3.09	3.09		4.53	4.53		6.65	6.65		9.76	9.76				
1.45			2.13			3.12			4.59			6.73			9.88					
												Values per decade			192	96	48	24	12	6

* ±0.1, ±0.25, and ±0.5%.

according to service variability will thus decrease stock varieties without significantly impairing ultimate selection utility, despite initial impressions to the contrary. This technique provides a major standardization benefit. One approach to estimation of service variability consists of performing a root-sum-square of the value change expected with purchase tolerance, plus that for each environmental extreme expected for the part in its application. Except for the initial tolerance, the most significant contributor is frequently that of long-term drift, which unfortunately is usually ignored by the procurement specification and must be assessed by experience.

Part-type selection Application analysis is an important part of standardization, and standard part selection listings for designers should be chosen to fit the needs of particular equipment function and use. Resistors, except for high-precision types, are chosen within general-purpose, medium-power, and high-power dissipation categories to provide the desired service variability, functional adequacy, packaging utility, and cost.

General usage parts should be selected to provide the fewest power and voltage ratings and the lowest acceptable precision that can be tolerated for the application. If, in a given piece of equipment, a significant quantity of applications require higher precision, lower drift, lower noise, or better RF performance than a lower-precision norm, it might be possible to show that a small quantity of low-precision parts needed will cost more than just increasing the already large quantity of more expensive close-tolerance parts—especially if procurement and stocking costs are considered.

Identification and marking Standardized marking schemes have been devised for resistors and capacitors using shape conventions, and alphanumeric designations. In addition to value and tolerance marking, some methods also label part type, "failure rate level," lot control number, and date code along with the electrical polarity and function markings for electrical terminals. This chapter presents resistor information; refer also to the chapter on capacitors. Table 1.2[1] and Figures 1.1[1] and 1.2[1] show marking examples and terminal identification information.

Resistor color codes Color coding for resistors is subject to damage from high-dissipation temperatures. Also, color bands do not adhere well to some encasing materials, although large improvements have been made in both respects by current materials technology. Their use to designate value and tolerance therefore finds greatest utility in low-wattage general-purpose types with alphanumeric methods being used by others. Three-color value codes are easier to memorize to a degree of one-look recognition, but are limited to two significant figures and cannot adequately identify decade values more finely graduated than 5 percent (see Table 1.1[1]). Four-band value codes are used in some military specifications and manufacturers.

Alphanumeric labeling Part-numbering systems incorporating functional values and other information, or "significant part numbers," are in almost universal use in both industry and military standards. In some instances, however, sequential nonsignificant dash numbers are used, and a value and type listing must be provided to all who have a need to identify parts. These lists are frequently keyed to a specification that must be referred to for value information. Although sequentially ordered dash number listings make for orderly arrays of categories in ascending value or size without disrupting the value progression with decimal multipliers, it does present user difficulties.

Value designators are almost universally a three- or four-digit sequence denoting, respectively, two or three significant figures, followed by a decimal multiplier number. Resistor values are given in ohms. For resistor values with a value order less than the number of significant figures, a letter R is sometimes inserted to depict the decimal point position. Thus, 1R1 signifies 1.1 ohms. Initial value tolerance at the time of purchase ("purchase tolerance") is stated as part of the value code by means of well-standardized

TABLE 1.2 General Color-Number Significance for Resistors and Capacitors and Approved Standard Abbreviations for Color Names

Color	Number	Multiplier, EIA-MIL Resistor	Multiplier, EIA-MIL Capacitor	Resistor value[d] tolerances MIL resistor, (±)%	Resistor value[d] tolerances EIA resistor, (±)%	Abbreviations MIL-STD-12	Abbreviations EIA 3-letter	Abbreviations EIA alternate	MIL-STD-1285 part-type identifier	MIL-STD-1285 failure rate[e]
Black	0	1	1	20		BLK	Blk	BK	Capacitor	L ( )
Brown	1	10	10	1	1	BRN	Brn	BR		M (10^4)
Red	2	10^2	10^2	2	2	RED	Red	R,RD		P (10^3)
Orange	3	10^3	10^3			ORN	Orn	O,OR		R (10^2)
Yellow	4	10^4	10^{4c}			YEL	Yel	Y		S (10)
Green	5	10^5			0.5	GRN	Grn	GN,G		
Blue	6	10^6			0.25	BLU	Blu	BL		
Violet	7	10^7			0.10	VIO	Vio	V		
Gray	8				0.05	GY	Gra	GY		
White	9					WHT	Wht	WH,W		
Gold[a]		10^{-1}	10^{-2f}	5	5	(a)	Gld			
Silver[b]		10^{-2}	10^{-1f} 10^{-2g}	10	10	SIL	Sil		Inductor	

[a] MIL-STD-12B uses the chemical symbol Au for gold.
[b] Metallic colors do not have EIA color standards.
[c] Not included in MIL-C-20.
[d] Capacitor value tolerance and stability characteristic designators vary widely and are covered in separate tables to avoid confusion.
[e] Failure rates are given in failures per 10^9 part-hours (FITS), symbol L has its value assigned by the individual part specification.
[f] EIA RS 198A ceramic TC and EIA RS 335-A composition capacitors.
[g] EIA RS 153B mica capacitors.

First
significant
figure

Second
significant
figure

Multiplier

Failure-rate
level (established
reliability types
only)

Tolerance

(a)

Equal width

Approx. 1.5 times
width of remaining
bands

A B C D E

First
significant
digit

Second
significant
digit

Multiplier

Terminal

Tolerance

(b)

0.020*
Typical ref

0.031*
Typical ref

Visibly wider spacing
between bands to
identify direction of
reading (from left to
right)

First
significant
digit

Second
significant
digit

Third significant digit

Tolerance

Decimal
multiplier

(c)

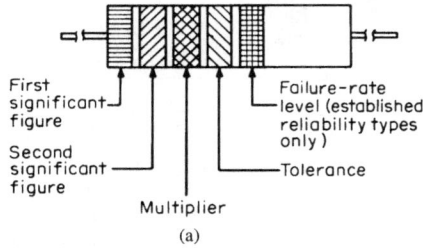

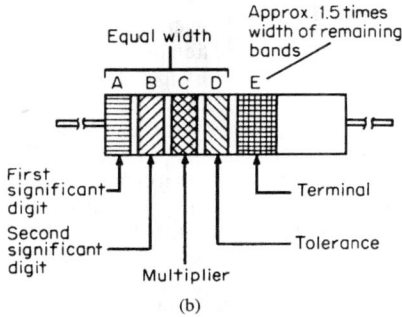

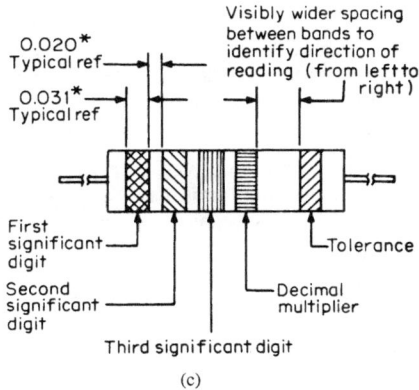

FIGURE 1.1 Color coding of resistors.

letter designators. Table 1.2[1] shows value tolerance letters, and examples of value and tolerance marking.

Date coding A standard convention for date coding is four-digit year/week designators, which use the last two digits of the calendar year followed by a two digit week of the year number. For example, date code 9608 would be the eighth week of 1996. MIL-STD-1285 has provided useful guidelines for part marking with the date code and other information.

(a)

(b)

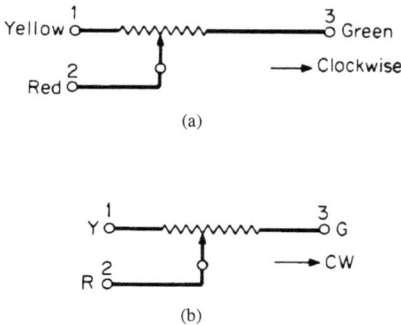

FIGURE 1.2 Circuit diagrams and terminal identification for variable resistors.

1.2 FIXED RESISTORS

1.2.1 General Information

Resistance is the scalar property of an electric circuit, which determines the rate at which electrical energy is converted to thermal energy while a given electric current is flowing. The property is analogous to viscous friction losses in mechanical systems. In its mathematical generalization, resistance is a function of the electric current, being equal to the dissipated thermal power divided by the square of the current. In practical usage, however, the resistance property is considered to be independent of the current flow. Fundamentally, a potential difference of 1 volt across a resistance of one absolute ohm is associated with a charge flow of 1 coulomb per second (1 ampere) and a thermal dissipation on one watt (1 joule per second). The international ohm, used prior to 1948, is 1.00049 absolute ohms—the absolute ohm being unity by definition (now almost universally used). The unit is designated in honor of Dr. Georg Simon Ohm, who, in 1827, demonstrated the electromotive force-resistance current relationship.

Basic design considerations are convenience of installation, stability of properties over a desired range of electrical and environmental exposures, and functional longevity. Further desired properties include small physical size, good mechanical strength, stability of resistance value, efficient thermal dissipation, ease and reliability of circuit connection, purity of resistive function over broad electrical frequency range, wide range of available resistance values, good dielectric strength, nonflammability, low-noise production, good producibility, avoidance of catastrophic failure mechanisms, and low relative cost.

General construction   Changes in resistance value are associated with physical, mechanical, and chemical changes in the functioning structure and materials of a resistor. Designs should therefore utilize stress-relieved mechanical systems, low-operating hot-spot temperatures, and chemically and galvanically compatible stable materials in low-energy states. Resistivity is the material property that summarizes many of these contributions, as defined in Figure 1.3[3]. Table 1.3[3] has a brief table of resistivities for resistance materials.

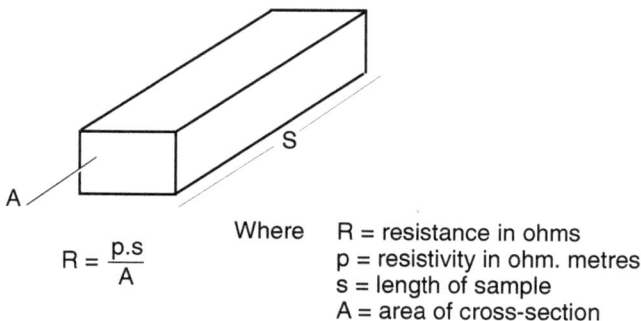

$$R = \frac{p.s}{A}$$

Where   R = resistance in ohms
p = resistivity in ohm. metres
s = length of sample
A = area of cross-section

FIGURE 1.3   General definition of resistivity.

TABLE 1.3 Resistivities by
Material Type

Values are in units of nano-ohm-
meters. To find absolute values,
multiply by 10^9. Values for pure
metals are shown first,
followed by values for alloys.

Metal	Resistivity
Aluminum	22.7
Copper	17.0
Gold	23.0
Iron	105.0
Nickel	78.0
Platinum	106
Silver	16.0
Tin	115
Tungsten	55.0
Zinc	62.0
Carbon-steel	180
Brass	60
Constantan	450
Invar	100
Manganin	430
Nichroma	1105
Nickel-silver	272
Monel metal	473
Kovar	483
Phosphor-bronze	93
18/8 stainless steel	897

The functional element of fixed and variable resistors is generally a core of electrically nonconductive material with a resistance wire wound onto or a film of resistance material bonded to the surface of the core supporting the terminals or termination methods, a solid pellet of resistance material formed around suitable terminations, or a film sputtered, deposited, or otherwise situated on a heat-dissipating substrate, such as beryllium oxide, aluminum oxide, or another ceramic material.

Various shapes, sizes, terminal, and core materials are used in the manufacture of resistors. Cylindrical shapes, often with axial leads, are commonly used for resistors, such as military types RLR and RNC. Coaxial disks are used in radio-frequency applications. Flat and cylindrical chips are used in hybrid and surface-mount situations. Threaded studs, chassis mounts, and integral heatsinks are provided to dissipate heat. Figure 1.4[1] shows an assortment of fixed resistor configurations. In general, the power rating of a given resistor is associated with its physical size and its ability to dissipate heat, resulting from its elec-

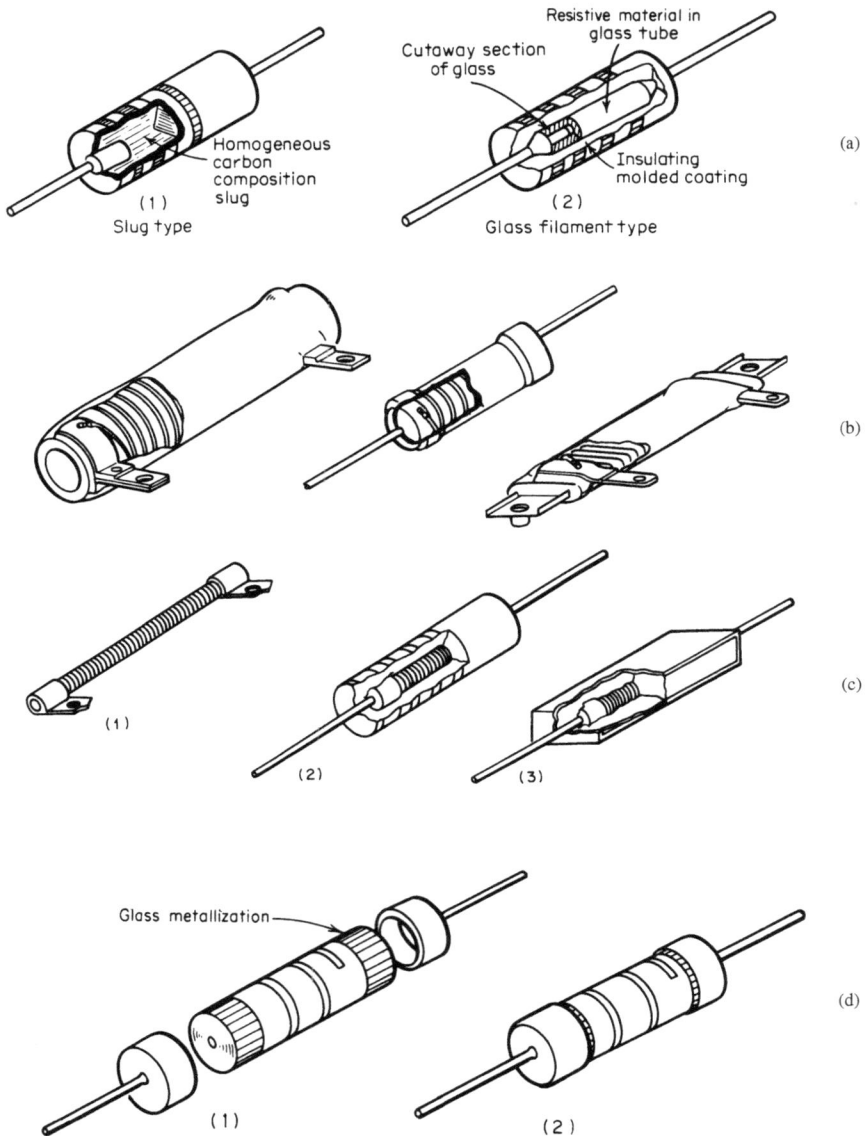

FIGURE 1.4 Common resistor configurations. (a) Carbon composition resistor construction. (1) molded composition pellet with phenolic outer layer. (2) Composition film on glass-tube filament with molded encasement, resistive material on glass tube. (b) Conformally coated wirewound power resistor construction. This basic shape runs from about 1¼ to 12 in. in length, with diameter and terminals scaled up in proportion, except that screw and nut terminals are furnished for 20 W and up. (c) Rod-type wirewound resistors, wound on resin-impregnated glass-fiber cores, chopped to length/value, compression-capped. (1) Automotive/appliance bare wire. (2) General-purpose molded case (note wide first color band). (3) Encapsulated in ceramic case with refractory cement. (d) End cap and lead to core assembly. Typical of most thin-film types, some thick films. (1) Core with spiraling ground in film, metallized termination areas. (2) End cap-lead assemblies pressed on, ready for coating, molding, or other encasement.

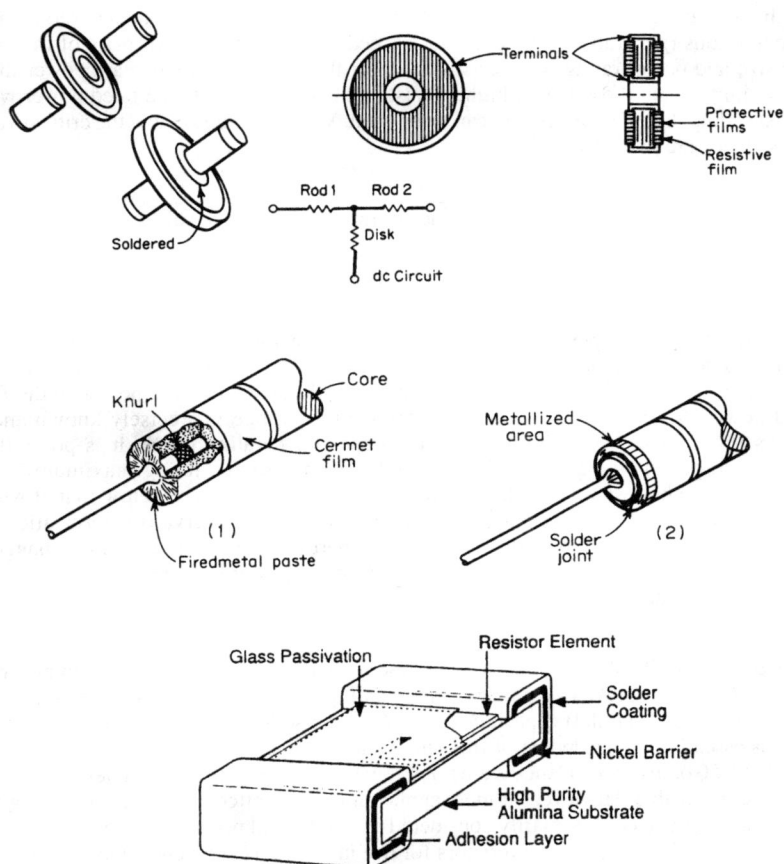

▲ Thick Film Chip Resistor Construction

FIGURE 1.4 *(Continued)* (e) RF tee pad assembled from rods and disk. Pyrolytic-carbon deposition on ceramic with metallized terminal areas. (*KDI Pyrofilm.*) (f) Examples of film resistor lead attachments not using end caps. (1) Headed and knurled lead is pressed into alumina core center. Firing of metal paste bonds lead to core and forms conductive termination for cermet element. (2) Flat-headed lead is attached with high-temperature solder to copper plated on termination area at end of core and resistor element (g) Typical thick film surface mount (chip) construction.

trical to energy conversion at a given rate or current. Material and thermal design for a given resistor material property also influence a resistor's power rating. Table 1.4 (Dale Electronics) summarizes the common characteristics of major types of resistors.

Critical resistance value   A maximum terminal to terminal voltage limitation is reached for a given case size and resistance element design. This limit is expressed as a maximum voltage rating and is considered to be dc or rms at the line frequency. Turn-to-turn and end-cap-to-end-cap dielectric properties under worst-case environmental conditions establish the part reliability. This limitation leads to the definition of a "critical value of resistance" for each resistor size and power rating, being the highest resistance value at which with voltage applied equal to rated voltage, power being dissipated at room ambient conditions (generally 25°C) does not exceed the rated power. All parts of the same case size and power rating with resistance lower than the critical value are power-dissipation limited, and those with a higher resistance cannot dissipate the rated power without exceeding the safe voltage that can be applied. Voltage ratings below the critical value of resistance are given by:

$$\text{Voltage max} = \sqrt{\text{power rating} \times \text{resistance value}}$$

$$= \sqrt{PR} \qquad\qquad (1.2^1)$$

If the ambient temperature in the area of the resistor under consideration is high enough to necessitate derating of the power dissipation, the voltage should be derated to a value determined by the square root of the ratio of the derated power to the full-rated power. If operating environmental temperature range is precisely known and is always lower than the inflection point on a specific derating curve, it is possible to modify the power ratings upward, provided that the voltage rating maximum is observed and the hottest spot on the resistor is held to a temperature below that at which the resistor is rated to be capable of zero dissipation. Conservative application, of course, will stop well short of that temperature, particularly if low resistance change in service is desired. Dissipation of one half of the rated power is often used as a standard rule for conservative practice.

Equivalent circuit   A resistor cannot be described as being purely resistive in electrical circuits. A typical equivalent circuit is shown in Figure1.5[1]. R_s is the series resistance, C_p is the distributed (parallel) capacitance, and L_s is the series inductance. The series resistance is essentially independent of frequency into the gigahertz range for most film resistors above 50 ohms. If lead lengths are relatively long in low-resistance values, skin effects could be noticeable. However, in most applications, skin-effect problems are negligible. The series inductance is primarily considered in wirewound resistors without inductance-canceling windings and in film resistors for use in gigahertz frequency applications. The parallel capacitance tends to dominate the effects of the inductance in most resistor types, except for the wirewound resistors without noninductive winding.

The impedance of the equivalent circuit of Figure 1.5[1] can be written as:

$$Z = \frac{R_s X_c^2 + j X_c [X_L X_C - (R_S^2 + X_L^2)]}{R_S^2 + (X_L - X_C)^2} \qquad\qquad (1.3^1)$$

where Rs is the true resistance, XL is the reactance of the true series inductance, and Xc is the reactance of the parallel capacitance. Resistance might change slightly with frequency, as mentioned previously, because of skin effect and dielectric losses in the equivalent capacitance might introduce small variations, but neither effect is considered to be of major proportions, so Rs is assumed to be constant. True inductance is also considered to be constant. Capacitance will vary slightly as dielectric constants change with frequency and voltage, but again, the effect is ignored here.

$$Z = \frac{R_S}{(R_S^2/X_C^2 + [(X_L - X_C)^2/X_C^2]} + j\frac{X_L - [(R_S^2 + X_L^2)/X_C]}{(R_S^2/X_C^2 + [(X_L - X_C)^2/X_C^2} \qquad\qquad (1.4^1)$$

TABLE 1.4 Parametric Comparison of Resistor Types

Resistor technology	Resistance range (ohms)	Voltage range (volts)	Tolerance range (± %)	Temperature coefficient (± PPM/°C)	Power rating range (watts)
Wirewound (power)	0.1 to 1 Meg	< 1000	≥ 1	≥ ± 50	< 300
Wirewound (general purpose)	0.1 to 1 Meg	< 300	≥ 0.005	≥ ± 3	< 10
Wirewound (surface mount)	0.1 to 10 K	< 50	.05 to 5	≥ ± 90	< 2
Film (power)	10 – 100 K	< 250	1 to 5	≥ ± 300	
Thick film (general purpose)	.01 to 10^{12}	< 250	1 to 5	≥ ± 300	< 8
Film (high voltage)	10^{11}	< 50 kV	1 to 5	≥ ± 50	< 50
Precision hybrid	4.7 to 1M	< 100	1 to 5	≥ ± 50	< 0.25
Carbon film	1 to 47 M	< 1000	> 5	> – 250	General purpose < 2 W other applications available
Carbon comp	1 to 100 M	< 1000	> 10	≥ ± 1200	< 4 W for general purpose, other applications available

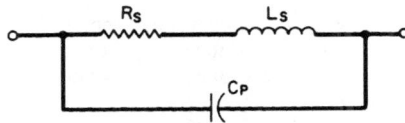

FIGURE 1.5 Equivalent circuit of a fixed resistor at high frequency.

Allows analysis of the effects of frequency. At very low frequencies, for instance, the Xc is very high, XL is very low, the denominators of the real and reactive terms approach unity. Although the real part approaches the value of Rs, the imaginary part is nearly zero because of the extreme values of Xc and XL. As frequency is increased, both real and imaginary portions are affected as the series and shunting effects interplay. Eventually, the lowered capacitive reactance will dominate. The real part becomes reduced in value as the denominator increases. The imaginary portion is also reduced, because of signs and because of the arrangement of first-degree denominator terms above the line, as opposed to second-degree denominator terms below the line. For resistors with low values of Rs, XL will play a more significant part in the impedance determination. After a "self-resonant point" is passed, the impedance always becomes capacitive and starts to decrease with frequency.

Temperature coefficient of resistance   Because resistors are heat-producing devices and can be expected to perform over a wide ambient temperature range, their ability to remain stable in resistance value over a wide range of temperatures is important. The term *coefficient* is used to denote a variation that is small or essentially linear, and temperature characteristic is used to define a wide nonlinear variation, confined mostly to carbon-composition resistors. Coefficients given in parts per million per degree C (ppm/°C), percent per degree C (which is parts per hundred per degree C), percent per degree C (which is parts per hundred per degree C), or sometimes just a decimal coefficient number per degree C are valid only within the temperature range specified and are defined for the temperature average of the resistor body, rather than the surrounding air temperature. Self-heating caused by the power being dissipated thus must be included in the determinations.

Power derating curves provide maximum hot-spot temperature information for free-air mountings, but almost always will give a very conservative estimate. This is apparently because limitation on resistance changes in operating load life is a more compelling design constraint than hot-spot temperature limits.

Resistors are designed to operate at given maximum temperatures dependent on the materials used. At full rated power, therefore, a high-temperature resistor will experience more resistance change for a given temperature coefficient of resistance than will a lower-temperature part or one that is operated at a fraction of its rated power. This consideration should be made when selecting resistor types. Resistance variation as a proportion of the total value tends to be greater with very low values of resistance, approximately 5 ohms or less. This is because lead-wire resistance, thermal elongations, and variables, such as end-cap contact resistance, are insignificant at higher resistances. Standards and specifications generally recognize this by allowing wider variations for low values. A means of calculating the temperature coefficient is given by Eq. 1-5.

$$R_2 = R_1 \left(\frac{1 + \alpha\theta2}{1 + \alpha\theta1} \right) \tag{1.5^3}$$

α = temperature coefficient (ppm/°C)
R_1 = known resistance at temperature T_1
R_2 = known resistance at temperature T_2
O_2 = temp 2 (°C)
O_1 = temp 1 (°C)

Voltage coefficient of resistance   Changes in conductivity caused by higher potential gradients across molecular interfaces cause resistivity to vary slightly with applied voltage. This is expressed as a coefficient of the nominal resistance (percent or parts per million) per volt. This quantity is specified to be independent of effect because of self heating, and measurement is thus difficult. The effect varies from −700 ppm/V for the higher resistance values of carbon composition through about 5 to 30 ppm/V for carbon film and cermet, and from 10 to 0.005 ppm/V for metal film and oxide films, although some thick-film types go as high as 400 ppm/V. The voltage coefficient is not usually of consequence for wirewound resistors.

Resistor noise   The noise output of a given resistor depends on its thermal "white" noise, plus a noise output from the applied current. The latter portion depends on the resistor design, and the former (Johnson noise) depends on the resistance and its temperature:

$$\text{(Johnson) } E_{rms} = 7.4 \sqrt{RT \Delta f} \times 10^{-12} \tag{1.6^1}$$

where E is the noise resistor voltage, R is resistance, T is absolute temperature in Kelvin, and Δf is the frequency bandwidth of consideration.

A noise index is conventionally specified for a given resistor type. This is the ratio of rms noise voltage (caused by a specified current flow through a resistor) to the average dc voltage across the resistor (measured over one bandwidth decade at one specific hot-spot temperature). The units are microvolts per volt or voltage ratio decibels, where 0 dB is 1 microvolt:

$$\text{Noise index} = 20 \log_{10} \frac{\text{noise voltage, } \mu V}{\text{dc voltage}} \tag{1.7[1]}$$

The current-noise voltage from frequency f_1 to frequency f_2 is given by:

$$\text{(current) } E_{rms} = V_{dc} \times 10 \left(\frac{\text{noise index, dB}}{20} \right) \times \sqrt{\log \frac{f_2}{f_1}} \tag{1.8[1]}$$

Where more than one frequency decade is of interest, the noise voltages add as the square root of the quantity of decades. 10 Hz to 100 kHz thus has an rms amplitude content twice that of 10 to 100 Hz (four decades have twice the noise of one decade). Except for possible noisy terminations and wire imperfections, wire-wound resistors produce only the inevitable thermal noise. Metal-film resistors have about –30- to –40-dB indexes, with lower resistances having lower noise. Carbon film ranges from about +10 down to –30 dB, with low values also having lower noise. Carbon composition ranges from about +40 dB for the highest values of 0.1 watt parts down to about 0 dB maximum for the lowest values of 2-watt parts. Fired thick films range from 0 dB down to about –20 db. For current-induced noise in the carbon types, the larger physical sizes have lower current densities for a given load, and are consequently less noisy. Figure 1.7[1] shows relative noise indexes for metal and carbon films. Total noise voltage is, of course, the sum of the thermal Johnson noise and the current noise.

1.2.2 Resistor Specifications and Standards

MIL-STD-199 and MIL-HDBK-978 can provide valuable assistance in the selection and use of military resistors, with much of their notes valid for similar commercial types as well. Established reliability specifications are as noted.

Typical power ratings versus temperature are shown in Figure 1.6[1]. Most, but not all (particularly high-power resistors), are rated to full power at 70°C, derating linearly to the zero-power derating temperature (typical values are shown in Table 1.5[2]). Manufacturers and reliability engineers should be consulted for more detailed information in particular applications.

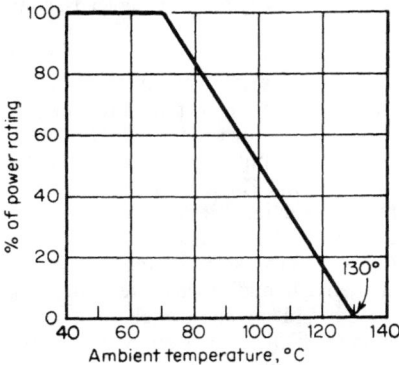

FIGURE 1.6 Typical power-derating curve.

1.2.3 General-Purpose Resistors

General-purpose resistors are those with power dissipation ratings of 2 W or less, and that have resistance stability characteristics slightly worse than or equal to semi-precision. This classification of resistors includes thick film, some thin film, wirewound, and carbon composition.

Carbon composition Carbon-composition resistors are widely used because of their low

TABLE 1.5 Typical Zero Power Derating Temperatures by Resistor Type

Type	Temperature (°C)
Wire wound (power-alumina type former)	275
Wire wound (power-beryllia former)	350
Wire wound (precision)	175
Metal foil	125
Metal film (power)	170
Metal film (precision)	175*
Nickel film	150
Tantalum nitride film	150
Cermet thin film	200
Metal oxide film	125
Carbon film (power)	100
Carbon film (precision)	120
Metal glaze (power)	200
Metal glaze (precision)	200
Thick film—metal oxide	150
Thick film—metal oxide (power)	275
Conducting plastic (power)	120
Conducting plastic (precision)	120
Carbon composition	130
Carbon—ceramic	220

cost (commercially), general lack of unwanted inductance and wide range of resistance values. Their drawbacks include service variability with temperature and humidity conditions.

Construction Two basic types are furnished, one having carbon granules molded with an organic binder into the proper pellet shape around the terminal leads, and another with a carbon resin paint deposition on a glass tube (called a *filament*), attached to the terminal leads by conductive cement, with a thick organic resin core molded around the assembly (Figure 1.6[1]). Both types use organic exterior coatings and are supplied in standardized values and tolerance painted on. Standards leads are solderable, but weldable versions can be supplied. Using weldable leads in such applications might require a derating of the power rating because of the lower thermal conductivity of weldable leads, compared to copper leads.

Resistance variability The chief disadvantage of carbon resistors is their variability of shelf and service life. Water absorption from exposure to humid atmospheres in nonoperating storage changes the resistance values dramatically; sometimes to the point of out-of-tolerance resistance values, when inspected by the purchasing user. Moisture causes resistance generally to increase by as much as several percent. The moisture can be safely removed and a partial recovery effected by baking or operational temperature rise, provided that the temperatures used are no greater than 100°C. Permanent effects on the organic structure are accelerated at higher temperatures. The recovery is temporary unless subsequent moisture is avoided. In service, normal relative humidities and irreversible operational aging can cause resistance value to vary well outside purchase tolerances, particularly for the 5 and 10 percent tolerances. Temperature characteristics are nonlinear and vary with resistance value. Because the slope of the temperature-resistance curve is not constant, the property is called a *characteristic*, rather than a *coefficient*. Noise is high, compared to that when using film and wirewound types. "Carbon noise", which is proportional to the load current, adds to the Johnson noise. Availability of parts with significantly different constructional character under the same standardized part designations requires evaluation in applications involving radio frequencies. Testing and circuit characterization is useful for carbon resistors considered for critical RF applications.

Above the critical resistance value (as defined earlier), the maximum rated voltage becomes also the power dissipation limitation. Both specifications warn that nonsinusoidal voltages with peaks exceeding 1.4 times the rms rated voltage require careful evaluation and that broad tolerance on resistance values (service variability, as well as purchase tolerance) should be considered to prevent possible dissipative overloads as resistance changes in service. Following general derating guidelines for resistors is again helpful to avoid such problems.

Another application for carbon resistors is the pulse performance. Table 1.6[1] shows pulse-performance capabilities for various rated carbon resistor families. MIL-R-11 and MIL-R-39008 are two commonly used specifications for this type of resistor.

Film resistors   Film resistors have become the most commonly used of fixed types, with applications in general-purpose, precision, and high-performance categories. Their configurations include axial, radial, surface-mount, networks, and potentiometers (the latter two are covered later in the chapter. Many terms common to film resistors are presented in the network discussion.

Thick- and *thin-film resistors* are terms commonly used to separate film resistors into two broad subcategories. The dividing point between the two types is film thickness—5 micrometers or greater are thick film and less than is thin film. Of more use, however, than terms, such as "thick" or "thin" are performance considerations, such as temperature stability, purchase tolerance, noise, and power-handling capability. Thick-film types are generally less precise than thin film, but can handle greater power and voltage and are not considered to be ESD (electrostatic discharge) sensitive. Thin film are generally more precise, handle less power, and are considered to be ESD sensitive, usually at voltage levels greater than 2000 volts. A standard, such as DOD-STD-1686, is useful for general guidance on ESD vulnerability.

Most film resistors, such as nickel chromium thin films and carbon films are vulnerable to mechanical damage and to atmospheric moisture and must be protected immediately after deposition. Varnish of the silicone variety are used directly over the film, being essentially inert and having sufficient high temperature capability. Some films, such as tin oxide, are mechanically hard, relatively inert chemically, and do not need this treatment. One type of thin-film resistor uses a metal film deposited on the inside of a ceramic tube. Helical grooving for value adjustment is followed by application of terminal metallization and sol-

TABLE 1.6 Pulse Performance Range for Allen Bradley Composition Resistors

Rated watts	Single pulse energy capability, W-s	Equivalent energy source
⅛	0.45	2 µF at 670 V
¼	1.8	10 µF at 600 V
½	6.4	32 µF at 630 V
1	16	32 µF at 1,000 V
2	44	32 µF at 1,650 V

SOURCE: Allen Bradley Company.

der sealed pressed on end caps to produce a self-sealing package, which is then epoxy encapsulated. The resulting product is marketed as a hermetically sealed metal-film resistor.

Carbon film   Pyrolitic carbon film depositions on ceramic and glass substrates have been manufactured for several decades and, when used within their limitations, provide better stability, lower noise, and better high-frequency performance than carbon-composition resistors. However, improvements in the performance, range, cost and popularity of other film types have lessened the use of carbon film in recent years.

For pyrolitic carbon deposition, the substrates are heated in a furnace and a carbon bearing gas such as methane is metered into the enclosure, where the red heat pyrolyzes the gas. Substrates are rotated or agitated to obtain equal exposures. Film uniformity, thickness, hardness, and resulting characteristics all depend on the properties of the substrate, mechanical features of the deposition exposure, control of the atmosphere in the deposition furnace, time of exposure, and similar considerations.

Examples of carbon film designs includes resistors specified by MIL-R-10509, MIL-R-11, and IRC type CF.

Carbon film resistors can be expected to remain within about 2.5 to 3 percent of their initial values over a life of 5 years, if conservatively applied. Derating curves are based on an expected resistance change of 1 percent per 1000 hours of load life. High-frequency characteristics, as with other film resistors, are fairly good. The shunt capacitance of 0.2 to 0.6 picofarads from end terminal to end terminal is the primary contributor to a roll-off in impedance for higher resistance values at higher frequencies. Lead inductance is the major inductance consideration.

Coatings and encasements available provide varying degrees of mechanical strength and environmental protection. Moisture is damaging to carbon film resistors and can cause failures if liquid accumulates on film surfaces. Changes in resistance value because of moisture are only partially recoverable; they amount to 1.0 to 1.5 percent in humid exposures. Coatings should therefore be mechanically strong, reasonably impermeable to moisture, compatible to solvents (particularly those generally used in board cleaning operations), inflammable, and nontoxic under temperature extremes (nontoxic smoke and fumes). Sealed construction, if used, must be extremely dependable because the reliability of such construction depends absolutely on the continued ability of the encasement to remain sealed. Small leaks in such cases tends to trap moisture and other contaminants in the very areas where they can do the most harm, whereas slightly pervious coatings can expel damaging fluids over time.

Tin oxide   Depositions of tin oxide into glass rod surfaces were introduced by Corning Glass Works in the mid 1940s. The process continuously extrudes nearly molten glass rods through an atmosphere containing the required gaseous compounds. Tin and/or antimony are necessary constituents of the gaseous medium. The result is a nominal tin oxide film diffused minutely and onto the glass rod surface. The film is under 5 micrometers in thickness, is smooth and glossy, and is on the order of tool steel hardness. Because it is a relatively stable oxide in a low-energy state, chemical activity is minimal in most commonly encountered ionic combinations. The glass substrate is not as mechanically strong as some ceramic types. The glass-tin oxide has good drift characteristics and high temperature-life capability, moisture resistance, along with low noise production and high frequency performance common to most metal film types. Tin-oxide resistors are available under military specifications MIL-R-22684, MIL-R-39017, MIL-R-55182, various DESC drawings, and many commercially comparable types.

Thick film   Thick-film configurations have become among the most popular and widely used fixed resistor configurations. These films are composed of stable combinations of metals and metal derivatives and other material combinations, usually in vitrified glass frit suspensions. Metals are usually gold, silver, palladium, ruthenium, iridium, and platinum for cermets, and tantalum, tungsten, or titanium for other types of film. After processing, these films are in stable states, with oxides, nitrides, and carbides being commonly formed, and are usually self-protecting by virtue of a passivating surface coating conversion. In glaze-type thick-film construction, finely divided metal particles are mixed with glass powder and a volatile vehicle. The mixture is painted on, rolled on, screened on, dipped, or otherwise applied to a ceramic substrate. The resulting combination is fired at temperatures of about 1000°C, melting the glass powder, suspending the metal particles therein, and bonding the resulting film to the ceramic substrate. Atmospheric controls are provided to control oxidation and reduction during the firing while the glass is molten. Metallized bands are applied to the ends for connection terminal leads or termination finish in surface-mount designs. Other thick-film types use anodized, oxidized, or otherwise modified applied films.

A range of resistance value is generally available for ±100 ppm/°C with ±50 ppm/°C available. Current-generated noise is generally a few decibels higher than for comparable-sized metal films, but still well below carbon composition. Size, type of film, and resistance value strongly influence the noise index. Stability varies with the film used. The voltage coefficient is likely to be higher for some thick films than for thin-film units. Thick-film resistors are available under military specifications MIL-R-22684, MIL-R-39017, MIL-R-55182, MIL-R-55342, various DESC drawings, and many commercially comparable types.

Thin films Initial thin films were platinum iridium or palladium silver, but most are nickel chromium or tantalum nitride, with silicon-oxide methods being developed. Partial reduction, oxidation, passivation, and other custom treatments modify characteristics to achieve desired traits. Most are produced by vacuum vapor deposition at high temperatures onto ceramic substrates, but sputtered tantalum nitride is also widely used.

Bulk temperature coefficients of resistance for pure metals are almost exclusively positive. In thin films, however, surface-conduction modes that are essentially independent of temperature become a large factor, and together with alloying and custom treatments, temperature coefficients very near zero are achieved. Resistivities depend on film thickness primarily, and as in other film types, the best characteristics are obtained in the middle of the practical range. Nickel-chromium films of 2000 to 3000 ohms per square provide for good stability, but higher sheet resistivities need thicknesses of film of less than 20 Angstroms, and drift stability begins to fall off rapidly. The addition of nonconductor materials to chromium in evaporation processes has allowed achievement of reasonably stable devices of up to 20 kilohm per square. Chromium-silicon monoxide film evaporated film, considered to be a "cermet" construction, has been successfully used in this regard.

Resistors are finished by grinding grooves in the film to produce a helixed conduction path. This is done usually on a Wheatstone bridge against a predetermined value, with the grinding wheel automatically lifted as the bridge balance is approached. Mild surface abrasion is sometimes used as a final trim. Most good, precision nickel-chromium film resistors, including hermetic versions, that are procured at 1 percent or better purchase tolerance and ±50 ppm/°C temperature coefficient or better can be expected to remain within about 1.5 to 2.0 percent of their value as received, for a normal service life of 5 to 10 years of normal operation and storage. For a 1-percent procurement tolerance, therefore, 2.5 percent is a practical end of life tolerance. Comparable performance is expected of tantalum nitride films. A generalized comparison between nickel chromium and tantalum nitride is shown in Table 1.7[6]. Changes tend to be greatest at installation and shortly thereafter, but cannot be depended on to track from part to part, unless they are empirically verified.

Metal films are able to remain stable at higher operating temperatures than carbon films and, along with other film types, have completely replaced carbon films in military standards. Various sealed designs have been developed. These are intended to exclude environmental conditions, and some have been very successful in use. One glass-metal sealed design is capable of direct transfer from 125°C oil to −180°C liquid without fracture or damage.

TABLE 1.7 Comparison of Tantalum Nitride and Nickel-Chrome Thin Films (Courtesy IRC)

Property	Tantalum nitride	Nickel chrome
Absolute TCR [1]	10 PPM/°C	10 PPM/°C
Tracking TCR [2]	2 PPM/°C	2 PPM/°C
TCR Slope [3]	Approx 0	10 PPM
Sheet resistance	4 to 150 ohms/sq	4 to 300 ohms/sq
Deposition process	Reactive sputtering	Sputtering
Life stability 1000 hours @ 125°C	0.03%	0.03%
Biased humidity 1000 hours 85/85 [4]	<0.05%	1 to > 10%
Aging mechanism [5]	Diffusion controlled	Complex
Overload capability [6]	Excellent pulse and ESD performance	?

Notes:
1. The TCR of tantalum nitride is adjusted by vacuum annealing which is a one-time batch process. After annealing at a high temperature, the TCR is not readily adjusted further. The TCR of nickel-chrome films is adjusted using a heat soak in air. The TCR will be adjusted further to some extent by continued heat soaking.
2. TCR tracking is influenced by:
   2.1 Design—TCR of conductors; location of resistors
   2.2 Sputtering—cathode area; in-line vs. batch; composition uniformity
   2.3 Annealing process—temperature uniformity
3. The TCR slope (the difference between $1/R * dR/dT$ at 125°C and −55°C) of tantalum nitride is essentially 0 as verified down to about 20 degrees absolute (liquid hydrogen temperature). Nichrome films typically have a more positive TCR at higher temperatures.
4. Nichrome films dissolve in the presence of humidity and a potential making them a reliability risk in harsh environments.
5. The aging of nickel-chrome films is by a combination of oxidation, grain growth, and interdiffusion of nickel and chromium. Both positive and negative shifts are observed.
6. Tantalum nitride performs well due to its high melting point of 3000°C (refractory metal). Resistor layout is also important.

Nickel-chromium thin films cannot tolerate water in liquid form if ionizable salts are present. Electrolytic couples that otherwise would be of minor significance become formidable in terms of the thin cross sections of metal films. Hermetically sealed designs, therefore, must be extremely good to avoid entrapment of contaminants. Coated and molded encasements provide mechanical strength and moisture protection in proportion to their particular capabilities; each must be measured in terms of cost, character, and performance. The high-frequency performance of metal films is extremely good, compared with that of wirewound or composition types, particularly in resistance values greater than 1000 ohms.

Helical spiraling has negligible inductive effect in lower resistances on the high-frequency characteristics. End-cap-to-end-cap and distributed capacitances dominate the performance in the higher values and frequencies, producing a shunting effect. Low-resis-

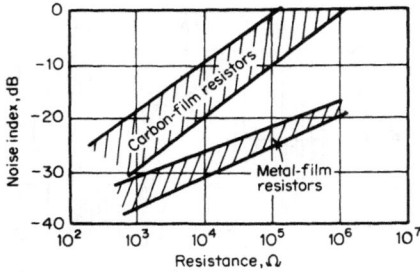

FIGURE 1.7 Relative noise characteristics of carbon-film and metal-film resistors.

tance values remain very close to labeled value. Very high frequency performance on low values can be improved by use of coaxial mounting to eliminate lead inductance. Nominal noise characteristics are shown in Figure 1.7[1]. Comparison of tantalum nitride performance (give Bos reference) shows improved performance in humidity testing when compared to nickel chromium films. Figure 1.8[5] shows results of tests, which compare change in resistance of 85 percent relative humidity/85°C testing. Because less environmental protection might be required or desired for tantalum nitride, greater design flexibility can be afforded for many applications.

Nickel-chromium and tantalum nitride thin films will provide the stabilities previously mentioned, if they are conservatively applied. Extreme conditions or less-refined production processes will, of course, result in degradation of stability. MIL-R-55182 includes a high-temperature exposure test, wherein the test sample is heated to the zero-power rating for 2000 hours without power applied. The allowed resistance change is controlled, thus providing a degree of assurance for use of the derating curve. Half power derating is still recommended. Chemical inertness, caused by the low energy states of the constituent materials, enhances stability by reducing the rate of change in character, and systems using such materials in stress-relieved systems with other materials of carefully matched compatibility enjoy the maximum stability and reliability. Electrolytic similarity and matched thermal expansion rates are equally important considerations. Conductor cross sections are extremely minute, and small changes tend to produce relatively large effects.

Temperature coefficients of resistance of ±25 ppm/°C are routinely furnished and ±5 ppm/°C are achievable.

Reliability figures stated for military uses are based on a predefined point of failure, usually some amount of change in resistance. These "failures" might not be catastrophic at all; in many circuit applications, they are negligible. Catastrophic failure modes, such as cracked cores, defective lead-to-end-cap connections, defective coatings, electrolysis, poor lead solderability and actual breakage are generally addressed by quality-assurance controls. Screening tests devised to locate film defects generally require overloading of parts up to about 5 times the rated power for a short time, the effect being the creation of localized hot spots at high current density points, resulting in either open circuiting or a large measurable change in resistance value.

Metal film resistors offer the highest temperature stability of any nonwirewound resistor. Their high-frequency and low-noise performances are excellent, and if they are well made and conservatively used, they can be expected to provide at least 10 years of reliable service. Although few specifications specifically address thin-film resistors, specifications such as MIL-R-55182 and MIL-R-55342 are generally used as a base line.

1.2.4 Wirewound and Medium-Power Resistors

Low-power wirewound Low-power wirewound resistors are offered in molded or conformally coated cases or in chip form with resistances as low as 0.1 ohm. Generally, these are used when custom values, greater precision, or slightly greater power are required and a film resistor will not suffice.

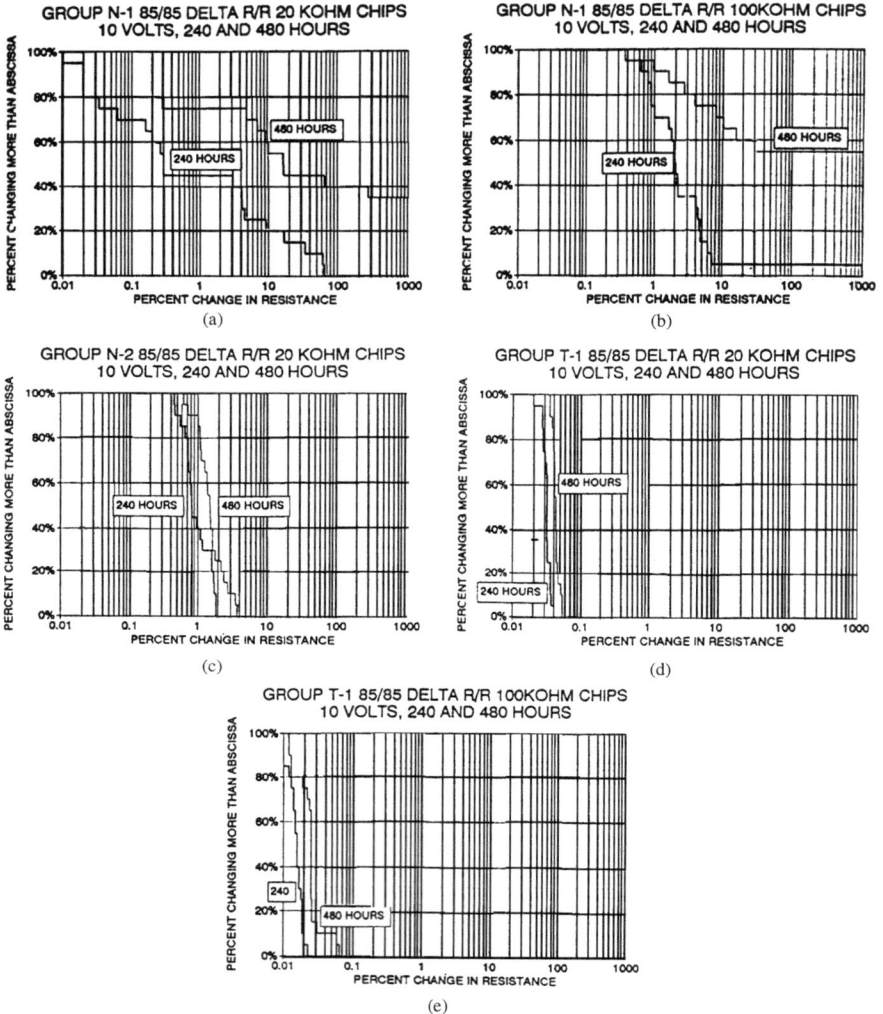

GROUP N-1 85/85 DELTA R/R 20 KOHM CHIPS
10 VOLTS, 240 AND 480 HOURS

(a)

GROUP N-1 85/85 DELTA R/R 100KOHM CHIPS
10 VOLTS, 240 AND 480 HOURS

(b)

GROUP N-2 85/85 DELTA R/R 20 KOHM CHIPS
10 VOLTS, 240 AND 480 HOURS

(c)

GROUP T-1 85/85 DELTA R/R 20 KOHM CHIPS
10 VOLTS, 240 AND 480 HOURS

(d)

GROUP T-1 85/85 DELTA R/R 100KOHM CHIPS
10 VOLTS, 240 AND 480 HOURS

(e)

FIGURE 1.8   Tantalum nitride versus nickel chromium humidity testing.

Noninductive windings   Reducing the effects of inductance in resistors, particularly wirewound resistors, is important to reduce undesired reactances. Noninductive windings and inductance-canceling techniques are sometimes used. However, as covered earlier, distributed capacitance, leakage flux, and other residual inductances remain.

One technique is the Ayrton-Perry winding. Two wires are wound in opposite pitches and the two are connected in parallel. At the wire crossing points, 180 degrees apart on the core, the potentials are essentially equal. This low potential decreases the effect of the distributed capacitance and allows the resistor, if desired, to be wound with uninsulated wire.

A disadvantage of this technique is that with the smallest usable resistance wire diameter of the highest obtainable resistivity, the highest resistance value that can be obtained is halved because of the parallel connection.

Another technique is the Chapron winding. Its advantage is that the resistance is not halved. This technique reverses the winding pitch in the center of the mandrel.

Bifilar winding is one of the two most-popular noninductive winding methods for precision wirewound multilayer resistors, but it is seldom used in single-layer power units. For this method, the length of wire is bent in half, and the resulting doubled wire is wound on the core. Each of the free ends then becomes a terminal, with one end being brought over the wound core opposite that where the winding started. The result is that two series-connected windings with the circuit current flowing in opposite directions in each, with respect to the core. The potential difference between adjacent turns at start ends is high, however, and the effects of distributed capacitance are correspondingly increased.

Another popular winding method for multilayer resistors is the reverse pi method. Here, the resistance winding is divided into an even number of approximately equal segments on a bobbin that is constructed with the needed segments separated by insulation barriers. One segment is wound, and the wire running end is dressed into a slot to the next segment. The next segment is wound in a reversed direction. The process is repeated so that the end result is a series of oppositely wound segments. Leakage flux and distributed capacitance are not completely eliminated, however, and results are about equivalent to that of bifilar winding. This technique is not easily adapted to power units.

Medium-power wirewound   Resistors having power dissipation ratings greater than 2 Watts are categorized as medium power. Wirewound resistors, due to their ability to withstand greater heat than other types, and film resistors (on substrates or with integral heatsinks) comprise the majority of this resistor class.

Application techniques for wirewound resistors in the range of 2 to 6 W must allow for their higher operating temperatures by arranging to remove the heat generated in dissipating the greater power to protect adjacent parts from damage. Although resistance-wire alloys have melting points as high as 1500°C, conservative practice and most industry and military standards specify a maximum hot-spot temperature of 275°C for resistors with power dissipation ratings less than 10 W.

Core material is a significant part of the thermal system, being in intimate contact with the resistance wire and usually well connected to the metal lead wires. Porcelain and pottery formulations are widely used. High alumina ceramics are generally used because of their high thermal conductivity and mechanical strength. Beryllium oxide is also used because of its even-better thermal conductivity, but extreme care must be exercised because it is toxic. Both materials are also used extensively in surface-mount and chip configurations. The effect of good core thermal conductivity is to lower the operating temperature gradient between the center of the resistor body and its ends, thus reducing the hot-spot temperature. Other materials, such as mineral filled plastics, impregnated glass fibers, and aluminum are also used to varying degrees. Glass fiber parts, despite a lower thermal conductivity, are used in high-volume applications, such as in automotives, because of their relatively low cost to manufacture.

Resistance wire   Resistance wire must be capable of withstanding high firing temperatures if vitreous coatings are used, and must withstand repeated thermal cycling and long exposures to elevated temperatures, while remaining stable and inert. Table 1.8[1] shows characteristics of various resistance wire materials.

When temperature changes of resistance value are assessed in comparison with film types or precision wirewound parts, the increased temperature permitted for a given power rating should be noted.

Reliability considerations of wirewound resistors   Good-quality wirewound resistors have windings evenly spaced on the core and will have well-made termination welds. Coatings should be capable if withstanding the specified high potential tests when the resistor body is intimately wrapped in foil with voltage from foil to winding. Fillers for silicone and cement coatings sometimes are sources of ionic salts and, in the presence of environmental moisture, can cause galvanic corrosion. Resistance alloys and terminal hardware are in themselves noncorrosive, but galvanic dissimilarities can cause corrosion problems in the presence of ionizable salts. The resistance wire joint at the termination is probably the most vulnerable to corrosion. Most standardized resistors in the medium-power range are of the axial-wire lead type, having conformally coated cylindrical bodies. Some designs have been encapsulated into metal tubes to allow efficient use of clip-type connections to the chassis for better heat removal. The bare and molded automotive types mentioned previously are also popular. One available commercial design uses a hollow ceramic tube or lidless rectangular box with an automotive-type resistor element intimately encased inside a ceramic-filled cement.

Heavy-bodied designs should be mounted by means other than their leads if shock and vibration environments are to be encountered, but care must be exercised to prevent breakage of ceramic and glazed parts by stress concentrations or tensile forces.

Power film   Film resistors provide can provide better RF performance over wirewound types and, in addition, allow a wider choice of resistance values. Tin oxide on glass has been available in large physical sizes for many years, and varnish-coated carbon-composition film and carbon-film types are used where sizes can be large enough to keep temperatures down and where environments permit.

1.2.5 High-Power Resistors

High-power resistors have power dissipation ratings greater than 6 W, except for devices designed specifically to be electrical heating elements. Usages of high-power resistors are more frequent in heavy-duty power equipment applications than in electronics applications.

Resistor types used in high-power requirements include wirewound, carbon composition, tin oxide, and thick film. Wirewound types have historically been the dominant choice, except for high-frequency applications. The high-temperature capability of resistance wire alloys provide a practical advantage for the wirewound, except as noted.

Terminals and mountings   The large physical sizes and masses require considerable attention to mounting details, and the familiar self-supporting axial-lead-mounted cylinder is little used above 10-W ratings. Radial tabs that are extensions of metal bands welded around flat oval or cylindrical core ends leave the hollow-core centers open for various mounting methods and are the most widely used termination method. The resistance wire is welded or hard soldered to the band, and the radial tab can be fitted with a variety of connection accommodations, such as an appropriately sized screw and washer terminals, long multistrand flexible, or short solid copper lead wires crimped and soldered to the tabs, or most frequently, a hole made in the tab to accommodate a wire for soft soldering. MIL-R-26, for example, requires that this hole be at least large enough to permit a fully affixed and crimp 14 AWG wire which has a 0.065 diameter.

Lugs and terminal connections are sized larger, proportional to the scale of the resistor and the current-carrying needs. Support of the resistor body by lead wires alone is increasingly less desirable up to about 20-W sizes, beyond which even the most benign mechanical and environmental considerations require body mounting. Some part designs require hard solid mounting to a metal chassis of a specified area to achieve their labeled power rating. Besides the radial tab terminal, parts are supplied with fuse clip ferrules, flat blades for pressure connections, long flexible insulated wires, and various standard Edison-type lamp screw bases. Lead wires are not as significant in the thermal dissipation process as lower-power resistors. Figures 1.6[1] and 1.16[2] show some typical resistor designs.

Thermal considerations   For conventional free-air-mounted 2 W and smaller resistors, lead wires account for about 50 to 75 percent of heat removal, but above 2 to 5 W, this fraction diminishes rapidly and radiation quickly gains significance for free-air designs, particularly for the high temperatures of wound wire and strip units operating at rated capacity. Convection is also prominent, retaining about half the significance of radiation at maximum-rated dissipation for wirewound styles. Resistor shape and surface finish variations have large effects on the convection rate, and the dissipation-mode proportions vary as the operating temperature achieved by a particular resistor changes. The relatively high operating temperature of wirewound units is a strong factor in the predominance of the radiation mode because the rate of energy dissipation by radiation from a body is proportional to the fourth power of its absolute temperature, whereas thermal conduction and the related convection are essentially linear or to a second degree with temperature.

Material types   Materials must remain stable in high-temperature service and should be well matched, with respect to thermal expansion characteristics to avoid physical damage in repeated thermal cycling. The finished item should be capable of withstanding the effects of atmospheric humidity and condensed moisture along with moderate industrial atmospheric contaminants if it is intended for use in unprotected applications. Galvanic metal couples should be carefully avoided by use of compatible platings and materials, and critical circuit points should be well protected to impede corrosion. If wirewound, resistance wire of the largest possible cross-sectional size should be used, and the heat-producing winding should be evenly distributed over the physical body of the device to avoid hot spots.

Coatings   Precision coatings and encasements available are of three basic types: filled silicone resins, vitreous enamels, and ceramic cement using binders other than silicone resin. Silicone-coated parts are used where precision resistance stability is needed because curing temperatures are low compared with those needed for vitrified materials, and wound resistance wire is less likely to be stressed and moved about during its curing process than is the case for the glossy vitreous enamels. Generally, the maximum operating temperature for silicone-coated units is 275°C. Detrimental effects from silicone coatings include outgassing, which can start to occur at 200°C, and vulnerability to industrial and circuit cleaning solvents.

Vitreous enamels are formulated to match the thermal expansion coefficients or wire, core, and terminals and, in general, are in a surface stress condition after firing so that the resistor wire and core are in a degree of compression. Vitreous enamel and silicone coatings provide handling and installation protection as well as improving turn-to-turn and element-to-mounting high-voltage capabilities. Protection from inadvertent short circuiting is provided by those types having completely embedded windings, and surface contaminants are prevented from accumulating across wire turns, as can happen in exposed types. Although high surface temperatures for parts operating at near-rated power are a factor in personnel safety, exposed conductors in accessible areas can be a much more serious hazard. Electrical shock dangers are greatly reduced by use of fully coated windings. All coatings

TABLE 1.8 Characteristics of Resistance Wire

Alloy type[a]	Resistivity, Ω cmil ft⁻¹ 20 °C	Resistivity,[b] Ω cm² cm⁻¹ × 10⁶ 20 °C	Resistance temp coefficient,[c] ppm °C⁻¹	Linear expansion thermal coefficient, cm/cm/°C × 10⁶, 20–100 °C	Min tensile strength lb in⁻² 25 °C[d]	Melting temp (approx), °C	Relative magnetic attraction	Density, g cm⁻³, 20 °C	Heat capacity,[e] J g⁻¹ °C⁻¹	Thermoelectric potential to copper[f] (approx), V °C⁻¹ × 10⁶
80–20 Ni-Cr	650–675	108–112	+60 to +90 ± 20	12–18	100,000	1400	None	8.41	0.435	+6.0
Constantan	294–300	49–50	0 ± 20	14.5	60,000	1350	None	8.90	0.393	-45
Manganin	230–290	38–48	0 ± 15[c]	18.7	40,000	1020	None	8.192–8.41	0.406	-3.0, +1
Alloy 180	180	29.9	+180 ± 30	15.7–17.5	50,000	1100	None	8.90	0.385	-37
Alloy 90	90	14.9	+450 ± 50	16–17.5	35,000	1100	None	8.90	0.385	-26
Alloy 60	60	9.97	+500 to +800 ± 200	16.2–16.3	50,000	1100	None	8.90	0.385	-22
Alloy 30	30	4.99	+1400 to +1500 ± 300	16.4–16.5	30,000	1100	None		0.385	-14
Linear TC[a]	120	19.9	+4500 ± 400	12–15	70,000	1100	Strong	8.46	0.523	-40
Nickel A	60	9.97	+4800	13	60,000	1450	Strong	8.90	0.544	-22
High purity Ni[a]	50	8.31	+6000	13.3–15	50,000	1400	Strong	8.90	0.544	-22
Iron	61.1	10.15	+5000 to +6200	11.7 (20 °C)	50,000	1535	Strong	7.86	0.445	+12.2
Copper	10.37	1.72	+3900 to +4300	16.5 (20 °C)	35,000	1083	None	8.90	0.385	0
Evanohm[g]	800	133	0 ± 5	12.6	100,000	1350	None	8.10	0.448	+3.0
Karma[h]	800	133	0 ± 20	13.3	180,000	1400	None	8.10	0.435	+3.0
Alloy 800[i]	800	133	0 ± 5	15	150,000	1260	None	7.95		+2.5
Chromel R[j]	800	133	0 ± 10	13.5	95,000	1398	None	8.1	0.448	~+1.0
Moleculoy[k]	800	133	0 ± 5	13.3	130,000	1395	None	8.12	0.435	+3.0
Nikrothal LX[l]	800	133	0 ± 5	12.6	150,000	1410	None	8.1	0.460	+2.0
Kanthal DR[l]	812	135	0 ± 20	11.9	100,000	1505	Strong	7.2	0.494	-3.5
Mesoloy[k]	825	137.2	0 ± 10	13.5	100,000	1500	Strong	7.15	0.481	-3.3
Alloy 815[j]	815	135.5	+82	15.9	~115,000	1520	Strong	7.25	0.460	~-3.7
Alloy K-20[j]	815	135.5	0 ± 20	13	100,000	1530	Strong	7.25	~0.460	-3.5
Evanohm S[g]	825	137	0 ± 5	13	100,000	1350	None	7.13	0.460	+0.2

[a] Refer to Table 9 for alloy composition and suppliers.

[b] Microhms per cm length of a 1-cm²-section was obtained by dividing the ohm–circular mil per foot value by 6.015.

[c] Various suppliers adjust minor constituents and processing to provide selected temperature coefficients or slightly different ranges. Values given cover generally a range of 0 to 100 °C. Nonlinearity of the curve will cause slight deviations outside this range. Manganin, in particular, has a parabolic TC curve whose peak can be adjusted for the desired operating temperature. TC stated for 20 to 35 °C for manganin. Values for 800 Ω cmil ft^{-1} and higher resistivity alloys cover a range of generally –55 to +150 °C.

[d] Tensile strength varies considerably with sample shape and size. Values given are advertised minimum values and are stated primarily for comparison purposes.

[e] Heat capacity in joules per gram for a 1.0 °C rise differs from "specific heat" referenced to calories per gram for a 1.0 °C rise, or an identical numerical quantity referenced to Btu per pound for a 1.0° Fahrenheit rise. The conversion factor for any given specific heat to joules per gram for 1 °C rise is therefore a multiplier of 4.186 J cal^{-1}.

[f] Thermoelectric-potential values given by different sources vary; also the alloy may be used with metals other than copper. Range of temperature for the values given is, in general 0 to 100 °C. (See "Metals Handbook," vol. I, 8th ed. American Society for Metals, 1961.)

[g] Registered trademark, Wilbur B. Driver Company.

[h] Registered trademark, Driver Harris Company.

[i] Registered trademark, C. O. Jelliff Corporation.

[j] Registered trademark, Hoskins Mfg. Company.

[k] Registered trademark, Molecu–wire Corporation.

[l] Registered trademark, Kanthal Corporation.

interpose a thermal impedance between the resistance element and the medium surrounding the resistor; under powered conditions, the resistance element will operate at a higher temperature than that measured on the outside of the coating. Military specifications, such as MIL-R-26 and other standards, set conditions under which exposed wires are not permitted and requires the minimization of voids in coatings. Film high-power resistors are sometimes furnished uncoated, or might have only a thin high-temperature varnish-type coating. Fired thick-film types in lower power ratings are usually heavily coated, but the carbon-composition film and tin-oxide-film types are used primarily in protected installations and are lightly protected.

Cores for some film types are glass tubes, but most other high-power units use ceramic cores. The relative size of core needed to provide required surface dissipative area makes it practical to use hollow cores, which also allows for ease of mounting by several means. The tubular film units can be fitted with ferrule terminals or fired colloidal metallization in a ferrule shape, which allows design of fittings that can be used to circulate fluid coolants through the core. Ceramic core materials are a variety of pottery formulations, most being steatite or electrical porcelain. Cores are popularly cylindrical, flat ovals, cylindrical with flat sides to allow precision in Ayrton-Perry noninductive windings, or helically grooved cylinders and flat ovals. Glass-fiber core units continuously wound in long thin rods, then chopped and terminated to size by length (automotive types) as described in the medium-power category above, are also produced in units having ratings greater than 6 W. They are furnished as bare windings for use in automotive and appliance applications, where they are used in protected enclosures and are rated in watts per inch. This type of element is also furnished as encapsulated unit, frequently square or rectangular in cross section. Because the diameter-to-length ratio is small and the surface area is therefore limited, a thermal advantage is achieved when the unit is intimately embedded in a relatively thermally conductive ceramic cement. Allowed surface operating temperature of these units is naturally much less than that of the large-diameter windings with thin coatings, and the technique is used only in lower-power ratings. These parts are inexpensive, suitable for many applications.

Some large, low-resistance-value parts are wound from heavy wire or strips and are designed to be supported by ribs or bars so that an open-wound unit is constructed. Those parts are usually not insulated or coated, and are used in protective enclosures, often with forced-air cooling. Several designs, as shown in Figure 1.9[1], are available.

A variety of embedded or tub-type resistor units are available. These parts use zig-zag noninductive windings on flat cards or ribs that are subsequently placed into a "tub" or depression in a fabricated base of ceramic or high-temperature molded material, there to be embedded in a refractory cement with terminals protruding from the cured embedment. These are convenient for mounting and require little height. One design shown in Figure 1.4[1] is that of a flat disk, having a mounting hole in its center with terminals brought out radially. Units of this configuration can be stacked on a long threaded rod, observing power derating, or can be chassis mounted, using a screw and washer combination through the center hole.

Related to these types are the military chassis-mounted units that are cylindrical wire-wound elements molded with thermoplastic into extruded aluminum cases. Chassis feed-through, standoff, and horizontal mountings are available, and the technique allows quite large power ratings to be furnished in relatively small physical sizes, provided that enough chassis area is available for heat dissipation. These units are designed to remove as much heat as possible by conduction. Even this convection-radiation efficiency depends on getting generated heat at the winding center to the outside case efficiently. Beryllium oxide solid cores are much used for that reason in these designs, being much better heat conductors than other ceramic materials; in fact, most power resistors that are designed to provide the smallest possible sizes use beryllia cores. The effect is to reduce the thermal gradient in a given resistor size under a given thermal operating condition or to "spread out" the hot

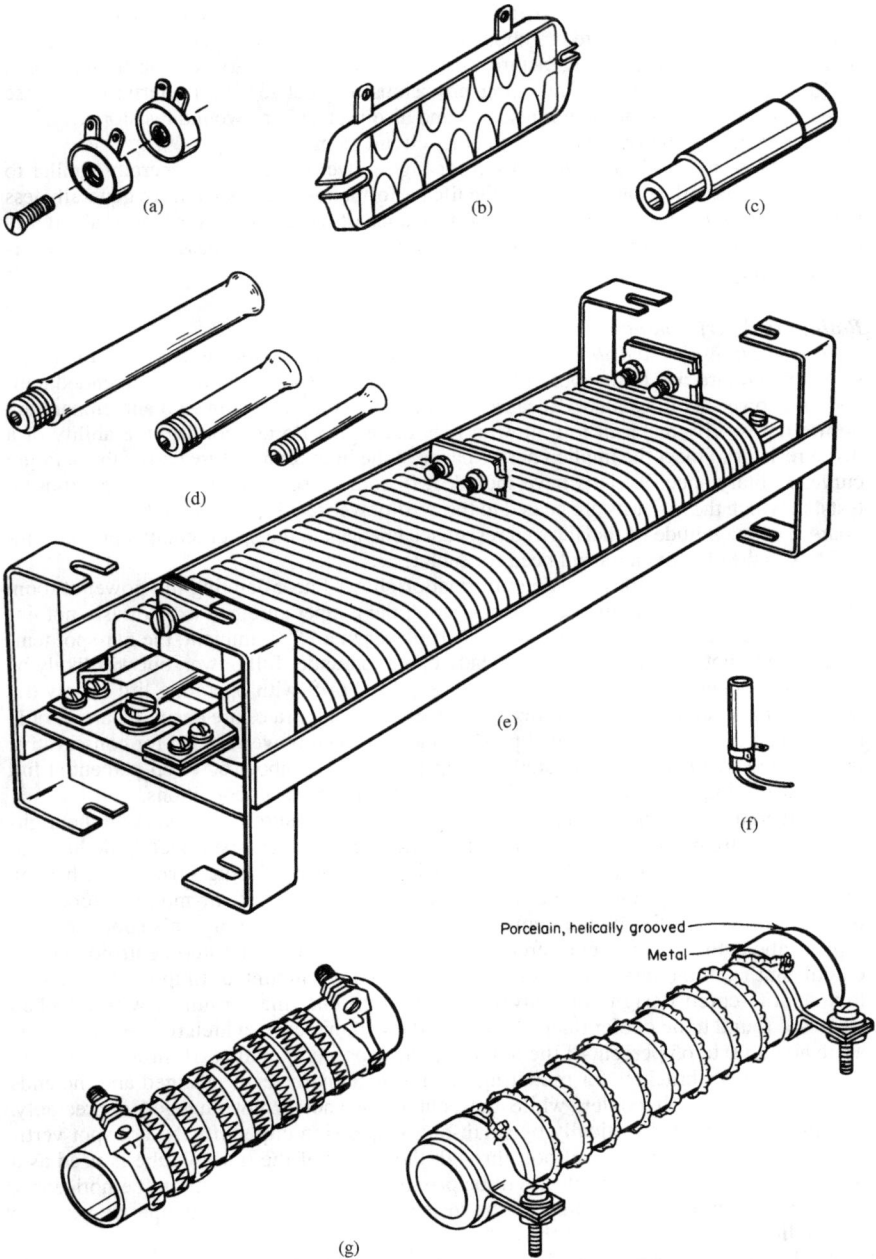

FIGURE 1.9 High-power resistor designs. (a) Tub-type disk. (Ward Leonard.) (b) Tub-type resistor. (Ward Leonard.) (c) Ferrule fuse-clip vitreous-enamel-coated resistor. (Ohmite.) (d) typical Edison-base resistors. (e) Edge-wound bare ribbon. (Ward Leonard.) (f) Appliance -0motor control. (g) Corrugated edge-wound ribbon. This is one type. There is another type in which the space between the ribbon is filled with vitreous enamel, leaving just the ribbon edges exposed.

spot. Some power metal-film resistors are supplied in heat-dissipating case designs, but currently availability is limited to about 12 W maximum. In addition to these configurations, most power resistor manufacturers list in their catalogs a variety of perforated metal enclosures that have resistor-mounting provisions and outside wiring terminals. These cases provide a shield for the hot resistors and also protect bare-wound resistors from accidental contact with personnel or with other circuit elements.

Resistance materials and alloys used for high-power designs are generally similar to other power classifications. However, the life performance and precision are generally less than that of the other types. Generally, performance of these resistors is that of about a 1- or 2-percent change in resistance value over service life with purchase tolerances generally around 5 percent.

Ratings and performance

Temperature rise and drift   Most military and industry standards do not explicitly state temperature rise limits, but instead imply limits by specifying a recommended temperature power-derating curve. Because full-load power tests at room ambient temperature do not necessarily result in full allowed rise, these tests do not confirm the ability of a given resistor to meet the implied performance at the high-temperature end of the derating curve. Military specifications have included, in some cases, high-temperature exposure tests, in which the resistors are heated to the zero-power rated temperature for 250 to 2500 hours. The magnitude of the allowed resistance changes in the specifications provides insight as to the effectiveness and need for such tests.

Although specifications derating information shows linear derating full power at room ambient to zero at maximum body temperature, the temperature rise is obviously not linear; if a given resistor is operated at one half power in a 25°C ambient, the hot-spot temperature will not be one half that for steady operation under full power, but will likely be somewhat higher than half. The amount of expected drift with operating life largely depends on the operating temperature; if low drift and failure rates are desired, parts should be operated at lower than the rated power. One half rated power is used for conservative practice and will result in an operating temperature rise of about 60 to 65 percent at full power, assuming a free-air single-unit mounting under sea-level conditions.

Temperature-rise ratings are all based on hot-spot measurements made on the highest-temperature portion of the resistor that is exposed. This can be logically deduced to be that portion of the assumedly evenly distributed heat-producing surface that has the highest thermal impedance to the surrounding media. For a free-air-mounted tube supported at its extremities by a mounting of low thermal conductivity, this spot occurs in a band about the outer circumference of the cylinder at its thermal centroid—about equidistant from the terminals if the unit is horizontally mounted. Empirically, on hollow steatite-cored designs of conventional proportions, the terminals with attached leads are found to be cooler than the ends of the ceramic core, which reaches a temperature about 60 to 65 percent of the maximum hot-spot temperature. If through bolts are used with heavy brackets for mounting, the thermal character is changed and the ends might be quite a bit cooler, while the center hot spot is not changed appreciably. Mounting the resistor vertically by small-end suspension clip shifts the hot spot vertically upward, and if a vertical bolt through the center of the hollow tube is used as a mounting to the chassis, the shift will be pronounced and the spot might be noticeably lower in temperature, depending on the resistor size, the operating temperature rise in the application, and the size of the bolt.

Each resistor size and design in a given application presents an individual set of conditions about which only qualitative data can be stated. Altitude, forced air cooling, intermittent operation and multiple stacking all affect ratings. For example, taps on resistors reduce the total power rating. MIL-R-26 resistors can be procured with single-center taps, but the power rating is cut by 10 percent. For movable taps and off-center taps, the power rating of a segment is proportional to its fraction of the total power rating. It is expedient to calculate the maximum allowed for full power, then use the value as a limit for any segment.

Voltage ratings The ability of power resistors to remain electrically isolated from their mounting or from adjacent conductors is evaluated in specification by standard tests that apply potential between the windings and body wrappings of metallic foil or a metal V block, or just to the mounting hardware, if it is integral. The coated tubular parts generally depend on spacing from the conductive chassis and the addition of electrical insulation in the mounting for high-voltage isolation. Test voltages are applied for one minute. For real-life usage, the applied voltages should be held well below the high-potential test levels. A maximum of one half the test voltage is suggested as a practice, unless extra insulation is used. The resistor element voltage depends on the power and resistance up to the critical resistance value; they are limited primarily by turn-to-turn breakdown above the critical resistance value.

In accordance with most supplier recommendations and military specifications, voltage ratings for embedded designs follow a general rule of 500 V per inch of winding, extended to 1000 V per winding inch for units of 150 W and up. Voltage ratings might become a limitation if steep leading edges of pulses are applied because the inherent inductive reactance of the winding could cause high-voltage gradients. For bare round wires, such as those used on adjustable tap units, voltage should not exceed 495 V per inch of winding. Arcing conditions cause very high energy densities, and the concentrated stress can rapidly deteriorate resistance windings. It is thus important to observe voltage ratings.

High-frequency performance As in lower-watt sizes, power-film resistors have generally superior performance for high-frequency applications. Wirewound units can be procured in Ayrton-Perry noninductive configurations in either flat oval-cored units or round tubular-core units, but have effectiveness limitations similar to those already covered in the medium-power wirewound part section. If inductance is a difficulty in the application, ferromagnetic mounting hardware should be avoided. Large high-power resistors have higher inductances than smaller-wattage ratings owing to their geometrical dimensions.

Pulse applications Resistors can withstand many times their rated power for short-pulse durations. Application notes at the end of the chapter provide methods of estimating suitability of wirewound for very short pulses. Power resistor thermal masses are basically quite large and thermal time constants are long enough to allow intermittent overpower applications of relatively long durations without endangering part reliability, particularly for large stripwound units. Film parts are not as good in this respect as wirewound units, and small wire diameters require more care than large wire. Manufacturers are able to supply information on intermittent duty for their parts and ICS 170 NEMA standards define standardized duty cycles for motor starting and braking uses of high-power wire and strip-wound resistors. Intermittent application of inputs on the order of 10 times the rated power are allowed, provided that duty-cycle limitations are observed. Voltage ratings should never be exceeded.

Figure 1.10[1] illustrates the range of performance that can be expected on two size classes of vitreous enamel-coated wirewound resistors produced by one manufacturer. Heavy strip-wound resistors limit, in general, at longer duty cycles than those shown. Specific data should be requested from particular manufacturers or laboratory tests should be made for other suppliers' parts because this trait varies with resistor design. It is worthy of note that very high-current surges in helically wound solenoids can produce considerable magnetic force on resistance windings. The force is in directions that would increase the self inductance if the wire were free to move (i.e., a radially outward force places wire in tension, and also longitudinal forces that are directed axially from the ends of the winding toward the winding center). With well-constructed resistors, the effect should not be of major concern. In general, the force can be defined as:

$$f \text{ (newtons)} = \tfrac{1}{2} i^2 \frac{\partial L}{\partial S} S \qquad (1.9^1)$$

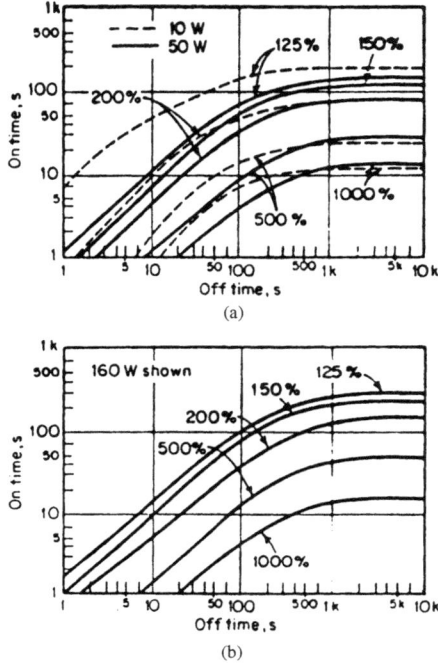

FIGURE 1.10 Percent of continuous-duty rating for pulse operation of vitreous-enameled wirewound resistors.

where i is the current in amperes, L inductance in henrys, and s is a unit vector in direction under consideration. A more detailed treatment is given in references 6 and 7.

Wire and strip-wound resistors

Air-mounted types　　Large suppliers of cylindrical high-power resistors list their product in categories of their standard core sizes. A listing of core inside diameters catalogs maximum available resistance values defined for the particular size. These cores are used alternatively to wind vitreous enamel-coated wire units or for parts wound with corrugated rectangular cross-section strip laid down on one of its narrow edges so that the remaining narrow edge projects radially from the core. A generous coating of vitreous enamel fills in between the helices and, in some units, coats the exposed edges. Some, however, are left intentionally bare to take advantage of the NEMA allowance of increased temperature rise for bared resistor conductors and to permit movable bands for taps. These edge-wound units in high-power ratings are rugged, reliable, and inexpensive.

Some embedded wire type resistors are furnished with silicone resin coatings usually at reduced allowable temperature rise limits, but with better resistance stability. Most high-power resistors, however, utilize some form of vitreous enamel coating. EIA specifications cover flat oval core wirewound parts, and the flatted packages with their integral mounting brackets furnished with spacers for stacking provide a convenient mounting and inherent economy of space. Several units can be easily stacked, provided that proper derating is observed.

Some styles in EIA specifications, called *miniature types*, require chassis mounting of a given area to reach the rated power with which they are labeled. Orientation of stacked parts

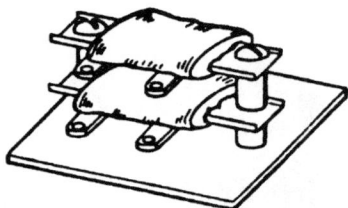

FIGURE 1.11   Stacked flat-strip resistors.

(see Figure 1.11[1]) should be ideally so that maximum air can flow across the flat surfaces without first being heated by neighboring units. Radiation from one unit to its neighbor is unfortunately higher than for comparably spaced round units with similar mounting and power-dissipation rates. Because of the dependence of electrical performance and reliability on the operating temperature, it is important to consider application conditions that will affect the heat dissipation process.

Chassis-mounted types   Metal heat-dissipating cases with mounting provisions are covered by MIL-R-39009 for parts from 5 W to 30 W and by MIL-R-18546 for 75-W and 120-W versions. These are all horizontally mounted parts, and all are available in noninductive windings. Beryllia cores are used on the established reliability parts, and the resistive element is molded into the extruded aluminum housing. Power ratings in MIL-R-18546 are stated for resistors mounted on a chassis of aluminum 0.040 inches thick and 4 by 6 by 2 inches for units up to 10 W, 0.040 inches thick and 5 by 7 by 2 inches for 20-W and 30-W units, and 0.125 inches thick and a 12-by-12-inch panel for the 75- and 120-W parts. All are rated at 275°C case temperature maximum. Steatite and alumina core are also used, depending on physical size. The same basic construction is used on parts designed for feed-through or standoff mounting in a chassis or panel hole large enough to accommodate the body of the resistor. A large nut threaded onto the body mounts the part. Various terminal configurations are available. Chassis-mounted resistors shown in Figures 1.12[1] and 1.13[1] are, as might be expected, more expensive than comparable power ratings in conventional types.

Mounting hardware and methods   Flat oval-core resistors have mounting brackets furnished as integral parts of the unit. Some, mostly the miniaturized designs, have a metal strap completely through the core, with the rods bent to form mounting feet, and others have push-in brackets that are held in the ends of the core by dimples or toothed projections until installation provides rigidity. Figure 1.11[1] shows resistors of this type mounted in a stacked configuration. The horizontal position shown is less preferred than a vertical arrangement, in which the flat sides are exposed to convected air that has not been preheated by its neighbor.

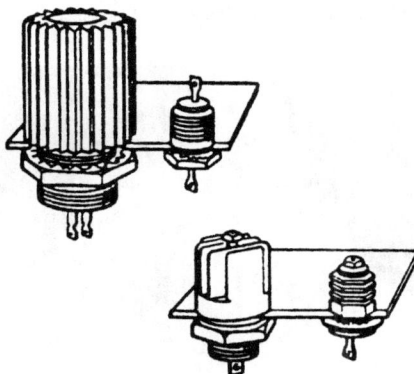

FIGURE 1.12   Chassis feedthrough and standoff aluminum-cased power resistors: grounded and ungrounded types.

FIGURE 1.13 Chassis-mount axial-lead power wirewound resistor.

For round tubular cores and those that have flat sides for Ayrton-Perry windings, the mounting hardware types can be divided into (1) those methods that require a long bolt through the hollow-core center, (2) those that insert short spring clips into the hollow ends of the core, and (3) ferrule-type terminals designed for fuse clips. Military drawing MS75009 covers threaded rod or long screws in 8-32 and 10-32 sizes and 3.375 to 12.750 inches in length, together with lock washer, centering washers of a cup type and an "eared" type, and right-angle brackets for single mounting of tubular resistors. Three sizes of brackets are depicted, ranging from a 1.0- to 1.5-inch center-hole height above the mounting plane. All threaded rod, screws, and brackets are zinc- or cadmium-plated steel. Centering washers can be brass or plated steel, and the split-type lock washers can be phosphor bronze or MS35338 corrosion-resisting steel. Four sizes of mica end washers are also provided, all of 0.031- to 0.094-inch thickness. These washers cushion the ceramic-steel junction at the ends of the core and also increase the high-potential surface creep distance to a grounded mounting bracket.

Figure 1.15c[1] illustrates the use of the mica washer. Although it is not covered by MS specifications, resistor manufacturers list a variety of multiple mounting brackets that are basically MS-type mountings, but with provisions for mounting two or more units on the same bracket (Figure 1.14[1]). Perforated metal protective housings for single and multiple resistors also use the through-bolt basic mounting. Noninductively wound resistors should not have steel screws through their center because of possible enhancement of the leakage

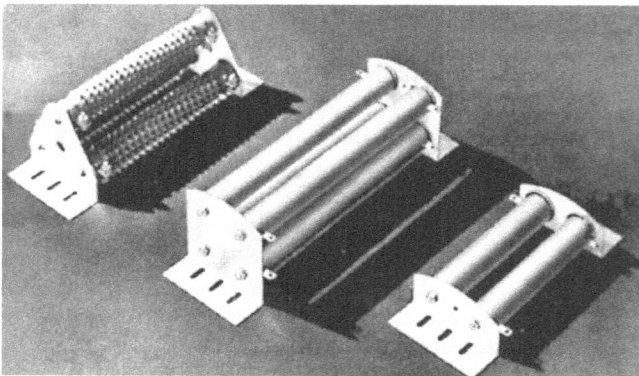

FIGURE 1.14 Multiple mounting of large power resistors.

flux effects. Also, if high voltages are present on the resistor, with respect to the mounting, supplier catalogs list porcelain end bushings that can be bonded into the ends of the core, or just used as loose bushings on through bolts to serve a dual function as additional insulation and as replacement for the normally used metal centering washer. When these bushings are used, the terminals are placed farther from the electrically conductive bracket. The through bolt must be lengthened accordingly. Figure 1.15d[1] illustrates the porcelain bushing use. The second most used mounting is the spring clip. This mounting method consists of inserting into the inside of the core end a clip that is configured to grip the inside of the tube

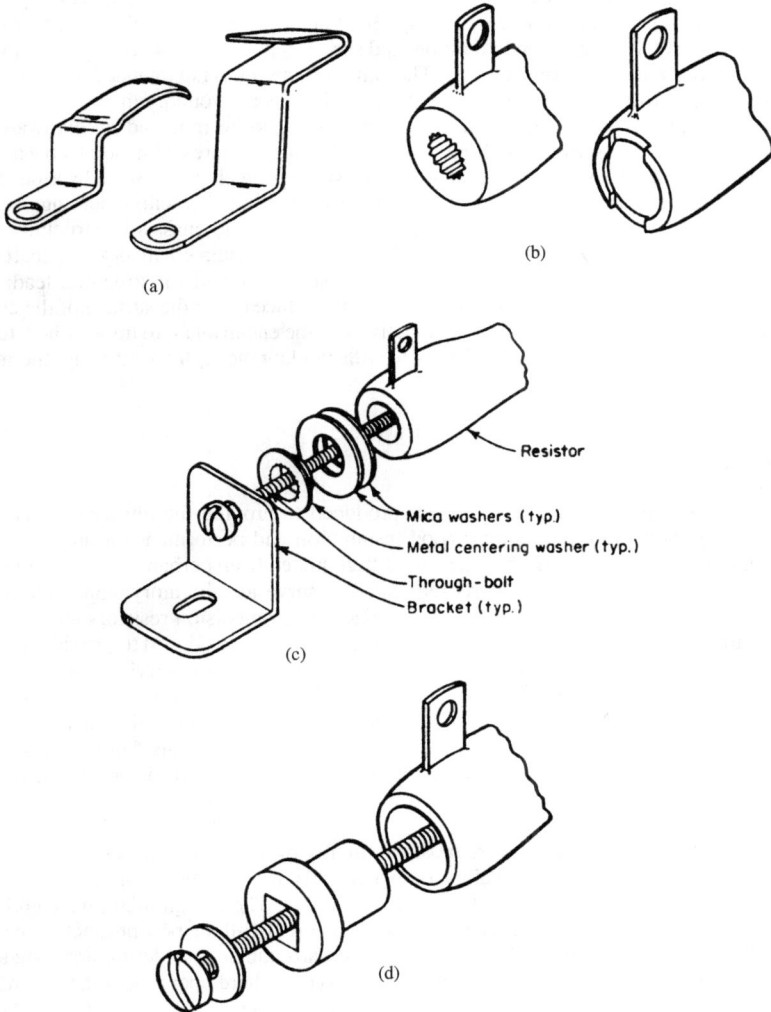

FIGURE 1.15 Power-resistor mounting accessories (a) Spring-clip types. (b) Nonturn features. (c) MS 75009-type mounting, opposite end similar but has a nut and a split lock washer. (d) Ceramic plug provides extra insulation to grounded mounting. (Ohmite Mfg. Co.)

by friction, brought to bear by spring pressure when the clip is deformed at insertion. These clips are available in brass, spring steel, and ordinary mild steel, with the brass parts useful for mounting noninductively wound resistors. Figure 1.15a[1] shows examples of spring clips.

Resistors are available with fluted interior or notches on the core ends to provide non-turn features (Figure 1.15[1]).

Ferrules for standard fuse-clip mountings are also supplied, with resistor terminal leads brought to the ferrules and attached by high-temperature soldering or welding. The ferrules are usually brass. Some types fit over an extension of the ceramic core, and others consist of cups attached to the ends by a through bolt. Fuse clips are of the standard electrical type. One other mounting for tubular resistors that is worthy of note is the Edison lamp base type. This is available in several standard sizes. Through bolts and fuse clips are obviously superior for applications involving shock and vibration, and spring clips are less likely to place strain on the resistor body during thermal cycling. The lamp base mount is not intended for high shock and vibration. Most other power resistor designs of rib open wound, tub potted units, and others have integral mounting means or are configured to be mounted by hardware and methods previously described. High-power film units on glass cores with radial tab leads can be mounted using the through-bolt method or by using spring clips. These and various high-power carbon-composition film parts frequently use the ferrule fuse-clip mounting.

High-power resistors should not be solidly bonded to the chassis or structure using resins or cements, and should be generally not be encapsulated in organic materials. Bonding materials can cause extremely high stresses in the resistor structure, leading to core breakages. Encapsulation moves the radiating surface out to the surface of the encapsulant, and it depends on the thermal conductivity of the encapsulant to transfer heat to that surface. Most encapsulants have difficulty in withstanding the high operating temperatures of power resistors.

1.2.6 Precision Resistors

Precision wirewound   Resistors that are produced to provide the ultimate in accuracy and stability while still maintaining good installation and performance utility are called *precision* or *"accurate" resistors*. They are differentiated from perhaps more accurate, but cumbersome and nonutilitarian parts fabricated to serve as laboratory standards, whose precision and stability they nevertheless approach. High-precision resistors are generally wirewound for the best achievable stability and are carefully produced to provide the lowest possible temperature coefficient of resistance and the minimum service resistance drift. Being wirewound, both the noise and voltage coefficient of resistance are of negligible consideration. The manufacturer must, however, avoid wire junctions that produce significant thermocouple-type potentials, must provide near-perfect joints from the resistance wire to the terminals, and must provide a design that avoids corrosion and disturbing stresses on the resistance wire.

Designs   The available designs are most commonly plastic-resin-encased elements having copper, Dumet, copper-nickel alloy, or nickel leads, with enamel or plastic-film-coated resistance wire wound on a molded, filled resin bobbin. The design might have end caps or it might just have leads embedded in the bobbin core. In all the best products, the resistance wire is joined to terminals by welding, but hard and soft soldering are sometimes used, with the latter being limited strictly to those types where operating temperatures are very low. Intermediate tabs are often used, with one end connected to lead wire and the other welded to the resistance wire. Various other designs are marketed, frequently with very little environmental protection, which is acceptable in many applications.

One frequently used design has a ceramic bobbin core on which the wire is wound, with radial wire leads soldered around end grooves. The element is protected by paper or plastic tape. Precision resistors are larger in comparable power ratings than are corresponding general-purpose parts. In addition to the temperature limitations of the resistance wire, insulation films, and served insulations, power ratings envision a minimum temperature rise to reduce temporary resistance changes because of the temperature coefficient, as well as reduce the probability of accelerating subtle drift mechanisms within the resistor that will result in permanent changes. Packages having internal cavities within cases intended to be sealed with epoxy or other resins avoid pressures on the resistance wire, but should be carefully evaluated in the intended environment. Some end-sealing encapsulation methods could leave undesired water or other residues inside the unit, or could leak at the lead exit in service as the unit is thermally cycled. The dissimilar metals at element to terminal welds are very vulnerable to corrosive attack by electrolytic processes. Systems that provide mechanical cushioning of the winding and then provide an essentially void-free dense encapsulation of the protected winding would appear to be better choices. Also, few materials are poorer thermal conductors than air, and elimination of interstitial spaces improves the thermal system. Avoidance of resistor wire stress is very difficult in solid constructions, however. Corrosiveness of fluxes and of encapsulants and their fillers and possible corrosive or aqueous by-products of the encapsulation process should be evaluated thoroughly. Although most resistance alloys and terminal materials are in themselves noncorrosive, it is easily possible to develop corrosion through dissimilar metal galvanic activity.

Care must be exercised in the measurement of low-value resistors to exclude the effects of lead length and contact resistance. Special techniques are usually necessary to resolve measurements closer than ±0.001 ohms. Temperature coefficients and drift mechanisms also have more effect on low-value parts when considered as a percentage of resistance values, and most specifications recognize this circumstance by providing slightly increased allowances for low values. For these low values, remember that copper circuit conductors have an approximate temperature coefficient of resistance of +3800 ppm/°C.

Considerable technical development and skill are required to wind precision resistors, particularly in fine wire sizes. Tensioning control is all important because the resulting system must be as stress free as possible to achieve in service stability. Just the act of bending wire around a mandrel (or core) produces stresses that affect stability, particularly for some alloys.

Wire sizes down to 0.0005 inches in diameter are used, but more conservative practice is to use wire no smaller than 0.001 inches in diameter. Special temperature stable alloys of about 800 ohms/cmil-ft are used almost exclusively for higher resistance values, and manganin or copper-nickel alloys are sometimes used for lower values. Wires are insulated with film coatings, like those used for transformer magnet wire and occasionally with glass fiber, silk, or other served fibers. Polyvinyl acetal is rated for 105°C and polyimide films such as DuPont Pyre M.L. are rated as high as 220°C.

Stability After winding adjustment and encapsulation, the resistors are stabilized by temperature cycling, baking, and high-power run-in. This step is particularly critical for manganin wire because of its tendency to change character significantly as winding strains are annealed out. After stress relief, well-made resistors can be expected to maintain value within ±0.1 percent and to retrace temperature coefficient of resistance within +10 to ±50 ppm/°C over the rated temperature range, depending on the resistance value. Fine trimming by the manufacturer is sometimes accomplished by mechanically scraping a series service loop. The manufacturer must assure that adjustment gradients are small and that the wire cross section is not seriously reduced in this operation. The ready availability of stable, precision wirewound resistors is, in large part, because of the development of special, stable, mechanically strong resistance alloys that have repeatable properties, and good handling and aging characteristics. The development of winding machinery capable of providing consistent controls has also been important, as well as the persistence of the relatively small number of producers specializing in this product in providing higher-quality, better-controlled products.

Power ratings   Power dissipation is low for physical sizes used. For best stability, temperatures should be held as low as possible.

Voltage rating   Voltage limitations are encountered because of the construction methods used. Turn-to-turn voltages must be limited. Consequently, critical resistance values are reached where a resistor of a given power rating and size becomes dissipation-limited by voltage.

1.2.7 Specialized Resistor Types

A brief mention is made here of resistors for specialized uses. Because of the number of such uses, only a few of the total variety are covered. Table 1.9[2] provides some general information.

TABLE 1.9   Types and Uses of Specialized Resistor Types

Variable	Name	Material	Uses
Temperature	Thermistors, N.T.C. or P.T.C.	N.T.C.—Semiconducting oxide	Temperature control Amplifier gain control Voltage regulation (N.T.C.) Current regulation (P.T.C.)
		P.T.C.—Barium Titanate based ceramic	Fluid flow control Control and alarm sets
Strain	Strain gauges	Metal wire or foil	Low pressure range sensing requirements
Voltage	Varistors, (See Figure 1.16 voltage dependent resistors (VDRs)	Metal oxide Silicon carbide	Transient protection Rectification of asymmetric pulses
Pressure	Transducers Microphones	Carbon, ceramic etc.	Numerous applications
Humidity	—	Carbon, metal films Thick film-metal oxide	Humidity sensors

High-frequency types   For RF devices and stripline assemblies, resistor shapes and designs are fabricated to provide broad useful high-frequency traits and ease of circuit attachment. Thin-film resistive elements perform well at high frequencies with tantalum nitride films providing superior environmental performance to nickel-chromium performance. Waveguide terminations, RF attenuators, power dividers, tee pads, and coaxial terminations all make use of flat, shaped cards, rod elements, coaxial disk elements, and other specialized shapes having deposited films. These can be thick- or thin-film deposited on mica, glass-fiber resin combinations, ceramic, glass, or other materials. Devices de-

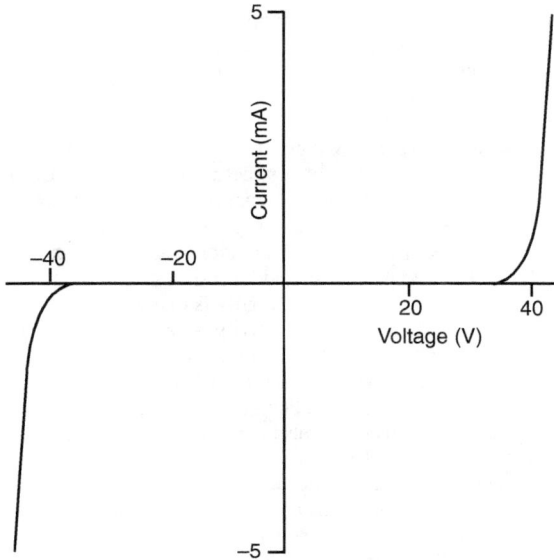

FIGURE 1.16 Typical *I* vs. *V* for a varistor.

signed using these basic elements must allow for the differences in thermal expansion rates, with respect to interfacing structures, usually by some type of spring pressure connection arrangement. Failure to consider this aspect of design will result in problems of cracked connection joints and damaged resistor elements. Various passivations are used. Thin quartz coatings have been available in the past, and high-quality silicone coatings are used. Coaxial disks allow fabrication of film resistors having very good characteristics well into the microwave range. Stripline components are also produced, in long, flat strips that can be cut and fitted as well as in pillbox types that fit in stripline sandwich constructions, between conductive pads. Chips are also produced for hybrid microcircuits. These are mostly metal thin films or fired cermet screenings that have metallized pads for soldering or wire bonding by semiconductor bonding techniques.

Meter multipliers and shunts   In addition to the precision wirewound resistors that find much use as meter multipliers and shunts, special resistor designs are manufactured for high-voltage meter multipliers. Also for high-current measurements, precision resistors of very low value are also available. MIL-R-29, listed in the specification discussion, covers large ferrule terminal metal-film resistors of value from 0.5 to 20 MΩ. These resistors are sealed in glass tubes and are said to provide long-term stability. Their power rating is that which is dissipated when a 1.0-mA current is induced through the element. Extra-long-bodied wirewound parts are also supplied for meter multipliers. They are wound with very-fine high-resistivity wire, and this, with the body length (to 10 inch on a nominal 1.5-inch diameter body), allows values up to 7.5 MΩ for a 1-mA rating. Low-resistance high-current meter shunts are made of shunt manganin or other resistive alloys in bars or heavy strips, mounted on a base having large, efficient screw-clamp terminals mounted directly onto the strip material. The strip element is calibrated by small grinding or filing adjustments to its cross section. Many sizes, shapes, and configurations of these high-current shunts are produced, many specially adapted to the mounting used in their application. Shunt manganium is a manganin alloy that has its minor constituents and metallurgical conditioning adjusted to move the characteristic parabolic temperature coefficient curve so that its low inflection point occurs at the operating temperature to be expected when the element carries heavy currents.

Power ceramic-composition types   A type of resistor that is not widely produced, but which has unique capabilities, consists of a structure in the form of a thick-walled tube, which is composed of a relatively strong homogeneous resistive material, and usually has ferrule-type terminals on the tube ends. The tube ends are metallized areas on the composition structure and serve as both electrical termination and mechanical amounting. As furnished by Cesiwid, two basic materials are said to be carbon-ceramic and silicon-ceramic compositions. A ceramic and carbon or silicon putty is extruded under pressure, and the resulting body is fired to sinter the mixture. Resistivity is varied by adjusting by materials and process controls. The result is a hard, rather durable structure. The ceramic carbons are said to be better for high-power pulse operation, and the silicon ceramics are superior for steady power application. Catalog listings show silicon ceramic units from 22.5 W at 1.0 to 150 Ω up to 1000 W at 1.0 to 500 Ω, with comparable sizes of carbon-ceramic resistors from 15 W, at 7.5 to 1000 Ω over a range up to 150 W at 50 to 10,000 Ω. Purchase tolerance tolerances are 5, 10, and 20 percent. Full power is rated at 40°C, derated to zero at 230 and 350°C, respectively, for carbon and silicon materials. The voltage coefficient of resistance is no greater than 1.0 percent per V per inch of resistive length for both, and temperature coefficient is said to be 750 ppm/°C maximum. These resistors have good high-frequency characteristics up to 50 MHz, and excellent survivability under high-power short pulses. They are widely used as RF loads, antenna termination resistors, and in high-power radar modulators. Epoxy coatings are applied to the carbon types if it is necessary to resist transformer oils.

High-resistance types   For high-voltage bleeders and dividers, and for other high-resistance uses, a class of resistors called "high-megohm" types is produced. These parts are almost all film types of various designs. The film types are made long in body and use either finely helixed thin-metal films, high-resistivity tin-oxide films, oxide-doped metal films, or high-resistivity metal frit-fired enamels. The latter have durability advantages, being relatively inert, as also is tin oxide. Metal and pyrolitic-carbon films should be carefully protected, and because of the high-resistance value, all the parts are vulnerable to surface leakage if moisture and contaminations are allowed to accumulate. Organic coatings must be sufficiently impervious to avoid volume resistivity decreases as a result of moisture absorption, at least during short-term exposures to condensing water. Temperature coefficient of resistance is large for those types which gain value by increasing film resistivities to values that are beyond the ranges needed to maintain optimum temperature stability. Some of the recent developments of thick films appear to be quite good in temperature stability and environmental endurance for this use. Various glass-cased designs are well developed and are available for this use.

Positive-temperature-coefficient resistors and ballasts   Ballast resistors are wound with wires having high positive temperature coefficients of resistance. They are used as voltage regulators and for related sensing functions. A resistor of this type in series with a power source to a load increases its resistance if the voltage increases and causes increased current flow and higher operating temperature, with the increase in resistance having a regulatory effect on the load voltage.

    Thermistors   Thermistors or temperature-variable resistors (Table 1.10[2]) are either NTC (negative temperature coefficient) constructed of semiconductor oxides or PTC (positive temperature coefficient) constructed of ceramics based on barium titanate. The thermal power dissipated in a thermistor is governed by the following equation:

$$W = IV = K \, (T - T_a) \qquad (1.10^2)$$

TABLE 1.10 Uses of Thermistors

Applications	Notes
NTC	
Temperature measurement. Temperature control.	Measured effect due to change of T_a
Liquid and gas flow measurement. Pressure measurement. Liquid level sensing.	Measured effect due to change of K
Voltage regulators	Measured effect due to existence of V_{max}
PTC	
Temperature measurement.	Only usable over very limited range—very sensitive. Usually used in alarm circuits.
Temperature control.	Measured effect due to change of T_a
Liquid level sensing.	Measured effect due to change in K

where W is the watts dissipated, I is the current, V is the voltage, T is the temperature of the thermistor, T_a is the ambient temperature, and K is a constant known as the thermal conductance or dissipation constant in mW/°C. For $T - T_a < T_a$, K is independent of temperature to the first order. For a transient condition the equation becomes

$$W = C \frac{dT}{dt} + K (T - T_a) \tag{1.11^2}$$

where C is known as the thermal capacity of the thermistor. This equation can be written in the form:

$$\frac{dT}{dt} = \frac{W}{C} - \frac{(T - T_a)}{\tau} \tag{1.12^2}$$

where tau is the thermal time constant.

The variation of resistance with temperature is governed by the equation:

$$R = R_O \exp B/T \tag{1.13^2}$$

where B is the material constant and is negative for PTC thermistors and positive for NTC thermistors. Typical values for B for NTC thermistors ranges between 1.5×10^3 to 6×10^3 K. Manufacturers such as Keystone, Ketema, Raychem, and Dale Electronics are good sources of additional information.

1.2.7 Special Application Guidelines

Fixed resistor trimming Trimming adjustment of resistor values is often desired in situations where subsequent retrimming is not desirable, or where environmental circumstances make the use of variable resistors undesirable. This can be accomplished as a production operation using selected values of fixed resistors. The incremental adjustment can be made arbitrarily small by using simple techniques. The total resistance of two parallel-connected resistors R1 and R2 is given by:

$$RT = \frac{R_1 R_2}{R_1 + R_2} \tag{1.14^1}$$

By fixing R1 at a predetermined "coarse" value and selecting R2 to adjust the combination, the desired trimming is accomplished. Figure 1-17[1] shows R_t as a function of the ratios of R_1 and R_2. R_2 is noted as a "star" value, its schematic value being conventionally denoted by an asterisk referring to a footnote. First, install R1, then clip in a precision-switching-resistor decade standard for R2, switch until the desired circuit condition is obtained, read the switch positions, and select the nearest standard marked value for installation as R2. The procedure can be made relatively insensitive to moderate in-service variability of R2, by judicious selection of ratios. Also, R1 can be selected as a precision type and R2 can be made less critical owing to the relative insensitivity to its variation. Radio-frequency circuits will, of course, require considerations of lead lengths, and resistor decade reactances for the bench adjustment.

The kit of values from which to select R2 can be minimized by circuit analysis to determine the most-probable needed values and calculating from an appropriate distribution curve. If the relative probability of needing a value step can be estimated, a Monte Carlo technique can be used to reduce the needed inventory. The range of adjustment can be increased and the span of needed stock values for R2 selection can be decreased by making a substitution selection for R1 as well. A number of pre-packaged networks using incremental adjustment to value are available from various suppliers. These are usually networks prepared on substrates by thick- or thin-film depositions, with terminal junctions brought up for connections and adjustment by terminal interconnection patterns to be accomplished after the package has been installed in its circuit location.

Short-pulse rating estimates   In circuit applications, it is frequently necessary to determine the proper wattage size for resistors for uses involving pulses of short duration. It is evident that resistors should be able to withstand higher power than rated continuous power if the exposure is sufficiently brief. Film and composition resistors do not lend themselves to simple analysis in this regard, although data published for pellet-type composition resistors by one manufacturer are shown earlier in the paragraphs on carbon-composition resistors. Films become an intimate part of the thermal mass of their substrate and overcoatings, but the current-carrying film is so thin that all energy from a short pulse (less than 100 milliseconds) can be considered to be concentrated in the very thin film, which limits the very short pulse overload performance of film resistors. Thick-film types should be a little better than thin films, but the core, coating materials, and design probably are as significant variables as film thickness. Because of the lack of empirical data and the difficulty of analysis, pulse overload operation of film resistors is not covered here.

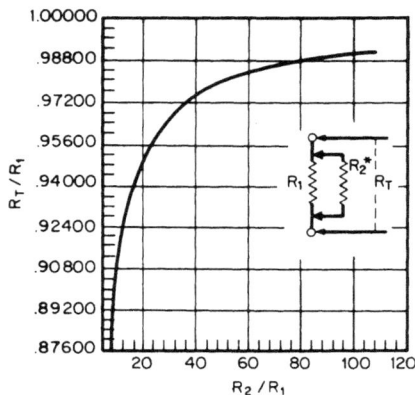

FIGURE 1.17   Resistor trimming.

Wirewound resistor construction is comparatively easy to treat, and most designs are capable of safe operation under large overloads for short pulses. Basically, the ratings of voltage and power must be addressed, and if after the calculations are made, the resulting safety margins are small, then methods should be explored for obtaining empirical data. One suggested method is to test a number of sample parts at stepped-pulse power levels until a significant change in resistance value or catastrophic failure occurs; then use the observed levels of significant effect to plot a distribution and calculate a standard deviation for an assumed normal distribution. If the 3 sigma point of damaging power-level distribution is outside the applied level, a case can be made for conservative adequacy.

For isolated pulses 100 ms or less in duration, it can be assumed that the pulse energy is absorbed completely by the resistance wire and that the wire temperature is raised to a level that is calculable if the mass and thermal capacity of the wire are known, the initial temperature is considered, and the total pulse energy is determined. A resistor can be disassembled to determine the diameter and length of the wire, or the manufacturer can be queried for the information. Heat capacities of most resistance alloys are given in Table 1.8[1]. If the exact alloy is not known, the heat capacities are sufficient to allow a good estimate as to whether the design application is marginal. First obtain the mass of the wire:

$$m = \pi \left(\frac{D}{2} \right)^2 Lp(2.54)^3 \times (10^{-2})^2 \qquad (1.15^1)$$

Where:

m	=	mass, g
D	=	diameter, mils
L	=	length, in
ρ	=	density, g cm^{-3}
2.54	=	conversion factor, inches to centimeters

Then compute pulse power:

$$P = \frac{V^2}{R} t \qquad (1.16^1)$$

Where:

P	=	pulse energy, J (or W–s)
V	=	pulse voltage, average
R	=	resistance value, Ω
t	=	pulse-time duration, g

The wire temperature can now be computed as a function of the initial wire temperature plus the incremental heat generated:

$$T = T_0 + \frac{P}{\theta m} \qquad (1.17^1)$$

where:
T and T_0 are total and initial temperatures, respectively,
I is the heat capacity of the wire in joules per gram $-°C$,
m and P are as defined above.

Combining these expressions using the defined quantities, the temperature of the wire is:

$$T = T_0 + \frac{4V^2 l \times 10^6}{\pi \rho \theta RD^2 L(2.54)^3} \qquad (1.18^1)$$

If the wire diameter is given in circular mils, it is proportional to cross-sectional area, being the diameter in mils squared; and it is necessary only to substitute D in circular mils for D^2 in the previous expression. When the temperature is obtained, it should not exceed 275°C for conservative uses that require resistance stability or a 350°C absolute maximum. T_0 must be based on the maximum ambient temperature for the resistor in its application. Note that heat capacity in joules per gram –°C is not equivalent to the specific heat, which is like specific gravity, is used to measure the material property against that of pure water at room temperature. Specific heat is given in calories per gram –°C or Btu per pound –°F. Both are the same number and can be converted to watt-seconds (or joules) per gram - C by the use of a constant multiplier of 4.186 J-cal^{-1}. Specific gravity (more properly, density), when expressed in grams per cubic centimeter, provides convenient units for this purpose; the SI units for density are kilograms per cubic meter.

If inductance-resistance or resistance-capacitance discharge pulses are involved and meet the 100-ms criterion, the pulse energy is obtained by assuming that all the energy stored in the capacitor or inductor is dissipated. This is:

$$E = \tfrac{1}{2}CV^2 \quad \text{for capacitors} \qquad\qquad (1.19^1)$$
$$E = \tfrac{1}{2}LI^2 \quad \text{for inductors}$$

Where:

E = energy, J
C = capacitance, F
V = voltage on the capacitor, V
L = inductance, H
I = inductor current, A

For a sinusoid, the average voltage of one-half cycle is 0.637 times the peak. Other pulse shapes are treated similarly, with the object being to obtain the total pulse energy. If trains of pulses are applied, it is necessary to determine the average hot-spot temperature for use as the initial temperature in Equation 1-17[1]. This can be done empirically using 30 AWG thermocouples on the center of the resistor body, or a worst-case estimate can be made by using the temperature-derating curves specified for the part in question. To determine worst-case temperature rise per watt, use the slope of the power-derating curve and express the result in degrees per watt. Using the calculated average power for the pulse train, multiply by this temperature rise per watt, then add to the expected maximum ambient environmental temperature to obtain the value of T_0 in Equations 1-17[1] and 1-18[1]. Use the pulse-power calculation technique for a single pulse, and add to T_0 to arrive at the total estimated temperature of the resistance wire. Temperature rise estimation using the power-derating curve is not dependable for situations where safety margins are slightly exceeded because the results are often unnecessarily pessimistic. In such cases, measurements should be taken, or suppliers should be consulted for information on their particular product.

Voltage limitation for short pulses is not established, although it is known that good coated wirewound designs will withstand several times their steady-state voltage rating for millisecond-range pulses, with higher power ratings being somewhat better in this respect than smaller units. One manufacturer recommends a design figure of the square root of 10 times rated voltage as a maximum for 4 W and larger resistors if pulses are 100 ms to 5 s and the square root of 5 times the rated voltage for the same pulse-width ranges on units smaller than 4 W. For shorter pulses, it is stated that work as shown that 20-ms pulses of 20,000 volts per inch of resistor can be applied if pulse-energy temperature-rise guidelines are observed. Obviously, resistor designs having large-resistance wire thermal mass are better for short-pulse applications from the standpoint of power capability.

For pulses longer than 100 ms, the momentary overload rating of the resistor can be utilized to access the design margin. Military and EIA specifications for resistors have momentary overload tests of 5 to 10 times the rated power for 5- or 3-s duration, depending

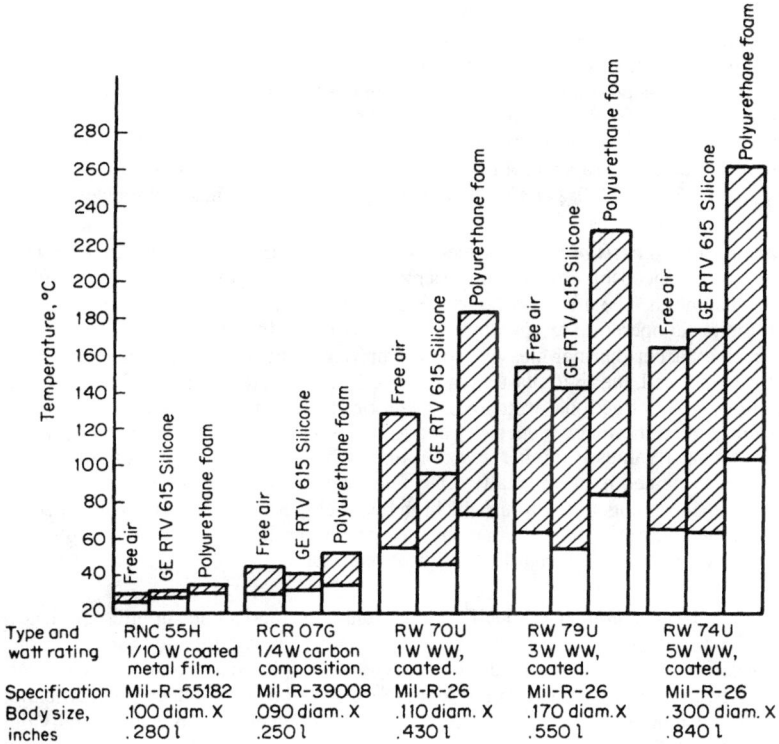

FIGURE 1.18 Experimental data showing the effect of encapsulants on resistor body temperatures.

on the resistor type and size. Resistance value shifts of varying amounts are allowed in the test. Allowed changes for wirewound resistors are no more than 2 percent or ±0.05 Ω; some premium performance styles are only 0.2 percent. If changes of about 1 percent or more are observed in iterations of the momentary overload test, the reliability of the part under those conditions is questionable. The specification test is an intentional stress on a sample of parts; thus, a derating factor of at least half the watt-second maximum test value is advised as a practice. For pulse widths less than 3, 5, or 10 s, whichever test is specified, the power overrating can be proportionally higher. A 1-s pulse would therefore be allowed to have an average power rate of five times that for a 5-s pulse. The longer pulses allow the resistor thermal system of core, leads, coating, etc., to begin functioning. From 100 ms to 1 s, the 1-s proportion should be used. For pulses longer than 5 seconds on parts of 20 W and higher, the intermittent service ratings for NEMA and UL requirements that are tabulated by some suppliers can be used as a guideline. The large parts possess longer thermal time lags. Short-pulse ratings on large parts, however, can be developed as outlined previously. Additional insulation to ground might be needed on large power resistors in pulse applications.

Specialized resistors are available for high-power pulse applications. Varieties that are, by experience, superior in ability to survive high-energy pulses, while also maintaining good high-frequency characteristics, are types of ceramic-composition units made by

Cesiwid. These parts, described in the specialized-resistor section, probably represent the best overall compromise for high-power short pulses. Both carbon-ceramic and silicon-ceramic types are furnished. The carbon-ceramic type is rated higher for pulse energy, but lower for steady-state dissipation than the silicon-ceramic type for comparable physical sizes. The 15- to 150-W and 40°C steady-power rated range in the carbon-ceramic type, for instance, is given a corresponding single-pulse energy range of 2500 to 50,000 J and a standard peak voltage range of 4 to 60 kV over the same range of sizes. Resistance values range from 7.5 to 10,000Ω at ±5-, ±10-, and ±20-percent purchase tolerances.

Encapsulation of resistors   Many electronic packages are designed to embed fixed resistors in various encapsulating systems. As might be expected, temperature rise is affected by the encapsulation. The effect varies, depending on the thermal conductivity of the embedment, the basic operating temperature of the resistor, and the location of the resistor, with respect to adjacent thermal masses. Some empirical data is presented in Figure 1.18[1]. As could be predicted, the rigid plastic foam caused temperatures to rise in all watt ratings, but only slightly so in fractional watt sizes. The room-temperature-vulcanizing silicone rubber has a fairly good thermal conductivity and lowered operating temperatures in low-watt ratings, but raised values in higher-power ratings. Each situation is an individual case and should be evaluated experimentally. One further consideration that must be made is the ability of encapsulants to withstand the high body temperatures of resistors—especially the small wirewound units, which often surprise packaging designers by their high operating temperatures and ability to char circuit board materials.

Resistor service variability   Specifications, standards, and some supplier application literature have taken responsible recognition of resistor in service-value changes. In the past, only occasional efforts were made to characterize service performance beyond the purchase inspection by supplier and user. With few exceptions allowances in specifications do not distinguish between possible test result situations that could show all the individuals in a test sample to be barely within defined change limits, and results from the same test on a different set of samples that could show all individuals tightly distributed about the initial value. Both situations are equivalent in the scheme of attributes testing. It would, however, be unlikely for the first situation to occur unless some pre-selection process were used either deliberately or inadvertently. Both military and EIA specifications require resistors to be subject to a series of tests, and the limits of success must be defined by the specifications for each test. When circuits are designed to use resistors, allowances must obviously be made for the purchase tolerance and temperature-resistance coefficient. Less apparent, however, are the variational allowances that must be made for lead pull and twist, soldering, load life, humidity, temperature cycling, and other expected exposures. The circuit designer must recognize that any one resistor can have performance that falls anywhere within the defined limits and still be within specifications. Fortunately, however, most conditions of use are not as severe as the tests, nor is it usual for particular resistor to vary the full allowed amount in the same direction for each test in a series.

An example using tolerance analysis is shown in Figure 1.19[1], as applied to general-purpose film resistor specification MIL-R-39017.

For MIL-R-39017, purchase tolerance (2 or 5 percent) and temperature coefficient of resistance (0 ±200 ppm/°C) are needed to reach the total service variability estimate. If the application is benign and avoids stresses of the types covered by the tests (high potential, for instance, was omitted in the example), it might be desired to omit the variation in the analysis, but shelf drift is not included, which is often significant. Also, in the case, life testing is limited to 2000 hours at rated load, and the 10,000-hour continuation allows an additional ±2-percent change. Note that purchase tolerance distributions might not be normal, particularly if close tolerance parts have been screened from the population.

Test	% allowed T_j	$(T_j)^2$
Power conditioning	0.5	0.25
Temperature cycling	0.25	0.0625
Low-temp. storage	0.25	0.0625
Low-temp. operation	0.25	0.0625
Short-time overload	0.5	0.25
Terminal strength	0.25	0.0625
Resistance to soldering	0.25	0.0625
Moisture resistance	1.0	1.0
Shock and vibration	0.5	0.25
Life, 2000 h	2.0	4.0
High-temperature exposure*	2.0	4.0
TOTALS	$T_j = 7.75\%$	$(T_j)^2 = 10.0625\ (\%)^2$
Total tolerance		$\sqrt{10.0625\ (\%)^2} = \pm 3.16\%$

* 2000 h at 150°C, not powered. Probably a good accelerated shelf-aging test.

FIGURE 1.19   Design tolerance determination of MIL-R-39017 resistors.

Tolerance analysis is a useful technique and should be used as a realistic evaluation of expected resistor performance. Military and EIA specifications provide test stress limits for most part types available domestically and should be used to evaluate service performance instead of the unrealistic practice of expecting resistors to maintain purchase tolerances in service life.

Power ratings   Industry standards, military specifications, and supplier data furnish power rating information on resistor designs. Except for the highest power types, these ratings are based upon a specified change in resistance value with time at rated power dissipation. It is generally established that drift rates for resistors are increased with increased temperature, following the expectations for increased chemical and physical activity with temperature. Power rating, temperature rise, designs, practical compromises between temperature maximums and drift rates have been established to define power ratings. A maximum ambient temperature of 85°C will include almost all severe environment applications, including most under hood automotive levels and military equipment. Some of the most severe military and space environments or exceptionally severe industrial applications require a 125°C maximum temperature, but few are higher.

Almost all resistors classed as power types are rated for full-labeled power dissipation only up to ambient temperatures of 25°C or sometimes 40°C. At higher ambient temperatures, the dissipated power must be reduced. Some types also must be mounted on chassis of specified sizes in order to realize even the full 25°C rating. General-purpose types, by contrast, are rated for full power up to some reasonably high temperature, 70 to 125°C, then derated linearly to zero at maximum-temperature resistor specifications, as recommended practice, not all test routines specified will determine the ability of a given resistor design to remain stable under high-temperature conditions. Only the more recent military specifications include tests that measure the resistance drift at temperatures approaching the zero power-rated level. Most of the other specifications require life tests at full-rated power and room temperature, and the hot-spot temperature reached by the test articles might be much less than the recommended maximum application temperatures. Therefore, be careful when using resistors at temperatures covered by the sloped portion of the derating curves. Specifications that require high-temperature exposure tests make significant resistance-change allowances for the test, and it can be expected that similar drift rates will be experienced by those designs not so tested. High-power resistors present special-application rating situations. It is frequently necessary to mount several units together in a bank, and power ratings must be reduced to allow for mutual heating. Forced-air cooling is sometimes used, and allows still air-power ratings to be exceeded without danger. At high altitudes, air is less dense and convection cooling is less efficient. All these variables have complex effects, and the exact results for a given installation are only grossly predictable, barring a complete and detailed thermal modeling. General rules are published by suppliers, however, based on experience and empirical data.

Fusible resistors   Resistors are furnished by several suppliers for applications in which it is desired that the unit open circuit if overload conditions should occur. In general, resistors for this purpose are not catalogued as standard items because of the varied application circumstances that must be considered. Failure under moderate overloads is a trait usually to be avoided in resistor design, and deliberately designing for such a function involves a complete knowledge and consideration of the application situation. If the expected overload condition is precipitous and substantial, the problem is simplified, provided that a resistance value can be used that is high enough to allow small-diameter resistance wire to be used in a wirewound unit. Millisecond-range fusing is said to be available if the overload gradient is high enough. Reference to the tables on resistance alloys show that these alloys melt at temperatures in the 1000 to 1500°C range, and good design temperatures for stable performance do not usually exceed 350°C. A reasonably well-cased wirewound unit will therefore withstand several times the rated current for long periods without open circuiting. If an underrated unit is used, the continual high-temperature exposure is likely to lower reliability in the normal function. Fifteen or 20 times the rated power (or even more) should be providing for fusing, if possible. Even then, it might be necessary to allow several seconds for action. The most common design furnished for fusible application is the small-diameter glass-fiber-core continuous-wound type of resistor. It is furnished as a molded general-purpose wirewound type by several suppliers and also is often cemented into ceramic cases and tubes to increase thermal mass and, thereby, power ratings. A high-current impulse into these units creates a high thermal drop to the case surface for the transient condition, but under long-term conditions, the case cement thermal mass and surface area dissipate well.

Under extremely high currents, complex effects occur. If current is high enough, wire can even detonate. Conditions are seldom this severe, however, but one of the concerns must be that of providing a mechanical package that will contain the fusing action and the resulting molten metal debris. Estimates for short-pulse fusing can be made using the calculations described previously for pulse applications if the melting temperatures of the alloys are substituted for the safe-temperature limits. Fusible designs other than those mentioned here are available. Film resistors have a better potential for this use than do wirewound units, and several film-type suppliers have characterized their products for fusing functions.

1.3 THICK-FILM NETWORKS

1.3.1 Introduction

Basic technology   Thick-film networks are resistor and conductor materials comprised of precious metals in a glass-binding system that has been screened onto a ceramic substrate and fired at high temperatures. Originally developed in the late 1950s and early 1960s, thick films established a broad range of applications and exhibited a substantial growth in demand. This was made possible by exploiting the inherent capabilities of the thick-film technology, which includes high power, precision, high performance, packaging flexibility, and low cost. Thick-film networks can provide miniaturization at costs comparable with those of general-purpose discrete resistors and performance comparable with that of semi-precision types. They are inherently reliable because they typically exhibit predictable changes, have a rugged construction, and are usually not subject to catastrophic failure.

Substrates are used to provide a base or carrier for the thick-film network. Their selection is based on a number of important properties, including low electrical conductivity (insulator), high thermal conductivity (heat conductor), mechanical strength, ability to withstanding high firing temperatures, and the absence of free alkalies (chemically inert). Because ceramics are well suited for most of these requirements, substrate materials (such as alumina, steatite, barium titanite, and beryllia) are most often used. A completed network is usually encapsulated by means of molding, potting, or coating to provide moisture protection, a marking surface, a modular package, and cosmetic appearance.

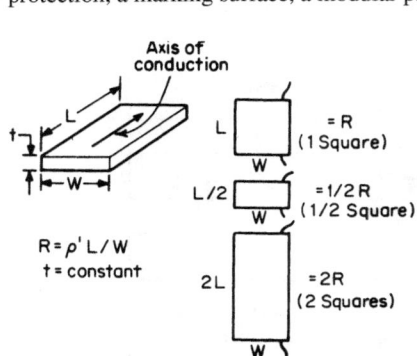

FIGURE 1.20 Aspect ratio for thick-film resistors.

The ohmic value of thick-film resistors is a function of sheet resistivity. Resistor formulations are available with resistivities from as low as 1 Ω per square to 1 MΩ per square. The desired nominal resistor values are obtained by varying resistor geometry by way of the screen printing process. Resistor values after firing can be tailored to precise values or tolerances by means of the abrading or trimming process, which increases the ohmic value by removing some of the resistor material. The trimming operation changes the length-to-width ratio, which results in an effective number of squares (Figures 1.20[1] and 1.21[1]).

A typical thick-film network manufacturing process therefore comprises screening, firing, and packaging steps, such as those outlined in Figure 1.22[1]. Such a process can be assembled for low cost to fabricate prototype networks or it can be developed into a highly sophisticated mechanized line for high-volume production.

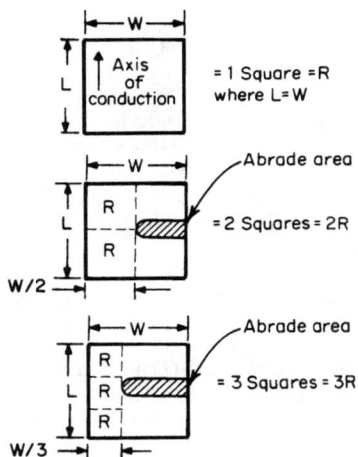

FIGURE 1.21 Thick-film resistor trimming geometry.

Scope of components and applications Thick-film technology can be used with a wide variety of applications, including discrete resistors, resistor chips, hybrid substrates, and RC networks, as well as resistor networks, which is the principle area covered here.

Resistor networks can be used in any resistor circuit application in which the thick-film technology offers some advantage in cost, performance, or size over alternative approaches. This, therefore, includes such common resistor functions as line termination, bias, bleeder, voltage divider, current sense, and load. Manufacturer's literature and application guides, such as that in *Bourns Soulutino Guide*, provide useful information. In most cases, economics dictate that as many resistors as possible be integrated into a network within the constraints of process yields, resistivity ranges, and package size. As microprocessor speeds have increased, so

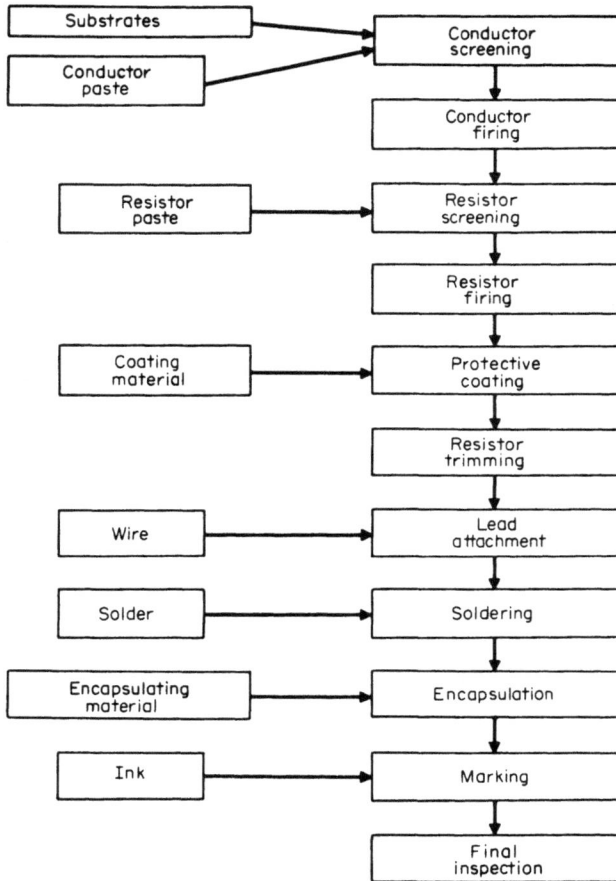

FIGURE 1.22 Typical thick-film process flow.

has the need for line termination to reduce undesired high-frequency effects (see Table 1.11[4]). Production devices have been made that incorporate from 1 to 50 resistor elements, but 5 to 15 is a typical range. The capabilities of thick-film networks, with respect to resistance values, tolerances, power ratings, and performance characteristics, are a function of the specific materials and package designs used, but capabilities indicative of the products generally available to industry are covered in the following sections.

General terms and definitions Following is a summary of several terms (in alphabetical order) with brief definitions that are commonly used in association with thick-film networks.

• *Aspect ratio* The geometrical relationship of the length and width (L/W) of a rectangular resistor element that is used in the layout of networks to establish "as-fired" resistance values (e.g., when $L = W$, the aspect ratio is 1:1, or 1 "square," and the resistor value is equal to the sheet resistivity).

- *DIP*   Dual-in-line package.
- *SIP*   Single-in-line package.
- *LCC*   Leadless chip carrier.
- *Formulations*   A specific mix of thick-film material, where the conductor, glass, and other additive ingredients are formulated to provide certain properties, such as sheet resistivity or TCR.
- *Paste/ink*   Screenable thick-film material comprising metals, oxides, and glasses in an organic vehicle, which, when fired, produces a circuit element, such as a resistor or conductor.
- *Power density*   The power dissipation per unit area of a resistor or substrate (in watts per square inch) used to determine the optimum layout design of a network.
- *Screen*   The process of printing a network pattern of thick-film ink or paste onto a substrate by means of a squeegee applied to a photoetched wire mesh "silk screen" or metal mask.
- *Sheet resistivity*   The nominal resistance per unit area of a thick-film ink or paste, which is usually expressed in Ω per square (assuming a constant thickness), where the design resistance value of the screened is determined by $R = pL/W$.
- *Substrate*   The base or carrier for the thick-film network, which is usually a ceramic plate.
- *Thick film/cermet/metal glaze*   Resistor and conductor materials comprising metals or metal oxides in a glass binding system, which can be screened onto a substrate and fired to provide circuit elements or networks.
- *Tracking*   The inherent capability of resistors that have been made from the same formulation and screened onto the same substrate to exhibit similar performance characteristics (e.g., drift and TCR).
- *Trim/abrade/adjust*   The process of tailoring a thick-film resistor element to specific value or tolerance by removing resistor material (via sandblasting or laser abrading), which increases the ohmic value.
- *Voltage gradient/field strength*   The linear voltage stress applied across a resistor element (in volts per inch), used to determine the optimum geometry of a high-value resistor.

Design considerations

Application requirements   Thick-film technology is extremely versatile and can be adapted to many applications if its characteristics are understood and its inherent capabilities

TABLE 1.11   Line Termination Demand in Microprocessor Applications

	Microprocessors				Microcontrollers	
	68000	68020	80286	80386	68HC11	8051
No. address lines	23	32	24	32		
Termination	0	32	0	32		
No. data lines	16	32	16	32		
Termination	0	32	0	32		
No. control lines	21	27	13	15		
Termination	0	27	0	15		
Total I/O lines	60	91	53	79	38	32
Termination	0	91	0	79	0	0

exploited. If a successful thick-film product is to be developed, the component design must be optimized to match the requirements of its intended application. Some of the principal application requirements that should be identified to permit effective component design include:

1. *Parametric requirements*   Resistor values, ratios, network configuration, and power ratings.
2. *Packaging constraints*   Size, shape, assembly, and power handling requirements.
3. *Performance requirements*   Initial tolerances, short-term stability (e.g., TCR, handling, and overload), drift, tracking, useful life, and failure rate.
4. *Cost constraints vs. alternatives (e.g., discrete resistors and ICs)*   Circuit cost, quantities, and packaging costs (e.g., handling assembly and space).
5. *Application environment*   Ambient temperature, humidity, and airflow.
6. *Process compatibility*   Handling, assembly, materials, and temperatures.
7. *Functional application characteristics*   Frequency, noise, duty cycle, and overload.

Packaging factors   Once the application requirements are determined, a number of basic design factors must be addressed to establish a component package compatible with the needs identified, including:

1. *Substrate material*   The principal factors are thermal conductivity, thick-film compatibility, cost, mechanical strength, insulating properties, and physical configuration. The most common substrates are ceramics, such as alumina, beryllia, steatite, and barium titanite.
2. *Resistor material*   The principal factors are resistivity, TCR, drift characteristics, and cost. Some of the most common systems are based on formulations of PdO, $RuO2$, $IrO2$, or AuO.
3. *Conductor and termination materials*   The principal factors are conductivity, solderability, resistor compatibility, and cost. Some of the most common conductors are based on formulations of Ag, Au, Pt-Au, Pd-Ag, and Pd-Au.
4. *Encapsulating materials and/or package*   The principal factors are environmental protection, physical configuration, mechanical integrity, process compatibility, cost, and cosmetic appearance. Some of the common packaging approaches are dipping, molding, potting, and roller coating.
5. *Interconnections or terminals*   The principal factors are the selection of materials (e.g., copper, nickel, and zirconium copper), design (e.g., lead frame, pins, and wires), and interconnection technique (e.g., staking, wire bonding, and soldering).
6. *Joining materials and process*   The principal factors are process and material compatibility, mechanical strength, cost, and reliability. Solders such as 60/40 or 10/90 Sn/Pb are by far the most common approach.
7. *Element and network layout geometry*   This is obviously influenced by resistor values, power dissipation, network configuration, substrate size, and the resistor materials available.

Typical component packages   Because such a wide variety of package styles, network configurations, material systems, and applications are feasible and available for thick-film resistor networks, it is obviously not possible to present and cover all the possible combinations. Three basic styles of packages that have been used widely for networks, however, can be used as a standard reference and, in fact, can satisfy most applications requirements. The most commonly used are the SIP (single-in-line package) and DIP (dual-in-line package) shown in Figures 1.23 (Vishay) and 1.24 (Vishay). Common schematics for these packages are shown in Figure 1.25 (Vishay). Other package styles include flatpacks and various surface-mount configurations (Figure 1.26 (Xicor Corporation)).

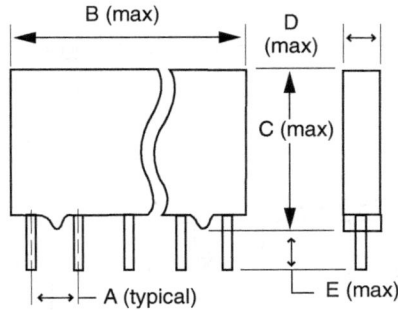

A	No. of leads	B	C	D	E
0.100	4	0.384	0.350	0.090	0.095
	6	0.584	0.350	0.090	0.095
	8	0.784	0.350	0.090	0.095
	10	0.984	0.350	0.090	0.095
0.125	2	0.234	0.350	0.110	0.095
	4	0.484	0.350	0.110	0.095
	6	0.734	0.350	0.110	0.095
	8	0.984	0.350	0.110	0.095
0.150	4	0.600	0.350	0.132	1.130
	6	0.900	0.350	0.132	0.130

FIGURE 1.23 Single-in-line package.

Package-variant characteristics The performance characteristics of thick-film components are obviously a function of such things as the application, package design, and material systems used. Therefore, generalizations about the performance of thick-film resistors can be made only in terms of their typical or relative stability because the specific characteristics unique to each package must be determined. These include the short-term permanent changes in resistance caused by handling and processing, and the effects of environment, as well as the resistor operating temperature under load. Some typical short-term performance characteristics as they relate to common specification tests (such as MIL-STD-202) are presented in Figure 1.27[1]. In general, thick-film resistors exhibit stability under such conditions. This stability is typically superior to carbon-composition and film-type resistors and comparable with that of general-purpose wirewound and metal-film devices.

Because operating temperature is typically the major influence on resistor degradation, the temperature rise characteristics of resistor networks under load conditions must be understood. Operating temperature characteristics are obviously a function of package construction and materials and therefore can be related only to specific designs. Using the three standard network package styles described previously as examples (i.e., SIP, DIP, and square), Figure 1.28[1] illustrates the average temperature rise for some standard package sizes and uniform network geometries in a natural convection environment. Moving air ambient will provide a reduction in operating temperature, as illustrated in Figure 1.29[1]. These figures, because of their general "ideal design" nature, can be considered indicative of maxi-

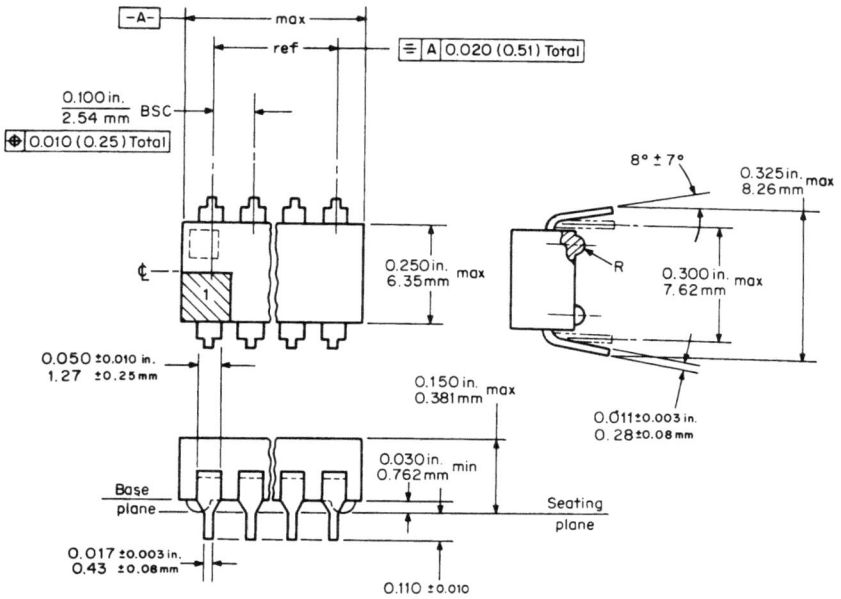

	A dimension	
No. of pins	inch	millimeters
8	0.450	11.4
14	0.750	19.1
16	0.850	21.6
18	0.945	24.0

FIGURE 1.24   Dual-in-line (DIP) package.

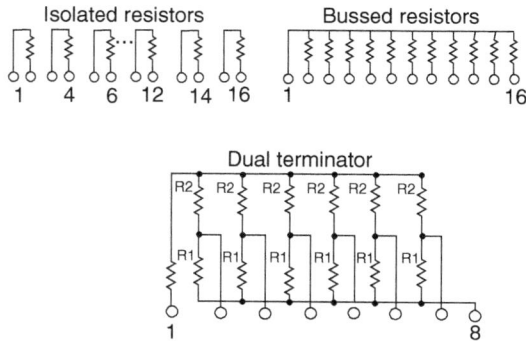

FIGURE 1.25   Typical resistors network schematics.

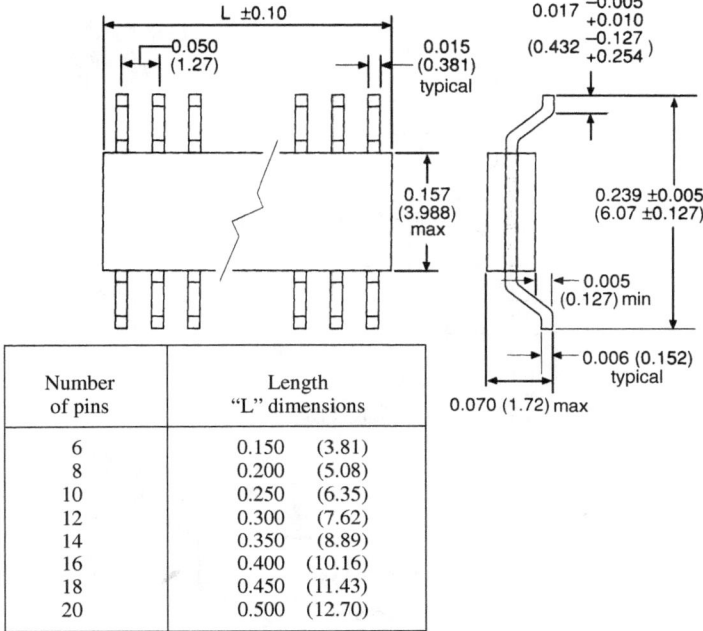

L ±0.10

0.050
(1.27)

0.015
(0.381)
typical

0.157
(3.988)
max

$0.017 \begin{smallmatrix} -0.005 \\ +0.010 \end{smallmatrix}$

$(0.432 \begin{smallmatrix} -0.127 \\ +0.254 \end{smallmatrix})$

0.239 ±0.005
(6.07 ±0.127)

0.005
(0.127) min

0.006 (0.152)
typical

0.070 (1.72) max

Number of pins	Length "L" dimensions	
6	0.150	(3.81)
8	0.200	(5.08)
10	0.250	(6.35)
12	0.300	(7.62)
14	0.350	(8.89)
16	0.400	(10.16)
18	0.450	(11.43)
20	0.500	(12.70)

(a)

B

D Max.

G Typ.

C

Terminal #1
Indicator

A Typ.

Terminal #1

0.008"
(0.203)
R Typ.

E Typ.

F

H

Index
Corner

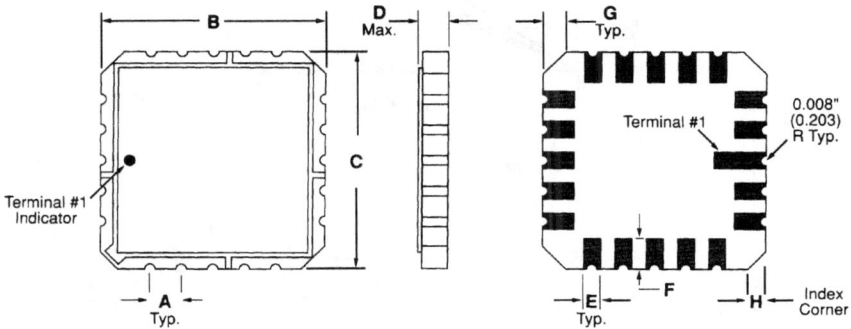

	A	B	C	D	E	F	G	H
16	0.050" (1.27)	0.300" (7.62)	0.300" (7.62)	0.077" (1.96)	0.025" (0.635)	0.050" (1.27)	0.040" (1.02)	0.020" (0.508)
20	0.050" (1.27)	0.350" (8.89)	0.350" (8.89)	0.077" (1.96)	0.025" (0.635)	0.050" (1.27)	0.040" (1.02)	0.020" (0.508)
24	0.050" (1.27)	0.400" (10.16)	0.400" (10.16)	0.077" (1.96)	0.025" (0.635)	0.050" (1.27)	0.040" (1.02)	0.020" (0.508)

(b)

FIGURE 1.26 Other network package configurations.

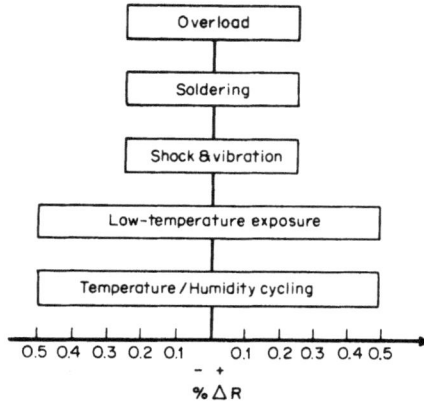

FIGURE 1.27 Typical short-term test performance of thick-film networks.

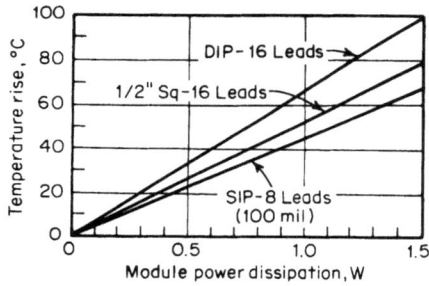

FIGURE 1.28 Temperature-rise characteristics of typical resistor networks (natural convection).

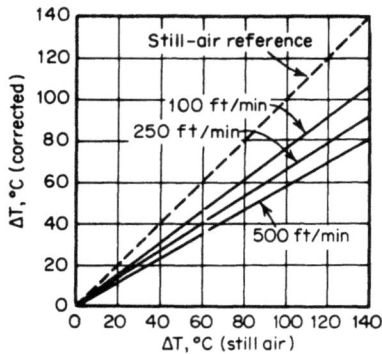

FIGURE 1.29 Temperature-rise correction for resistor networks in moving air ambients.

mum power-handling capability. A number of other variables, however, even within these standard packages, could have a significant effect on operating temperature and must therefore be compensated for, such as resistor size, resistor geometry, number of resistor elements, type of printed circuit board, and application environment (particularly with respect to neighboring heat dissipators). These factors all tend to reduce power-handling capability from the ideal data presented and are unique to each specific network design and application.

1.4 TRIMMING POTENTIOMETERS

1.4.1 Scope of Components and Applications

Trimming potentiometers, which are also known as *resistance trimmers* and *variable resistors*, are adjustable three-terminal resistor elements commonly used as voltage dividers or rheostats. Trimming potentiometers are available in wide a variety of styles, sizes, and constructions.

Five basic resistor technologies are used for potentiometers, with the wirewound and cermet (thick film) types being the most popular for a broad range of applications and the carbon-composition, carbon-film, and metal-film types used for selective markets. Potentiometers are available in both single and multi-turn adjustment styles and are typically compatible with printed circuit board mounting.

Trimmers are used in a wide variety of applications where precise resistance values or voltage levels are required or where circuit functions must be set up or adjusted.

Some typical applications for potentiometers include setting biases for transistors, adjusting time constants for RC networks, setting reference voltages for control circuits, adjusting the gain of an amplifier, and varying or limiting current in a bleeder circuit. The large number of styles, technologies, and constructions of trimming potentiometers available permit the user to select a device that optimizes cost, performance, and size for an application. The capabilities and characteristics of potentiometers in general, as well as some of the most common styles, is covered in the following sections.

As an additional reference and for specified detail relating to standard specifications, the following documents are applicable to potentiometers:

Applicable military specifications:

- *MIL-R-39015* Resistor, variable, wirewound (lead-screw actuated)
- *MIL-R-22097* General specification for resistors, variable, nonwirewound (adjustment type)

Applicable EIA standards:

- *RS-345* Resistors, variable, wirewound (lead-screw actuated)
- *RS-345* Resistors, variable, nonwirewound (lead-screw actuated)

1.4.2 Basic Technologies

Two basic types of resistor elements are primarily used in potentiometers: one is incremental (i.e., wirewound), and the other is continuous (i.e., composition or film). A third type using integrated-circuit technology has begun to appear in the literature (Figure 1.30). The resistor can be fabricated as either a linear or circular element, depending on the type of potentiometer desired. Although many package styles exist, there are only two basic approaches to trimmer construction (i.e., single or multi-turn design). Multi-turn units typi-

PIN Configuration

DIP
SOIC

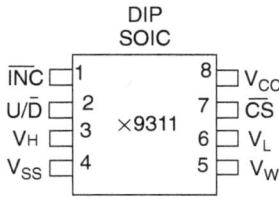

PIN names

Symbol	Description
V_H	High terminal
V_W	Wiper terminal
V_L	Low terminal
V_{SS}	Ground
V_{CC}	Supply voltage
$U/\overline{D}$	Up/down input
$\overline{INC}$	Increment input
$\overline{CS}$	Chip select input

Functional Diagram

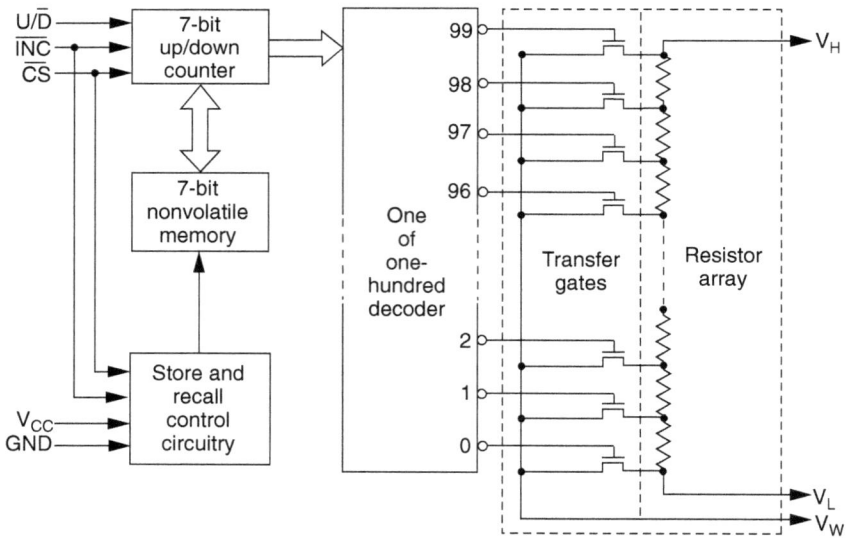

FIGURE 1.30   Digitally controlled potentiometer.

cally use lead-screw actuated sliding contacts and might have either circular or rectilinear elements. In the single-turn units, however, the wiper is usually driven directly by the screwdriver slot or shaft on a circular element. Following is a brief description of the five basic resistor technologies used for potentiometers:

Wirewound The resistor element is fabricated by winding fine-resistance wire around an insulating bobbin or core, which is circular in cross-section and either rectilinear or circular in shape. The lead terminations are typically welded to the ends of the resistor element. Wirewound potentiometers are the most common type and usually exhibit low TCR and good stability, but by nature of the noncontinuous element construction, are limited in the resolution and noise performance they can provide.

Cermet The resistor element is fabricated by screening a thick-film composition of precious metal and ceramic materials on a ceramic substrate. The element can be either rectangular or circular in shape and is typically terminated by soldering thick-film conductor pads to pins staked through the substrate. Cermet potentiometers provide a continuous element for high resolution as well as the capability for excellent performance.

Carbon composition The resistor element is comprised of carbon molded with an organic binder (similar to the discrete axial-leaded carbon-composition resistor element) as an integral part of the potentiometer baseplate subassembly. This provides a relatively inexpensive potentiometer with a continuous element, but its performance is generally the least stable of the five technologies and is intended primarily for general-purpose applications.

Carbon film The resistor element is fabricated by depositing a film of carbon (similar to the discrete axial-leaded carbon-film resistor) on a substrate or baseplate. This approach can provide an economical continuous-element potentiometer with improved performance over the composition type.

Metal film The resistor element is fabricated by depositing a thin metal film (similar to the discrete axial-leaded metal-film resistor) on a ceramic substrate with similar configurations and termination techniques to those used for cermet types. This approach can provide a precision, high-performance potentiometer, and perhaps offers the greatest potential to achieve ideal characteristics (in terms of resolution, linearity, noise, and stability), but it is generally more expensive than the conventional types.

1.4.3 General Terms and Definitions

Following are brief definitions of several general terms normally used in association with trimming potentiometers.

Total resistance The resistance measured between the two end terminals of a potentiometer.

End resistance The resistance measured between the wiper terminal and an end terminal, with the wiper element positioned at the corresponding end of its mechanical travel.

Travel The clockwise or counterclockwise rotation of the wiper along the resistor element. *Mechanical travel* is the total rotation of the wiper between end-stop positions. *Electrical travel* is the total rotation between maximum and minimum resistance values. These are not identical, owing to discontinuities at the end positions.

Resolution or settability The ability of an operator to set the potentiometer to a predetermined ohmic value, voltage, or current. This is a measure of the sensitivity or degree of accuracy to which a potentiometer can be set. Wirewound potentiometers, having noncontinuous elements, typically are referred to in terms of a theoretical resolution, which is the reciprocal of the number of turns of wire. Settability is affected by the material and the uniformity of the resistor and wiper elements, the length of the resistor element, and the design of the adjustment mechanism.

Noise   The effective contact resistance introduced in the wiper arm while rotating the wiper across the resistor element. Wirewound potentiometers are normally referred to in terms of equivalent noise resistance (ENR), which is caused primarily by variations in wiper contact during travel along the wire element. Although nonwirewound potentiometers have a continuous resistor element, there is usually a built-in dc offset caused by a measurable contact resistance between the wiper and the resistor. Their noise is therefore normally referred to in terms of contact resistance variation (CRV), which is caused primarily by the changes in contact resistance between the wiper and the resistor element during rotation.

Torque   The mechanical moment of force applied to a potentiometer shaft. *Starting torque* is the maximum torque required to initiate shaft rotation. *Stop torque* is the maximum torque that can be applied to the adjustment shaft at a mechanical end-stop position.

Setting stability   The ability of a potentiometer to maintain its initial setting during mechanical and environmental stresses, normally expressed as a percentage change in output voltage, with respect to the total applied voltage.

Rotational life   The number of cycles of rotation that can be attained at certain operating conditions while remaining within specified allowable parametric criteria. A cycle comprises the travel of the wiper along the total resistor element in both directions.

1.4.4 Potentiometer Styles

Design considerations   When selecting a potentiometer, you should attempt to match and optimize the design factors that satisfy each of the basic requirements of the application. The principal application requirements which should be considered when specifying a potentiometer are:

- *Parametric requirements*: total resistance, initial resistance tolerance, and power rating.
- *Performance requirements*: TCR, resolution, noise setting stability, drift, and rotational life.
- *Cost constraints*: cost of component, assembly, and adjustment.
- *Application conditions*: assembly process, operating environment, and circuit characteristics.

The major design factors to be considered in the selection of a suitable potentiometer are:

- *Resistor technology*: wirewound, composition, and film.
- *Adjustment style*: single-turn (and mechanical rotation), multi-turn (and number of turns).
- *Package design*: shape (round, square, rectangular), size, top or side adjust, slot size, and seal.

1.4.5 Military Designations

Trimming potentiometers that have been qualified for military applications are classified by a standard type-designation system. The type number identifies the style, size, performance characteristics, terminals, resistance, and failure rate, which are applicable to the designated potentiometer. Figure 1-31[1] presents a summary outline and explanation of some of the common designations for military-grade trimmers.

Package designs   Many potentiometer styles and sizes exist even beyond those identified in the military designation system. Only two basic kinds of packages are available, however (i.e., the single-turn and multi-turn), from which many variations have developed, with respect to size, shape, adjustment position (i.e., top- or side-adjust), lead configuration, and resistor technology. Figure 1-31a[1] illustrates several common single-turn poten-

Numerical Code Format

XXX– XX– X – X –XXX– X
Style Size Char. Term. Res. F/R

Styles
RT–Wirewound
RJ–Non-wirewound
XXR–Established reliability

Sizes
10–1" rectilinear, multi-turn, 3/4W
11–1 1/4" rectilinear, multi-turn, 1W (0.280" wide)
12–1 1/4" rectilinear, multi-turn, 3/4W (0.190" wide)
22–1/2" square, multi-turn, 3/4W
24–3/8" square, multi-turn, 1/2W
26–1/4" square, multi-turn, 1/4W
50–1/4" round, single-turn, 1/4W

Characteristics

Designation	TCR (+ppm/°C)	Max. temp. (°C) at RP†	Max. temp. (°C) at NL‡
B	600	70	125
C	250	85	150
F	100	85	150
H	50	85	150
J	10	85	150

Terminals*
L–Flexible, insulated wire lead
P–Printed circuit pin (base mount)
W–Printed circuit pin (edge mount–top adjust)
X–Printed circuit pin (edge mount–side adjust)
Y–Printed circuit pin (staggered)

Resistance

Designation	Value, Ω
101	100
201	200
102	1,000
202	2,000
103	10,000
etc.	

Failure rate (F/R)

Designation	Failure rate (%/K hrs.)
M	1.0
P	0.1
R	0.01
S	0.001

* Color code, terminal identification, and schematic diagram conventions are shown in Fig. 1-2.
† Rated power
‡ No load

FIGURE 1.31 Military-type designations for trimmers potentiometers.

tiometer styles, including top- and side-adjust types, which are most often selected when low cost and small size are required and resolution is not critical. Some of the more popular styles of multi-turn potentiometers are also illustrated in Figure 1.31b[1], including rectilinear and circular element types, which are usually used when high resolution is required.

Performance characteristics A potentiometer's performance is comprised of many elements, the importance of which should be assessed for each application (Figure 1.32[1]). The principal performance characteristics can be categorized into four groups, which are covered in the following sections: initial parameters, functional performance, package-variant characteristics, and drift.

Initial parameters Only one basic initial electrical parameter is of significance for a potentiometer, total resistance. Each resistor technology has its own inherent capabilities, with respect to resistance values, tolerances, and TCR characteristics, which are summarized in Table 1.12[1]. Total resistance is a function of the resistivity and length of the resistor element. Therefore, the wirewound and metal-film technologies, which are restricted to

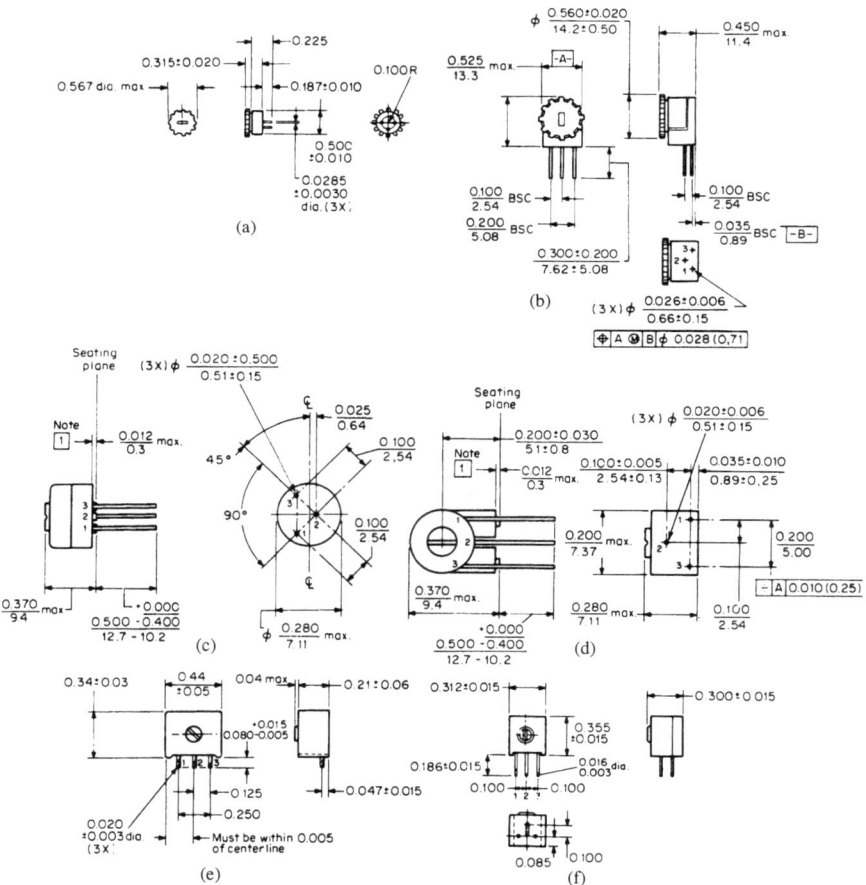

FIGURE 1.32 Typical potentiometer configurations.

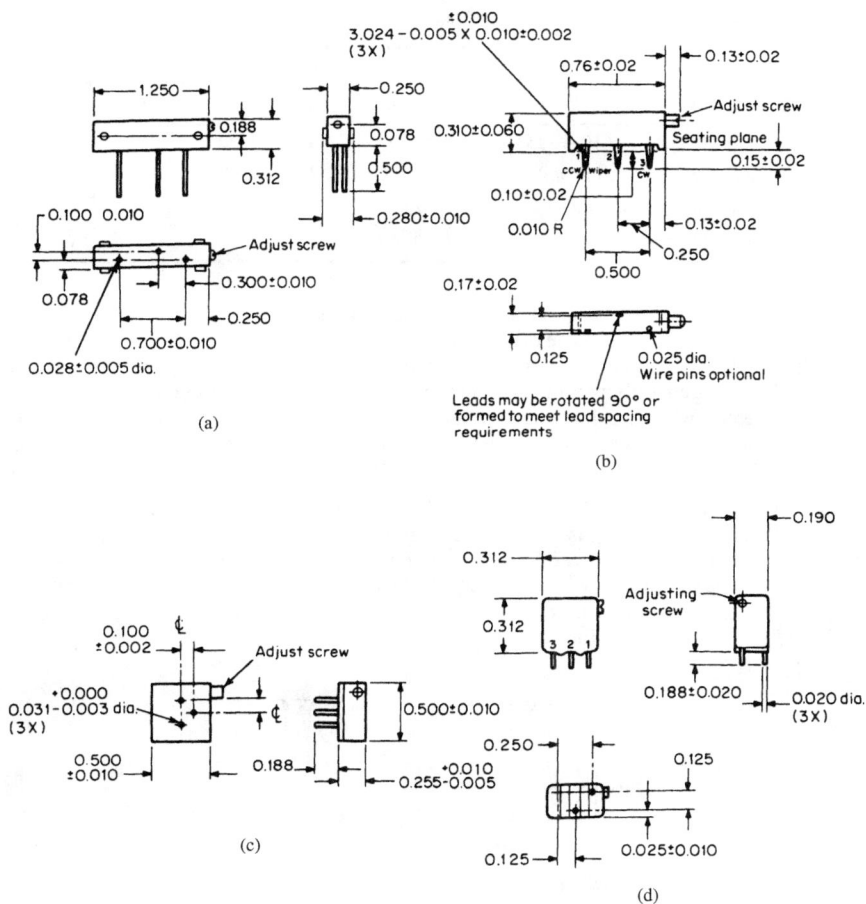

FIGURE 1.32 Typical potentiometer configurations. *(Continued)*

TABLE 1.12 Initial Parameter Capabilities of Trimming Potentiometers

Parameter	Resistor technology			
	Wire	Carbon	Metal	Cermet
Total resistance:				
Typical range	100–10 kΩ	1 kΩ–1 MΩ	100–1 kΩ	100–100 kΩ
Max range	10–100 kΩ	100–5 MΩ	10–10 kΩ	10–1 MΩ
Resistance tolerance, ±%:				
Typical	10	20	10	20
Min	5	10	1	10
TCR, ± ppm/°C:				
Typical range	50–100	500–1,000	20–50	200–500
Min	20	500	10	100

TABLE 1.13 Functional Performance Characteristics of Trimming Potentiometers

Characteristic	Resistor technology			
	Wire	Carbon	Metal	Cermet
ENR	10–100 Ω	2% or 10 Ω	2% or 10 Ω	3% or 20 Ω
End resistance	1% or 2 Ω	1% or 5 Ω	2.5% or 5 Ω	1% or 5 Ω
Setting stability, %	0.5–1.5	2	1	1
Rotational life, %	2	5	2	2

low resistivities and simple geometric constraints, provide the most limited range of resistance values. The initial tolerance capability of a potentiometer is determined primarily by the nature and economics of resistor-element fabrication process. Although the carbon and cermet types are generally the most economical to fabricate, their inherent process distribution is greater, which prohibits tight tolerances. The TCR characteristics of each technology are similar to those available with discrete resistors so that the wirewound and metal-film types can normally be expected to be the most stable.

Functional performance   A number of performance characteristics of potentiometers are functional in nature; that is, they relate directly to the adjustment of a potentiometer, and their relative importance varies according to the application. The principal functional performance characteristics are noise, end resistance, setting stability, rotational life, and resolution. Typical examples of trimmer performance are presented in Table 1.13[1]. Figures 1.33[1] and 1.34[1] are covered further:

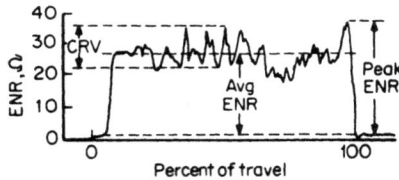

FIGURE 1.33   Typical contact resistance variation.

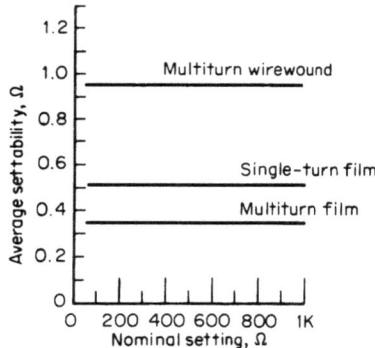

FIGURE 1.34   Potentiometer settability data.

- *Noise*   The equivalent noise resistance of a wirewound potentiometer is influenced by the wire size and wiper design. Nonwirewound potentiometers exhibit a contact resistance during rotation, which is primarily a function of the surface characteristics of the resistor element and the wiper design.
- *End resistance*   The end resistance is primarily a function of the termination technique and materials. It is generally highest in the film-type potentiometers owing to the relatively high resistance of the termination metallurgy interface.
- *Setting stability*   This is basically a function of potentiometer construction, and is influenced by the torque of the wiper and the geometry of the resistor element. In wirewound potentiometers, setting stability generally improves with the number of windings and finer wire size (i.e., the higher-resistance values are more stable).
- *Rotational life*   A common measure of rotational-life performance for trimmers is a 100-cycle test. Total resistance and noise typically decrease during early rotation, owing to the break-in of the wiper and resistor-element contact. Eventually, however, wearout and a loss of lubrication begin, which will tend to reverse this process. Trimmers are generally not subjected to many rotations. Therefore, wearout is not expected to occur.
- *Resolution*   The settability of a potentiometer is determined primarily by the design of the adjustment and the nature of the resistor element. As summarized in Table 1.14[1], the theoretical resolution of wirewound potentiometers is a function of wire size and the number of windings. Therefore, it improves dramatically for higher resistance values. Nonwirewound potentiometers have an infinite theoretical resolution because their resistor elements are continuous, but actual settability is limited by the mechanics of the adjustment technique and the contact resistance variation. Operational settability (i.e., actual setting trials) demonstrates that multi-turn potentiometers and film-type elements provide a significant improvement in resolution capabilities. It should also be noted that resolution is generally limited by a fixed resistance, which means that higher resistor values should be expected to have a better resolution capability on a percentage basis.

Package-variant characteristics   Certain characteristics of the performance of potentiometers are a function solely of the package style and construction. When these performance characteristics are critical to an application, therefore, the selection of a suitable potentiometer must include an assessment of the appropriate package design factors. The

TABLE 1.14   Theoretical Resolution of Wirewound Trimmers (Typical for ½-in-square, 20 Turn Style)

Total resistance	Max resolution, %
100	0.51
200	0.42
500	0.40
1,000	0.36
2,000	0.29
5,000	0.26
10,000	0.14
20,000	0.11

major package-variant characteristics are power-handling capability, environmental performance, solvent resistance, and short-term effects each covered briefly:

Power-handling capability   The power-handling capability of a potentiometer, as represented by its power rating, is based on an assumed level of performance (with respect to resistor drift) at an assumed maximum resistor operating temperature. Each resistor technology exhibits its own inherent drift characteristics as a function of operating temperature rise caused by power dissipation. In most cases, trimmers are used at low enough power levels that the power rating is not a significant concern. When trimmers are used as rheostats (i.e., as a two-terminal variable resistor), it is recommended that the nominal power rating be reduced by 50 percent, in addition to other deratings, because only a portion of the resistance element will be powered and the full current will be drawn through the wiper.

Environmental performance   The principal concern in relation to the environmental performance of a potentiometer is the effects of moisture during shipping, storage, and operation. This is obviously a function of the package materials and construction, as well as the resistor technology selected. However, you can expect, in general, that under standard humidity exposure test conditions (i.e., 40°C/90% RH), wirewound devices will typically exhibit the most stable performance and carbon types the least.

Solvent resistance   A potentiometer's performance, when exposed to solvents and cleaning processes, is dictated solely by the package construction. In general, the resistor and wiper elements cannot be expected to provide acceptable stability and noise characteristics if they are exposed to water, solvents, or fluxes. Only sealed potentiometers, therefore, such as those having O-rings as part of the lead screw or slot assembly, are recommended for applications that require exposure to washing or cleaning processes. Unsealed and open-construction units should, therefore, normally be considered nonwettable and should be assembled to printed-circuit boards manually.

Short-term effects   As with any other component, potentiometers are subjected to a great deal of handling prior to actually performing in the circuit application. This includes shipping, storage, and assembly, which expose the potentiometer to environmental, mechanical, and thermal stresses. A number of common tests are used to evaluate these effects, such as those included in Table 1.15[1] (e.g., temperature cycling, resistance to soldering, vibration). Wirewound and film-type units generally exhibit similar performance characteristics under these conditions, which (as would be expected) are typically superior to the carbon types.

Drift   Once a potentiometer is performing in its application, the only performance characteristic of major concern (aside from the stability of the setting) is the drift or aging of the resistor element. A small change in total resistance is generally not critical, but there should be an awareness of the expected drift characteristics, particularly under load. Potentiometer elements perform as discrete resistors, which would utilize the same tech-

TABLE 1.15 Standard Test Performance for Trimmers
(Maximum % Change in Total Resistance)

		Resistor technology		
Test	Wire	Carbon	Metal	Cermet
Humidity exposure	1	10	2	2
Temperature cycling	0.5–1.5	2	2	2
Soldering	1	2	1	1
Vibration	1	2	1	1
Load life	2	10	1	3

nology and are usually subjected to standard load-life tests to evaluate their long-term drift characteristics. As presented in Table 1.15[1], the metal-film devices are typically expected to be the most stable, followed closely by the wirewound and cermet types.

1.5 ACKNOWLEDGMENTS

The author of this chapter expresses sincere appreciation for the assistance of the editor of this book for his encouragement and to the previous authors of this chapter for their time-worthy efforts.

1.6 REFERENCES

1. E. L. Hierholzer, H. B. Drexler, and J. H. Powers, Resistors and Passive-parts Standardization in *Handbook of Components for Electronics* (Edited by C. A. Harper), McGraw-Hill, New York, pp. 7-1 to 7-102, 1977.
2. D. S. Campbell and J. A. Hayes, *Capacitive and Resistive Electronic Components*, Gordon and Breach Science Publishers S. A., Yverdon, Switzerland, pp. 153-264, 1994.
3. Ian Robertson Sinclair, *Passive Components: A User's Guide*, Butterworth-Heineman Ltd., Oxford, England, pp. 33-85, 1991.
4. Bourns Corporation, *Solutions Guide: Electronic Components RC4*, pp. 138-154, 1995.
5. Larry Bos, "Performance of Thin Film Chip Resistors," *IEEE Transactions on Components, Packaging, and Manufacturing Technology*, Part A, Vol. 17, No. 3, pp. 359-365, 1994.
6. Larry Bos, *IRC Application note LWB-10-17-95*, IRC, Inc., Corpus Christi, TX, 1995.

Catalog and application data from the following companies:
Allen Bradley Co., Milwaukee, WI
Bourns, Inc., Riverside, CA
Caddock Electronics, Riverside, CA
Cesiwid, Inc., Niagara Falls, NY
CTS Corp., Berne, IN
Dale Electronics, Columbus, NE, Subs. of Vishay Electronics
IRC, Inc., Boone, NC
IRC, Inc., Corpus Christi, TX
KDI/Triangle Electronics, East Hanover, NJ
Vishay Electronics, Malvern, PA

CHAPTER 2
CAPACITORS

H. Bennett Drexler

2.1 DEFINITIONS AND
GENERAL CHARACTERISTICS

2.1.1 Capacitance

Experimentation in the mid- and late-1700s demonstrated the ability to store a charge in capacitive devices. Experimenters were able to build stored charge by small increments until a desired amount was obtained, then to discharge the accumulated potential energy over a relatively short time period to achieve a short-term high-current result. Experiments in The Netherlands at the University of Leiden (Leyden) in 1747 used "Leyden Jars," later called *condensers*, to accumulate charge.

In mathematical derivation, quantifying ability to store charge can be expressed in physical volume units, leading, no doubt, to the term *capacity*, later modified to the present designation of *capacitance*.

Capacitance is a physical property possessed by two electrically conductive surfaces separated by a dielectric material, such as air, vacuum, or any material of suitably high resistivity. Applying an electrical potential across the conductive plates drains the charge carriers from one conductive surface and causes them to accumulate on the opposite conductive surface to a degree that depends on the time and amount of electromotive force (voltage) applied, the limitation or lack thereof for charge flow (resistance, inductance, amount of charge available), and the capacitance of the system. The physical explanation of the processes involved were poorly understood initially. In the mid-1800s, Michael Faraday devised a visualization of magnetic and electric fields using force-field lines, and did much toward the development of the concept of capacitance. Although quantum mechanics concepts have since supplanted Faraday's force field theories at the atomic level, we still use his field lines as visualization aids, and they are helpful in visualizing cause and effect at the macroscopic level. Even if field lines don't exist, the concept produces accurate descriptions of results. Faraday's work with capacitors is remembered by the unit name for capacitance, the farad.

Many fundamental physics textbooks, for instance, Halliday and Resnik[1], begin with Gauss's law and provide complete derivation of the capacitance property, capacitance units and the relationships between materials and physical dimensions of capacitors. The reader is referred to these texts for the derivations. Here, we only state some of the resulting relationship formulas for ready reference.

The unit of capacitance, the farad, is defined as the amount of capacitance that will produce or accept a current of one ampere during a voltage change of one volt per second or, removing the time element:

$$1 \text{ farad} = 1 \text{ coulomb per volt}$$

because 1 ampere is equivalent to 1 coulomb of charge per second. For practical purposes, the farad is much too large for convenience, and the microfarad and picofarad, being 10^{-6} and 10^{-12} farads, respectively, are most often used. Also, the nanofarad (10^{-9} farads) is frequently used. To provide a measure of the size of the farad, Halliday and Resnik[1] demonstrate that to have one farad of capacitance, two square parallel plates separated by 1 millimeter of air must be more than six miles on edge.

From the above definition:

$$C = \frac{q}{V} \tag{2.1}$$

where C is capacitance in farads, q is charge in coulombs, and V is potential (electromotive force) in volts. The derived formula for the capacitance of two parallel plates is:

$$C = \frac{\epsilon A}{d} \tag{2.2}$$

where capacitance, C, is in farads, A is the overlapping plate area in square meters, d is separating dielectric thickness in meters, and ϵ is the permittivity of the separating dielectric in coulombs squared per newton-meter squared. The permittivity of isotropic, linear, homogeneous space can be deduced as a proportionality quantity in using Coulomb's law relating mechanical force, charge, and charge separation distance, but is also accurately derived as one of the triumphant results in using Maxwell's third equation to describe electromagnetic wave propagation. The Maxwell result is:

$$c^2 = \frac{1}{\mu_0 \epsilon_0} \tag{2.3}$$

Note that the c is lower preferred case in this expression, and it represents the speed of light in a vacuum, 2.9979×10^8 m/s $\approx 3 \times 10^8$ m/s. Units of measure at this point are in meters, seconds, amperes, coulombs, and newtons, c is in meters per second (and is capable of very accurate measurement), μ_0 is the magnetic permeability of isotropic free space measured in a defined proportionality term, webers per ampere•meter, and ϵ_0 is in coulombs squared per newton•meter squared. The units of expression come from defining the Maxwell results in terms of the derivations of static magnetic and electric field effects using Gauss's law. Rewriting the preceding equation:

$$\epsilon_0 = \frac{1}{\mu_0 c^2} \tag{2.4}$$

where μ_0 has the value:

$$\mu_0 = \frac{4\pi}{10^7} \; webers/amp \bullet m \tag{2.4.1}$$

and:

$$\epsilon_0 = \frac{10^7}{4\pi c^2} \ coulomb^2/nt \cdot m^2 \tag{2.4.2}$$

Using this quantity for ϵ_0, the formula for this vacuum capacitor becomes:

$$(meters) \ C \approx \frac{A}{36\pi d} \times 10^{-9} \ farads \tag{2.4.3}$$

where d and A are measured in meters. Combining constants, this becomes:

$$(meters) \ C \approx \frac{8.854A}{d} \ picofarads \tag{2.4.4}$$

with A and d again in meters. If the dielectric is other than free space with a permittivity of ϵ_0, the permittivity of the separating material must be substituted for ϵ_0 in the expression to calculate the capacitance. A measure of the permittivities of materials other than space, referenced to ϵ_0 by a ratio independent of units, allows practical length measurement units other than meters to be used in calculations by just calculating and re-defining the units of ϵ_0 to provide a numerical value consistent with the selected units of measure. Convention thus defines a relative dielectric constant as the ratio of the permittivity of the material, ϵ_d to that of free space,

$$k = \frac{\epsilon_d}{\epsilon_0}. \ Re\text{-}writing: \epsilon_d = k\epsilon_0 \tag{2.5}$$

with k being a dimensionless ratio number. We must admit that the relative dielectric constant definition is not quite so simple as this. Actual dielectrics do not return exactly all the energy used to energize them, and material permittivities must contain a quadrature term to define this trait of dielectric loss, usually represented by the imaginary part of a complex number representation of the permittivity. We follow convention here in disregarding the loss term and using only the real part of the permittivity, an omission that is almost never significant at frequencies of excitation below the megahertz domain. The relative dielectric constant would be more accurately titled *the relative dielectric coefficient* because with most materials, it is not quite constant, varying with frequency, temperature, and field intensity. Interestingly, the relative dielectric constant of a given material also turns out to be its electromagnetic wave refractive index. We hereinafter compound the slight mistitling of the property as a constant by dropping the "relative" and follow common usage by calling k just the *dielectric constant*. To see how the definition as a relative dielectric permittivity ratio tracks units of measure, we combine the constants in ϵ_0, writing the speed of light and the permeability of space in the length units by which we wish to define plate area and dielectric thickness. If we remain with meters and assume two parallel plates with overlapping area, A, in square meters, separated d meters by a dielectric of material having dielectric constant, k, the capacitance formula becomes:

$$(meters) \ C \approx \frac{8.854kA}{d} \ picofarads \tag{2.6}$$

transforming all length dimensions in the definition of ϵ_0 (including the value of μ_0), the formula becomes:

$$(inches) \ C \approx \frac{0.2249kA}{d} \ picofarads \tag{2.7}$$

This formula is valid for capacitors having very large plate areas relative to separation distance. But for parallel plates, where d is greater than about $10^{-2} A$, electric field nonuniformity caused by "fringing" or edge effect, shown graphically by Figure 2.1, will cause capacitance to be slightly higher than that predicted by the formula. For such capacitors having circular plates, add a factor of $0.11d$ to the radius of the plates when calculating A, and for those small-area capacitors with straight edges, $0.44d$ should be added to the dimension of each side.

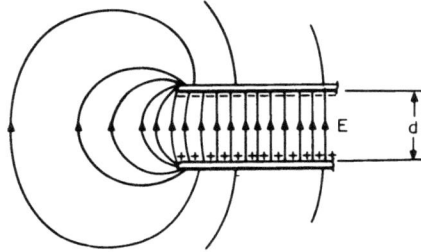

FIGURE 2.1 Electrostatic field line depiction, showing edge fringing.

Several other useful formulas are presented here for ready reference:

Energy Electrical energy stored in a capacitor is contained within the dielectric and, when capacitance is in farads and voltage is in volts, is given in joules as:

$$E = \frac{CV^2}{2} \tag{2.8}$$

Power Because 1 watt is 1 joule per second, power (or the average rate of energy transfer) in watts is:

$$P_{av} = \frac{CV^2}{2t}. \tag{2.9}$$

when time, t, is in seconds.

Ac energy transfer
For average rate of energy transfer in a sinusoidal ac circuit, this becomes:

$$P_{av} = 2\pi f\, CV^2 \tag{2.9.1}$$

where V is rms voltage, f is in hertz, and power is in watts.

Mechanical analogy The analogous equivalency of mechanical and electrical systems has long been an item of interest. If the differential equation for response of a series circuit containing resistance, inductance, and capacitance is written:

$$V(t) = R\frac{dq}{dt} + \frac{q}{C} + L\frac{d^2q}{dt^2} \tag{2.10}$$

(a)

(b)

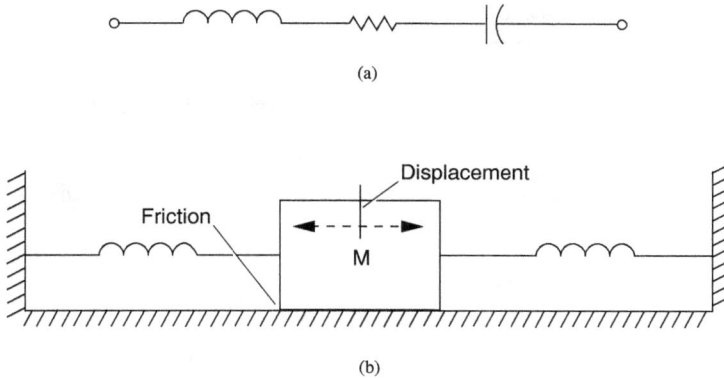

FIGURE 2.2 Models for analogy between electrical and mechanical systems: (a) 2-terminal series capacitor-inductor-resistor circuit (b) series double-acting spring-mass-friction system.

and compared to the equation for the response of a mechanical system (Figure 2.2), it can readily be seen that resistance, R, is analogous to friction; inductance, L, corresponds to mass; capacitance, C, is analogous to the reciprocal of mechanical spring rate; and voltage, V, corresponds to impressed force. Displacement corresponds to charge, q, and velocity corresponds to current, i. The spring constant analogy can be seen more easily, perhaps, if one uses electrical current, i as dq/dt, and rewrites the equation:

$$V(t) = iR + \frac{1}{C}\int i\,dt + L\frac{di}{dt} \qquad (2.10.1)$$

This compares to a mechanical system containing a mass on a spring with friction involved:

$$F(t) = \phi v + k\int v\,dt + m\frac{dv}{dt} \qquad (2.10.2)$$

where v is velocity (dx/dt), ϕ is friction coefficient, k is spring constant, m is mass, and F is force.

If a steady-state sine function $V(t) = V_0 \sin 2\pi ft$ is substituted into the electrical equation and the resulting differential equation is manipulated, a process too lengthy for this chapter, but available in many texts (e.g., Wylie[2]) the formulas for impedance, capacitive reactance, instantaneous voltage, and inductive reactance can be derived. If a step-function voltage is used for $V(t)$, and L is assumed to be zero, solution of the equation yields definition of the familiar RC time-discharge curve. Table 2.1 presents a list of additional formulas useful in applying capacitors.

Capacitor equivalent circuit Lumped-impedance equivalent circuits for practical capacitors are shown in Figure 2.3. (The circuit shown in Figure 2.3a is that most commonly shown in the literature, although it contains some lumping itself.) For convenience, any of the equivalent circuits can be mathematically transformed into a two terminal network having an equivalent value of capacitance in series with a resistance, called, appropriately, the equivalent series resistance (or, in EIA publications, the effective series resistance), commonly abbreviated ESR. Such an equivalent is shown as Figure 2.3b. One transformation technique is demonstrated by Equation 2.12.

TABLE 2.1 Common Formulas Used in Defining and Applying Capacitors

Capacitance, C, of parallel plates of overlapping area, A, separated by dielectric of dielectric constant k and thickness, d, with dimensions in inches	(inch) $C(\mu F) = \dfrac{0.2249 \bullet 10^{-6} \, kA}{d}$
Capacitance, C, of parallel plates of overlapping area, A, separated by dielectric of dielectric constant k and thickness, d, with dimensions in centimeters	(cm) $C(\mu F) = \dfrac{0.8854 \bullet 10^{-7} \, kA}{d}$
Total capacitance, capacitors connected in parallel	$C_{total} = C_1 + C_2 + C_3 + \cdots + C_n$
Total capacitance, capacitors connected in series	$\dfrac{1}{C_{total}} = \dfrac{1}{C_1} + \dfrac{1}{C_2} + \dfrac{1}{C_3} + \cdots + \dfrac{1}{C_n}$
Total capacitance, 2 capacitors in parallel	$C_{total} = \dfrac{C_1 C_2}{C_1 \times C_2}$
Capacitive reactance in ohms, when capacitance is in μF, f in Hz	$X_C = \dfrac{10^6}{2 \pi f C}$
Capacitive-inductive circuit resonant frequency in Hz, with capacitance in μF, inductance in H	$f_r = \dfrac{10^3}{2 \pi \sqrt{LC}}$
Dissipation factor for known equivalent series resistance in ohms, C in μF, f in Hz ($R = $ ESR)	$DF = \dfrac{R}{X_C} = 2 \pi f RC \times 10^{-6}$
Figure of merit for known equivalent series resistance and X_C in ohms ($R = $ ESR)	$Q_M = \dfrac{X_C}{R} = \dfrac{1}{DF}$
Equivalent series resistance in ohms (R) for known power factor or for known power loss in W and leakage current in A. f in Hz, C in μF, ($P_L = $ power loss)	$R = \dfrac{PF}{2 \pi f C} \times 10^{-6} = \dfrac{P_L}{I^2}$
Impedance of series circuit with L, C, and R, all values in like units (e.g., ohms)	$Z = \sqrt{R^2 + (X_L - X_C)^2}$
Power factor ($R = $ ESR, $Z = $ two-terminal impedance)	$PF = \dfrac{R}{Z}$
Capacitor charge amount in coulombs, capacitance in μF, potential in V	$q = CV \times 10^{-6}$
Stored energy in joules (W-s) contained in charged capacitor, capacitance in μF, voltage in V	$E = \dfrac{CV^2 \times 10^{-6}}{2}$
Capacitor heating, total watts within the capacitor, known equivalent series resistance at frequency of interest	$W_{total} = W_{dc} + W_{ac}$ $W_{dc} = V_{dc} \times leakage\ current$ $W_{ac} = I^2_{acrms} \times ESR$
Capacitor voltage during charge through resistance R in ohms, C in μF. e is the base of natural logarithms	$V_C = V_0 \left(1 - e^{-\frac{10^6 t}{RC}}\right)$
Capacitor voltage during discharge through R in ohms, C in μF. e is the base of natural logarithms	$V_C = V_0 e^{-\frac{10^6 t}{RC}}$

TABLE 2.1 Common Formulas Used in Defining and Applying Capacitors *(Continued)*

Temperature coefficient of capacitance for linear
curve segment spanning temperatures T_1 and
T_2, °C

$$TC(PPM/° C) = \frac{C_1-C_2}{(T_1-T_2) C_1} \times 10^6$$

Time constant, t in seconds (time required for
63.7% of charge to transfer from or into
capacitance, C, in μF, through series
resistance, R in ohms)

$$t = 10^{-6} RC$$

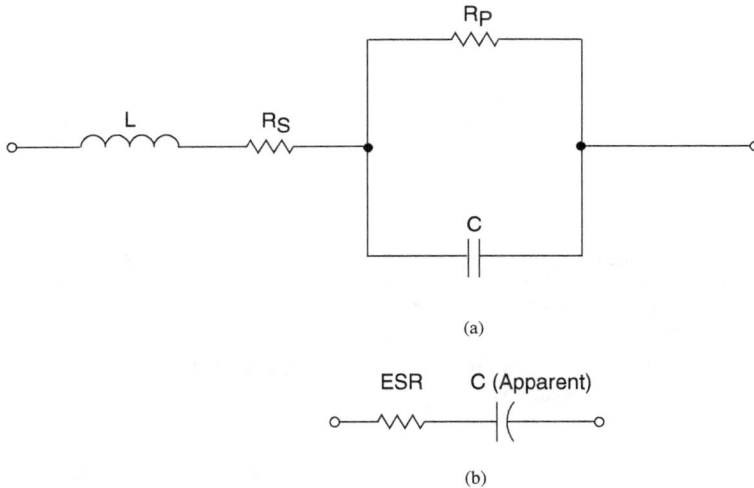

(a)

(b)

FIGURE 2.3 Simplified capacitor equivalent circuits: (a) The most common equivalent circuit showing true capacitance, C, equivalent series inductance, L, a lumped series resistance, R_S, and a lumped capacitor parallel resistance, R_P. (b) The two-terminal resolution of (a) with apparent capacitance that includes the effects of the equivalent series inductance.

To gain insight into what happens to practical capacitors as frequency, temperature, and voltage are varied, and how design options can influence these properties, it is instructive to further examine equivalent circuit representation. The equivalent series circuit of Figure 2.3b is that used to define specification values and is the simplest depiction, but the quantities of resistance and capacitance are valid only at a given set of conditions. It is evident from examining Figure 2.3a that as the frequency increases, capacitive reactance decreases, and the equivalent inductive reactance increases, reaching a frequency where the two are equivalent, defined as self-resonance, where each cancels the other to leave the equivalent series resistance as the total impedance. Exactly analytically determining that frequency and the ESR value is not a trivial endeavor, however. It is interesting that as frequency increases past self-resonance, the total impedance becomes increasingly inductive, but without the direct-current connection afforded by all coil-type inductors.

The complex impedance of the circuit shown by Figure 2.3b is represented as:

$$Z_{EQV} = R_S - jX_{CS} \qquad (2.11)$$

Using the conventional algebraic complex number representation, where R_S is the ESR and X_{CS} is the capacitive reactance of the equivalent series capacitance, sometimes represented as $1/\omega C$, equivalent to $\dfrac{1}{2\pi f C}$ where f is the frequency of interest.

The circuit of Figure 2.3a is, of course, somewhat more detailed, showing the series-inductance equivalent of leadwires, terminations, and plate conductors, with a resistance, dominated by the dielectric insulation resistance, across the capacitive element.

To discourage oversimplification, however, a slightly more detailed capacitor representation, better suited to radio frequencies, is shown by Figure 2.4.

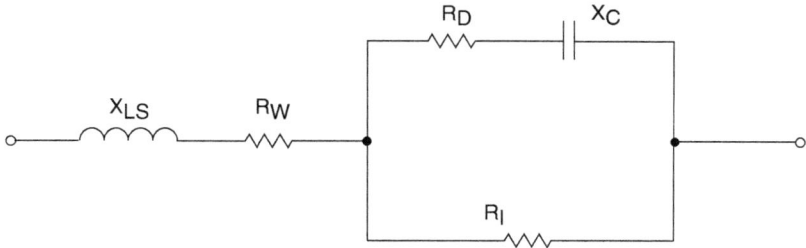

FIGURE 2.4 Capacitor equivalent circuit showing dielectric loss: X_{LS} = lumped series inductance, R_W = lumped wire, plate resistance, R_I = insulation resistance, R_D = dielectric loss equivalent resistance, and X_C = true capacitive reactance.

In this, the dielectric losses are separated from the equivalent resistances and reactances, and are shown as an effective resistance in series with the capacitive circuit element. This circuit can be represented algebraically and transformed into its two-terminal lumped series-circuit equivalent by using conventional impedance-combining techniques, then separating real and complex parts. The algebraic expression for this circuit is:

$$Z_{EQV} = R_W + jX_{LS} + \frac{R_I(R_D - jX_C)}{R_I + R_D - jX_C} \tag{2.12}$$

Where:

R_W means leadwire and terminal resistance

X_{LS} means inductive reactance of series equivalent inductance

R_I is a parallel resistance dominated by the dielectric insulation resistance

R_D means a resistance equivalent in series with the capacitive element and containing the dielectric losses and plate dc resistance

X_C means the capacitive reactance of the capacitive element

After manipulation, the foregoing becomes:

$$Z_{EQV} = R_W + \frac{R_D^2 R_I^2 + X_C^2 R_I}{(R_D + R_I)^2 + x_C^2} - j\left[\frac{X_C R_I^2}{(R_D + R_I)^2 + X_C^2} - X_{LS}\right] \tag{2.12.1}$$

Which does not simplify much more. Without taking this farther, we can examine what we have. The real or resistive term has a component containing the capacitive reactance, which varies with frequency. The complex portion or reactive term contains portions of the dielectric losses and the insulation resistance. So long as the denominator of the resistive term remains large, with high-insulation resistance, each term of the equivalent resistance will remain small, except when the dielectric loss term becomes large (as at higher radio frequencies). The wire and terminal resistances and inductance in this representation have direct first-order effects. If frequency is held constant and temperature is varied, a dielectric constant change causes the capacitive reactance to change. Insulation resistance changing as a function of temperature will cause apparent capacitance to change a bit as well as ESR. Dissipation factor, which is a ratio of resistance and reactance, will change as the two values vary. Holding temperature constant and varying frequency causes capacitive reactance to change in response, which has possible effects on equivalent resistance and apparent capacitance that are quite independent of any changes in the dielectric properties. A capacitor design that attains increased capacitance by adding more plate area while maintaining dielectric thickness will behave differently from one that reduces dielectric thickness to increase capacitance.

Frequency, design, and temperature effects are not simple. If a critical application is involved and one of the highly stable low-loss dielectrics cannot be used, consider making performance measurements on candidate capacitors at circuit-use conditions.

For capacitors with high-insulation resistance, at frequencies where dielectric losses are very low, conditions that prevail for most capacitors at non-RF inputs, the expression is much simplified and accurate predictions are possible. Specifying the frequency and signal levels when specifying capacitance value and ESR is highly important. The realized capacitance is essentially that determined by the previously stated relations of plate area, dielectric thickness (plate spacing), and dielectric constant, reduced by the effective series inductance caused by terminal leads and plate configuration, but perturbed by insulation resistance and dielectric losses if conditions make their effects significant. The two-terminal impedance equivalent can be conveniently measured over frequency with a suitable swept-network analyzer or impedance meter, and is the most useful representation of practical capacitors up to the frequency limits of available instruments. As frequency increases and series resonance is passed, the circuit impedance becomes inductive, but the series capacitor is still there from a dc viewpoint, its dc resistance being determined by the insulation resistance, and, so long as the ac impedance is still low enough to be useful, capacitors can be effectively used as bypass elements well into their inductive range (see Figure 2.13 and its accompanying text).

Capacitors can be connected in parallel for circuit applications that require higher capacitance values than available in a single unit, to obtain a capacitance value equal to the sum of the paralleled parts, with the voltage rating of the combination being that of the lowest-rated part. In instances, where a higher voltage rating than that of available capacitors is desired, capacitors can be connected in series, but with caution. When a high voltage is applied across the series-connected parts, a leakage current will flow, determined by the insulation resistances of the capacitors in series. The total voltage will divide across the series combination with each capacitor having an applied voltage equal to the total leakage current multiplied by its particular insulation resistance; thus capacitors with higher insulation resistance will have higher voltage applied, which can encroach on voltage-rating limits if the companion part has much lower insulation resistance. If the application circuit characteristics permit, this problem can be allayed by connecting equal high-value resistors across each capacitor, with the resistance values being as high as possible, but at least as low as one tenth of the lowest-rated capacitor insulation resistance. The series-combination capacitance value is determined by the formula shown in Table 2.1: the reciprocal of the total capacitance equals the sum of the reciprocals of capacitances of each of the series capacitors, and the combination always will have a capacitance lower than the lowest of the individual capacitances. Parallel combinations of capacitors are effectively used to broaden capacitor frequency-performance characteristics in bypass and decoupling applications. The rising impedance of high-capacitance paper, film, or electrolytic units as the applied frequency passes self-resonance, can be mitigated by shunting with a capacitor of a different type and lower capacitance, but of a value that presents a low impedance above the large capacitor self-resonance.

2.1.2 Practical Capacitor Considerations

The balance between ESR, capacitance, physical size, achievable voltage rating, frequency sensitivity of materials, temperature stability of characteristics, and cost have fostered development of a wide variety of capacitor types and designs to suit various application needs. Each capacitor type and design is often more suitable for particular classes of circuit applications than for others by virtue its specialized design refinements, materials, and packaging, although many instances exist where two or more types perform equally well. Despite a great deal of voluntary industry standardization and incentive standardization by military users, a relatively large variety of capacitors is available. An understanding of the performance-design relationship is thus necessary for proper capacitor selection and circuit application.

An ideal capacitor would be purely capacitive (without inductance and resistance), extremely small, insensitive to temperature and other environmental conditions, and would endure forever in the harshest of environments. In practice, the ideal cannot be achieved, of course, and enhancement of one or more traits to suit a particular application need usually results in compromise or trade-off of other desirable characteristics.

2.1.3 Capacitor Categories

Of the ways in which capacitors can be categorized, a major division can be made between those designed to be fixed in capacitance value and those designed to be variable. The variable capacitor definition is commonly taken to mean those whose capacitance is variable by mechanical means, but sometimes the term includes those intended to change capacitance as a function of the voltage applied. The latter are usually semiconductor devices. Specialized capacitors that are designed to change capacitance predictably as a function of temperature are included as part of the fixed capacitor domain because they are usually just fixed capacitor designs with carefully constituted dielectrics.

Fixed capacitors represent the largest category and, for this purpose, they are separated by dielectric types. Other logical separations are possible: For instance, categories by application could be made such as "power supply filter," "RF coupling," "energy storage," etc., but this would cause duplication in many cases. Comparisons that are shown between different capacitor types categorized by dielectric will provide suitable information to guide selection for various applications. The technology associated with producing and applying practical capacitors of each dielectric type is diverse and extensive, and sees constantly added knowledge of materials, techniques, and capabilities.

Information about dielectrics includes accumulated data published as technical papers, materials texts, material supplier specifications, and industry materials standards documents. Also included is a large amount of special production technique and process knowledge that is available only as released by producers in technical papers in various industry technical data exchange forums. The scope of this presentation will necessarily be limited, being directed primarily toward identification of parts available, with their application characteristics and a few aspects of design and production that might alter application decisions.

Dielectric types   An electric field can, of course, exist in a vacuum between two charged surfaces, but the introduction of a dielectric material provides an improved medium in which electrical energy can be stored. Dielectrics can be amorphous materials (such as glass and some polymers), liquids (such as oils), polycrystalline materials (such as ceramics), single crystalline materials (such as mica sheets), mixtures of crystalline and amorphous materials (like many plastic films), or even gases.

Many solid-state physics and electromagnetic-theory physics texts[4] cover the mechanisms of electrical energy storage in dielectrics. The treatments are much too lengthy to include here, but energy storage is shown to be by means of displacements and the alignment of electrical charge dipoles within the materials in response to an applied electric field. In a normal condition, electrical charges distribute themselves within a material so that a charge-balanced equilibrium is attained.

Crystal lattices, for instance, have positive atoms or ions balanced by adjacent negative atoms or ions, placed in a periodic-arrayed arrangement. The internal electrical attractive forces in this arrangement bind the material together. Exposure to an electric field alters the neutrality of the equilibrium and forces the lattice to a new equilibrium that includes the influence of the applied field. The lattice is forced into an elastic distortion that creates dipoles, which are aligned with the applied field. The displacement from equilibrium requires energy, which is released when the field is removed, and the energy transfer is manifested by charge flow in the capacitor discharge circuit. Noncrystalline dielectrics all have some form of charge distribution of ions or molecules forming dipoles or dipole moments that are positioned in an equilibrium state in the absence of an applied field. Applying a field realigns the dipoles into positions that form an equilibrium including the effect of the field applied, and the result is similar to that for crystalline dielectrics. In a charged capacitor, the energy is thus said to be stored in the dielectric. Note that poor conductors or dielectrics are characterized by structures that contain few free electrons, most being tightly bound into the material makeup. Tables 2.2 and 2.3 list some characteristics of commonly used dielectrics.

TABLE 2.2 Some Fundamental Characteristics of Commonly-Used Dielectrics

Material	Approximate dielectric constant, k	Approximate upper practical temperature limit for capacitors (°C)
Inorganic		
Vacuum	1 (by definition)	none
Air	1.0006	none
Ruby Mica	6.5–8.7	150
Flint Glass	10	150
Fused quartz	4.4	150
Class I ceramic	5–450	125
Class II ceramic	200–2400	125
Class III ceramic	2200–12,000	85
Porcelain	5–7	150
Aluminum oxide (Alumina)	2.5–2.8	150
Tantalum pentoxide	8.4	125
Titanium dioxide	28	125
Silicon nitride	80	150
Silicon dioxide	7.5	150

TABLE 2.2 Some Fundamental Characteristics of Commonly-Used Dielectrics *(Continued)*

Material	Approximate dielectric constant, k	Approximate upper practical temperature limit for capacitors (°C)
Organic		
Kraft paper	4.4	125
Mineral oil	2.23	85
Castor oil	4.7	85
Halowax	5.2	70
Epoxy, various	3.2–3.5	125, no melt
Polyisobutylene	2.2	125
Polytetrafluoroethylene (PTFE)	2.1	150, melts 315
Polyethylene terepthalate (PET)	3.0–3.2	125, melts 260
Polystyrene (PS)	2.5	70, melts 160
Polycarbonate (PC)	2.8–3.0	125, melts 230
Polysulfone (PSU)	2.8–3.2	150, melts 280
Polyvinylidene floride (PVDF)	6–10	100, melts 180
Polypropylene (PP)	2.2	90, melts 165
Polyphenylene sulfide (PPS)	3	125, melts 285
Poly-para-xlylene (Parylene)	2.7	125, no melt

Dielectric constant values are commonly at 60 Hz and 1000 Hz, with some materials being measured at higher frequencies. For polar plastics, values are generally between 3 and 7, and decrease with frequency increase. Polarization below T_g (glass transition temperature) is lower than above T_g. Nonpolar plastics dielectric constants are generally less than 3.

Dielectric breakdown strength is not shown because capabilities are influenced greatly by the conditions of test such as length of exposure to test voltage, shape of electrodes, temperature, rate of voltage increase, voltage-frequency (if alternating) purity, thickness, and processing history of the specimen. As specimen thickness is decreased, dielectric strength expressed as voltage-per-unit-thickness increases, dramatically in some cases. Increased temperature generally decreases breakdown by significant factors many of the values given in the literature are for material specimens of ¼-inch thickness. Thin-film values might be higher by half an order of magnitude. Permissible stress in a practical capacitor might be another matter altogether, because time of exposure must be much longer than the commonly specified short-term values, the materials are stressed in capacitor fabrication, and spacing, which might not be uniform, must be set to allow design margins at the weakest point.

2.1.4 Capacitor Design Considerations

Good dielectric materials should have a high dielectric constant to reduce capacitor size and weight and have a high insulation resistance to provide low charge-leakage rates and low ESR. Good dielectrics should have high-voltage breakdown per unit thickness to reduce capacitor size, and the dielectric constant and insulation resistance should be stable over the temperature range to provide end-item capacitor stability. For some capacitor applications, the dielectric should be easily chargeable to reduce dielectric losses as the frequency increases. Finally, and very important, the material must be mechanically and chemically compatible with the other capacitor materials and with the fabrication processes used, should be low cost, easily fabricated into capacitors, and should remain electrically and physically stable under years of use.

Dielectric materials   The characteristics of some common dielectric materials, shown by Tables 2.2 and 2.3, are nominal values and can be expected to vary somewhat over temperature, frequency at which measured, manner in which the specimen is processed (e.g., crystal size within the material, presence of impurities, etc.), and even geometry of the sample. For most electronic applications, the struggle to provide larger capacitance $\times$ working voltage product (CV product) in ever-smaller volumes to fit the shrinking sizes of functional electronic packages dominates capacitor development efforts. Using high k (dielectric constant) materials is an obvious approach, but reaches serious limitations when the highest k materials are relatively unstable with temperature, and with increased frequency and voltage, and might suffer from greater dielectric losses than other candidate materials. Dielectric losses are analogous to friction or viscous losses in the dielectric when aligning and relaxing dipoles, and cause the apparent series resistance to increase with frequency. These limitations are noted for various capacitor types (following). Production techniques have not yet been found to produce very small low-voltage capacitors with some dielectrics, such as mica, for instance, which has excellent stability, low losses, and high dielectric strength, but has a relatively low dielectric constant, and must use comparatively critical fabrication methods. Figure 2.5 shows approximate ranges of practical capacitors for some dielectrics.

TABLE 2.3   Dielectric Absorption Values of Some Common Dielectrics

Dielectric material	Approximate dielectric absorption (percent)
Polycarbonate (PC)	0.1–0.25
Polytetrafluoroethylene (PTFE)	0.015–0.02
Polystyrene (PS)	0.02–0.05
Polyethylene terepthalate (PETP)	0.25–0.3
Flint Glass	0.011–0.012
Mica	0.5–0.8
NPO Ceramic	0.1–0.8
General Purpose Ceramic (EIA X7R)	1.0–3.0
Unstable Ceramic (EIA Z5U)	1.0–3.5

Capacitor physical size   In addition to finding high dielectric-constant, stable, low-loss materials, another obvious size-reducing design approach is to find ways to increase plate and dielectric area in a small capacitor volume, while reducing dielectric thickness to a minimum consistent with reliable margins of voltage breakdown. As various capacitor types and designs are covered, the size-versus-performance design driver is a recurrent theme. However, for many applications where physical volume is not a primary consideration by reason of abundant application space, and where low-application frequencies make the increased inductances of massive plate and termination structures less important, capacitor designs have been made about as small as they need to be, and large sizes persist in the available types. Large physical size is an advantage where large capacitor currents occur and heat-dissipation surface is needed for the I^2R losses. Examples of these application types are motor-run phase shifters, power-factor correctors, and many power filters (Figure 2.6).

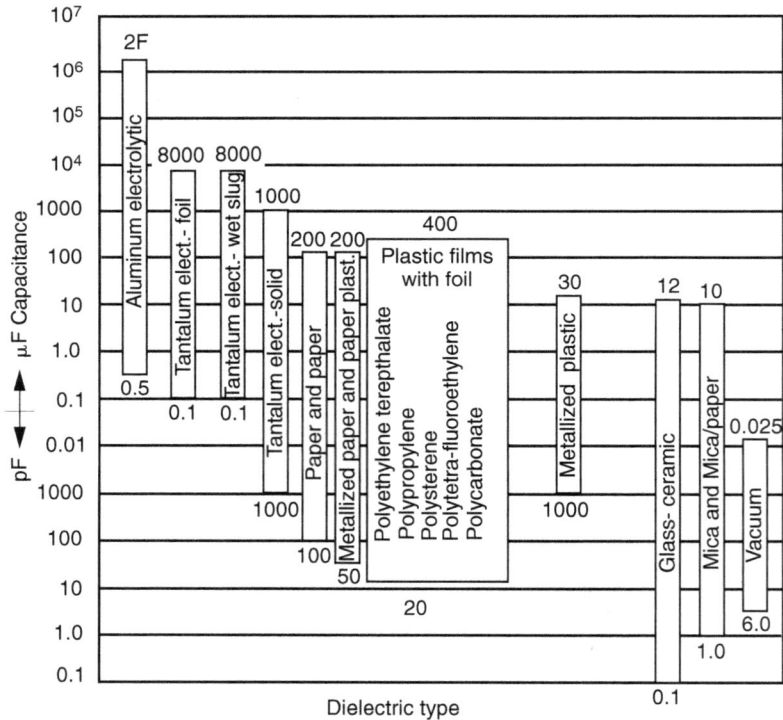

FIGURE 2.5   Approximate ranges of capacitor performance for some typical dielectrics: (a) capacitance values.

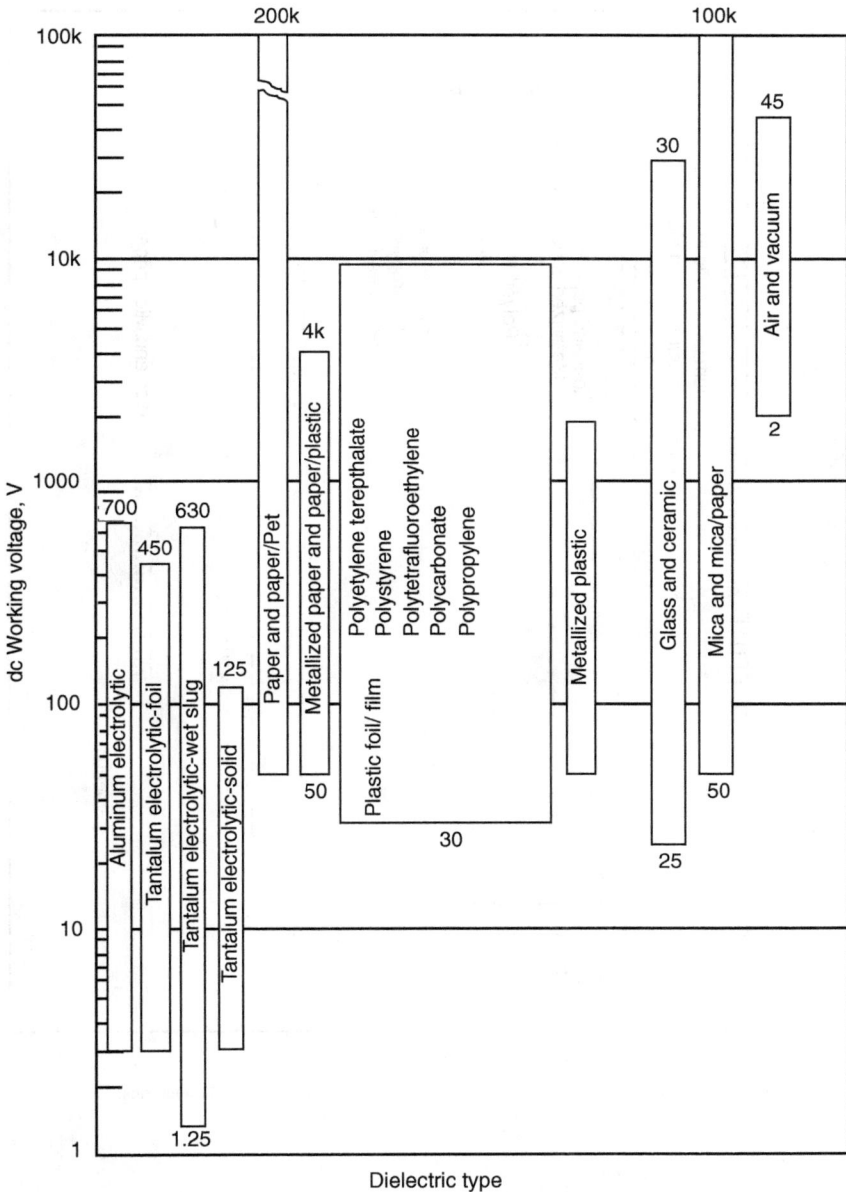

FIGURE 2.5 Approximate ranges of capacitor performance for some typical dielectrics: (b) dc working voltages. *(Continued)*

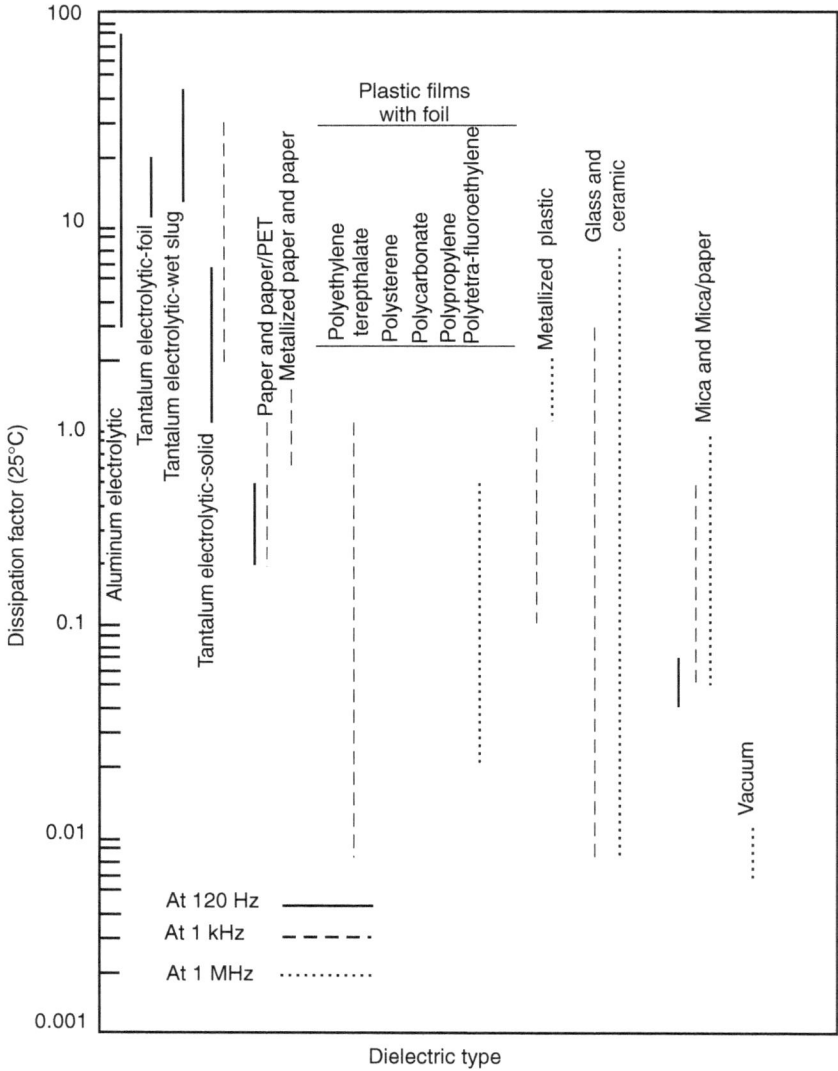

FIGURE 2.5 Approximate ranges of capacitor performance for some typical dielectrics: (c) dissipation factor. *(Continued)*

2.1.5 Capacitor Electrical Performance Traits in Specifications

Capacitor electrical-performance specifications must always include the capacitance value and the maximum voltage that can be safely applied. Also essential is some representation of the equivalent series resistance (ESR), which can be expressed by one of several measurable properties. Insulation resistance or, for some capacitor types, dc leakage current, can be an important application property if long-term energy storage is involved, but is also a primary figure-of-merit indicator for individual capacitors within a batch. Low insulation resistance can indicate internal (or external) contamination or a faulty dielectric that might lead to early in-service failure. External contamination can usually be corrected by cleaning.

Capacitance and dissipation factor In practice, capacitance and dissipation factor (DF) or power factor are commonly obtained in a single measurement, with capacitance representing the reactive component between the two measurement terminals at the frequency and voltage of measurement, and with a dissipation factor or power factor that represents the resistive component of the two-terminal series impedance (see Figure 2.4b and the sec-

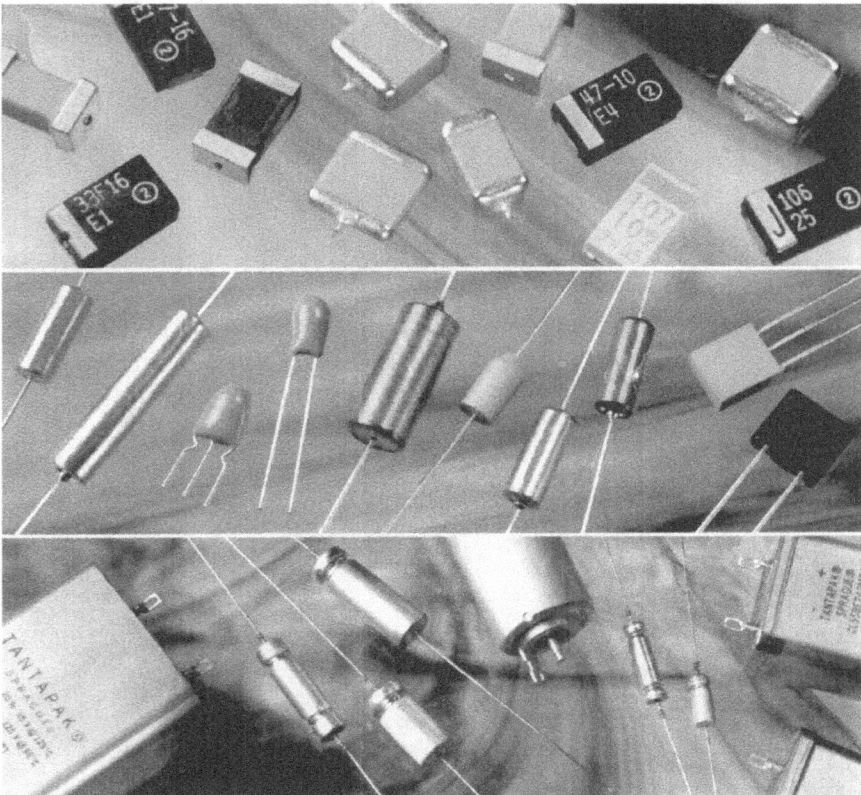

FIGURE 2.6 A variety of capacitor types and sizes. Photographs courtesy of Vishay-Sprague, a Vishay Intertechnology Company

FIGURE 2.6   A variety of capacitor types and sizes. *(Continued)* (Photograph courtesy of Vishay-Sprague, a Vishay Intertechnology Company.)

tion "Current ratings"). Both reactive and resistive components of the impedance are commonly needed to balance the measurement instrument for obtaining a reading. The ESR is easily obtainable from the DF or power factor (see Table 2.1).

Current ratings   Capacitors intended for carrying high currents at RF or power-line frequencies will have specified current ratings. For general-purpose capacitors not having explicitly specified current ratings, users might find it necessary to decide whether currents in the application are safe. This is easiest done by calculating the heat dissipation using the ESR, and looking to the manufacturer's information for determining the heat-dissipating capabilities of a given capacitor design. A prediction method is given, for instance, by MIL-STD-198 for large electrolytic capacitors using a technique based on a given heat-dissipation rate in watts per square inch of case area (see Section 2.8.3). High temperatures are enemies of capacitor reliability, and should be carefully considered for all capacitor types (for further information, see the following section). Some capacitors used primarily for filtering will have ripple voltage at a specified frequency stated as an indirect current rating, and, if so stated, this measure can be used in lieu of calculating ESR heating. Users should be especially wary of high ripple and other ac currents in high CV-product per unit volume ceramic and electrolytic capacitors, which inherently have reduced heat-dissipating area.

One other type of current rating must be mentioned—that of feedthrough capacitors designed to provide noise or high-frequency interference filtering for power circuit leads that pass through the capacitor structure. This feedthrough current rating mostly depends on the feedthrough conductor size, however, deducing the internal temperature of a power feedthrough might not be simple. Consider that if a high-value capacitor is used to bypass relatively low frequencies and the power source is ac, enough bypass current might exist to significantly raise the capacitor temperature. Low-impedance interference sources are not only more difficult to filter, but they can also produce dissipating capacitive currents for which allowance must be made.

Temperature rise estimation The following is an example of a temperature-rise estimation procedure for a military CKR06 BX-characteristic ceramic capacitor (see the section on ceramic capacitors for a description). The capacitor temperature is determined by the rate of heat generation and the thermal impedance of the capacitor body to the ambient conditions. Heat generation is determined by the real (resistive) component of the capacitor current multiplied by its ESR under the circuit conditions; in this case, the thermal impedance to ambient is overwhelmingly dominated by the capacitor mounting configuration. For this design, the external case area has little effect within several multiplying factors because thermal conductivity of the plastic case is so low, compared to that of the internal ceramic body and the attached leadwires—almost all of the dissipated heat leaves the capacitor body through the leadwires.

Empirical results for CKR05 and CKRO6 capacitors cited in the literature[3], despite their difference in case sizes, are about equal and state an impedance of about 90°C per watt for 1-inch leads into an infinite heatsink, 55°C per watt for 0.25-inch leads into an infinite heatsink, about 77°C per watt for a part flushmounted to an 0.062-inch thick glass-epoxy board with small copper circuit traces, and about 53°C per watt for the same mounting with 4 square inches of copper trace. ESR in X7R (and BX) capacitor dielectrics decreases dramatically with temperature, but increases somewhat with an applied ac voltage. Leadwire and capacitor plate conductors have positive temperature-resistance variations, as do most metals, partially compensating for the ESR drop with temperature.

A sample calculation for a 1.0-μF, 50-V capacitor with 10 Vrms at 20 kHz is shown, based on manufacturer nominal capacitor characteristics[3]:

- The capacitor is assumed to be mounted to a glass-epoxy 0.062-inches thick board with small copper circuit traces (Figure 2.7).

- Ambient temperature in this case is given as +30°C.

- Manufacturer curves show an impedance of 9.0 Ω and an ESR of 0.16 Ω at 20 kHz and 25°C.

- Delta ESR versus temperature curves show ESR at 30°C to be 0.92 times that at 25°C, at which temperature it is specified.

- Effects of dc bias below 60 percent of rated voltage are negligible.

- Manufacturer curves show that ESR goes up by a factor of $1.55x$ when ac rms voltage increases from 0 to 10 V.

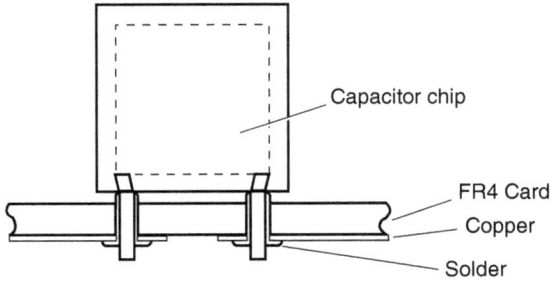

FIGURE 2.7 Model of circuit card mounting for temperature-rise example: A CKRO6 capacitor is mounted on an epoxy-glass laminate circuit card.

Using these factors, ac current, I_Z, is 10 volts rms divided by 9.0 Ω impedance, or 1.11 A. The ESR at 0 Vac and +30°C is $0.92 \times 0.16 = 0.15$ Ω. Using the 1.55 ESR multiplier for 10 Vrms yields $0.15 \times 1.55 = 0.23$ Ω. In this case, the temperature coefficient of terminations and plates is included in the manufacturer's graphic data for ESR versus temperature. Real power dissipation by this 0.23 Ω at 1.11 A rms is 1.11 squared times the 0.23, or 0.28 watts. Using 77°C per watt thermal impedance shows a temperature rise of $0.28 \times 77 = 22$°C. Adding this to the 30°C ambient predicts a surface temperature of +52°C. Actually, the internal rise will depress the ESR more than the 30°C value that is used as a calculation basis, so the estimate is conservative. Manufacturer curves are often given as nominal, rather than limit, values, and there are part-to-part variations, so conservatism is not misplaced. For these capacitors, internal temperatures should be held below +125°C in the worst case. Also, peak ac voltage plus dc bias must never exceed the rated voltage. Ceramic capacitor ESR behaves differently with temperature as the capacitance values vary, caused by the differences in ratios of leadwire and plate resistances versus dielectric losses that are its constituents, also because of differences in dielectric volume with capacitance. Similarly, the effects of the frequency vary with the capacitance value as self resonance and other factors enter into the picture. To follow this procedure, you must have manufacturer characteristic curves on hand or generate data in the laboratory.

2.1.6 Capacitance Measurement Conditions

Knowing the conditions at which specified capacitor characteristics are measured is sometimes essential to proper application. Industry standards, military specifications, and supplier information are all sources of this important information.

Frequency and voltage in capacitance measurements The user must be aware that capacitance and equivalent resistance values are all sensitive to frequency and other conditions in practical capacitors, and have limited accuracy of meaning unless at least the frequency and temperature of measurement are specified. The measured or apparent capacitance and ESR variations stem from both the changes in dielectric properties with voltage, temperature, and frequency, and from the variation caused by effects of inductance and stray capacitances. Some capacitor dielectrics have relatively high sensitivity to applied dc and signal voltages, thus the signal level during the measurement is of prime importance in those cases. Other capacitors have relatively high inductances, which have increased effects as the frequency increases. The labeled capacitance and DF or ESR are those values measured under the conditions stated in the specifications for the capacitor, not necessarily the frequency and level in the user's application.

Fortunately, capacitor dielectrics that are most often used in applications requiring precision capacitance value, such as mica, temperature-stable and temperature-compensating ceramic, and some polymer-film types are relatively insensitive to frequency and voltage over fairly broad ranges because of their dielectric properties and construction. Fairly well-standardized measurement conditions are followed by industrial and military suppliers, primarily driven by the needs to promote in-circuit interchangeability of capacitors. Standardized frequencies and voltages are selected to be high enough to provide measurement sensitivity in the capacitance range being measured, but low enough in frequency to minimize the effects of self-inductance, keeping the "true" capacitance as near as possible to the "apparent" or measured capacitance at the measurement condition.

Applied signal voltages used for capacitance and loss-parameter measurements are as low as practicable, consistent with avoiding ambiguity due to ambient noise. Standard measurement voltages are lower for those dielectric types having high-voltage sensitivity. Electronic Industries Association (EIA) standard 198 for ceramic capacitors and various military and commercial standards state recommended measurement frequencies and voltages, some of which are summarized in Table 2.6. Capacitors designed and intended for use at high radio frequencies, such as radar microstrip ceramic units and high-current transmitting types have their own detailed measurement conditions of voltage and frequency and for mounting configurations, which often involve mounting in transmission-line fixtures. EIA 483, MIL-C-55681 (Appendix A), and MIL-C-49464 cover RF test methods, and material has been published by various manufacturers of microwave capacitors[4,5,6]. Polarized capacitors, particularly some types of wet tantalum electrolytics, might require a polarizing dc voltage to prevent the signal from reversing polarity on the capacitor being measured. Most polarized units will withstand the small reversal, however, and can be measured without using a dc bias. Dissipation factor figures measured in this fashion should be used with caution.

Environmental conditions for measurements   The temperature at which capacitance and loss parameters are specified is almost universally 23 to 25°C, unless the measurements are being made for temperature-variation determination. The reference point for specifying temperature-capacitance variance is +25°C. Altitude is not important unless voltages above about 500 V are involved. Relative humidity should not exceed about 70 percent, but, if it is below 40 percent, care should be taken with handling low-capacitance, low-voltage parts to prevent damage from electrostatic buildup on tools and handling surfaces.

There are many exceptions to the general-measurement conditions guidelines, so look to manufacturer specifications for specific values and conditions.

Loss angle, power and dissipation factors   Figure 2.8 shows a conventional means of representing the vector impedance of an ac circuit containing resistance in series with reactance. The total reactance magnitude is represented by the length of a line placed at a right angle to the line representing the applied voltage and the resistive impedance component, thereby depicting the phase angle lead/lag for pure reactance. The length of the right-angle reactance line is just the properly scaled arithmetic difference between the inductive and capacitive reactances because one is 90 degrees leading, the other is 90 degrees lagging (180 degrees apart). In the example, the capacitor's inductive reactance is small at our frequency of investigation, and when subtracted, shortens the capacitive reactance line only slightly. The resistance, too, is small, relative to the

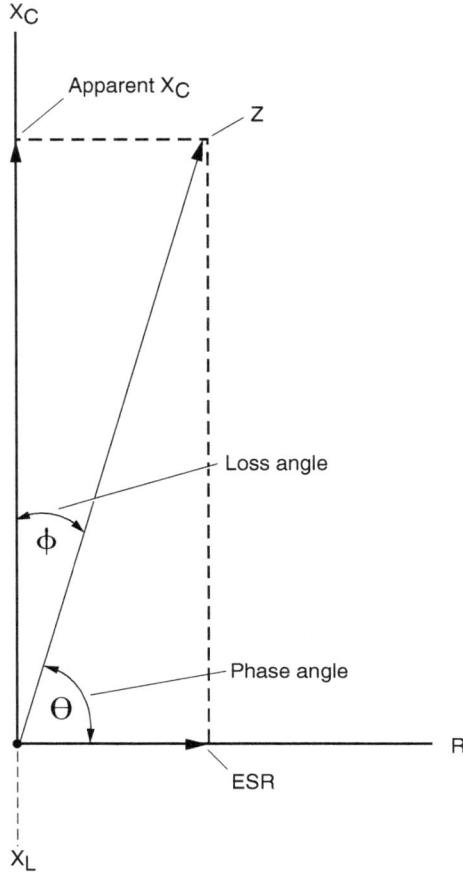

$$\text{Cos } \theta = Power\,factor = \frac{ESR}{Z}$$

$$\text{Tan } \phi = Dissipation\,factor = \frac{ESR}{X_C}$$

Both are usually multiplied by 100
and expressed as a percentage.

FIGURE 2.8   A graphic vector representation of 2-terminal equivalent capacitor circuit showing phase angles and definitions of measured quantities. Apparent X_C is the vector sum of X_L and X_C (true) at the frequency of measurement. ESR is the real component of the total two-terminal series impedance.

capacitive reactance (the line should be actually much shorter proportionately for practical capacitors than that shown). The total impedance, as shown, is represented by the vector sum of the resistive and reactive components, with the vectors represented by length and directions of the lines. Purely reactive circuit elements do not dissipate power; all circuit power dissipation is performed by the resistance. With application of Ohm's law and basic trigonometry, the total circuit power dissipation is just:

$$P_{Total} = E_Z I_Z \cos \theta \qquad (2.13)$$

The ratio of ESR to total impedance is cos θ, which is seen to be the multiplying factor for converting apparent power to real power in a series ac circuit. Cos θ is called, appropriately, the power factor, and is often multiplied by 100 and expressed as a percentage. Because the power factor provides an indirect measure of the ratio between pure capacitance and ESR in a capacitor, it is sometimes used in capacitor specifications. Used more often, however, is another term that is derived from the same vector diagram. The geometric complement of the phase angle, θ, labeled ϕ, is called the *loss angle*. (In EIA and many other standards publications, the loss angle is identified with the letter δ.) Its trigonometric tangent, seen to be the ESR divided by the capacitive reactance, is called the *dissipation factor (DF)*, and now has almost universal use in electronic capacitor specifications. It is also often multiplied by 100 and expressed as a percentage. Actually, if one compares the trigonometric function curves for the cosine near 90 degrees with that of the tangent near 0 degrees, the two are seen to be almost coincident for nearly a 20° span, and the values for power factor and dissipation factor are thus almost the same numbers for low ESR. It must be stated that dissipation factor is sometimes called the *loss factor* or *loss tangent*, and is sometimes defined by cot θ, in this diagram, seen to be equivalent to tan ϕ. See Table 2.1 for relationship formulas. One other quantity sometimes used in materials specifications is the loss factor, defined as the power factor times the dielectric constant.

Quality factor (Q) Another term used for radio-frequency capacitors to express a measure of the ESR at a frequency of interest is Q, the quality factor. This is just the ratio of the capacitive reactance to the series resistance at the measurement frequency, which is either cot ϕ or tan θ in our series equivalent circuit representation. At the risk of injecting confusion, it must be admitted that it is often more convenient to think of a high-frequency capacitor as a pure capacitor shunted by a resistance representing the resistive losses, disregarding, but not forgetting, the inductance. In this circuit, the Q representation is inverted; it becomes the ratio of parallel loss resistance to capacitive reactance in a parallel-circuit vector diagram different from that shown. It sometimes becomes easier to rationalize parallel-circuit representations if you use conductances rather than resistances and reactances, where the desired high-parallel resistance becomes a desired low resistive conductivity.

Equivalent series resistance (ESR) Electronic Industries Association (EIA) and Institute of Electrical and Electronics Engineering (IEEE) standards refer to ESR as "effective series resistance." Some specifications, usually for electrolytic capacitors, in lieu of specifying DF, state a value for ESR, as more convenient for use in assessing comparative part performance. Electrolytic capacitors have relatively high ESR values, in part because of their usually high capacitance values, thus their applications frequently make ESR a matter of primary interest and a convenient property for use.

Dielectric loss factor *Material loss factor* is defined as power factor times dielectric constant.

Insulation resistance (IR) Suitable dielectric materials, as has been stated, are classed as nonconductors or electrical insulators, having almost no free mobile electrons to accommodate charge transfer or current flow through their bulk. Their tightly bound electrons are not easily available to act as charge carriers. All materials, however, contain impurities that might have mobile charge carriers, and some materials have charge carriers less tightly bound than others. Thus, the term *nonconductor* is not absolutely accurate. Dielectric materials are *poor* electrical conductors, having very high resistances, but are not precisely "nonconductors."

A capacitor's insulation resistance is an important property, partially because its value, when measured against the known dielectric material capability in a given capacitor design, provides a "goodness" indicator for the individual capacitor. External contamination, such as fingerprints across capacitor terminals or moisture absorbed into external encasement material surfaces, can cause low insulation resistance, and should be considered, particularly if a high-voltage measurement is involved. IR for capacitors is measured using a dc voltage lower than or equal to the capacitor rated voltage. Because of small measurement currents, the low-voltage ratings of some miniature ceramic and film capacitors make measurement difficult, but the measurement should still be made. As a high-impedance voltage source is applied, the capacitor charges: a high-current spike occurs first, tailing off exponentially in response to the RC charging time. The measurement circuit must contain sufficient series resistance to limit current spikes to no more than about 50 mA for most capacitors. As the capacitor charges, the decreasing current is indicated as increasing resistance, which asymptotically approaches an end value. An "electrification" time limit should be observed in reaching the specified minimum value to avoid accepting marginal parts.

IR depends heavily on the area and thickness of the dielectric material because bulk resistance of any material depends on the cross-sectional area and the length between the points of measurement. As expected, therefore, large capacitance values have lower IR than small values, and low-voltage ratings in electrostatic capacitors have lower IR than higher-voltage ratings if the dielectric thickness is less. IR is commonly expressed in megohm-microfarads for higher capacitance value capacitor families, meaning that the expected resistance value of a given capacitor is the stated value in megohm-microfarads, divided by the number of microfarads in its capacitance. It should be noted that IR is temperature sensitive for most materials. As temperature rises, the normal electron random motion in materials is increased, molecular bonds become modified, and the IR decreases, usually by orders of magnitude. Users of capacitors in applications that are sensitive to reduced IR should be aware of this circumstance. IR is not often directly stated for electrolytic capacitors, rather, the dc leakage current at a rated voltage is specified, and is even more important as a figure-of-merit indicator.

Dielectric withstanding voltage (DWV) The limitation in using thinner dielectrics to make higher capacitance values in small volumes is reached when the dielectric becomes too thin to reliably withstand the desired voltage. Breakdown occurs when the dielectric material has its structure disrupted by the voltage stress, causing a usually localized bridging of the insulation and current flow between the capacitor plates. If, in the application, a low-impedance current source is connected across the capacitor, the result is almost invariably a catastrophic and, sometimes, spectacular capacitor failure. Some capacitor designs have built-in self-healing features, with the self-healing action depending somewhat on limiting the input current at the breakdown site. DWV is an important capacitor property from a reliability viewpoint. It is common for manufacturers to test each nonelectrolytic capacitor at 200 or 300 percent or more of rated voltage before delivery, using the overstress to eliminate marginal units and reduce early in-service failures. The voltage overstress magnitude in the practice is not overly conservative when it is considered that the test is at room ambient temperature on new capacitors that have not been subjected to years of cumulative voltage stress and thermal cycling. Small capacitors designed for voltages of over 500 volts are often tested at lower over-voltage margins for practical reasons, but with increased risk of future breakdown.

Do not depend on the insulation resistance measurement to satisfy a DWV requirement; the IR test does not impose the full test voltage on a capacitor. Many commercial high-potential test instruments use alternating current at line frequency as a high-voltage source. This is, in general, less suitable for higher-value capacitance testing because the current through the capacitor is "seen" by the instrument as leakage current. Dc testing becomes necessary in those instances, and it is very important that sufficient series resistance be used in dc testing to limit current spikes to no more than about 50 mA.

Capacitor current limitations must be observed when using sinusoidal ac. Repeated exposures to voltages above the capacitor rated voltage should be avoided because, by various mechanisms, each exposure tends to weaken a dielectric further at its already weakest points. Electrolytic capacitors cannot be subjected to over-voltage stresses of the aforesaid magnitudes, but moderate overvoltage (measured with respect to rated voltage) application, usually at elevated temperature, for 100 or more hours is used as an effective screen for infant mortality and characteristics stabilization. Electrolytic capacitor manufacturers have knowledge of dielectric-forming voltages, and are better equipped to select over-voltage screening conditions for those parts.

Another important use of capacitor high-potential testing is that of measuring the effectiveness and integrity of insulating sleeving, wraps, cases, and dipped coatings placed over capacitor bodies to prevent exposed or thinly insulated electrically conductive portions of a capacitor structure from short-circuiting to exposed adjacent electrical conductors in its installation. Metal-cased capacitors often are internally connected to the metal case directly and, for some electrolytic types not connected directly, the case might be connected to the capacitor element through an effective resistance. Good, durable case insulation, therefore, is essential. Many conformally coated capacitors, particularly in element-to-leadwire-termination areas, are only thinly protected from outside conductors that might contact the body, and might need DWV case testing. DWV tests can be conducted by closely wrapping a thin foil conductor around the capacitor body insulation, and conducting the test between the foil and both capacitor leadwires short-circuited together. The foil wrapping must be intimately applied, but kept well clear of the terminations. This test can be used to demonstrate a large design margin for coverings, which most have, and then be subsequently used as a sample test, unlike the 100-percent high-voltage testing needed for population conditioning and screening of capacitor elements.

Dielectric absorption When a capacitor is quickly discharged by connecting its plates through a low impedance or a short circuit, and the low impedance is removed after a short period, capacitors might still show residual charge. This is a dielectric-related property, said to be caused by a space charge trapped in the dielectric during the rapid discharge[7] This reluctance of the dielectric to return immediately to its uncharged equilibrium state is called *dielectric absorption*, and is a matter of vital concern in some applications.

Standardized tests are specified for those capacitors having particular susceptibility if they are types likely to be used in applications where this is critical. In the tests, the capacitors are fully charged, then abruptly discharged through a low impedance for a specified time period, then open-circuited for voltage measurement with an electrometer or other high-impedance voltmeter. A low-value resistor, specified at some number of ohms per volt is used to limit charge and discharge currents. The ratio of recovered voltage to charging voltage, usually expressed as a percent is called the *dielectric absorption*, with the quantity being valid at room ambient temperature, where the measurement is commonly made. Table 2.3 presents dielectric absorption values for some common dielectrics. Manufacturer data and parts specifications should be consulted for accurate information on specific capacitor designs.

Corona In capacitors operating above about 300 to 400 volts, the ionization of air or other gases that exist between areas or points of high potential or high charge density can become a problem. The ionized gas becomes a leakage path for the charge and, worse, the chemical products of the ionization process, such as ozone (O_3) and various oxides of nitrogen, can degrade the capacitor insulation system. Polymers are attacked and deteriorated by ozone, and the oxides of nitrogen can mix with any existing water to become acids that will attack metal plates and terminations and could embrittle dielectrics. The high-velocity electrons and ions in the discharge can bombard and erode insulation surfaces over time, particularly at localized high-temperature spots that are associated with high-current-density ionic leakages. Oils can be converted to semi-solid paraffins and thus be prevented from performing their void-filling purpose.

To discourage corona, capacitor designers and manufacturers must exclude air in areas of high potential. Oil or fluid polymer impregnation is an obviously effective measure. Some materials have a physical propensity toward maintaining a very thin layer of air at their surfaces, unless wetted by an impregnant—even if vacuum-pressure techniques are used to reduce interstitial air—and, as a result are of reduced utility at high voltage. With dry polyethylene terepthalate (the E.I. Dupont trademark is Mylar), for instance, it is very difficult, if not impossible, to eliminate corona at an elevated voltage, although the material withstands the damaging effects of the corona much better than many others. Conversely, mica or a reconstituted mica sheet, thoroughly impregnated under vacuum with a suitable resin or other material, then mechanically compressed while the impregnant polymerizes, can be made corona-free for quite high voltage. Corona formation is exacerbated by reduced air or gas pressures and by the existence of sharp points or dielectric discontinuities that cause charge concentrations, and can be addressed by careful capacitor design and manufacture for those applications where important.

Manufacturers of corona-measurement instruments, such as Hipotronics and AVO Instruments (Biddle), are good sources of information about corona-measurement techniques. The corona start and stop voltages are determined by monitoring the radio noise in the circuit being tested. Both corona start and corona stop voltages should be specified for high-voltage capacitors, with corona stop being the voltage at which the corona, once started, is extinguished.

High-rate discharge For some energy-storage applications, it is desired that charge stored in a capacitor be discharged as rapidly as possible or, perhaps, just at a known rapid rate. Such applications arise when explosive actuated switches, detonators, and many other impulse-actuated or pulsed devices are served. The discharge rate is governed by the ESR (RC time constant) and the series inductance of the capacitor and the discharge circuit. When a low-inductance testing fixture has been designed to charge and discharge the capacitor under test, residual inductance cannot be totally eliminated, and the rise time of the discharge current impulse will be basically sinusoidal in shape. The time to maximum current is limited by the capacitor and circuit inductance, particularly if the capacitor is an electrostatic type of relatively high capacitance and voltage. The maximum-current rise rate is closely related to the equivalent series self-resonance of the capacitor with its series inductance. Capacitors for this type of application must therefore be selected for low inductance, test fixtures must be carefully designed, and application circuit runs should be short and direct, having low resistance and inductance.

2.2 APPLICATION STRESSES ON CAPACITORS AND GENERAL RELIABILITY CONSIDERATIONS

Depending on design and materials, temperature, applied voltage, and applied frequency have various degrees of temporary effects on capacitance, dissipation factor, and insulation resistance of capacitors. Figures 2.9a and 2.9b show the relative sensitivities of some dielectrics. In addition to the immediate temporary effects, elevated temperature and high-er applied voltages exert time-integrated stresses on capacitor insulation systems that permanently consume usable service life. If capacitors are not conservatively designed and applied, capacitor life-span reduction can begin to encroach on the expected service life span of the equipment in which they are used. Service life end for capacitors can be defined as the point at which the capacitor no longer performs satisfactorily in its application. This definition is a bit more liberal than those of part specifications, wherein point of failure in life

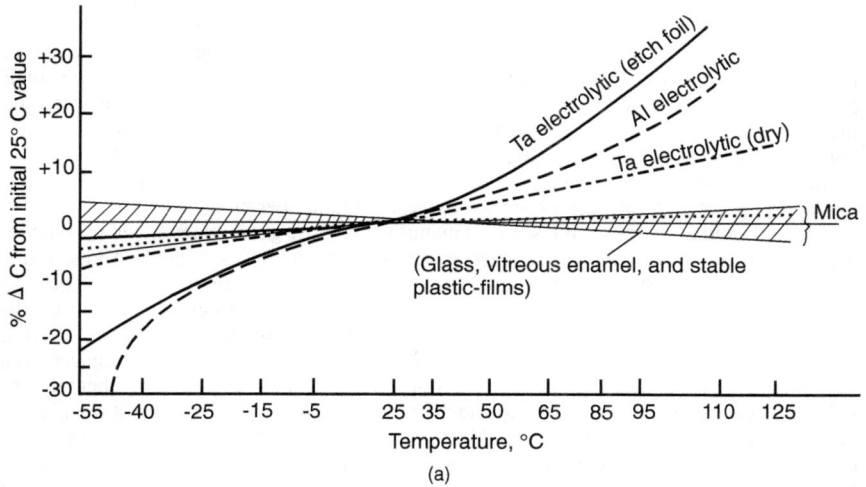

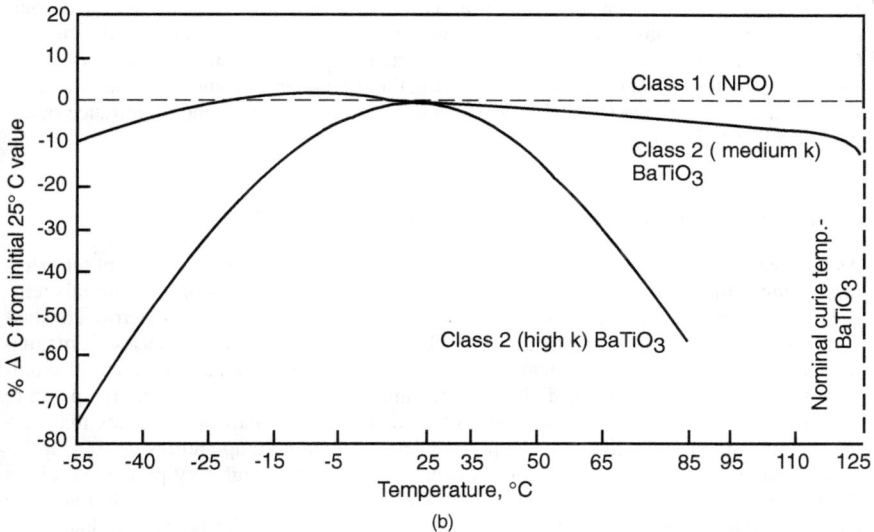

FIGURE 2.9   Comparative temporary effects of temperature on capacitance value for some capacitor types (a) electrolytics, stable plastic films, mica; (b) ceramics.

tests must be defined as the capacitance change or leakage current past some preselected points that are generally less-forgiving than in-circuit performance failure point.

Circuit and equipment package designers must always be aware that part reliability depends on the demands made on the part in its application, both in terms of performance limits that can be tolerated and applied stresses that must be withstood. Unless capabilities are stretched by performance demands, capacitor failure is seldom defined by excessive capacitance-value change, although such a change might portend a more serious event. More often, capacitor failure is characterized by short circuiting or lowered insulation resistance to a degree that causes a malfunction or degraded function in the using assembly. Less often, failure is caused by increased ESR or open-circuiting. In general, the accepted approach to capacitor reliability enhancement entails initially selecting capacitors with conservative design margins and wide-margin over-voltage screening and conditioning of finished capacitors by the manufacturer; then applying the capacitors well within their electrical performance capabilities, derating voltages and currents where possible, and making sure that maximum operating temperatures are held as low as possible and well within dielectric and encasement material capabilities. Good manufacturer screening and voltage conditioning removes infant mortality after capacitors have been baked, thermally cycled, or subjected to other stress-relief and stabilization operations.

Users should determine if the selected manufacturers use good material and production process controls for reducing incidence of random defects and substandard lots. Users must carefully select the correct capacitor design for the application, practice voltage and current derating, and be certain that capacitors are not exposed to environmental conditions that are beyond the part capabilities—either during equipment manufacture or during service life. One condition to avoid, for instance, is placement of a capacitor adjacent to a heat-producing part, such as a power resistor or power transistor. Even a fundamentally reliable capacitor cannot achieve its expected life in an application where it must remain stable beyond its defined performance ability or where it is subjected to environments or other stresses beyond its defined capability. Proper and conservative circuit application is thus a basic factor in providing reliable capacitor service. Most electrostatic capacitor and many electrolytic capacitor failures appear as sudden short-circuiting through the capacitor to become a serious threat for equipment permanent damage from excessive current. Corollary damage risk is high because many capacitor applications involve connecting the capacitor directly across the power bus, making the full power-current capability available for diversion through the short circuit. This circumstance increases the importance of capacitor reliability to major proportions.

2.2.1 Temperature Stress

As the internal temperature of a capacitor changes, whether because of ambient temperature changes or internal heating, the capabilities and characteristics of its materials temporarily change in response. It has already been mentioned that dielectric constant changes, and that insulation resistance goes down as the temperature increases. Some materials undergo more dramatic changes than others, but, in general, all voltage withstanding features of insulators and dielectrics become temporarily degraded as temperature increases. Permanent chemical decomposition and material character changes that normally occur slowly over time are sharply accelerated by high temperatures.

In general, it is accepted that chemical-physical property degratory processes follow the Arrhenius model, named for Svante A. Arrhenius, a very productive early twentieth century Swedish physical chemist, who introduced it. The model predicts reduction in time to failure at an elevated temperature versus time to failure at a reference temperature, based on the activation energy threshold for the dominant deteriorating mechanism for a particular dielectric system. The principle can be extended to cover voltage stress as well because stored charge is also energy, but many reliability models just assign an exponent to the ratio of applied to rated voltage, using the Arrhenius prediction to define the effects of temperature, multiplying the two together to obtain the total prediction. EIA 521 and military handbook 217 use an exponent of 3 for the voltage application to ceramic capacitors, meaning that the expected life changes at a rate defined as the ratio of applied to rated voltage raised to the third power.

Arrhenius equation   The Arrhenius equation is, in one form:

$$\frac{t_2}{t_1} = e^{\frac{ea}{K}\left(\frac{1}{T_2 - 1/T_1}\right)}$$

(2.14.1)

where t_1 and t_2 are, respectively, time spans at reference and investigated conditions, T_1 and T_2 are the absolute temperatures at the reference and investigated conditions (respectively), e is the base of natural logarithms, K is the Boltzman constant (here usually expressed as 8.62×10^{-5} electron volts per Kelvin), and ϵ_a is the threshold activation energy for the physical-chemical process under investigation, generally taken to be 1.0 eV for ceramic capacitors. (Note that the upper case K is used for Boltzman's constant to avoid confusion with the lower case dielectric constant, k, and we cannot dare to use the international symbol for absolute temperature degrees, Kelvins, which is an upper case K.) For materials that exist in already low-energy stable states, such as mica and glass, temperature-induced chemical changes and material phase alterations require higher activation energies, and elevated temperatures are of less concern, but organic polymers and high-energy-state oxides need allowances for high-temperature life reduction. EIA-521 states a time-temperature acceleration factor for ceramic capacitors as:

$$Time\ Acceleration\ Factor = \left[\frac{V_2}{V_1}\right]^3 X^{\frac{10^5}{8.2}\left(\frac{1}{T_1} - \frac{1}{T_2}\right)}$$

(2.14.2)

Where V_1 is the reference voltage, V_2 is the proposed voltage, T_1 and T_2 are the respective reference and proposed absolute temperatures in Kelvins, and $10^5/8.2$ is the ceramic-associated activation energy of 1 eV divided by Boltzman's constant (8.2×10^{-5} eV/kelvin). Other types of capacitors have similar equations defined in military handbook 217, in EIA 521, in manufacturer literature, and in other reliability papers and publications. Exponents for voltage acceleration and failure-mechanism activation energies are based on reported testing results. Operating temperature and voltage limitations are the most significant techniques available for improving well-made capacitor reliability in terms of catastrophic failure. Capacitor manufacturers and researchers have derived time-voltage-temperature exponential-acceleration-factor equations similar to those previous for most capacitor types, wherein constants have been chosen to fit empirical data accumulated for the particular capacitor design being addressed. If available in the manufacturer's data, these expressions are quite useful when the reliability in a particular application must be quantitatively addressed. A frequently overlooked application aspect is the already-mentioned capacitor internal heating because of large ripple or alternating-current conduction. Measurements and analysis should be conducted if this condition is present, particularly for small solid electrolyte tantalum and ceramic parts whose heat-dissipating abilities in relation to capacitance value are limited by their small physical sizes. Most polymers and other organic materials have limits of elevated temperature at which physical state changes, such as elastic modulus and tensile and compressive yields start to occur. If these temperatures are exceeded during equipment production or in service, the capacitor structure can be altered, causing the capacitor to change value, open-circuit because of structural failure at terminations, or to become susceptible to dielectric voltage breakdown at reduced voltage. Where alternating currents through capacitors are significant, even for physically large parts, the effects of internal heating from dissipation by the ESR of the capacitor must be determined and added to the capacitor localized ambient temperature maximum. Internal heating under high current has a degree of self-moderation for electrolytic capacitors, where, because of increased electrolyte efficiency, ESR generally decreases with temperature, causing dissipation and temperature rise to be somewhat reduced. Note that the electrical polarization nature of organic polymers often undergoes an abrupt change at the material glass transition temperature, causing the dielectric constant to change. This is not an important consideration so long as the capacitor operating temperature range does not include the glass transition point, whether the material's normal use is in its glass or in its plastic range. For structural reasons, installation soldering temperatures should be diligently assessed against dielectric material capabilities as a possible source of jeopardy for plastic film-type capacitors. (Note that the heat-distortion temperature and glass-transition temperature are not interchangeable or even synonymous terms.)

Another danger associated with heating capacitors during solder installation occurs when certain capacitor types, notably ceramic dielectric types, are thermally shocked by subjecting them to soldering heat without preheating the capacitor bodies. A high thermal gradient at leadwire or other termination attachments can crack the adjacent ceramic during rapid heating and cooling, creating structural damage that destroys the termination immediately or becomes a latent condition awaiting in-service thermal cycling to fail completely. Such failures have been produced during leadwire solder dipping by users who did not use preheating and slow-cooling techniques. Capacitors assembled with soft solder can experience problems from solder reflow during installation, a particular problem with some solder-in feedthrough radio frequency interference (RFI) filters, where heatsink attachments and special soldering techniques might be needed. Where possible, capacitor manufacturers use solder alloys with higher liquidus temperature than that of the 63-37, 60-40 tin-lead alloys, or 62-36-2 tin-lead-silver, commonly used for circuit installations, but temperature controls and time-exposure limits must often be used to prevent reflowing capacitor construction solders.

Voltage failure	Applied voltage is associated with capacitor failure in at least two ways: high-voltage disruptive breakdown, either by material displacement stress or thermally induced bridging, can occur at the weakest point of the dielectric or at a point of maximum charge concentration under any condition (if the voltage is high enough). Secondly, high voltage can decrease the time to long-term wearout failure, particularly when combined with elevated temperatures. Most capacitor reliability models combine the effects of voltage and temperature for predicting life degradation. Figures 2.10a, 2.10b, and 2.10c show generally predicted life reliability numbers, as related to voltage and temperature. These charts are representative examples of those derived from curves based on the previous models. Degraded dielectrics will fail sooner with a higher (than with lower) applied voltage for the same reasons that weak spots in dielectric systems will fail first as voltage is elevated. Also, while at high temperature, insulation resistance is reduced, leakage currents are increased, and the chance for a localized high current-density leakage fault is sharply increased. Small localized high temperatures caused by high current-density spots can bridge the dielectric, permitting even higher current and rapid failure or, for certain capacitor designs and types, a self-healing mechanism might be activated, provided that the current is sufficiently limited to prevent regenerative destruction, but high enough to activate the healing system.

Polarized terminal electrolytic capacitors should, in general, not be exposed to voltage polarity reversal, even if the reversal is just tips of a reverse polarity ac waveform. Some electrolytic capacitors can tolerate limited voltage reversal, but perform differently in the

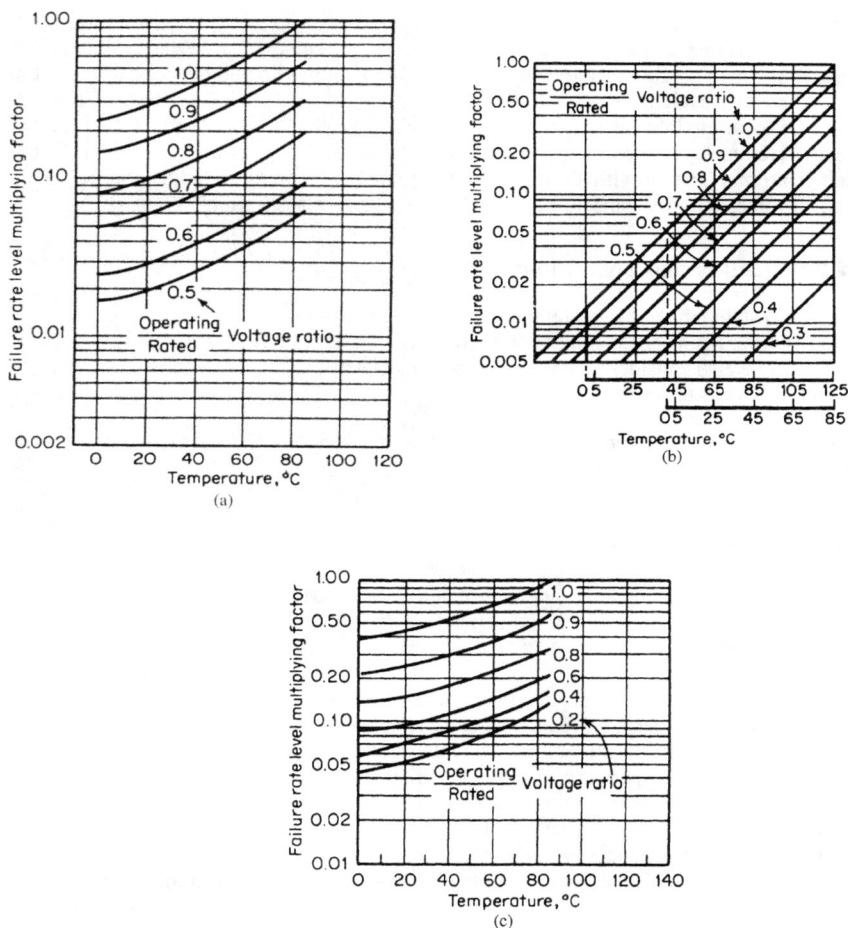

FIGURE 2.10 Sample of predicted operating-life degradation charts based on empirical data and curve-fitting (see MIL-HDBK-217 for additional data). (a) Solid electrolyte tantalum; (b) General purpose ceramic; (c) Non-solid electrolyte tantalum.

reverse direction, and will usually have stringent manufacturer limitations on the magnitude and conditions of the reversed polarity. Part specifications must be followed, and manufacturer information should be obtained if polarity reversal is an unavoidable possibility in an application. As follows, polarized electrolytic capacitors can be connected back-to-back in series for some ac applications, but the most frequently used technique is to bias the polarized part with sufficient dc voltage to prevent polarity reversal by applied ac. The combination of dc level and ac voltage peaks should not be permitted to exceed the capacitor rated voltage with the capacitor in the circuit for both electrostatic and electrolytic capacitors. As a general rule for conservative design, voltages applied to capacitors should be half of their rated values. This rule should be diligently followed for physically small capacitors in applications where appreciable ac or ripple currents are carried and heating occurs, remembering that high temperatures and high voltage work together in reducing capacitor life. Voltage-derating guidelines can be safely moderated somewhat where conditions are benign, but the rated voltage should never be exceeded—even if the duty cycle is very small. Reduced voltage ratings for elevated temperature are based on local environment at the capacitor site, and must be observed. Manufacturer-specified reduced voltage ratings for applied alternating or ripple current are based on capacitor internal heating with alternating current and are essential for capacitor longevity. Peak ac values with the capacitor in the circuit are used to assess applied voltage.

2.2.2 Reliability and Environments

It has been stated that elevated temperatures shorten capacitor life, whether high temperatures resulting from self-heating or from ambient conditions. Other environmental conditions must be also considered when selecting capacitors for particular applications if long, reliable service is to be expected.

Low temperature and thermal cycling effects   Very low temperatures can be detrimental to capacitors, although for reasons different from high-temperature concerns. The effects of low temperatures cause capacitor values to temporarily change, often dramatically, such as for many electrolytic capacitors. Electrolytes and fluid impregnants are susceptible to degraded performance at lowered temperatures, but by using premium materials, acceptable performance of electrolytic capacitors can usually be obtained down to −55°C, but at reduced capacitance or another tradeoff. Effects are generally more pronounced for aluminum types, but depend on design. Some aluminum electrolytic designs are limited to −35°C or −40°C (or even warmer limits), depending on the capacitance loss that can be tolerated.

Problems can arise with packaging or encasement of capacitors at low temperatures when differences in thermal expansion coefficients and decreased tensile strengths of materials become significant limitations. The effects of material mismatch can appear particularly when sharp temperature changes occur. If an installed capacitor is stabilized at −40°C or lower, and power is abruptly applied to the equipment, rapid temperature changes can occur, causing thermal gradients between portions of the total mass, resulting in stresses. Production practices typically cure encapsulating and molding plastics at temperatures of 125 to 150°C, and the materials polymerize or solidify with all components at that temperature, producing a structure that is stress-free at that temperature. As the encasing structure cools, the differences in thermal expansion characteristics cause stresses at the material interfaces, with maximum stress occurring at the lowest temperature, or that temperature which is farthest from the stress-free polymerization temperature. If molding and encapsulating materials bond to capacitor surfaces (such as ceramic, for instance, which has a very low expansion coefficient compared to most polymers), stresses can be damaging, particularly if the encasement is nonsymmetrical. Ceramics withstand uniformly applied compressive forces very well, but nonsymmetry can cause stress risers and ceramic fracture at material mass interfaces. Users should keep thermal stresses in mind when applying coatings or adhesives to installed thinly encased ceramic or glass capacitor bodies. Circuit conformal coatings should ordinarily be kept as thin as possible and be applied uniformly, with respect to capacitor bodies, avoiding thick buildups in isolated areas.

Thermal cycling in atmospheric conditions will cause condensing moisture and subsequent freezing as the dew point and freezing points are traversed. The cycling process might permit water penetration into small openings and cracks in capacitor-element encasements, expanding the openings as the water freezes, and degrading the encasement effectiveness in preserving the capacitor performance. Encasement systems can be designed to resist these conditions.

Leaded capacitors are typically mounted so that leadwires provide a relatively flexible member between the mounting points and the capacitor body. With temperature excursions, when the circuit card or chassis distance between the leadwire mounting points expands and contracts at a different rate than does the capacitor body, the difference is absorbed by the flexible leadwires, and no damage is done. Chip-type ceramic capacitors, however, when mounted normally, have their bodies rigidly attached to termination pads by soft solder or by a metal-loaded electrically conductive polymer. If the circuit card to which capacitors are attached is a polymer resin-based material that is not one of the few available low-expansion types, a significant mismatch will exist between the linear thermal expansion coefficients of the capacitor body and the circuit card. Thermally cycling the described assembly produces stresses in the capacitor bodies and in the attachment joints, and, if the attachments are rigid and the thermal excursions are wide range, capacitor bodies or attachment joints can be cracked. If solder is used for attachment, continued working of the solder by the cycled stresses can cause grain growth and weakening of the joint. Large capacitor bodies having greater distances between termination pads are more susceptible to stress because stress is proportional to the amount of expansion displacement, and expansion coefficients are stated in length per unit length per degree (viz., parts per million per °C). This problem is a major concern for chip-type capacitors that are mounted by termination pads; the situation should be carefully evaluated when utilizing chip-type capacitors.

Low temperature and thermal cycling concerns are effectively addressed by the design of the capacitor for use in the applications where this environment is to be encountered, and by awareness of the phenomena when selecting circuit card materials and attachment methods. Table 2.4 shows thermal expansion properties of some capacitor-related materials.

Humidity effects Intrusive moisture has long been recognized as an enemy of capacitors, and many design measures have been taken to combat it. Moisture is a greater problem if voltages are high or if moisture-sensitive dielectrics (such as paper) are used. Impregnation with wax in a paper case, and impregnation with oil and other fluids in gasket-sealed or hermetic metal cases or enclosure in hermetic cases without impregnation are still successfully used where environments and available spaces permit. Wax-impregnated paper casing and wax impregnation of porous resin coatings are effective, but suffer from upper-temperature limitations and their use has declined. The availability of dielectric films with low moisture sensitivity, and high-performance dipping, molding, and impregnating materials have reduced, but not eliminated, the need for capacitor hermetic sealing. For ultimate survivability in harshest environments, hermetic sealing and part designs that are inherently sealed are still the best choice.

At the present, all available organic polymer materials permit eventual moisture intrusion through their molecular structure, with some, however, being more resistant than others. For short-term exposures, such as diurnal cycling with brief periods of condensation followed by lower humidity and normal drying, polymer encasements are quite effective, particularly if the capacitor structure is monolithic with no internal cavities. In practice, well-designed and constructed plastic impregnated, molded, or wrapped capacitors have immediate survivability in moist environments that is comparable to the polymer circuit cards and other plastic-encased and structured components used in typical circuit card assemblies and, unlike hermetically sealed parts that might develop defects and permit water intrusion through small leaks, can be restored by warm, dry conditioning after significant time periods of exposure to high humidity if the exposure has not been long enough to permit internal corrosion or metal migrations to occur.

Not to mislead, however, capacitors are much more sensitive to the effects of moisture intrusion than many other circuit components. Water effects can be seen as prompt change

TABLE 2.4 Linear Thermal Expansion Rates of Some
Representative Materials That May be Associated with
Capacitors

Material	Thermal Conductivity (W/m•K)	Linear Thermal Expansion Coefficient (PPM/K)
Alloy 42	17.3	5.3
Aluminum Oxide Al_2O_3	34.6	6–7
Barium Titanate	4–5	9–11.5
Brass (70–30 Cu–Zn)	105–108	22
Conductive Epoxy	1–4	50–100
Copper	390	17
Cyanate ester-glass laminate	0.4	8.7–12.5[1]
Epoxy-glass laminate	0.18–0.30	11–15[2]
Filled Epoxy	≈0.5–0.9	15–28
Gold	314	14.2
Nickel	86	12.8
Polyimide-glass laminate	0.39	16–19[3]
PTFE	2.5	55
Silicon Dioxide (glass)	3.4	3.4
Silver	419	20
Steel	≈50	15
Tantalum	55	6.5
Tin	230	16.5
Titanium Dioxide	6.7	8.2–10.3
Tin-lead	34	24

1 Figure given is linear axis, Z axis is 33–40 ppm/K.
2 Figure given is linear axis, Z axis is 60 ppm/K.
3 Figure given is linear axis, Z axis is 52 ppm/K.

in insulation resistance and, depending on circuit voltages and impedances, might result in
an immediate failure of the capacitor to perform its intended function, or might cause a high
current-density insulation leakage that progresses under power to permanently short-circuit
the capacitor plates. Longer-term effects can be short-circuiting from metal migration
across dielectric surfaces when voltage is applied, or internal corrosion to reduce insulation
resistance or open-circuit current paths. Galvanic potentials are present between various
metals (such as aluminum, copper, and tin) that are used inside capacitors, and exposure to
moisture can cause rapid corrosion of metals that are, by themselves, inherently corrosion
resistant. Metallized film and paper capacitors are particularly vulnerable to electrolysis
because of the small amount of metal movement needed to seriously alter their plate and
termination structure.

Mechanical shock and vibration Depending on the design, capacitors can be susceptible
to a number of problems in mechanical shock and vibration environments:

- the capacitor mass, suspended by its leadwires, might respond to the applied mechanical
  forces, causing the leadwires to yield,

- the internal structure of the capacitor might be moved or damaged by the mechanical forces

- the capacitor element structure can be distorted temporarily or permanently, causing a transient or permanent capacitance shift or, in the case of dielectrics, such as high-k ceramics having piezoelectric properties, unwanted "microphonic" electrical outputs might be produced.

All of these effects can be addressed by proper capacitor selection and proper installation design. Many capacitors can be eliminated from consideration for applications involving shock and vibration by simple examination of their construction and mounting means. For others, the susceptibility might not be obvious, and testing will be required. Military specifications require demonstration of vibration and shock durability by test, but even this does not ensure immunity in a given circuit installation. Leadwire failures in high-vibration environments (such as found in automotive, railroad, aircraft, and helicopter installations) are more likely to be associated with heavy capacitors that are not provided with hold-down clips or other restraints, but can occur on relatively small parts if the chassis or circuit-card structure resonates at some frequency within an applied vibration spectrum. A structural resonance at a capacitor mounting site can apply a very high vibration input to the capacitor at that frequency, and rapid flexing of capacitor leadwires can occur when the capacitor body mass, which they suspend, is accelerated. Metal fatigue of the leadwires by back-and-forth bending causes fracture failure. Military capacitor leadwires are typically copper-clad steel, which resists metal fatigue better than annealed copper, but also functions better as a spring.

Internal shock and vibration damage to capacitors is almost always associated with designs that have a capacitor element supported within a hollow case by its terminating leadwires or with capacitors that have little or no element restraints within a hollow case. These design types are produced in hermetically sealed paper and film types, and in electrolytics of end-terminal and axial-leadwire types. For uses in high-shock and high-vibration environments, selected capacitor designs should be examined for adequate internal structure. Mechanical responses of internal structures usually cause degradation or disruption of the capacitor element termination, a condition that is detectable by an increase in DF and ESR or open-circuiting. Leadwire flexing at capacitor seal interfaces can also damage the seal, causing case leaks and resulting latent failures.

Troublesome microphonic effects during shock and vibration are fortunately rare; most such transient electrical effects being absorbed by the circuit characteristics. If microphonics become a problem, however, a type of capacitor that does not exhibit the trait can usually be found to replace the offending part in the sensitive application. High-k ceramic capacitors (using ferro-electric dielectrics), which can produce piezoelectric impulses, are the worst offenders, but capacitors whose structure is loose enough to be distorted during shock and vibration can change capacitance in response to vibration and shock. Chip capacitors, used in hybrid and microwave assemblies, often use microcircuit-type wire bonding to make electrical connection. Like the one-mil, more or less, diameter bondwires used for microcircuits, these wires are subject to resonance if dynamic shock inputs have significant energy in the frequency range from about 5000 Hz to 20,000 Hz. The resonance amplification will move the relatively low-mass wires and if inputs are repetitive or prolonged, wires will be embrittled by the repeated flexure, and might fail. Ultrasonic cleaning of this type of assembly should not be done unless first investigated thoroughly, and if equipments are subjected to pyrotechnic or other shock spectra containing significant high-frequency components, careful testing must be done.

Altitude stress   The operating altitude for capacitors is an important specification for parts that have operating voltages of 400 to 500 volts (and upward). Probabilities of arc-over and charge leakage across insulator surfaces, through case sleevings and coatings, and through air that separates points of high charge concentration (such as sharp conductor projections) are increased in accordance with the Paschen altitude curve. Designs must provide adequate spacing and means for reducing charge concentrations at vulnerable points. (For those not familiar with the Paschen altitude curve, it shows the steadily decreasing dielectric strength between two conductors separated by air as the air pressure is decreased until near vacuum is approached. The dielectric strength begins to rise as very high vacuum conditions begin to prevail, whereby conditions are measured by electron mean-free path.)

Questionable features should be performance-verified with a generous design margin by high-potential testing at reduced pressure. Nonhermetic oil-filled parts might exude oil at reduced pressures, and reduced pressures will stress gasket and polymer seals on liquid-filled parts, such as aluminum electrolytic capacitors. Reduced pressure over long periods can increase aluminum electrolytic capacitor electrolyte water vaporization if water is used in part designs that are exposed to the low atmospheric pressure and if seals are not sufficiently robust. Electrolyte dryout is probably the most significant long-term failure mechanism for aqueous electrolyte aluminum electrolytic capacitors.

Handling and mounting Damage from mishandling and installation are common sources of capacitor failures. Damage often does not cause immediate failure, but creates conditions that ultimately lead to failure. Understanding the vulnerabilities of different capacitor designs can help you to avoid these problems. Listed here are some of the more common difficulties, but by no means is this a complete listing.

Cracks and chips   Cracks and chips in ceramic capacitor bodies result from many causes. These openings into the ceramic plate structure permit moisture to intrude into the plate area, which causes silver migration, high rated-voltage leakage current at the damage site, high leakage at low voltage, and eventual hard failure. Cracks can also propagate, ultimately fracturing the body, with the plate displacement causing a short circuit between plates or, if the crack is at a termination, the part might fail open-circuit.

Cracks and chips can be caused by mishandling stresses in the soldering operation or in handling and mounting of circuit cards on which the capacitors are mounted. Any time uncased ceramic capacitors are permitted to contact each other or other sharp-cornered hard objects, chipping can occur. Sharp points concentrate force so that a few ounces of force can become thousands of pounds per square inch at the point of contact. This frequently happens with automatic pick and place machinery in placing ceramic-chip surface-mount capacitors. The centering jaws can become worn by the strong abrasive action of the ceramic material and, instead of a flat contact surface, might present a wedge shape that contacts the capacitor body only on a corner edge, thus multiplying the relatively gentle jaw centering force many times. Figure 2.11 is an example of subtle-appearing damage to a ceramic capacitor that resulted in catastrophic failure. Forcing the capacitor center downward against the mounting surface while it is supported by its termination pads can also crack the part. Soldering without preheating or improper rapid cooling after soldering can cause chip-capacitor body-stress cracking or can cause leaded capacitors to crack at the terminations. If circuit cards are warped after soldering, and the warpage is mechanically forced flat when mounting the card to another structure, surface-mounted chip capacitors (or their attachment joints) and even radial-leaded capacitors mounted in contact with the board can be broken as the straightening commences.

Excessive or too-long-duration heating during soldering   Excessive or too-long-duration heating during soldering can damage hermetic seals by reflowing solder at capacitor solder seals, thus creating a latent failure. Small cases are more vulnerable because of their lower thermal mass, and their small spacing between axial terminal wires and case is

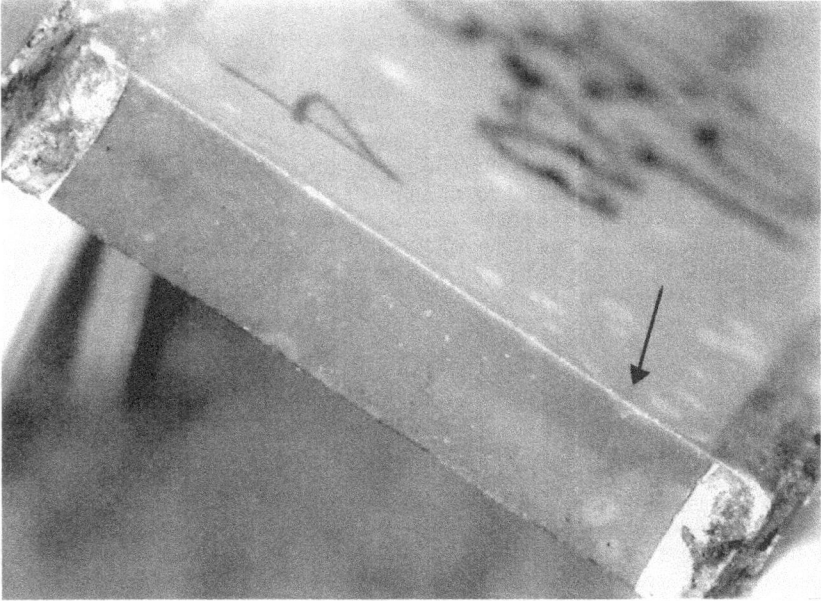

(a)

(b)

FIGURE 2.11 Example of fatal handling damage to a ceramic chip capacitor. (a) The arrow identifies small surface crack near a corner. (b) The same area after rated voltage had been applied. The capacitor short-circuited within the crack and the dissipated energy displaced the chip, leaving the capacitor short-circuited. (*Photographs by Lockheed Martin Electronics and Missiles Failure Analysis Laboratory.*)

easy to bridge and short-circuit with the solder reflow. Leaded ceramic parts are fairly well protected from solder reflow by manufacturer use of slightly higher melting-point solder than the near-eutectic alloys used in mounting, but if conditions become excessive, reflow can occur. Both thermoplastic and thermoset plastic films used to make capacitors have temperature limitations. Above these limitations, they might distort and change the capacitor structure or deteriorate the capacitor end-metal to film joints. These temperatures are often well below the melting point of the mounting solder, and the thermal mass of the capacitor and case must be depended on, together with the brief soldering-heat exposure to keep the temperature down. This is a critical balancing act, and it sometimes becomes necessary to remove film capacitors from automated soldering, adding them later manually. Soldering temperatures can deteriorate or shrink plastic sleeving over metal capacitor cases, baring the case ends to short-circuit adjacent conductors. If sleeves shrink excessively, it is a fair indicator that time-temperature exposure is excessive because sleeves are selected to be capable of withstanding safe high-temperature exposures. Case coverings are particularly vulnerable to penetration if they are subjected to a point force during soldering. The softening plastic might flow away from the point stress, permitting the point to contact the metal case.

Damage by cleaning solvents   Electrolytic capacitors with elastomer gland seals can be damaged by cleaning solvents. Aluminum electrolytic capacitors can be ruined if tiny amounts of chlorine get inside, and chlorinated hydrocarbon solvents can penetrate elastomeric seals—even if the seals are gas-tight. A latent failure is created, to be manifest in the future. Table 2.5 lists some acceptable and unacceptable solvents for use with aluminum electrolytic capacitors. One manufacturer's literature cites internal chlorine contamination as the primary cause of field failures that he has been called in to investigate.

Storage damage   Capacitors to be soldered must be protected during storage to maintain solderability. Using equipment manufacturers commonly limit fluxes to only rosin or mildly activated rosin, and solderability must be established and maintained to ensure that reliable circuit joining is possible. Pure tin, pure nickel, and very thin coatings of tin-lead are generally unsatisfactory finishes—even when well-protected in storage. Storage in very dry oxygen-free conditions in closed, dark containers might permit satisfactory use of these finishes after limited-time storage, but the best practical finish for solderable terminations at this writing is still a sufficiently thick tin-lead solder coating applied over a clean, active surface, which is then heat-fused to form a nonporous protective surface over the intermetallic zone that forms in the surface of the underlying metal. The intermetallic

TABLE 2.5 Acceptable and Unacceptable Solvents to be Used Adjacent to Aluminum Electrolytic Capacitors

Safe	Unsafe
Xylene	Freon TF, TE, TMC
Ethyl alcohol	Carbon tetrachloride
Butyl alcohol	Chloroform
Methyl alcohol	Trichloroethylene
Isopropyl alcohol	Trichloroethane
Calgonite (detergent)	ALL (detergent)
Naptha	Methylene chloride
Water	Methyl ethyl ketone
Toluene	
Alkinox	

TABLE 2.6 Recommended Standard Frequency and Voltage Measurement Conditions for Some Capacitor Types.

Capacitor type	Measurement frequency	Measurement voltage	Remarks
EIA class-I ceramic ≤1000 pF	1 MHz	1.2 V rms max	Includes Mil-C-20, some MIL-C-55681
EIA class-I ceramic >1000 pF	1 kHz	1.2 V rms max	Includes MIL-C-20
EIA class-II ceramic	1 kHz	1.0 V ± 0.2 V rms	Mildly voltage-sensitive, includes MIL-C-39014, some MIL-C-55681
EIA class-III ceramic	1 kHz	0.5 + 0.1 V rms	Voltage-sensitive, includes some MIL-C-39014, some MIL-C-55681
EIA class-IV ceramic	1 kHz	0.1 V rms max	Highly voltage-sensitive
RF ceramic 0-1 nf	130 MHz to 1.25 gHz	See EIA 483 and Appendix A to MIL-C-55681	Uses a resonant 50-Ohm transmission line
Aluminum electrolytic for electronics	120 Hz, 1.0 V rms, dc biased to 2.2 V dc	EIA 395 specification sheets	Not significantly voltage sensitive except for polarity
Aluminum electrolytic for motor starting	60 Hz	EIA 463 specifications	Not significantly voltage sensitive
Paper, paper-plastic and plastic for electronics	<10 μF: 1.0 kHz ≥10 μF: 60 Hz	Not specified	Not significantly voltage sensitive
Paper, paper-plastic and plastic for power applications	60 Hz	Not specified	Not significantly voltage sensitive
Glass	≤1000 pF: 1.0 MHz >1000 pF: 1.0 kHz	Not specified	Not significantly voltage sensitive
Mica	≤1000 pF: 1.0 MHz >1000 pF: 1.0 kHz	Not specified	Not significantly voltage sensitive
Solid tantalum	1.0 kHz CSR 21 only. All others 120 Hz. 1.0 V rms, 2.2 Vdc	120 Hz. 1.0 V rms, 2.2 V dc	Not significantly voltage sensitive

formations continue to form with time, and the tin-lead must be sufficiently thick to maintain protection: at least 200 to 500 micro-inches. Even with this finish, storage should be dry, preferably not over 40-percent relative humidity. EIA 534 provides guidelines for maintaining solderability.

2.2.3 Reliability Enhancement and Assurance Through Part Testing

Like many other electronic parts that are produced at high production rates, capacitor quality and uniformity off the production line depends strongly on the producer controls exercised on the materials and processes during production. The effectiveness of these controls depends on the capacitor design and the production tools and methods used and their controllability. Highly developed designs and processes available today permit a more uniform and higher overall quality than ever before, but no matter how good, the product off the assembly line will have a range of variation. The many random variables in materials and processing result in normal (Gaussian) distributions of properties about average values, with a tight grouping for well-controlled properties and wider variations for less easily controlled traits. In both tight and broader distributions, there will also be some number of "outliers" or parts that, for random causes, fall well outside the usual population distribution.

Each measurable feature of each production lot or batch of materials will have its own average, its own distribution of variation, and its own proportion of outliers. For well-controlled products, the lot-to-lot variation is small and the distributions within each lot are comparatively narrow. Other part features and possibilities can affect reliability: It is possible to have hidden or latent problems that cannot be easily measured or detected. For instance, a metal-cased capacitor might have a leaking seal that, over diurnal cycling, will accumulate water inside to deteriorate insulation resistance and corrode conductors. This leak will probably not be detected unless a specific test is conducted on the particular leaking part. Also, it is the nature of structures just fabricated to undergo a stress-relieving equalization and relaxation in which materials move, stretch, compress, bend, and otherwise change to a balanced condition and, in so doing, might cause an otherwise marginal condition to become critical.

Like many other complex products, it is an accepted practice to design capacitors with a safety margin, that is, make the product somewhat better than it apparently has to be, to allow for the variations in properties, the initial stress relaxations, minor latent interactions, and degradation of properties in service. The amount of the design margin is a cost driver and, in the case of capacitors, often a marketability factor, because larger design margins typically translate into lower CV product per unit volume, making a higher design margin part physically larger than one of the same rating and value with a lesser margin.

Recognizing that even the most stringent quality controls are not infallible in avoiding the production of outliers and parts with undesirably low performance margins, it is accepted practice to perform conditioning and overstress screening on capacitors at the assembly-line end. The conditioning screen can be grouped with the manufacturer's quality-assurance tests (as opposed to quality-control measures) if one should desire, but, in the writer's opinion, conditioning screens are more properly deemed part of the production process and its quality controls—even if some threshold of total lot acceptability is observed based on fallout proportion from a given lot. For types of capacitors with dielectric self-healing features, initial voltage conditioning is clearly part of manufacturing, being essential to producing functional capacitors.

In terms of reliability, end-of-the-line stress screens must be designed and implemented to remove the infant mortality portion of the well-known reliability bathtub curve (see Figure 2.12) without damaging the normal population. It is important that 100 percent of the parts be screened, that the voltage stress be greater than rated voltage by as high margin as possible without endangering good parts, and that an elevated temperature be used in combination with the over-voltage—both as a production stress relief accelerator and to enhance the screen effectiveness in defect detection. It is generally conceded that most infant-mortality defects will manifest themselves within 200 hours at maximum-rated conditions of temperature and voltage, and conditioning screens are often based on that assumption. Test times

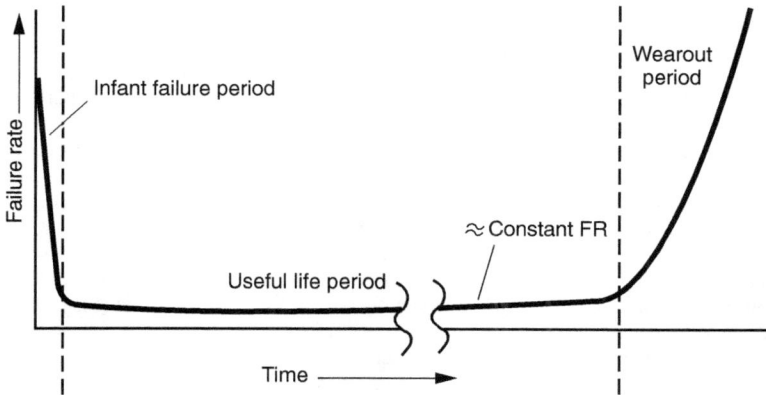

FIGURE 2.12   A classic reliability "bath tub" curve.

are frequently shortened by applying acceleration factors of over-voltage and temperature, based on an established acceleration model for the capacitor type (see the previously stated Arrhenius-based equations). A thermal cycling step prior to voltage-temperature conditioning might be helpful for some capacitor designs to accelerate stress relief. Despite normally used infant mortality assumptions, screening exposure time is best selected by experience, based on fallout history as a function of time. Many capacitor performance features can be satisfactorily controlled by statistical methods in the production processes, but, if military or other user performance specifications are involved, sample assurance testing of the finished product is usually imposed. Some critical features may need 100 percent measurement. For instance, case leakage of hermetically sealed and fluid electrolytic capacitors should always be checked 100 percent because each part's individual reliability depends absolutely on its seal integrity, and short-term side-effect indicators of leakers are few. Other variables likely to be individual part traits, rather than group distributions, might need scrutiny, but the 100-percent stress conditioning before final electrical performance measurement is an essential step if early in-service failures are to be reduced to insignificance.

Recognize the distinction between reliability enhancement and reliability assurance, in terms of the rationale for parts testing. Part failure rate predictions, such as those covered by MIL-HDBK-217, assume that outliers and early failures have been diligently removed, that no undue damage was caused when installing the parts, and that parts start service on a constant failure rate curve, based on wearout mechanisms. Military Handbook 217 assigns failure rate prediction acceleration factors to recognize probabilities that this might not be totally the case; parts that have had less stringent end-of-the-line quality-assurance testing are assumed to have higher probability of defective escapees. It is a fact that almost all in-service failures actually seen are caused by one or a combination of:

- Undetected part defect
- Part damage incurred during part installation
- Transitory or long-term part overstress beyond its design capability.

Note that less end-of-the-line quality-assurance testing does not necessarily equate to lower reliability because reliability and quality are actually provided by the manufacturer's material and process quality controls, with the end-of-the-line tests (except for 100-percent screening), providing only some assurance that the quality controls were effective.

Handbook 217 also assigns factors based on general-rule application stresses (e.g., the voltage derating factor used), which are legitimate wearout accelerators. Understand that reliability failure-rate prediction using handbook values is somewhat related to, but is a distinctly different activity from that of dealing with the practical aspects of part reliability enhancement through design, selection, testing, and application.

Manufacturers have in the past recognized that solid electrolyte tantalum capacitors (and other electrolytic capacitors, as well) are particularly sensitive to stringency of material and process controls. Military specifications for established reliability solid electrolyte tantalums were among the first for electronic parts to implement a modified sampling system based on a Weibull distribution-life test-data analysis, rather than the long-used exponential distribution. The Weibull analysis technique for evaluating and extrapolating lot capability based on fallout experience in sample test under accelerated stress is particularly suited to parts that have a generally decreasing failure rate over time in normal service, such as demonstrated for solid electrolyte tantalums. Correct use of the method emphasizes product fabrication controls, and has apparently worked very well. Unfortunately, however, the cost of assurance testing is about the same as for the exponential system. Details of the Weibull sampling program can be found in MIL-e-39013.

2.3 MILITARY AND INDUSTRY STANDARDS DOCUMENTS

The U.S. military has been a strong proponent for improved reliability in all electronic components and has funded a very strong standardization and specification development and maintenance effort from the 1960s through the mid-1990s. Military needs and those of the U.S. space program for fairly large quantities of reliable parts at competitive prices have driven product and process development efforts to meet better and more stringent specifications for capacitors and other parts as well. The array of all-service part specifications that were begun during the unprecedented mobilization of World War II, were used into the 1970s, but are now, with some exceptions, supplanted by a later generation of "established reliability" (ER) specifications.

The ER specifications, apart from being more detailed and definitive than their predecessors, all require that parts delivered undergo some type of 100-percent timed overstress conditioning followed by 100-percent measurement of key electrical characteristics. This is sometimes apart from the manufacturer's usual quality-control measures that might implement the same sort of process, but often it is just a definitive requirement for what responsible manufacturers do anyway, with the advantage for the user being that all his bidders are quoting to the same set of minimum screening requirements, not to what a particular cost-conscious supplier might think is enough.

The ER parts specifications all have one other significant requirement—that for sampling from parts ready for shipment and placing the sample on long-term accelerated life testing. A life-test failure doesn't usually cause delivered parts recalls, but success in meeting a threshold failure rate over time is a condition for remaining on the acceptable supplier list for the demonstrated reliability level that the manufacturer has achieved by accumulating numbers of successful part-hour testing.

Do not use the ER specification "demonstrated failure rate" numbers in predicting in-service part-hour failure rates. The numbers are based on a very narrow set of accelerated failure mechanisms, as measured by observed performance parameter changes, based on a predetermined threshold that often would not constitute a failure to function in an equipment application. Nevertheless, the tests serve as standardized comparison yardsticks to assess whether production controls are producing an acceptable result.

In a few cases, ER military specifications are just upgrades of older specifications, and parts kept their existing identification numbers while being upgraded to specify ER requirements. In most cases though, a new number was assigned, leaving the former specifications, often for the same physical part, in place. Some of the non-ER specifications have now been retired, but some are still used. Older military specifications for products that are not used at all or are little used in new designs are retained as valid, so long as active procurement for spares and maintenance occurs—even when identical parts produced under a newer specification are available. This presumably is because the military inventory system is based on part numbers, and the newer parts are often at least slightly more costly, making it unwieldy to publish and observe interchangeablity lists. Some obsolete specifications bear a title page label prohibiting their use for new equipment design, but design contracts for new military equipment generally require use of the newer established reliability parts—even when the older documents are still valid. Some military specifications also do not, at this writing, have qualified suppliers in their qualified products listings, so the user should habitually check qualified products lists before assuming that a specification-listed part is available. This occurs with increased frequency as the military market shrinks in relation to commercial procurement, and military users are abandoning the qualified products listings in favor of "qualified manufacturer lists," which are listings of recommended manufacturers for the products covered by the associated specification, presumably having been surveyed and found satisfactory, and having produced successful parts in the past.

In years past, the relative size of the military and space market made the considerable effort and expense of maintaining the required testing and record keeping worthwhile for manufacturers, and users thus obtained parts that were much better than in the past at prices that were only initially somewhat higher than the older parts. As industry processes developed, and improved production and material-control techniques were implemented to compete effectively, relative costs actually decreased for most parts. At this writing, however, the military and space market has declined mightily in proportion to the commercial communications, computer, entertainment, and automotive markets.

The U.S. military has decided that it can no longer maintain its strong efforts in fostering product improvement and in maintaining specifications and standards. Many manufacturers, faced with a small and dwindling military market, are being forced to forego the expenses of maintaining a qualified military status, ending a 50-year trend. The results of this convolution are yet to be seen, but it is apparent that military users will be forced to depend more on industrial and commercial voluntary standards and, if a military (or other) user needs to reduce risk that a part might be unreliable, he will be forced to write his own specifications and to conduct reliability-assurance testing and conditioning himself, or pay a much higher part cost to have the supplier do it. Parts manufacturers, losing the military market resource that was willing to pay a higher up-front price for improved products, might find it more difficult to economically justify high-risk product-improvement initiatives. Whatever the outcome, the passive electronic parts industry, its standards, and its practices will be decidedly altered during the end years of the 1990's and the first decade of the 21st century.

Commercial standards are published by several industry organizations, and instances exist in which two or more organizations have identical standards that are identified by different document numbers. Unfortunately, the identification numbering methods are not uniform, even within the same organization, and several numbering schemes exist for both document identifiers and part number designations, mostly as a function of the chronological vintage of the documents. Revision is also constantly in process, with older specifications being superseded by new ones, and existing documents being revised.

The Electronic Industries Association (EIA) is the most active of the U.S. standards groups insofar as capacitors are concerned. EIA standards and military specifications are often closely parallel, as they should be, but there are also diversities. Some EIA standards are adopted, as written, as international standards, and many are included in the American National Standards Institute (ANSI) listings. The EIA "RS" designator for specification numbers has been dropped in listings, but still appears on older documents. The RS designator is dropped in this chapter.

Tables 2.7, 2.8, and 2.9 summarize current military and industry standard documents for electronic capacitors. Note that commercial and industry standards are almost invariably "voluntary" standards, in which manufacturers and furnishers of the product covered cooperatively publish a document that annunciates their agreed-upon collective position as to what a product marketed as meeting that standard should be. Generally, new and revised standards are published when individuals active in the industry (users, manufacturers, or a combination of the two) perceive a need and convince enough other industry-active individuals to support the action. For instance, at one time there was a producer perception that many user-reported multilayer capacitor defects were not capacitor defects at all, but were actually anomalies induced by the methods used to prepare capacitor cross-sections for viewing. Accordingly, a standardized method and materials document for resin imbedding and polishing ceramic capacitor cross-sections was prepared, circulated for approval and inputs, and published as EIA 469-B, "Standard Test Method for Destructive Physical Analysis of High Reliability Monolithic Ceramic Capacitors."

In parts specifications and standards, the part descriptions are usually quite complete, stating form, fit, and expected performance based on ability to pass tests that are described in detail. Contrary to military specifications, which are "mandatory" standards, the voluntary standards do not mandate required test sample sizes, testing frequency, and other conditions of acceptance that are *required* by military specifications, but sometimes state *recommended* standard sampling and testing. Other than this, many industrial standards are very similar to their military counterparts, serving very well to moderate, if not eliminate, the proliferation of part package types, sizes, capacitance values and voltage ratings, and to set standards of expected performance. Standardization of part types, sizes, test methods, identification marking, and packaging provide immeasurable benefits to producer and user alike. This is seldom valued highly enough, and when quality standards are added to the list of benefits, fruits of standardization activities become enormous.

In addition to those military specifications listed, standardized test and inspection methods are published in MIL-STD-202, Standard Test Methods for Electronics, MIL-PRF-38534, Hybrid Microcircuits, General Specification for, and other related documents. EIA documents sometimes contain standard environmental test methods as well as electrical test standards. EIA 186 contains standards for environmental testing of passive electronic parts. Defense Electronics Supply Center Standardized Military Drawings (DESC SMD) also cover a number of capacitor types and designs that are not specifically covered by military specification listings. SMD drawings exist for packaged capacitor arrays, extended capacitance-voltage parts, special high-voltage capacitors, and for form factors not covered by the military specifications. Table 2.9 is a partial listing of SMD's that are current as of this writing. Addresses of standards organizations are given in the footnotes for Tables 2.7 through 2.10.

2.4 CAPACITOR TYPES AND RANGES OF PERFORMANCE

Figure 2.5 shows capacitor dielectric types and the ranges of capacitance values and working voltages for practical capacitors. Aspects of capacitor performance other than capacitance value and size enter into a selection decision, however, to a greater degree than for most

TABLE 2.7 Military and Commercial Capacitor Standards and Specifications.

Capacitor type	Commercial			Military		
	Specification	Style	Remarks	Specification	Style	Remarks
Electrolytic						
Solid-electrolyte tantalum w/leadwires	EIA 535AAAA, EIA 535AAAB-, EIA 535AAARF			MIL-C-39003[1], MIL-C-49137	CSR, CS[2], CX	Hermetic, metal case; Dipped, molded
Solid-electrolyte tantalum, chip	EIA 535BAAA- EIA 535BAAD			MIL-C-55365[1]	CWR	Coated, molded
Wet-foil, wet-slug tantalum	EIA 535ABAA- EIA 535ABAE, EIA 535ACAA- EIA 535ACAD			MIL-C-3965, MIL-C39006[1]	CL, CLR, CLS[2]	Foils, slugs; Foils, slugs
Aluminum electrolytic	EIA 395 (long life), EIA 395-1 (gen purp use), EIA 463 (ac motor hvy dty), (ac motor lt dty)	Type 1, Type 2, Type 1, Type 2		MIL-C-62, MIL-C-39018[3]	CE, CU, CUR	Axial and single-ended; Axial and single-ended
Double layer				DOD-C-29501		Molded plastic, metal can
Ceramic/mica/glass Ceramic, leaded	EIA 198 (CL I, II, III, IV, incl HV)	CC		MIL-C-20[3]	CC, CCR	Temperature stable and compensating, tubular, disc, molded, resin coated
	EIA IS-39 (glass-encased axial ld)	AG	Axial ld	MIL-C-123[1]	CKS[2]	Temperature stable and general purpose, molded resin coated

TABLE 2.7 Military and Commercial Capacitor Standards and Specifications. (*Continued*)

Capacitor type	Commercial			Military		
	Specification	Style	Remarks	Specification	Style	Remarks
				MIL-C-11015	CK	General purpose, molded, resin coated
				MIL-C-39014[1]	CKR	General purpose, molded, resin coated, multilayer high voltage
				MIL-C-49467[1]	HVR	All ceramic types, commercial item description
Ceramic, surface mount	EIA IS-36 (MLC chips)			A-A-55089	---	
	EIA IS-37 HV MLC chips)					
	EIA 595 (visual insp. of MLC chips)			MIL-C-55681[1]	CWR	Temperature stable and general-purpose MLC, has high-frequency types
Ceramic, chip, high frequency				MIL-C-49464[1]	CPCR	Single-layer, high-frequency microstrip parallel plates, stable, and general purpose
Mica, leaded	EIA 153			MIL-C-5	CM	Resin coated, molded
				MIL-C-39001[1]	CMR$_2$	Resin coated
				MIL-C-87164[1]	CB	Resin coated
				MIL-C-11015		Feedbthrough, standoff, resin sealed, metal-

Type	EIA	MIL-C	Code	Description
Glass		MIL-C-11272	CY	Glass, porcelain
		MIL-C-23269[1]	CYR	Hermetic glass w/leads
Film/Paper				
Paper/film-foil, paper-foil	EIA 361 (feed-through)	MIL-C-19978[3]	CQ, CQR	Hermetic glass, ceramic, metal
	EIA 376 (dc)	MIL-C-27297	CTM	Plastic, dc
	EIA 401 (pwr semi cond. use)	MIL-C-25	CP	Hermetic, metal, dc
	EIA 454 (ac)	MIL-C-11693[3]	CZ, CZR	Hermetic metal feedthrough, dc, ac, RFI reduction
Metallized paper, film, or paper and film	EIA 377 (dc)	MIL-C-83439	CA	Hermetic, metal, EMI feedthrough
	EIA 356 (ac)	MIL-C-12889	CH	Hermetic, metal, bypass, dc, ac, RFI reduction
		MIL-C-18312	CHR	Hermetic, metal, dc
	EIA 530A000 (chip, metallized PETP)	MIL-C-39022[1]	CFR, CRH, CRS[2]	Hermetic metal or ceramic, dc, ac
	EIA 495 (metallized paper for ac)	MIL-C-55514[1]	CHS2	Nonmetal, dc, ac
		MIL-C-83421[1]		Hermetic metal or ceramic, dc or dc/ac
		MIL-C-87217[1]		Hermetic, metal dc, low energy, high impedance

TABLE 2.7 Military and Commercial Capacitor Standards and Specifications. *(Continued)*

Capacitor type	Commercial			Military		
	Specification	Style	Remarks	Specification	Style	Remarks
Trimmer/variable						
Ceramic				MIL-C-81	CV	Class-I ceramic
Glass, quartz, air, gas				MIL-C-92	CT	Air dielectric trimmer
				MIL-C-14409	PC	Air, glass, quartz piston-types
				MIL-C-2318[3]		High-voltage, high-current air, gas dielectric

[1]Established reliability military specification and part style. Demonstrated failure rates: M - less than 1.0% per 1000 hours, P - less than 0.1% per 10000 hours, R = less than 0.01% per 1000 hours, S - less than 0.001% per 1000 hours.
[2]Military high reliability specification or part style. S-level or better.
[3]Specification contains both nonestablished reliability and established reliability styles.

TABLE 2.8 Military and Commercial Standard Information, Testing, and Applications Instructions

Subject/Title	Identifying Number
Ceramic capacitors	
Application Guide for Multilayer Ceramic Capacitors	EIA 521[1]
Hybrid Microcircuit Ceramic Chip Capacitors Specification Guidelines	ISHM SP 004[2]
Surface Mounting of Multilayer Chip Capacitors, Guidelines	EIA CB11[1]
Visual Inspection of MLC Chips	EIA 5951
Decoupling Capacitor DIP Sockets for Electronics Equipment	EIA 540DBAA[1]
Paper, film, paper-film and power capacitors	
Power Line Carrier Coupling Capacitors and Coupling Capacitor Voltage Transformers	ANSI C93.1[3]
Standard for Series Capacitors in Power Systems	IEEE 824[4]
Standard for Shunt Power Capacitors	IEEE 18[4]
Guide for Application of Shunt Power Capacitors	IEEE 103[4]6
External Fuses for Shunt Power Capacitors	NEMA CP-9[5]
Guide for Protection of Shunt Capacitor Banks	IEEE C37.99[4]
Electrolytic capacitors	
Storage Shelf Life and Reforming Procedures for Aluminum Electrolytic Fixed Capacitors	MIL-STD-1131
General capacitor subjects	
Selection and Use of Capacitors	MIL-STD-198[6]
Effective Series Resistance and Capacitance Measurements at High Frequencies	EIA 483[1]
Applications Guide for Soldering and Solderability Maintenance of Leaded Electronic Components	EIA 534[1]
Design Guide-Components and Their Mounting Characteristics	IPC D-30, sec 5-92[7]
Standard for Safety Capacitors	UL810[8]
Underwriters Laboratories Standard for Safety for Across-the-Line, Antenna Coupling, and Line Bypass Capacitors for Radio and TV-Type Appliances	UL1414[8]
Lead Taping of Axial Lead Components	EIA 296[1]

[1] Electronic Industries Association: 2500 Wilson Boulevard, Arlington, VA 22201-3834, Telephone 703-907-7500.

[2] International Society for Hybrid Electronics: 1850 Centennial Park Drive, Suite 105, Reston, VA 22091, Telephone 703-758-1060.

[3] American Society of Mechanical Engineers: (American National Standards Institute) 348 East 47th Street, New York, NY 10017, Telephone 212-705-7722.

[4] Institute of Electrical and Electronics Engineers, Inc.: 445 Hose Lane, PO Box, 1331 Piscataway, NJ 08855-1331, Telephone 800-678-4333.

TABLE 2.8 Military and Commercial Standard Information, Testing, and Applications Instructions *(Continued)*

[5] National Electrical Manufacturers Association: 1300 N. 17th Street Suite 1847, Rosslyn, VA 22209, Telephone 703-841-3200.

[6] U.S. Department of Defense. See footnotes to Table 2.9.

[7] The Institute for Interconnecting and Packaging Electronic Circuits: 2215 Sanders Road, Northbrook, IL 60062-6135, Telephone 708-509-9700.

[8] Underwriters Laboratories, Inc.: 333 Pfingsten Road, Northbrook, IL 60062, Telephone 847-272-8800.

TABLE 2.9 Current Standardized Military Drawings for Capacitors

Ceramic, Porcelain, Mica

DESC* SM Drawings

87040	MLC High Voltage X7R 2 kV dc
87043	MLC High Voltage X7R 1 kV dc
87046	MLC High Voltage CG 1 kV dc
87047	MLC High Voltage X7R 3 kV dc
87052	Assembly
87070	MLC High Voltage X7R 5 kV dc
87076	MLC High Voltage CG 4 kV dc
87077	MLC High Voltage CG 5 kV dc
87081	MLC High Voltage X7R 10 kV dc
87106	Ceramic, Switch Mode Power Supply
87107	Ceramic Feedthrough
87112	Ceramic BX 7-Section Assy SIP Package
87114	MLC High Voltage CG 3 kV dc
87116	Ceramic CG 7-Section Assy SIP Package
87119	Ceramic BX 8-Section Assy SIP Package
87120	Ceramic CG 8-Section Assy SIP Package
87122	Ceramic BX 4-Section Assy SIP Package
87125	Ceramic Disc High Voltage 15 kV dc
88011	Ceramic Switch Mode Power Supply CG
88019	Ceramic BX 9-Section Assy SIP Package
88035	Mica High Voltage
88037	Ceramic Chip High Voltage
89044	MLC High Voltage X7R 4 kV dc
89045	Porcelain Chip Ultra High Q
89086	Ceramic CG 4-Section Assy SIP Package
89087	Ceramic Disc High Voltage 20 kV dc
89089	Ceramic Chip High Voltage
91019	Ceramic Chip .56 – 1.0 μF
94006	Ceramic Chip

Marshal Space Flight Center

MSFC-SPEC-85M03931	Ceramic High Voltage

Goddard Space Flight Center

GSFC-SPEC-S-311-P-15	Ceramic High Voltage
GSFC-SPEC-S-311-P-15/1	Ceramic Disc HV Monolayer
GSFC-PROC-S-311-S-536	Screening Procedure for Tusonix HV Ceramic

TABLE 2.9 Current Standardized Military Drawings for Capacitors *(Continued)*

Paper, Film and Paper-Film

DESC* SM Drawings

88012	Metallized Teflon Hermetic
88028	Paper or Paper-Plastic
88029	Metallized Polypropylene for AC Application
89028	Metallized Film or Metallized Paper-Film Hermetic
89041	Metallized PTFE Hermetic
89062	Metallized Polycarbonate Non-hermetic
89115	Metallized Polyester Film Surface Mtg Non-Hermetic
92019	Metallized Polyethylene Terepthalate DC Hermetic
92020	Metallized Plastic DC Hermetically Sealed

Electrolytic

DESC* SM Drawings

88022	Polarized Al Electrolytic Sgl End Mtg
89012	Polarized Al Electrolytic Low Screw Insert Mtg
89021	Polarized Al Electrolytic Ultra Low ESR
89022	Ta Module
89085	Polarized Al Electrolytic Low Profile Snap Mtg
91024	Assemblies Hermetic All-Ta Wet Ta Slug
93026	Polarized Wet Slug Ta
93026	Polarized Wet Slug Ta

Marshal Space Flight Center

MSFC-SPEC-85M03923   Wet Ta Hermetically Sealed

Goddard Space Flight Center

GSFC-SPEC-S-311-P-17   Solid Ta for Space Flight

Kennedy Space Center

KSC-SPEC-0009	Wet Ta Etched Foil Non-Polar KCL23
KSC-C-133	Polarized Wet Ta Slug KCL65

Double Layer

DESC* SM Drawing

92001   Double Layer (classed as electrolytic)

---

*Defense Electronic Supply Center. Formerly located in Dayton, Ohio, now consolidated into Defense Supply Center Columbus (DSCC) at 3990 East Broad Street, Columbus, OH 43216-5000, document control department is VA, Telephone (614) 692-7603. Government documents are available from Standardization Document Order Desk, Building 4D, 700 Robbins, Ave, Philadelphia, PA 19111-5094, Telephone (215) 697-3321.

other passive-circuit component types. The overlapping ranges in the figures might not necessarily provide the range of type choices that appears to be available for given applications.

Figure 2.13 shows laboratory frequency-sweep comparisons of impedance and apparent series capacitance for certain solid tantalum, ceramic, and film capacitors. Special designs for particular applications also are available. Higher-than-normal-value ceramic capacitors have been made for switching regulator power supplies, for instance. Prior to 1965, vacuum tube circuits predominated, and available capacitors were largely those suitable for high-impedance, relatively high-voltage circuits: low capacitance, high-Q mica and ceramics, many low-capacitance high-voltage paper types, and high-voltage electrolytics.

Solid-state circuits operate at lower impedances and require much lower voltages, and their increased complexity and smaller active-element size generate a constant drive for ever higher stable-capacitance values at low voltage in ever smaller volumes. Some of the older capacitor types survive with much-reduced usage, many have virtually disappeared, and some have been adapted to the changing needs by new production techniques and expanded capabilities. Many specialized capacitor designs are produced, with some degree of voluntary standardization, to serve particular usage niches.

Electrical power applications are probably the largest usage area of specialized standard designs, parts associated with the power broadcasting and large radars are a smaller, but significant, market for high-quality RF power capacitors.

A large and growing usage area is that of small radar and communications assemblies designed for use at frequencies that range upward into the millimeter wavelengths. This equipment uses microstrip RF substrates upon which active- and passive-component chips are mounted and connected using microcircuit wire-bonding techniques, microcircuit soldering and conductive polymer techniques. The needed parts range from surface-mount conventional types for power decoupling to tiny parts having capacitances in the single-digit picofarad range.

Measurement techniques and specifications for performance at the frequency of use for the highest-frequency parts must use waveguide transmission-line techniques, making a very specialized market. Part designs use low-loss materials in mostly single-layer construction, with porcelain, NPO and other ceramics, and MOS construction dominating. In MOS construction, a single crystal substrate, usually doped silicon, has a silicon dioxide (glass) or silicon nitride layer deposited for its dielectric, followed by a metallization layer for the opposite plate, with a final passivation layer, all deposited using microcircuit techniques. The silicon backside is metallized for circuit connection. A few present standards exist for these capacitor types.

2.5 PAPER AND PLASTIC FILM CAPACITORS

Tubular paper capacitors are one of the oldest capacitor designs still available. Large numbers have been produced since the early days of telephone, telegraph, and radio, and many are still being made. When properly constructed with the high-grade kraft material now obtainable, and when protected from moisture, the rolled paper, foil, metallized paper capacitors now available are a reliable relatively low-cost selection for many uses, and are superior to polymer films for some applications, offering long, reliable life in current-impulse service, for example.

Film capacitors are better choices for most applications in small-scale solid-state functional electronics constructions, but paper dielectric capacitors have many practical uses as motor start and run capacitors, power-supply filtering and decoupling, across the line-power filtering and decoupling, automotive power decoupling, high-power energy storage and pulsing, and as switching contact protectors. Mineral oil, mineral wax, or other im-

TABLE 2.10 Examples of Capacitor Part Number Identifiers.

Non-significant identification examples

M39018/01-1075R

M is military identifier, 39018 is the basic specification number, 01 is the detail specification sheet ("slash number"), which covers the style or case and leadwire type, polarization, and element design, 1075 is the dash number, which in this instance, specifies detail case size, capacitance, voltage rating, and capacitance tolerance, and R is the demonstrated failure rate. MIL-C-39014, however, includes the demonstrated failure rate in the dash number (e.g., M39014/02-1320V specifies a 470,000 pF ±10%, 100-volt BX characteristic MLC capacitor with demonstrated failure rate R, solderable leadwires, and a molded-in V-shaped case standoff feature).

Significant identification type (a) example

CDR32BP110BFWR

CDR32 identifies the style, which in this instance, is a MIL-C-55681 MLC chip 1.6 mm × 3.2 mm × 1.3 mm max height, with wrap-around terminal metallization. The BP specifies the temperature range, B denotes –55°C to +125°C and P, the voltage-temperature characteristic limits, with P denoting 0 ± 30 ppm/°C. 110 is the capacitance –11.0 pF, B is the rated voltage of 100 V, F is the capacitance tolerance of ±1 percent, W is the termination finish (barrier metal with solder finish) in this instance, and R is the demonstrated failure-rate level.

Significant identifier type (b)

CFR02AMC682JM

MIL-C-55514 non-hermetic film capacitor case marking identifier breaks down as follows. CFR02 is the military identifier, next letter may be R for radial leadwires or A for axial, M is the dielectric material and high temperature limit identifier: M = PET at 85°C, N = PET at 85°C, Q = PC, 85°C voltage de-rated to 67% at 125°C, R = PC, 85°C voltage de-rated to 50% at 125°C, S = Poly-para-xylylene, 85°C voltage de-rated to 50% at 125°C. Third letter following CFR02 is the operating rated voltage, A is 50, B is 100, C is 200, D is 300, E is 400, F is 600, G is 75, H is 150, J is 25, K is 250. 682 in this case is the capacitance value in pF (6800), J is the tolerance (±5%), and M is the demonstrated failure rate number. (Note that the rated voltage symbology does not completely follow the standard of Table 2-10.)

Significant identification type (c)

M49464F01A001PN

This number signifies military specification MIL-C-49464 high frequency parallel plate ceramic capacitor, F temperature characteristic (0 ± 15 ppm/°C), detail specification sheet number 1, case size A (must consult specification), dash number 001 (must consult specification detail sheet to obtain capacitance, rated voltage, and demonstrated failure rate), P is the capacitance tolerance, (±0.01 pF in this instance), and N is the termination finish (consult specification).

pregnation is a necessary part of the construction, with the porous paper acting as a good matrix for impregnants. Some constructions have plastic film interleaved with a double-metallized impregnated-paper "soggy foil" to provide a self-healing effect. Plastic film capacitors began as better-performance substitutes for paper dielectric parts, and the familiar rolled tubular construction that predominated for paper capacitors was adapted for film types. Most film capacitors are still of rolled-layer construction, but some have departed

A: |Z| B: Cs   o MKR   3 019 951.720 Hz
A MAX   5.000   Ω   MAG   1.05970   Ω
B MAX   5.000   μF   Cs   -271.012   nF

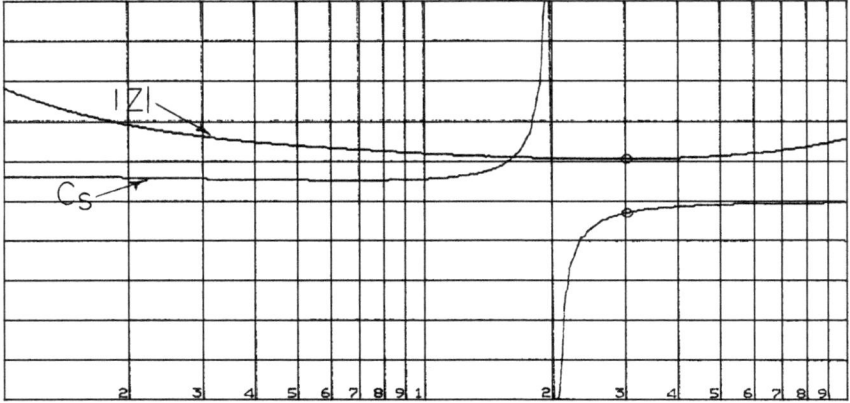

A/DIV   1.000   Ω   START   100 000.000 Hz
B/DIV   1.000   μF   STOP   10 000 000.000 Hz
OSC= 5.00E-01 V

0.68μF 50V MILITARY CSR 13 SOL TA |Z| & Cs

A: |Z| B: Cs   o MKR   3 019 951.720 Hz
A MAX   5.000   Ω   MAG   270.005   mΩ
B MAX   5.000   μF   Cs   -196.364   nF

A/DIV   1.000   Ω   START   100 000.000 Hz
B/DIV   1.000   μF   STOP   10 000 000.000 Hz
OSC= 5.00E-01 V

~0.68μF 50V BX CERAMIC |Z| & Cs

FIGURE 2.13 Comparative frequency-swept performance data for several capacitors. The data was taken with Hewlett-Packard 4194 Gain-Phase Analyzer using approximately 1-inch total leadwire length. Discontinuities on capacitance value plots occur as self-resonance is passed and arise because the instrument measures the capacitive reactance and calculates the value of capacitance from $C = (2\pi f X_C)^{-1}$. As the true X_C = apparent X_L at self-resonance, the apparent reactive value tends to zero and the calculated C tends to infinity. Past self-resonance, the reactance is inductive, as indicated by an indicated negative capacitance value. On the R-X plots, capacitive reactance is shown as a negative quantity, and inductive reactance is depicted as positive. *(Courtesy of Lockheed Martin Electronics and Missiles Engineering Laboratories.)*

```
A:   |Z|      B:  Cs      o  MKR      3 019 951.720 Hz
A MAX   5.000    Ω        MAG           738.906   mΩ
B MAX   5.000    µF       Cs          -165.141    nF
```

Cₛᐧᐧᐧᐧᐧᐧᐧᐧᐧ↘

|Z|ᐧᐧ→

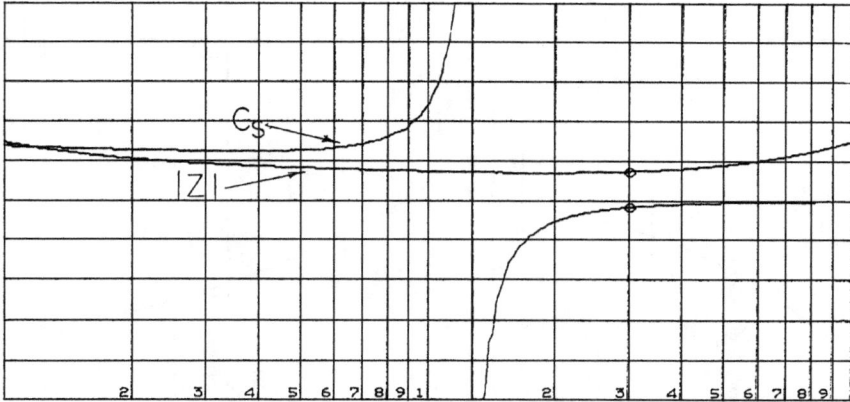

```
A/DIV   1.000    Ω      START      100 000.000 Hz
B/DIV   1.000    µF     STOP    10 000 000.000 Hz
OSC= 5.00E-01 V
```

1·5 µF 50V MILITARY CSR 13 SOL TA IZI & Cₛ

```
A:   |Z|      B:  Cs      o  MKR      3 019 951.720 Hz
A MAX   5.000    Ω        MAG           169.765   mΩ
B MAX   2.000    µF       Cs          -313.279    nF
```

|Z|↘

Cₛ↘

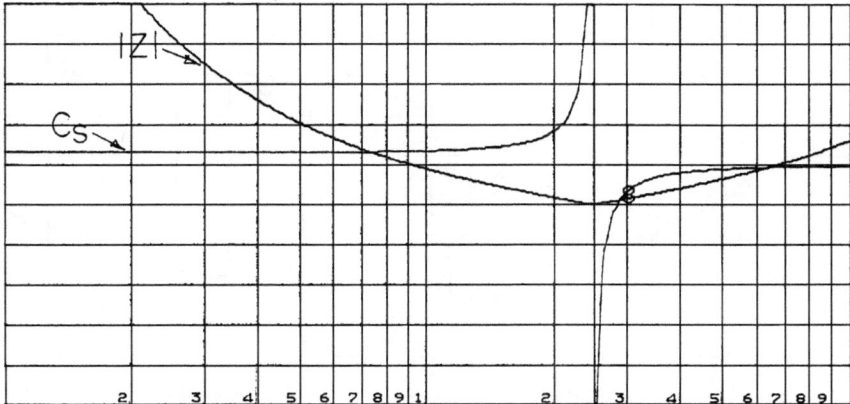

```
A/DIV   1.000    Ω      START      100 000.000 Hz
B/DIV   500.0    nF     STOP    10 000 000.000 Hz
OSC= 5.00E-01 V
```

0.15 µF 100V MTLZD PET IZI & Cₛ

FIGURE 2.13 Comparative frequency-swept performance data for several capacitors. The data was taken with Hewlett-Packard 4194 Gain-Phase Analyzer using approximately 1-inch total leadwire length. Discontinuities on capacitance value plots occur as self-resonance is passed and arise because the instrument measures the capacitive reactance and calculates the value of capacitance from $C = (2\pi f X c)^{-1}$. As the true X_C = apparent X_L at self-resonance, the apparent reactive value tends to zero and the calculated C tends to infinity. Past self-resonance, the reactance is inductive, as indicated by an indicated negative capacitance value. On the R-X plots, capacitive reactance is shown as a negative quantity, and inductive reactance is depicted as positive. *(Courtesy of Lockheed Martin Electronics and Missiles Engineering Laboratories.) (Continued)*

```
A:  R         B:  X          o MKR     3 019 951.720 Hz
A MAX    1.450      Ω        REAL           1.09438      Ω
B MAX    2.000      Ω        IMAG         175.020       mΩ
```

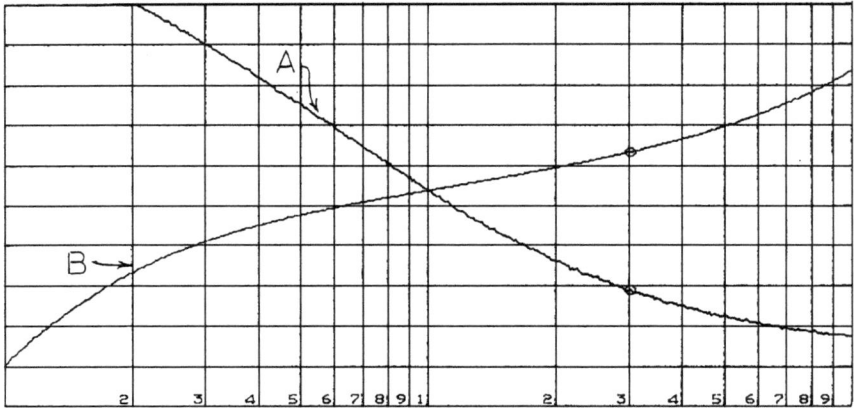

```
A/DIV   50.00    mΩ      START          100 000.000 Hz
B/DIV   500.0    mΩ      STOP       10 000 000.000 Hz
OSC=  5.00E-01   V
```

0.68 μF 50V MILITARY CSR 13 SOL TA   R&X

```
A:  R         B:  X          o MKR     3 019 951.720 Hz
A MAX   100.0     mΩ         REAL          29.3115      mΩ
B MAX    2.000     Ω         IMAG         267.600       mΩ
```

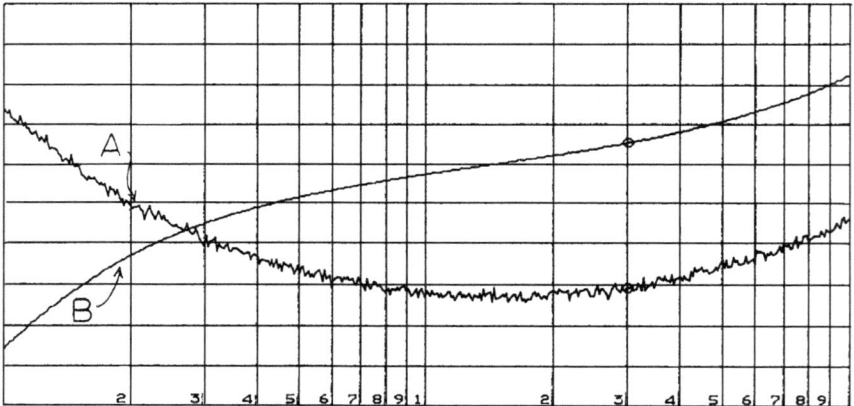

```
A/DIV   10.00    mΩ      START          100 000.000 Hz
B/DIV   500.0    mΩ      STOP       10 000 000.000 Hz
OSC=  5.00E-01   V
```

0.68 μF 50 V BX CERAMIC   R&X

FIGURE 2.13 Comparative frequency-swept performance data for several capacitors. The data was taken with Hewlett-Packard 4194 Gain-Phase Analyzer using approximately 1-inch total leadwire length. Discontinuities on capacitance value plots occur as self-resonance is passed and arise because the instrument measures the capacitive reactance and calculates the value of capacitance from $C = (2\pi f X_C)^{-1}$. As the true $X_C =$ apparent X_L at self-resonance, the apparent reactive value tends to zero and the calculated C tends to infinity. Past self-resonance, the reactance is inductive, as indicated by an indicated negative capacitance value. On the R-X plots, capacitive reactance is shown as a negative quantity, and inductive reactance is depicted as positive. *(Courtesy of Lockheed Martin Electronics and Missiles Engineering Laboratories.) (Continued)*

```
A:  R          B:  X            O  MKR       3  019  951.720  Hz
A  MAX     180.0    mΩ          REAL              25.0729    mΩ
B  MAX     2.000     Ω          IMAG             184.259    mΩ
```

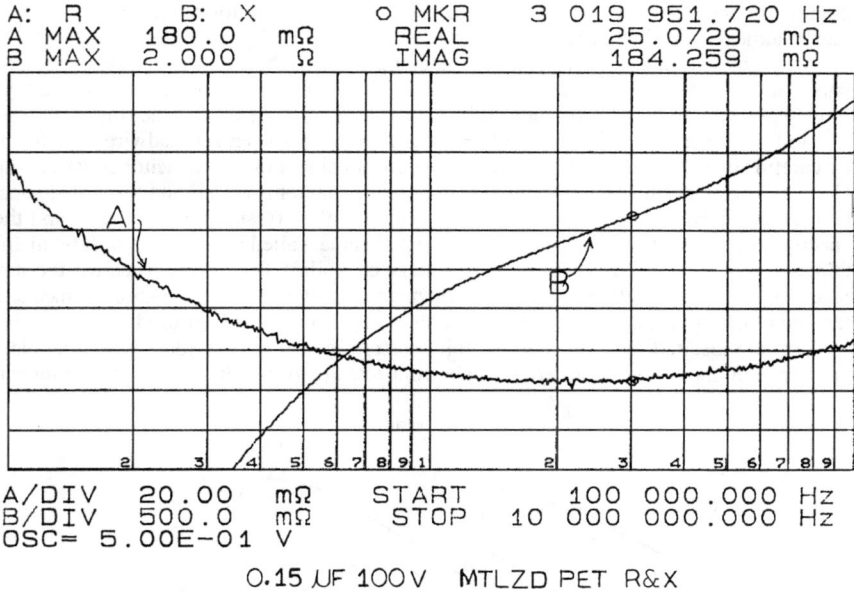

```
A/DIV    20.00    mΩ        START                100 000.000  Hz
B/DIV    500.0    mΩ        STOP       10 000 000.000  Hz
OSC=  5.00E-01  V
```

0.15 UF 100 V   MTLZD PET R&X

FIGURE 2.13 Comparative frequency-swept performance data for several capacitors. The data was taken with Hewlett-Packard 4194 Gain-Phase Analyzer using approximately 1-inch total leadwire length. Discontinuities on capacitance value plots occur as self-resonance is passed and arise because the instrument measures the capacitive reactance and calculates the value of capacitance from $C = (2\pi f X_C)^{-1}$. As the true X_C = apparent X_L at self-resonance, the apparent reactive value tends to zero and the calculated C tends to infinity. Past self-resonance, the reactance is inductive, as indicated by an indicated negative capacitance value. On the R-X plots, capacitive reactance is shown as a negative quantity, and inductive reactance is depicted as positive. *(Courtesy of Lockheed Martin Electronics and Missiles Engineering Laboratories.) (Continued)*

from that norm to achieve improved case geometry and enhanced performance by stacking layers of metallized films or film and foil layers.

2.5.1 Rolled Paper and Film Construction

Basic rolled construction is shown in Figure 2.14. Inserted-tab termination, once used extensively for termination of rolled-construction capacitors has largely given way to the electrically superior extended-foil configuration. Inserted-tab termination components use foil strips that are periodically layered into the winding layer structure, forming pressure electrical contacts with the capacitor plate material. The tabs, with those contacting separate plates being extended from opposing ends of the roll, are collected and consolidated on each end of the structure for the attachment of terminating leadwires.

In extended-foil construction, the two plate-conductor foils or metallized dielectrics are layered with conductive edges of opposing plates situated at opposing ends of the roll and with the conductive foil or metallized area recessed from the roll end, opposite that where its edge is exposed. Figure 2.15 shows this and other capacitor-winding details. Layers of film, foil, or other plates and dielectrics are wound around a mandrel. The mandrel void is later plugged or the element is pressed to close it. Extended-foil construction has advantages in that it does not depend on pressure contact, and because it reduces the capacitor series resistance and inductance over that of inserted tab. Extended foils are commonly pressed together on each end of the capacitor roll to form flat surfaces for further treatment as leadwire attachment surfaces.

After press-forming the capacitor or plugging to close the mandrel cavity, metallized paper and metallized films have finely divided hot zinc, tin, or other metals or alloys flame-sprayed into the layered ends. This makes electrical contact with the film metallization to form a surface, referred to in Europe as a "Schoop" layer, for leadwire attachment by careful soldering. Termination formation is a critical part of the capacitor construction. Good contact with film or paper metallization is essential for low ESR and low inductance. Good mechanical strength must be attained, often with very small contact area, and the spacing from contact to the recessed opposite plate area, called the *margin*, must be maintained to provide dielectric-withstanding voltage capability. This margin is across the surface of the dielectric with no solid dielectric material between the termination surface and the end of the opposing plate conductors, unless impregnation is later used.

For both inserted-tab and extended-foil constructions, if the outside layer of the plate conductor on the roll is identified, it can be connected to ground in the circuit, forming an

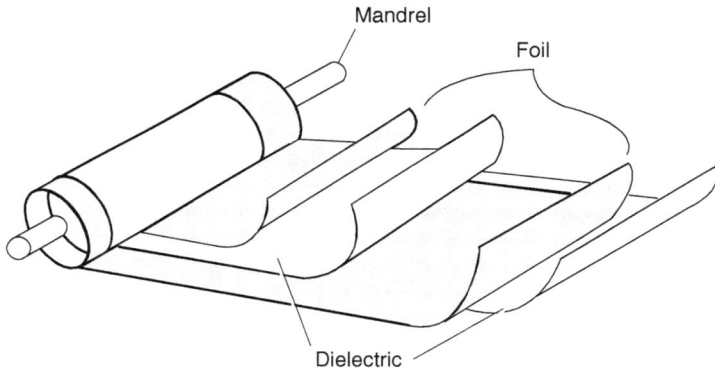

(a)

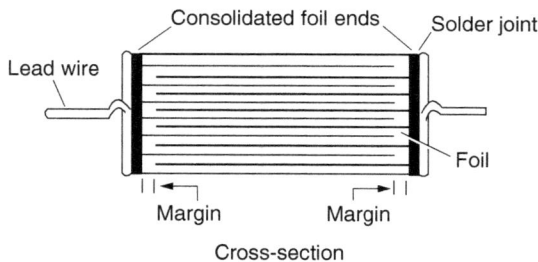

(b)

FIGURE 2.14   Rolled paper and film element construction: (a) basic extended foil construction, (b) extended foil cross-section.

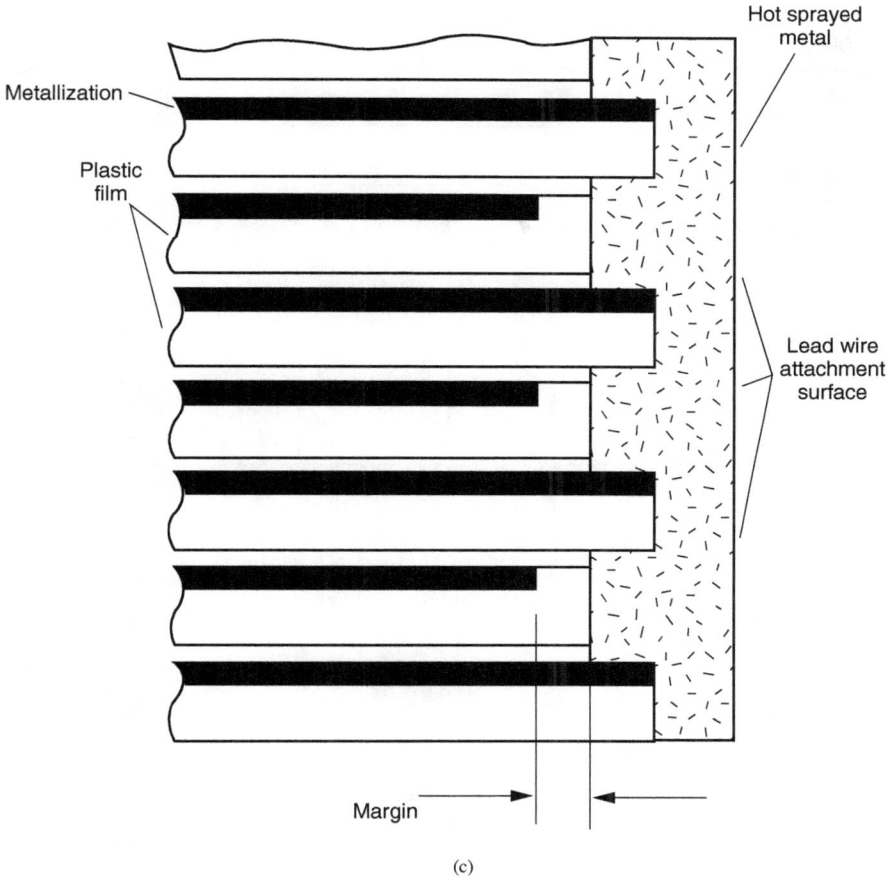

Hot sprayed metal

Metallization

Plastic film

Lead wire attachment surface

Margin

(c)

FIGURE 2.14 Rolled paper and film element construction: (c) exaggerated cross-section of end termination for metallized dielectric-film constructions. *(Continued)*

electrostatic shield around the capacitor body. In metal-cased capacitors, this outside foil is normally connected to the case and one leadwire, unless the intended circuit use makes isolation from the case desirable. Capacitors with metal cases that are electrically connected to the capacitor inside have insulating sleeving placed over the case outside to avoid short-circuiting to adjacent conductors (see DWV testing). Rolled capacitors not in metal cases have their outside foil identified by a mark on the case consisting of a band or segment of a band around the case near the end.

After rolling, many film capacitors are pressed to form oval, rectangular, or even triangular cross-sections to provide for efficiencies in application circuit placement. Manufacturers, depending on the film used, heat and vacuum-condition the rolled capacitors to remove interstitial moisture and to consolidate and tighten the structure as stresses are relieved and films shrink. Note that the ESR and inductance of rolled-capacitor construction can be reduced by making the roll shorter with a larger diameter. The volumetric efficiency might suffer because of the relatively larger volume required for terminations and case clo-

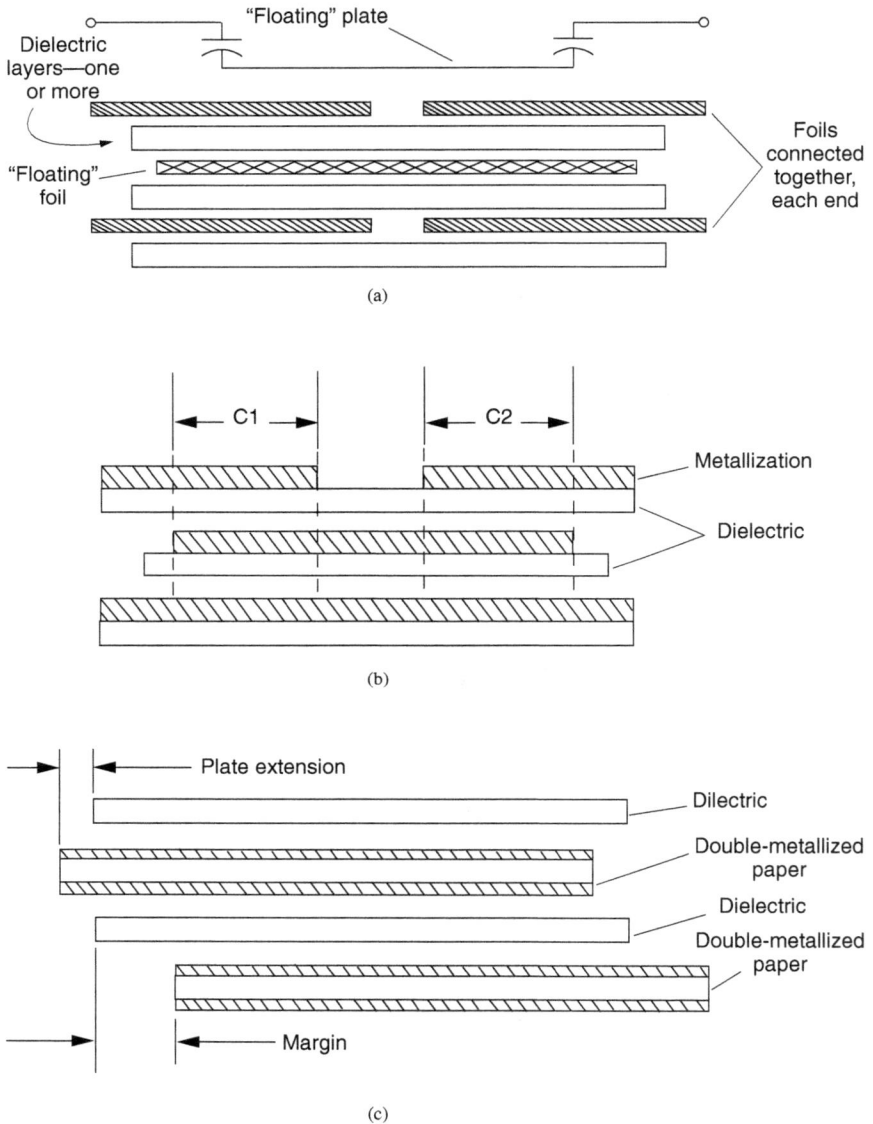

(a)

(b)

(c)

FIGURE 2.15   Some film-capacitor construction variations: (a) Series-wound foil construction, (b) Series-wound metallized film winding. Metallization patterning can be used to construct a number of effective series-connected capacitors between the end terminations. (c) The "soggy plate" construction uses impregnated double-metallized paper as a plate, taking advantage of the self-healing clearing capability. Any number of dielectric layers can be used in any of the designs. The "soggy plate" construction can also be series wound.

sure. Long, slender cylinders are more efficient volumetrically, but have higher ESR and inductances.

2.5.2 Series-Wound Film Capacitors

To gain higher-rated voltages using thinner film, some capacitors are series wound. In this configuration, two series-connected capacitors are constructed by interleaving with a film dielectric to separate a double metallized (both sides) kraft paper from the two foils. The electrically floating metallized paper plate has metallization margins on both ends, and the foils are recessed on opposing ends to form dielectric margins. The metallized-paper floating plate is impregnated, its porous nature readily accepting the impregnant, forming a "soggy plate" with the result being two series-connected capacitors of approximately double the voltage rating of either, but half the capacitance value of each, and with the self-healing capability of the common metallized paper plate. Add the corona-formation resistance of an impregnated construction, and a serviceable high-voltage capacitor is the result. Figure 2.15 shows a series-winding schematic.

2.5.3 Stacked-Plate Film Capacitors

In the efforts to provide producible smaller size, better-performing, competitively priced film capacitors, many construction and packaging techniques have been developed, among them, the small stacked-plate metallized-film capacitor. Most stacked-plate constructions use spool-winding of specially patterned metallized films, but on large spools— about 1 meter in diameter. After winding, the resulting layered metallized film cylinder is cut into layered strips that, because of the large diameter of the winding and heat tacking or other means used to fix layer-relative positions, are virtually flat. Termination edges of the layered patterned metallization are treated by the aforementioned hot-metal spraying to form termination surfaces. The layered strips are cut again, into flat capacitor elements ready for lead attachment and encasement.

The large-roll stacking technique is adaptable to automated processing of multiple "sticks" of capacitors attached to one another, which can later be cut apart. A similar-shaped capacitor element can be made by conventional winding on a flat mandrel, but is less adaptable to high-rate production. The thin, flat capacitors (see Figure 2.16) are typically encased in close-fitting heat-closed polyester cases or are processed into plastic molded units. Ceramic-cased hermetic parts are made, with leaded metal end caps welded to a foil tab that is attached to the capacitor end metallization, then high-temperature soldered to metallized portions of the encasing ceramic to complete the seal. Flat-film capacitors using chip-type surface-mount terminations have been proposed at this writing and without doubt will see increasing use as the problems of high temperatures in surface-mount soldering and their effects on the film dielectrics are solved.

2.5.4 Encasement of Film and Paper Capacitors

Available encasement configurations for rolled capacitors with leadwires attached are adapted to particular expected service environments and installation methods used. If an impregnant is used, the structure must be dried by heating under a vacuum to remove air and water, with the vacuum being broken while the capacitor is immersed in the liquid impregnant, and then, if the impregnant remains liquid, it must be encased in an outside structure that can be sealed well enough to contain the liquid at all the expected operating temperatures. Figure 2.17 shows some common constructions.

Sprayed-on
metal termination

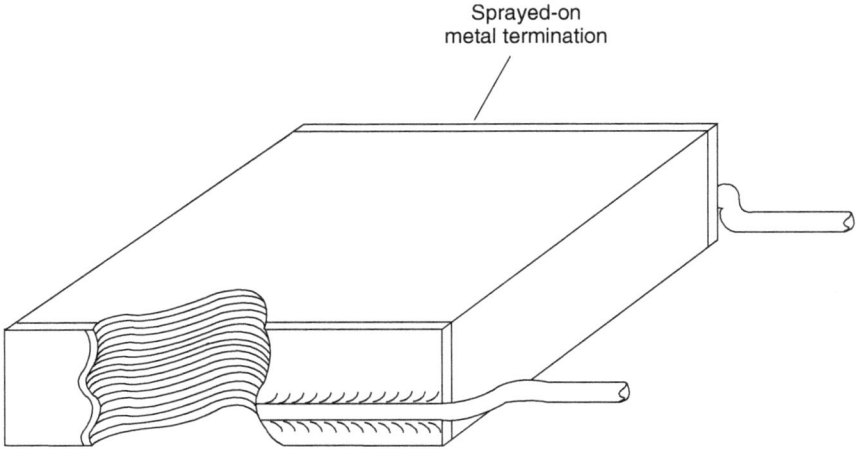

FIGURE 2.16   Stacked film construction.

Hermetic metal cases commonly have tin-lead solder-attached glass-bead or ceramic feedthroughs to the case. Tubular cases have glass-metal seals with a tubular through-glass conductor, into which the leadwire is inserted after the capacitor body is in place in its support spacers inside the case; the opposite leadwire being extended through a hole in the opposite case end (a hole in the case if the leadwire is to be attached to the case, a previously installed glass seal with tubular feedthrough, if not). After solder-sealing the case-attached or opposite feedthrough leadwire, impregnation is completed, if applicable, and the top glass seal is soldered to the case and its leadwire tube is filled with solder, completing the seal.

Cases are typically tin- or tin-lead-coated brass; steel is avoided because of its magnetic properties. Differences in thermal expansion properties of the capacitor roll and the case material make it necessary to provide some form of strain relief between the leadwire, now rigidly affixed to the case, and the mechanically weak joint between leadwire and capacitor roll, lest thermal cycling degrade or destroy the electrical connection to the foil or end metallization. A flexible relief is provided usually by forming the leadwire into a helix, hairpin, or other nonrigid shape between the capacitor end and the inside of the end seal. Large capacitors with post terminals on a single end can be constructed on the terminating structural surface, which then becomes the lid to be soldered or welded onto a drawn metal case or resin-sealed into a plastic case. For many commercial applications, resin-sealing, plastic feedthrough terminals, gasket sealing, roll crimping, and other nonhermetic sealing techniques provide a sufficiently serviceable container for impregnants, and see frequent use. In other encasements, impregnants might be a polymerizing resin, forming a structure that can be subsequently dipped into a further outside coating to form a conforming case. For those film capacitors not using impregnants, the leaded capacitor structure can be overwrapped with pressure-sensitive plastic tape, such as polyethylene terepthalate polyester having a heat-curable adhesive, with the wrap extended beyond the roll termination area. The pockets thus formed around the leadwires by the extended tape are then filled with a thermoset resin, such as filled epoxy, to make what is commonly called *wrap-and-fill construction*, a lightweight construction that provides some short-term moisture protection.

The strength of wrap-and-fill construction depends on the materials used and the physical size of the capacitor. Some capacitors made in this manner are made with the aforementioned oval-shaped, flat, or rectangular prismatic cross section to permit favorable orientation in cir-

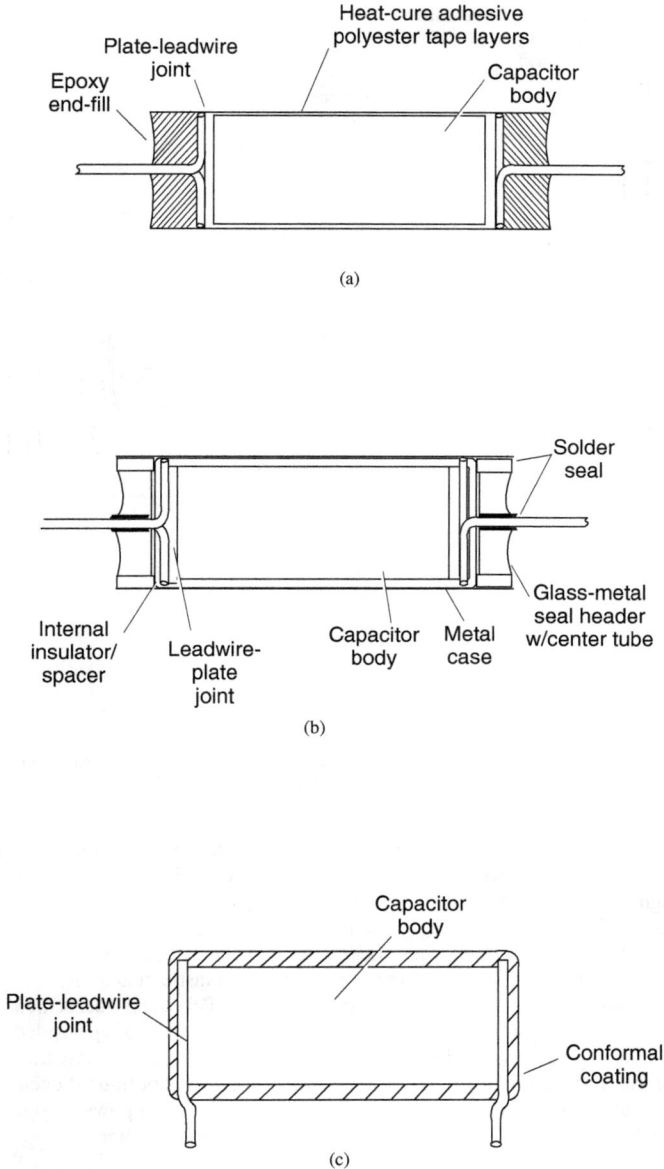

FIGURE 2.17 Axial-lead rolled film capacitor encasements: (a) wrap and fill (cylindrical shape shown). Oval, rectangular, and other cross-sectional shapes are similar, (b) metal hermetic seal, (shown is 2-seal case-isolated, may be one glass seal with outside foil leadwire soldered to metal case) (c) conformal coating on radial lead cylinder. Conformal coating is used for stacked film and other prism-shaped parts, and shrink film covering is used on commercial parts of cylinder and prism shapes.

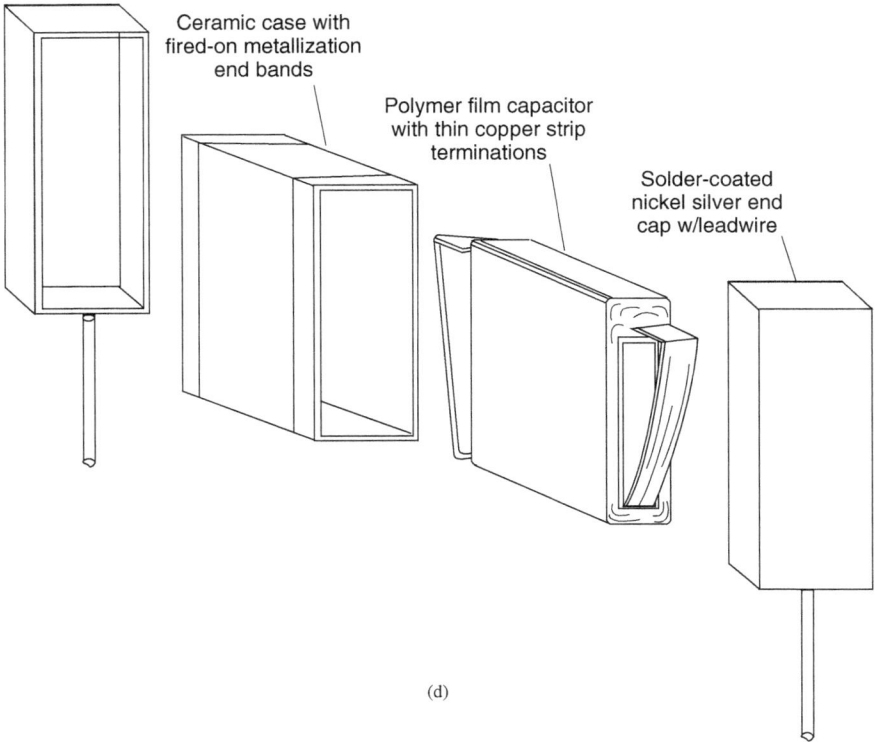

(d)

FIGURE 2.17   Axial-lead rolled film capacitor encasements: (d) one type of ceramic hermetic seal encasement, for a rectangular prism, in this instance. *(Continued)*

cuit locations where area or height are limited. Some case constructions use a conformal coating of thixotropic material, which, after curing, is relatively thick and porous, and which is then impregnated to seal the pores. If a polymerizing impregnant is used, the coating is strengthened. In the past, wax impregnation of porous coatings was a common moisture barrier encasement for this and other types of capacitors. Other encasement methods include injection and transfer molding with thermoset or thermoplastic materials, and potting in pre-molded plastic cases using filled polymerizing resins. This latter encasement method is extensively used for commercial power-circuit capacitors with single-ended terminals. Polymer coatings applied in powder form, then heated to coalesce and flow the powder can be used to form a relatively thick conformal coating. Specially configured encasements are produced for low-inductance connections across power busses and power-supply switching components. Figure 2.17 depicts some types of film capacitor encasements.

2.5.5 Foil Versus Metallized-Dielectric Construction

Paper dielectric with thin vacuum-vapor deposited metallization was originally introduced as a construction material in reduced size alternates for paper and foil capacitors, primarily for applications not involving heavy transient or alternating currents. The metallization used is typically aluminum, but might be zinc or other metal, deposited at about 1000-angstroms thickness. Good paper or film-and-foil capacitor construction, with the foil being typically aluminum, must use more than one thickness of dielectric to avoid the effects of inevitable pinholes and thin spots. Using at least two dielectric layers reduces probability of two random defects aligning to cause failure. Using thin metal coatings on the dielectric itself for the plates permits a significant size reduction, not only in the number of interleaves needed, but also in mostly eliminating the foil thickness from the capacitor volume needed. Moreover, the thin metal, it was found, vaporizes at the site of a plate-to-plate short circuit in response to the heating effect of the current flowing through the defect. Removing the metal at the localized spot electrically isolates the defect, effectively self-healing the fault.

The metallized-paper technique was extended to plastic-film capacitors with great success and is now common. For a plastic-film dielectric fault, not only does the automatic metallization removal isolate the defect, the film can flow and change to form a tough scar, which further neutralizes the spot. Metallized-dielectric capacitors are practical for applications where capacitive currents are low or nonexistent. Current-carrying ability is limited because of the small plate-conductor cross section for the thin metallization. Both the ESR and DF are higher than for foil and film parts, hence self-heating under current-flow conditions is significantly greater. In service, dissipation factor and ESR increases are sometimes associated with degradation of the plate metallization-to-termination transition, which can result from continued severe thermal cycling or repeated mechanical shock and vibration, both conditions more often seen in metal-cased hermetically sealed units.

2.5.6 Special-Purpose Paper and Film Capacitor Constructions

As mentioned, film and paper capacitor construction can be varied to enhance performance in particular applications. A shorter, but larger-diameter roll, for instance, will reduce inductance at the expense of volumetric efficiency. Performance in radio noise suppression is greatly improved if rolled capacitors are made in a feedthrough configuration, with one foil connected to a grounded outside metal case. In this construction, the power or other conductor to be bypassed passes through the capacitor roll center via an insulated tube or wrapping, and is electrically connected to the ungrounded foil or end metallization at one end of the roll. The feedthrough lead must be carefully insulated where it passes through the margin area on the opposite end of the roll. Here, that end's foil or metallization is connected to the outside metal case.

Figure 2.18 shows schematically this type of construction. The capacitor metal case is designed to be mounted on a metal bulkhead that shields input from output, and provides a very low inductance path from conductor wire to the capacitor connection and from the outside foil to ground. Other radio frequency interference (RFI) constructions are similar, providing a feedthrough conductor, but with the outside foil leadwire also fed through and brought out to both ends of the capacitor for grounding, in a four-terminal configuration. Four-terminal RFI capacitors might have nonmetallic-case construction with some obvious disadvantage in shielding and mounting. Grounding leads should be kept as short as possible. Figure 2.18 shows a performance comparison between bulkhead-mounted RFI and nonfeedthrough film-capacitor constructions.

2.5.7 Film Dielectric Comparisons

Figure 2.19 shows the comparative characteristics of commonly used film dielectrics. The production of capacitor dielectric films is a specialized endeavor that requires attention to details of material purity and achievement of tightly controlled thicknesses with few defects and thin spots. In the past, some films, such as polytetrafluorethelene (PTFE is

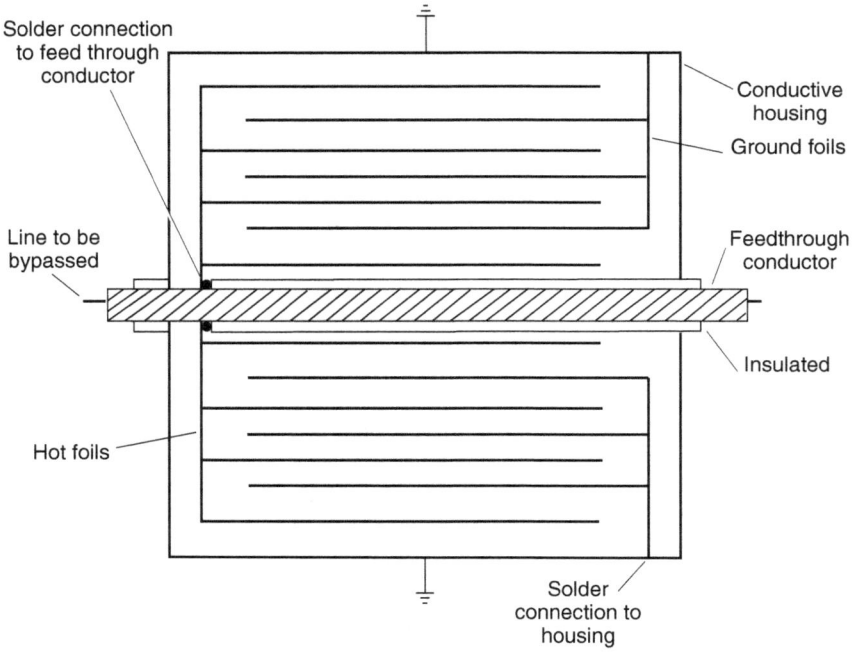

(a)

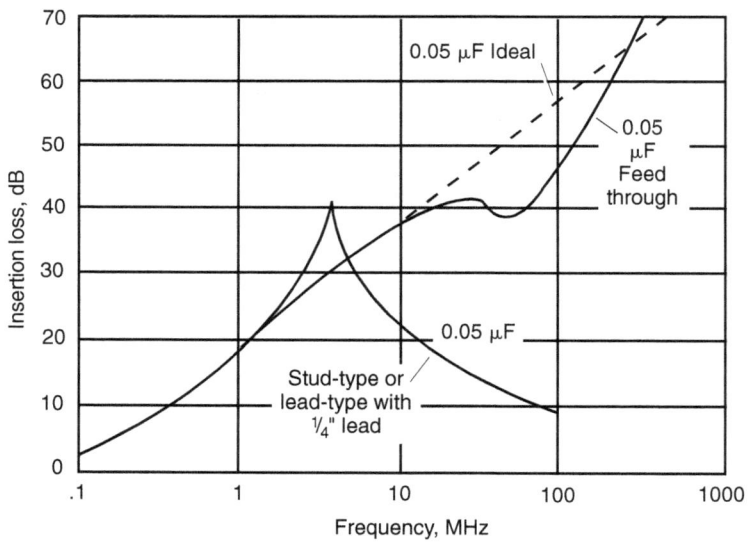

(b)

FIGURE 2.18 RFI feedthrough construction variant: (a) cross-sectional schematic. (b) filtering performance compared to non-feedthrough configuration.

Dupont's trademark Teflon) were skived or shaved from a solid material. Today, most films are either extruded or solvent cast. Extrusion involves forcing the fluid material under carefully controlled conditions through a properly sized die slit.

In solvent casting, the plastic material mixed with a solvent is fed through a dispensing arm to a large heat-controlled rotating drum, where, because of the drum motion and the design of the liquid material dispenser arm and its positioning relationship to the drum sur-

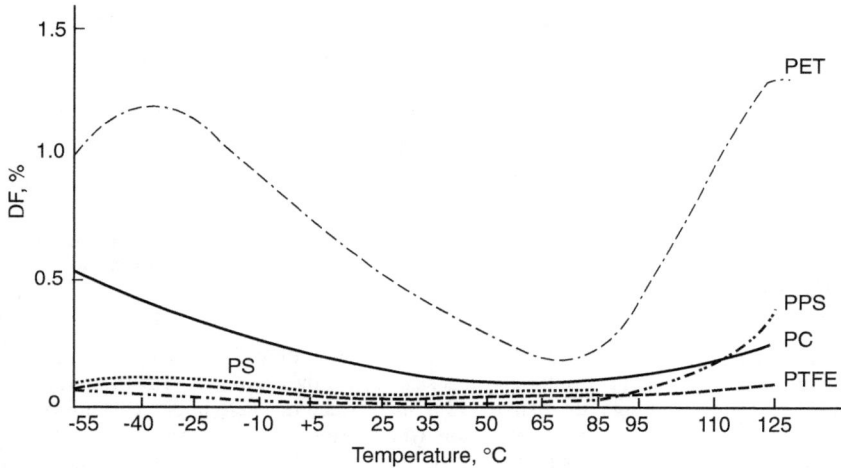

(a)

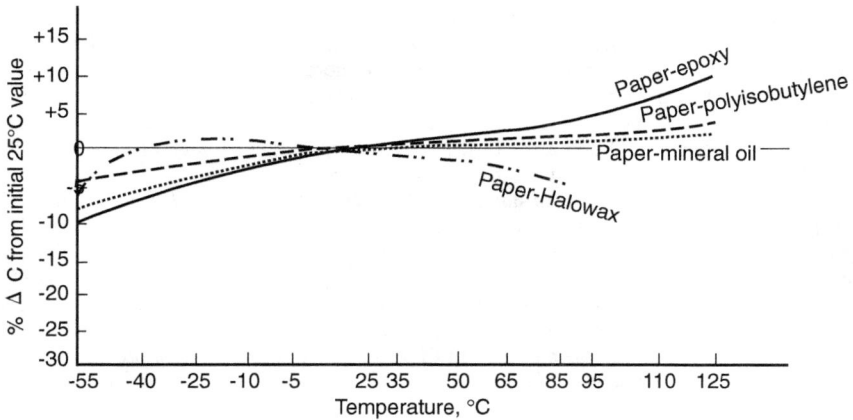

(b)

FIGURE 2.19   Some compared-performance traits of film dielectric types. The performances are generalized. Capacitor design, whether foil or metallized construction, dielectric thickness used, etc. will affect detail performance of individual specimens.

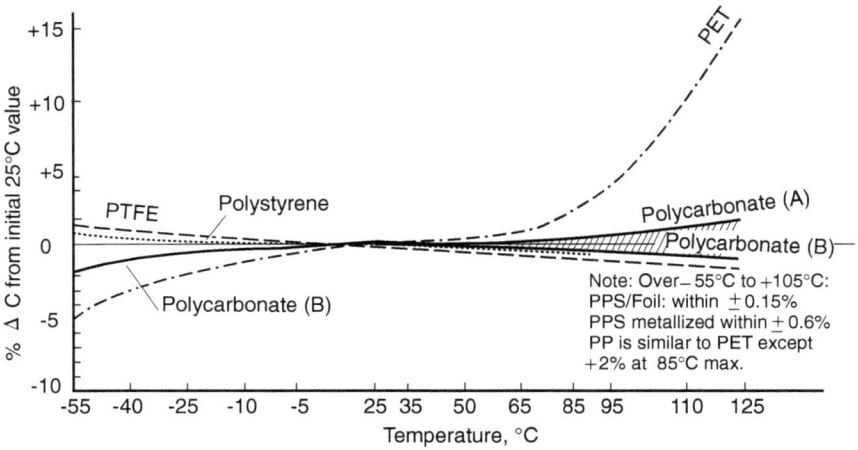

(c)

FIGURE 2.19   Some compared-performance traits of film dielectric types. The performances are generalized. Capacitor design, whether foil or metallized construction, dielectric thickness used, etc. will affect detail performance of individual specimens. *(Continued)*

face, it becomes deposited uniformly over the internal drum wall. The solvent is then driven off by heating, and the resulting film is stripped off the drum wall. The film is further conditioned by stretching it to orient the film microstructure and to improve its utility and stability in the capacitor-winding operation. Comparison of film materials shows that no single one is ideal for every purpose. Our limited material-properties charts do not show all the important properties, some of which are corona resistance, moisture absorption, dielectric absorption, and frequency sensitivity, but all are considered in choosing films for practical production and application. At the glass-transition temperature, Tg, polymers undergo a change: below that temperature, they retain dimensional stability, and are more elastic, above Tg, they begin to become rubbery and more fluid. For those materials whose Tg is near usual upper operating or capacitor-installation temperatures, the Tg becomes a condition that cannot usually be exceeded without the expectation of changes in capacitors made from the material. For a few films, Tg is at very low temperature, and the normally seen state of the material is in its plastic-transitioned condition.

The most widely used film is said to be polypropylene (PP), which has excellent dielectric strength, very low dielectric loss, low cost, and good dielectric constant stability, but is limited in application temperature to about +105°C. Polystyrene (PS) also has attractive stability and low losses, but is even more temperature limited—to about +85°C. Also, it is said that PS capacitor film is presently not obtainable as new material. The bulk of PP uses are in commercial entertainment, motor run, appliance, and consumer electronics. Polyethylene terepthalate (PET or PETP) films, often called just *polyester* in advertising literature, presently dominate military and industrial uses, with polycarbonate being a distant second.

One promising relatively new film for capacitor applications is polyphenylene sulphide (PPS), which has temperature endurance properties much like PET, but with better volume resistivity and dielectric constant stability over temperature. New films are constantly being investigated, and better new materials or improved existing ones will no doubt come into use.

2.5.8 Film-Capacitor Specifications and Standards

Physical sizes, termination configurations, and encasement methods for paper and film capacitors intended for electronics applications have seen only loose standardization outside the military specifications. The logarithmic decade system of assigning capacitance values is a well-established standard for film and paper and for other capacitor types as well, with 12 values per decade plus numerical values of 1, 1.5, 2, 3, and 5 being frequently used. The 24-per-decade values can be used for precision capacitor values, but are not frequently seen.

Table 2.17 in the ceramic-capacitor section shows the 12-value decade. Marking conventions for capacitance value and tolerance are well-followed for both commercial and military types: unless a capacitance is simply stated on the part body, and sometimes in addition to that, value in pF is stated using two significant value digits followed by a powers of ten multiplier digit, followed then by a letter indicating the tolerance. Figure 2.20 shows typical alphanumeric-value marking, also see the section on ceramic capacitors for further marking-standards information. Standard letter-tolerance correspondence for military and industrial identification is M: ±20 percent, K: ±10 percent, J: ±5 percent, and F: ±1 percent, D is ±0.1 percent. Special tolerances for values too small to have measurable tolerance values when expressed as percentages are also standardized, using the letters A through C.

Beyond value, tolerance, and outside foil identification marking, designators are not well-standardized for paper and film capacitors. Voltage-letter correspondence in Table 2.18 is not totally followed, other than over the 50- to 500-volt range. An array of case style, dielectric identifiers, temperature range, and other designators are used by both military and commercial organizations. The EIA 4-number case-size identifier is not presently used by military specifications for paper and film capacitors, but is increasingly used by commercial suppliers. Examples of military-identifier numbers are shown in Table 2.10. Color codes, once extensively applied, are seldom used today on paper and film parts. As mentioned, the outside foil of rolled capacitors is identified by a band or a band segment around the roll circumference, near the proper end of the case. Military date coding is a 4-digit number identifying the year and week the part completed its acceptance testing, and, in accordance with MIL-STD-1285, must be the last two digits of the calendar year, followed by two digits that represent the week of the year. Commercial capacitors for motor start and run, energy storage, power-factor correction and other power applications have a degree of case size, shape and terminal standardization driven by the need for replacement interchangeability.

Tables 2.7 and 2.8 show industry- and government-type designations and specification listings. The very useful EIA practice of designating capacitor case sizes in the type number by using a four-digit number associated with length × width has been extended to film and other capacitors. For cylinders, the length and width are just those of the widest plane through the body, but for prismatic shapes, the plane of the largest area is generally used. Four numerals identify length and width in tens of mils (tens of thousandths of an inch), or more recently, in tens of millimeters. Presently, you must consult the manufacturer's catalogs or part specifications to discover whether dimensions are metric or in inches.

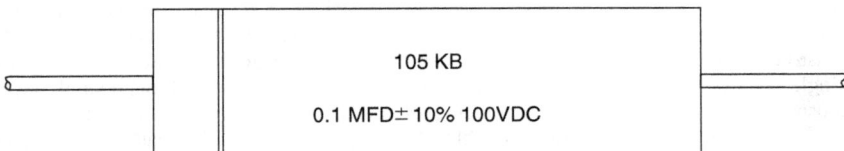

105 KB

0.1 MFD± 10% 100VDC

FIGURE 2.20 The alpha-numeric value-marking convention for film and paper capacitors.

2.6 CERAMIC AND GLASS DIELECTRIC CAPACITORS

Ceramic capacitors, once confined to applications requiring small capacitance values at relatively high operating voltages, now provide CV product per unit volume, surpassed only by small electrolytic types, with which they, for some uses, share an overlapping area of application choice. In their upper range of capacitance values, ceramic capacitors also serve many uses that were formerly reserved almost exclusively for the larger-sized paper and film types. In the past few years, however, advances in thin-film fabrication and packaging have provided competing equivalent CV product-per-unit-volume plastic-film parts. To compete equally for most uses, higher-temperature thin films or other means must be found to avoid the difficulties caused by high temperatures seen in automated soldering of film capacitors.

Glass dielectric capacitors, once produced in volume as multilayer leadwire parts of high performance and stability are now viewed mostly as specially configured parts for high-voltage and high-current RF applications, where the attainable high Q and stability are essential. Multilayer glass dielectric capacitors are constructed with alternating layers of specially formulated glass ribbon and aluminum foil, which, after layup, leadwire attachment (welding), and enclosure in cover glass halves, are fired under controlled conditions of pressure to flow the glass and produce a voidless multilayer capacitor structure sealed in glass. Glass is also used as a single-layer dielectric for some mechanically variable trimming capacitors because of its dielectric stability, thermal stability, and high Q. Note that various glass formulations provide differing properties of dielectric constant, insulation resistance, dielectric withstanding voltage, and long-term stability of properties under applied voltage and environmental conditions, and the production technology is both material and process intensive. Multilayer glass production was originally developed by Corning Glass Works and the product is currently manufactured and marketed by Kyocera International/AVX Corporation.

Porcelain enamel, usually magnesium titanate Mg_2TiO_3, also of low loss and high stability, but with a relatively low dielectric constant of about 15, was once used to produce multilayer-construction low-capacitance, high Q, very stable parts for general use, but is now confined mostly to specialized high-frequency, low-value, high-Q chip parts that are very useful at high frequencies into the microwave range. Several manufacturers produce this type of porcelain capacitor. Porcelain dielectric properties are very comparable to those of glass and mica (see Figure 2.26). Porcelain requires a much higher firing temperature than does most other ceramic capacitor materials, which limits the plate metallization material choices and makes the parts generally much more difficult to produce than titanate formulations. Both glass and porcelain enamel multilayer capacitors were developed in the United States as substitutes for mica capacitors near the end of World War II, when the availability of high-quality muscovite mica from India was reduced by the war activities. The same need fueled the development of techniques for producing multilayer ceramic (MLC) capacitors, using some of the same basic process technology as that for the multilayer porcelain enamel and glass products.

The market demand generated by the advent of solid-state devices and integrated circuits, with needed higher capacitance in smaller low-voltage packages, drove process technology refinements that permit today's wide availability of high-reliability, relatively low-cost MLC capacitors. Single-layer ceramic capacitors in disc and tubular configurations and in mechanically variable trimming capacitors were extensively used before MLC construction development extended the capacitance range of temperature-stable parts. Single-layer ceramic construction still is widely produced and used. Although single-layer construction requires careful production, the refinement of processes and the meticulous attention to detail needed to ensure reliable MLC capacitors is much greater. Single-layer parts are generally lower-priced, and can serve quite well in many applications. Single-layer ceramic capacitors are also used in microstrip RF constructions as very small chip-type devices using the same mounting and connecting techniques as used for solid-state devices. As might be suspected, capacitance values for these high-frequency applications are likely to be quite low, and the dielectrics are frequently the stable, high-Q materials, such as glass (silicon dioxide), NPO ceramic, aluminum oxide, or porcelain. Tubular single-layer ceramic capacitors, in addition to being used along with multilayer tubular-leaded configurations as circuit elements, have wide usage as feedthrough RF filters. Their shape is nearly ideal: the

TABLE 2.11 EIA Standard Ceramic Dielectric Performance Identifiers

Class-I dielectrics					
TC symbol	Significant figure, TC of capacitance in PPM/°C	TC multiplier symbol	TC decimal multiplier (PPM/°C)	TC tolerance symbol	Tolerance of TC in (PPM)
C	0	0	−1	G	±30
B	0.3	1	−10	H	±60
U	0.8	2	−100	J	±120
A	0.9	3	−1000	K	±250
M	1.0	4	−10000	L	±500
P	1.5	5	±1	M	±1000
R	2.2	6	±10	N	±2500
S	3.3	7	±100		
T	4.7	8	±1000		
U	7.5	9	±10000		

Class-II, -III, and -IV dielectrics					
Low temperature symbol	Low temperature (°C)	High temperature symbol	High temperature (°C)	Capacitance range symbol	Maximum capacitance change over temperature range (% of +25° value)
Z	+10	2	+45	A	+1.0
Y	−30	4	+65	B	±1.5
X	−55	5	+85	C	±2.2
		6	+105	D	±3.3
		7	+125	E	±4.7
		8	+150	F	±7.5
		9	+200	P	±10
				R	±15
				S	+22
				T	+22, −33
				U	+22, −56
				V	+22, −82

inner metallization is connected to the feedthrough conductor, and the outside metallization is connected to a ground plane, through which the filtered conductor passes.

2.6.1 Ceramic Dielectric Types and Characteristics

The general stability and high-Q performance of glass and porcelain have already been mentioned. Although glass is an amorphous material that resembles a viscous liquid, ceramic materials are basically polycrystalline. Ceramics used for capacitor dielectrics are mixtures of compounds, almost always containing titanium compounds: barium titanate (Ba_2TiO_3) and titanium dioxide (TiO_2), but with other materials as well. Metallic silver is the plate material of choice, with added palladium or platinum to discourage migration and solder scavenging and to improve firing characteristics, but other additions and combinations are also used, including pure palladium. Stable temperature-compensating capacitor types (EIA class I) are made without or with very little Ba_2TiO_3, but are basically TiO_2 or Ca_2TiO_3 with additive materials, some of which are magnesium titanate ($MgTiO_3$), and strontium titanate ($SrTiO_3$), neodymium (Nd) compounds, magnesia (MgO_2), alumina (Al_2O_3), bismuth stannate ($Bi_2Sn_3O_9$), or Manganese titanate (Mn_2TiO_3) with proportions to produce the desired temperature coefficients when mixed and fired in accordance with the manufacturer's developed processes.

Dielectric constants for temperature-compensating capacitor materials are relatively low, ranging from about 6 to as high as 500, with the DF generally from 0.0001 to 0.004. EIA class-II general-purpose less-stable capacitors use Ba_2TiO_3 as a base because of its high dielectric constant, adding stabilizing materials, which can be the same as listed for temperature stable (class I) materials, including TiO_2 and $CaTiO_3$, plus calcium zirconate ($CaZrO_3$), niobium pentoxide (Nb_2O_5), and others. A ceramic dielectric based on lanthanum modified-lead zirconate titanate or "PLZT" has been favorably reported in the literature, having general-purpose Class-II ("stable") stability characteristics, but with capacitance-stability high-temperature limits extended beyond +125°C to about +145°C.

Fired ceramic materials have high insulation resistance, high dielectric strength per unit thickness, and neither absorb moisture nor do they permit it to pass through. Moreover, ceramics are not physically degraded by uniformly applied temperatures (without accompanying electrical stress) well beyond those applied for any soft soldering processes, though encasing and terminal attachment can limit temperature capabilities. Producing predictable and reliable electrical performance and strong, voidless MLC structures with well-defined plate patterns and uniform dielectric thicknesses requires stringent control of raw material purity, and of firing conditions and processes.

2.6.2 Ceramic Material Identification Conventions and Standards

Ceramic materials are standardized by performance characteristics that do not specify the mixtures and processes to be used in attaining the particular performances. Conventional colloquial names of one type for temperature stable and compensating no longer have official standards backing, but still are widely used. This simple nomenclature just uses the letter "P" for positive and the letter "N" for negative, with the approximate slope of the capacitance-temperature curve in parts per million per degree celsius (ppm/°C) following the letter. Thus, "P100" means positive 100 ppm/°C, N750 means negative 750 ppm/°C, and (the one exception) "NPO" means stable with temperature.

Electronic Industries Association (EIA) and military designations with corresponding temperature performances stated in EIA 198 and in pertinent military specifications are shown in Tables 2.11 and 2.12. These tables require some care in interpretation because they are based on practical capacitors constructed with the materials, rather than characteristics of the materials themselves. The characteristics are defined under the performance measurement methods in specifications for those capacitors. The three EIA and two military alphanumeric identification characters specify the low and high temperatures for defining the performance range, and the limits of capacitance variation within that temperature range expressed as percentages (or parts per million) of the initial room temperature value.

The U.S. military specification two-letter system of dielectric identification (see Table 2.12) is undergoing diminishing use. Its initial letter defines the part temperature range within which stability is defined, and the second designates the limits of the change over the temperature range, including applied voltage, if so stated, defined as percentages of the initial room temperature value. In the temperature-compensating capacitor range, the EIA 198 and the military definitions covered in MIL-C-20, MIL-C-55681, MIL-C-81, and MIL-C-49464 are nearly identical, except that more-recent MIL-C-20 documents have eliminated the spectrum of temperature-compensating characteristics, retaining only temperature-stable types. The performance standard for the most widely used military general-purpose ceramic capacitors in MIL-C-39017, MIL-C-11015, and MIL-C-55681, designated BX characteristic, differs slightly from the similar EIA X7R characteristic.

The military BX designation includes a voltage instability limit in combination with its temperature instability limits, not required for EIA X7R. This can permit a higher room temperature CV product per unit volume for the X7R (generally of higher k), at the expense of the voltage stability. Note that the military and EIA specifications both define specific sequenced stability measurement procedures for high-dielectric-constant ceramic capacitors, which must be followed to produce repeatable results because recent voltage and temperature history and order of temperature and voltage application affect the results obtained when using ferroelectric dielectric constituents.

TABLE 2.12 Military Standard Ceramic Dielectric Performance Identifiers

Temperature stable and compensating: Capacitance changes referenced to value at +25°C			
Characteristic	Over temperature at 0 V dc	Over temperature w/rated voltage	Dissipation factor max (percent)
Temperature stable types			
CG	0 ± 30 ppm/°C	0 ± 30 ppm/°C	0.15
CH	0 ± 60 ppm/°C	0 ± 60 ppm/°C	0.15

A long list of temperature compensating characteristics covered by MIL-C-20 have two-letter designations with initial letters: A = 100 ppm/°C, C = 0 ppm/°C, H = –30 ppm/°C, L = –80 ppm/°C, P = –150 ppm/°C, R = –20 ppm/°C, S = –330 ppm/°C, T = –470 ppm/°C, U = –750 ppm/°C. The second letter denotes the tolerance for temperature characteristic in ± ppm/°C: F = 15, G = 30, H = 60, J = 120, K = 250, X ≈ 0.

General purpose types			
BX	± 15 percent	+ 15, –25 percent (± 15 percent at 0 dc voltage)	2.5
BR	± 15 percent	+ 15, –40 percent	3.0

Initial letter B denotes operating temperature range of –55° C to +125° C, letter C denotes –55°C to +150°C, initial letter A means –55°C to +85° C.

In addition to the standardized terminology, other popular names have also been used with some imprecision for dielectric combinations as well: "stable k" refers to a material of dielectric constant about 1100 to 2400, with the approximate characteristics of X7R. This stems from a previous situation where EIA had only three ceramic classes (class II was split into "stable" and "unstable" groupings). You might also hear a material identified by its nominal dielectric constant (e.g., "K1200").

EIA 198 and EIA 521 now categorize ceramic dielectrics as Class I, II, III, and IV. Class I uses the most controlled temperature-performance materials. Class II contains those with higher dielectric constant and lesser, but moderately good, temperature stability. Class III uses still-higher dielectric-constant materials with large temperature and voltage instabilities. Class IV is a special category for barrier-layer or semiconductor-layer ceramic parts of generally very high CV product per unit volume, but wide instability.

Formerly, the EIA used only three classifications (as stated), with class II containing the present classes II and III, and the former class III corresponding to the present class IV. Class II now defines capacitors with a maximum total capacitance variance of about ±15 percent over −55°C to +125°C, having relative dielectric constants ranging from about 250 to 2900. "Unstable" class III now means all less stable, except for barrier-layer types, with a dielectric constant ranging from 3000 to about 10,000. Class-II and class-III have DF ranges from about 0.004 to 0.04, measured under established specification conditions.

The class-IV barrier-layer ceramic capacitors have present usage, but were formerly more popular. This type capacitor is made by heating a ceramic substrate in a reducing atmosphere until its resistivity is lowered to about 10 ohm-cm, then changing to an oxidizing atmosphere to form a thin-oxidized high-resistivity layer, which is polarized. Metallization is then applied over the oxidized layer to form capacitor plates. The resulting structure uses the now low-resistivity ceramic substrate as a common plate to form two capacitors in series between the metallized sides. The high-resistivity layer has polarity-sensitive resistivity that is somewhat similar to a diode; this type capacitor is polarized unless two series capacitors are formed back to back. The advantage of the technique is that very high capacitance values, albeit of relatively unstable character and low-voltage capability, can be attained in very small sizes without using MLC production methods.

2.6.3 Ceramic Performance Stability

Comparative capacitance variations over temperature are shown graphically in Figure 2.21. These curves represent essentially the variation of dielectric constant over temperature. The class-I (temperature-stable) materials have much lower dielectric constants, and generally very low dielectric losses. Their temperature characteristics are fairly linear within the range of usually encountered temperatures and the change can be stated with acceptable accuracy as a "coefficient" defining the slope of the change per unit temperature (with a tolerance, of course). Actual determination for specification purposes uses a two-point measurement at 25°C and +85°C. More nonlinearity is said to occur in the lower temperature range. Class II, III, and IV ceramics have temperature "characteristics" specified in lieu of temperature coefficients: their variations over temperature are markedly nonlinear, and are bounded at the extremes of their specified temperature range by specifying a total-capacitance range tolerance about their +25°C value when measured under specified conditions of input signal voltage, dc bias, and frequency.

As ceramic material mixes are varied to obtain higher dielectric constants, basically by adding higher proportions of ferro-electric Ba_2TiO_3 or other materials to the formulation, temperature variation and nonlinearity within the range of interest increases. Not only does the temperature variance become greater with increasing dielectric constant, but the charge accumulation as a function of applied voltage becomes increasingly nonlinear, making dielectric constant more sensitive to applied voltage. The direction of the nonlinearity is such that dielectric constant is driven downward with increasing voltage. The material becomes lossier with increased frequency, but dissipation factor or Q is fairly stable with temperature, with DF generally decreasing at higher temperatures. Dormant aging capacitance changes, covered later, also increase with dielectric constant of the material formulation.

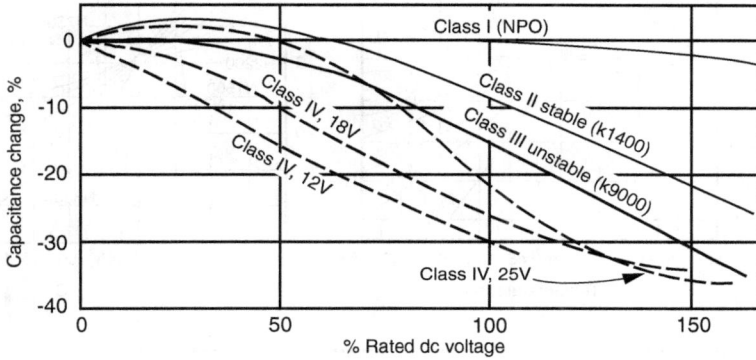

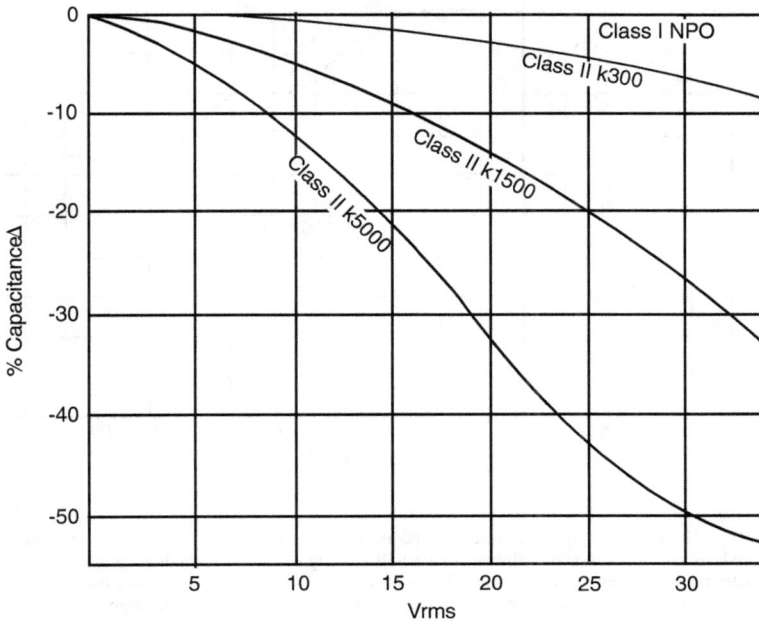

FIGURE 2.21 Ceramic capacitor typical temperature-capacitance characteristics for commonly encountered designs (individual part characteristics will vary significantly, depending on dielectric thickness, plate area, plus details of material mix and processes): (a) Capacitance dc voltage sensitivity referenced to 1.0 V rms, 1 kHz, and 0 V dc; (b) Capacitance ac-signal voltage sensitivity with 0 V dc.

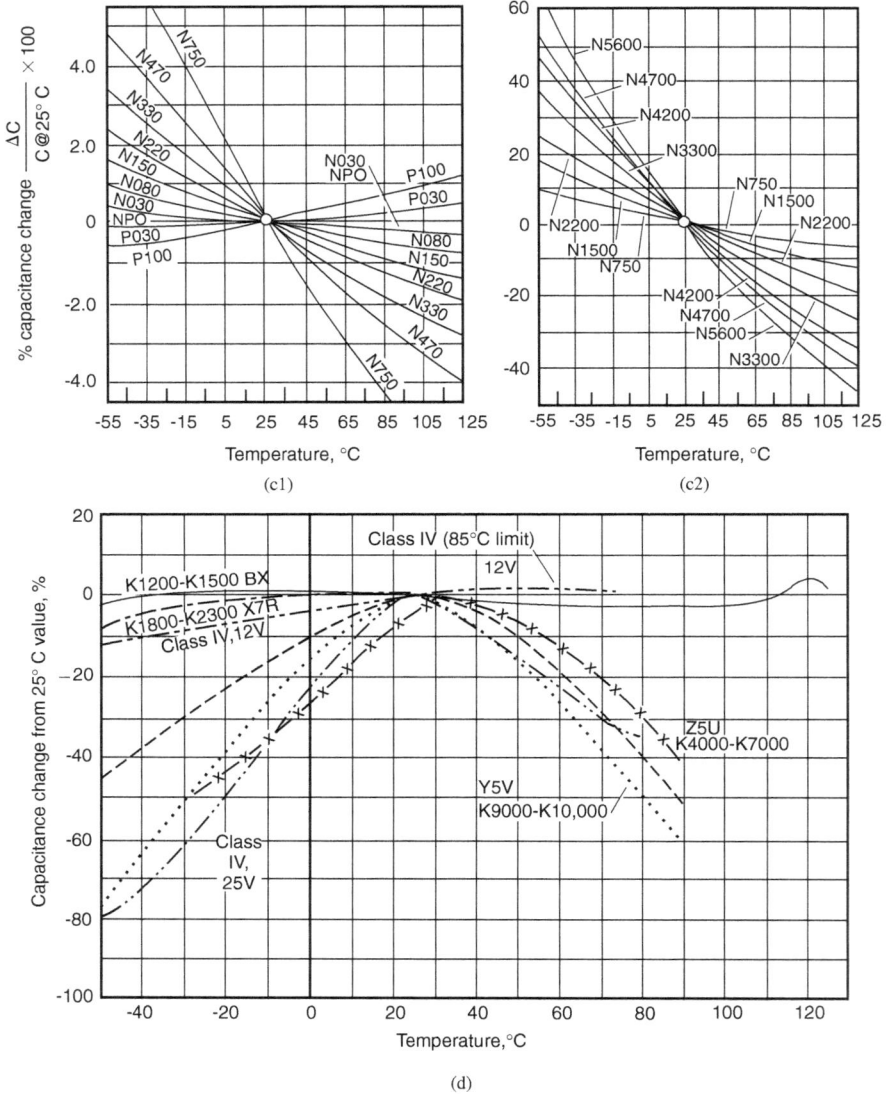

FIGURE 2.21 Ceramic capacitor typical temperature-capacitance characteristics for commonly encountered designs (individual part characteristics will vary significantly, depending on dielectric thickness, plate area, plus details of material mix and processes): (c) Low-range (c1) and high-range (c2) temperature-compensating capacitor nominal capacitance variation with temperature. Tolerance envelopes not shown. (d) Typical capacitance variation with temperature for capacitors made with various ceramic formulations. *(Continued)*

Pure titanium dioxide has a temperature coefficient of about −750 parts per million per degree celsius or per kelvin (PPM/°C or PPM/K) with a dielectric constant of about 85, and is mixed with materials having differing coefficients to reach various temperature-stability

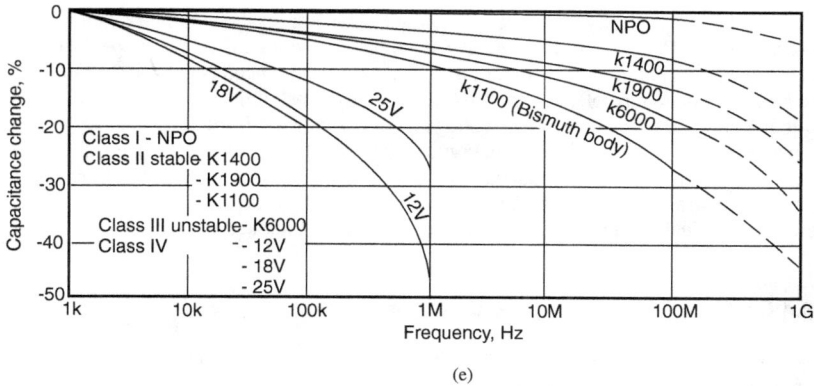

(e)

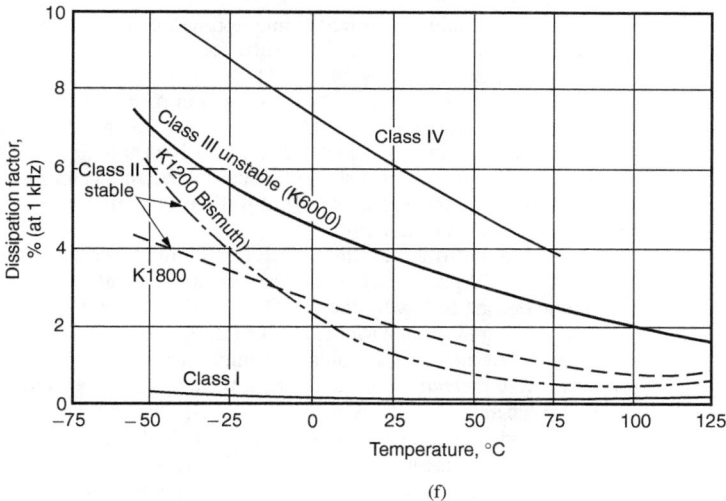

(f)

FIGURE 2.21 Ceramic capacitor typical temperature-capacitance characteristics for commonly encountered designs (individual part characteristics will vary significantly, depending on dielectric thickness, plate area, plus details of material mix and processes): (e) Frequency sensitivity of capacitors made with various dielectric formulations. (f) Dissipation factor variation with temperature for capacitors made with various ceramic formulations. *(Continued)*

characteristics. Barium titanate is classed as a ferroelectric material, responding electrostatically in the presence of an electric field much as soft iron behaves magnetically in a magnetic field, including the presence of an electrical hysteresis characteristic. The crystalline structure is highly polarized and changes its nature as the temperature increases past its Curie point of 121°C. The Curie point is associated with the peak in the capacitance-temperature characteristic at that temperature. Ba_2TiO_3 crystals change from cubic to tetragonal unit cell structure during exposure to temperature above the Curie point, and slowly revert to cubic structure over time when temperature is lowered. This is believed to account for the time-aging characteristic. Actually, the material undergoes several crystal symmetry changes within its −55°C to +125° application range. These points can be identified by the peaks in the dielectric constant changes over temperature. The most pronounced peak is at the Curie point. The barium titanate ferroelectric property accounts for most of the nonlinear decreases in ceramic material dielectric constant as voltage is increased. Graphs of various ceramic capacitor performance traits are shown in Figure 2.21. The actual characteristics of individual capacitor designs depend on nuances of dielectric mixtures, processing procedures used, volts per mil stress in the design, geometry, and many other factors. Envelopes of limits are therefore practical for designs other than the temperature-stable types.

2.6.4 Ceramic Capacitor Aging Changes

Depending on the detailed makeup and processing of a particular capacitor containing barium titanate, a dormant aging characteristic of capacitance value occurs. For EIA X7R and military BX dielectrics, this aging rate is downward in capacitance value and amounts to a 1 to 2 percent per time decade in hours measured from the time that the material was last exposed to a temperature above the Curie point for a sufficiently long time period to "de-age" the part. Capacitance thus drifts downward 1 to 2 percent in the first hour, another 1 to 2 percent the next 10 hours, and so on. When the manufacturer measures the part for his acceptance, its age should have passed the 1-, 10-, and 100-hour points, but unless the manufacturer makes special efforts, the 1000- (six weeks), and 10,000-hour points (a little over a year) can transpire before the part is installed in equipment, and a capacitor on the low side of tolerance at production measurement might well be several percentage points below its "purchase" tolerance at its installation.

To avoid incoming inspection difficulties, manufacturers should "guard band" capacitance value to allow for the initial part of aging, and usually do so, aiming usually for the 1000-hour point. The time decade following the 10,000-hour decade, of course, amounts to more than 10 years, and aging in service for the types of uses for which X7R capacitors are used is not a problem. Other dielectric formulations might have higher or lower aging rates, generally proportional to dielectric constant, which the user might need to consider in his choice of capacitor type and value. NPO or COG dielectrics can be assumed to have zero aging, while Y5U has about 2 to 3 percent per decade, Z5U is about 4.5 to 5.5 percent per decade, and less-stable dielectrics might go up to about 7 percent per decade. Exposure to high temperature for a time resets the aging process, with amount of upward capacitance change dependent on the exposure time, the exposure temperature, and the aging decade in which the observed part initially exists.

2.6.5 Ceramic Capacitor Construction

As previously mentioned, ceramic capacitors are seen in three basic constructions, but with many variations to suit particular uses. The class-IV barrier-layer construction has been described, and for other parts, single-layer construction is the simplest, with forms that vary from the familiar disc with leadwires soldered onto metallizations on opposite sides, to tubular forms with metallizations on the outside and inside of the tube. As mentioned, the tubular form is also readily adaptable to feedthrough bypass applications when the outside metallization is electrically connected to a grounded bulkhead. The disk is also adaptable to feedthrough use when a center feedthrough hole is added with a leadwire margin in the opposite-side metallization. The discoidal configuration provides a short low-inductance path to ground when the outside plate is connected to conductive ground around its total periphery. Sketches of some ceramic capacitor constructions are shown in Figure 2.22.

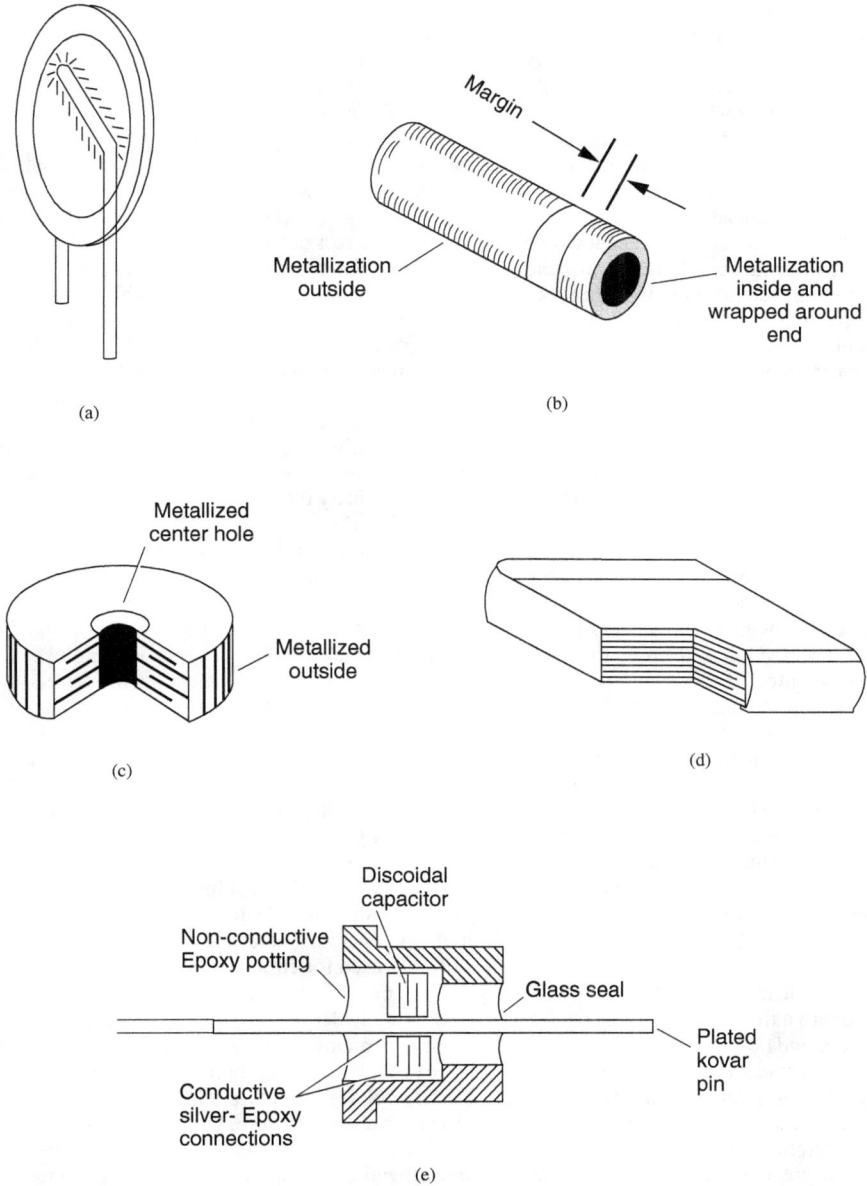

(a)

(b)

(c)

(d)

(e)

FIGURE 2.22 Ceramic capacitor construction: (a) single-layer disc, (b) single layer tubular, (c) multilayer discoidal chip, (d) multilayer chip with metal terminations. It can be made into radial or axial leadwire type by attaching suitable leadwires to the terminations. (e) A cross section of a hermetic seal feedthrough bypass using a discoidal chip.

Silver and silver-alloy metallization has a tendency to migrate across ceramic surfaces when exposed to moisture and voltage, and should be well-protected from humid conditions in service. Fired ceramics are impervious to moisture, but if internal voids exist in multilayer ceramic (MLC) construction and are accessible by means of cracks or other imperfections, the capacitor is vulnerable to failure because of moisture in very small amounts. Multilayer ceramic (MLC) construction permits higher-capacitance values using stable dielectrics and, by its nature, provides a self-sealed structure with only fired ceramic and the plate end connections exposed to atmospheric environments. Figure 2.22 shows a breakaway of MLC construction for a chip-type package. In the chip-type configuration, the plate ends are connected by a fired-on silver-glass frit mixture, often alloyed with palladium, which is then plated and tin-lead solder coated for attachment into the using circuit. Addition of palladium or platinum reduces the tendency for silver to migrate across ceramic surfaces.

Silver with its additions, suspended in a volatile fluid vehicle is usually applied by screening techniques to thin sheets of flexible green ceramic, with the paint vehicle being evaporated away, leaving the metallic coating on the ceramic. Another thin green ceramic sheet is layered over the previously metallized sheet and similarly screen-coated, with metal painting being patterned to bring alternating metal layers to the opposite edges of the future structure, leaving an unmetallized margin on the opposing unterminated ends. After the desired number of layers are constructed, green ceramic covering sheets are laid on, the green ceramic-sandwiched structures are cut into strips, the terminating plate ends are painted over with a silver alloy-glass frit mixture. In careful processes designed to prevent voids within the structure, the construction is fired to vitrify the ceramic and form a monolithic capacitor with silver-alloy terminations.

Silver easily goes into solution in molten tin, causing termination silver to be rapidly scavenged away when tin-lead solder is used to attach terminal leadwires or to coat terminations for ease in surface-mount chip solder mounting. The scavenging problem is allayed somewhat if a silver-bearing solder (usually about 2 percent) is used for soldering. Better protection is afforded for surface-mount paste soldering if a nickel barrier plating is applied before solder coating or other plating. Nickel plating must, in turn, have a protective plating or solder dip if it is to be soldered or installed using conductive thermoset plastics because it rapidly forms a durable unsolderable oxide upon exposure to humid atmosphere, with the oxide also interfering with good low-resistance conductive-polymer bonding as well. Tin plating suffers from the same difficulty, with its oxide being even more intractable than that of nickel, plus an additional ability to grow dendritic filaments in moist environments. Tin dendrites have been known to cause short circuiting. The addition of a small amount of alloying lead to the tin plate is said to eliminate the dendritic growth problem. If the reflow of the solder during capacitor installation is to be avoided, soldering leadwires to the capacitor termination areas to make leadwire-terminated parts requires a solder alloy having a higher liquidus temperature than the eutectic 63-37 or near-eutectic 60-40 tin-lead solder, which is usually used for circuit installation.

After leadwire attachment and cleaning, capacitors can be conformally coated by dipping into a thixotropic polymer or by coating with a powder that is subsequently heated to liquefy and polymerize it, can be enclosed by forming an injection-molded case around the element, or can be placed into premolded plastic cases with a fluid thermoset potting resin, which is subsequently cured to form an essentially voidless structure. Plastic coatings and encasements provide protection from short-term moisture exposure, but function primarily to strengthen and protect the leadwire and capacitor structure against handling stresses, and to prevent short-circuiting of the end metallizations to adjacent conducting circuit structures. Coatings and cases also add some amount of thermal mass to lessen probability of end-solder reflow during installation. Plastic materials have much higher linear expansion coefficients than the metals and ceramic materials of capacitor bodies, and coating and encasement systems must be carefully chosen and implemented to keep stresses on capacitor bodies within capabilities during in-service thermal cycling and during installation stresses. Thinly coated ceramic capacitors should not be thickly overcoated by the user with circuit-card plastic coatings unless the effects are first carefully evaluated, lest temperature cycling of the equipment cause the ceramic capacitor bodies to fracture.

2.6.6 Ceramic Capacitor Applications

Class-I dielectric capacitors can be successfully used for temperature compensation to stabilize frequency-determining tuned circuits, for RF coupling, sample-and-hold functions, RF filters, delay-line construction, and for many other uses that require very stable and predictable capacitance with high Q and high insulation resistance, under conditions including radio frequencies into the microwave range. These high-quality capacitors can be used in practically any capacitor application where their relatively large size versus capacitance-voltage and resulting limited capacitance values can be tolerated. Package configurations available include bare chip types with terminals for solder and conductive polymer mounting and for microcircuit wirebonding. There are axial and radial leadwire tubular types sealed in glass, parts painted with enamel, conformally resin coated, molded in plastic with leadwires, with dual in-line terminals on molded plastic cases, conformally coated discs, and bare chips for surface mounting with wire bonding or active terminal attachment with solder or conductive polymers. Both single-layer and multilayer constructions are furnished in a multitude of shapes.

Class-II dielectric capacitors of the X7R or the similar military BX, and class III dielectric parts of Y5U, Z5U, and Y5V stability characteristics have the widest use of any ceramic capacitors. They are available in package and mounting configurations even more varied than those of class-I dielectrics, including all those previously mentioned, plus a variety of feedthrough and standoff packages and a significant number of special custom packages for higher-voltage energy storage and power-supply filtering, where size is at a premium. All are widely used for decoupling, bypassing, feedthrough filtering, and some coupling applications. Good acquaintance with the stability and self-resonance performance of the available packages and performance characteristic classifications will permit reliable choices of value and type.

One general rule is to provide a bypass impedance of no greater than 0.1 times the load impedance at the lowest frequency to be bypassed or decoupled. This rule should be extended by defining this minimum condition at the worst case of localized temperature, widest capacitance tolerance, highest circuit voltage, and maximum aging tolerance, to set the minimum capacitance value that can be used. The capacitance value that can be used in the particular package style selected is determined by the highest frequency of concern and the self-resonance and Q of the bypass capacitor. If the self-resonance or impedance versus frequency is not given in the manufacturer's data, and is critical to the application, a swept impedance meter of the proper frequency range with properly sized leadwires makes it easy to see what is going on with the capacitor over frequency.

It is not unusual to discover that a lower capacitance value will do a better job than one selected for low frequency overkill. However, carefully consider the performance changes because of the applied ac and dc voltage, and changes caused by temperature and by aging, particularly in the less-stable classifications. Most such changes drive capacitance downward. If frequencies are very low, it might be better, of course, to use a large capacitor of another type to respond to low frequencies, shunted by a lower-valued ceramic to take care of the high frequencies. Figure 2.13 shows swept-frequency data for impedance and apparent series equivalent capacitance for some selected capacitors. For high ac current components, the power dissipated by the capacitor ESR might require calculation. Capacitor body temperature should not exceed +125°C or the limitations of the capacitor packaging materials, whichever is less. Capacitor reliability is greatly enhanced by limiting high-temperature exposure (see the previous reliability paragraphs).

Less-stable class-III dielectrics, such as those used in the relatively popular Z5U characteristic are applied in bypass and decoupling applications where their wide value swings over time, temperature, and voltage are partially compensated by their ability to contain much higher capacitance per unit volume. Their selection process is similar to that for the general-purpose X7R or BX characteristic with the difference that much larger capacitance swings must be considered. As stated, most of the value changes are downward in capacitance, making it necessary to select a somewhat higher capacitance value than would be necessary if the selection were an X7R part, remembering that standard specifications specify capacitance value under very low-signal levels and with no dc bias. Temperature rise can also be a factor for selection of this type of dielectric. The same temperature limitations apply as for stable class-II dielectrics, but dissipation is somewhat higher for the same ac conditions because of higher dissipation factor.

2.6.7 Ceramic Capacitor Specifications and Standards

Tables 2.7 and 2.8 show military and industrial standards for ceramic capacitors. Materials for ceramic and for other types of capacitors also have a number of voluntary electronics industry standards in active use.

2.6.8 Ceramic Capacitor Identification Standards

Tables 2.13 through 2.19 show alphanumeric identifications for military- and industrial-standard ceramic capacitors. Color codes for ceramic capacitors can become elaborate because of the many features that need to be identified: capacitance value, capacitance tolerance, temperature characteristic, tolerance on temperature characteristic, working voltage, etc. Because color code use has declined to a very low level, only limited tables are included here. If color code information is required, the part manufacturer or the appropriate industry standard should be consulted. When stated as capacitor coded marking, capacitance is always expressed in picofarads, whether for alphanumeric or color coding.

Capacitance tolerances that are equilateral plus-minus numerals can be alphanumerically coded with the standardized military and commercial letters: F, G, J, K, M, etc., corresponding to percentages of labeled value as shown in Table 2.13 through 2.17. Very small capacitance values in close tolerances can reach resolution limitations for measurement equipment, and require that tolerance be stated as a maximum-value variation, rather than as a percentage of the labeled value. EIA letter designations for these and the standard tolerances are shown in Table 2.15. Color codes for tolerances vary, and might require reference to part specifications. Introduced by the EIA, identification of capacitor surface-mount chip nominal body sizes by using a 4-digit number in the type designation has proven to be most useful. It is quite widely used by commercial suppliers, and some newer military specifications have used the method as a style identifier and part marking, but not in part-identification numbering.

The convention states the length and width of the largest prismatic surface or, if cylindrical, the body outline profile in tens of mils or hundredths of an inch (e.g., 0402 means 0.040×0.020 inches, and 1204 means 0.120×0.040 inches). Recently, the practice has been extended to state the dimensions in tens of millimeters if the capacitor is designated as a metric style. Supplier's catalogs or the appropriate military or commercial standard will show which measurement system is used. Disc-capacitor type designations have three numerals only, which identifies them as disc-shaped, with the numerals disclosing disc diameter, to be deciphered. Some established reliability military specifications, unfortunately,

TABLE 2.13 EIA Ceramic Capacitor Type Designations

Type designator:

Example (1) CC068COG120M500
Example (2) CC1515X7R103K101

Example	EIA identifier	Style (Size)	TC code	Capacitance	Capacitance tolerance	Rated voltage
(1)	CC	068	COG	120	M	500
(2)	CC	1515	X7R	103	K	101

Style: Largest body dimensions in tens of mils (or in millimeters) L × W or, for 3 numerals: disc diameter. TC Code: Table 2.11. Capacitance: in picofarads, 2 significant numbers and decimal multiplier, use R as a decimal point. Capacitance tolerance: See Table 2.17. Voltage code: 2 significant numbers followed by a decimal multiplier.

TABLE 2.14 EIA Standard Color Code for Ceramic Capacitor Value and Tolerance Marking

	Capacitance			Tolerance		
First significant figure	Second significant figure	Decimal multiplier	Color dot or stripe	≤ 10 pf	> 10 pf	Color dot or stripe
0	0	X1	Black	± 2 pF	± 20%	Black
1	1	X10	Brown	± 0.1 pF	± 1%	Brown
2	2	X100	Red		± 2%	Red
3	3	X1,000	Orange		± 3%	Orange
4	4	X10,000	Yellow		+ 100%, − 0%	Yellow
5	5	X100,000	Green	± 0.5 pF	± 5 %	Green
6	6	X1,000,000	Blue			Blue
7	7	X10,000,000	Violet			Violet
8	8	X.01	Gray	± 0.25 pF	+ 80%, − 20%	Gray
9	9	X0.1	White	± 1.0 pF	± 10%	White

TABLE 2.15 EIA Standard Ceramic Capacitor Marking for Temperature
Characteristic (EIA 198[1])

	Class I Dielectrics	
Temperature coefficient[2]	Color code dot(s)/stripe(s)	Alpha marking code
P100	Red + Violet	C
P090	Green + Blue	G
NP0	Black	A
N030	Brown	H
N080	Red	L
N150	Orange	P
N220	Yellow	R
N330	Green	S
N470	Blue	T
N750	Violet	U
N1000	Red + Orange	V
N1500	Orange + Orange	W
N2200	Yellow + Orange	X
N3300	Green + Orange	Y
N4700	Blue + Orange	Z

		Class II Dielectrics		
Temperature range code (°C)	Temperature range color dot/stripe	Capacitance range over temperature	Range color dot/stripe	Alpha marking code
Z5 (+10 +85)	Brown	D (± 3.3%)	Brown	B
Y5 (−30 +85)	Silver	E (± 4.7%)	Red	B
X5 (−55 +85)	Gold	F (± 7.5%)	Orange	B
X7 (−55 +125)	Red	P (± 10%)	Yellow	C
X8 (−55 +150)	Unassigned	R ± 15%)	Green	C
X9 (−55 +200)	Violet	S (± 22%)	Blue	C
		T (+22, −33%)	Violet	D
		U (+22, −56%)	Gray	E
		V (+22, −82%)	White	F

1 Military specifications have largely abandoned color code markings for capacitors. Also, temperature compensating capacitors are specified in newer military specifications only for microwave chip capacitors. MIL-C-55681 covers only stable military G = 90 ± 20 ppm/°C and P = 0 ± 30 ppm/°C, but MIL-C-20 lists ratings down to −750 ppm/°C. These letters, preceded in military marking by the temperature range letter, usually B for −55°C to +125°C should not be mistaken for the EIA one-letter temperature characteristic marking code.

2 Temperature characteristics of stable ceramic capacitors and of temperature compensating capacitors are actually not linear within their tolerance envelopes, and the slope cannot be exactly defined by a single "coefficient", but the prevailing practice is to refer to the stable and temperature compensating characteristics as coefficients (with only small error).

TABLE 2.16A
Standard Letter Marking
for All Capacitors:
Rated Voltage (EIA
and Most Military)

Symbol	Rated voltage
A	50
B	100
K	150
C	200
D	300
E	500
F	1000
G	2000
H	3000
J	4000

TABLE 2.16B EIA and Military
Capacitor Standardized Capacitance
Tolerances and Identifications.

Capacitance value ≤ 10 pF		Capacitance value > 10 pF	
Delta from 25°C value (pF)	Letter symbol	Percent of 25°C value	Letter symbol
±0.01[1]	P	±1.0[1]	F
±0.05[1]	A	±2.0	G
±0.1[1]	D	±5.0	J
±0.25	C	±10	K
±0.5	D	±20	M
		+100/−0	V
		+80/−20	Z

[1] Many specifications do not require measurement resolution lower than 0.25 pF.

TABLE 2.17 EIA and Military Capacitor Preferred Standardized Capacitance Values vs. Tolerances. (Although given here for ceramic capacitors, the preferred values apply to all capacitor types. Except for military standards, not as well-followed for capacitors as for resistor values and voltage regulator diodes, the logarithmic decade system of values nevertheless is a valuable tool for reducing the number of values needed to cover a decade while avoiding overlapping values with individual part value tolerance range. The formula for the ratios of succeeding values is: $r = (10^{1/m})$ with n being the desired number of values per decade. Selecting an initial term and multiplying each successive term by the ratio corresponding to the selected number of values in the decade provides a sequence of values. Standard numbers of values per decade are 3, 6, 12, 48, 96, and 192. Capacitors seldom use more than 12 values per decade.)

Tolerances			Tolerances			Tolerances		
±5%[1]	±10%	±20%, +80/−20%	±5%[1]	±10%	±20%, +80/−20%	5%	±10%	±20%, +80/−20%
10	10	10	22	22	22	47	47	47
11			24			51		
12	12		27	27		56	56	
13			30			62		
15	15	15	33	33	33	68	68	68
16			36			75		
18	18		39	39		82	82	
20			43			91		

[1] Tolerances tighter than 5 percent and values below 25 pF for class I capacitors may use 24 values per decade. Values: 25, 35, 40, 45, 50, 60, 70, 80, and 90 (X10E-01) are frequently used. Generally, specifications limit tolerances to a measurement resolution of ±0.25 pF or higher. EIA standard 395 for aluminum electrolytic capacitors states the following values, with a recommendation that the logarithmic value sequence tabulated above be used in the future; 1, 2, 3, 4, 5, 8, 10, 20, 30, 40, 50, 60, 70, 80, 90, 100, 200, 300, 400, 500, 600, 700, 800, 900, 1000, 1500, 2000, 3000, 4000, 5000.

TABLE 2.18 Chip Ceramic Limited Space Capacitance
Value Marking. (EIA 198 and MIL-C-55681)

First character (alphabetic)		Second character (numeric)	
Significant figures	Character	Decimal multiplier	Character
1.0	A	$\times 10^0$	0
1.1	B	$\times 10^1$	1
1.2	C	$\times 10^2$	2
1.3	D	$\times 10^3$	3
1.5	E	$\times 10^4$	4
1.6	F	$\times 10^5$	5
1.8	G	$\times 10^6$	6
2.0	H	$\times 10^7$	7
2.2	J	$\times 10^8$	8
2.4	K	$\times 10^{-1}$	9
2.7	L		
3.0	M	Example:	
3.3	N	$K3 = 2.4 \times 10^3$ pF	
3.6	P	$= 2400$ pF	
3.9	Q		
4.3	R		
4.7	S		
5.1	T		
5.6	U		
6.2	V		
6.8	W		
7.5	X		
8.2	Y		
9.1	Z		
2.5	a		
3.5	b		
4.0	d		
4.5	e		
5.0	f		
6.0	m		
7.0	n		
8.0	t		
9.0	y		

K3

TABLE 2.19 EIA Identification Codes for Ceramic Capacitor Values and Tolerances (at 25°C)

Capacitance value ≤10 pF		Capacitance value >10 pF	
Delta from value (pF)	Letter symbol	Percent of value	Letter symbol
±0.01[1]	P	±1.0[1]	F
±0.05[1]	A	±2.0	G
±0.1[1]	D	±5.0	J
±0.25	C	±10	K
±0.5	D	±20	M
+100/–0	V	+80/–20	Z

[1] Many specifications do not require measurement resolution lower than 0.25 pF.

have dual part-number designation systems. One is a significant type-designation number having the part type, value, and voltage coded in. The actual part number is a nonsignificant dash number to the military specification number that requires reference to the specification listing to identify the characteristics. Some military part numbers are a mixture of significant and nonsignificant characters. Most manufacturers obligingly mark the value, voltage, tolerance, and characteristic identifier on parts, space permitting, even if nonsignificant part number marking is required, but part listings of nonsignificant part numbers are inscrutable without a specification in hand. Military parts also must have a four-digit year-followed-by-week date code, space permitting, which is the most useful if a batch of problem parts should be identified. At least two marking schemes have been proposed in EIA circles for value marking where space is limited, such as on chip-type capacitors. One method uses a single letter or numeral whose color designates value multiplier, with the letter or numeral designating a two-digit value in a 5-percent decade (see Table 2.18). This single letter-digit with color has apparently lost favor and the most widely limited-space marking used is a proposed EIA standard that uses one letter and a number for value and decimal multiplier (see Table 2.18). Figure 2.23 shows an assortment of ceramic capacitor designs.

2.7 MICA CAPACITORS

Natural mica is one of the oldest and, for some capacitor applications, one of the best dielectric choices. Natural mica includes several complex silicates of aluminum with many other contained materials, principally potassium, magnesium, iron, sodium, and lithium. The crystalline structure is well-bound along one plane, but with almost no binding energy perpendicular to the bound plane, hence the layered structure. The various mica chemical compositions have differing mechanical, thermal, and electrical properties, but the material is electrically very stable. Muscovite mica has the best combination of mechanical and electrical properties for capacitor use, being usable up to 500°C, and has high dielectric strength. The dielectric constant of capacitor-grade muscovite mica ranges from 6.5 to 8.7. Being a natural material, quality and properties are variable, and capacitor grades must undergo selection and grading from the candidate mined stock, with virtually flawless muscovite being required for most capacitor uses.

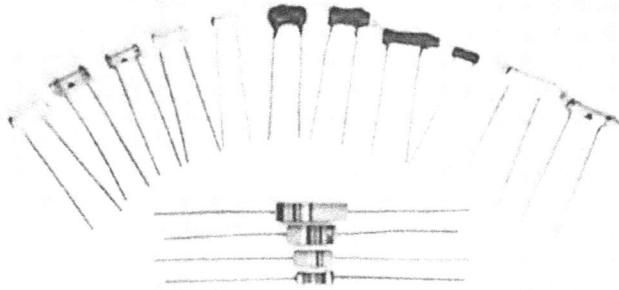

(a)

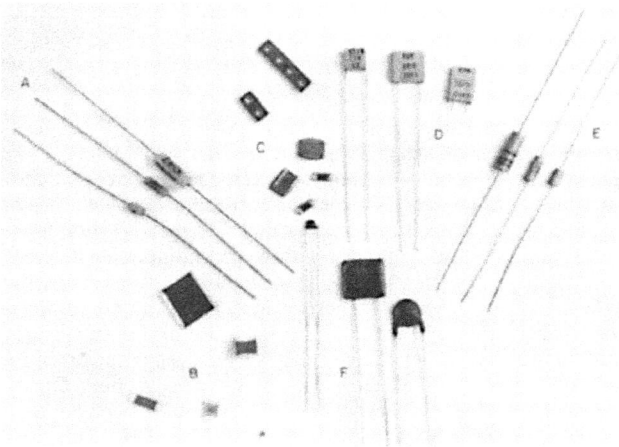

(b)

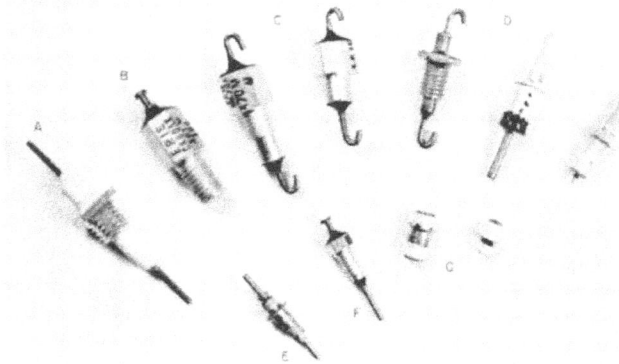

(c)

FIGURE 2.23   Typical ceramic-capacitor configurations. Others abound: (a) tubular shapes, (b) multilayer chips and packaged parts, (c) some of the available feedthrough designs.

Phlogopite mica, a softer material with less desirable dielectric performance, but temperature capability up to about 900°C, is used for some applications. Muscovite mica is termed a potassium aluminum silicate, $KAl_2Si_3O_{10}(OH)_2$; phlogopite is a potassium magnesium silicate, $KMg_3AlSi_3O_{10}(OH)_2$, both having also varying amounts of iron, manganese, copper, and chromium. For capacitor use, sheets are cleaved to 0.005 to 0.001 inches (or thinner). Theoretically, cleavage could produce sheets of a single molecule of thickness, but obviously unusable mechanical strength.

2.7.1 Reconstituted Mica

A paper-like material made from shredded mica particles, often manufactured mica rather than processed from natural mica, is used to produce high-quality capacitors for some specialized applications. This material is prepared by subjecting mica material that would not ordinarily be mechanically suitable as split plates for capacitor use to high-pressure water jets, producing a slurry of small platelet particles. This slurry is processed much like cellulose slurry is processed to produce paper, and the resulting material is a matrix of clean mica particles held together by their natural cohesive forces. The unimpregnated material has about 800 volts per mil dielectric strength, and the dielectric properties are comparable to good natural sheet mica. This material can be used to produce rolled or stacked plate capacitor bodies and, when vacuum impregnated with a material of choice (such as polystyrene, other polyesters, or silicones), it can have dielectric strength up to 2000 volts per mil. Temperature capability and temperature stability of capacitance reflect the capabilities of the impregnant materials as modified by the presence of the stable mica. The resulting capacitors can be made to have corona-free high-voltage capability and excellent stability.

2.7.2 Mica Capacitor Construction

For low-power applications, cleaved and gaged mica sheets of thickness of about 0.001 inches (or thicker), depending on the desired voltage rating and mechanical strength, have a paste of finely divided silver with a volatile fluid vehicle applied in patterns using screening techniques. Screening permits controlled edge margins and termination areas. Firing the silvered mica in an oxidizing atmosphere drives off the vehicle and permanently bonds the silver and mica. The resulting silvered wafers can be stacked and clamped together with inserted foil tabs to produce capacitors of the desired ratings.

Terminations are made by metal-forming clamping terminal forms around the mica stacks. The terminated capacitor is then molded or conformally coated with multiple layers of a durable resin to protect the element from moisture and handling and to insulate it. Present techniques mostly use resin powder coatings that are coalesced by heat. Figure 2.24 shows "dipped mica" construction.

High-power and high-voltage mica capacitors used extensively in RF transmitting equipment are made by stacking thickness-gaged cut-to-size mica leaves with interleaved foils of tin-lead or other material. The stacks are clamped together, the foil end tabs are metallurgically bonded together by soldering or welding, and the capacitor structure is vacuum impregnated with one of a variety of materials, depending on the design. Insulating oil, polyester resins, and silicones are among the materials used. Air must be excluded to prevent corona, and the material must provide the needed mechanical and electrical stabilities at the rated operating temperatures and voltages, and must not absorb water. Capacitor element stacks can be connected in series, parallel, or combinations to obtain the needed properties. Terminations and interconnections must be of large cross-

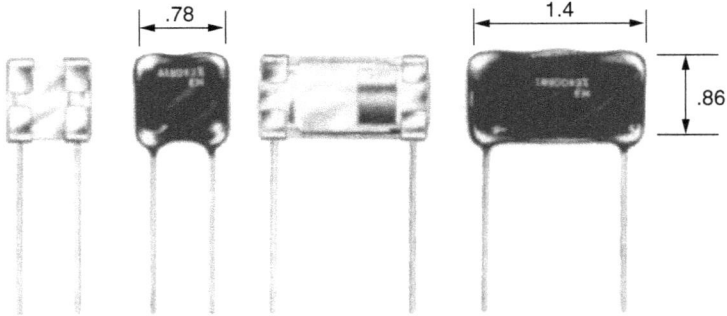

FIGURE 2.24 Dipped-mica capacitor construction (dimensions approximate in inches).

section to provide the surface area for large RF currents, but also must be configured for minimum inductance.

Most encasements for high-power mica capacitors have provisions for maintaining clamping pressure on stacks. For lower power uses, stacked-foil mica capacitors can be supplied with molded-resin cases containing the stack with screw or tab terminals that are attached to the internal foils and stack structure by metal-formed pressure clips. Button mica capacitors, widely used in the past but now seldom seen, are configured for efficiency in low-inductance coupling to a grounded case for decoupling or, with a feedthrough terminal to provide efficient filtering of a conductor passing through a grounded bulkhead. Button capacitors use disc-shaped silvered mica wafers with metallization brought alternately to the outer edge and to a center hole through the discs. The center hole has an eyelet with terminal attached installed by metal forming, and the outer terminal consists of a rolled metal rim, also formed to apply pressure. The whole is commonly oil- or wax-impregnated to fill voids. Solder tab projections are often provided on the outer metal rim for mounting. In the past, some cases were designed to be completely soldered into a chassis hole, with the center terminal then offering a standoff-type terminal. Later versions provide hermetic sealing of the capacitor element into a threaded feedthrough or standoff bushing, thus avoiding some difficulties of high temperatures needed for solder mounting. Some versions have even used glass-metal seal headers welded into the feedthrough bushing to avoid even the danger of reflowing the header sealing solder during wire terminal soldering. Figure 2.25 illustrates some button mica capacitors.

2.7.3 Mica Capacitor Performance

Silvered mica capacitors are among the most stable-with-temperature types, being comparable to stable ceramic, porcelain, and glass, and having very low loss with increasing frequency. Well-made mica capacitors also have excellent reliability, high insulation resistance, and excellent high-frequency characteristics. At this writing, however, they have been little adapted to surface mounting and miniaturization. Surface-mount chip-types are made, but despite their excellent stability and Q, their generally higher production costs have not yet been able to compete with the high-usage ceramic parts. Their physical size for the relatively large capacitance values needed for low-impedance solid-state applications, apart from high radio-frequency applications, makes them difficult to adapt to high-volume

FIGURE 2.25   Some types of button mica capacitors (Murata-Erie).

uses in digital and video applications. In their niche as a stable, high-Q capacitor for many applications, however, they are sometimes indispensable, having largely outlived the glass and porcelain parts that were designed to replace them. Operating characteristics, in comparison to other dielectrics, are shown by Figure 2.26. Capacitance-temperature stability for dipped parts ranges from about 0 ± 500 PPM/°C to as good as 0, +70, –0 PPM/°C.

2.7.4 Mica Capacitor Specifications

The older military specification, MIL-C-5, is still a valid document, but has been eclipsed by the MIL-C-39001 for leadwire-type mica capacitors. EIA 153 covers the same wire-lead types, which are, at this writing, all conformally coated radial-lead types, although the older molded-case capacitors with their six-dot color codes are still shown in the EIA documents. MIL-C-87164 covers conformally coated radial-lead types with specially implemented supplier controls and testing to reach highest reliability. MIL-C-11015 is a nonestablished reliability specification, but is still valid. It covers the button-type feed-through and stand-off designs, once extensively used, but now in declining application.

Special high-voltage and high-current transmitting types, like those shown in Figure 2.27, have no surviving standard specifications known to the writer, but still are used in their specialized applications. Newer designs in molded and coated encasements are also available for high-current, high-voltage applications.

2.8 ELECTROLYTIC CAPACITORS

With the exception of some special-purpose "double layer" capacitors, the highest CV product densities (capacitance × voltage per unit volume) are achieved by electrolytic capacitors. A large variety of electrolytic capacitor designs are produced using two fundamental materials: aluminum and its oxides, and tantalum and its oxides. (Capacitors made with niobium (columbium) were once briefly available commercially, but not at present.

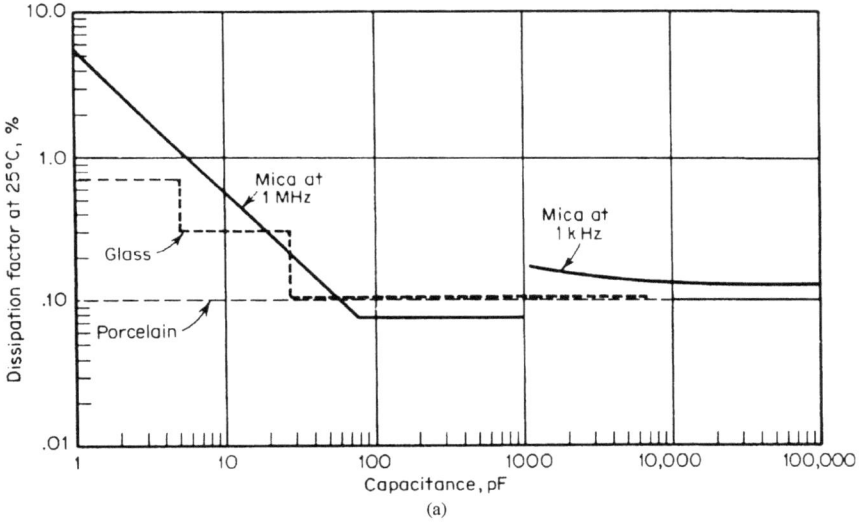

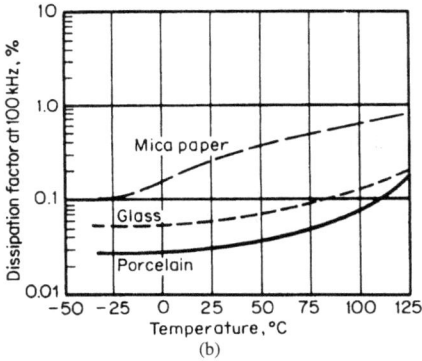

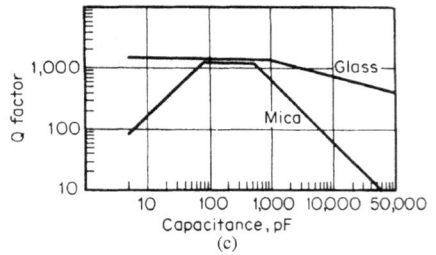

FIGURE 2.26   Some typical mica-capacitor performance curves: (a) dissipation limits over capacitance range, (b) dissipation factor-temperature stability, (c) Q factor versus capacitance at 1 MHz.

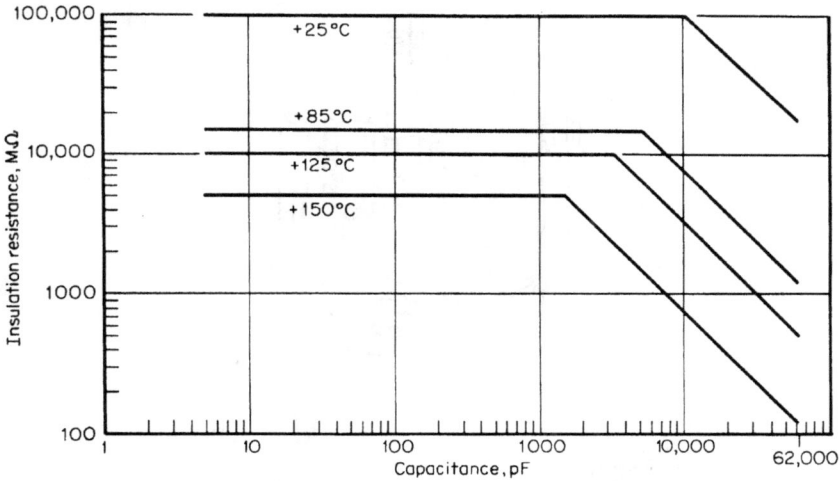

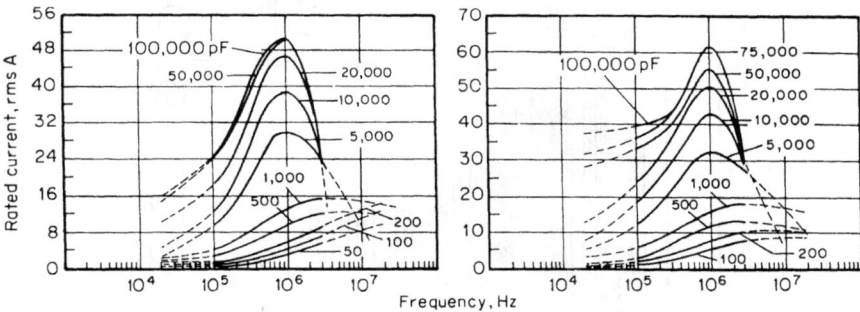

FIGURE 2.26 Some typical mica-capacitor performance curves: , (d) minimum insulation resistance values over temperature, (e) representatrive RF current capabilities for some typical mica power transmitting capacitors. *(Continued)*

Niobium is very similar to tantalum, and occurs naturally in the commonest tantalum ore. Titanium and zirconium also have been used in a very limited manner.) Aluminum and tantalum and their oxides have characteristics that are particularly well-suited for use in electrolytic capacitors, with the variety in design being primarily caused by the need for enhancement of one or more properties for better function in particular types of applications, to reduce cost, or to improve reliability.

The dielectric in aluminum electrolytics is amorphous aluminum oxide, (alumina) Al_2O_3, which has a very good dielectric strength in thin sections, is very stable chemically and physically, and has a dielectric constant of about 8.4. The dielectric in tantalum electrolytic capacitors is amorphous tantalum pentoxide, Ta_2O_5, which also has good dielectric strength, a dielectric constant of about 28, and a stable, inert nature. Producing high capacitance values in small volume is possible because of an ability to provide dielectric self-healing mechanisms for the thin, uniform, dielectric layers that can be formed onto irregular surfaces by electrochemical surface conversions and deposition. Using irregular surfacing allows a large

(a)

(b)

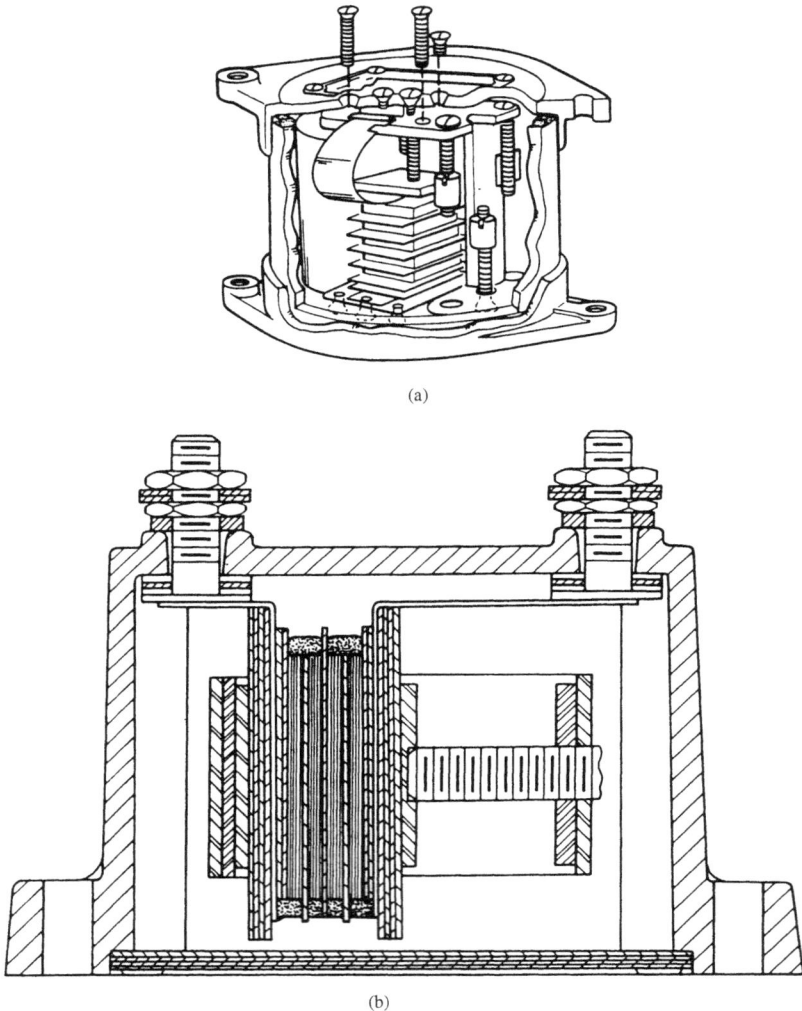

FIGURE 2.27 Some established designs of high-voltage transmitting mica capacitors.

plate area to be provided on a small plate length × width. After dielectric forming and structuring, some electrolytic capacitor element types are then placed in their container encasement with an electrically conductive liquid or semiliquid electrolyte, which provides the electrical conductor for one of the capacitor plates, but others, after forming and structuring, are cleansed of forming electrolyte and encased with solid materials that are electrically conductive and capable of forming a plate conductor in intimate contact with the film.

Electrolyte materials can be dry, liquids, gelled liquids, or liquids impregnated into porous media in contact with the dielectric. Some aluminum capacitors labeled as "dry" electrolytes might have gelled or porous media-absorbed electrolyte, but solid electrolyte tantalum sintered anode and parts labeled as solid aluminum types have truly dry electrolytes.

2.8.1 Basic Foil Electrolytic Construction

Figure 2.28 shows the basic construction schematics for polar and nonpolar wet electrolyte tantalum and aluminum foil capacitors. Termination-lead attachment to the foils is a critical step and is typically performed by welding the flat-formed same-metal lead to the foil ma-

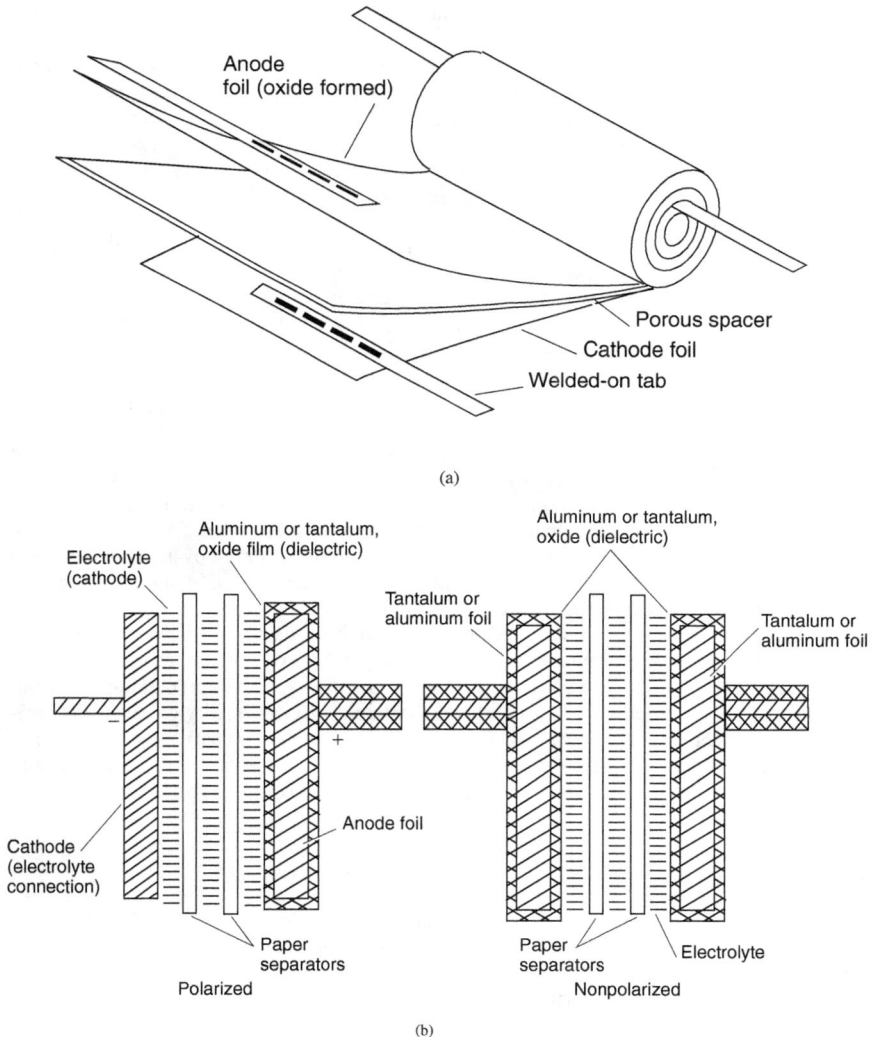

(a)

(b)

FIGURE 2.28 Tubular foil tantalum and aluminum electrolytic capacitor construction: (a) Foil and separator roll. The number of termination tabs can be more or less. In some designs, foil extensions can be used instead of welded-on tabs. (b) Schematic of construction cross section. In most cases, though not shown here, the cathode foil will undergo some oxide formation because of unintended or deliberate polarity reversals. The oxide formed will be thin if the reverse voltage is small, and the resulting high-value low-voltage capacitor will act in series with the forward capacitor, reducing the total capacitance value somewhat.

terial. Dielectric forming requires the use of phosphate or borate solutions for aluminum, and usually sulfuric acid for tantalum. Cathodes for formation can be any suitable material, such as copper or nickel. The forming voltage applied is related linearly to the desired oxide thickness, which then determines the working voltage of the finished product. For aluminum systems, the oxide is about 11 to 15 angstroms per volt applied; for tantalum, it is 16 to 20 angstroms per volt, depending on other variables (such as forming temperature).

The formed oxide is highly polarized, being electrically conductive, although with some resistance, in the "reverse" direction, but having a high insulation resistance in the "forward" direction. In the finished capacitor, with application of potential in the "forward" direction, the high insulation resistance is seen, and the film functions as a dielectric with a small leakage current, caused primarily by impurities and minute imperfections in the film. If potential is reversed, the film appears as a resistive impedance of fairly low value and generally nonlinear characteristics. Moderate amounts of reverse current can be tolerated in most electrolytic capacitors, with the chief danger being that of overheating because of the power dissipated within the oxide. A few capacitor systems, however, should never be reversed, even for a short impulse. These are fortunately few, and are wet electrolytic capacitors, in which the formed dielectric system is vulnerable to "poisoning" by ions from the case material. Alternating current reverse peaks and reverse peaks in damped oscillations from switching transients are some sources of reverse currents that might need to be evaluated.

Formation voltage, related to working voltage, has a practical maximum for aluminum of about 760 V, for tantalum, about 500 V and generally determines the oxide dielectric thickness. Dielectric thickness can be traded against plate area to reach optimized performance in a given volume. Some electrolytic capacitors have constructions other than foil, but the basic oxide formation systems are the same. It is possible to construct two or more capacitors of differing capacitance value and operating voltage in a common structure with a common electrolyte if a common cathode, usually at ground potential, is used. Multiple electrolytic capacitors are essentially aluminum types, made for commercial applications, where the cost of a special space-saving design can be easily absorbed by a high-production volume and reduced cost of one capacitor body versus two or more. EIA standard 395 contains standard requirements and measurement methods for the crossover coupling impedances for multi-section aluminum electrolytic capacitors using a common electrolyte. Electrolytic capacitor performance and reliability depend heavily on the purity of the basic materials used, exclusion of contaminants from the processing, care in processing, and good seal designs and materials for the liquid types. Various capacitor-quality grades using various electrolyte choices and mechanical designs (all strongly tied to cost) are purveyed, particularly in the aluminum electrolytic types. Extreme care with purity of the materials and control of processes produces capacitors capable of 20 years (or more) of reliable service.

2.8.2 Polarization Options

The most common electrolytic capacitors are polarized, made with a single dielectric film. If such a wet electrolyte capacitor uses an unformed (unanodized) foil or plate of aluminum or tantalum, as applicable, as its cathode, low-voltage reversals of polarity can form an oxide film on the hitherto unformed plate, which will reduce the effective forward capacitance. Some nonpolarized electrolytic capacitors intentionally use this technique to produce the nonpolar effect, with a corresponding sacrifice of CV product per unit volume. The effect is that of two capacitors in series, in which, by the rule of combining capacitors in series, will always have total capacitance less than the least of the two. Films formed from low-voltage polarity reversals will be thin, and the small dielectric spacing to the conductive electrolyte acting as the opposing plate makes its capacitance larger than that of the intentionally formed thicker main dielectric, assuming equal plate areas. Because the cathode capacitor has less reducing effect on the main capacitor value in the series combination if it is a high value, the cathode area is often enhanced by etching the cathode foil to increase its effective area.

As noted previously, polarity reversals with large accompanying currents are not well tolerated by polarized electrolytic capacitors, primarily because of internal heating, and should be addressed with caution. Semipolarized aluminum electrolytic capacitors are available, in which the capacitors are supplied with a deliberate thin cathode film to permit low-voltage reversals in service with greatly reduced risk of damage. Others are supplied with fully formed cathode films, are nonpolar, and can be used for applications, such as ac motor starting. Power factor (or dissipation factor) is specified as the chief figure of merit for nonpolar electrolytic capacitors, being a measure of the internal resistance of the system under ac conditions, but nonpolar dc capacitors use dc leakage current as a measure of the capacitor system insulation resistance. Dissipation factor is also often specified for polarized capacitors and should be measured using dc bias or a very low ac voltage.

2.8.3 Current Ratings

Because of the dc leakage current and resistance of the electrolyte conductor, power losses in electrolytic capacitors tend to be significant, and temperature rise under high currents is an application concern, particularly if a capacitor is used for high-current charge and discharge, motor starting, high-current power-supply filtering, or another high-current application. Current ratings for these types of uses are important specifications. High-current energy storage and high ripple-current capacitors (such as motor-starting units) should ideally be specially designed for such service. Heavy-duty plate terminations with large-area foil attachments, heavy case configurations, and plate aspect ratios can be adjusted to enhance capabilities. Commercial aluminum motor-start capacitors are, for instance, commonly rated for 20 starts per hour at 3 seconds per start. Lowering case temperatures dramatically improves service-life expectancy and, as a general practice, aluminum electrolytic capacitor case temperatures should not be permitted to exceed +60°C.

A single valid routine for calculating capacitor case temperature versus capacitor current cannot be stated because of the wide differences in capacitor type and design. Manufacturer information is the most dependable source of this information. Estimating techniques for a given capacitor utilize the worst case specified dissipation factor or ESR and dc leakage current along with circuit values of voltage and ripple current to arrive at a dissipated power for worst-case conditions. The rate at which a capacitor case dissipates heat in free space depends on its surface area, the nature of its surface, whether forced air flow is present, air density (for high altitude), the difference in temperature between case and air, and a few other variables. Modifying the condition from free space to that of a mounted configuration with thermal flow paths affects heat loss rate even more. A capacitor's resistive components are the only heat dissipating portion of its impedance, with the reactive portions being nondissipative. Because electrolytic capacitor ESR typically decreases as the temperature rises, using the room temperature value in calculations provides some conservatism. The insulation resistance dc leakage current is often a small enough perturbation to be ignored, but some manufacturers include it in calculations. Once the resistive dissipated power is determined, the rate of heat dissipation from the capacitor must be estimated, based on its temperature difference with the surrounding air and the nature of the capacitor case contact with the air (*e.g.*, whether or not a plastic sleeve is present). If forced-air movement is present, the heat transfer rate is much higher, of course, and that must be factored into the estimate. MIL-STD-198 and most manufacturer literature provide multiplying factors for calculated capacitor case areas, but most do not provide correction factors for capacitor length-vs-diameter ratios, which would allow for differences in thermal impedance from the internal core hot spot to case surface. Because internal temperature is the governing limit, the internal thermal impedance is a significant factor.

The MIL-STD-198 equation is:

$$T_{case} - T_{ambient} = 100 \left[\frac{Power\ (watts)}{Case\ area\ (inches2)} \right]^{.82} \tag{2.15}$$

where temperatures are measured in degrees C. The equation is said to apply for cylindrical aluminum capacitors and allows dissipation of about 0.06 W/sq in of case surface, not including one end, for a case-temperature rise of 10°C above ambient at sea level with no forced air. Elevated temperature shortens capacitor expected lifetime, primarily because of electrolyte dryout, and increases the probability of spontaneous early failure—even when kept within the manufacturer's recommended maximum. Heat production should always be investigated in applications where there is concern that temperature rise might approach the rated maximum.

2.8.4 Temperature-Temporary Effects on Capacitance

Typical capacitance-temperature curves for a number of electrolytic capacitor types and values are shown in juxtaposition by Figure 2.29. These curves are not absolute because temperature effects are greatly influenced by the specific electrolyte used, and by the specific capacitance value and dielectric thickness for the part being investigated, but it is apparent that the greatest concern is the dramatic loss of capacitance at low temperature,

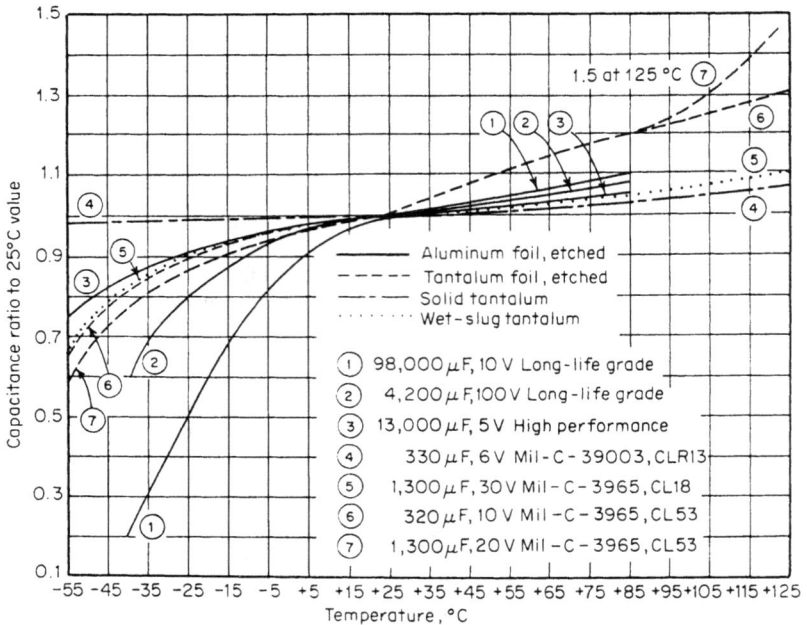

FIGURE 2.29 Capacitance-temperature performance of some electrolytic capacitor types (capacitance changes depend primarily on electrolyte used and the capacitance value. Small values are more stable.

experienced by the high-capacitance wet electrolyte types. Aluminum systems using ethylene glycol-based electrolyte with included water do not fare as well as those with dimethyl formamide (DMF)(not shown by Figure 2.29).

Low-temperature capacitance-value decrease is not quite as severe with wet tantalum
capacitors as with aluminum types. Many aluminum electrolytics are not usable as capacitors at –55°C. By careful processing and use of premium electrolytes, however, aluminum
electrolytic operation range is extended to –55°C, with less than a 40-percent loss at the
25°C capacitance value. Large capacitance values lose a higher percentage of their value
than do smaller values. Wet tantalum electrolytics can be expected to lose about 20 percent
of their 25°C capacitance at –55°C. Low-temperature limitations are caused primarily by
low-temperature changes in and freezing of the electrolyte used, with better-performing
capacitors using electrolytes that retain their efficiency at lower temperatures.

2.8.5 Electrolytic Capacitor Shelf Life

The oxide film of many wet aluminum electrolytic capacitors undergoes as a degree of deforming of over time in storage if no voltage is applied. This effect is greatly reduced for
premium-quality capacitors, made better by selecting and optimizing electrolyte and by
carefully controlling purity of materials and quality of processing. It has been demonstrated that high-quality aluminum electrolytic capacitors can be expected to retain function with acceptable temporary loss of performance after storage under reasonable
conditions for over 10 years and up to 20 years for most parts in such long-life configuration populations. Parts might be degraded somewhat, but quickly recover in service if the
electrolyte has not dried out.

Parts produced under less-controlled conditions of purity and processing can be well-
served by subjection to a current-limited reforming ritual after storage for 1.5 to 2 years or
more. A reforming procedure can be obtained from the part manufacturer, or one is published in MIL-STD-1131. The procedure, usually a current-limited small over-voltage conditioning, is followed by careful leakage-current screening to identify and eliminate
permanently affected parts. Drying out or loss of electrolyte through defective or poorly performing seals will cause loss of capacitance and elevated ESR during storage, and also constitutes the major wearout mechanism seen in electrolytic capacitors in operation under rated
conditions. Wet tantalum capacitors do not have any concerns associated with deforming
during dormancy, primarily because of the better stability of their basic materials. Solid electrolyte tantalums not only do not deform, but they have no electrolyte loss concerns.

2.8.6 Electrolytic Capacitor Leakage Current

Electrolytic capacitors have lower insulation resistances than electrostatic types, partly
because electrolytics have much higher capacitance values and therefore have a much
higher dielectric cross-sectional area with their thin dielectrics, but the nature of the capacitor system also influences the IR value, generally specified as a dc leakage current at
the rated dc voltage. As the temperature increases, the oxide-film insulation resistance
decreases and the electrolytic capacitor dc leakage current increases. Figure 2.30 shows
the temperature effects on leakage currents for several electrolytic capacitor types, although this particular comparison suffers because the capacitance values in the figure are
different for each type. Leakage currents raise capacitor internal temperatures and decrease charge retention time, in addition to being the primary quality and health indicator
for electrolytic capacitors.

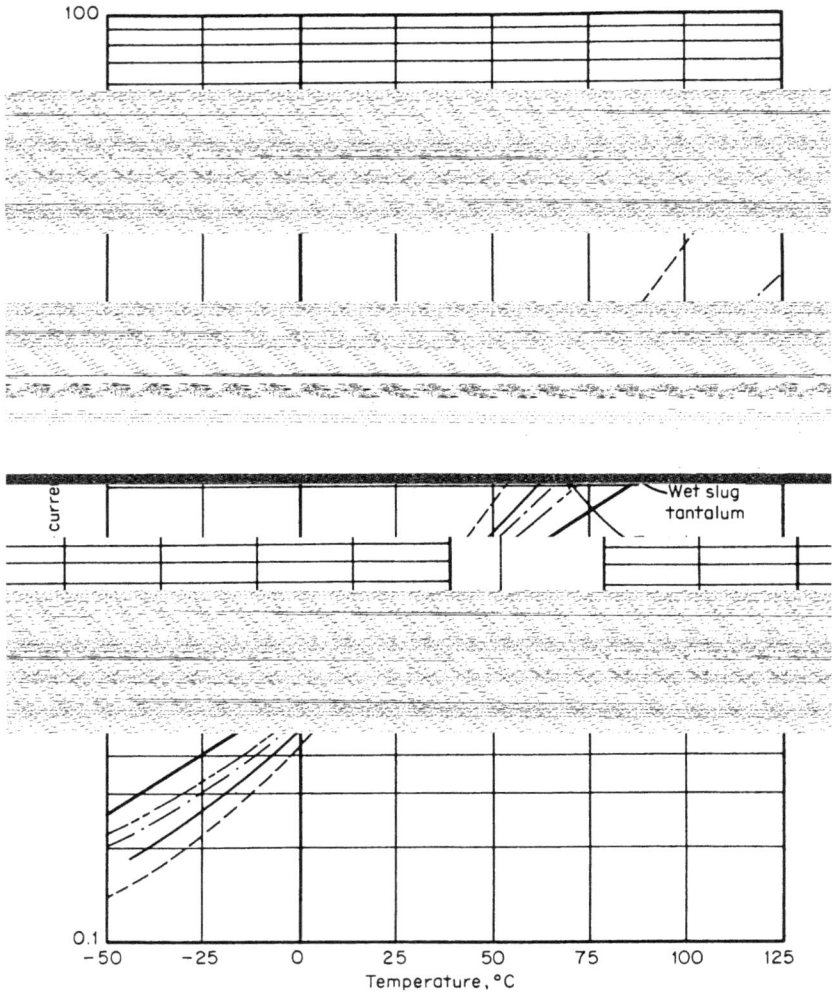

FIGURE 2.30 Electrolytic capacitor leakage currents versus temperature: typical ranges.

2.8.7 Aluminum Electrolytic Capacitors

See also the previous section on electrolytic capacitors, in general. An impressive array of aluminum capacitor sizes, shapes, and specialized designs are available. In addition to various design options that enhance performance for specific kinds of applications, manufacturers make available various quality grades, whereby, for differences in cost, you can select parts with lesser or greater temperature range or whose population can be expected to have more or fewer in-use failures after shorter or longer time periods. Parts are available with epoxy over-caps to discourage the introduction of chlorinated hydrocarbon solvents during equipment production, which some manufacturers have cited as the major reason for aluminum electrolytic capacitor field problems. These resin over-seals do not appreciably improve electrolyte loss, however, and should not be depended upon for that purpose.

To the writer's knowledge, no hermetically sealed aluminum electrolytic capacitors are now available, but the elastomer pressure seals that are used can be very tight. However, even the best seal is vulnerable to the minute chlorine intrusion that is needed to destroy an aluminum electrolytic in service. Chlorine-bearing materials are best kept well away from aluminum electrolytic capacitors, and it is essential that this be emphatically communicated to equipment production operators.

To organize and arrange the aluminum electrolytic varieties, an informal system of "grading" has been used by some manufacturers for their commercial offerings, and the following terms are often used: "computer grade", "general-purpose grade", "premium grade", "extended-life grade", etc. The precise meanings are defined by manufacturers in catalog data. EIA 395 specifies 2 grades of aluminum electrolytic capacitors, and MIL-C-62 also specifies 2 grades. Selection of electrolytes, physical construction, purity of materials, care in production and testing, and, ultimately, the manufacturer's degree of individual part conditioning and screening make the differences in grading (Table 2.20).

The electrolyte most commonly used in aluminum capacitors is a combination of ethylene glycol and boric acid, mixed and reacted to form a pastelike material of medium resistivity. Lower resistivities, increased low temperature capability, and lower leakage currents are provided by other nonaqueous electrolyte combinations based on ethylene glycol and dimethyl formamide, at higher costs.

Aluminum electrolytic capacitor application in some types of military equipment has been discouraged by many military design authorities for the past 40 years, primarily because of the defined equipment needs for a –55°C low-temperature extreme for high-altitude airborne use and a –45°C low limit needed for worldwide ground-level battlefield performance. Also of concern, the lack of guaranteed instantly available full performance after long-dormant storage periods has caused prohibition for many military applications, but this concern pales to insignificance if high-grade capacitors are used. There are many military uses for which aluminum electrolytic capacitors are suitable and perform very well. Aluminum parts have much lower cost than tantalum types, are lighter weight, and are available in a wider diversity of capacitance values and voltage ratings. If selected for military applications, attention should be given to mechanical ruggedness of internal structure in addition to temperature range and life expectancy. Some aluminum capacitors have been designed with solid electrolytes, notably MnO_2. The aluminum oxide lower dielectric constant, as compared to tantalum pentoxide, makes for an aluminum volumetric disadvantage to counter the lower cost.

Aluminum electrolytic capacitor specifications and standards Electronic Industries Association voluntary standard EIA 395 covers two polarized aluminum electrolytic capacitor quality grades, primarily for use in electronic equipment. Other commercial standards (see Table 2.7) cover capacitors for power-system applications. Specifications MIL-C-62 and MIL-C-39018 cover military aluminum electrolytics, with MIL-C-62 being the less stringent of the two, and covering a wider diversity of types. MIL-C-39018 covers both polarized and nonpolarized capacitors having ruggedized construction, excellent elastomer seals, and the best available life and stability. Table 2.21 shows some capacitor features and their relative variations between performance grades.

Aluminum electrolytic capacitor case types Polarized aluminum electrolytics almost always use the case (can) as the cathode external terminal with a terminal wire attached to the can, or with a solder lug or sometimes a screw terminal electrically connected to the can and the cathode foil. Cylindrical parts, with their familiar circumferential furrow

TABLE 2.21 Graded Characteristics for Some Aluminum Electrolytic Capacitors.

Grade	Temperature range (°C)	Manufacturer life test (hours)	Relative highest CV product per volume[1]	Relative lowest ESR[1]	Relative highest ripple and rating[1]	Expected life (years)	Relative highest surge voltage[1]
Commercial Grade	−20/+85	500 or 1000	3	4	3	3–5	3
Computer Grade	−20/+85, also −40/+85	500 or 2000	1	3	3	5–10	2
Long Life Grade	−40/+85	1000 or 2000	2	2	2	10–20	1
Premium Computer Grade	−55/+105, also −55/+125	1000 or 2000	1	1	1	10–20	1

[1] Number 1–4 indicate best through worst, with 1 being best. Grades are arranged in order of ascending relative cost. Mounting and terminating hardware for higher cost grades is more rugged and durable, higher cost.

TABLE 2.20 Elevated Temperatures and Surge Voltage Ratings Related to
Room Temperature Voltage Ratings for Electrolytic Capacitors

Rated dc voltage	Aluminum foil surge[2]	Wet tanatalum, foil, slug surge	Solid electrolyte tantalum[3]		
			+85°C surge	DC rated at +125°C	Surge rating at +125°C
2			2.6	1.3	1.7
3	4	3.4	4.0	2.0	2.7
4			5.0	2.7	3.5
5	7				
6	8	6.9	8.0	4.0	5.0
7	10				
8		9.2			
10	13	11.5	13.0	7.0	9.0
15	20	17.2	20.0	10	12
20	25		26.0	13	16
25	30	28.8	32.0	17	21
30	40	34.5	39.0	20	26
35	45		46.0	23	28
40	50				
50	65	57.5	65.0	33	40
60	75	69.0			
75	95	86.2	98.0	50	64
100	125	115.0	130	67	86
125		144.0			
150	175	172.0			
175	200	230.0			
200	250	287.0			
250	300				
300	350	345.0			
350	400				
375		431.0			
400	475				
450	525	518.0			

[1] Ratings may vary with manufacturers and detail part designs. Consult manufacturers
for current information.

[2] Manufacturers commonly reduce surge ratings below these figures for their high-performance graded aluminum electrolytic capacitors.

[3] Military specifications for most, but not all, solid electrolyte tantalum capacitors require reduction of rated dc voltage to 66 percent of the –55°C – +85°C rated voltage
when operating at +125°C, with a linear rating gradient between +85 and +125°C.

where the can has been roll-formed into the terminal block sealing groove, are an almost universal configuration (see Figure 2.31), although the sealing furrow is not always present. Terminal leadwires can be axial, extending from alternate ends of the case, or single-ended, with both anode and cathode extending from the same end of the cylinder. Solder or screw terminals mounted on the case-end header are also common single-end termination types for very large size cases, and standardized blade connector contacts are provided on many capacitors intended for power electrical system use. Plated, flat solder terminals on single-end termination types are often bent over to contact printed-circuit termination pads for automated soldering. For small single-end cases with wire-type terminals, the terminations are made mechanically strong enough to support the capacitor when terminals are soldered directly to circuit pads. Can rim bend-over or twist tabs can be provided to provide extra strength for larger cases. Cans must be carefully designed so that bending and twisting terminal tabs does not distort the relatively soft aluminum can in a manner that degrades or destroys the rolled-metal seal and terminal block retention.

Single-end termination wires for small-size types intended for mounting by their wires onto circuit cards are commonly of different lengths. This feature serves a dual purpose, providing a visual and tactile polarity indicator for the capacitor installer and also making the insertion one hole at a time instead of requiring simultaneous alignment of two leadwires and their circuit board holes. Because the capacitor case is electrically connected to the capacitor cathode for most polarized electrolytic capacitors, it is practical to insulate the case from casual electrical contact by using a shrink plastic sleeve or, sometimes, a close-fitting paper tube. Some parts also can be molded into a plastic encasement.

For safety, aluminum electrolytic capacitors must be provided with some means of venting the case if the capacitor should become short-circuited while connected to a stiff power source. Unless vented, the pressure can increase to levels that cause damage and injury if the can ruptures. Various venting schemes have been used in the past, with some capacitors having rubber blowout plugs. Most capacitors now have deliberately weakened thin points in the aluminum can, formed by pressing grooves into the soft aluminum can before the capacitor is constructed. These weak points will rupture at a lower internal pressure than would the unaltered can, thus relieving the pressure and avoiding a more violent event. Unlimited high current into a failed, short-circuited capacitor can still generate enough sudden energy to overcome venting systems and cause capacitor explosion on occasion. Connecting high-value capacitors in parallel might create a high-current source when parallel-connected good capacitors discharge through a short-circuited companion. Capacitor users for high-energy applications should "anticipate" such an event, no matter how infrequent, and provide protective packaging to contain the event, and should provide over-current protective shutdown (e.g., fusing), if possible.

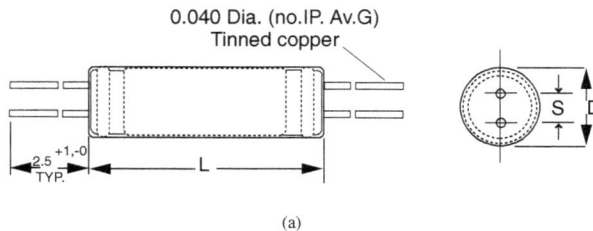

(a)

FIGURE 2.31   Aluminum capacitors: Some popular case configurations: (a) Four-wire with common ground and an anode on each end, (b) Several designs of commercial metal can and plastic-cased metal can parts.

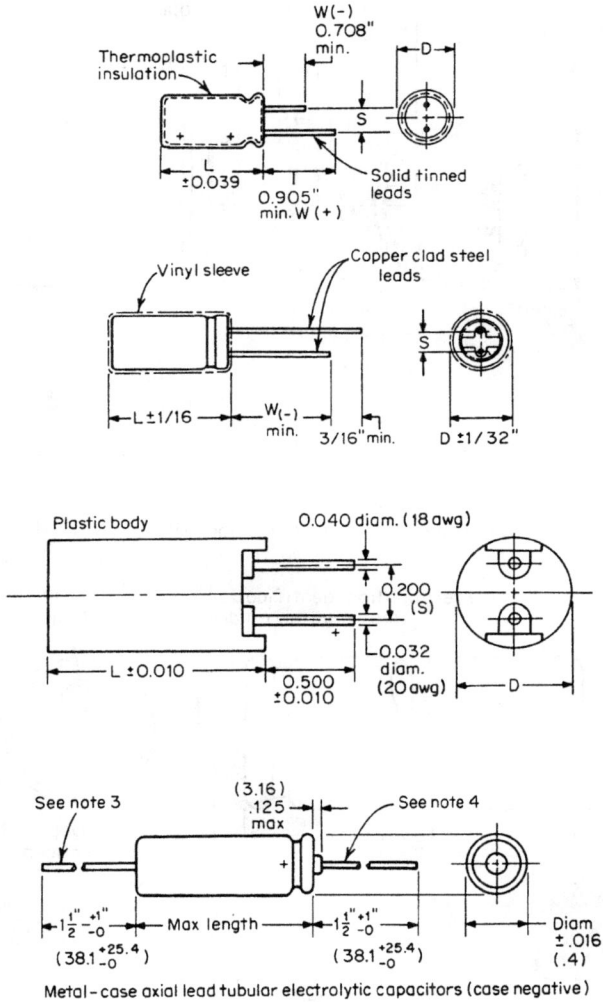

FIGURE 2.31 Aluminum capacitors: Some popular case configurations: (a) Four-wire with common ground and an anode on each end, (b) Several designs of commercial metal can and plastic-cased metal can parts.

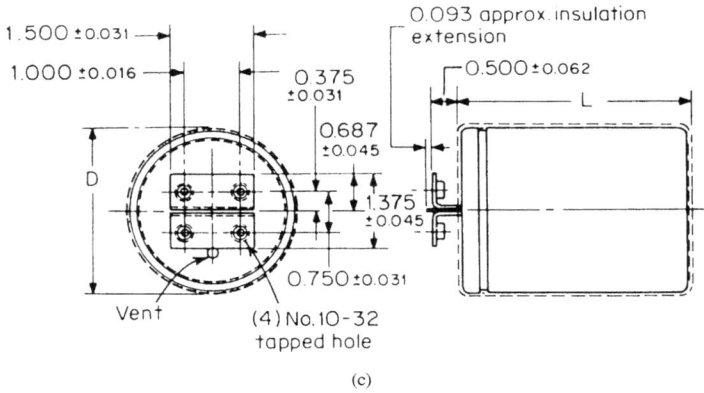

(c)

Metal-Case Round Aluminum Electrolytic Capacitors with Screw-Insert Terminations

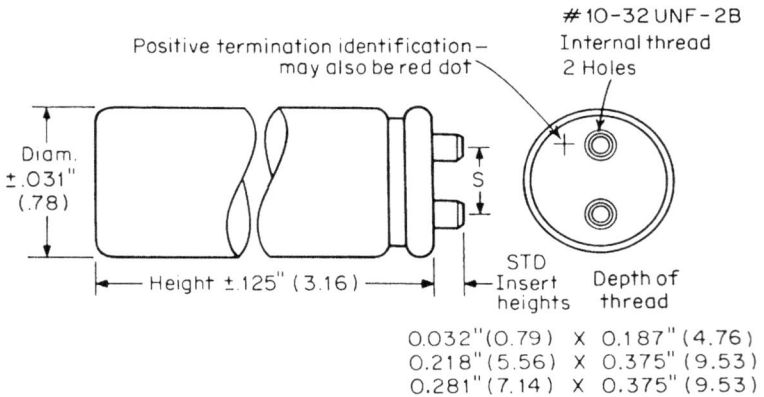

0.032"(0.79) X 0.187"(4.76)
0.218"(5.56) X 0.375"(9.53)
0.281"(7.14) X 0.375"(9.53)

(d)

FIGURE 2.31 Aluminum capacitors: Some popular case configurations: (a) Four-wire with common ground and an anode on each end, (b) Several designs of commercial metal can and plastic-cased metal can parts. *(Continued)*

				Standard Case Dimensions					
in	mm	in	mm	in	mm	in	mm	in	mm
				Case diameters					
1.375	34.93	1.75	44.45	2.0	50.8	2.5	63.5	3.0	76.2
				Insert spacing					
0.50	12.7	0.75	19.05	.875	22.23	1.125	28.58	1.25	31.75
				Heights					
2.125	53.98	2.125	53.98	2.125	53.98	3.125	79.38	3.625	92.08
2.625	66.68	2.625	66.68	2.625	66.68	3.625	92.08	4.125	104.78
3.125	79.38	3.125	79.38	3.125	79.38	4.125	104.78	4.625	117.48
3.625	92.08	3.625	92.08	3.625	92.08	4.625	117.48	5.125	130.18
4.125	104.78	4.125	104.78	4.125	104.78	5.125	130.18	5.625	142.88
4.625	117.48	4.625	117.48	4.625	117.48	5.625	142.88	6.625	168.29
5.125	130.18	5.125	130.18	5.125	130.18			8.625	219.09
5.625	142.88	5.625	142.88	5.625	142.88				

FIGURE 2.31 Aluminum capacitors: Some popular case configurations: (a) Four-wire with common ground and an anode on each end, (b) Several designs of commercial metal can and plastic-cased metal can parts. *(Continued)*

Capacitor body clamps and clips for mounting large capacitors or those to be subjected to mechanical shock and vibration must be carefully chosen and applied because the nearly pure aluminum case material is soft and subject to damage and distortion. Also, the venting mechanism must not be defeated by covering it with a clamp. Capacitor manufacturers provide guidance in selecting auxiliary mounting hardware and they should be consulted before departing from the recommended configurations. Metal clamps and clips that are plated with tin, zinc, nickel, or even cadmium should generally not be used in direct contact with the aluminum capacitor case if the equipment is to be subjected to high-humidity conditions. Condensed moisture can activate a galvanic corrosion process because of the dissimilarity of the metals. Insulating sleevings are effective in somewhat alleviating this problem.

Aluminum electrolytic capacitance tolerances   Because of the nature of the product, aluminum electrolytic capacitors should not be expected to provide precise and stable capacitance values. Rather, they are suitable for applications where it is desired to present a low impedance for bands of frequencies or for specific frequencies, and a variation in the impedance value can be tolerated up to some maximum value. They are also suitable for energy-storage applications, where large design margins are possible to allow for variations. Table 2.23 shows EIA-recommended capacitance tolerances. Low-temperature characteristics for capacitors should be investigated, and the role of the leakage current (insulation resistance) and dissipation factor in the capacitor impedance must be known. None of these characteristics improve with time in capacitor service, so design margins are important.

TABLE 2.22 EIA Recommended Capacitance
Tolerances for Aluminum Electrolytic Capacitors

Rated dc voltage	Tolerance for single section capacitor	Tolerance for multi-section capacitors
3 – 100 V	–10%, +100%	–10%, +150%
101 – 300 V	–10%, +75%	–10%, +100%
301 – 450 V	–10%, 50%	–10%, +50%

2.8.8 Tantalum Electrolytic Capacitors

Tantalum electrolytic capacitors are of three basic types: tantalum foil, wet sintered slug, and dry (or solid electrolyte) sintered slug. Some small-value, very small-size etched-wire types are made as well. All utilize tantalum pentoxide as their dielectric. Capacitance value and voltage range is much more limited and costs are much higher, when compared to aluminum electrolytics, but electrical performance and stability over environmental extremes is much better. Long-term reliability in service is generally better than for aluminums, there being no recognized wearout mechanisms in well-made capacitors, and the stability of the chemical system is much higher.

Tantalum-foil capacitors   As of this writing, tantalum-foil electrolytic capacitors are not being manufactured anywhere in the world. Nevertheless, in the interest of completeness, a brief discussion is presented because many are still in service. Tantalum foil electrolytic capacitors provide the stability and reliability of the tantalum-tantalum pentoxide system with a capacitance range up to 2500 µF. Operating voltages up to 300 volts can be made, and good stability over –55°C to +125°C is provided. Reversed voltage on polarized parts can be tolerated up to 3 V, surge current tolerance is good, and degraded performance after long-dormant shelf life is not a significant concern. Nonpolarized parts have been made. Dc leakage current is much lower than for aluminum electrolytics and is about equivalent to that of solid electrolyte sintered-slug tantalums, somewhat higher than wet-slug tantalums. The cost for foil tantalums had escalated when production stopped, to be the highest of the electrolytic capacitor types, slightly greater than for all-tantalum wet-slug types.

Tantalum foil construction   Tantalum foil capacitors are made in much the same way as aluminum foil types, except that the materials are different. Tantalum leadwires are welded to two high-purity tantalum foils, usually 0.5 mils in thickness. Amorphous tantalum pentoxide is formed on the anode foil surfaces by electrochemical processing. Cathode foils are also formed for nonpolarized capacitors, but are left plain for polarized types. Before forming, the foils can be etched to increase surface area, permitting higher CV product at the expense of increased ESR and lowered temperature-capacitance stability. Cathode foils in polarized parts are not usually etched. The case is not directly connected to either foil other than by the electrolyte, and it does not intentionally enter into the functional system. The foils are interlayered with porous separator materials of paper, glass fiber, or other material, wound into a cylindrical form with the tantalum leadwires extending from opposite ends.

The structure is placed into a close-fitting case with fluorocarbon spacer blocks, as needed. A circumferentially grooved fluorocarbon header with silicone elastomer gaskets is put into place with one tantalum wire protruding through its center axial hole, and the case is metal-formed to compress the elastomer around the center wire and the case to header-block interface, and to form the case into the header-block groove. The capacitor structure is then impregnated with the electrolyte, which is dimethyl formamide with dissolved salts, lithium chloride, sulfuric acid, or another material selected for low resistivity, enhancement of temperature stability, and support of oxide-film self-healing. The remaining header end block and seal is put into place over the opposite tantalum leadwire, and the case is metal-formed to seal the unit. Solderable copper, copper sheathed steel or nickel leadwires are then butt-welded to the tantalum wire terminals. See Figure 2.32 for a cross-section schematic of a typical tantalum-foil capacitor.

Tantalum-foil electrolytic capacitor specifications At present, MIL-C-39006 and the older MIL-C-3965 covers polar and nonpolar etched and plain foil tantalum foil capacitors but, at this writing, no qualified sources exist. EIA voluntary standard EIA 228 covers the same products, but there are no active domestic manufacturers.

2.8.9 Wet and Solid Electrolyte Sintered-Tantalum Slug Capacitors

Constructing a capacitor plate by using a porous sintered tantalum slug or pellet produces a very high surface-area-to-volume ratio, in the order of 1000 to 1 and, when combined with the ability to electrolytically form uniform thin films of stable dielectric oxides, permits production of very high CV product capacitors. Wet slug tantalum electrolytic capacitors have the highest CV product per unit volume of any capacitor type covered thus far, and the ESR achieved is the lowest of any conventional electrolytic capacitor type. Because the pellet or slug construction and dielectric formation is identical for both wet-slug and solid electrolyte parts, that portion of the construction is presented together.

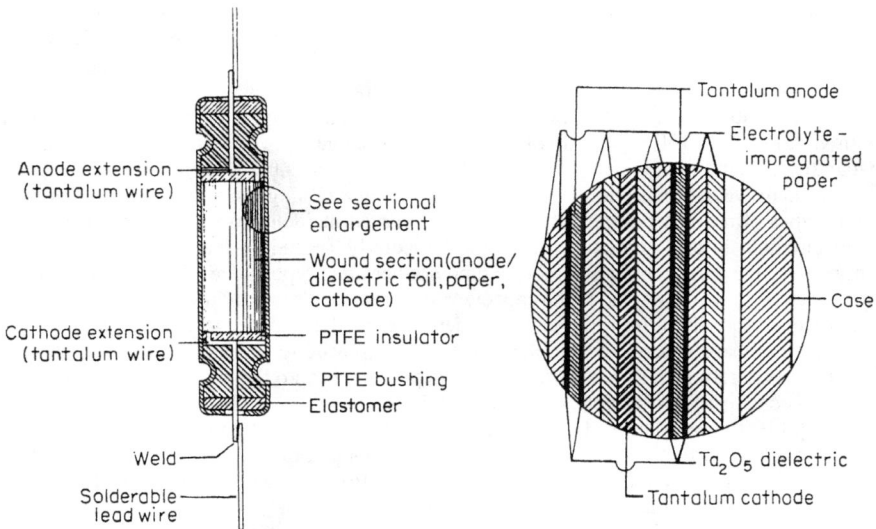

FIGURE 2.32 Typical wet tantalum foil cross-section.

Sintered tantalum slug construction for wet and solid electrolyte capacitors Solid-electrolyte porous-slug capacitors begin the same way as do the wet-slug variety. Finely divided, very pure tantalum metal powder (sometimes with a small amount of binder material) is pressed to a desired size and shape and sintered under high vacuum at approximately 2000°C to clear the binder material and weld the tantalum particles together, forming a spongy structure with a large surface area. The combination of high vacuum and high temperature cleans the surface and removes many of the trace impurities that could otherwise introduce difficulties into further processing and degrade quality of the final product. A tantalum leadwire must be provided to make electrical connection to the pellet, and can be attached by pressing the tantalum powder around the leadwire during the pellet pressing, driving it, later in the pellet processing, into a hole left in the pellet for that purpose, or the wire can be welded to the pellet after sintering.

Each of these options have advantages and disadvantages: the welded-on connection is said to provide the lowest ESR in the finished capacitor, but must be processed to remove welding stresses and damage to the structure. The porous pellet with anode wire is immersed into a liquid-forming bath, and a forming voltage is applied to the anode wire with circuit return through the forming electrolyte. An amorphous coating of tantalum pentoxide, Ta_2O_5, typically less than 500 angstroms thick, is thus formed on the large tantalum surface area of the porous sintered pellet. The bath temperature, run time, and current flow are carefully controlled to ensure that the coating is uniform and amorphous, rather than crystalline.

When the oxide is formed to the desired thickness, as determined by the final forming voltage, which is typically two to three times the intended capacitor working voltage, the tantalum metal will have a color when viewed in white light that is characteristic of its thickness, usually shades of red, green, or blue. Uniformity of color also indicates uniformity of oxide thickness. At this point, processing for wet-slug capacitors deviates from that for solid electrolyte parts, with the electrically conductive materials used to form the cathode plate being different. Figure 2.33 shows the formed dielectric within a sintered slug viewed by means of a scanning electron microscope.

Wet-slug tantalum electrolytic capacitors Except for solid electrolyte tantalums, wet-slug tantalum capacitors, as stated previously, have lower ESR than other electrolytics and have the lowest dc leakage current of any electrolytics covered here. The wet-slug tantalum has good capacitance stability over its normal operating range of –55°C to +125°C, and single units are made with extended-range capacitance values up to about 2200 µF with voltages up to 150 V. Wet-slug tantalum capacitors have a very efficient self-healing capability, which forms dielectric over small breakdown areas in the dielectric.

For many years, restrictions were placed on using wet-slug tantalum electrolytics in many types of military equipment because of observed incidents in which the capacitors short circuited in use, sometimes failing spectacularly, spreading their unfriendly electrolyte over the surrounding assembly. Some nonhermetic types were also prone to electrolyte leakage, which degraded the capacitors and damaged surrounding circuits. A major part of the short-circuiting problem was identified as dendritic growth from the silver cases then in use and a related problem caused by exposure to reverse-polarity potential, which quickly removed silver from the case, depositing it onto the slug (pellet) and initiating a mechanism to inevitable short-circuit failure.

The problem was addressed by developing a design having a tantalum metal case and a hermetic seal. The true hermetic seal required development of headers using either compression glass around the tantalum riser wire with enclosure by a tantalum annulus, or one in which the glass and tantalum are closely matched in thermal expansion characteristics and the molten glass is induced to wet the tantalum center wire or tubulation and the tantalum annulus, fusing to them. The compression glass seal depends on thermal-expansion co-

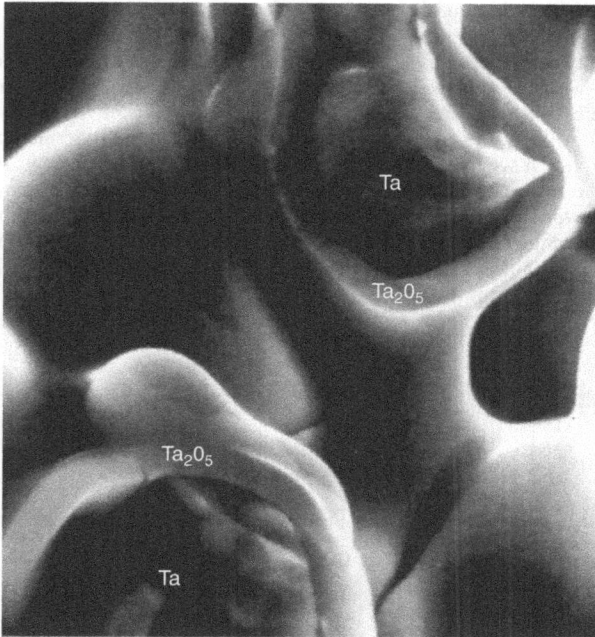

FIGURE 2.33 Scanning electroni microscope view of a typical formed sintered-tantalum pellet. *(Lockheed Martin Electronics and Missiles Failure Analysis Laboratory.)*

efficients of glass and tantalum, being such that the glass-metal interfaces are held in compression over the temperature range of exposure. The seal outer annulus of either type is laser welded to the tantalum case, and a weld is used to close the header tubulation around the tantalum anode wire if that design is used, thus completing a hermetic seal. Internal gasket sealing is necessary to retain the electrolyte, keeping it away from the header internal surface, where it might provide an electrical leakage path from the anode to the case.

There is no military restriction for using wet-slug tantalums of this design, and it now has wide application. Other case types with various internal platings have been tried, with some success, to replace the relatively expensive tantalum case. Never, even for an instant, subject a silver-cased wet-slug tantalum capacitor to a reverse polarity potential. Silver-cased wet-slug tantalum capacitors are thus not suitable for back-to-back connection to achieve a nonpolar capacitor effect.

Wet-slug tantalum capacitor construction The pellet or slug, with its formed dielectric has PTFE fluorocarbon support blocks affixed for support within the case to avoid failure because of mechanical shock and vibration, and is inserted into its closed-end case, which, as previously stated, can be silver, tantalum, or other metals that are specially plated. An electrolyte, commonly 40 percent by volume sulfuric acid to which fine silica powder can be added to produce a thixotropic gel, is introduced into the case in a manner that ensures complete impregnation of the pellet. Ungelled 40-percent sulfuric acid is also used. Other PTFE vibration supports having elastomeric gaskets are added to seal the electrolyte into its intended volume, and, if so designed, the case is metal formed to compress the gasket seals and retain the header plug.

If a hermetically sealed design is being constructed, the glass-to-metal header is then installed by laser welding. Butt-welding a solderable nickel leadwire to the tantalum anode wire completes the structure, the drawn case having a nickel leadwire previously welded or swaged to it for the cathode connection. Two all-tantalum hermetically sealed constructions are shown schematically in Figure 2.34. The gelled 40-percent (by volume) sulfuric acid is the most common electrolyte, although lithium chloride has been used. Using the silica powder gelling agent avoids having the capacitor performance become position-sensitive when the electrolyte fill has a void that might move to an unfavorable location for certain sustained capacitor orientations, but it is said that ungelled liquid electrolyte provides a lower ESR capacitor.

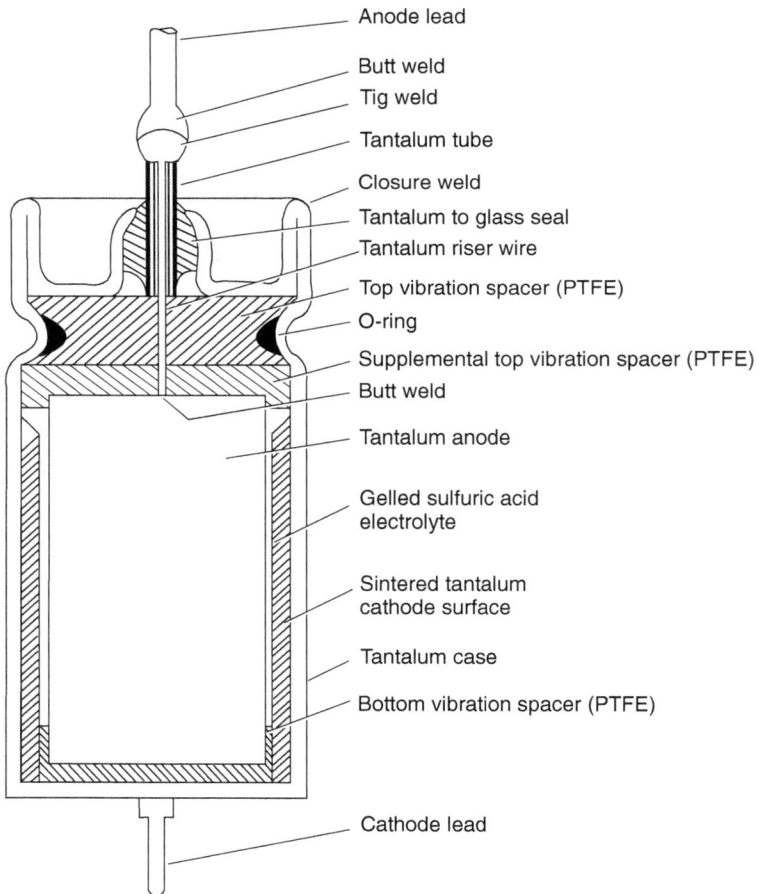

FIGURE 2.34 Two Designs for all-tantalum hermetically sealed wet-slug capacitor construction. *(Vishay-Sprague.)*

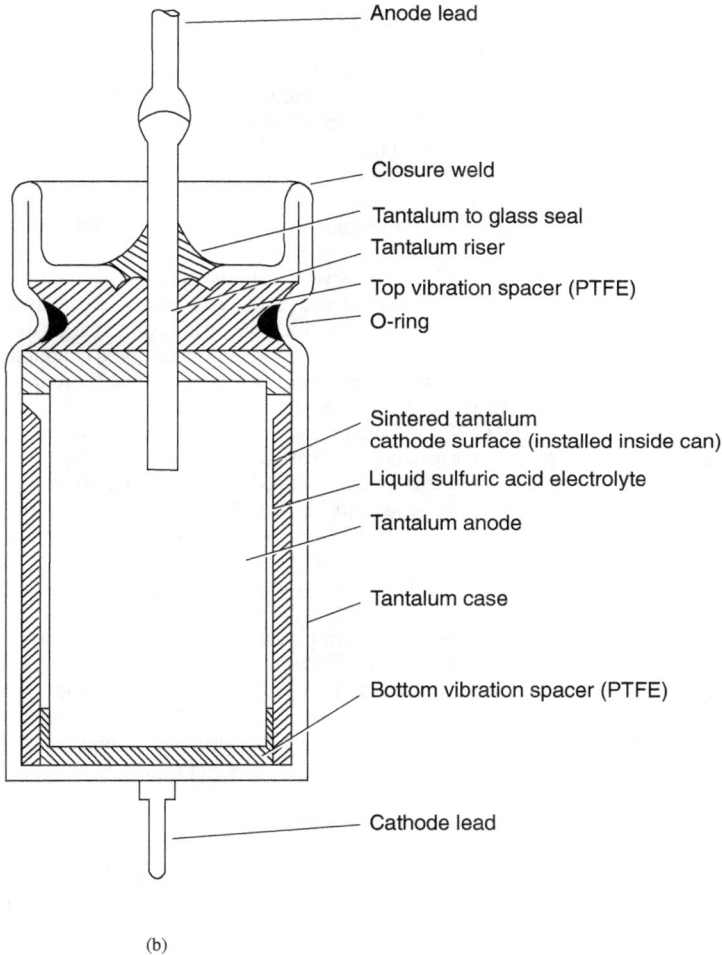

(b)

FIGURE 2.34 Two Designs for all-tantalum hermetically sealed wet-slug capacitor construction. *(Vishay-Sprague.) (Continued)*

The tantalum case is the cathode connection, and the capacitor is improved by altering or sleeving the inside of the case to provide a sintered tantalum surface with large surface area. A plating is commonly used on the tantalum interior surface to prevent the formation of a counter electrode oxide under reverse polarity. Some manufacturers use sputtered or electrolytically deposited palladium with high-temperature treatment to alloy it with the tantalum, which then inhibits the tantalum oxide-forming ability. An oxide layer that forms on the cathode during reversed polarity places a capacitor in series with the pellet oxide capacitor, lowering the total capacitance.

Other capacitors use silver-plated cases with palladium or platinum overplating and have elastomer, rather than hermetic, seals. Using silver cases leaves the parts vulnerable to the previously mentioned silver-induced failure mechanisms after polarity reversal, unless extraordinary steps are taken to isolate the silver case from the functional system. Wet-slug tantalums with their efficient self-healing capability have a good tolerance for the high surge currents that are experienced in applications across switched power busses. This type of capacitor has long been available in multiple-packaged units, in which two to four complete capacitor units are series connected within a hermetically sealed case to produce a capacitor package with a voltage rating two to four times that of the individual capacitors at, of course, reduced capacitance (in accordance with the properties of capacitors connected in series). Other multiple packages are produced to provide very large capacitance values by parallel connections. Capacitance values of up to 22,000 microfarads at low voltages are advertised.

One other vintage type of porous electrode wet tantalum capacitor is commonly called a *button type*. These units have disc-shaped sintered tantalum electrodes with tantalum pentoxide dielectric formed on one, enclosed in a flat disc-shaped elastomer gasket-sealed case with a sulfuric-acid electrolyte. These parts are configured to mount on a conductive metal surface that is most negative, with a solder tab or pin positive terminal presented for connection. MIL-C-83500 (USAF) values range from 6 V, 1200 µF to 125 V, 47 µF with CV product at approximately 6000 to 7000. The parts experience a severe loss of capacitance at –55°C, but are still usable for their primary application of power filtering at that temperature. Specification values for 120-Hz impedance at –55°C range from 20 to 60 Ω maximum. These parts exhibit the same good ripple current and surge capability of other wet-slug tantalums, and are subject to the same limitations. Polarity must never be reversed, and 125°C rated voltages are 66 percent of the labeled rated voltage, which holds to +85°C. All-tantalum parts of this type are available, but not in hermetically sealed configurations.

Tantalum wet-slug capacitor performance   Ripple current ratings for wet-slug tantalum capacitors range up to well above 1 A rms at 40 kHz in larger case sizes. Dc leakage current is very low, ranging from 20 µA at rated voltage and +125°C for the highest CV-product-per-unit-volume designs downward to less than 1 µA at room ambient temperature for lower CV products. Capacitance variation over temperature, as with many wet electrolytics, is rather dramatic, particularly for higher capacitance values and higher CV-product-per-unit-volume parts. At –55°C, these worst-case parts lose from 40 to 75 percent of their room-temperature capacitance, and those in the mid and low ranges of capacitance, depending on value and voltage ratings, might lose only 30 to 40 percent in the mid-range and 10 percent or less for low CV per volume parts. At a higher temperature of –25°C, capacitance loss is less than 40 percent of the +25°C amount for all values, and can be in the range of 5 percent for low CV per volume. Capacitance increases above 25°C, but not as dramatically, being no greater than about 25 percent up to +125°C for most values. ESR generally decreases with temperature increase, but might start a moderate increase for some case sizes and values above about +60°C. Self-resonance is about 100 kHz for capacitors in the 200-µF range, and can approach 1 MHz for values below 3 µF. Impedance is, of course, sharply affected by capacitance loss at low temperatures, but for most capacitance values shunting effectiveness is below 10 Ω up to 5 MHz, with capacitance values of about 5 µF being well below that. Below 5 MHz, impedance is determined by capacitance value and ESR. If a 5-Ω cutoff up to 5 MHz is desired over the full –55°C range, CV product must be held low. ESR decreases with frequency up to 1 MHz. These capacitors excel in power-supply filter input applications and other circuit locations, where low-impedance decoupling, involving heavy surge and pulse currents, are involved.

Wet-slug tantalum capacitor specifications   MIL-C-39006 and the obsolescent, but still valid, MIL-C-3965 cover wet-slug tantalum capacitors for military applications, and EIA standard EIA 535 has coverage of essentially all available case types.

2.8.10 Solid-Electrolyte Sintered-Slug Tantalum Capacitors

Solid-electrolyte porous-slug tantalum capacitors were invented at Bell Telephone Laboratories and have been developed and refined extensively to provide consistent and reliable products since initial commercial production in the late 1950s. It is now the most extensively used tantalum electrolytic capacitor type, having many applications in both military and commercial equipment. Well-made parts have firmly established reliability, moderate cost, and wide availability of several package styles from both U.S. and international manufacturers. Unlike the wet electrolytic capacitors, the solid electrolyte types are relatively easily adapted to miniature surface-mount package configurations, which accounts for a substantial portion of present usage.

The electrical performance characteristics outstrip wet electrolyte types in many areas: solid tantalums have lower ESR, higher-capacitance stability over time and temperature, and a higher, but still-low dc leakage current, when compared to wet tantalum electrolytic types. Solid-electrolyte tantalum capacitors have a self-healing ability, although not quite as efficient as that of their wet-slug cousins, and cannot tolerate current surges as well as wet tantalum foils and slugs or as well as aluminum electrolytics. Solid electrolyte tantalums can tolerate current-limited polarity reversals without damage, and will self-limit current at low-voltage polarity reversal. Manufacturer specifications and industry standards provide recommendations for polarity-reversal limits (Figure 2.37).

The ability to tolerate limited polarity reversals makes it possible to connect two capacitors back-to-back, producing a nonpolar capacitor with a voltage rating equal to that of one of the single capacitors, and capacitance equal to the series combination of the two. It is not intuitively easy to see how this is possible (see the next section). Success of the sintered-slug solid-electrolyte design has depended greatly on the effectiveness of the manganese dioxide solid electrolyte used and its ability to modify chemically when heated, changing from MnO_2 to other oxides of manganese, chiefly Mn_2O_3, which have higher resistivity. When impending dielectric failure is presaged by an increase in leakage current at the weak site, usually quite a small area, the resultant joule heating caused by the locally high current density effects this chemical change, isolating the potential defect with higher resistivity oxide, which interrupts the dry manganese-dioxide counterelectrode contact with the anode through the defect site. This electrical isolation, called a *self-healing effect* is not perfect; the lower manganese oxides not being perfect insulators; and capacitor dc leakage current permanently increases at the site. A disproportionately large number of these isolation spots can cause capacitor leakage current to increase past its acceptance threshold limit. Incidence of these isolation events is sometimes called *scintillation* because the dc leakage current will appear to "flicker" or "scintillate" during the action. Electrical screening near the end of a capacitor production cycle is designed to force this process, clearing weak spots before final leakage current measurement.

Solid-electrolyte sintered-slug tantalum leadwire-capacitor construction High-purity sintered-tantalum pellets with an oxide layer formed, as described, are cleaned and dried. To form the solid electrolyte, the porous capacitor pellets are impregnated with an aqueous solution of manganous nitrate, $Mn(NO_3)$. The pellets are then heated above 200°C to drive off the water and to chemically convert the manganese nitrate to manganese dioxide (MnO_2) crystals, deposited over the interstitial surface area of the pellets. The gas driven off in the process prevents complete layer coverage, and the process of impregnation and heating must be iterated until the oxide is fully coated. Also, during the manganous nitrate conversion, tantalum pentoxide defects are often incurred, and the pellets are electrolytically reformed at lower-than-original forming voltage between steps to heal any defects.

After the manganese dioxide coating is complete, a coating of graphite is introduced by impregnating the structure with a water suspension of graphite powder, then driving off the water. The pellet outside is then coated over with finely divided silver, suspended in a organic vehicle, which, when cured, forms a silver-cathode contact surface that, with proper technique, can be soldered to a terminating metal shape or plated and treated for surface-mount soldering. See Figure 2.33 for a scanning electron-microscope view of a solid-electrolyte sintered-tantalum slug.

For metal-can hermetically sealed capacitors, a cylinder-shaped silver-coated pellet is preheated and forced into a metered amount of molten soft solder at the bottom of a plated closed-end drawn brass case. Upon cooling, the rough-surfaced pellet becomes mechanically secured and electrically connected to the case, which, with its pre-attached leadwire, becomes the cathode terminal. The tantalum riser wire is butt or lap welded to a nickel leadwire, usually before the pellet is soldered into the case, and a glass-to-metal seal header with a metal-tube feedthrough is threaded over the leadwire and put into place at the open case end. Soldering the header outside ring to the case and filling the leadwire tube with high-temperature soft solder completes the structure ready for conditioning, testing, marking, and application of an outside shrink plastic covering to protect the cathode-connected case from unwanted electrical contact.

Figure 2.35 shows a cut-away of a typical solid-electrolyte hermetic metal-can capacitor and the outlines of several typical leadwire case configurations. If nonmetallic cased capacitors are being constructed, the silvered cathode surface is soldered to a metal terminal form with attached leadwire and the anode tantalum wire is welded to its terminal leadwire. The resulting capacitor is encased variously by transfer molding, powder coating in several layers, dipping into thixotropic resin mixtures or combinations thereof. One common configuration uses three radial leadwires with the anode wire at the center, between two outer cathode wires. This arrangement makes it impossible to install the capacitor with the polarity reversed.

Solid electrolyte sintered tantalum slug surface-mount capacitor construction For surface mounting, depending on the package design, the cathode terminal, formed as de-

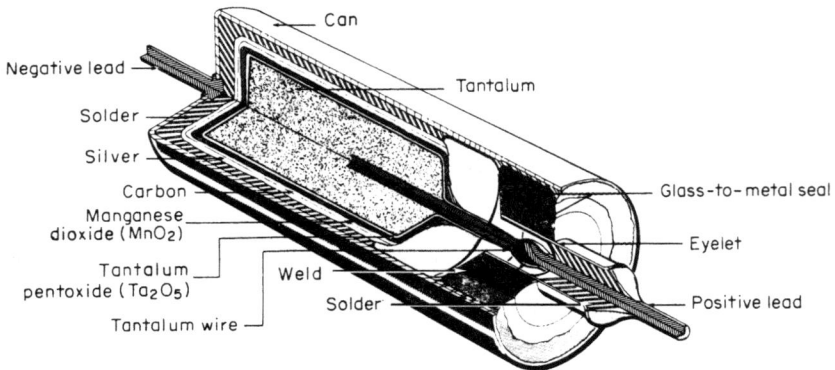

FIGURE 2.35   (a) Solid-electrolyte hermetically sealed sintered-tantalum slug capacitor: typical construction. *(Kemet Electronics.)*

tailed previously, can be plated and conditioned for direct soldering to circuit board pads, or it can be soldered to a flat terminal form that can be soldered to a circuit pad. The tantalum wire is not solderable, however, and must always be welded to a solderable terminal form. Some surface-mount package forms use an "A"-shaped anode termination, with the apex of the "A" welded to the riser wire and with the outward-formed feet of the "A" providing the solderable terminal. Others use a flat, thin metal tab that is folded back under the anode terminal end (Figure 2.36). Some surface-mount packages are conformally coated using powder or liquid-coating techniques, and others use very thin molded encasements. The conformally coated configurations have been known to trouble some types of automated pick and place machinery, which is unable to correctly handle the resulting nonuniform shapes, a condition that is avoided by the uniformity of the molded case designs. The molded case packages now available have terminations that provide a measure of strain relief for the surface-mounting joints as circuit boards and capacitor bodies expand and contract at different rates, and provide uniform shapes for automated handling.

Well-made plastic coated parts will withstand brief water immersion and relatively short term saturated atmosphere humidity without detriment. Longer exposures can be tolerated if a drying period is experienced before circuit power is applied. No plastic coating or covering provides a long-term protection from moisture, but well-designed coatings generally provide protection adequate for most uses. Military specifications quantify protection ability by specifying well-defined humidity tests which the capacitors must survive, as do the voluntary requirements of commercial standards.

Terminal finishes for surface-mount parts that do not have attached terminals present somewhat of a dilemma for solder mounting: The cathode silver coating is highly susceptible to scavenging by molten tin-lead solder, rapidly entering into solution with tin, and depleting the coating over the graphite layer. To defeat or reduce the scavenging process, manufacturers plate the silver with a barrier of nickel, which then must be immediately protected with another plating to preserve its solderability, tin-lead plating being frequently used. Solder dipping can be used, but it adds a difficult-to-control amount of solder, whose physical nonuniformity sometimes causes pick-and-place problems. Tin-lead plating effectively protects nickel solderability only if it is thick enough (greater than 100 to 300 microinches) and heat fused to form a nickel intermetallic and consolidated-alloy outer surface. This creates a somewhat complex terminal treatment that is still not totally immune from solder scavenging. Some suppliers prefer to plate pure tin over the silver, and the tin remains solderable in unprotected storage slightly longer than does nickel, which rapidly deteriorates, however, in humid air and exposure to light. Pure unalloyed tin also has a reputation for growing dendritic whiskers than can produce short-circuiting in circuit applications. Gold plating is available as a coating over the nickel, but is considered unsuitable for soldering because the gold is rapidly dissolved into tin-lead mounting solder and weakens the solder joint. However, gold plating is the preferred finish for conductive polymer attachment, avoiding both the possibility of tin oxide on tin plate and reduced adhesion to tin-lead. With proper attention to termination finish and part-storage conditions, reliable attachment for both conductive polymer and soldering can be achieved. If solder is used for attachment, it is recommended that a tin-lead-silver eutectic combination be used, containing about 2 percent silver, military designation Sn62.

Solid electrolyte tantalum capacitor performance Solid-electrolyte tantalum capacitors are useful in a wide range of coupling, decoupling, filtering, and energy-storage applications, but have some limitations that should be considered. Their relatively small physical size limits heat-dissipating ability, and presents a small thermal mass for absorbing heat transients. Applications where high ripple currents or repeated heavy charge-discharge surges occur should be carefully examined for temperature rise caused by internal dissipation. Internal temperatures above 125°C begin to deteriorate the oxide dielectric structure chiefly by causing tantalum pentoxide crystal formation in the otherwise amorphous dielectric, and constitute an overstress on the capacitors. Depending on capacitance value and case size, solid-electrolyte sintered-slug tantalum capacitors have impedances less than 1 Ω up to about 100 kHz, with lower capacitance values exceeding that frequency somewhat for their less-than-1-Ω threshold. The curves of Figure 2.37 show general expected impedance over frequency.

Capacitance values under the specified measurement frequency conditions can be expected to remain within about ±15 percent over –55°C to +125°C; the more-stable solid

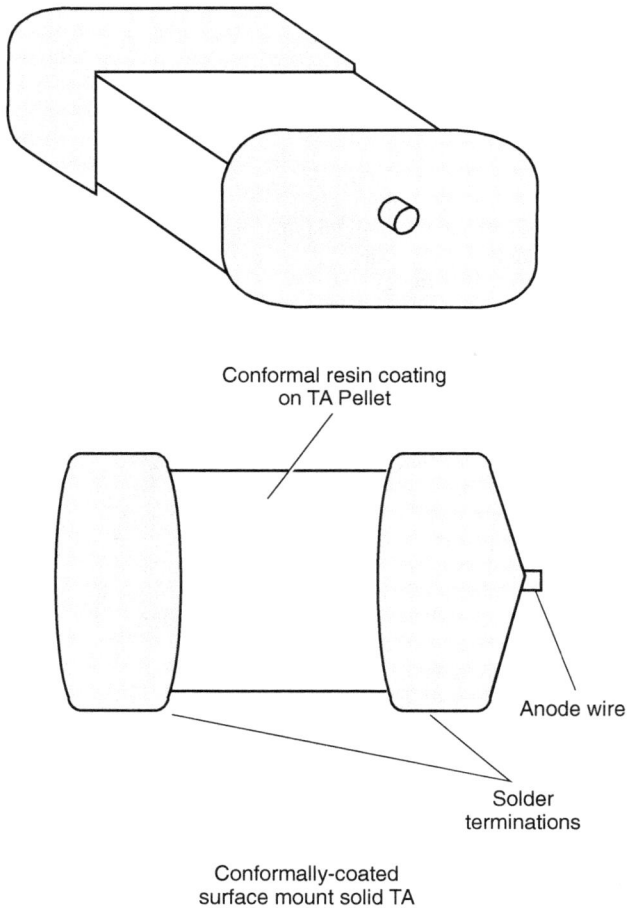

Conformal resin coating
on TA Pellet

Anode wire

Solder
terminations

Conformally-coated
surface mount solid TA

FIGURE 2.36 Solid-electrolyte sintered-tantalum slug capacitors for surface mounting: (a) Conformally coated, direct attachment.

Cathode terminal

Tantalum pellet-silvered

Tantalum riser wire

Weld

Molded plastic case

Anode terminal

Top

105
20

Contrasting stripe identifies positive terminal

Capacitance

Rated voltage

Side

Mounting surface

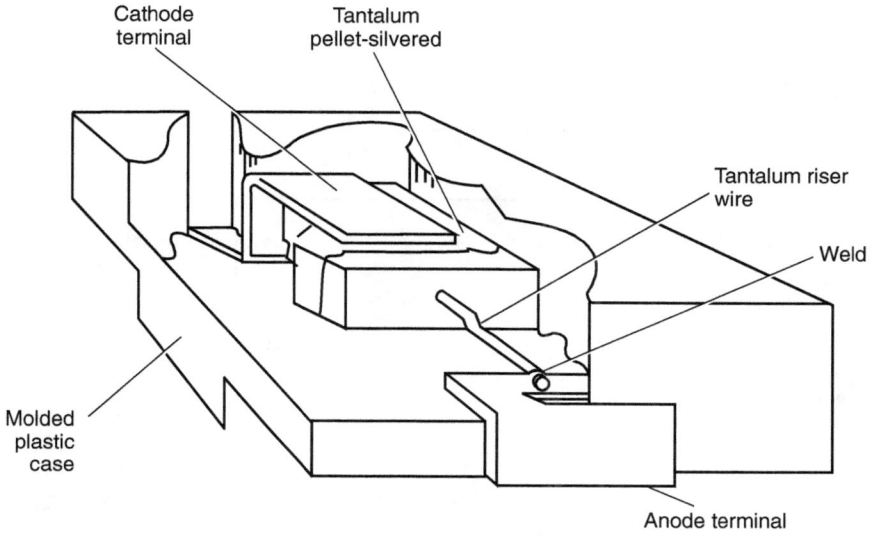

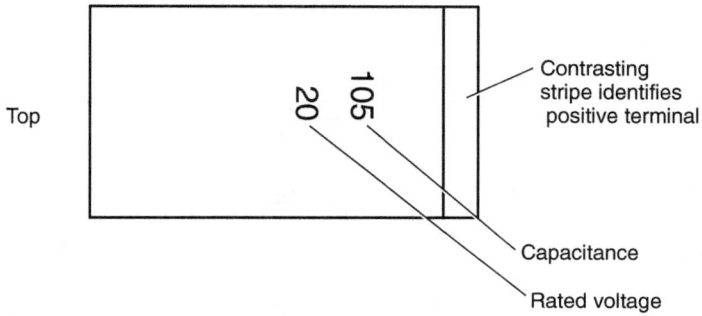

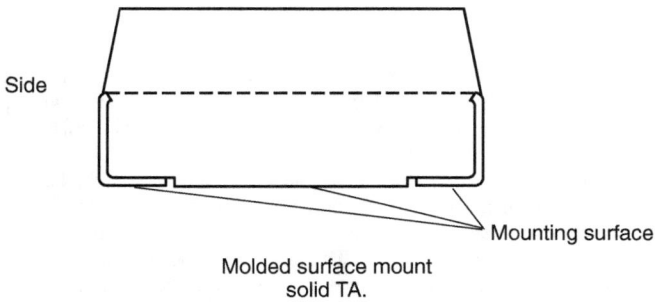

Molded surface mount solid TA.

FIGURE 2.36 Solid-electrolyte sintered-tantalum slug capacitors for surface mounting: (b) molded case, flexible metal tab attachment. (*Continued*)

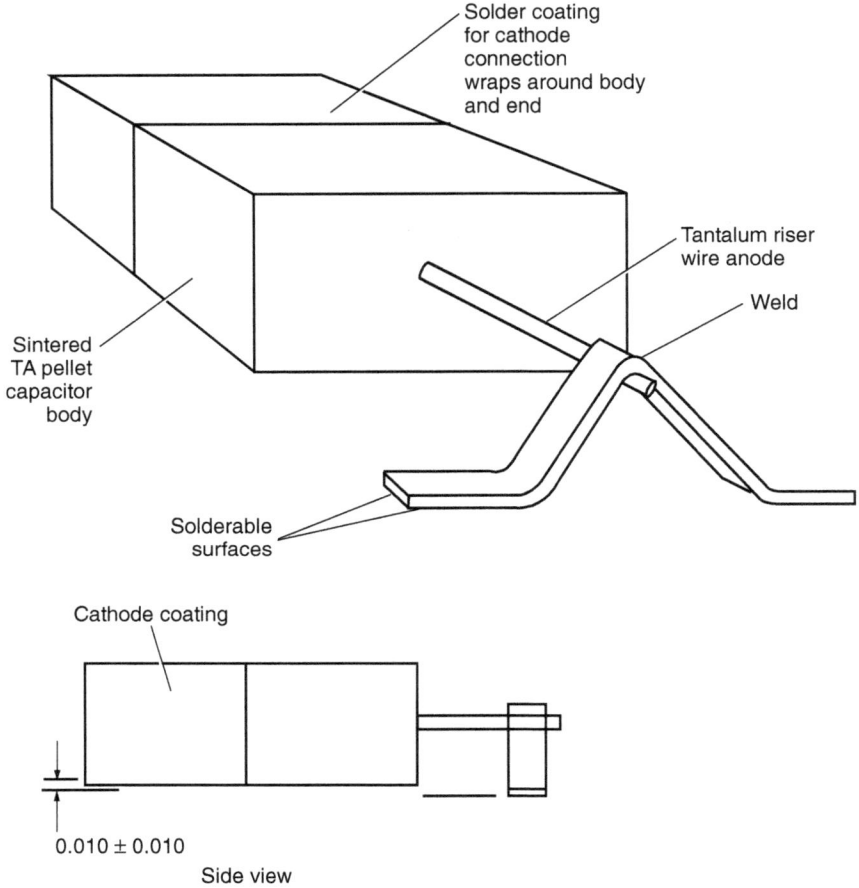

FIGURE 2.36   Solid-electrolyte sintered-tantalum slug capacitors for surface mounting: (c) extended riser wire welded tab attachment, conformally coated. (*Continued*)

electrolyte provides an advantage over the wet electrolytes. Although many thousands are successfully used in such applications, solid-electrolyte tantalums sometimes fail by spontaneous short-circuiting when subjected to heavy current surges, as might be encountered in unlimited charging current when a power bus is turned on across the capacitor, in short-circuit discharges, or when used in parallel-connected applications where a charged companion capacitor might act as a low-impedance current source. The exact cause is not fully understood, but manufacturers have greatly reduced the problem by improving material purities and varying processes to provide better products. The failures can occur as a scintillation defect isolation event (see 2.8.10) begins, and the spot current density is not suffciently limited to prevent localized melting of the tantalum material at the spot. When the metal bridges the dielectric, larger-area melting occurs, eventually causing a short circuit with high-current ca-

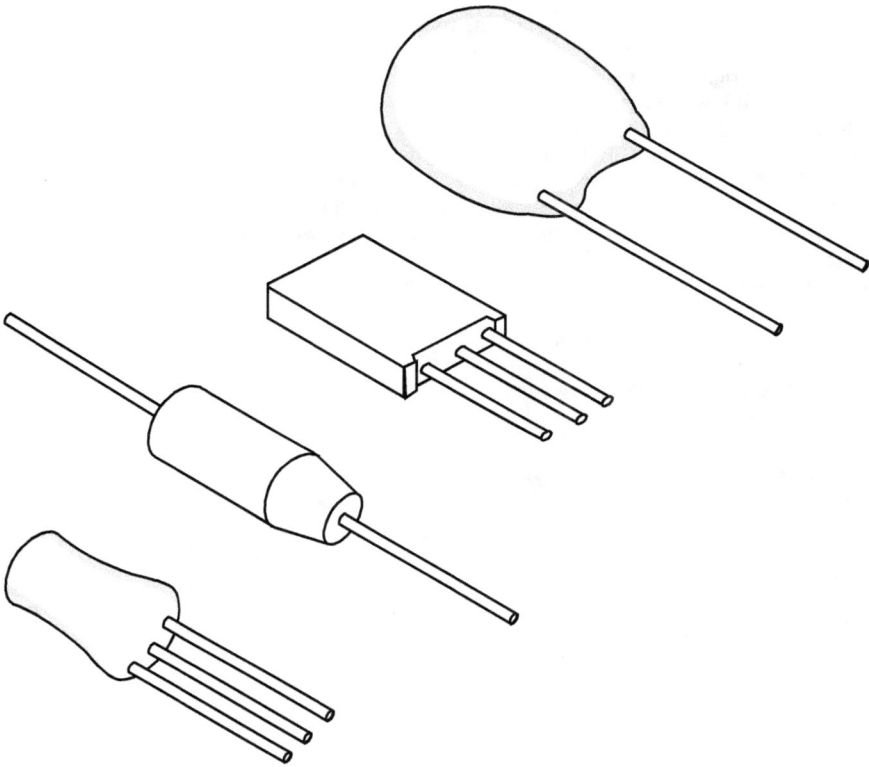

FIGURE 2.37 Solid-electrolyte sintered-slug tantalum capacitors: Typical impedance values versus frequency.

pability. One investigation finding associates the occurrence of many scintillation sites with unwanted formations of the crystalline form of tantalum pentoxide, rather than the desired amorphous form, said by some to be exacerbated by material impurities.

One theory examines whether damage sites at dielectric weak spots are worsened by the electrostatic forces present across the dielectric. These electrostatic forces are relatively small until their total pounds-per-square-inch values are applied to areas that are reduced to microscopic point size, as at points where manganese-dioxide crystals contact the pentoxide dielectric. Resolved to a point, the forces can be hundreds of pounds per square inch. Another theory views thermal cycling the capacitor structure, particularly the metal cased designs, which might have some failure cause-enhancing effects, as the multi-material composite structure undergoes stresses because of differences in material thermal-expansion coefficients. Suffice it to say, however, failure incidents in low-impedance surge applications are greatly reduced in present-day solid tantalum capacitors, particularly those with a CV-product-versus-case-size lower than about 75 percent of maximum in a given case size. At one time, high-reliability practice recommended a series impedance of at least 3 ohms per volt for solid tantalums, but product improvements have reduced concerns.

Much data has been presented showing that failures can be reduced to nearly zero by conducting surge-current screening of all capacitors to be used in low-impedance circuits. Personal experience seems to verify this, provided that the surge currents are applied at both high- and low-temperature extremes. The capacitors, with external fuses in series, are subjected to a number (five has been shown to be sufficient) of essentially unlimited current charges to rated voltage from a charged low-inductance, low-resistance capacitance bank of several hundred thousand μF, using a bounceless switch, such as a mercury relay. Each charge is followed by a low-resistance, low-inductance short-circuit discharge. Those capacitors that survive without blowing their fuse wire and that still have leakage current within their specification value have been shown to have very low risk of spontaneous short-circuit failure in current-surge applications. The surge-voltage test of MIL-C-39003 should not be confused with this surge-current screen. Most solid tantalum capacitor manufacturers are well equipped to conduct this screen, and will do so for a small price increase. Any high-CV-product solid tantalums to be used in parallel connection with other high-energy capacitors or across switched low-impedance circuits should have this screen (or its equivalent) if high reliability is sought.

Back-to-back nonpolar solid-electrolyte tantalum capacitors   Two approximately equal-value polarized solid-tantalum capacitors can be connected cathode to cathode, making a capacitor that, for periodic waveforms, behaves as a nonpolar capacitor with half of the capacitance value of either half, and with the same dc or dc plus peak ac voltage rating as one of the capacitors (Table 2.24). Capacitor units made in this fashion in a single case are marketed as nonpolar capacitors. Back-to-back capacitors that are matched in characteristics have also been used to resist the effects transient gamma-radiation effects because their two radiation-induced discharges are in opposing directions and balance out.

TABLE 2.24 EIA Standard Value and Tolerance Color Coding for Solid Electrolyte Tantalum Capacitors

Color	Significant figures for capacitance in pF	Multiplier	Tolerance (at 25°C)
Black	0		
Brown	1		
Red	2	10^2	
Orange	3	10^3	
Yellow	4	10^4	
Green	5	10^5	
Blue	6	10^6	
Violet	7	—	
Gray	8	—	
White	9	—	
No Color	—	—	±20%
Silver	—	—	±10%
Gold	—	—	±5%

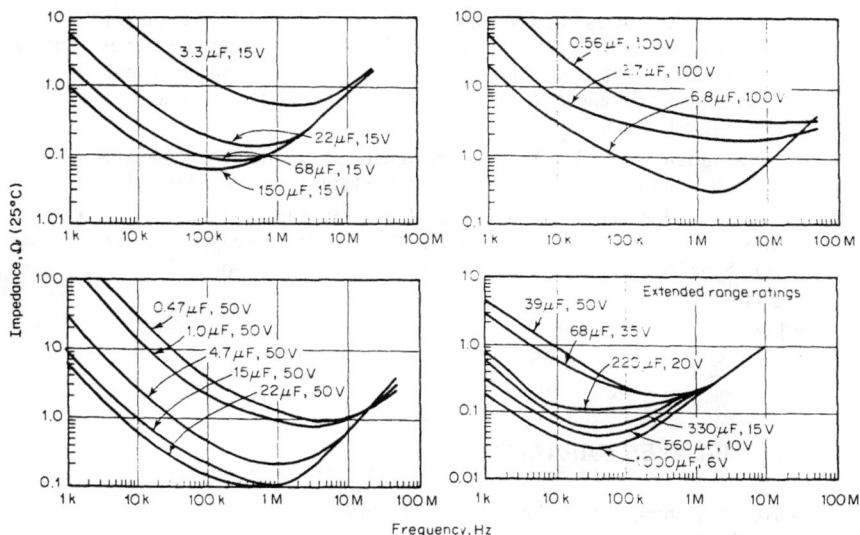

FIGURE 2.38 Equivalent circuit model for back-to-back solid-electrolyte sintered-tantalum slug capacitors.

Because, with an ohmmeter, solid- and wet-electrolyte capacitors appear to have capacitive properties in their "forward" direction, but appear to be just resistive in the reverse-polarity direction, it is not easy to deduce how this back-to-back configuration functions. The following is extracted from the first edition of this handbook, written by Joseph F. Rhodes, formerly of Martin Marietta Corporation.

Ignoring the resistive and inductive properties, a functional solid electrolyte tantalum capacitor model can be depicted by a perfect capacitor shunted by a diode having a reverse voltage-current characteristic with breakdown in the order of 3 to 4 volts, depending on the capacitor (see Figures 2.39 and 2.40). Connecting two such capacitor models in series, back to back, makes a configuration as shown in Figure 2.39. Assume that the two capacitors are initially discharged and all voltages are zero. Applying a positive step voltage, V_1 of, for example, 1 volt, causes C1 to charge through the forward-biased diode in C2 and through R back to the assumed low-impedance grounded source. Assuming perfect diodes, the time constant for this charge is $R \times C_1$. When the assumed zero-impedance voltage source now drops to zero, C1 discharges through R, through the power-source zero impedance into C2, charging C2. Neither diode conducts during this discharge-charge, and its time constant is $R \times$ the series combination of C_1 and C_2. Assuming that C1 and C2 are equal, after several time constants, the voltage across each capacitor will be equal and one-half that of the initial 1-volt V_1. If V_1 drops to −1 volt, C1 will discharge, and C2 will charge, still with time-constant $R \times$ the series combination of the two capacitors (one-half $R \times C_1$, in this case) until C2 becomes charged to −1 volt as seen at V_2 through the zero-biased perfect C1 diode. When neither diode is conducting, the time constant is one-half the $R \times C_1$ (or C_2) value, but when either diode conducts, the time constant is $R \times C_1$ (or C_2). Capacitors and diodes, particularly these, are never perfect, having leakages and both forward and reverse voltage drops and breakdown characteristics, and the model is not exact. Nevertheless, the following empirical results apply:

1. For initially applied cycles of an ac signal up to about 10 percent of rated voltage for a single one of the capacitors, little difference will be seen between the effects of using two back-to-back polarized capacitors and two equivalent electrostatic capacitors, that is, the capacitance will appear to be one half the value of a single one of two equal-valued back-to-back capacitors.

2. After a few initializing cycles, a sinusoidal signal sees an effective capacitance almost precisely as that defined in 1, regardless of applied signal level (within ratings).

3. As the signal level is increased from about 10 percent of rated voltage to rated voltage and before reaching the stabilized condition of 2, the effective capacitance seen for the initial 3 to 5 cycles only will increase to a maximum limit of about 1.5 to 1.8 times the low-signal value or steady-state value.

4. Very low frequency sinusoidal signals might appear to "see" increased capacitance, thought to be associated with dielectric absorption characteristics.

2.9 DOUBLE-LAYER CAPACITORS

Double-layer capacitors, called "SuperCapacitor" by NEC, "Ultracapacitor" by Pinnacle Research, and "Dynacap" by Elna were introduced relatively recently. In construction

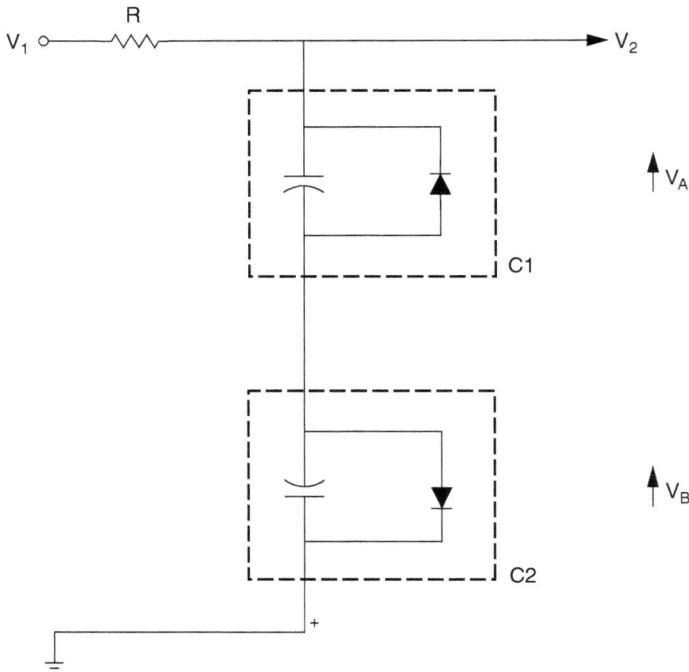

FIGURE 2.39   Equivalent circuit model for back-to-back solid-electrolyte sintered-tantalum slug capacitors.

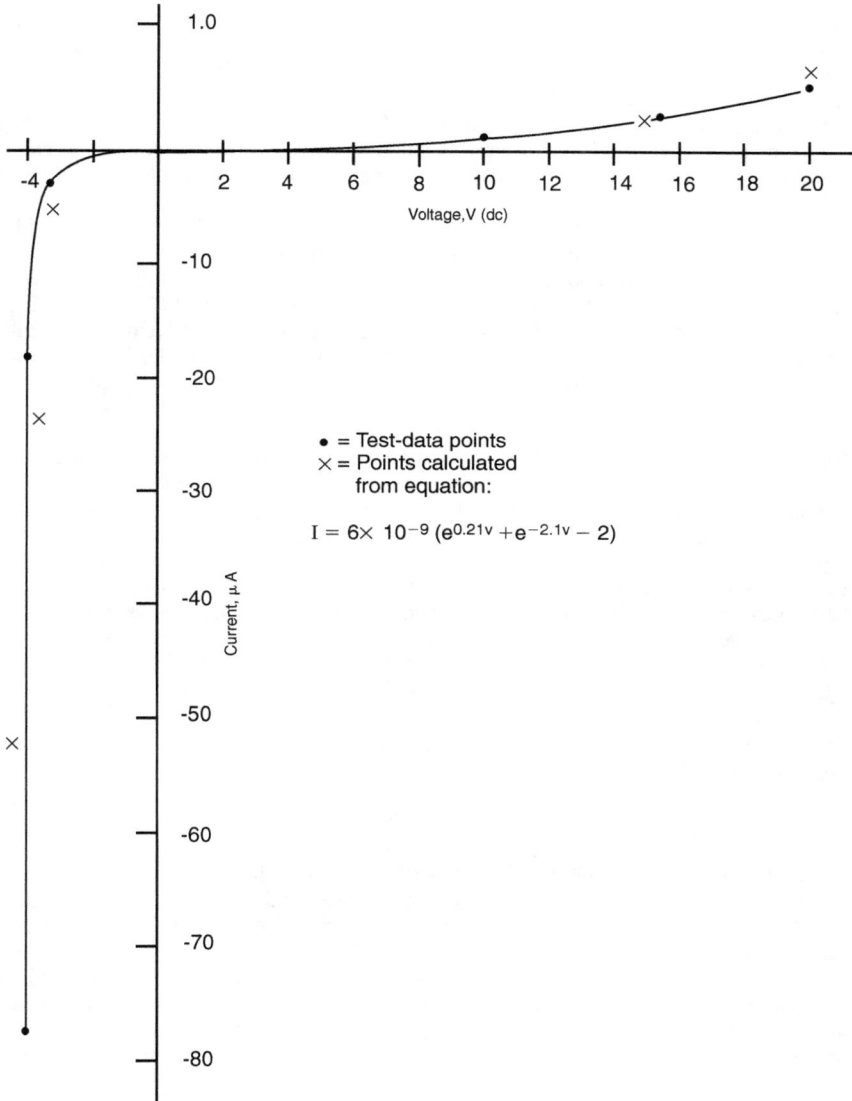

FIGURE 2.40 Voltage-current plot for a typical solid-electrolyte tantalum-slug capacitor. The pilot is for a 47-μF, 20-Vdc CSR13 military part.

only, those designs now available are somewhat akin to the wet electrolytic types, but are actually fundamentally different in function and makeup. Very large values of capacitance, up to several farads at 6.5 volts (and lower) are achieved in very small volumes. Cylindrical metal can units are offered at about 150 or more µF per cubic centimeter at 5.5 volts rating. At this writing, application is limited essentially to that of energy storage for supplying data retention during power interruptions for low-current-demand digital systems, such as complementary metal-oxide semiconductor (CMOS) logic, but device designs can be optimized to supply multiple-millisecond pulses of several amperes.

The temperature range for presently available parts is rated at –40°C to +85°C, with a sharp decrease in life expectancy for elevated temperatures. Limitation to about 50°C or lower appears to provide good probability of 10-year life expectancy, with end of life defined as decrease in capacitance to 70 percent of the initial. For the NEC 5.5-volt ratings, the discharge time from 5.0 V to 2.0 V varies with the current demand and the particular design (size), with 100-µA demand being furnished for 0.5 to 30 or 40 hours, and 1.0 µA available for 100 to about 2000 hours, depending on design and size. Leakage current is obviously of great importance. These capacitors are presently not suitable for filtering or similar uses because of their relatively high ESR (hundreds of Ω in values below about 0.1 F, tens of Ω down to 20 to 30 Ω at 1 F), their small relative heat-dissipating mass, and their limited high-temperature capabilities. Their failure mode is said to be open-circuiting, and NEC claims a 0.4 FIT (failures per 10E09 part-hours) at room temperature.

Double-layer capacitor construction utilizes finely divided inert particle electrodes, typically porous activated carbon, saturated with (and suspended in) dilute sulfuric acid. In the NEC design, electrical contact to the electrode is provided by a conductive elastomer portion of an overall enclosing elastomer container capsule having two symmetrical regions of electrode-electrolyte mixtures. The mixtures are insulated from each other by nonconductive portions of the enclosing elastomer capsule, but are separated internally by an ion-permeable membrane, which permits ion transport between the two electrodes within the electrolyte. Applying a potential below the electrolysis decomposition value of water (range of 1 volt) across the two opposing sections for a sufficient time period causes opposing ion concentrations at the interfaces between the electrolyte and the extremely high-surface-area granular electrodes.

The opposing polarity ion concentrations at the potential interfaces, called *double layer* (Maxwell-Wagner layer) formation constitute the capacitor-like function. The time required for the initial charging is as much as 100 hours to realize full backup power capabilities. Capacitor sections are symmetrical and, consequently, nonpolar before charging, but, once charged and subsequently applied with polarity reversed, will require a substantially longer time period to remove the existing charge completely and establish the opposite polarity function. The voltage rating is increased by serially stacking capsules within a two-terminal package. The parts can be used in parallel to increase effective capacitance and the resulting stored energy. Plastic encapsulated and metal-can-packaged parts with unsealed roll-crimping to just retain capsules or resin-sealed versions of the crimped-can configuration are furnished by NEC and Elna. The production cost is higher, compared to other capacitor types, because of the specialized techniques necessary. The unique nature of the product and its applications requires unconventional (but straightforward) measurement techniques, descriptions of which can be found in manufacturer catalogs and in military specification DOD-C-29501. The fullest capabilities of this capacitor type have yet to be explored and developed. Manufacturer data will have current information. Figure 2.41 shows an available configuration.

Plastic
sleeve

Lead connected
to case
⊖ Negative polarity

Negative polarity
identification mark

H max.

L min.

0.3 min.

Insulation
sleeve

P±0.5

ΦD±0.5

d1±0.1

d2±0.1

Lead
connected
to case
⊖ Negative polarity

LEAD

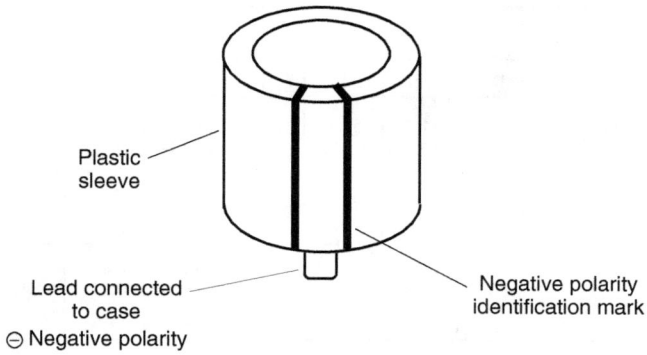

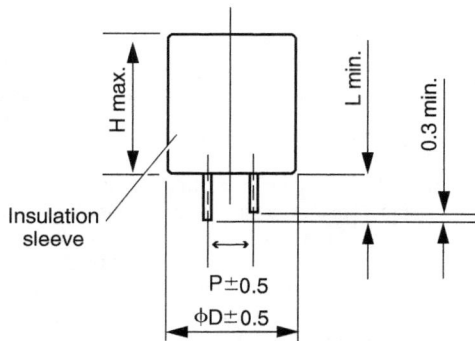

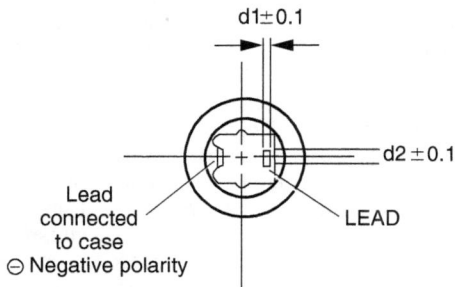

FIGURE 2.41   Double-layer capacitors. *(NEC Corporation.)*

Capacitance value	Dimensions $\frac{mm}{(inch)}$						Weight
	D	H	P	d_1	d_2	L	g (oz)
0.022 F	11.5 (0.453)	14.0 (0.551)	5.08 (0.200)	0.4 (0.016)	1.2 (0.047)	2.7 (0.106)	2.3 (0.081)
0.1 F	14.5 (0.571)	15.5 (0.610)	5.08 (0.200)	0.4 (0.016)	1.2 (0.047)	2.4 (0.095)	4.3 (0.152)
1.0 F	21.5 (0.850)	22.0 (0.866)	7.62 (0.200)	0.6 (0.024)	1.2 (0.047)	3.0 (0.118)	13.3 (0.470)

Note: Weight is typical

FIGURE 2.41   Double-layer capacitors. *(NEC Corporation.) (Continued)*

2.10 MECHANICALLY VARIABLE CAPACITORS

2.10.1 Variable Capacitor General Properties

For certain types of circuit designs, capacitors that can be mechanically adjusted over a small range are indispensable. Capacitors that can be set mechanically to an optimum value and that will remain stably set at that value over time and environmental exposures are used to "trim" capacitance at a critical circuit point. They are called, aptly, *trimmer capacitors.*

In the days of vacuum tube radio, many relatively physically large air-variable capacitors were used to tune radio receiver local oscillators and RF input stages and transmitter oscillators and amplifiers. The most common designs consisted of a series of metal plates attached to a rotatable shaft in a manner that permitted shaft rotation to intermesh the plates with mating plates affixed to a stator. Air spacing was maintained to provide a dielectric. The stator was usually insulated from ground, with the rotor being grounded through its bearings and often a slip-ring arrangement. This basic design is still used for tuning RF circuits, largely for transmitters. The air dielectric and good mechanical design make it quite stable. Shaping the plates permits a desired nonlinear capacitance variation with shaft rotation. Each section of these large air variables for broadcast receivers had attached small adjustable capacitors, usually ceramic-insulated mica-sheet spring-compression types with a screwdriver adjustment. Two small adjustables were usually used for each section: one in series with the large variable and its companion fixed inductor, called a *padder,* and the other connected in parallel across the inductor and large capacitor, and called a *trimmer.* Adjusting the padder set the low-frequency resonance limit, and the trimmer was used to set the high-frequency limit, thus the origin of names "trimmer" and *padder.* Alternate ways of tuning oscillators and RF amplifiers have evolved, and the high vacuum-tube plate voltages that gave an advantage to the air-gap spacing of the rotating plate variable capacitors are largely in the past, except for high-power broadcast equipment, where variations of these large air-variable capacitors are still found.

Remembering the fundamental capacitor equation from the chapter introduction, varying the capacitance of two conductor plates separated by a dielectric can be accomplished by changing the relative overlapping area of the plates, the spacing between the plates (dielectric thickness), or by varying the dielectric constant of the dielectric. Most available trimming capacitors accomplish capacitance change by mechanically varying the overlapping plate area. Many use air as their dielectric, and some use dielectric materials, such as temperature-stable ceramic or glass. Metal parts in the most precision types must be low-linear thermal-expansion-coefficient alloys (such as invar and kovar), not only to enhance mechanical and, hence, capacitance stability over temperature variation, but to permit the low expansion-coefficient glasses and ceramics to be reliably attached to metal portions by welding and soldering to metallized portions or, in some cases, allowing the glass to fuse to metal.

Some mica-dielectric spring-compression trimmers do not require the thermally stable metals. In this type, usually inexpensive, thin sheets of mica are placed loose on a small metal plate, mounted on an insulator base structure. The opposite plate is a flat metal spring of matching area, also attached to the base insulator. A metal screw passes through a hole in the spring plate, through a clearance hole in the mica dielectric, through clearance holes in the base plate and insulator base into a captive nut or other threaded receptacle. Tightening the screw compresses the spring plate, varying the dielectric spacing and the dielectric constant separating the two plates by changing the proportions of low-dielectric-constant air and higher-dielectric-constant mica. Metal tab extensions of the two plates provide terminals.

Tubular trimmers sometimes have a leadscrew-driven piston inside a glass tube that is metallized on its exterior surface to form a plate, or it might use the lead screw to move an open cylindrical cup-shaped member attached to the screw inside a glass or ceramic tube so that the movable cup meshes with a mating metal member using air spacing as the dielectric. The cup shapes can have two or more concentric walls on one or both members so that as the two intermesh, a larger plate area is achieved. The most common designs use high-alumina tubes metallized to accept high-temperature solder for the assembly of the screw bushing and stator assemblies. Most available designs of tubular trimmers control screw rotational torque by using split-threaded bushings with a C-spring in a groove around the bushing to provide a hoop force. This, combined with gold or other durable high-conductivity plating, also provides good electrical contact with the movable plate through the screw threads.

Most flat ceramic trimmers use a lapped ceramic base with a silver or other metal deposited in a film, shaped to the desired plate area. The mating plate is film-deposited on a class-I ceramic rotor that is affixed to a metal adjusting screw. The screw is often connected electrically to the movable plate with a slip-ring electrical connection to a terminal. The rotor and stator are assembled with a pressure spring to maintain a constant pressure, and rotating the adjustment screw varies the overlapping plate areas separated by the ceramic dielectric. This basic design is made in many variations and sizes, and the tiniest are suitable for surface mounting in hybrid microcircuit assemblies. Except for compression mica and some plastic-dielectric types, ceramic-dielectric trimmers provide higher capacitance than most others because the dielectric constant is relatively higher than the air and glass used by other types. Plastic materials also permit higher capacitances. Capacitance for all but the compression-mica types is limited to about 150 pF, with most being much lower. Stabilities are quite good for the better designs, being ±100 PPM/°C (or less), except for mica, which can be ±500 PPM/°C over -30°C to $+85$°C. Consult the manufacturers for current information. Figure 2.42 illustrates some common trimmer designs.

Two other types of variable capacitors must be mentioned. Some single-layer chip capacitors on ceramic substrates, primarily for microwave applications, are designed to permit selective removal of the top metal by laser scribing, thus reducing the capacitance. Another type of single-layer microwave chip-ceramic capacitor has separated top-metal patterns of varying area, permitting wire bonding to a selected metal pattern to obtain the capacitance value to the common back plate that is associated with that plate (Figure 2.43).

(a)

(b)

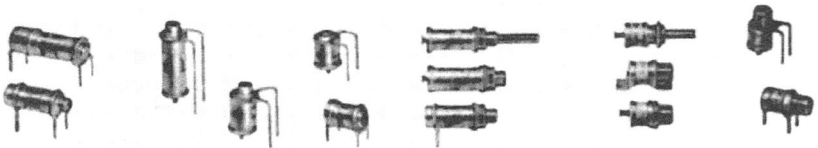

(c)

FIGURE 2.42 Commonly available trimmer capacitors: (a) Air dielectric types, typically with high aluminia insulation; (b) ceramic types; (c) glass dielectric types.

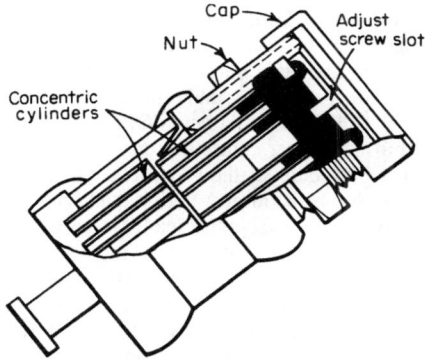

(d)

FIGURE 2.42 Commonly available variable trimmer capacitors: (d) cross section of a tubular air variable. *(Continued)*

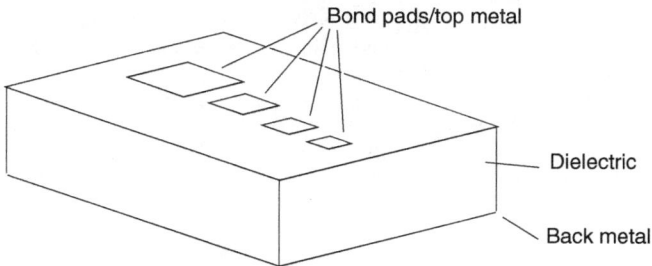

FIGURE 2.43 Stepped-value microwave trimmer.

2.11 CAPACITOR PACKAGING FOR AUTOMATED USE

Handling difficulties for small surface-mount capacitors and the need to automate pick-and-place installation for these, and similar considerations for leaded capacitors that are used in high-rate production operations has generated a need for standardized packaging that can be accommodated by automated production-handling equipment. Most of the standards directed toward automated handling involve some manner of tape-and-reel configuration, although a few involve magazine-type loading. Chip-type parts are furnished by manufacturers on reels, resembling motion-picture film reels with the capacitors residing in the embossed recesses of paper or plastic tape. This section is covered with a thin tape, usually having a pressure-sensitive adhesive, overlaid to complete the encasement, or the tape might have punched-out recesses in thick paper or plastic tape, with overlaid thin tapes on top and bottom to form a 3-layer structure.

The carrier tapes have punched-in sprocket holes for machine feeding. These tape-and-reel packages are in standard sizes of 8-, 12-, 16-, and 24-mm, width tapes for ceramic, solid tantalum, and stacked film and other chips. Dimensions, polarity orientation for polarized parts, drive sprocket hole dimensions, and other characteristics are covered by EIA 481. Reel diameters are 7 or 13 inches in diameter. Other tape and reel specifications of different dimensions for chips and other configurations are covered by EIA 296. Leaded parts are commonly taped at both ends of their leadwires, if of axial lead cylindrical or prism configuration, but they are also (rarely) body-taped with the leadwires extended. Tape configurations are also available for axial leadwire parts that have leadwires preformed to symmetrical radial form or for stand-on-end mounting with the top-end hairpin loop. These special tape configurations commonly have the parts secured by both leadwires to a sprocket-holed backing tape using a pressure-sensitive adhesive over-tape. Reels for these items and the lead-end taped axial leadwire parts are, of course, much wider than those mentioned previously, the width being governed by the overall length of part and leadwires. These wider reels are usually constructed of cardboard.

EIA 296 covers axial-leaded part taping and reeling. Radial-lead parts can be taped at their leadwire ends on 18-mm-wide paper sprocket-holed carrier tape using a narrow pressure-sensitive adhesive tape in accordance with EIA 468. Capacitors that are configured as two-pin dual in-line (DIP) molded units (presently limited to ceramic capacitors and some film types) or capacitor arrays that use DIP packages are commonly packaged for delivery in the plastic tubes that were originally designed for DIP microcircuit devices. EIA standards for these parts show them both with leads extended and formed like microcircuit DIPs, and also annunciate packaging details. Correct packaging of good quality is of profound importance to the high-volume capacitor user, not only to protect parts from handling damage, but to ensure that polarized parts are correctly placed, and correct capacitance values and voltage ratings end up in their proper circuit locations. When high-rate machinery errs, the quantity of incorrect or damaged parts is likely to be nontrivial.

Packaging standards and guidelines are also published by the Aerospace Industries Association: NAS 3414, NAS 3429, and NAS 3430. The Institute for Interconnecting and Packaging Electronic Circuits (IPC) also publishes pertinent design guidelines in IPC D-317A. The International Electrotechnical Commission (IEC) also has published standards. IEC 286 covers tape and reel packaging. Figures 2.44 and 2.45 show some standardized tape-and-reel designs.

2.12 CONCLUSION

The subject of capacitors is truly diverse and interesting. This chapter merely brushes the surface of the technology that has been developed to supply the needs of our expanding worldwide electrical and electronic industries. Though lacking the glamour and much of the "edge of technical knowledge" aspects of solid-state device technology, many facets of capacitor design and production technology are fields of technology in themselves. For instance, production of thin polymer films that are suitable for high-rate production of quality capacitors is a challenging specialized endeavor. Design and production of equipment for producing such films and the machinery for handling thin films through metallization and winding is, likewise, a technically demanding undertaking.

Ceramics technology and production processing, optimization of electrolytic capacitor materials and production processing, and many other similar specialized disciplines are continually being investigated, improved, and reinvented with new knowledge by a cadre of specialists, many of whom have been dedicated to the capacitor industry throughout their working lives. The electronics industry and the millions of users of the ever-more-re-

Top cover
tape
thickness

Carrier

Embossment

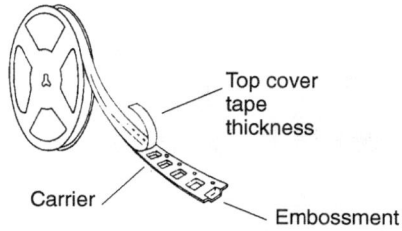

**Standard orientation is with the cathode (-) nearest to
the sprocket holes per EIA-481-1 and IEC 286-3.**

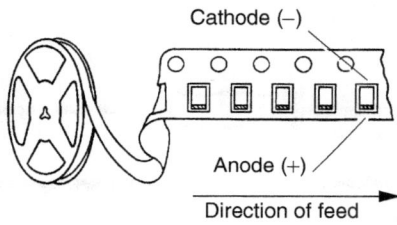

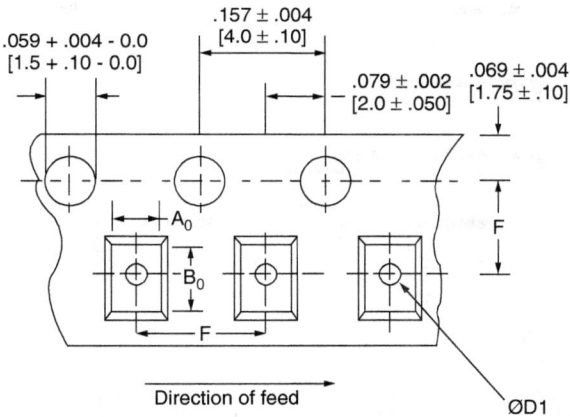

Cathode (−)

Anode (+)

Direction of feed

$.157 \pm .004$
$[4.0 \pm .10]$

$.059 + .004 - 0.0$
$[1.5 + .10 - 0.0]$

$.079 \pm .002$
$[2.0 \pm .050]$

$.069 \pm .004$
$[1.75 \pm .10]$

A_0

B_0

F

F

Direction of feed

ØD1

Tape size	B1 (Max.)	D1 (Min)	F	K (Max.)	P	W
8mm	.165 [4.2]	.039 [1.0]	-138 ± .002 [3.5 ± .05]	.094 [2.4]	.157 ± .004 [4.0 ± .10]	.315 ± .012 [8.0 ± .30]
12mm	.323 [8.2]	.059 [1.5]	.217 ± .002 [5.5 ± .05]	.177 [4.5]	.315 ± .004 [8.0 ± .10]	.472 ± .012 [12.0 ± .30]

FIGURE 2.44 Typical standard carrier tape for chip-type capacitors. *(Vishay Sprague)*

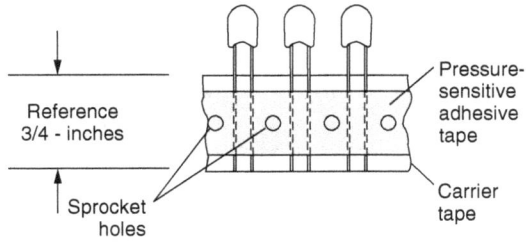

(a) Typical radial leadwire tape and reel configuration

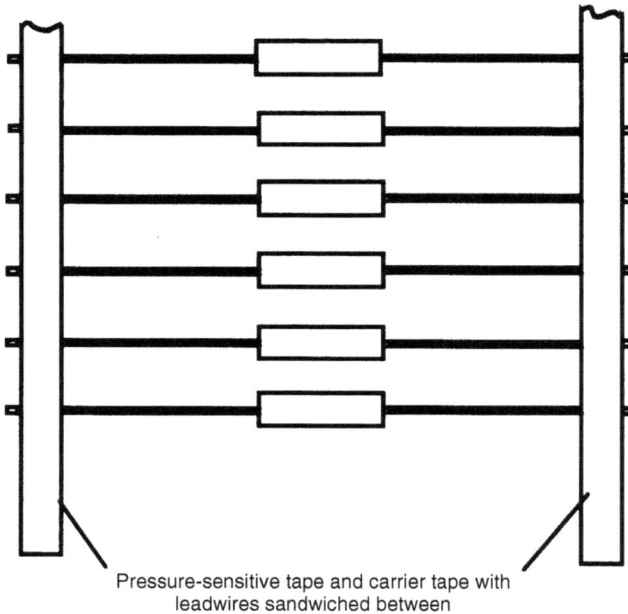

Pressure-sensitive tape and carrier tape with
leadwires sandwiched between

(b) Typical axial leadwire tape- and -reel configuration

FIGURE 2.45   Taped and reeled leadwire capacitors: Typical methods.

liable electronic and electrical products owe a debt of gratitude to these dedicated and largely unsung individuals. Often, in miniaturization of electronic circuits, physical sizes of capacitors needed for particular circuit functions become limiting features, despite the great progress that has been made to date, so needs still remain.

The author is grateful to the many who have given information and advice, particularly to Messrs. Jerry S. Sellers, James R. Dupree, and Timothy Shipe, Electronics Engineers at Lockheed Martin Electronics and Missiles.

2.13 REFERENCES

1. Halliday, David and Resnick, Robert, *Fundamentals of Physics*, Chapters 23 and 35, Wiley, New York, 1970.

2. Wylie, C. Ray, *Differential Equations*, Chapter 4, McGraw-Hill, New York, 1979.

3. Reed, Erik K., *Power Dissipation Characteristics of Kemet C052 (CKR05), and C062 (CKR06) Monolithic Multilayer Ceramic Capacitors*, Proceedings of 4th Capacitor and Resistor Symposium by Components Technology Institute, Inc, Huntsville, AL, 1984.

4. AVX Corp., A Kyocera Group Company, Myrtle Beach, SC.

5. American Technical Ceramics Corporation, Huntington Station, NY.

6. Dover Corp./Dielectric Laboratories, Inc., Cazenovia, NY.

7. Condon and Odishaw: *Handbook of Physics*, 2nd Ed., New York, 1967; part 4 chapter 1.

CHAPTER 3

TRANSFORMERS AND INDUCTIVE DEVICES

Leo E. Wilson
Advisory Engineer, Retired, Westinghouse Electric Corporation

3.1 INTRODUCTION

In elementary terms, a transformer consists of two or more wirewound coils that are coupled inductively. When an alternating voltage is applied to one winding (usually called the *primary*), a corresponding voltage is induced in the other winding (usually called the *secondary*). The magnitude of this voltage is determined by the number of turns in each of the coils. If the coils are adequately coupled magnetically, it is expressed as:

$$\frac{E_1}{E_2} = \frac{N_1}{N_2} \tag{3.1}$$

or the winding voltages are directly proportional to the winding turns. This principle applies to all transformers, regardless of the application or the operating frequency. Figure 3.1a, 3.1b, and 3.1c shows a simple two-winding, single-phase transformer and its equivalent circuit.

3.1.1 Basic Relationships

If the voltage drops and various losses are considered to be negligible (a perfect transformer) the power received by the transformer will equal the power delivered to the load.

$$E_1 I_1 = E_2 I_2 \tag{3.2}$$

or

$$\frac{E_1}{E_2} = \frac{I_2}{I_1} \tag{3.3}$$

and from Eq. 3.1:

$$\frac{I_2}{I_1} = \frac{E_1}{E_2} = \frac{N_1}{N_2} \tag{3.4}$$

or the winding currents are inversely proportional to the winding turns.

Replacing the primary of the transformer with an equivalent impedance Z_1 and the secondary load resistance R_l with impedance Z_2, the primary and secondary currents may be expressed as

$$I_1 = \frac{E_1}{Z_1} \tag{3.5}$$

and

$$I_2 = \frac{E_2}{Z_2} \tag{3.6}$$

substituting these expressions in Eq. 3.3:

$$\frac{Z_1}{Z_2} = \left(\frac{E_1}{E_2}\right)^2 \tag{3.7}$$

and from Eq. 3.1:

$$\frac{Z_1}{Z_2} = \left(\frac{N_1}{N_2}\right)^2 \tag{3.8}$$

3.1.2 Transformer Parameters

The values of the various inductances, capacitances and resistances will vary with the size of the transformer, the materials of construction, the type of transformer and its application. It can be seen that the performance of the transformer as a circuit element will change as the operating frequency changes. Normally, the various secondary parameters are reflected to the primary by the ratio of the square of the turns of the two windings. Figure 3.1c reflects this. This ratio is reasonably accurate for most power, audio and pulse transformers; however, in IF and RF transformers where the magnetic path is largely through air, the equivalent circuit depends on the coupling of the windings. Table 3.1 shows the effect of each of these parameters on the operation of power, audio and pulse transformers for steady-state conditions, i.e., the condition of normal operation. The operation of a transformer or inductor during turn-on, turn-off, switching modes, circuit faults, or other transient-inducing modes of operation is covered elsewhere.

3.1.3 Transformer Operating-Characteristics Polarity

Figure 3.2 shows schematically the convention used to indicate transformer polarity. The dots establish the terminals of the same polarity. The + and − signs indicate instantaneous polarities (i.e., V_1 and V_2 reach positive maximums at the same time). The dots also show that current flows into the primary terminal and out of the secondary terminal of the same polarity.

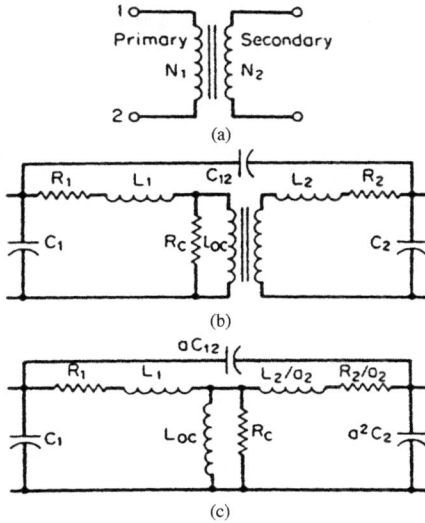

N_1 = primary turns
N_2 = secondary turns
a^2 = N_2/N_1
C_1 = primary capacitance
C_2 = secondary capacitance
C_{12} = primary-to-secondary capacitance
L_1 = primary leakage inductance
L_2 = secondary leakage inductance
L_{OC} = primary open-circuit inductance
R_1 = primary resistance
R_2 = secondary resistance
R_C = equivalent core-loss resistance (shunt)

FIGURE 3.1 Transformer equivalent circuit: (a) schematic diagram, (b) Equivalent circuit, (c) Equivalent circuit referred to primary.

Regulation Transformer regulation is the ratio of the differences in secondary voltage between no load and full load to the full-load voltage; it is usually expressed as a percentage:

$$Percent\ Regulation = \frac{100\ (V_{NL} - V_{FL})}{V_{FL}} \tag{3.9}$$

Efficiency Efficiency is the ratio of the power out of the transformer to the power into the transformer and is expressed:

$$\eta = \frac{output\ power}{input\ power} \tag{3.10}$$

The input power is equal to the output power plus the transformer losses.

TABLE 3.1 Effect of Transformer Parameters on Transformer Characteristics

Parameter	Transformer characteristic affected				
	Power transformer	Inverter transformer	Wideband transformer	Pulse transformer	Inductor
R_1	Regulation, efficiency, temperature rise	Regulation, efficiency, temperature rise	IF response, HF response, efficiency, temperature rise	Efficiency, temperature rise	Temperature rise, voltage drop, Q factor
L_1	Regulation, commutation	Efficiency, high-voltage spikes, HF response, primary balance, loss of inverter switches	HF response	Front-edge response	
C_1	Turn-off-pulse suppression	HF response	HF response	Front-edge response, trail-edge response	HF impedance
$C_{1\,2}$	EMI	EMI	EMI	Front-edge response	
R_c	Temperature rise, exciting current efficiency	Efficiency, temperature rise, exciting current	Efficiency, temperature rise	Efficiency, temperature rise	Temperature rise, Q factor
L_{OC}	Exciting current	Exciting current	IF response	Top response, pulse droop, trail-edge response	Inductance
R_2	Regulation, efficiency, temperature rise	Regulation, efficiency, temperature rise	IF response, HF response, efficiency, temperature rise	Efficiency, temperature rise	
L_2	Regulation, commutation	Efficiency, high-voltage spikes, HF response, secondary balance, loss of inverter switches	HF response	Front-edge response	
C_2	Turn-of-pulse suppression	HF response	HF response	Front-edge response, trail-edge response	

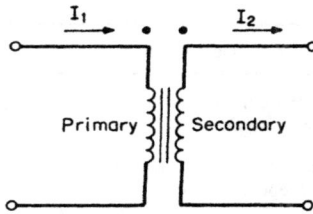

FIGURE 3.2 Transformer polarity.

Power factor The power factor is the ratio of the transformer input power to the input volt-amperes and is expressed:

$$Power\ Factor = \frac{input\ power}{input\ volt\text{-}amperes} \tag{3.11}$$

3.1.4 Wave Shapes*

Current wave shapes Transformers in electronic circuits can be subjected to alternating and direct currents simultaneously, to modified sine waves, or to other nonsinusoidal waves. Although there is a relation between current and voltage shapes in a transformer, the two frequently are not the same. Dc components of primary voltage are not transformed; only the varying ac component is transformed. Secondary current can be determined by the connection to the load. For example, if the load is a rectifier, the current will be some form of rectified wave; if the load is a modulator, the secondary current will be the superposition of two waves. If the primary voltage is nonsinusoidal, the secondary current will almost certainly be nonsinusoidal.

If the primary voltage comes from an alternating source only and the load is a half-wave rectifier, the secondary current has a dc component, except under changing conditions. That is, in the steady state, there is no primary dc component resulting from secondary dc components alone. This is true because any direct current in the primary requires a dc source. But by the initial assumption, no direct current is present in the primary. Under these conditions, the core flux will be very distorted because the flux excursions go into saturation in one direction only.

Generally, two values of current are of interest in circuits with nonsinusoidal waves: the average and the rms. Average current causes core saturation unless an air gap is in the magnetic circuit. Rms current determines the heating of the windings and is limited by the permissible temperature rise. Common current waveforms are shown in Table 3.2.

Voltage wave shapes Several different voltage waveforms are used in the design and specification of transformers covered in this section. Figure 3.3 shows three of the most commonly used of these waveforms. The equations showing the voltage-flux-turns relationships have the constants corrected so that turns and flux density can be calculated using the normally stated values of voltage, time, and frequency.

---

* With permission, John Wiley & Sons, *Electronic Transformers and Circuits, Third Edition*; R. Lee, L. Wilson, and C. Carter, 1988.

TABLE 3.2 Nonsinusoidal Current Waveforms

Current Wave Shape	Description	I_{rms}	I_{av}
	Direct current with superposed sine wave	$I_{dc}\sqrt{1+\dfrac{M^2}{2}}$	I_{dc}
	Half-sine loops of T duration and f repetition rate	$I_{pk}\sqrt{\dfrac{fT}{2}}$	$\dfrac{2I_{pk}fT}{\pi}$
	Square waves of T duration and f repetition frequency	$I_{pk}\sqrt{fT}$	$I_{pk}fT$
	Sawtooth wave of T duration and f repetition frequency	$I_{pk}\sqrt{\dfrac{fT}{3}}$	$\dfrac{I_{pk}fT}{2}$
	Trapezoidal wave of f repetition frequency	$I_{pk}\sqrt{\dfrac{f(2\delta+3T)}{3}}$	$I_{pk}f(\delta+T)$

Source: Electronic Transformers and Circuits, Third Edition, Lee, Wilson and Carter, John Wiley, New York, 1988.

3.2 ELECTRONIC TRANSFORMERS

The use of alternating current for the generation, transmission, and distribution of electrical energy is largely caused by the reliability, efficiency, and convenience of the static transformer. In the application of electronic transformers, it is found that they often perform many functions besides the basic changing of voltage and/or current values. The development by design of the various characteristics is not discussed, except to bring out the application and to show the effect on size and complexity of manufacture.

3.2.1 Power Conversion

Perhaps the best-known use of transformers is for the conversion of ac power from one voltage to another, generally at relatively low frequencies. The generation frequency has been chosen primarily for transmission considerations and to a lesser extent for reduced losses in the transformers and equipment both in the magnetic material and in the windings. Commercial power in the United States is almost entirely 60 Hz, but 50 Hz is used extensively in Europe. Where transmission over great distances is not required (as in aircraft), 400 Hz (and even higher frequencies) permit a reduction in the size of devices that use magnetic materials, such as generators, motors, and transformers.

Improved magnetic materials, semiconductors, and coil construction have enabled power transformers to operate at considerably higher frequencies when the power is converted to direct current close to the transformer. Modern practice using inverters transforms power in kilovolt-ampere blocks at frequencies greater than 100 kHz for airborne equipment. As more power is required in airborne or space applications, the limitations on size and weight are leading to the development of conversion techniques at even higher frequencies and power levels.

The conversion of power from one voltage to another has many applications in operation of ac equipment, but in electronics applications, conversion to direct current at various voltages is common. In general, older equipment, which uses vacuum tubes, requires higher voltages than does more modern solid-state and digital circuitry. This use of lower voltages naturally means higher currents for equivalent power.

SINE WAVE VOLTAGE

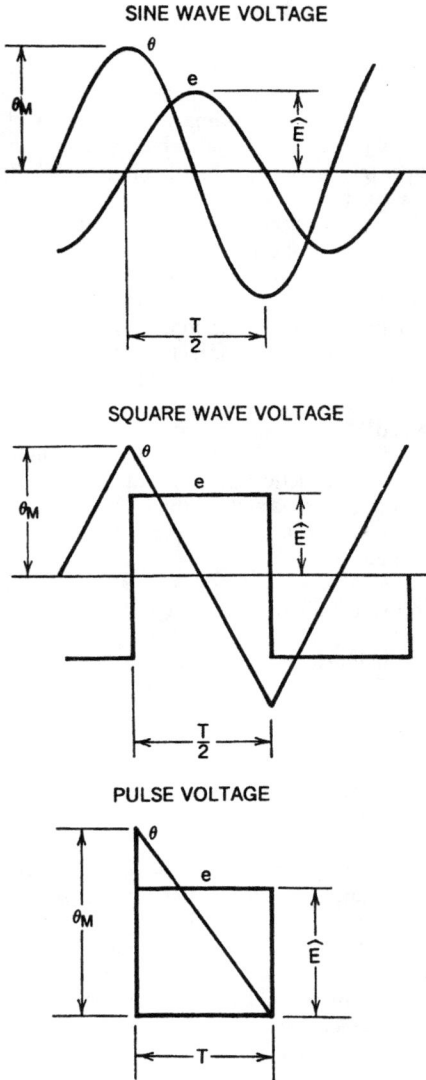

$$(1)\quad e' = -N\frac{d\phi}{dt} \times 10^{-8} = -N\frac{\Delta\phi}{\Delta t} \times 10^{-8}$$

$$(2)\quad e' = -N\frac{2\phi_m}{T/2} \times 10^{-8} = -N\frac{2\phi_m}{1/2f} \times 10^{-8}$$

$$(3)\quad e' = 4fN\theta m \times 10^{-8} = 4fNAcBm \times 10^{-8}$$
Where B is in Maxwells per In^2
For B in Gauss

$$(4)\quad e' = 25.8fNAcBm \times 10^{-8}$$
Where $e' = $ average induced volts

$$(5)\quad \text{Erms} = 28.6fNAcBm \times 10^{-8}$$

$$(6)\quad N = \frac{E \times 10^8}{28.6fAcBm} = \frac{3.49E \times 10^6}{fAcBm}$$

SQUARE WAVE VOLTAGE

EQUATIONS (1) THRU (4) ABOVE

$$E(pk) = e'$$

$$(5)\quad E = 25.8fNAcBm \times 10^{-8}$$

$$(6)\quad N = \frac{E \times 10^8}{25.8fAcBm} = \frac{3.88E \times 10^6}{fAcBm}$$

PULSE VOLTAGE

EQUATION (1) ABOVE

$$(2)\quad e' = -N\frac{\phi_m}{T} \times 10^{-8} = -N\frac{AcBm}{T} \times 10^{-8}$$

$$E(pk) = e' \text{ and for } B \text{ in Gauss}$$

$$(3)\quad E = \frac{6.45NAcBm}{T} \times 10^{-8}$$

$$(4)\quad N = \frac{ET \times 10^8}{6.45AcBm}$$

FIGURE 3.3 Voltage-flux-turns relationship. *(Reprinted with permission, John Wiley and Sons,* Electronic Transformers and Circuits, Third Edition, *R. Lee, L. Wilson, C. Carter, copyright 1988.)*

3.2.2 Power Sources

Commercial power is standardized into insulation classes, each of which is assigned a basic impulse level (BIL). The BIL is primarily the ability to withstand lightning surges, which are a transmission hazard. It follows that the BIL will apply to the primary or ac winding of transformers supplied directly from the commercial power lines. Transformers supplied for local generators or for in-plant distribution sources might not require BIL ratings.

When local generation is used, the frequency can be subject to choice. Rotary generators usually operate at frequencies below 2000 Hz, but oscillators or inverters can generate frequencies that are limited primarily by the switching speeds obtainable.

The number of phases will also require consideration. Many applications use polyphase power to reduce the size of equipment for equivalent power, as compared to a single-phase transformer. This is especially important for rectifier operation, where the ripple frequency and magnitude are directly related to number of phases, rectifier circuit used, and primary frequency. Overall performance, including filtering and regulation, must also be considered.

Alternating current at high frequencies obtained from inverters introduces transmission limitations. The waveshape, especially when combined with pulse-width regulators, might introduce transformer problems. Increased losses in magnetic materials and in windings from eddy currents and skin effect might result. If transformed and rectified at the inverter, this method of producing dc power at various voltages has proven to be desirable—especially when size and weight are primary considerations.

Tolerances   Two types of tolerances must be provided. Short-term voltage and/or current transients usually present a hazard, rather than out-of-specification performance, although it might affect performance, too. Voltage surges might cause insulation breakdown in the transformer, as well as in other locations in the circuit. Frequently, protective devices or circuits are used to limit such voltage transients.

Inrush current at turn-on can reach several times the normal load current and can operate overload devices, such as fuses and circuit breakers. This hazard can be eliminated by gradual turn-on, instead of switching on at full voltage. Another method is to place a current-limiting impedance in series with the primary, which is removed from the circuit after steady-state conditions have been reached.

Improper performance can result from unusual voltage variations, source impedance, transformer regulation, auxiliary equipment impedance, etc. In rectifier circuits at light loads, filter capacitors tend to charge to peak voltage. Thus, for zero-load, a single-phase full-wave rectifier with filter capacitor will reach 1.57 times the average voltage. The effect is less with polyphase rectifiers and two-way rectifier circuits, but it is present to some extent in all rectifier circuits. Peak charging can be eliminated by using a bleeder, which prevents the load current from dropping into the critical region. Transformer regulation can be limited by design. Other voltage changes can be provided for with transformer taps.

3.3 TRANSFORMER CONSTRUCTION, MATERIALS, AND RATINGS

3.3.1 Configuration

Usually, the configuration of a transformer used in an electronic circuit is determined by the materials used in its fabrication. The selection of these materials can be influenced by cost, environment, circuit application, available space, weight limitations, etc. In most instances, the electronic circuit designer is not particularly concerned with what materials are in the transformer. The primary concern is that the device meets a set of requirements and works in the circuit in the manner for which it was designed. The ultimate selection of a part is often made on the basis of its cost.

However, the circuit designer should have some knowledge of the materials used in the construction of transformers and of the various configurations of transformers with the advantages and disadvantages of each. Usually, no specific configuration is indigenous to a particular type of magnetic device. Figure 3.4 shows various configurations of transformers and inductors. Table 3.3 lists the usual applications of each type, together with advantages and disadvantages of each.

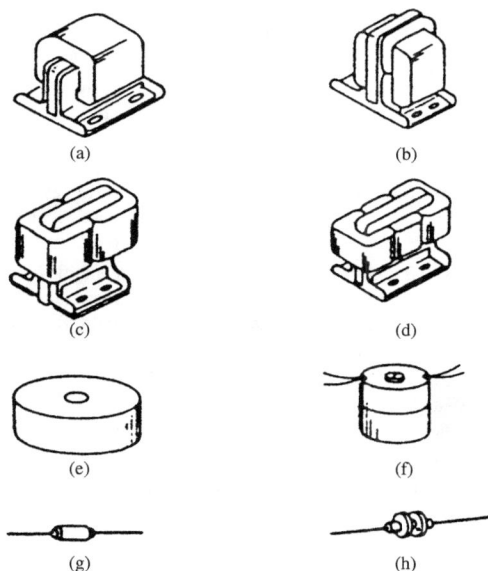

FIGURE 3.4   Transformer configurations: (a) Simple (one-phase), (b) Shell (one-phase), (c) Core (one-phase), (d) Three-phase, (e) Toroidal, (f) Cup core, (g) Single-layer solenoid, (h) Pie.

An electronic transformer can be considered to consist of several physical systems—each of which has an effect on the performance of the device.

Magnetic circuit   The magnetic circuit consists of the core, which can be any type of magnetic material, and the magnetic path, which also includes air spaces, gaps, and leakage paths. In some applications, usually at high frequencies, the magnetic path can include little or no magnetic material.

Transformer cores   The component most responsible for the shape, size, and weight of transformers and inductors, except air-core transformers, is the magnetic core. Transformer cores are basically iron, sometimes with nickel or cobalt in various proportions, and usually with small amounts of other elements. Power, audio, and large pulse-transformer cores are usually wound from thin sheet or strips of magnetic tape. Often, the

TABLE 3.3 Application of Transformer Types

Type	Uses	Comment
Simple	Power transformers, inductors, pulse transformers	Only one coil to wind, low leakage inductance, large mean turn
Core	Power transformers, inductors, wide-band transformers, pulse transformers, inverter transformers	Permits winding balance in push-pull application, minimizes external magnetic field, permits lower winding capacitance in high-voltage transformer
Shell	Power transformers, inductors, wide-band transformers	Better winding efficiency than core type, smaller mean turn than simple type
Three-phase	Power transformers	E core not affected by unbalanced currents in primary
Toroid	Power transformers, pulse transformers, inductors, wide-band transformers, inverter transformers	Minimum leakage inductance, more difficult to wind, usually degraded by direct currents in windings, may be difficult to cool, can function over wide frequency ranges
Cup core	Pulse transformers, inductors, wide-band transformers, pulse transformers, inverter transformers	Available only in ferrite-type core materials. Low external magnetic field. Relatively fragile. Can adjust to given inductance. Difficult to wind with larger magnet wire
Single-layer solenoid	IF and RF transformers, IF and RF inductors	Can be wound on magnetic or nonmagnetic form
Pi winding	IF and RF inductors	Lowest interwinding capacitance

cores are assembled from the laminations. Adjacent layers of the tape or adjacent laminations must be insulated from each other to minimize eddy currents, which represent losses. As the operating frequency of the transformer increases, eddy current losses increase so that the thickness of the magnetic tape must be reduced to control the eddy currents.

Laminated cores Laminated cores are fabricated in many shapes, the simplest being a ring. The ratio of core area to winding area can be varied over a limited range, and the cross-sectional shape of the iron circuit can also be varied a limited amount. Constrictions on transformer shape usually affect the core configuration and result in a reduction of efficiency and/or an increased cost.

Ferrites and powdered iron cores In addition to laminations and strip wound cores, cores can be made of powdered metal bonded with a resin or of ferromagnetic oxides in a ceramic structure called a *ferrite*. Powdered metal cores can be constructed as toroids, solenoids, or formed as an adjustable slug. Ferrite cores can be of many shapes, including toroids and pot cores, which completely surround the windings. Powdered iron and ferrites are usually used for the higher-frequency applications; however, they are often used at the power and audio frequencies.

Table 3.4 lists several standards for cores and core materials issued by various technical, trade, national, international, and other associations. Figure 3.5 shows the types of core configuration often used. Table 3.5 lists many of the types of core material used.

Electric circuit The electric circuit consists of successive turns of wire, usually single strands of copper or aluminum wire; the wire can be round or, for higher currents, square or rectangular. As the operating frequency of the transformer increases, skin effect can affect losses and regulation. To reduce these effects, Litz wire can be used. Litz wire is made of multiple strands of individually insulated wire. The need for Litz wire is determined by the current being conducted and the operating frequency.

Insulation system The insulation system consists of the insulation on the magnet wire; sheet insulation, which can be used between layers of windings and between windings themselves; various plastic impregnants used to provide mechanical support, heat transfer, and/or environmental protection; various gaseous or liquid dielectrics; and miscellaneous tapes and adhesives. Table 3.6 lists various insulating materials with recommended voltage stresses and temperature limits for each. These materials can be organic or inorganic and are generally selected for one or more of the following: their compatibility with each other, the expected life of the transformer, the allowable hot-spot temperature, their compatibility with the anticipated environment, the magnitude and type of voltage stress, their effect on size and weight, and their cost. It is also important to recognize that many factors affect both the temperature and the voltage at which an insulation system is used.

Voltage considerations In a transformer, the voltage to which a combination of insulating materials can be stressed is not the summation of the voltages at which a given thickness of each material breaks down. Rather it is usually a substantially reduced voltage that is determined by the dielectric constants of the materials, the geometry of the windings, the geometry of terminals and associated ground planes, the atmospheric pres-

TABLE 3.4 Standards for Magnetic Cores and Core Materials

Electronic Industries Association, (EIA)

EIA 217-83	*Wound Cut Cores*
EIA 260-83	*Tape-Wound Toroidal Cores*
EIA 393-83	*Core Laminations, Vertical and Horizontal Channel Frames for Transformers and Radio Receivers*

Magnetic Materials Producers Association, (MMPA)

PC110	*Standard Specifications for Ferrite Pot Cores*
TC200	*Standard Specifications for Ferrite Threaded Cores*
UEI310	*Standard Specifications for Ferrite U, E, & I Cores*
SFG-89	*Soft Ferrites, a Users Guide*

International Electrotechnical Commission, (IEC)

113	*Dimensions for pot cores made of ferromagnetic oxides and associated parts*
205	*Calculation of the effective parameters of magnetic piece parts*
226	*Dimensions of cross cores (X-cores) made of ferromagnetic oxides and associated parts*
367	*Cores for inductors and transformers for telecommunications*
401	*Information on ferrite materials appearing in manufacturers catalogues of transformer and inductor cores*
431	*Dimensions of square cores (RM cores) made of ferromagnetic oxides and associated parts*
525	*Dimensions of toroids made from magnetic oxides and associated parts*
647	*Dimensions for magnetic oxide cores intended for use in power supplies (EC cores)*
701	*Axial lead cores made of magnetic oxides or iron powder*
723	*Inductor and transformer cores for telecommunications*

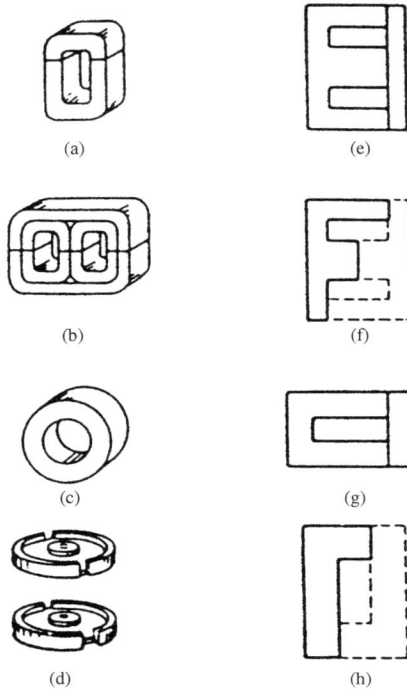

FIGURE 3.5 Transformer cores: (a) Tape-wound cut, C core (one-phase applications); (b) tape-wound cut E core (three-phase applications); (c) toroid, either tape wound or solid; (d) cup core (used in pairs); (e) E-I laminations (used in sets); (f) F lamination (used in pairs); (g) U-I laminations (used in sets); (h) L lamination (used in pairs).

sure, and other factors. It is also affected by the nature of the voltage stress: alternating current, direct current, impulse, or combinations of all of these.

Thermal considerations   In a transformer, the temperature at which a combination of insulating materials can be used is determined by the compatibility of the various insulations and the anticipated life of the transformer. Transformer life is essentially temperature-related, being determined by the rate of thermal decomposition (within reasonable bounds). The Arrhenius equation is often used to relate the life of a transformer manufactured with organic insulation to temperature:

$$\log{(rate)} = \log A - \frac{E}{RT} \tag{3.12}$$

where A, E, and R are constants based on the materials, and T is the temperature in degrees Kelvin.

The corresponding graph of log life vs. $1/T$ (Kelvin) results in a straight line. The slope of this line leads to the often-used 10° rule (i.e., that the life of an insulation system is reduced 50 percent for each 10°C increase in temperature). Modern insulation systems often have life plots with slopes providing 8 to 14° rules. The graph in Figure 3.6 depicts the anticipated life of a transformer insulation system, which consists of organic materials and is impregnated with an organic resin.

Thermal circuit The thermal circuit consists of:

- All elements that generate heat in the transformer: winding losses, core losses, and possibly dielectric losses.
- All elements that conduct heat out of and away from the transformer: the windings, the core, the mounting structure, the various insulating materials.
- All elements that create thermal barriers: the insulation materials and the transformer geometry.
- The environment.

TABLE 3.5 Core Materials

Material	B_{max} (T)	Suggested frequency range	Uses	Forms
Silicon-iron alloy:				
Oriented, Hipersil	17.6	To 20 kHz and pulse	A, B, C, D, E, H, J	1, 2, 3
Nonoriented, AISI M-19	15.0	To 400 Hz	A, B, D, I	3
Nickel-iron alloys:				
50% Ni, square loop	16.0	To 50 kHz and pulse	B, C, E, H, J	1, 2, 3
48% Ni, round loop	15.0	To 20 kHz	A, D, F	3
79% Ni, square loop	8.0	To 50 kHz and pulse	B, C, E, H, J	1, 2, 3
79% Ni, round loop	8.0	To 20 kHz	A, D, F	2, 3, 7
Supermalloy	7.0	To 50 kHz and pulse	A, C, E, F	2
Cobalt-iron alloy,				
Supermendur	21.0	To 5 kHz	A, B, D	1, 2
Powdered moly permalloy	2.0	To 100 kHz	A, D	4, 7
Ferrites	5.0	To 300 MHz	A, B, C, D, F, G, H	4, 5, 6
Iron powders	5.0	2 kHz to 300 MHz	B, G	7, 8, 9, 10

Uses:
A. Audio	F. Wideband
B. Power	G. RF and IF
C. Pulse	H. Inverter
D. Inductor	I. Communications
E. Saturable reactors	J. Instrument transformer

Forms:
1. Tape-wound cut cores, C and E	6. Cast or machined shapes
2. Tape-wound cores, toroidal and rectangular	7. Bars
3. Laminations	8. Rods
4. Toroidal	9. Threaded rods
5. Cup cores	10. Sleeves

TABLE 3.6 Recommended Voltage Stress and Temperature Limits for Various Insulations

Insulation		Dielectric constant	Dielectric strength	Max stress	Hot-spot
Type	Material	100 Hz, ε	V(rms)/0.001	V(rms)/0.001	max, °C
Gases	Air	1.0	See Fig. 3.22	See Fig. 3.22	
	Sulfur hexafluoride (SF$_6$)	1.0	See Fig. 3.22	See Fig. 3.22	180
Liquids	FC75 or FC77*	1.9	350	150	105
	Coolanol 20⁺	2.5	350	150	105
	Askarel	5.5	350	150	105
Solids	Epoxy resin	4.2	>500	200	130
	Polyester resin	4.0	550	150	130
	RTV	3.0 – 3.5	>300	150	200
	Laminate, epoxy glass	5.2	500	150	130
	Diallyl phthalate (DAP)	4.2	>300	150	130
Films	Mylar† (0.001-in. thick)	3.3	>3,000	200	130
	Kapton† (0.001-in. thick)	3.5	>5,000	250	200
	Teflon (0.002-in. thick)	4.2	>900	200	180
Sheet	Kraft paper, unimpregnated	2.0	>250	75	105
	Kraft paper, resin-impregnated	3.5	>500	200	130
	Nomex†, unimpregnated	2.7	150	75	180
	Nomex†, liquid-impregnated	2.8	>850	200	180
	Nomex†, resin-impregnated	2.9	>350	150	155
	Pressboard	2.0	>200	100	130
Other	Mica	5.8	3,000	250	200
	Wire enamel	3.2	>800	200	105

*3M Co. trademarks for fluorocarbon liquids.
⁺Monsanto Co. trademark for silicate-ester-base fluid.
†Trademark of E.I. du Pont de Nemours & Co., Inc.

Mechanical system The mechanical system consists of all elements that support the transformer and maintain its physical integrity. It can include parts of the insulation system, the windings, the core, and additional hardware used for support or protection.

Transformer and inductor standards Some standards that pertain to various types of transformers and inductors have been promulgated by various technical, trade, national, international, and other associations. Table 3.7 lists several transformer and inductor standards by the organization that issued them; this list is not intended to be inclusive, but rather is indicative of the types of devices that are covered by standards. No international standards are listed. However many countries have standards issued by their national commissions that are similar to those issued by ANSI, the American National Standards Institute.

3.4 TRANSFORMER APPLICATIONS

3.4.1 Power Transformers

Power transformers are used primarily to change the magnitude of the voltage and current of the source of the electric energy. Additional uses might include isolation of circuits for safety or insulation purposes, change in the number of phases, or change of phase angle (including polarity reversal). Special-purpose transformers (such as current-limiting, constant current, regulating or power-factor-correcting) can also be classified as power transformers. By far, the most common use of electronic power transformers is in rectifier circuits for the conversion of an ac voltage to a dc voltage.

Transformer connections Power transformers can be single or polyphase with three-phase the most common poly-phase connection. Systems beyond three phases are in multiples of three. The principal use of these 6-, 12-, and 24-phase arrangements is in rectifying circuits to reduce the average current in the rectifiers and/or to reduce the size of filter components and the magnitude of harmonic voltages.

Three-phase Figure 3.7 shows several different configurations of polyphase transformations. There are many more combinations of 3-, 6-, 12-, and even 24-phase transformer configurations, whose selection depends on the rectifier circuit being used. Figure 3.7 contains only delta primary windings; however, each can be replaced with a wye winding.

Delta windings The main advantages of a delta winding, either as a primary or a secondary, are that it provides a circulating path for the third-harmonic currents, which are generated by the non-linearity of the transformer core and it also requires a smaller size conductor for given line current than does a wye winding for the same line current. The main disadvantages of a delta winding are that it must hold off the full rms line-to-line voltage and it is more difficult to change taps in a tapped delta than in a tapped wye.

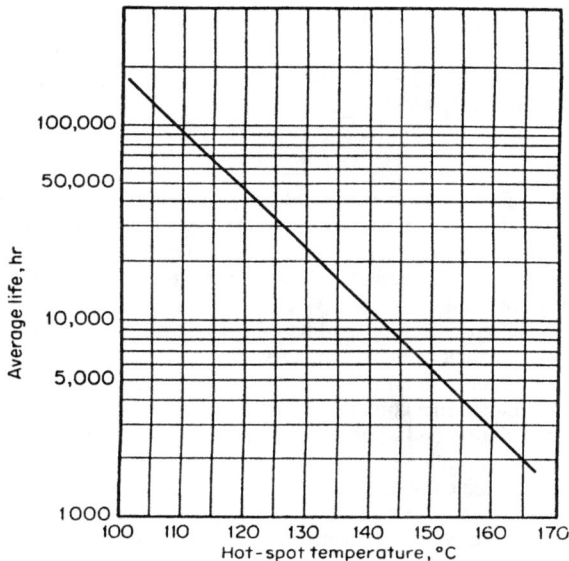

FIGURE 3.6   Transformer life.

TABLE 3.7 Standards for Magnetic Devices

Electronic Industries Association, (EIA)

EIA 174-82	*Audio Transformers for Electronic Equipment*
EIA 180-82	*Power Transformers for Electronic Equipment*
EIA 175-82	*Audio Inductors*
EIA 197-A-86	*Power Filter Inductors for Electronic Equipment*
EIA 175-82	*Pulse Transformers for Radar Equipment*
EIA/IS-48	*Axial Lead Fixed Radio Frequency (RF) Coils*

Institute of Electrical and Electronic Equipment, (IEEE)

IEEE 295-69	*Standard for Electronic Power Transformers*
IEEE 111-84	*Standard for Wide-Band Transformers*
IEEE 264-77	*Standard for High-Power Wide-Band Transformers*

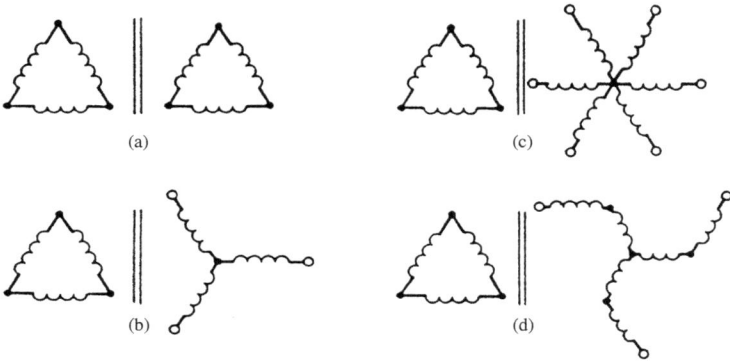

FIGURE 3.7 Three-phase transformer connections: (a) Delta-delta, (b) Delta-wye, (c) Delta-doublewye, (d) Delta-wye zigzag.

Wye windings The main advantages of a wye winding are that it must sustain only $1/\sqrt{3}$ of the line-to-line voltage, the neutral might have graded insulation, and it is easier to change taps in a tapped wye than in a tapped delta. The main disadvantage of the wye is that it must carry full line current and, unless the neutral is grounded, it does not provide a path for third-harmonic currents, thereby creating a possible line-to-line voltage unbalance. The wye-wye transformations, while offering some advantages, are not generally recommended for electronic circuits, primarily because if a neutral ground is not provided in the primary, there is no path for the flow of the third-harmonic currents generated by the nonlinearity of the transformer core. If the primary neutral is grounded, the third-harmonic currents flowing in the ground might create an EMI problem. However, a delta tertiary winding will provide the path for the circulation of the third-harmonic currents and eliminate the need for a grounded neutral.

Three-phase/two-phase The main transformation configurations for converting from three-phase to two-phase or vice versa are the Scott and LeBlanc connections (Figure 3.8). In general, the LeBlanc connection will result in a smaller transformer. However, when the LeBlanc connection is used to convert from two phase to three phase, it performs essen-

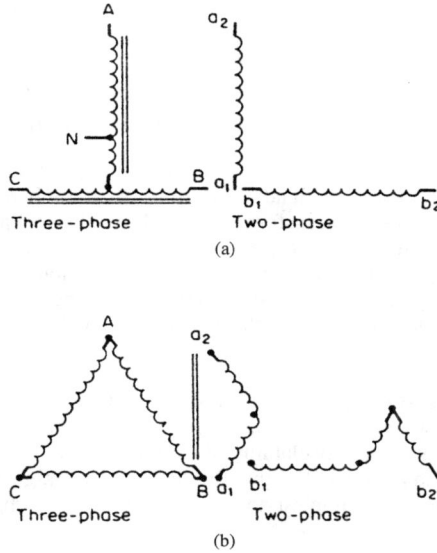

FIGURE 3.8 Three-phase-two-phase connections:
(a) Scott T, (b) LeBlanc.

tially as a three-phase wye-wye connection and might therefore require a delta tertiary winding to eliminate possible line unbalance.

Single phase When either a single-phase or three-phase source of power is available, single-phase transformers are normally used only to supply relatively low power loads, usually less than 200 W. Large single-phase loads can unbalance the phase voltages of a three-phase power system. Virtually all transformers used in home entertainment equipment and general low-power commercial applications are single-phase.

Tap changing The adjustment of voltages in a transformer is accomplished by arranging windings in parallel or in series or by taps, which add or eliminate portions of windings. Changing taps or winding connections is usually done with the power off; however, tap changing under load is possible with proper equipment. The most common example of this is the continuously variable autotransformer.

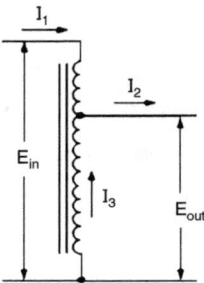

FIGURE 3.9 Single-phase autotransformer.

Autotransformer An autotransformer can be either single-phase or polyphase; their principles of operation are identical. The connection of a single-phase autotransformer is shown in Figure 3.9. The transformer has a single winding tapped in such a way that a fraction of the primary voltage is across the secondary load. Figure 3.9 shows a step-down transformer; a step-up transformer would have reversed connections.

The volt-ampere rating of autotransformers depends on the ratio of the primary and secondary voltages. In Figure 3.9, the ratio of $100 \times E_{out}/E_{in}$ equals the percent tap. If p = percent tap divided by 100 then $I_2 = I/p$ and $I_3 = (I/p - 1) I_1$.

Then

$$VA \ above \ the \ tap = (1 - p)E_{in} I_1 \qquad (3.13)$$

$$VA \ below \ the \ tap = PE_1 I_3 = (1 - p) \ E_{in} I_1 \qquad (3.14)$$

When ratio p is close to 1, the VA rating is small, resulting in a small transformer. When the ratio p is small, there is not much advantage in size, as compared with a two-winding transformer. In practice, the autotransformer offers little physical advantage over a two-winding transformer when $100XE_{out}/E_{in}$ is less than 50 percent. The main advantage of the autotransformer is the possible reduction in size with a reduction in regulation and leakage inductances, when compared with similarly rated two-winding transformers. A major disadvantage of the autotransformer is that there is no isolation between the primary and the secondary.

Power transformer specification The electronic circuit designer must, when selecting a power transformer for a given application, consider many factors, all of which will have an affect on the rating, size, and availability of the transformer.

The block diagram in Figure 3.10 shows the elements that might be included in power-transformer applications. The simplest application is that in which a transformer is connected directly to an ac load.

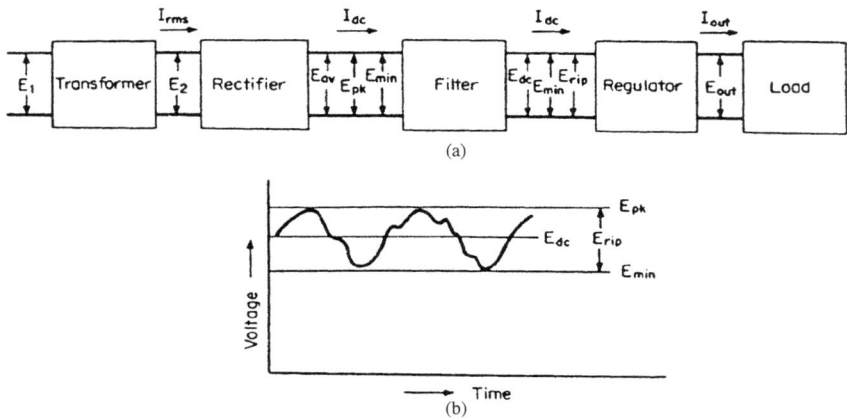

(a)

(b)

E_1 = rms input voltage; may be single-phase or three-phase
E_2 = rms output voltage; may be single-phase or three-phase
E_{av} = average rectified voltage
E_{pk} = peak rectified voltage
E_{min} = minimum rectified voltage (also called valley voltage)
E_{dc} = rectified voltage with filtering
E_{rip} = ripple voltage, specified as a percentage of E_{dc} in either rms or peak-to-peak
E_{out} = voltage into the load
I_{rms} = rms current in the output of the transformer
I_{dc} = average current in the output of the transformer
I_{out} = average load current

FIGURE 3.10 Electronic power-transformer application: (a) block diagram; (b) output voltage waveshape (symbolic), where transformer can be single-phase or three-phase, rectifier can be single-phase or three-phase one-way or two-way; filter can be any type; regulator can be any type.

The requirements for power transformers can be established in two ways. First, the transformer output can be specified in rms volts and amperes for a given input voltage and frequency. Second, in applications in which the load is direct current, the circuit designer might specify the dc voltage and current required at the load or input to a regulator and the rectifier circuit; this permits the transformer designer to establish the transformer voltage and perhaps the filter component values.

Practical limitations The following practical limitations in the design and operation of power transformers must be considered by the electronic circuit designer when specifying power-transformer requirements.

- The normal tolerance on the absolute output voltage of a single-output, low-voltage transformer is ± 3 percent.
- The normal tolerance on the absolute output voltage of a multiple-output low-voltage transformer is ± 5 percent.
- The normal regulation of a single-output, low-voltage transformer, no load to full load, at nominal input voltage is 5 percent or less.
- Normal regulation of a multiple-output, low-voltage transformer, no load to full load, at nominal input voltage is a function of the percent volt-amperes the secondary winding being considered is of the total output volt-amperes. A heavily loaded winding, greater than 30 percent of the total volt-amperes would be about 5 percent. More lightly loaded windings would have less regulation.
- Compensation for line-voltage drop, rectifying-diode drop, and filter inductor drop must be designed into the transformer.
- The maximum differential between the minimum of the rectified voltage at full load (E_{min}) and the peak of the rectified voltage at minimum load (E_{pk}) might be 15 percent of E_{dc}. The effect of input-voltage variation must be added to this variation to determine the maximum possible variation in output voltage.
- Volt-ampere safety factors are not normally designed into a transformer; the load currents specified are those used to establish core and wire sizes.

Transformer requirements To approach the optimum transformer, the electronic circuit designer must specify:

- *Power source* Voltage with limits of deviation (including surges, number of phases, and phase unbalance) and frequency (including tolerances, line impedance, turn-on provisions, and protective devices).
- *Output* Output voltages and load currents, load type, power factor, rectifier circuit, filtering, regulation limits, duty cycle, and tapped voltages.
- *Loads may be continuous, intermittent, or variable* Intermittent and variable loads permit duty-cycle rating so that more output volt-amperes can be realized than are available for continuous operation of the same size of transformer. The power rating of transformers with duty less than continuous is:

$$P = \sqrt{\frac{H_1^2 S_1 + 1/3[(H_2^2 - H_3^3)/(H_2 - H_3)]S_2 + \text{etc.}}{S}} \tag{3.15}$$

where
   P = equivalent rating in current or volt-amps
   H_1, H_2, H_3, etc. = various current, etc. levels during
   S_1, S_2, etc. = *time*
   S = total time of one cycle

In this formula, H_2 and H_3 represent the maximum and minimum power level, which changes at essentially a constant rate during time S_2.

Environment   Ambient temperature limits (including cooling available, operating temperature limits, protection from weather and/or contamination) and altitude limits.

Other   Size and weight limitations, special mounting, shock and/or vibration, interface requirements, auxiliary equipment, corona limits and electrostatic or magnetic shielding required.

3.4.2 Rectified Loads and Circuits

Most electronic systems require a power supply to provide the various needed dc voltages. These dc voltages are created either by the direct conversion of the prime power to a dc voltage without any frequency conversion or by converting a dc voltage to a higher-frequency alternating voltage with subsequent transformation, rectification, and regulation if required; the dc voltage can be obtained by rectifying the alternating prime power source or by using a battery or other dc power source. The first of these are called *direct-conversion power supplies* and the second *switching power supplies*.

Direct conversion, ac source to dc use   *Rectified loads and circuits*   A selection of electronic power transformers must also include information about rectifier circuits and their impact on the design and operation of the transformer in the circuit. Although switching power supplies usually operate from a dc source, either a battery or the ac line-voltage rectified, the majority of rectifier applications must have the transformer designed to match the circuit for the desired dc output.

The single-phase half-wave rectifier is the simplest possible rectifier circuit, consisting of a transformer, a rectifying element, and a load. The transformer in this arrangement has a utility factor (UF) of only 0.287, meaning that the transformer is large in proportion to the power output. The utility factor is defined as the ratio of the dc power output to the volt-ampere rating or size of the transformer.

The ripple factor is the ratio of the amplitude of the first harmonic of ripple to the dc output voltage; in the half-wave rectifier circuit, it is 1.57. In contrast, the full-wave, two-way or bridge rectifier circuit, which requires four rectifying elements, has a UF of 0.90 and a ripple factor of 0.67. This means a much smaller transformer and less filtering are required.

The single-way rectifier circuit, as the name implies, causes current to flow in one direction in the secondary of the transformer. This means that the current flows only during the part of the cycle when the voltage causes a rectifier to conduct. Thereby, it creates a duty cycle of less than unity. It also means that the current, when it is flowing, must be greater than it would be with uninterrupted current flow, the increase being equal to the reciprocal of the duty cycle. The transformer secondary size increase is proportional to the current increase multiplied by the square root of the duty cycle:

$$p = \sqrt{I^2 \frac{t}{T}} = I \sqrt{\frac{t}{T}} \tag{3.16}$$

where

    P = transformer volt-ampere rating
    I = current in winding during time t
    t/T = fraction of cycle of current flow = duty cycle

The transformer primary current and voltage cannot contain a dc component because it is connected to an ac line. Therefore, a magnetizing current flows in the primary when no-load current is flowing, which has an average value equal and opposite to the load current. The primary of the transformer must be designed to handle the load current plus the magnetizing current.

ANSI standard C57-18 recognizes more than 60 different rectifier circuits. Table 3.8 contains the more frequently used of these circuits together with some of the more important constants, which might influence the selection of a suitable circuit.

Dc source to dc use

Switching power supplies A major application of power transformers is in switch-mode power supplies, wherein the basic conversion of voltage and the transfer of energy occurs at a frequency higher than that of the prime power source. In some applications, the power source is dc, a battery, for instance. The methods of generating these higher frequencies are described in many different books on power supplies. However, the topology of the switching power supply has a definite bearing on the transformer requirements and its subsequent design. For that reason, the type of switching circuit must be considered when the transformer is specified. The three main categories of converter circuit and the functioning of the transformer in these circuits follows.

In one type of converter, the transformer functions as an inductor and stores energy for transfer to the load. In the other, it couples the source of energy directly to the load by transformer action; there is no energy storage. In each of these modes of operation, the transformer is driven differently and operates over different parts of the core flux-current loop; this has a direct affect on the size of the transformer.

Flyback converter The most commonly used circuit in which energy is stored in the transformer before being transferred to the load is the flyback converter (Figure 3.11a). The flyback transformer, actually a coupled inductor, accomplishes the transfer of energy in the following manner. When the transistor is "on," the current in the primary (the inductor) is increasing and energy is being stored in the core gap; this energy cannot be transferred to the secondary because the output rectifier is reverse biased. When the transistor turns "off," the polarity of the voltage in the primary reverses and the ampere turns stored in the core and core gap are transferred to the secondary, through the rectifier to the load. Figure 3.11b shows a typical flux-current loop for a flyback transformer, notice that the flux swing is from B_r to B_{max} in the first quadrant and that the loop is narrower and less square than that of the forward converter in Figure 3.12; this is caused by the relatively large gap in the core magnetic path in which energy is stored.

Forward converter In the single-ended circuit, such as the forward converter shown in Figure 3.12a, the flux-current loop is completely within the first quadrant of the B-H loop (Figure 3.12b). When the switch and the transistor are "on," a square-wave voltage is applied to the transformer primary, the core is driven from B_r to B_{max}, and energy is transferred through the secondary to the load. When the switch is turned "off," the voltage drops to zero and the core returns to its T_r state. This limits the theoretical maximum flux swing in the core, ΔB, to $(B_{max} - B_r)$.

Push-pull converter In most bridges, half-bridge, and full-wave center-tap circuits, the transformer core is driven symmetrically so that the flux current loop is symmetrical about the B and H axes. Figure 3.13 shows a typical push-pull switch-mode power supply

TABLE 3.8 Rectifier Circuits with Transformer Connections

Name (Single way)		Single-phase half-wave	Single-phase full-wave	Three-phase delta-wye	Three-phase delta-zigzag	Three-phase delta-diametric
Number phases primary		1	1	3	3	3
Number rectifiers		1	2	3	3	6
Circuit						
E_{dc}		$0.45\,E_s$	$0.90\,E_s$	$1.17\,E_s$	$1.17\,E_s$	$1.35\,E_s$
$I_{rectifier}$	AV	I_{dc}	$0.5\,I_{dc}$	$0.33\,I_{dc}$	$0.33\,I_{dc}$	$0.167\,I_{dc}$
I_s	RMS	$1.57\,I_{dc}$	$0.707\,I_{dc}$	$0.577\,I_{dc}$	$0.557\,I_{dc}$	$0.408\,I_{dc}$
Secondary U.F.		0.287	0.636	0.675	0.585	0.552
I_p	RMS	$1.21\,I_{dc}$	I_{dc}	$0.471\,I_{dc}$	$0.471\,I_{dc}$	$0.578\,I_{dc}$
Primary U.F.		0.373	0.90	0.827	0.95	0.78
Ripple factor		1.57	0.667	0.25	0.25	0.057
Ripple frequency		f	$2f$	$3f$	$3f$	$6f$

Name (Double way)	Single-phase full-wave (bridge)	Three-phase delta-wye	Three-phase wye-delta	Three-phase delta-zigzag
Number phases primary	1	3	3	3
Number rectifiers	4	6	6	6
Circuit				
E_{dc}	0.90 E_s	2.34 E_s	2.34 E_s	2.34 E_s
$I_{rectifier}$ AV	0.5 I_{dc}	0.33 I_{dc}	0.33 I_{dc}	0.33 I_{dc}
I_s RMS	I_{dc}	0.816 I_{dc}	0.472 I_{dc}	0.816 I_{dc}
Secondary U.F.	0.90	0.95	0.95	0.95
I_p RMS	I_{dc}	0.816 I_{dc}	0.141 I_{dc}	0.816 I_{dc}
Primary U.F.	0.90	0.95	0.95	0.95
Ripple factor	0.667	0.057	0.057	0.057
Ripple frequency	2f	6f	6f	6f

(a)

Hysteresis loop of magnetic
core in flyback circuit

(b)

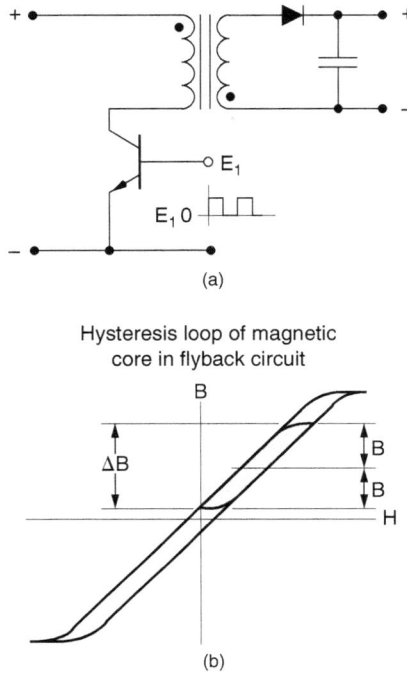

FIGURE 3.11 Flyback converter: (a) schematic
diagram, (b) hysteresis loop of magnetic core in flyback circuit.

with its associated flux current loop. The push-pull circuit is an arrangement of two forward converters operating out of phase. The switches in the input are alternately "on" and
"off," thereby applying alternating square waves to each half of the primary. Energy is
transferred to the load through the transformer secondaries and the full-wave bridge rectifier circuit. This provides a theoretical maximum ΔB of $2B$ max.

Table 3.9 is a comparison of the three switching power-supply topologies and a resonant power supply, their advantages and disadvantages. These are the basic forms of the
three switching power supplies. The many variations of these are described in published
volumes and in technical publications.

Series-resonant converter[*] The series-resonant converter (Figure 3.14) uses a resonant
circuit to produce sinusoidal waveforms instead of the rectangular voltages and triangular
current waveforms of the switching supplies. When S_1 closes, a sinusoidal current is formed
by the transformer leakage inductance, L, and the distributed capacitance, C. In addition to
the leakage inductance, intrinsic inductances in the circuit and any finite external inductance
must be added to the leakage inductance. Similarly, any other capacitance in the series circuit must be included with the distributed capacitance. After the current reaches a peak, it
falls to zero. Then S_1 opens and S_2 closes and the current flows in the opposite direction, returning to zero when the peak is reached. The cycle is then repeated. The flow of current in

---

* With permission, McGraw-Hill Companies, "Handbook of Transformer Design and Application, Second
Edition", W.M. Flanagan, 1993.

the primary induces a sinusoidal voltage in the secondary, which is connected to the load. Voltage regulation is achieved by delaying the opening of the switches.

The operating frequency is the series-resonant frequency of the circuit. For optimum operation, the impedance of the resonant circuit should equal the load impedance.

3.4.3 Pulse Transformers

In analog circuits, pulse transformers generally perform a coupling function or are used in a circuit that generates pulses.

Pulse-coupling transformers In these applications, the transformer must pass a square wave or pulse, which approaches a square wave. The square wave differs from the sine wave in that the rise and fall of the pulse is very steep. Applications involving pulses that are not square are not covered because a transformer that can pass a square wave can also pass pulses with a sloping rise or fall and with nonflat tops. The standard pulse waveform is shown in Figure 3.15.

Figure 3.15 shows that the pulse contains a front edge, a top, and a trailing edge. By adapting the transformer-equivalent circuit (Figure 3.1c) to pulse transformers, the three equivalent circuits in Figure 3.16 for front edge, top, and trailing edge are developed.

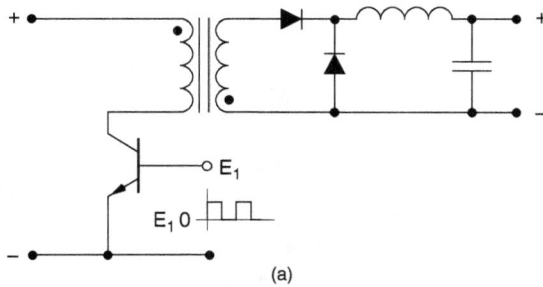

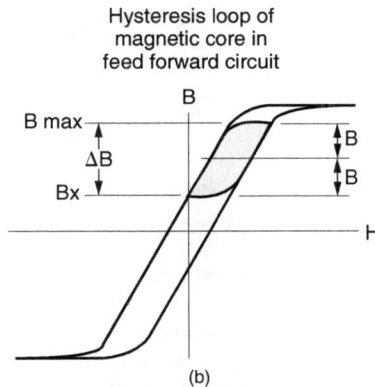

FIGURE 3.12 Feed-forward converter: (a) schematic diagram, (b) hysteresis loop of magnetic core in feed-forward circuit.

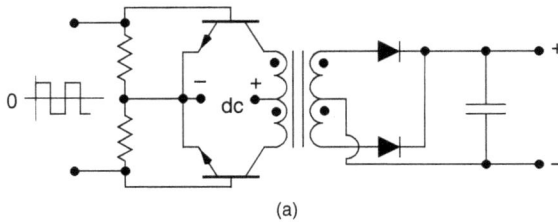

(a)

Hysteresis loop of magnetic
core in push-pull circuit

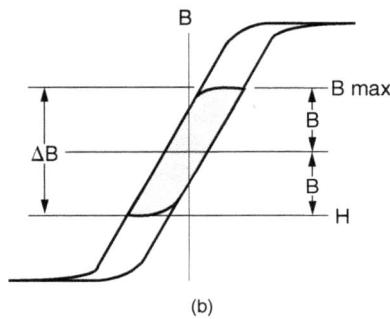

(b)

FIGURE 3.13 Push-pull converter. (a) Schematic diagram, (b) Hysteresis loop of magnetic core in push-pull circuit.

TABLE 3.9 Comparison of Switching Power Supplies

	Flyback
Advantages	Simple topology, lowest parts count, and multiple outputs possible.
Disadvantages	Transformer core utilization poor, transformer design critical, leakage inductance high. Large core gap can increase core loss. Output ripple high.
	Feed forward
Advantages	Simple topology, multiple outputs possible, and low output ripple.
Disadvantages	Transformer core utilization poor, poor transient response, and low output ripple.
	Push-pull
Advantages	Simple but has higher parts count, good transformer core utilization, low ripple, and noise.
Disadvantages	Transformer design critical, core saturation possible, high parts count, poor dynamic range, and transient response.
	Series resonant
Advantages	Low switching losses, leakage inductance and capacitance are part of the resonating circuit, good transformer core utilization.
Disadvantages	Transformer design is critical.

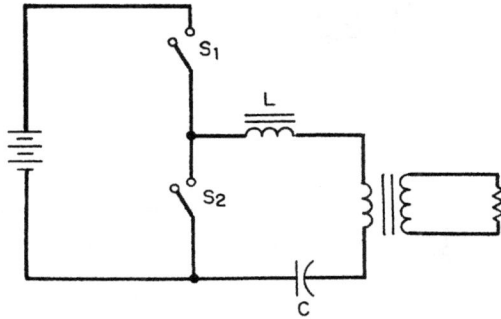

FIGURE 3.14 Simplified circuit of a series-resonant inverter. *(Reprinted with permission, McGraw-Hill Companies,* Handbook of Transformer Design and Applications, *Second Edition, W.M. Flanagan, copyright 1993.)*

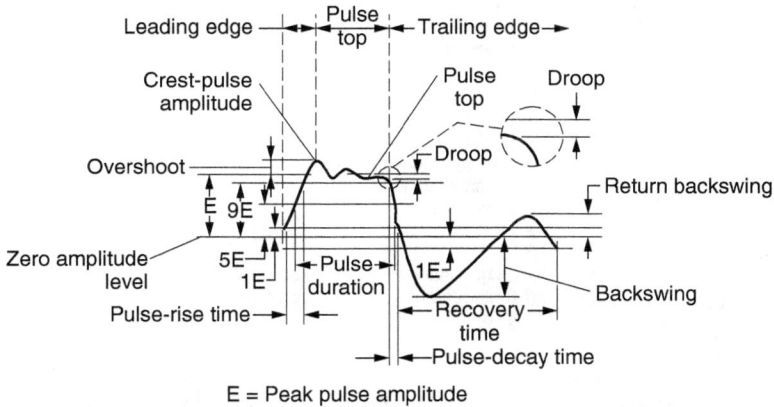

E = Peak pulse amplitude

FIGURE 3.15 Pulse waveform.

As in wideband transformers, the pulse transformer is affected by factors external to the transformer. These include:

- The source and load impedances
- The linearity of these impedances
- The pulse width
- The repetition rate
- The magnitude of the pulse voltage

In addition to the foregoing external factors, it can be determined from Table 3.1 and from an examination of the equivalent circuits in Figure 3.16 which transformer characteristics affect pulse response.

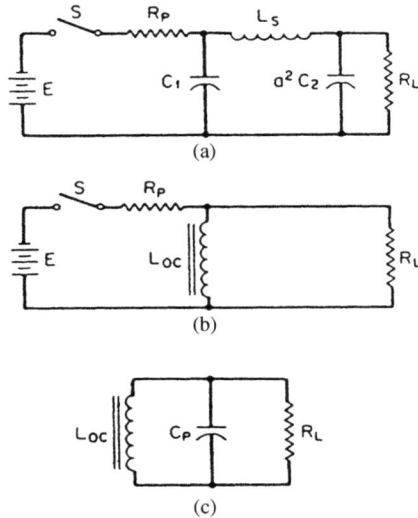

(a)

(b)

(c)

E = voltages
S = switch
R_P = R_1 + source impedance
P_L = R_2 + load impedance
L_{OC} = primary open-circuit inductance
L_S = leakage inductance referred to primary
C_1 = primary capacitance
C_2 = secondary capacitance
C_P = primary equivalent capacitance

FIGURE 3.16 Pulse transformer equivalent circuits: (a) front edge of pulse, (b) top of pulse, (c) trailing edge of pulse.

Pulse formation Transformers and inductors are often used in circuits that generate pulses. These include blocking oscillators, pulse modulators, and sweep generators. The transformers and inductors in these circuits perform pulse-forming, pulse-coupling, charging, and switching functions.

3.4.4 Higher-Frequency Transformers

Wideband transformers Included in this classification of transformers are the types listed in Table 3.10. The primary difference between low-frequency power transformers and wideband transformers is that the latter operate over a range of frequencies where leakage inductance and winding capacitances affect performance while these parameters have little effect in properly designed low-frequency power transformers. There is a similarity between the transformers in switching power supplies and wideband transformers— even though the switching transformers usually operate at a single frequency or over a narrow band of frequencies. The proper operation of the switching transformer is also affected by leakage inductance and winding capacitances. Pulse transformers are a special case of wideband transformer that require the passage of a broad spectrum of harmonics.

TABLE 3.10 Wideband Transformer Types

1. Audio-frequency transformers that operate from vacuum-tube voltage source. Full-power frequency response normally ranges from 20 Hz to 20 kHz.

2. Audio-frequency transformers that operate from solid-state current source. Full-power frequency response normally ranges from 20 Hz to 20 kHz.

3. Modulation transformers for amplitude-modulated transmitters. This type of transformer is usually designed for specific application and not for general use.

4. Driver transformers used to supply power to the grids of class-AB and class-B amplifier tubes.

5. Line-matching transformers that receive power from one impedance and deliver it to another impedance.

6. Load-matching autotransformers that provide impedance matching with taps.

7. Control-system transformers used in open or closed system feedback. These transformers operate on a carrier frequency modulated by an error signal. Either amplitude or phase modulation can be used.

8. Transducer service transformer that are usually step-down ratio and in which the load impedance can vary with frequency.

9. Random-noise transformers used in vibration-machine applications. These operate on random frequencies of varying amplitude and duration.

10. Ultrasonic transformers used in various communication and industrial applications, where the lowest frequency is higher than 10 kHz.

11. Carrier transformers which operate at a specific frequency and transmit intelligence by amplitude, frequency, or phase modulation.

12. Video transformers that are exceptionally wideband and can operate over part or all of the range from 10 Hz to 10 MHz.

13. Baluns are used to provide balanced-to-unbalanced impedance transformations, usually in the ratios 1:1 and 4:1. Typically, they are used to drive balanced antennas and other balanced loads.

Wideband transformers operate over a frequency range of less than 1 Hz to greater than 100 MHz. The ratio of the highest frequency to the lowest frequency in the more common audio-frequency transformers is about 2000:1. Most RF transformers operate over relatively narrow bandwidths.

In a wideband transformer, for a given source impedance and core material, the product of the bandwidth and the turns ratio is an approximation of the size. By adapting the transformer-equivalent circuit in Figure 3.1c to wideband transformers, the two circuits in Figure 3.17 for low- and high-frequency response are developed.

To obtain proper operation of wideband transformers, factors external to the transformer must be considered. These include:

- Frequency
- Source and load impedances
- The linearity of these impedances

In addition to these external factors, you can see from Table 3.1 and from the equivalent circuits in Figure 3.17 which transformer parameters affect performance.

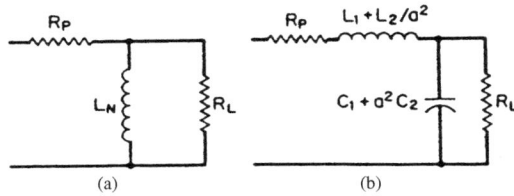

FIGURE 3.17   Wideband transformer equivalent circuits: (a) low-frequency equivalent circuit, (b) high-frequency equivalent circuit.

Impedance matching   Because wideband transformer operation usually requires the matching of source and load impedance, the ratio of the input and output voltages is usually expressed in decibels (dB), according to the expression:

$$dB = 20 \log_{10} \frac{E_1}{E_2} \tag{3.17}$$

However, to be meaningful, voltage and power must be related to a reference level. The standard reference level is 1 mW and is expressed as zero dBm. The voltage for 0 dBm across 50 Ω is $\sqrt{50 \times 0.001} = 0.224$ V.

Power level   The power level of a wideband transformer is expressed in dB as:

$$dB = 20 \log_{10} \sqrt{\frac{P_1}{P_2}} \tag{3.18}$$

A transformer whose maximum power level is expressed as 200 dBm can deliver 100 mW.

Balun transformers   The balun is a special form of wideband transformer, used to connect a balanced load to an unbalanced source. Figure 3.18 is the circuit for a balanced-unbalanced 4:1 impedance transformer: it is commonly used to match a 300-Ω twin-lead transmission line to a 75-Ω television receiver. Baluns used in electronic circuits are usually wound on low-loss, high-permeability ferrite toroids, which are intended for high-frequency operation. Although the balun is considered a wideband transformer, it is designed as a transmission line with the coils arranged so that the interwinding capacitance becomes a part of the impedance of the line. In this way, the bandwidth is not limited by the resonance of the leakage inductance and the interwinding capacitance. Leakage inductance is usually kept to a minimum by using twisted bifilar windings; this ensures close coupling and good high-frequency response. Wide-band transformers, including the balun, were covered in detail in a paper by Ruthroff (1959) (Ref. 1).

Air-core transformers   In higher-frequency applications, transformers without iron cores or with small slugs of powdered iron or ferrite are widely used to couple circuits. In transformers with magnetic cores, the exciting current is a small percentage of the total primary current. In air-core transformers, all primary current is exciting current, which induces a secondary voltage proportional to the mutual inductance.

The coefficient of coupling of the primary to the secondary, k, is the ratio of the mutual inductance to the geometric mean of the self inductance of the primary and secondary windings.

$$k = \frac{L_m}{\sqrt{L_p L_s}} \tag{3.19}$$

The value of k is never greater than unity and can be as small as 0.01 (or lower) at high frequencies.

The maximum power transfer from the primary to the secondary occurs when the reactance caused by mutual inductance equals the geometric mean of the primary and secondary circuit resistances. The primary and secondary can be tuned to appear pure resistance.

$$X_m = \sqrt{R_p R_s} \tag{3.20}$$

3.4.5 Regulating Applications

Ferroresonant transformers A ferroresonant transformer normally regulates for changes in the input voltage. The equivalent circuit of a ferroresonant transformer is shown in Figure 3.19. The inductance, L_{sat}, saturates at the low end of the range of the input voltage to be regulated. L_{sat} resonates with the capacitor, C, and the leakage inductance, L_l. Above the frequency where resonance occurs, the impedance of L_l and C in parallel decreases as the input voltage increases, resulting in an increase in the voltage drop across the leakage inductance, this provides the regulation. The average voltage across L_{sat} remains essentially constant because the flux density is equal to the saturating flux density of the core. However, the average voltage is also proportional to the frequency. Hence, the basic ferroresonant transformer provides a constant voltage only at a single frequency. Furthermore, the ferroresonant transformer functions as a conventional transformer with changes in load current, it does not regulate for changes in load current.

Although the basic ferroresonant transformer regulates for changes in input voltage only, ferroresonant voltage regulators can be produced which provide dc output voltages with regulation better than 0.5% for input voltage, load current, frequency, and temperature changes.

Magnetic amplifiers Magnetic amplifiers are saturable inductors arranged to control large amounts of power with small power inputs. Power gains in excess of 10^6 can be

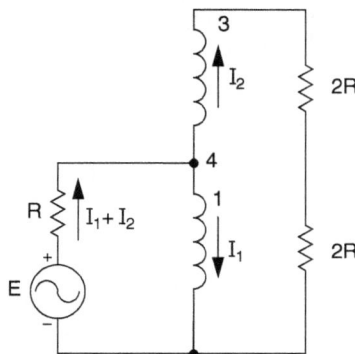

FIGURE 3.18 Balanced-unbalanced 4:1 impedance transformer, equivalent circuit.

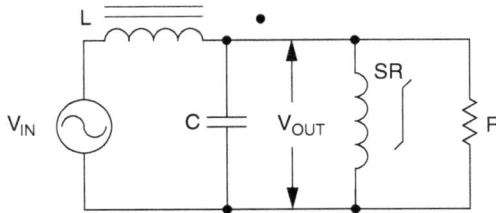

FIGURE 3.19   Ferroresonant-transformer equivalent circuit.

achieved. Feedback is often used to obtain automatic control. Magnetic amplifiers are used in industrial-control applications—especially when large power handling is involved. Their use in electronic circuits is gradually giving way to solid-state amplifiers.

3.5 INDUCTORS

Inductors are circuit elements of lumped inductance that are normally designed so that their impedance at a specified frequency or over a specified frequency range is predominately inductively reactive. They are considered with transformers because the same theory is used in design and the same general construction methods and materials are used for both. Some electronic transformers have inductance requirements and some inductors actually perform voltage, current or impedance transformations. The inductance of an iron-cored inductor may be calculated from the following expression

$$L = \frac{3.19 \, N^2 A_c \times 10^{-8}}{1_g + 1_c/\mu_\Delta} \tag{3.21}$$

where

L =   Inductance in Henries
N =   Total number of winding turns
A_c =   Net core cross-sectional area
l_g =   Length of the core gap
l_c =   Length of the core magnetic path
μ_Δ =   Incremental permeability

3.5.1 Inductor Requirements

Table 3.11 lists the electrical, environmental, and other requirements that should be specified for various iron-core inductor applications.

Inductor Q   The equivalent circuit of an inductor can be derived from that of the transformer (Figure 3.1). The parameters which apply to the inductor are R_1, L_N, and R_N. In this circuit, R_N depicts a resistance equivalent to the core losses and is in shunt with the winding as shown in Figure 3.20 and is expressed:

$$R_{SER} + jX_L = \frac{jX_L R_{SH}}{R_{SH} + jX_L} \tag{3.22}$$

The Q of the coil is the ratio of the coil reactance to ac resistance. For values of $Q > 5$,

$$R_{SH} = \frac{X_L}{R_{SER}}$$

(3.23)

TABLE 3.11 Inductor Requirements

Type	Specify
1. Linear inductor for ripple reduction	Minimum inductance Type of rectifier circuit Fundamental frequency Rectified voltage Direct current Maximum working voltage Transient voltages Maximum dc resistance Thermal and mechanical environment
2. Swinging inductor for ripple reduction	Same as type 1, except maximum and minimum inductance must be specified together with range of direct currents with correspond to inductances
3. Charging inductor	Inductance with tolerance Type of charging circuit Dc charging voltage Average current Peak current Pulse-repetition rate Dc resistance Inductor Q Thermal and mechanical environment
4. Current-limiting inductor	Inductance and tolerances Maximum alternating current Nominal alternating current Maximum voltage drop Nominal voltage drop Frequency Thermal and mechanical environment

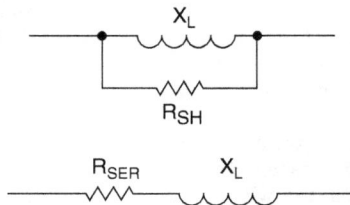

FIGURE 3.20 Equivalent inductor core-loss resistance.

Because the winding resistances are not negligible, the expression for Q becomes, with R_1 equal to winding resistance:

$$Q = \frac{X_L}{R_1 + R_{SER}} \qquad (3.24)$$

If the value of R_{SER}, obtained from Eq. 21, is substituted, the equation becomes approximately:

$$Q = \frac{1}{R_1/X_L + X_L/R_{SH}} \qquad (3.25)$$

This equation relates Q to winding resistance, open-circuit inductance, core losses, and operating frequency.

3.5.2 Saturable Inductors

Saturable inductors are iron-core inductors in which the impedance is varied over a wide range by varying the flux density in the magnetic core with a bias current in an auxiliary winding. The bias current is usually direct current and bias winding contains many turns of small magnet wire, making ampere-turn control possible with a small dc current.

Most iron-core transformers are designed to operate at a flux density in the magnetic circuit which does not reach saturation. Under this condition the open circuit impedance is virtually all inductive reactance with a very small resistance component. If operating conditions are changed so that the flux density is increased to the point of core saturation for a part of each cycle, the inductive reactance will decrease. If the core is completely saturated, the inductance becomes virtually the same as if no core were present, and the impedance becomes the resistance of the winding.

A saturable inductor is similar to a transformer with an ac winding connected in series with the line which is adequate to carry the full line current and a dc winding with which the impedance of the ac winding can be controlled over a wide range by means of dc-flux saturation of the magnetic core.

3.5.3 Air-Core Inductors

Air-core inductors, used primarily in RF applications, are available in a variety of configurations. Fixed inductors are available for surface mounting in chip and toroidal forms, as axial-lead chokes and as solenoid units for printed circuit or surface mounting. Although these devices have been identified as "air-core inductors," they often are wound on an iron-powder core—sometimes with a slug for tuning purposes. For each style of RF coil, manufacturers normally provide a range of inductances with minimum SRF, maximum R_{DC}, and maximum dc current for each inductance value. EIA 48-88 listed in Table 3.7 is an interim standard that covers "Axial Lead Fixed Radio Frequency (RF) Coils." These industrial-grade coils are intended for machine packaging, insertion, and soldering on printed wiring boards.

Fixed high-frequency inductors consist of a coil of insulated magnet wire wound on a coil form. Two-terminal wire leads are anchored to the coil form and the wire-coil leads are soldered to the terminal end wires. This internal assembly is encapsulated with a molded jacket of plastic insulating material, or is conformally coated with a protective coating, such as epoxy.

3.5.4 Shielding Beads

Shielding beads usually are small toroidal cores used to attenuate unwanted high-frequency noise or oscillations on lines in electronic and digital systems. The beads, normally a ferrite, are strung on the lines or are wired into the circuit. The resulting inductor possesses low impedance at low frequencies and high impedance over a wide band of high frequencies. Thus, the ferrite bead is a broadband lowpass filter. The effectiveness of the filter depends on the impedance of the source, the bead, and the load.

The attenuation at a given frequency can be determined from the following expression:

$$Attenuation = 20 \log \frac{Z_S + Z_{FS} + Z_L}{Z_S + Z_L}, dB \qquad (3.26)$$

where
Z_S = Source impedance
Z_L = Load impedance
Z_{FB} = Ferrite bead impedance

Z_{FB} is normally much greater than the source and load impedances in the frequency of interest. For example:

$$Z_S, Z_L = 2\Omega \text{ each}$$

$$Z_{FB} = 500\Omega$$

Then:

$$Attenuation = 20 \log \frac{2 + 2 + 500}{2 + 2} = 42 \, dB \qquad (3.27)$$

Choosing a bead   The basic objective is to choose a bead that provides high impedance at noise frequencies, but low impedance at the desired-signal frequencies. The attenuation provided by a bead can be determined from Eq. 3.26. However, because the impedances might not always be known exactly, an empirical approach might be required to determine which bead or combination of beads are needed to achieve the attenuation desired. Ferrite bead manufacturers sell kits that contain a variety of bead sizes and materials for this purpose. The magnetic properties of the beads will be affected by temperature and by flux density: the impedance of the bead will decrease at elevated temperatures or if the bead is saturated. When the temperature is reduced, the impedance will return to its initial value; however, some ferrite materials are permanently degraded if the bead is saturated.

3.6 HIGH VOLTAGES

High voltage is a relative term and is often considered to be any voltage which is a safety hazard. Is the voltage a danger to a human being? From the standpoint of transformer insulation and electrical clearances, it can be considered any voltage at which partial discharge can occur. In air at one atmosphere, this is about 250 V rms in a uniform field. However, it has been shown that partial discharge can initiate at less than 50 V (Ref. 2).

The effect of corona or partial discharge in transformers can be twofold: it might be destructive to the insulation system and it might create undesirable electromagnetic interference. Partial discharge is not a phenomenon indigenous to transformers. Any device, circuit element, wire, cable, etc., could be a source of and could be damaged by its existence.

The destructive effects of partial discharge under ac stress are accelerated as the frequency of the voltage is increased. While under dc stress, the occurrence of discharge is intermittent, the resistivity of the insulation, which dissipates the charge developed by the discharge being the limiting factor.

Flashover is arcing in gas between points of differing voltage. Creepage is flashover across the surface of insulation between points of differing voltage. Figure 3.21 provides design limits for flashover and creepage in air. Table 3.12 provides a comparison of flashover and creepage in air with those in sulfur hexafluoride, SF_6.

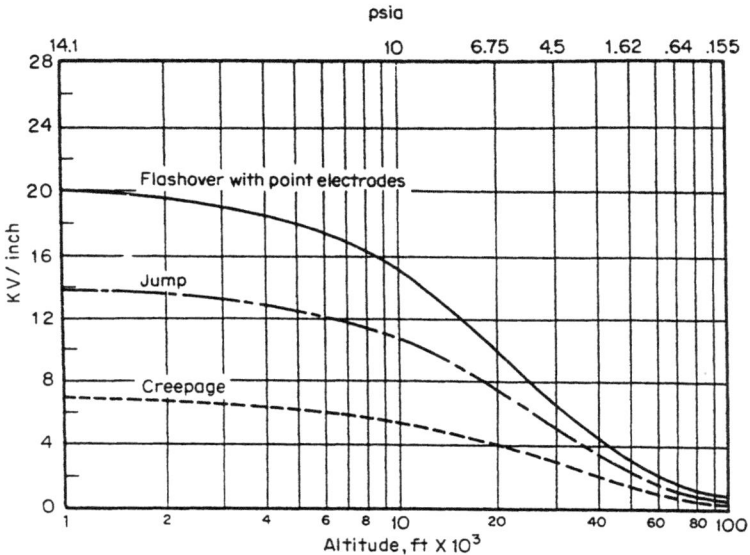

FIGURE 3.21   Crest-voltage design guide. Jump clearances and creepage over clean, dry surfaces; maximum temperature, 125°C. Data from literature sources.

The dielectric strengths of typical insulating materials are given in Table 3.6. This table also provides recommended maximum stresses for these insulating materials as they apply to transformers.

Solids typically have much higher dielectric strength than liquids or gases. The utilization of their inherent strength is usually not possible because of the difficulty in obtaining and maintaining void-free impregnation, particularly adjacent to metallic electrodes and other materials within the transformer. The dielectric strength of solids, as well as the gaseous and liquid dielectrics, decreases with increasing thickness (i.e., the strength in volts/mil decreases as the insulation thickness is increased). The inherent dielectric strength of solids can be realized only by achieving a uniform field and avoiding series gas gaps between electrodes.

Liquids are very effective for filling the spaces within the transformer, which are not filled with solid insulation or other materials. When the transformer container is properly filled, liquids effectively impregnate the porous and permeable insulating materials. This increases the dielectric strength and corona inception voltage of the system. The presence of gas bubbles, entrapped air, in liquids reduces the dielectric strength drastically; liquids

TABLE 3.12 Flashover and Creepage in Air and SF$_6$

1. Creepage is a function of the dielectric constant of the insulation; the higher the dielectric constant, the lower the voltage to flashover. Creepage has the least effect (as compared with flashover) when the dielectric constant is low and when the surface is parallel to the electric field

2. Small series of air gaps (cracks or surface irregularities) reduce the ac flashover in air, but have little effect in SF$_6$ below 2 atm

3. The ratio of impulse (unidirectional pulse) creep over glass in SF$_6$ to air is approximately 2.00

4. The ratio of ac creep over glass in SF$_6$ to air at 1 atm ≈2.25; at 2 atm ≈ 2.75

5. Impulse ratio (unidirectional pulses to ac crest) in air and in SF$_6$ (approximately—varies with spacing):

	Air		SF$_6$	
	1 atm	2 atm	1 atm	2 atm
Flashover	1.19–1.45	1.20	1.2–1.4	1.1
Creep (glass)	1.21–1.65	1.16–1.57	1.0–1.28	0.99

saturated with gas will break down at lower stresses when subjected to lower pressure. The container must be designed to accommodate changes in fluid volume with changes in temperature so that a partial vacuum will not exist when the temperature is lowered.

Gases are normally used as the primary dielectric medium around barrier insulation and the structural parts of the transformer. Consequently, dielectric strength is their most important property. The breakdown voltage varies as the product of the gas density (pressure) and the electrode spacing as shown in the Paschen's curves in Figure 3.22. The gradient, in volts per mil, decreases with increasing spacing. All gases exhibit a minimum voltage below which breakdown will not occur in a uniform field. At pressure-spacing values below the Paschen's minimum, larger gaps will break down at lower voltages than do smaller gaps; this must be considered in very low pressure applications.

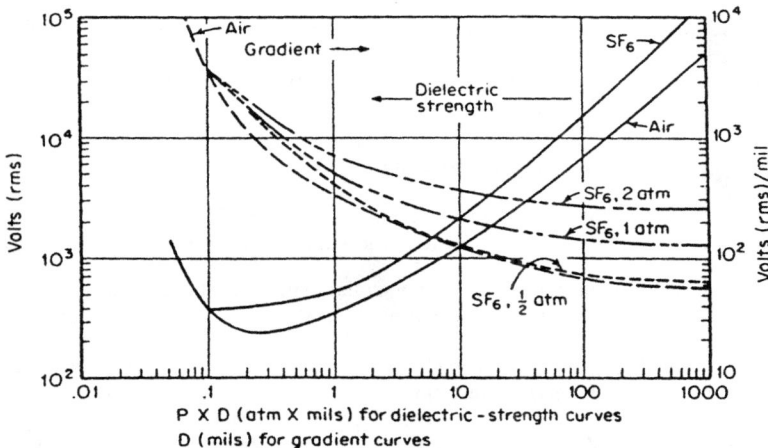

FIGURE 3.22 Paschen's-law curve for air and SF$_6$; at 25°C.

3.6.1 High-Voltage Magnetic Devices

High-voltage electronic circuits might require a variety of magnetic devices. A transformer can be the source of the high voltage or it can isolate high-voltage circuitry from lower-voltage circuits or ground. It might even be required to couple high voltages at different impedance levels. Regardless of the circuit function, all must withstand ac, dc, pulse, or transient voltages individually or in combination.

Power transformer   A power transformer is usually the source of the high voltage in electronic circuits; transforming a low voltage to a high ac voltage, which can then be rectified, filtered and regulated to a usable dc voltage. The transformer can be operated at the nominal system frequency—either single or polyphase, or it can be operated at a higher frequency that has been generated by an inverter circuit. The power transformer must be designed to withstand the high induced voltages: the high dc voltage, which is determined by the rectifier circuit, and various high peak voltages, which might be generated during system operation. A three-phase step-up transformer is usually connected in a delta-wye configuration to reduce the induced winding voltages.

Filter inductor   An inductor that filters a high dc voltage can be connected in two ways. In the first, the winding and the core are isolated from ground. In the second, the winding must provide isolation between the winding and ground. Each connection requires a different method of insulation.

In the first case, one end of the coil is connected to the core so that the core and the mounting bracket are always at the same voltage as the coil. Normally, the voltage across the coil is that developed by the resistive drop of the inductor. However, under fault or switching conditions, the load end of the winding might be instantaneously grounded and, because current through the coil cannot change instantaneously, at least the full dc voltage will be placed on the input of the inductor. This voltage does not distribute equally across the coil and it might require special winding and insulating methods.

In the second case, the core is grounded and the coil must be insulated to withstand the full dc voltage to ground. The winding must also be designed and insulated to withstand transient voltages during power supply turn-on, turn-off, and fault conditions.

Voltage isolation   Frequently, low-voltage ac circuits must operate at a high dc voltage. A transformer is commonly used to provide the electrical isolation between the high voltage and ground, often with very low capacitance to the ground and the primary.

3.6.2 High-Voltage Insulation Systems

Some considerations in selecting an insulation system for a high-voltage transformer and in applying a high-voltage transformer to an electronic circuit are:

1. The shape of elements that are at high potential, with respect to ground, and other elements on the breakdown voltage of insulation between these points. The Paschen's curve in Figure 3.22 is for uniform field. A utilization factor of 0.33 is sometimes used for point electrodes, but as indicated in Figure 3.23, at small spacings the utilization factor approaches unity. The effect of electrode shape is most pronounced in gases, although it is obvious that stress concentrations will occur at point or sharp-edge electrodes in liquids and solids. In this connection, remember that a small wire is a sharp edge.

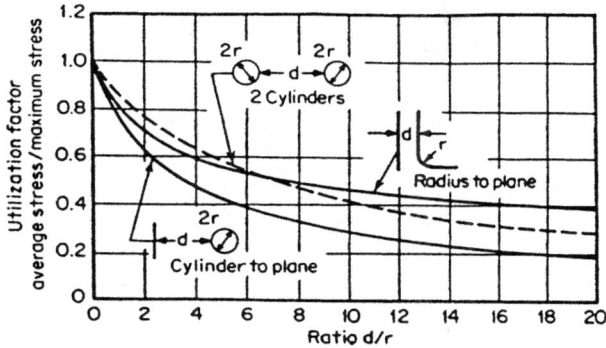

FIGURE 3.23 Utilization factor as a function of spacing-to-radius ratio.

2. When more than one insulating material is used in series between circuit elements of ac potential difference, the stress in each is inversely proportional to dielectric constant ε of each material. Thus, the insulation with the lowest ε has the highest stress. Often this is air or gas, which usually has the lowest dielectric strength.

When the voltage is dc, the voltage drop is simply *IR*. Therefore, the material with the highest resistance will have the highest stress. Don't forget that the total resistance is composed of both volume and surface resistances.

When the stress is unidirectional pulses, an ac-voltage component is present that could be the source of corona. Even a large ripple on dc voltages might produce ac stresses sufficient to require investigation; it might be a corona source.

With unidirectional pulses above the corona-inception level, a charge is deposited on the surface of the dielectric. These will increase the next pulse-breakdown level. With ac voltage, the effect is reversed (i.e., the corona-inception voltage will be lowered).

3. Great care must be taken to ensure that liquid and gaseous dielectrics are kept free of contaminants.

4. In processing rigid or flexible plastic dielectric materials, great care must be taken to ensure that no voids, cracks, or physical interfaces exist at locales of high stress, above the Paschen's minimum for air, within the plastic castings. This is very difficult to accomplish!

5. When possible, electrical stresses within a transformer, between layers, and between turns should be kept below the Paschen's minimum for air.

Voltage stresses on insulations in series Usually in high-voltage transformers, several different insulating materials are required. Often, more than one of these materials intervene between electrodes of differing electrical potential. The manner in which the different insulations share the electrical stress is determined by the voltage across the insulation. If the voltage is ac, the stress distribution is capacitive; if it is dc, it is resistive[3].

Ac voltage stress When more than one insulation intervenes between electrodes at different potentials, the electric stress in each is proportional to the ratio of the insulation thickness to the dielectric constant. Thus, the insulation with the lowest ε has the highest stress if all insulation thicknesses are the same. This is often air or an insulating gas, which would probably have the lowest dielectric constant of the several insulations.

The gradient (electric stress) on each insulation element in series between two electrodes of different potential can be calculated by:

$$G_x = \frac{E}{\varepsilon_x(T_1/\varepsilon_1 + T_2/\varepsilon_2 + \ldots + T_n/\varepsilon_n)} \tag{3.28}$$

where

E = Total voltage across all insulation
G_x = Gradient (electric stress) in V/mil
T = Thickness in mils
ε = Dielectric constant
Subscript $x = 1, 2, \ldots, n$

Dc voltage stress When more than one insulation intervenes between electrodes at a different dc potential, the dc voltage stress in each is proportional to the resistivity-thickness products (provided that the electric field passes through the same area of each). Thus, the insulation with the highest ρ will have the highest stress, all other things being equal.

The gradient (dc stress) on each insulation element in series can be calculated by:

$$G_x = \frac{\rho_x E}{(T_1\rho_1 + T_2\rho_2 + \ldots + T_n\rho_n)} \tag{3.29}$$

where

E = Total voltage across all insulation
G_x = Gradient (electric stress) in V/mil
T = Insulation thickness in mils
ρ = Volume resistivity
Subscript $x = 1, 2, \ldots, n$

3.7 SHIELDING

Shielding in transformers can be of two types, electrostatic and electromagnetic.

3.7.1 Electrostatic Shielding

Electrostatic shielding prevents or reduces the transfer of voltage (noise) through interwinding capacitances. It is usually required to isolate the power-input circuit to the transformer from transient voltages or high-frequency noise, which might appear on the transformer secondary windings. An electrostatic shield is usually placed between the primary and the secondary windings, with the shield being suitably grounded.

3.7.2 Electromagnetic Shielding

Electromagnetic shielding can be of two types, that which is required to attenuate the magnetic field formed as a part of the power transformer magnetic field and that which is required to shield a low-level transformer from magnetic fields created by power transformers or other circuit elements. For instance, these fields can induce small voltages in amplifier input transformers, which are then, in turn, amplified by high-gain circuitry. Because the main purpose of shielding is to reduce the effect of stray magnetic fields, several techniques can be used.

1. Separate the input to the high-gain circuit from the power transformer as much as possible. This is most effective because the field varies as the reciprocal of the distance cubed.
2. Use the core-type construction in the power transformer. This is most effective in single-phase transformers because the leakage fluxes generated in the two legs effectively cancel at a distance from the transformer. The three-phase transformer wound on the E-type core generates a third harmonic field because the third harmonic flux cannot circulate in the core.
3. Orient the power transformer and/or the input circuitry for minimum pickup. The axis of the input transformer coil should be 90° to the stray field.
4. Provide magnetic shielding around the component or circuitry being affected by the stray field magnetic field.

Placing the shield around the power transformer usually is not too effective because many of the lines of the flux that originate at the transformer would be perpendicular to the shield and would not be attenuated.

Shielding usually consists of ferromagnetic material or layers of high-permeability magnetic material normally interleaved with sheets of nonmagnetic material, such as copper.

Cooling All transformers have power losses that raise the temperature of the various parts until reaching the point of thermal equilibrium (where the losses generated equal the power dissipated).

The normal transformer is manufactured from many different materials, most of which have different coefficients of thermal conductivity. Generally, the materials used with the best thermal conductivity also generate the bulk of the losses, although the electrical insulators are usually thermal insulators. Transformer cooling is usually accomplished by providing a combination of solid insulation (varnishes or solventless-resin systems) with a mechanical structure that will permit the most efficient conduction of the winding and core losses to the surface of the transformer (if it is to be cooled by convection) or to a mounting surface (if it is to be cooled by conduction).

There are approximations that permit a reasonably accurate determination of the temperature rise of transformers cooled by convection. Figure 3.24, which is based on these

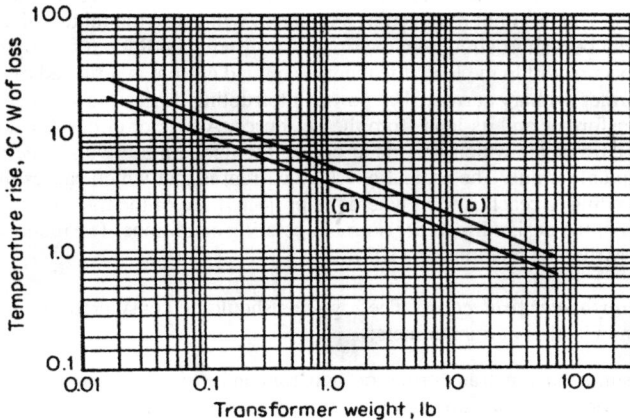

FIGURE 3.24 Temperature rise of transformers with convection cooling: (a) 70°C rise at 95°C ambient, (b) 50°C rise at 20°C ambient.

approximations, relates temperature rise to transformer weight and loss per unit weight. Table 3.6 lists the hottest-spot temperature limitation for various insulation materials.

The determination of temperature rise in transformers cooled by conduction only is more complex and should be determined by more rigorous thermal analyses.

Transformer reliability   The concept of reliability in electronic components is based on the establishment of failure rates for these components as a number of failures during a finite number of operating hours. Magnetic devices are considered to be reasonably reliable; however, to realize this reliability, the part must be properly designed, manufactured, and applied.

Failure occurs in a transformer when the materials used are overstressed electrically, mechanically or thermally, and individually or in combination. This overstress can be caused by improper design, misapplication of the part, improper specification of requirements, improper manufacture, mishandling, defective materials or the failure of related circuit elements.

The end of life should not be confused with reliability; electronic transformers that are designed properly, manufactured properly, and which are properly applied, will have an anticipated life (covered in Section 3.3.1).

Power-transformer applications and protection   Most power transformers and inductors are operated from low-impedance sources. These sources will continue to supply power to the transformers until the circuit is opened by either the opening of a switch or the operation of some protective device.

Usually a transformer fails because it is subjected to localized or general overheating. The conditions that usually cause the device to overheat are component failures or other conditions, which place a short circuit or overload on the transformer or inductor. There are some instances where dielectric failure occurs within the transformer or inductor, but these usually happen in a device that is improperly designed for the application or that is improperly applied.

The most common approach to protecting transformers is to prevent the overheating of the insulation. This can be done either by removing the source of power to the fault or by isolating the fault itself. A fault can be any condition of operation that imposes a current, voltage, or temperature in excess of the safe limit for the parts or circuitry that is affected by the fault and which, if not removed or prevented, will cause the failure of a part or circuit. A fault can be a short circuit, an overload, a transient condition, or an open circuit.

If a fault occurs within the load, the increase in load current is reflected by an increase in the transformer primary current. The power-interrupting device must be capable of isolating the transformer and its load from the ac source. A fuse or circuit breaker must respond to the current for which it is rated before the temperature within the transformer exceeds acceptable limits. The critical overload current is that which causes the temperature of the winding to reach a point where the projected life of the transformer is reduced to one-fiftieth of the design life under normal operating conditions. The reduction in life is based on the Arrhenius relationship (10° rule).

Guides to transformer protection   These guides should be considered by the circuit designer in the application of transformers:

1. Do not assume that any transformer or inductor can be operated under conditions other than those for which they are designed. Safety factors are not normally designed into transformers used in electronic equipment.

2. Do not assume that any transformer or inductor can operate in any environment other than that for which it was designed. A transformer designed to operate in air should not

be subsequently embedded in a plastic resin—even though the transformer losses are in the milliwatt range.

3. Do not assume that a fuse or circuit breaker in the primary of a single-phase transformer will operate for all types of faults on the transformer secondary or within the transformer. The impedances of the transformer windings and the related circuitry could limit currents to a level that might permit the transformer to overheat, but not cause a fuse to open or circuit breaker to function.

4. Do not assume that fuses in all three input lines to a three-phase transformer will isolate the transformer if a fault occurs in the load or within the transformer. It is likely that a three-phase transformer will operate in a single-phase mode for some time if only one input line is opened.

5. Do not assume that a "short-circuit"-proof circuit will prevent a transformer from overheating. The overload current might be below the cutoff threshold of the protective circuit.

6. Consider all plastic materials as potentially hazardous—even though they pass requisite military, ASTM, or other specification, with respect to flammability, dielectric strength, and arc resistance.

7. Perform a careful electrical-stress analysis on all insulation systems used. Remember that the distribution of stress within a system that utilizes several insulations with different dielectric constants will vary with the dielectric constants and the thickness of the materials. It is possible that although the average stress between two points of different electrical potential is well within maximum limits, the actual stress on the lower dielectric-strength material is much greater than is safe.

8. Perform a careful thermal analysis of all embedded assemblies, subassemblies, or parts, particularly those that contain heat-dissipative elements. Most materials that are good electrical insulators are also good thermal insulators.

3.8 REFERENCES

1. Ruthroff, C. L. 1959. "Some Broad-Band Transformers." *Proc. IRE*, 47, 1337–1342 (August).
2. Dakin, T. W., and Berg, O. 1962. "Theory of Gas Breakdown." *Progress in Dielectrics-4*, Heywood and Company Ltd., London.
3. Dakin, T. W. 1986. "Insulation Reliability-5." *IEEE Electrical Insulation Magazine*, Vol. 2. No. 4, p. 54, July.

CHAPTER 4
RELAYS AND SWITCHES

Earle R. Wright
Quality Improvement Manager, Leach International, Inc.

4.1 INTRODUCTION

An electrical switch is simply a device for making, breaking, or changing the connections in an electric circuit. A relay is an electrically operated switch. This definition is nonrestrictive enough to embrace both solid-state (semiconductor) relays and electromagnetic or electromechanical and hybrid types.

4.2 RELAYS

4.2.1 Definitions of a Relay

The National Association of Relay Manufacturers (NARM) defines a *relay* as an electrically controlled device that opens or closes electrical contacts to affect the operation of other devices in the same or another electric circuit. The reference to contacts obviously restricts this definition to only the most common form, the electromechanical relay. Webster's definition, "an electromagnetic device . . . actuated by a variation in conditions of an electric circuit and that operates in turn other devices (as switches, circuit breakers) in the same or a different circuit," can be made to suit today's needs just by replacing the word *electromagnetic* with *electric*.

A solid-state relay (SSR) is a device without any moving parts that performs a relaying or electrical-switching function. It uses semiconductors in both input and output electrical circuits to perform essentially the switching function normal to the simple electromagnetic relay. A hybrid relay is an electrical device or unit having a solid-state input and electromagnetic output, or vice versa, to perform an electrical-switching function.

Solid-state relays and hybrid relays are treated as special forms in this chapter.

The electromagnetic/electromechanical relay is still the form most often encountered, the simplest and most readily understood, and it demonstrates best the basic switching principles and problems. Unless otherwise noted, it is the type of relay covered here.

FIGURE 4.1   A common type of relay having a clapper armature and heavy-duty contacts. *(Magnecraft Electric Company.)*

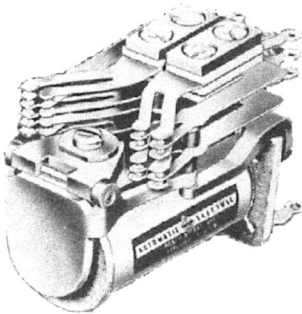

FIGURE 4.2   A long-frame telephone-type relay with medium-duty contacts. *(GTE Automatic Electric Company, Inc.)*

FIGURE 4.3   A dry-reed contact type of relay. *(C. P. Clare & Co.)*

Thermal relays, solid-state relays, hybrid relays, and such are treated separately later on and are clearly identified. Figures 4.1 through 4.4 illustrate some of the most common relay types.

Relays must be considered by the engineer as a basic component and not an accessory. The kind, overall shape, method of mounting, and aesthetics are somewhat a matter of choice, but usually the coil characteristics, kind and amount of power required, overall size, nature of the contacts required to handle the known load, etc., are largely dictated by the limitations of the job to be done. For example, it is unrealistic to expect a microminiature relay to handle a heavy-duty load requiring 25-A contacts.

Relay terms   As with many other electrical devices, custom (as much as usage) has resulted in the terms commonly applied to relays. Many are confusing and some are contradictory as commonly used. To some extent relay manufacturers have created and used terms peculiar to their own product, sometimes at variance or in conflict with those used by another manufacturer. The terminology common to one field of application might bear little resemblance to that of another field. In an attempt to improve this situation, The National Association of Relay Manufacturers (NARM) and American National Standards Institute (ANSI) agree upon some standardized definitions, which are listed in the Glossary. Use of these terms is helpful when discussing relays and relay problems with vendors and other engineers.

Relay classifications   Before a meaningful discussion of relays can be carried on with the vendor, broad classification is necessary. Good system design using relays necessitates a consideration of relay characteristics from the output-requirements end forward. That is to say, to ensure that the relay is large enough, sensitive enough, rugged enough, etc., the nature and requirements of the job the relay has to do must be first established and a choice made, working from the output or load-handling end forward to the input requirements. All too often, the tendency is to practically finish the design, with relays hung on almost as an afterthought. This can nullify an otherwise good design and lead to unjust criticism of relays in service.

Relays classified by output

- Low-wattage dc load
- Medium-wattage dc load
- High-wattage ("power") dc load
- Low-wattage ac load
- Medium-wattage ac load

FIGURE 4.4 Examples of a solid-state relay. *(Hamlin, Inc.)*

- High-wattage ("power") ac load
- Specialized loads (e.g., coaxial switching of high-frequency power)
- Specialized contacts (e.g., sealed contacts or solid-state switching)

Relays classified by input

- Direct current: neutral, polarized, or thermal
- Alternating current: commercial power or other low-frequency sources, frequency sensitive (e.g., tuned-resonant reed), or thermal

Relays classified by duty rating

- Contact performance: Because much of a relay's size, shape, method of terminating, method of mounting, coil power requirements, protection from ambients, etc., depends on the nature of the contact load, it is important that this matter receive early attention. Large electrical-contact loads usually require a larger relay of a particular type than small loads. This can be a factor in the method of mounting, as well as terminating. It is therefore essential that the vendor be made aware of all pertinent data regarding contact service requirements initially.
- Service life: Another early factor classifying the relay is service life. If the need is for longevity with a lot of operations, a certain relay type can be precluded from favorable consideration in favor of a type that is capable of providing the required life. On the other hand, if other factors are ruling, such as the need for small size or resistance to shock and vibration (as in airborne service), for example, then this needs to be recognized early so that a proper choice will be made.

Relays classified by custom and usage Relays are sometimes most importantly classified by one or more of the following:

- Commercial
- Industrial
- Military
- Communications
- Railway

Customer or user acceptance, training and experience of service personnel, field-operating extreme conditions, or safety might dictate that relays of a certain type, as specifically associated with one of the previous classifications, be chosen. An industrial application, for example, might dictate the use of a relay-type meeting National Electrical Manufacturers Association (NEMA) standards, and such a relay might be totally unsuited to a military or railway application.

Relays classified by performance Relays are classified by what they can do and are required to do. Three broad classifications are:

- General-purpose
- Special-purpose
- Definite-purpose

Specific-performance classifications are these:

- Marginal (with respect to pickup and/or dropout current)
- Timing (either or both allowable operate and/or release times, operate and/or release delays)
- Sensitivity (capability to recognize narrow pickup and/or dropout currents, or the ability to operate on much less than normal current)
- Latching
- Sequencing
- Frequency-sensitive
- Thermal-response
- Stepping

4.2.2 Relay Performance

Electromechanisms do not switch instantaneously. Relay performance presents a series of sequential events on both energization and deenergization. These events are identified as to sequence and defined in Figures 4.5 and 4.6. A look at these figures, when associated with a study of the most common terms applied to the relay from Table 4.1, provides an understanding of basic relay performance. The relay terms of Table 4.1 are defined in detail in the Glossary. Figure 4.7 is helpful in placing the events of Figures 4.5 and 4.6 in their proper time frame.

4.2.3 Relay Construction

All electromechanical relays consist of at least three basic elements: the actuating coil, linkage to transform coil energization into output, and a change of output conditions because of coil energization (switching).

It is customary to picture relay coil and contact details by symbols. Basic contact symbols are presented in Figures 4.8 and 4.9 and coil symbols in Figure 4.10.

In Figure 4.8, the heavy arrow indicates the direction of operation. The armature contact spring (indicated by the long spring in each example) moves downward. In forms D and I, some electrical discontinuity might be caused by contact chatter. The symbols are taken from ANSI C83.16-1959, Y32.2-1962, and Y32.2a-1964. When abbreviations in Figure 4.8 are used to designate a contact assembly, the following order is used:

1. poles
2. throws
3. normal position
4. double make or break (if applicable). Example: SPST NO DM refers to single-pole sin-glethrow, normally open, double-make contacts.

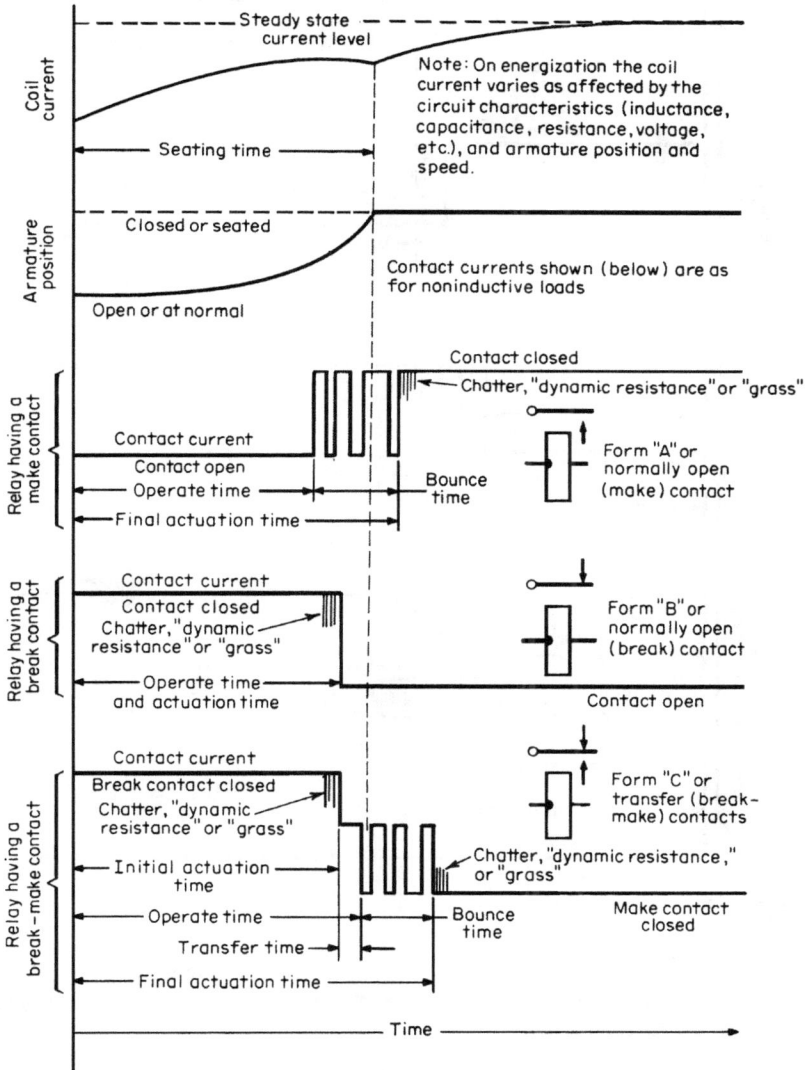

FIGURE 4.5 Time traces typical of relay pickup. [1]

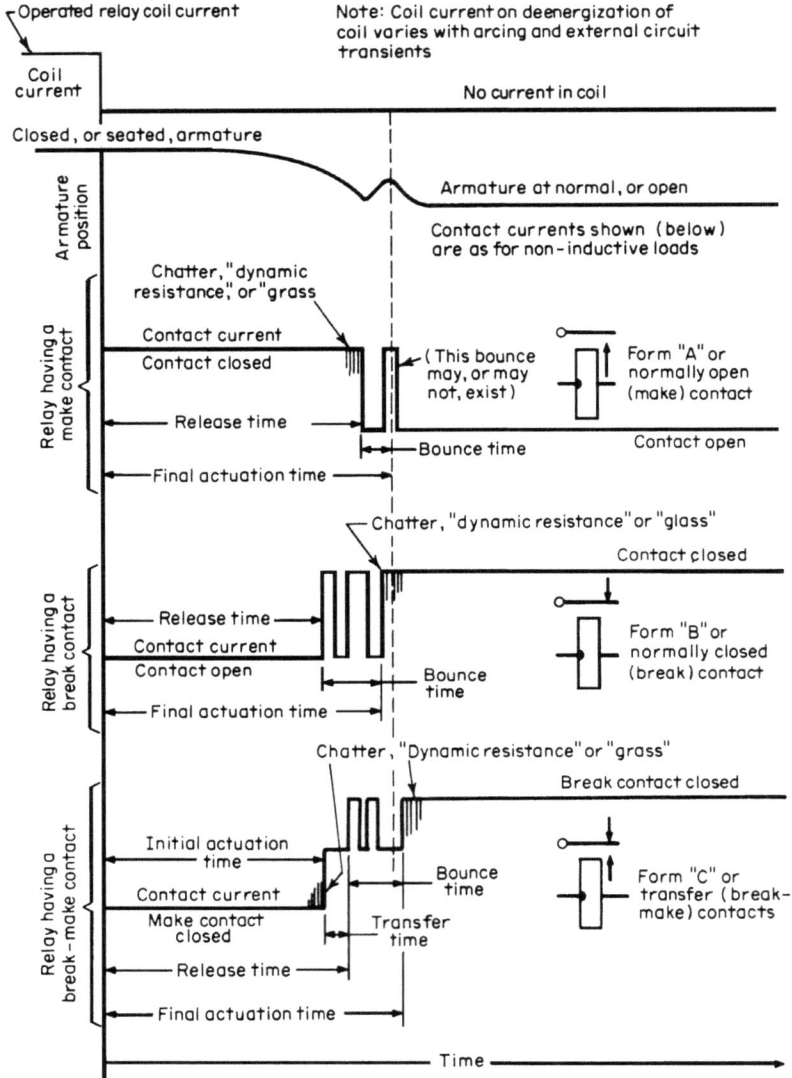

FIGURE 4.6 Time traces typical of relay dropout.[1]

It is significant in the symbolic presentation of the contact combinations in Figure 4.8 that, although Form A comes before Form B alphabetically, in a normal relay contact assembly, the closed contacts are closer to the armature than the open contacts (Figure 4.11b). This prevents any armature spring tension from going to waste by keeping the back contacts closed with as much pressure as possible. Thus, an order calling for a relay having 1A, 2B, 1C contact combinations will usually be arranged in the order of 2B, 1C, 1A, unless otherwise specified by the purchaser, and for a good reason. If an "early make" is

TABLE 4.1 Most Commonly Used Terms Relating to Relay Performances[1]

Preferred	Not preferred
Hold, measured	Nondropout, measured
	Nonrelease, measured
Hold, specified	Maximum dropout
	Nondropout, specified
	Nonrelease, specified
Nonpickup, measured	Nonoperate, measured
Nonpickup, specified	Minimum pickup
Pickup, measured	Operate, measured
	Pull-in (or pull-on) value, measured
	Operate value, just
Pickup, specified	Operate, specified
	Pull-in (or pull-on) value, specified
	Operate value, must
	Maximum pickup
Operate time	Pickup (or pull-in) time
Dropout, measured	Release, measured
Dropout, specified	Release, specified
	Minimum dropout
Release time	Dropout (or drop away) time
Transfer time	

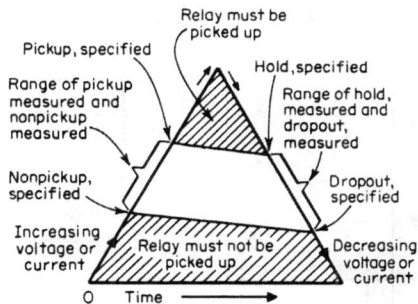

FIGURE 4.7 Relationship of relay performance to definitions.[1]

required, it must be specified (as in Figure 4.11c, where the X associated with a make combination indicates that the circuit requires one A combination to be preliminary). In Figure 4.11c, the contacts are drawn in vertical alignment with the coil symbol. Movable, armature, or lever contacts are drawn as if attracted to the coil on energization.

The contacts are numbered in sequence from the mounting surface outward; No. 1 is closest to the mounting and No. 10 is the farthest away. X contacts are preliminary and operate first. Y contacts are break contacts and operate last. As make contacts normally operate last, they are not so identified. Battery and ground symbols are shown connected to the coil (negative battery to inside terminal, positive battery grounded) to minimize electrolysis in permanent installations. Switching ground is, by telephone practice, at 48 Vdc, nominal. This is not permitted above 50 Vdc by National Electrical Code.

Form	Description	USASI Symbol
A	Make or SPSTNO	
B	Break or SPSTNC	
C	Break, Make, or SPDT (B-M), or Transfer	
D	Make, Break or Make-Before-Break, or SPDT (M-B), or "Continuity transfer"	
E	Break, Make, Break, or Break-Make-Before-Break, or SPDT (B-M-B)	
F	Make, Make , or SPST (M-M)	
G	Break, Break or SPST (B-B)	
H	Break, Break, Make, or SPDT (B-B-M)	
I	Make, Break, Make, or SPDT (M-B-M)	
J	Make, Make, Break, or SPDT (M-M-B)	
K	Single pole, Double throw Center off, or SPDTNO	

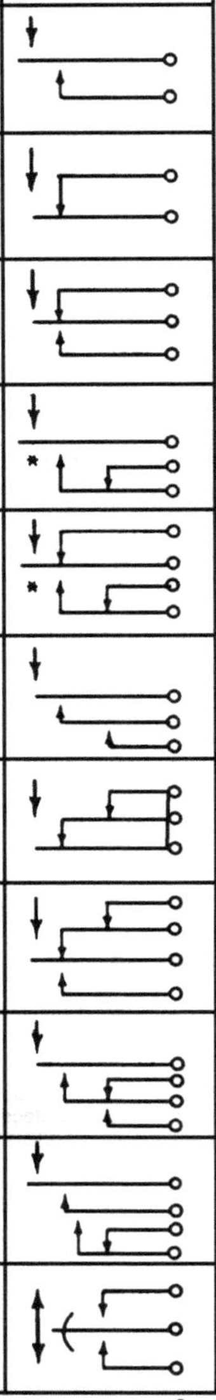

Form	Description	USASI Symbol
L	Break, Make, Make, or SPDT (B-M-M)	
M	Single pole, Double throw, Closed Neutral. or SP DT NC (This is peculiar to MIL-SPECS.)	
U	Double make, Contact on Arm., or SP ST NO DM	
V	Double break, Contact on Arm., or SP ST NC DB	
W	Double break, Double make, Contact on Arm., or ST DT NC-NO (DB-DM)	
X*	Double make or SP ST NO DM	
Y**	Double break or SP ST NC DB	
Z	Double break, Double make , or SP DT NC-NO (DB-DM)	

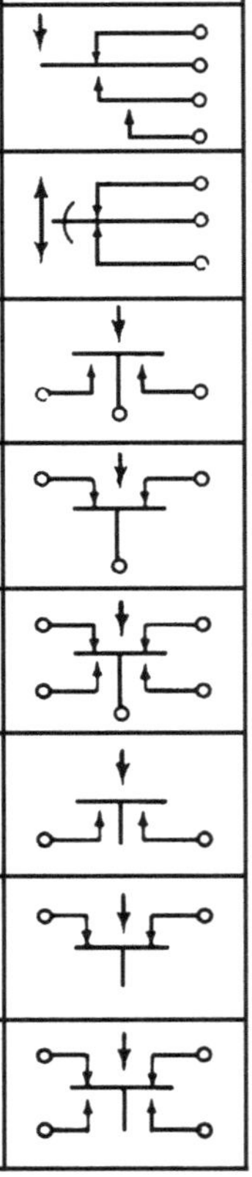

* Not to be confused with preliminary ("X") make
** Not to be confused with a late ("Y") break

Special A	Timed close	T.C. OR
Special B	Timed open	T.O. OR

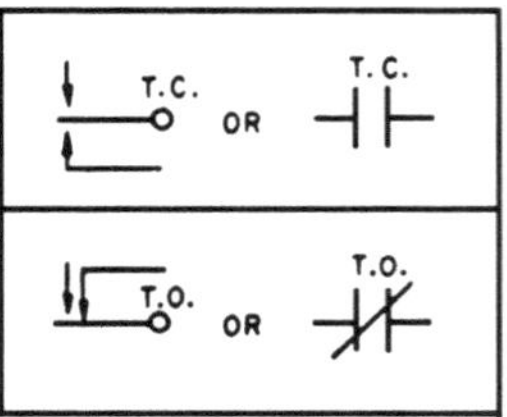

Multi-point selector switch   — or —

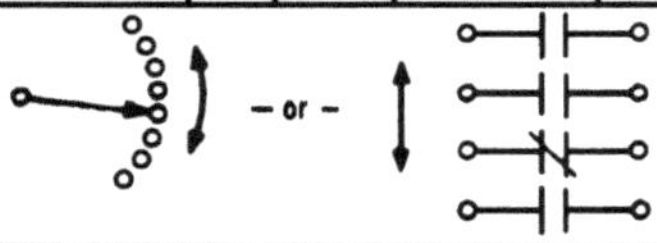

FIGURE 4.8 Symbols for relay-contact combinations established by American National Standards Institute *(ANSI)*.[1]

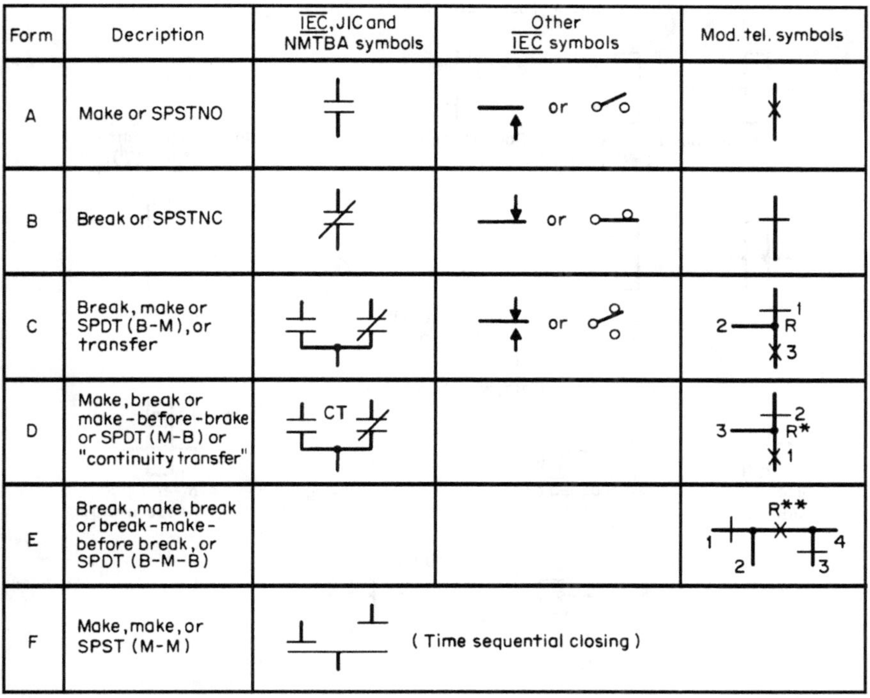

Form	Decription	IEC, JIC and NMTBA symbols	Other IEC symbols	Mod. tel. symbols
A	Make or SPSTNO	(symbol)	(symbol) or (symbol)	(symbol)
B	Break or SPSTNC	(symbol)	(symbol) or (symbol)	(symbol)
C	Break, make or SPDT (B-M), or transfer	(symbol)	(symbol) or (symbol)	(symbol)
D	Make, break or make-before-brake or SPDT (M-B) or "continuity transfer"	CT (symbol)		(symbol)
E	Break, make, break or break-make-before break, or SPDT (B-M-B)			(symbol)
F	Make, make, or SPST (M-M)	(symbol) (Time sequential closing)		

FIGURE 4.9 Alternative symbols for relay-contact combinations. Sources of symbols: IEC (International Electrotechnical Commission), JIC (Joint Industry Conference Electrical Standards for Industrial Equipment), NMTBA (National Machine Tool Builders Association, Electrical Standards); Mod. Tel. (Modern Telephone practice). CT indicates continuity transfer, one asterisk denotes make before break, and two asterisks denote break make before break.[1]

It is notable that the 2B, 1C, 1A contact-assembly designation appears to be simpler and less likely to be misunderstood than the equivalent "one DPSTNC, one SPDT, and one SPSTNO."

The association of coil and contact symbols in a simple relay as frequently seen is covered in Figure 4.11a. There is ordinarily no picturing of the linkage between actuating coil and contacts, but the assumption is made that the movable contact is attracted toward the coil on energization, returning to the pictured position on deenergization, unless otherwise stated.

Joint Industry Conference and National Machine Tool Builders Association have adopted relay coil and contact symbols as seen in Figure 4.11d. Symbols for thermal relays, covered in a later section, are pictured in Figure 4.11e. Symbols used in motor-control circuits (JIC/NMTBA) are pictured in Figure 4.11f. An explanation of the method of operation for this slightly more complicated diagram follows:

Two-wire control is generally thought of in relation to a pilot device, such as a thermostat or pressure switch, or to a simple, maintained SPDT toggle or pushbutton switch. As the term implies, these devices require the use of only two wires between the control unit and the starter. The device is connected in series with the main contactor coil of the starter, and the opening or closing of the pilot device directly controls the deenergizing or ener-

Common, fast-acting, neutral types

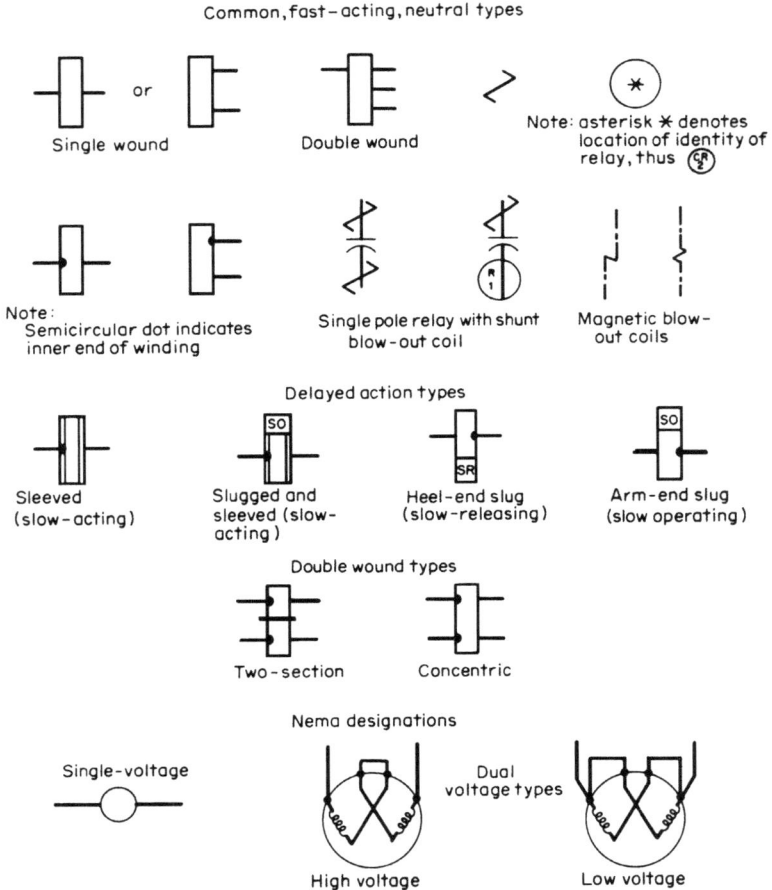

Single wound

Double wound

Note: asterisk ✳ denotes location of identity of relay, thus

Note:
  Semicircular dot indicates inner end of winding

Single pole relay with shunt blow-out coil

Magnetic blow-out coils

Delayed action types

Sleeved (slow-acting)

Slugged and sleeved (slow-acting)

Heel-end slug (slow-releasing)

Arm-end slug (slow operating)

Double wound types

Two-section

Concentric

Nema designations

Single-voltage

Dual voltage types

High voltage

Low voltage

FIGURE 4.10   Commonly used symbols for relay coils.[1]

gizing of the starter. The major feature of a two-wire control system is the low-voltage release. The starter drops in the event of a power failure, but operates automatically when the power is restored.

In three-wire control systems, the main contactor coil of the starter is wired in series with its own no auxiliary contacts. The start-stop push-button station, which requires the use of three wires between the control and the starter, is connected in parallel with the coil. In the event of a power failure, the starter will drop out and remain deenergized until the start button is pressed. Because the starter drops out when there is a power failure and will not operate again until the start button is pressed, this control system provides low-voltage protection.

Two of the most common forms of the relay are demonstrated by Figures 4.12 and 4.13. The linkage that converts coil energization to contact switching is obvious. The force acting on the armature when the coil is energized is directly applied to cause contact switching, and its effectiveness is somewhat proportional to the degree of coil energization.

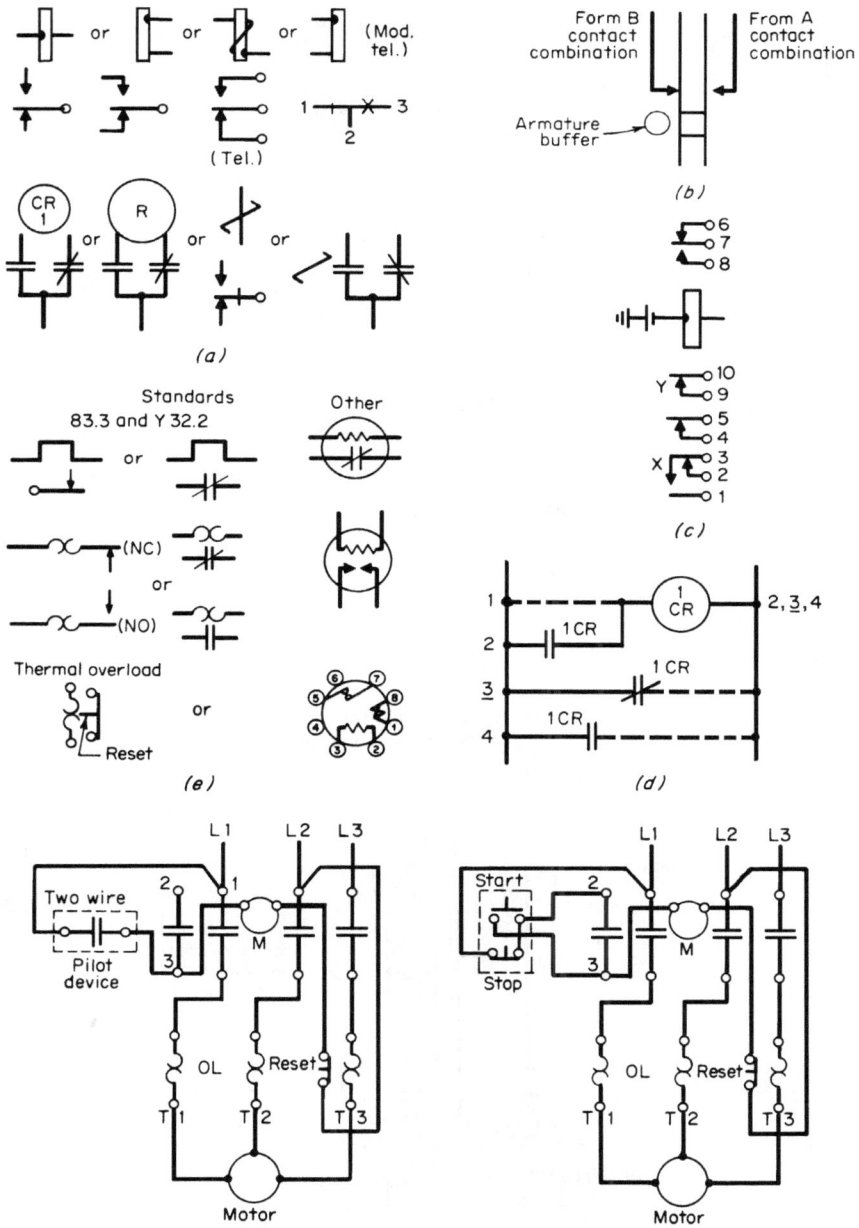

FIGURE 4.11 Common symbols used for relay coils and contacts in communications and general systems. (a) Common symbols used to show a relay with a form C contact combination. (b) The preferred order of contact arrangement in a simple relay pileup. (c) A relay with a large contact-spring pileup showing preferred order of arrangement. (d) Location of contact symbols with respect to coil on JIC/NMTBA simplified diagrams. Sequence of numbers at right of coil symbol locate associated contacts in lines numbered in left column. An underscored number signifies a normally closed contact. (e) Symbols for thermal relays. (f) Symbols used in motor-control relay circuits (JIC/NMTBA); 1. two-wire control; 2. three-wire control.[1]

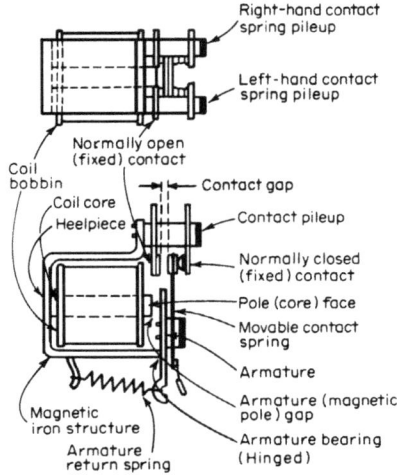

FIGURE 4.12 A relay commonly referred to as general purpose.[1]

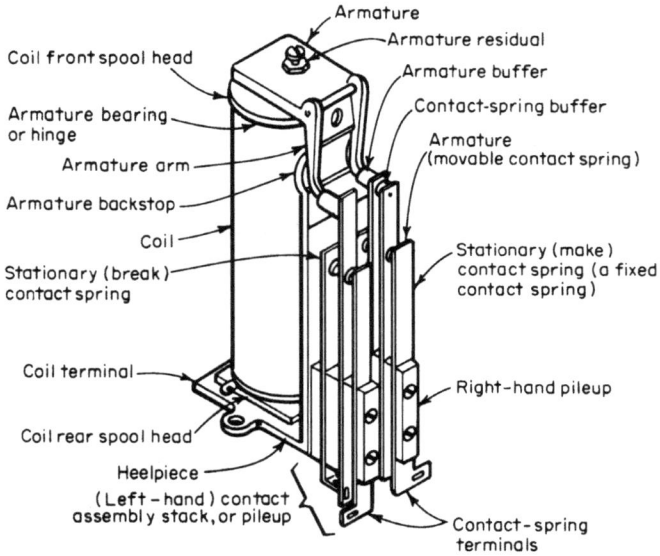

FIGURE 4.13 A typical relay of the long-frame and leaf-spring type of construction.[1]

A design rapidly increasing in use is that called *permissive make*, as demonstrated in Figure 4.14. In this type of relay, contact switching occurs when the energized coil provides sufficient force to overpower a pretensioned spring that held the contacts in a biased (unoperated or normal) position. When the biasing force is overcome by sufficient armature pull owing to coil energization, switching of the contacts takes place. When the coil is deenergized, the contact springs return to their unoperated position because the biasing force of the restoring

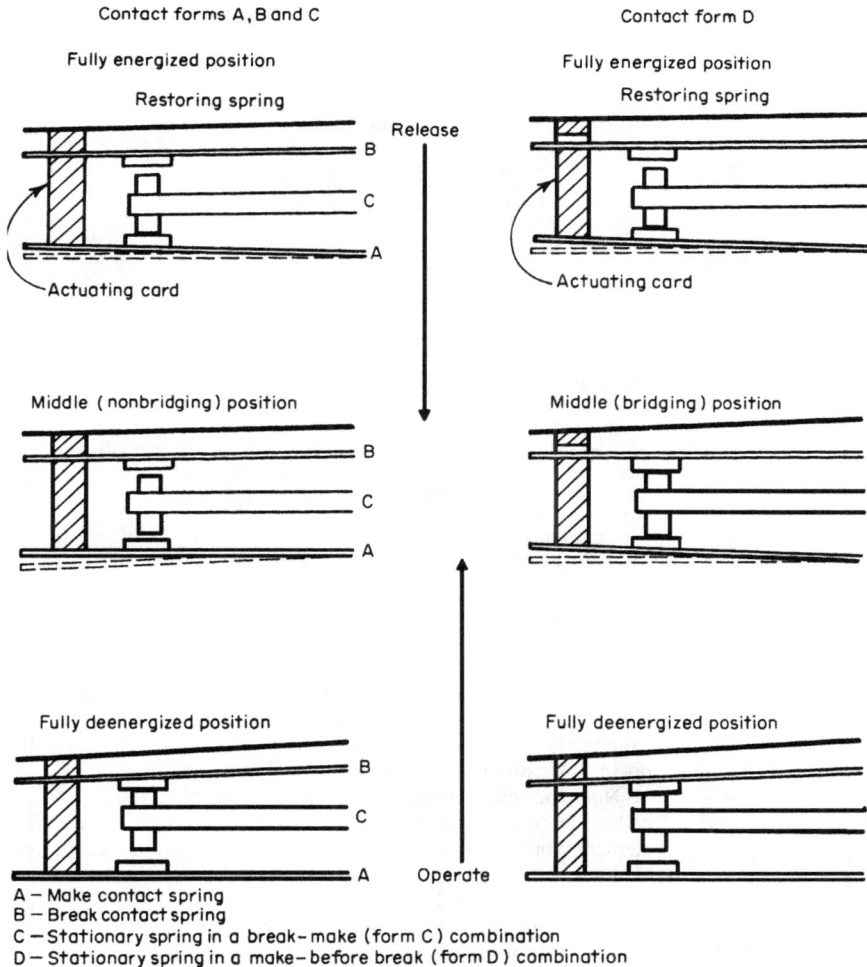

A – Make contact spring
B – Break contact spring
C – Stationary spring in a break-make (form C) combination
D – Stationary spring in a make-before break (form D) combination

FIGURE 4.14 Permissive make contacting. Contact forms A, B, and C: When the coil is energized, the armature operates the actuating card, moves spring 1 into an open *(break)* position, and then permits spring 2 to close *(make)* with the stationary contact 3. On deenergization, the contacting sequence is in reverse. Contact form D: When the coil is energized, the armature operates the actuating card, which is shaped to cause spring 2 to close *(make)* with stationary contact 4 before spring 1 opens *(breaks)* its contact with stationary contact 4. In the fully operated position, spring 1 does break contact with 4. On deenergization, the contacting sequence is reversed.

spring is now unopposed. Advantages of this design are many, but principally they are relative freedom from the need of readjustment, long life, predictably adequate and uniform contact pressures, a capacity for handling a large number of contacts with a minimum expenditure of consumed power, and independence from the effects of gravity because of mounting positions.

An increasingly popular relay design is called "dry reed." No mechanical linkage is used. The contacts are switched as a result of being placed directly into the magnetic field of the coil. Figure 4.15 shows in simplified form how this is accomplished. Much more complicated relays of this type are possible, wherein one or more coil windings controls a battery of reed-type contacts. These relays, too, are explained in more detail later.

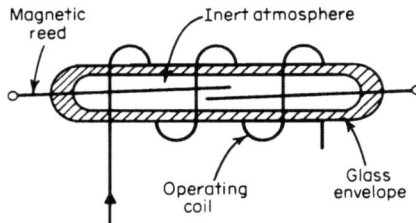

FIGURE 4.15   Basic parts of a dry-reed contact relay. *(C. P. Claire & Co.)* Basically, the switch is composed of only three parts, a glass envelope and two reeds. The reeds are flat and made of a magnetically soft material. The glass seals at the ends of the envelope support the reeds as cantilevers so that their free ends overlap and are separated by a small gap. If the capsule is now enclosed in an operating coil and coil current is applied, a magnetic flux will be present in the reed gap. This flux causes an attractive magnetic force to act on the reeds and pull them together.

FIGURE 4.16   A cutaway view of mercury-wetted contact relay. *(GTE Automatic Electric Company Inc.)*

Specialized forms of the reed-contact relay use a mercury coating of the reeds to increase contact life and rating. These differ little from the dry reeds in size and method of operation, but are usually position-sensitive. They are covered in more detail later.

Not to be confused with the previous reed-contact relay is the so-called mercury-contact relay. It has a reed as one of the contacting elements, but the size and design are considerably different. The cutaway view in Figure 4.16 shows much of the design detail. This relay, too, is examined in greater detail later.

The use of mercury as a wetting element of the contacting surfaces in both these cases reduces and stabilizes contact resistance and increases load-carrying capabilities.

4.2.4 Classes of Service

Some relay characteristics must be considered early owing to the nature of the service.

- *Aircraft, commercial*   The environmental requirements for relays in this category are now rather well understood, and features are designed into the relay to make it compatible with the service needs. Danger exists when the vendor is not made aware of all the facts. For example, if a vendor was aware that the relay was to mount in communications equipment, but was unaware that said equipment was to be airborne, some deficiency might exist.

- *Aircraft, military*   In general, the relays required for this kind of application are of special design, required to meet a particular military specification referred to in the government contract, which must be made known to the vendor at time of ordering.

- *Air conditioning (and heating)*   Standards that apply for air conditioning and heating are under the cognizance of the Underwriters' Laboratories (UL) or a different national equivalent (such as Canadian Standards Association). This must be made known to the vendor initially.

- *Appliances, household electrical*   Frequently Underwriters' Laboratories (UL) approval of relays will be required, with respect to not only flame prevention but also safety for the operator of such equipment. One approach is to find out, quite early in the design, what the UL ruling will be. Submission of the entire device so that the relay gets approved as a portion of the whole is a desirable approach, where a UL-listed relay of the kind needed is not readily available. The use of an unlisted relay will always require adherence to recognized UL standards of insulation, crepe path, and flameproof materials before eventual approval can be obtained. It is best to start in this direction to save time and frustration. Contact life is usually not required to be of a high order, but its magnitude should be expressed.

- *Automobiles and trucks*   Environmental requirements are usually quite severe. The worst of these in respect to shock, vibration, dust, temperature, humidity, etc., should be made known to the vendor at once. Contact life is not usually of a high order, but it needs to be carefully established.

- *Computer input-output devices*   Owing to the heavy-duty requirements, a maximum life expectancy with utmost reliability is needed. Quick-disconnect mounting so that a unit in need of attention can be instantly replaced by a standby, is frequently a requirement. Low coil power consumption is usually specified, but high-order electrical-interference suppression must be provided and carefully chosen. The environment is usually not a problem.

- *Electric-power control*   In this field, relays are used principally in supervisory equipment. Long life, reliability, and freedom from too-frequent maintenance are important. The environment is usually favorable to relays.

- *Electronic data processing*   Process control is a frequent user of relays in this rather broad field. Gathering of data from sensors and the reduction of such to usable form frequently requires relays. The most troublesome area is usually contact selection to accommodate a high variation in contact load. For example, very low-level signals as generated by thermocouples, strain gauges, etc. make necessary the use of especially reliable contacts (usually gold-plated or gold alloy), but the other functions scanned by even the same relay might be of a destructive type of load. Great care must therefore be taken in describing all contact-load functions to the vendor. Environment can be a great variable in this field, varying from an air-conditioned room to a highly contaminating manufacturing-plant area. It should be fully discussed with the vendor.

- *Laboratory test instruments*   Maximum reliability with good life and freedom from too-frequent servicing is the ruling requirement for relays in this field. Environment is normally of no consequence.

- *Machine-tool control* In general, the relays used in this field are of a particular design, in accordance with NMTBA recommendations and to meet NEMA and UL standards. Some of the associated equipment is exempt from these requirements. It is therefore necessary to have a clear understanding with the vendor on this matter. Environment is frequently such that maximum protection from the ambient extremes is necessary.

- *Production test equipment* Cable testers, circuit checkers, automatic continuity, voltage circuit testers, etc., make rather extensive use of relays. The required properties are that the relays not introduce any faults of their own. This means a high order of insulation resistance and dielectric withstanding voltage with low contact resistance.

- *Military applications* This is a highly varied category. The requirements are ordinarily so specific to the job that only by taking the vendor entirely into the designer's confidence can an adequate choice of relay be made. Because the military requirements are so often quite stringent and a ruling military performance specification for the whole job frequently (and too often) is extended to the relay maker, the best relay for the job is not always obtained. The question must be asked early whether the relay is to function or only survive the worst of the specification. In meeting a performance requirement of a specification, a relay is sometimes not as reliable, in general, as would have been the case had it only been required to survive maximum temperature, shock, vibration, etc. The four main military and space applications in which relays find usage are: military aircraft, land-based equipment, missiles and aerospace, and naval shipboard. The logic on which specification writing is based is quite different for each, although some overlap is frequent.

  In many military applications, environments are not very friendly to relays. Several relay manufacturers build relays to meet the stringent requirements of well-known specifications MIL-PRF-6106 and MIL-PRF-83536. These specifications do contain electrical loads and environmental requirements that will satisfy most applications for the military and commercial aircraft.

  These specifications contain requirements for a wide range of currents and voltages. The general range spans from 5 amperes up to 25 amperes for a variety of loads, which include lamp, motor, inductive, and resistive loads for both ac and dc voltages. Some examples of these relays are shown in Figures 4.17 through 4.20.

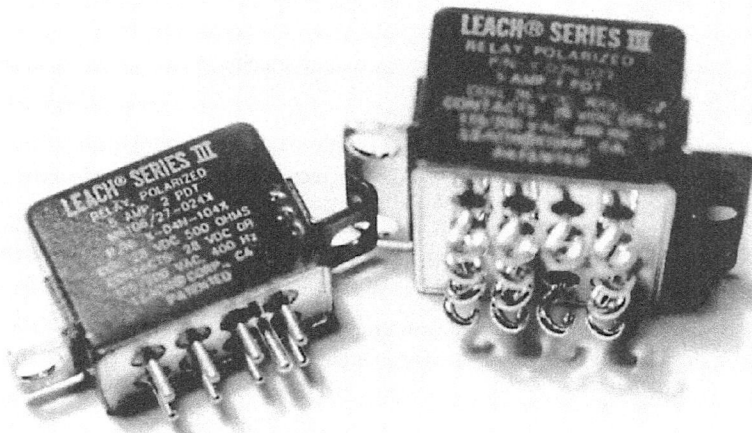

FIGURE 4.17   A common 5-ampere *(dc or ac)* 2 PDT/4 PDT military relay utilizing permanent magnets to assist in surviving high temperature, shock, and vibration environments. *(Leach International, Inc.)*

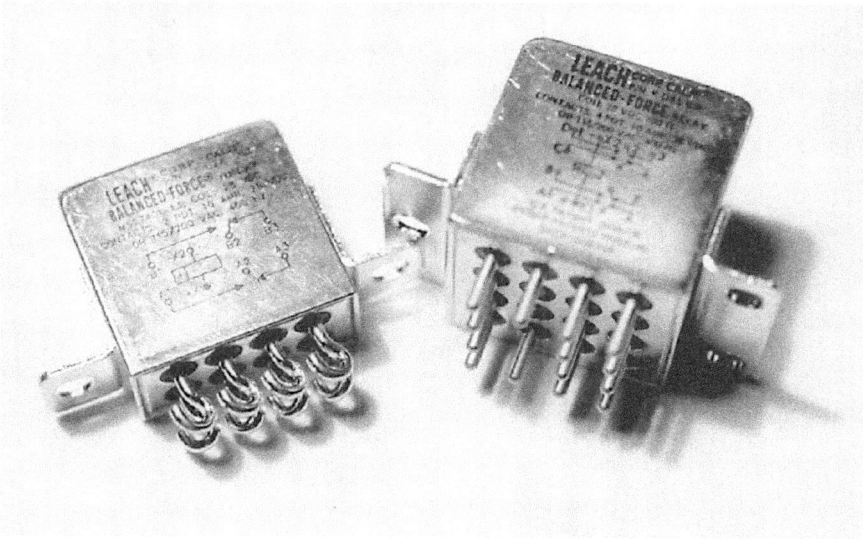

FIGURE 4.18   A common 10-ampere *(dc or ac)* 2 PDT/4 PDT military type relay using a balanced - force design to survive high temperature, shock, and vibration environment. *(Leach International, Inc.)*

FIGURE 4.19   A 25-ampere *(dc or ac)* relay with 3 PNO main contacts and a 2-ampere 1 PDT auxiliary contact for military application. *(Leach International, Inc.)*

FIGURE 4.20   A common 10-ampere
(dc or ac) 4 PDT military-type relay using
a balanced armature design to survive military-type environments. *(Leach International, Inc.)*

4.2.5 Factors in Relay Selection

Power input, ac or dc   As pointed out, relays are not something to be added on after all
design work is finished. A suitable power supply should be a part of the basic design. A decision on choice of operation from alternating or direct current is required early in the design. Commercial alternating current usually offers economic advantages, but direct
current is most often used, for the following reasons.

If the only requirement is that the relay simply shall operate when a switch to it is
closed and release when that switch is opened, it probably matters little with respect to performance in the circuit whether it is powered by alternating or direct current. But many relay applications are so complex that dc power in some form is required. The bulk of this
chapter, therefore, relates almost entirely to dc-operated relays. In addition to the usual inductance and induced noise problems that a supply of ac power is likely to produce, there
are a number of specific reasons for going the dc route.

Dc relays usually have longer life. The contacts of ac relays flatten prematurely as a result of wear because of noticeable ac vibrations during their closing and opening, as always occurs to some degree just before the armature has sealed in. There is also some
perceptible light chatter of the armature for relays operated on alternating current—even
while fully operated. Adding to longer life of dc relays is the reduced bearing wear that results from the absence of ac vibration at this point.

Dc relays usually have greater sensitivity. Because there is no tendency to chatter,
lighter energizing forces can be used than with alternating current, where early saturation
of the shading coil is required and where a poor power factor might cause the coil current
to be large. In other words, the power that is adequate on direct current to start armature
movement is usually more than adequate to maintain a securely operated position.

The heat loss of dc coils is usually noticeably lower. There are both fewer iron losses (no hysteresis on direct current) and fewer copper losses (because usually the required holding power is less).

On a grams-per-contact operated-pressure basis, dc relays for the same contact load, can be kept smaller than ac relays. This reduced size can be quite important at times—especially when mounting on a printed-circuit (PC) board.

Costs should favor dc relays. Dc coils are less expensive to make than ac (a solid-coil core is cheaper than laminated, and no shading coil is required for direct current).

Dc relays, especially if heavily loaded, can accommodate to a wider voltage range than ac relays.

Desired timing variations are almost impossible of achievement when operating conventional relays on alternating current.

Reliability is usually in the dc relay's favor because no compromises in contact pressure, number of contact springs, or adjustment refinement need be made in the interest of quieting the relay.

If the battery type of dc operation is economically feasible, the relay equipment is practically independent of commercial power failure. This kind of reliability, as would exist for battery-powered relays with the battery and rectifier floated across the alternating current, is required in many cases.

If alternating current is chosen, there is usually no concern as to the adequacy of the power source to maintain the required operating voltage and wattage. If direct current is to be provided, however, there is frequently a need to determine carefully in advance that there is voltage stability—especially if the power-supply source is a rectifier. Most dc relays will function well on rather poorly filtered rectified power supply, but this must be determined well in advance of the final design.

Rectification of ac to provide dc-relay operation Some simple arrangements to permit operating dc relays from an ac power supply are shown in Figure 4.21. The following paragraphs present various schemes for providing relays with a variety of rectification circuits for permitting the use of the relay directly on an ac line.

Figure 4.21a shows a full-wave bridge rectifier circuit using semiconductor diodes as the rectifying elements. To improve the quality of the output, a suitable filter is sometimes placed at position C.

Figure 4.21b represents a more common arrangement for producing satisfactory dc power for relays. Capacitor C is a large electrolytic (at 115 Vac applied power, a commonly employed capacitor is 40 µF, 230 V test with a protective resistor in series with the diode of approximately 30 to 40 Ω). The economy of this arrangement is apparent.

Figure 4.21c shows a circuit requiring a relay with two equal windings, one of which functions on one half-cycle, the other on the other half-cycle. The diodes alternately block and conduct, with respect to each other, to maintain unidirectional flux.

A modification of this circuit is shown in Figure 4.21d. Here connection A ties a diode directly across each winding, providing a low-impedance path for circulating current.

Figure 4.21e is essentially the same as Figure 4.21c, except that a varistor has been used to protect the diodes from line surges.

Figure 4.21f is a half-wave, inexpensive circuit found satisfactory for relays inherently not too fast to release. Flux established during the current on half-cycle remains in force owing to the effect of the counter emf during the current off half-cycle. The current freewheels through VAR and K1 in series during the half-cycle when no external voltage is be-

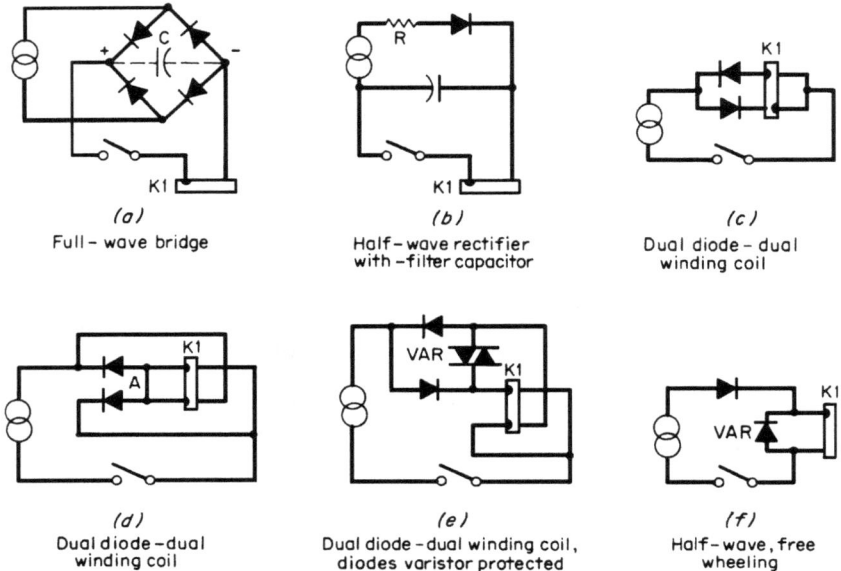

(a)
Full – wave bridge

(b)
Half – wave rectifier
with –filter capacitor

(c)
Dual diode – dual
winding coil

(d)
Dual diode – dual
winding coil

(e)
Dual diode – dual winding coil,
diodes varistor protected

(f)
Half – wave, free
wheeling

FIGURE 4.21   Rectification of alternating current to provide dc-relay operation.[2]

ing applied, hence the name that has been given to this technique. Relays that have a tendency to release fast might not perform satisfactorily because of a tendency to chatter.

4.2.6 Inductive Transient Suppression of Relay Coils

Many methods of coil transient suppression are in use today—a few are effective and some are actually harmful to the successful operation of relay. This section covers the merits of seven such approaches, however, the final selection must be an economic as well as a technical judgment. As such, the choice of method used will depend on the meters of the specific application.

Relay coils are inductive and when the coil is turned off, the inductive energy can create undesirable surge voltages on the dc power line. It might not be apparent, but transients of 1500 volts or more can be generated by this inductive energy.

Prior to the use of solid-state devices, this surge voltage was not as critical. However, this surge voltage can be disastrous to today's solid-state systems. In many applications, the surge values must be reduced to a more acceptable level (40 to 80 volts).

The basic aim of suppression is to reduce the spikes, but if proper care is not taken, a very serious reduction in relay contact life can occur. Long drop out times, slow contact transfer and contact bounce on break are usually the results of extreme or improper transient suppression.

The bifilar winding suppression method causes the most complications. The proponents of the bifilar winding system claim it is the reliable system (no auxiliary solid-state components) and most trouble-free, but this claim bears further examination.

Bifilar suppression   When a relay is designed, the space for the coil winding is usually optimized for best economy of coil power, weight, size, and cost. To then use a bifilar winding, which takes considerable space away from the primary winding, poses a serious problem. A bifilar coil has two windings, (see Figure 4.22a), a power winding, and a shorted winding to absorb the energy from the main winding when the coil circuit is opened.

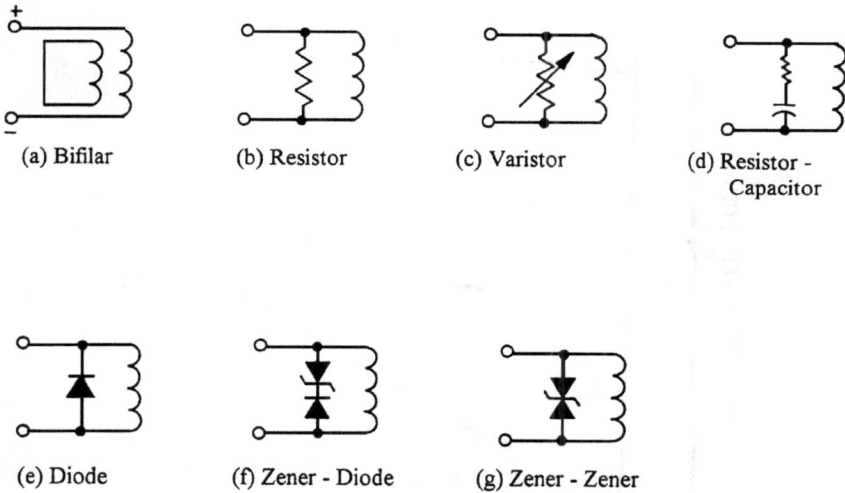

(a) Bifilar   (b) Resistor   (c) Varistor   (d) Resistor - Capacitor

(e) Diode   (f) Zener - Diode   (g) Zener - Zener

FIGURE 4.22   Types of coil-suppression circuits.

The shorted winding reduces the surge voltage that is generated in the power winding when the circuit to the relay coil is opened and the magnetic field collapses. To reduce this surge to a minimum requires that the shorted winding of the bifilar coil absorb and dissipate as much of the power winding's energy as possible. To do this effectively (surge reduced to approximately 1½ times the nominal coil voltage), the shorted winding must be equal to the main winding in turns and approximately equal to 1½ times the resistance of the main winding. To accomplish this, a smaller wire size must be used. The small wire size coupled with the dual windings on the coil increase the cost by a considerable amount.

When coil power to the main winding is interrupted, the shorted winding absorbs the energy stored in the magnetic circuit. The energy circulates in the shorted winding and is slowly dissipated causing the dropout time of the relay to be extended. This energy is gradually dissipated in the resistance (I^{2R}) of the shorted winding, causing a gradual decreasing holding force on the relay's armature. The contacts are thus prevented from transferring rapidly from the closed to the open position. In fact, the transfer time of the relay contacts can increase as much as five times. In addition to the time increasing, two other very undesirable conditions can occur: break bounce and armature rebound. These two additional contact deterioration actions occur because of high energy in the shorted coil. This effect on the magnetic circuit causes a variable air gap as the armature opens and then recloses.

A typical curve of the current in the shorted winding during dropout conditions is shown in Figure 4.23. The curve shows quite dramatically what occurs when the relay is de-energized; it shows the current flow in the short circuited secondary bifilar winding. The ampere turns (NI) at the instant the main winding is interrupted is: $\times NI$. At this point, the armature begins to move, creating an air-gap variation—a rapid change in magnetic flux causing an increase in self-induced current in the secondary winding. The increase is NI (to as much as $0.3 \times NI$) increases the magnetic flux, arresting the motion of the arma-ture, and it can even reverse its direction. As can be seen from the curve, this oscillation occurs several times during dropout.

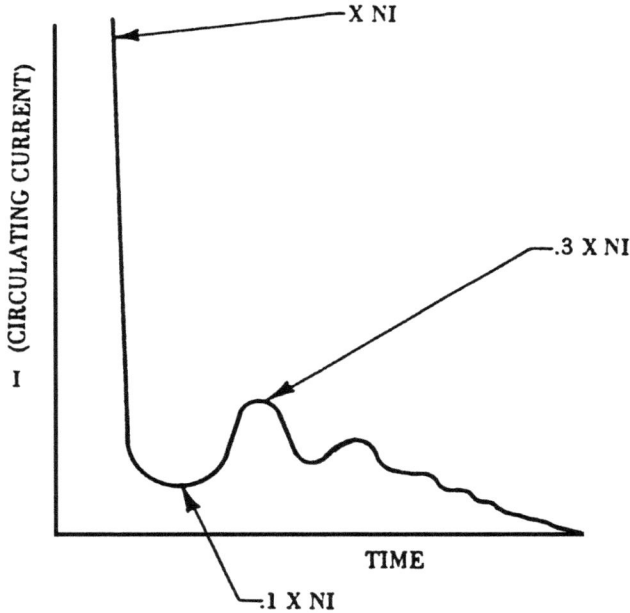

FIGURE 4.23 Typical curve of the current in a shorted winding *(bifilar)* dur-ing dropout conditions.

During this period, the armature motion is such that the contacts open a very short dis-tance and then can reclose, causing break bounce to occur—all depending on such things as armature inertia, simultaneity of contact adjustment, etc.

When the contacts are loaded, this hesitation, when the contact gap is still very small, causes excessive arcing. This resultant arcing severely damages the contacts and shortens the life. For example, on one relay type, contact load-life failures were experienced after 25,000 cycles. Without surge suppression, this same relay type exhibits failures usually af-ter 350,000 cycles.

To minimize the effects of the bifilar winding on relay performance, the following se-quence is required:

• Extensive adjustments of contact pressures and armature position.
• Very careful evaluation of relay function, as to transfer time, break bounce, and armature rebound.

In addition to these requirements, a much-higher pickup voltage allowance must be made to offset the loss of turns on the main winding or an allowance made for a reduced input power relay. The "fix" for the bifilar is further aggravated because each relay design reacts differently to the bifilar effect, depending on its magnetic circuit design and contact system. This means that a single approach cannot be used to solve the problems and each design must be carefully evaluated separately. The effects of operation during life can alter the relay adjustments and might cause problems later that did not exist during original operational checks.

Therefore, the use of bifilar windings necessitates either reasonable changes of the specifications or use of a lower powered relay. In the final analysis, it is still subject to all of the undesirable results of the unrestricted energy suppression of coils. As indicated later in the table, the excessive cost might become serious in some instances and the user must decide for himself the reliability of a relay using a double-wound coil.

Other types of inductive suppression   Solid-state and other types of suppression can be used in a variety of forms and an attempt is made to indicate the relative effects on such characteristics as transfer time, dropout delay, break bounce, and degree of reliability. A table follows the conclusion giving these relative merits.

Resistor   The simplest and oldest method of suppression is the shunt resistor (Figure 4.22b). This method will give adequate suppression if its value is properly chosen, but it does require extra power from the line and across control contacts. The better the suppression (lower resistance), the longer the dropout time of the relay becomes and the greater the change of causing dropout contact problems.

Varistor   Next to the use of a resistor, the next simplest device is a varistor (Figure 4.22c). This device, like the resistor, draws current continuously when energized. However, because it is voltage sensitive, it has a high resistance at low voltage. Therefore, it does not draw an appreciable amount of current until the inductive voltage appears across it. When properly chosen, this device allows near-normal relay operation during dropouts.

Resistor-capacitor   Next in simplicity and extensively used in the past is the resistor-capacitor combination (Figure 4.22d). This circuit eliminates power drain when the coil is on, but, like the resistor, causes some problems on relay dropout.

The proper size of capacitor to do the required job usually causes a space problem and is, therefore, impractical in some applications. Care must be taken in selecting the proper resistor to avoid damaging control contacts during capacitive inrush.

Diode   The shunted diode (Figure 4.22e), because of its small size, was the first system to be used that permitted its installation inside the can with a miniature relay.

It is an excellent voltage suppressor, but it affects the relay characteristics approximately the same as the bifilar winding, causing very great increases in both relay dropout time, transfer time, and quite often causes break bounce to occur. To reduce the effect on the dropout, a resistor of a proper value in series with the diode can be used and still provide adequate protection, but the resistor adds more packaging problems. It is also polarity sensitive and reverse polarity can destroy the diode.

Zener-diode and zener-zener   Next is the diode-zener (Figure 4.22f) or back-to-back zener combination (Figure 4.22g). This combination is small in size, provides excellent suppression, and has practically no effect on the relay's dropout and life characteristics. Of the two, only the diode-zener combination is polarity sensitive.

Conclusion   In all of the previously discussed methods of suppression, the degree to which they affect dropout conditions and subsequently relay life is the key to successful use.

Although Figure 4.23 (the circulating current during dropout) was covered only in the bifilar section, a similar condition is present in the other suppression techniques as well, but in lesser degrees. The one exception is the diode-only type; it and the bifilar type cause the greatest effect on dropout.

As mentioned earlier, it is possible in some cases to adjust sufficient factors so that a relay using extreme suppression (bifilar or diode only) can function near normal at the beginning of life. However, because all relays change somewhat during life, the dropout conditions could change radically because of the suppression and drastically reduce contact life.

For a relay to operate as near normal as possible over life, and supply the required suppression (40 to 80 volts), all the data points to the zener-diode or zener-zener type.

However, a final choice must be made by the user, who is fully aware of his needs and particular requirements, relative to all the data in the table, keeping in mind that the higher the suppression voltage allowable, the less the potential relay problems.

See Table 4.2 for a summary of additional information for different coil-suppression methods.

Coil resistance   In many discussions with engineers, the question is asked about calculating the coil resistance at various temperatures. A graph can be prepared by using the coil resistance measure value at +25°C as one point on the graph. Compute a second point, such as +100°C, and place the second point on the graph. Because the coil resistance changes linearly with temperature change, a straight line can be drawn between two points over any temperature range desired.

The formula used for computing the coil resistance values at various temperatures is as follows:

$$Rt = Rt_1 [1 + 2(t - t_1)]$$

Where
   Rt = Coil resistance at the desired temperature
   Rt_1 = Coil resistance at +25°C
   a = Temperature coefficient for type coil wire used

Note: Most relay coils are wound using annealed copper wire. (Temp. coefficient is 0.0040074 at +25°C).
   Example:

$$Rt_1 = 450 \ \Omega$$
$$t = +100°C$$
$$t_1 = +25°C$$
$$Rt = 450 [1 + a(t - t_1)]$$
$$= 450 [1 + 0.0040074(100 - 25)]$$
$$= 450 [1 + 0.0040074(75)]$$
$$= 450 [1 + 0.300555]$$
$$Rt = 585 \ \Omega + 100°C$$

Timing   Because of the exponential rise in current through the relay coil on circuit closure, there is considerable elapsed time between initial relay-coil energization and operation of the contacts (Figure 4.5). There is a similar delay in release time following circuit opening (Figure 4.6). The inherent operate and release-time delay might be of little or great importance. Usually operate and release timing should be as short as possible, but sometimes considerable delay on either pickup or dropout (or both) is needed. The operate and release times of relays are controllable to some degree, depending on the particular relay design and the constants of the circuit in which the relay is applied. If the relay coil is of sufficient size, delay factors can be built into the coil. However, where space is at a premium; as in very small relays, series connected or shunting capacitors, resistors, and diodes can be used to modify the relay operate and release times.

TABLE 4.2 Types of Relay Coil Transient Suppression.

Figures	Possible Increase Over Standard Cost	Degree of Possible Space Problem	Polarization Requirement	Possible Temperature Problems	Line Surge Sensitive	Possible Effect On D.O., Transfer Time and Bounce Before Complete Break	Effect On Power Relay	Possibility of Reducing Life (Reliability)
Standard Relay	-------	-------	Some	None	No	None	None	None
1. Bifilar	Great	Great	No	None	No	Excessive	Considerable Reduction	Excessive
2. Resistor	Minor	Resonable	No	Yes	No	Considerable	None	Considerable
3. Varistor	Minor	Resonable	No	None	No	Minor	None	Minor
4. Resistor-Capacitor	Resonable	Great	No	Yes	No	Considerable	None	Considerable
5. Diode & Diode-Res.	Minor	Minor	Yes	Yes	Yes	Excessive	None	Excessive
6. Zener-Diode	Resonable	Minor	Yes	Yes	Yes	Remote	None	Remote
7. Zener-Zener	Resonable	Minor	No	Yes	Yes	Remote	None	Remote

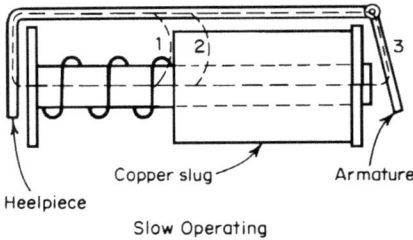

Heelpiece

Slow Operating

FIGURE 4.24 Effect of copper armature-end slug on relay operate time. 1. Magnetic-flux path immediately after coil-circuit closure. 2. Magnetic-flux path a moment later. 3. Final flux links armature to cause pickup.

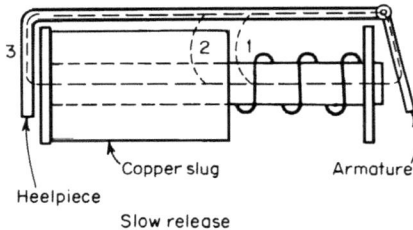

Heelpiece

Slow release

FIGURE 4.25 Effect of copper heel-end slug on relay release time. 1. Magnetic-flux path immediately after coil-circuit closure. 2. Magnetic-flux path a moment later. 3. Final flux links slug. Slug can now delay dropout on coil-circuit opening.

The following paragraphs are included to demonstrate how the operate and release times of medium- to large-sized relays are affected by coil design.

Slow-operating relays Slow-operating relays (Figure 4.24) use coils with a large copper collar or slug at the armature end of the core. The copper collar, acting as a short-circuited secondary winding, retards the building up of the magnetic field. These coils are most effective when used with a large armature stroke and a heavy spring load. The large armature stroke reduces magnetic pull on the armature, and the large spring load prevents pickup until the magnetic field has been built up to, or near, its full value, thus providing the maximum operate time delay. Relays with armature-end slugs are also somewhat slow to release.

Slow-releasing relays Slow-releasing relays (Figure 4.25) have a copper collar or slug around the heel end of the coil core, a small residual gap, and a light contact load. The copper collar retards the collapse of the magnetic field once it has been established. This, together with the small residual gap and the light contact load, permits the armature to remain in its operated position until the magnetic field has died down to a very low value. Where the greatest possible release delay is desired, or where permanence of adjustment is very important, a short-lever armature is helpful.

Slow-acting relays Slow-acting relays use coils with a full-length copper sleeve about the coil core to provide release-time delay. Because the sleeve extends full core length, it also causes some delay in operate time, as well as the sought-after release time delay. Sleeve coils are primarily used to save copper and reduce weight. Maximum release-time delay equal to that achieved by a full-sized slug is accomplished with about one-half the copper volume by using a sleeve.

Where small relay size makes the use of slugs and sleeves impractical, circuit elements (such as diodes, capacitors, resistors, and thermistors) are used to alter the relay's inherent timing. Examples of some of these techniques follow, together with the explanation of how the timing is affected by associated circuit components. More information is available from Ref. 2.

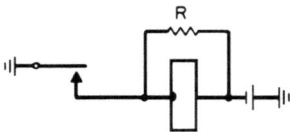

FIGURE 4.26 Slow release by use of noninductive shunt.[2]

Slow release by noninductive shunt This circuit provides for a slight increase in the normal release time of the relay by means of a noninductive resistor R connected across the relay coil (Figure 4.26). The shunt path provides means for the back emf (created by the decay of flux in the core when any relay coil is deenergized) to circulate and somewhat prolong the operated condition of the relay.

Slow release with shunt diode If the resistor of Figure 4.26 is replaced by a properly poled diode, a longer release time will be achieved than for the use of the resistor. Normally, for a dc electromagnetic relay, dropout will occur following a delay of several milliseconds after the coil is shunted with a diode, such as might be used for transient voltage suppression in transistor circuits or to protect electrical contacts from voltage surges from an inductive load. The delay can be lengthened to 20 to 30 ms (or longer). The delay is obtained by circulating the exponentially decaying current through the diode shunt after the coil circuit is opened.

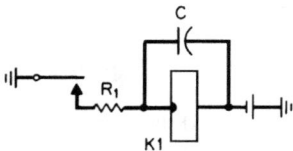

FIGURE 4.27 Slow operate and slow release by series resistor and shunt condenser. [2]

Slow action using series resistor and shunt capacitor A simple application of an RC circuit to provide a brief time delay in the operation of a relay is illustrated in the circuit of Figure 4.27. When the circuit is closed, capacitor C, uncharged, provides a direct shunt around the relay coil. As the voltage across C builds up and the current through it decays at the same rate, a design point is reached where the current diverted through the relay provides sufficient ampere-turns to operate it. Resistor R limits the current available for charging the capacitor and operating the relay, but it also makes the timing voltage-sensitive. Because the capacitor is charged when the relay circuit is opened, the capacitor discharges slowly through the relay coil, giving a release time delay.

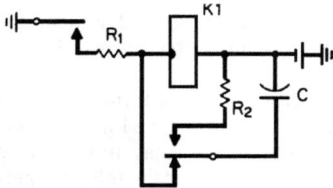

FIGURE 4.28 Slow operate by switched condenser shunt and series resistor. [2]

Slow operate by switched RC shunt The circuit of Figure 4.28 provides the same operate delay as for that shown in Figure 4.27, but once the relay is operated, the RC network is discharged locally so that no release delay is introduced. Operate delay will be a function of R, C, the applied voltage, and the operating characteristics of the relay. Because of the marginal arrangement of the relay coil and the resistor, this circuit is voltage-sensitive. Timing will change somewhat as the voltage changes.

Slow operate and slow release by coil-winding interaction Short delays in operate or release times are possible for multiwinding relay coils by making use of the circuit conditions of Figure 4.29.

Slow release by shorted heel-end winding The relay shown in Figure 4.29a has a two-section quick-acting coil, which is energized and deenergized on its No. 1 (armature-end) winding. When the relay is energized, a pair of auxiliary contacts will short the No. 2 heel winding, forming somewhat of a delay slug. This shorted winding is not nearly as efficient as a solid-copper slow-operate slug, but it provides some increase in release time. The connection between the in terminals of the two coil windings is provided so that both windings will have a negative potential standing on them when the relay is idle. This eliminates any possibility of electrolytic corrosion on the windings and associated wiring.

Slow operate by shorted armature-end winding The circuit of Figure 4.29b is the converse of Figure 4.29a. In this case, the relay is energized and deenergized on its No. 2 heel-end winding, and prior to operation, the No. 1 armature-end winding is shorted through a pair of normally closed contacts on the relay. This arrangement is not as efficient as a solid-copper slow-operate slug, but will introduce some operate delay. The connection shown between the in terminals of the two coil windings is provided as before so that any possibility of electrolytic corrosion on the windings and associated wiring is eliminated.

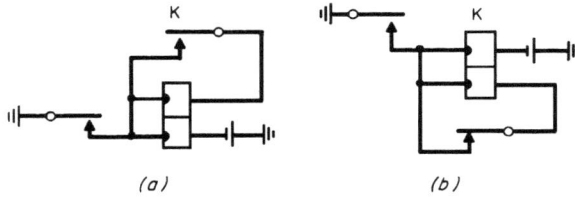

FIGURE 4.29   Slow operate and slow release by coil-winding interactions. (a) Slow release by shorted heel-end winding. (b) Slow operate by shorted armature-end winding.[2]

Circuit fundamentals   The requirements that a relay take the operated state when its operating-coil winding has power applied and restored when deenergized is basic and in need of no further explanation. A few other requirements are almost as basic, but do require recognition before a relay selection can be made. The most elementary of these circuit fundamentals follow. More sophisticated relay circuits can be found in many places, such as Refs. 1 and 2.

   Locking through own contact(s)   Figure 4.30a shows a basic locking circuit in which the relay is energized by the closure of S1 and is held energized, or is locked up electrically, through one of its own make contacts and S2. It is now operated independent of S1 and will release when S2 is opened after S1 is closed.

   A variation of this is shown in Figure 4.30b. Here, closures of S1 and S2 are required to operate the relay, which is then held through one of its own make contacts and S2. The relay restores to normal when S2 is opened.

   Another form of lockup utilizes a continuity (make-before-break, or form D) contact combination. In this circuit (Figure 4.30c), the relay is operated through S1 and the normally closed portion of the form D contact. After operation, the relay is locked up through the normally open portion of the form D contact and S2. After the relay is operated, S1 is free to control other circuits without being affected by the ground potential to the winding of relay K. A make-before-break contact is used in this case because a break-make would fail to secure the holding circuit and the relay would buzz in a partially operated position.

   Variations in continuity lockup circuits that require closure of both S1 and S2 before the relay will operate are illustrated in Figures 4.30d and 4.30e.

   In Figure 4.30f, a method is illustrated for isolating an operating and lockup circuit through use of two electrically isolated relay windings with a common source of power. Operation of S1 will energize the relay, but it can be locked in only if S2 is closed before S1 is released. Both S2 and S1 must be open to release the relay.

   Use of a preliminary make contact   A more desirable way of arranging the contacts to provide for a locking circuit is by using a form A make-type spring combination. However, if the operating signal is of extremely short duration, such that there is some doubt as to whether or not the relay will have time to operate fully, the locking contact can be provided as a preliminary make. Once these locking contacts have closed, complete operation of the relay is ensured. If the operating signal is still present, the armature will be driven by both the locking winding and the operating winding.

   To provide this preliminary make in a telephone-type relay, it is necessary that the relay be equipped with a heavy normally closed back contact. The balance of the spring pileup above the preliminary make is then supported in a position so that the preliminary make is free to

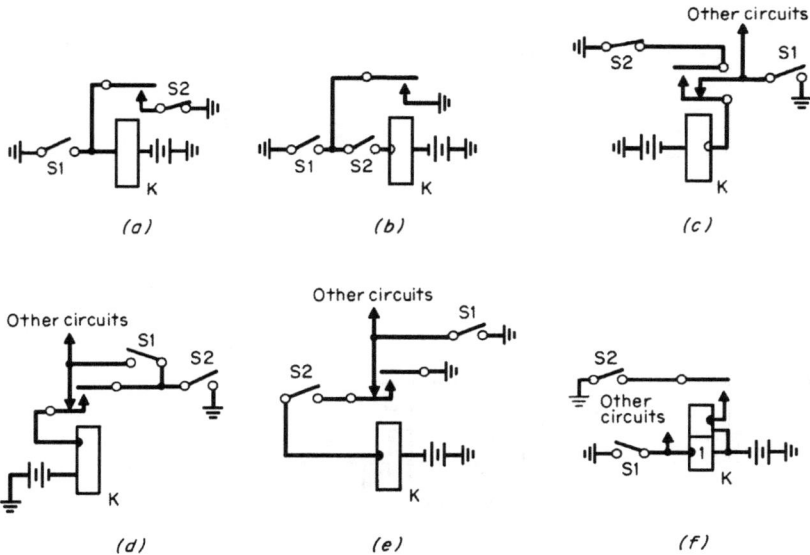

FIGURE 4.30 Six methods of electrically locking up a relay through one of its own contacts.[2]

move when the relay is energized. This heavy, normally closed, contact must be provided, even though the controlled circuit does not require a normally closed contact.

Avoiding malfunction from feedback If a relay locks itself in with the simple circuit of Figure 4.30a as a function of some remote signal, it is possible that the locking potential could be fed back over the operating lead and cause a malfunction in some other part of the circuit. These sneak or feedback circuits can be eliminated by the addition of a blocking diode in series with the operating lead.

A simpler, more direct means of eliminating this feedback problem is to use a dual-coil relay with windings aiding and to use an independent locking contact. In this circuit (Figure 4.31), the relay is operated on its No. 1 winding and locks itself in the operated condition on its No. 2 winding.

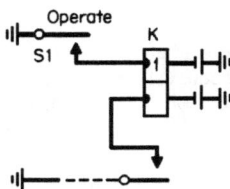

FIGURE 4.31 Locking up a relay on an auxiliary winding.[2]

Forced release of relays The holding circuit of Figure 4.32a uses dual coils, but differs from that of Figure 4.30f in that the two coil windings are connected in opposition and two Form A contacts are used. It provides lockup indefinitely until the release key is closed to energize the opposition winding. The only precaution essential to successful use of this dual-opposed winding relay is that the release circuit be connected to the most powerful winding (usually the armature-end section). This precaution is taken to ensure that the flux is driven through zero so that the relay will release positively.

A somewhat different circuit for forced release of a dual-opposed coil relay is shown in Figure 4.32b. This mode of operation assumes that the operating circuit was opened prior to the closure of the release circuit.

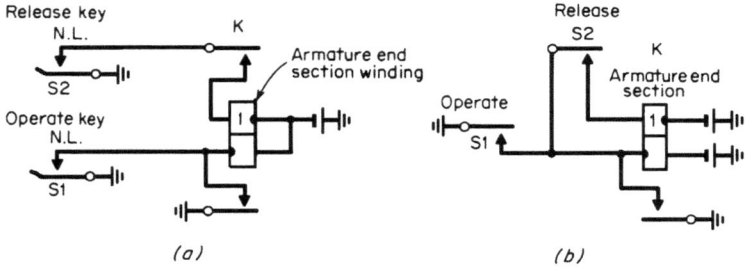

FIGURE 4.32   Two methods of forced release of a relay by use of an opposed winding.[2]

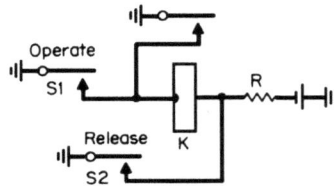

FIGURE 4.33   Release of a relay by shunting it down.[2]

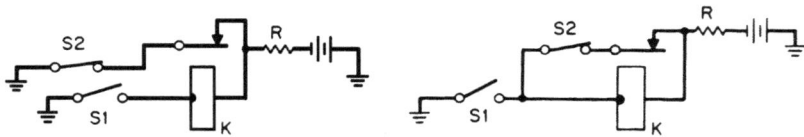

FIGURE 4.34   Two forms of lockdown circuit using a coil-shunting resistor. [2]

In either case, such forced release requires considerably more time to affect restoration of the relay to normal than would have been required had the relay been restored by merely opening the holding circuit. This increased release time is the sum of the time for the counter magnetic flux to build up to the point where the relay is caused to release plus the armature-movement time necessary to restore the relay.

Shunt control of relays   The shunt-release relay shown in Figure 4.33 is always operated in series with a resistor and is provided with a local locking circuit. This locking circuit is not necessary if the remote operating circuit is maintained. To release the relay, closure of the remote-release (RLS) contact effectively shorts out the relay coil and leaves the resistor in series with the power supply. If the RLS contact is maintained, this resistor must have sufficient wattage to operate satisfactorily under this condition.

When the relay is shorted out to affect its release, the low-resistance short across the relay coil provides means for the back emf to circulate for the complete exponential decay time of the flux in the magnetic circuit of the relay. The relay remains in the operated condition until the flux density has decayed to the point where the spring load, in combination with the residual gap, used can restore the relay. Normally, larger residual gaps are required for this type of service than would ordinarily be expected for a particular relay.

Generally speaking, shunt circuits of this type operate more satisfactorily on higher than on lower voltages. When the voltages are lower, the current, which must be handled by the operate and locking contacts (as well as the release contacts), must be higher to achieve the required relay wattage.

Lockdown operation by coil shunt A variation in shunt release of a relay is found in the lockdown circuit, in which a relay prevents itself from being operated through one circuit path until another circuit path has been opened. Two forms of relay lockdown circuits involving a shunt-resistor path around the relay coil are shown in Figure 4.34. In both circuits, the relay cannot be operated through switch S1 until S2 has been opened. However, after the relay has been operated through S1, opening and closing of S2 will not affect the operated state of the relay. Because the resistor remains in series with the relay winding during operation, it can have an appreciable effect on operate time. These circuits have the disadvantage of dissipating power in the shunt resistor during the lockdown state.

Contact load and contact choices Because the prime purpose of a relay is to establish the flow of electricity through a mating pair of contacts in the controlled circuit(s), the choice of materials ensuring that this is accomplished becomes a first priority. The contact must be large enough so that there is no deterioration from destructive melting or welding, yet not so large that the current density is below a critical value. The material must be highly conductive, yet hard enough to meet the required number of closures without excessive wear (flattening).

For each kind of load condition, there is a best choice of material, shape, size, and pressure when in contact. To determine what is best for the specific job, the vendor must be fully informed by having the purchaser include such information as ambient temperature, humidity, and degree of dust and dirt likely to reach the contacts. The required service life is also a factor in the choice to be made, also frequency of operation. Voltage and current to be handled are naturally of prime concern. The nature of the load with respect to whether it is resistive only, capacitive, inductive, or combinations of these must be stated. If resistive, is it fixed, lamp (surge), or subject to overload currents? If inductive or capacitive, what is the likelihood of surges? What is the chance of a flashover? What is the duty cycle? How many sets of contacts are required, and in what sequential order?

One common error made by relay users of little experience is to assume that if a contact is large enough, or rated above the known load, it will perform perfectly. This frequently leads to real trouble. Each contact material has a critical minimum or threshold voltage and current for its composition, size, and shape. The best contacting results when there is sufficient electrical pressure (voltage) and current, along with sufficient mechanical pressure on the contacts, to cause some fusing of contact surfaces on each operation. This fusion is of a low order, but it means that a very small amount of welding occurs on each operation, with a rupture of that weld each time the contacts are separated. If the critical or threshold voltage is not reached or exceeded, if the current density is inadequate to melt and weld even one tiny spot, if the contacting surfaces are too large and flat to give the critical current density necessary to get surface softening, or if the mechanical pressure is inadequate to push any insulating film aside that might be covering the surfaces, then contact failure results. This minimum electrical requirement is known as minimum (reliable) current.

Table 4.3 shows some critical values applying to the most commonly used contact materials (which presumes that there is sufficient mechanical pressure to make conducting spots on the contact faces touch through any insulating film that might exist). It is difficult to generalize on contact materials, as related to load current and voltage-handling capabilities. This is why the vendor's judgment must be relied upon. As an aid in understanding how a decision is reached, Table 4.4 indicates some generally accepted practices. From this, it might be assumed that gold contacts are not used often enough. Pure gold is not an

TABLE 4.3 Typical Softening, Melting, and Boiling Voltages of Commonly Used Contact Materials. *(Holm)*[1]

Material	Softening	Melting	Boiling
Silver	0.09	0.37	0.67
Gold	0.08	0.43	0.90
Palladium		0.57	1.30

TABLE 4.4 Considerations Affecting Choice of Material for Relay Contacts

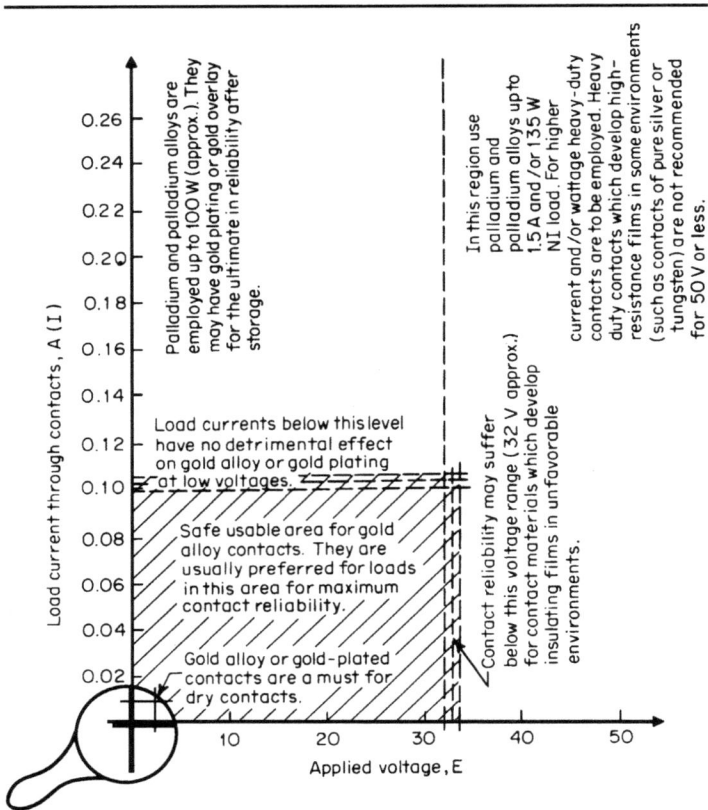

acceptable material because it cold-welds without even the presence of current. Impurities, to prevent cold welding and to add some mechanical hardness for improvement of life, are deliberately added to so-called gold contacts to make them acceptable for use in relay circuits. Even so, their mechanical life is not very great; for that reason, other noble metals of greater hardness are more often used. If conditions permit, an alloy of palladium and silver is probably the best practical contact.

Cost factors favor the use of silver and silver alloys wherever service requirements are favorable. Silver and some of its alloys tarnish readily in industrial atmospheres. The tarnish is of very high resistance, high enough in many cases to appear as an insulator. In general, silver is not satisfactory for use much below 50 V, or even above 50 V, if the current flow is light owing to large circuit resistance.

Understandably, most concern is normally shown by the user regarding maximum allowable current figures. Typical values are shown by Table 4.5. Because of maximum concern with what can be seen, rather than what is invisible, it must again be emphasized that failure to reach or exceed minimum (reliable) current has caused as many, or more, contact failures than overload. That is why the choice of kind of material, size, and shape of contact is important.

Contact characteristics   Although the final choice of contact is a matter best left to the vendor in the end, the user must give it some consideration before even a tentative choice of relay type and vendor can be made. In other words, kind of relay, relay size, and relay availability depend to a considerable degree on the nature of the contact load to be handled and the contact life required. The need to use a contact small enough that the minimum current is adequate for keeping the contacting faces in a conducting state has already been covered. The opposite situation is now to be examined. If such relays as dry reeds, mercury-wetted contacts, miniature size, or other kinds with specialized contacts are the choice, there is little latitude for the user to explore. However, if the more common general-purpose relay, telephone-type relay, or power-type relay is being investigated, the kinds, sizes, and shapes of contacts available vary extensively. As an example of what is available in the way of contacts on a large- to medium-sized telephone-type relay, one manufacturer offers as standard the contact variety of Table 4.5. All manufacturers have similar data applicable to the relays of their manufacture.

In interpreting Table 4.5, recognize that the advertised wattage ratings of relay contacts are based on fixed-resistance resistor (noninductive) loads, and care must be exercised to avoid introducing inductance or capacitive reactance accidentally, and to recognize its effect when intentional. When setting up ratings for inductive circuits, it is common practice to derate the contacts to one-half (or less) of the wattage for noninductive circuits. Adequate arc and spark suppression must also be provided with inductive loads; otherwise, the contacts should be derated even further.

Contamination by small amounts of airborne dust or fumes is another common cause of reduced contact life. Even accidental fingerprints on the contacts can accelerate their rate of deterioration.

More detailed information concerning contact capability for each of the relay types is contained in the section about individual relay characteristics. Table 4.4, for example, shows a relationship between contact load and kind of contact material that is applicable to many kinds of relays other than large- and medium-frame size telephone types, such as many kinds of general-purpose relays and some specialized relay types.

Contact protection   A relay operating near sensitive circuits might cause trouble in electronic equipment from arcs generated as the contacts function. Also the coil transients represent both radiated RF interference and conducted radio frequency back over the power leads. Some form of suppression must be applied as electrical protection of the controlling contacts. Ideally it can serve the purposes of both contact protection and interfer-

TABLE 4.5 Typical Data for User Guidance in Choosing Telephone-Type Relay Contacts

Code	Material	Break or make, load	Will carry load	Remarks	Diam, in	Height, in	Shape
0-20	Palladium-silver	50 W (max 1 A N.I.)* 135 W (max ½ A) inductive	135 W (max 2 A)	See Code 0-18	0.055	0.020	
3-18	Gold alloy			For low-level circuits			
0-18	Palladium-silver	135 W (max 3 A N.I.)	150 W (max 3 A)	Resistant to tarnish, and nonmicrophonic	0.067	0.020	
9-18	Platinum-ruthenium	Same as Code 4-18	Same as Code 4-18	More resistant to wear than Code 0 contacts			
4-18	Palladium	135 W (max 3 A N.I.)	150 W (max 3 A)	See Code 0-18			
4-14	Palladium	150 W (max 3 A N.I.)	150 W (max 3 A)	See Code 0-18			
9-14	Platinum-ruthenium	Same as Code 4-14	Same as Code 4-14	See Code 9-18	0.084	0.031	
0-14 Flat-or domed	Palladium-silver	150 W (max 3 A N.I.)	150 W (max 3 A)	See Code 0-18			
Tungsten	Tungsten	450 W (max 2 A N.I.)	450 W (max 2 A)	Use in highly inductive, low-current circuits	0.125	0.063	
Laminated silver	Laminated silver	450 W (max 4 A N.I.)	450 W (max 4 A)		3/16	3/64	
Tungsten silver	Tungsten silver alloy	575 W (max 5 A)	1,150 W (max 10 A)		¼	1/16	
Silver cadmium oxide	Silver cadmium oxide	Meets NARM Grade B life—100,000 operations N.I. load—20 A, 28 vdc; 10 A, 48 vdc; 1 A, 110 vdc; 5 A, 115 vac	Will handle 300-W lamp loads (allow for inrush currents) or the current drain of an average ¼ hp or a well-designed ½-hp motor	Especially applicable to circuits where high current density causes welding of other material	3/16	0.040	
Snap-action switch, ac		AC N.I. Loads Break 10 A, 115 V; Make 6 A, 115 V; Break 5 A, 230 V; Make 3 A, 230 V		Also available for loads up to 30 A, 220 V, ac on special order			
Snap-action switch, dc				On special order, magnetic blowout switches up to 15 A dc			

SOURCE: Industrial Products Division, GTE Automatic Electric Company, Inc.
* N.I. = noninductive.

282

ence suppression. To determine that both objects are achieved, some considerable experimental testing might be necessary. This subject is much too complex for complete analysis here, but from a practical standpoint, it can be rather easily solved by adjusting the contact protection for maximum effectiveness (see Refs. 2 and 4).

Closing contacts, dc and ac loads Contacts can be damaged on both closure and opening. Contact damage at closure is frequently caused by current surges because contact forces at this instant are light, permitting contact sliding and bouncing to occur. This is particularly bad because the load current is often many times the steady-state value at this instant. A microscopic weld or "bridge" will frequently form at the point of contact closure. In dc circuits, this bridge usually ruptures asymmetrically at the next contact opening, resulting in metal transfer. In ac circuits, there is usually a net loss of contact material. The metal vapor that condenses in the vicinity of the actual contact area is normally black and is frequently mistaken for carbon.

Loads that produce transients at contact closure are as follows:

- Tungsten lamps whose cold resistance is 7 to 10 percent of their hot resistance.
- Transformers and ballasts that cause transients 5 to 20 times their normal currents when switched in their inputs.
- Ac solenoids and some kinds of motors.
- Capacitors placed across contacts or loads with inadequate series-connected current-limiting resistance.

To meet these circuit conditions, the relay designer might elect to use:

- Heavy-duty contacts and a high contact force to minimize contact bounce on closure.
- Contact materials with the highest possible electrical and thermal conduction, usually silver or silver alloys.
- Contact-material additives to inhibit welding, such as cadmium or cadmium oxide.

The circuit designer can usually add small values of series resistance to the circuit to reduce current surges.

Opening contacts, dc load There are almost certain to be transients when the contact opens. When the circuit to an inductive dc load is opened, much of the energy stored in the load must be dissipated as arcing at the contacts unless some alternative means of energy absorption is provided. Some of the load energy is dissipated as heat in the load resistance, in eddy-current losses in its magnetic circuit, and in the distributed capacitance of the coil winding. For dc circuits, a number of simple solutions are available to lessen or inhibit contact arcing:

1. A semiconductor diode can be connected across the inductive load (see Figure 4.35) so that it blocks the applied voltage at contact closure, but allows the stored energy in the load to recalculate through it at contact opening. The time for the load current to decay to 37 percent of its steady-state value equals L/R. The blocking diode will prevent any inductive transients from appearing across the contacts during the switching operation. For loads below the minimum arcing current, the time required for load deenergization can be materially reduced by adding a Zener diode, resistor, or varistor in series with the blocking diode, thus increasing R. The rate of energy dissipation after the contacts are opened is thereby increased as the load current circulates back through this additional voltage drop. The instantaneous voltage plus the source voltage should not exceed 320 V.

2. Either the load or the contacts can be shunted with a resistor-capacitor combination (Figure 4.36). For load currents in the stable-arc range, the resistor R_c can be selected

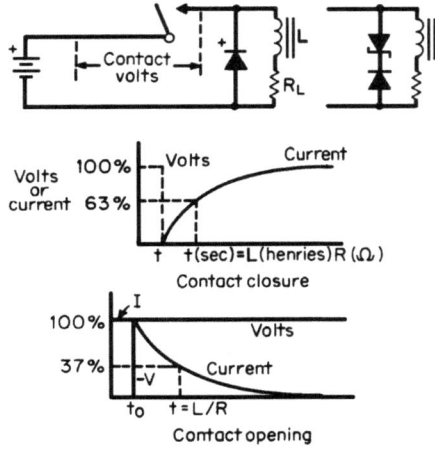

FIGURE 4.35 Method of using semiconductor diode to suppress voltage surge from inductive load at contact opening.[2]

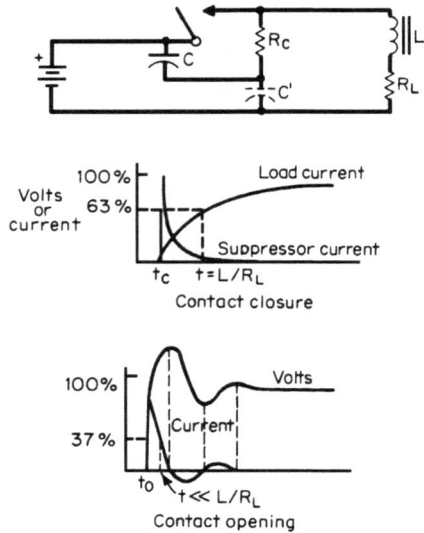

FIGURE 4.36 Use of capacitor-resistor combination to suppress surge from inductive load. The capacitor can be at either C or C'.[2]

to match the load resistance, or it can be ½ or 1 Ω V – 1 of the power source. For smaller load currents that can use a stable arc, the resistor can be higher in value.

A reasonable value is one that results in a voltage transient of less than 300 V for the sum of the source voltage and the instantaneous voltage generated by the load current in the resistor, calculated thus:

$$R = \frac{300 - E_{source}}{I_{load}}$$

The resistor is essential and must be large enough to limit the current transient from the capacitor discharge (or charge) on contact closure to prevent contact welding. The capacitor should be large enough to accept the stored energy of the load without permitting an electric breakdown of the contact gap, normally at greater than 320 V. An oscillograph is the best way to determine when these transients are adequately suppressed. In Figure 4.32, the capacitor can be connected at either C or C'. Connection at C is preferred, because it protects against source and line, as well as load, inductance.

3. A varistor (voltage-sensitive resistor) or thyristor can be used to shunt the load. If such a device carries 10 percent as much current as the load, the maximum switching transient will be about twice the source voltage. This method is also suitable for ac circuits.

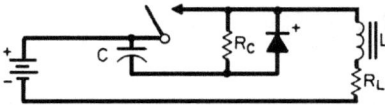

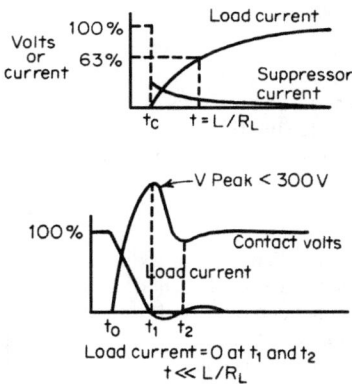

4. For extremely inductive loads, for the longest possible life, or for load power and contact-gap length above the minimums for a stable arc, the circuit of Figure 4.37 can be used. In this circuit, the capacitor is charged through the diode, but can discharge only through the resistor. This arrangement gives essentially zero contact-voltage drop at the instant of contact opening. The capacitor value should be such that when the energy transfer from the load is complete, the peak voltage to which it charges will not cause a breakdown of the diode, the contact gap, or itself. Usually, the peak voltage should not exceed 200 to 350 V. For dc inductive loads for which the conditions for a stable arc can be satisfied by the partially opened contacts, the circuit of Figure 4.38 permits the contact gap to be established without drawing an arc, and the stored-energy transfer is accomplished more quickly than would have been the case if the contacts had been allowed to arc. The reason for this is that the integrated inverse voltage to which the capacitor charges is greater than the voltage drop in an arc, were arcing permitted.

FIGURE 4.37 Use of capacitor-resistor-diode combination for arc suppression with a highly inductive load.[2]

5. Where the inductive load of a relay coil presents a hazard to transistor drive circuits, coils with dual windings (called *bifilar coils*) can be used with one winding shorted. This arrangement provides a pronounced damping effect on the rate of change of the

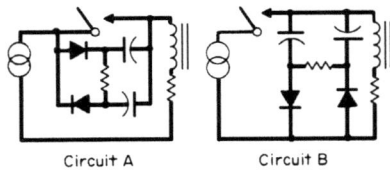

Circuit A   Circuit B

FIGURE 4.38 Use of capacitor-resistor-diode combination for suppressing contact arc on ac inductive load.[2]

magnetic flux in the iron. Hence, it provides a significant moderating effect on the induced voltage. It is, however, wasteful of coil winding space and increases relay cost.

Opening contacts, ac load Ac loads are most commonly treated in a different manner from dc loads because a stable arc will normally be terminated when the current passes through zero and reverses at the end of the first half-cycle following contact separation. It is fairly common to use arc-resistant contact material, preferably in a relay in which the contacts separate slowly, and let arcing be terminated by the reversal of the current, rather than by the continuing separation of the contacts. When load currents get too heavy for safe interruption by small relays (greater than 10 to 25 A), the current-reversal effect can be supplemented by magnetic or air blowout, multiple break contacts, or arc-gap cooling labyrinths, or by evacuating the contact chamber.

Under moderate arcing conditions, contact life can be greatly increased by shunting the load with a resistor-capacitor-diode combination whose time constant is equal to that of the load:

$$R_c C = \frac{L}{R_1}$$

or assuming R_c equals load resistance R_v:

$$C = \frac{L}{R_2}$$

This network makes the load characteristics essentially resistive. When the maximum possible contact life is required, either of the capacitor-diode combinations shown in Figure 4.38 can be justified. For 115-Vac service, the diodes should have a peak inverse voltage rating of 400, the capacitor should have a dc working voltage of 200 Vdc, and there should be a 100-kΩ resistor, which will dissipate nearly 1 W. The capacitor discharge time after a switch closure can be as long as 1 s.

Cautions The transient voltages developed when the contacts open the load circuit can exceed the dielectric withstanding voltage between contacts and another part of the relay. In some circuits, these voltages can be high enough to cause the breakdown of another circuit component. These transients often cause interference in adjacent or associated circuits. Usually, a resistor-capacitor network, as recommended by the relay vendor's literature, will reduce the voltage to a level that suitably protects the contacts and avoids dielectric breakdown. However, it is sometimes necessary to use diodes to eliminate radio interference from arcing. For the latter cases, no general rules can be formulated because the interference is closely associated with the particular circuits.

In general, careful attention to contact protection can increase life expectancy by as much as three orders of magnitude. System reliability can be greatly improved by the elimination of high-voltage transients, and the speed of response and its consistency are often substantially improved.

Cost Good engineering always necessitates much consideration of cost. If only one relay or a few are involved, this consideration might be slight. Where the quantities per device are great or the production heavy, this matter might justify a great deal of thought. Several items must be balanced against first cost before a decision to reduce relay purchase cost is made. These are cost of maintenance, cost of field replacement, customer satisfaction, and value of product enhancement accomplished by using the most reputable relay.

Mountings and enclosures Sometimes the method of mounting becomes of prime importance. For example, if all the other equipment is PC-board-mounted, the relays are probably most acceptable if they are mounted that way, too. In some cases, mounting preferences will limit the kind of relay types that can be used.

The least-expensive relay, with respect to original purchase price, is usually without extraneous materials, using only two or more screws into tapped holes and without dust covers. Simple angle-iron mounting brackets are usually the next least costly. Mounting on strips, with or without dust covers, usually represents modest added cost.

Enclosures are used mostly for protection from ambient dust, dirt, oil, metallic chips, etc. Enclosures can also be used with the intent to keep out prying fingers. Enclosures run the wide gamut from simple drawn metallic covers and molded-plastic covers to elaborately tailored enclosures and hermetically sealed enclosures. Hermetic sealing protects the contacts thoroughly from the environment, but it has several disadvantages that make its use less than ideal in all cases. For example, contact troubles develop in the confined area of the contacts that do not occur in air, and hermetic sealing is expensive and prevents inspection of the working elements of the relay when it is in service.

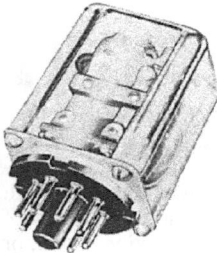

FIGURE 4.39 A covered general-purpose relay having an octal plug-in base. *(Potter & Brumfield, Div. of AMF, Inc.)*

Plug-in relays exist in many forms. A common variety that is very popular is demonstrated by Figure 4.39. The increase in the use of relays directly on PC boards has resulted in the development of the DIP method of mounting, as demonstrated by one form of this relay in Figure 4.40 (see also Figure 4.3).

Before the design is finalized, it is best to examine the matter of accessibility. The ideal arrangement in this regard is to have relays mounted so that they are accessible for service, but discourage tinkering. Use of a relay having a cover firmly in place and practically impossible to remove, as on the relay of Figure 4.41, is one solution. This relay mounts on a PC card, and the entire assembly is replaced when being serviced. It is discouraging (as well as expensive) for an authorized service person to have to dismantle a complicated device, partially just to see if the contacts on a relay are in physical contact or are in need of cleaning, for example.

FIGURE 4.40 A relay in a DIP enclosure for mounting on a PC board. *(Magnecraft Electric Co.)*

Terminals and connections A variety of wire terminations, as used for connecting the relay into a circuit, are available for many types and kinds of relays. Some have only one kind, possibly by decree of the governing industry association. Machine tool people, for example, prefer screw terminals. Connections to relays can be made as follows: lead wires (not usually a good idea); terminals shaped for insertion into a PC board and then to be soldered; solder terminals of the tab, tang, eyelet, or other types; screw terminals; quick connectors (AN and similar types); taper tabs; taper pins; solderless (gas-tight) wrap; Termi-point; plugs and sockets; and DIP. Some relays accommodate to several of these, some to only one or a few. The manufacturer's catalog is usually sufficient for finding out what is available in a particular relay type, but sometimes it is necessary to ask about specific needs.

FIGURE 4.41   A small telephone-type relay designed for insertion into and soldering directly to a PC board. *(GTE Automatic Electric Company, Inc.)*

Environment For some kinds of relays, any environment can be accommodated. Other designs might do a good job in only one kind of environment. Usually, however, there is one type of relay that functions best in any specific environment. It is not good engineering, therefore, to try to fit a personally preferred type of relay to an environment to which it is not ideally suited. For example, it is not good practice to use military-type relays of the hermetically sealed and shock-resistant variety in a stable communications setup, or conversely to use a telephone-type relay in a military application, where severe shock, vibration, excessive humidity, etc., are to be encountered.

The relay chosen should be just as accommodating to the environment as possible, but not over engineered. Some environmental extremes occasionally encountered for which relays might have to be specially engineered are the extremes of shock, vibration, humidity, radiation, temperature, etc.—especially as encountered in airborne service and space applications.

Circuit requirements Some circuit requirements can dictate the relay type or size. For example, operate and release times of some circuits might require the use of copper collars, slugs, and sleeves to provide the needed time delays, as covered elsewhere in this section. Such relays have to be as large as a certain minimum size for best economy because the timing that can be generated by slugs or collars and sleeves of copper requires a certain mass and volume to provide the desired results. This would preclude the use of a very small relay. But if size and weight are of prime importance, such as on some airborne or space applications, then the required timing delays in the performance of the relay will have to be accomplished by other means (such as capacitors, diodes, or thermistors). Also, larger conventional relays are normally slower than smaller conventional relays, or relays designed for fast operation, such as reed contact relays, mercury-wetted contact relays, and polarized relays. Timing can, therefore, require a great deal of consideration during the early design activities.

Sequence of relay operation, exclusive of the operate or release time of any individual relay, can dictate that a particular kind of circuit performance be provided. Not all kinds of relay designs permit just any kind of circuit feature, so this, too, must be taken into account during the relay-selection stages.

Vendor capability In the end, the vendor is probably the best able to determine whether relays can be provided that will do a required job. Some original search of catalogs, literature, and engineering aids will be necessary on the part of the design engineer, however, to determine if the kind of relays the job needs seem to be available and what and how many manufacturers can provide them. Also vigilance is required to make certain the vendor does not overlook an important requirement.

If a vendor is to be reasonably certain to have what is needed and will satisfy the job requirements, all information pertinent to the performance of the required relay must be available. Specifically regulatory codes, specifications, and practices are pertinent, and/or any prescribed performance abnormalities must be known. The designer should not over specify. This not only can forestall the use of the ideal relay, but can increase costs needlessly and cripple the performance of even an ideal relay. On the other hand, the vendor must know all the pertinent facts, such as voltage extremes, circuit peculiarities, shock and vibration extremes, humidity and temperature extremes, military specifications, sensitivity needs, power-supply limitations, contact performance, life, marginal operating conditions, timing required and/or delays in operation and release that cannot be tolerated, contact insulation withstanding voltage, and permissible insulation leakage.

Some commonly overlooked pitfalls are shunt circuits that were not noticed, lamps in series or parallel with the relay coil, other relay coils in series or parallel with the operating coil of the relay, capacitors that charge or are discharged in series or in shunt with the relay coil, and large inductances that affect the performance of the relay—either by induction, due to proximity, or through an electrical feedback into the relay coil.

The National Association of Relay Manufacturers has prepared several specification checklists oriented to specific industries. A checklist for airborne applications, one of the broadest, is reproduced in Table 4.6. The type of purchase specification that might result is given in Table 4.7.

A too-common mistake is that of a relay purchaser who attempts to apply a specification, such as one of the military specifications (see IV of Table 4.6), to a job for which it does not fit, is not needed, or is a handicap to obtaining the best-performing or most economical relay. To aid both vendor and buyer in this matter, NARM offers the use of a specification as shown by Table 4.8.

Incoming tests and inspections   It is advisable to make adequate tests and inspections of representative samples of the purchased relays as soon as incoming shipments are being received. All too often, just simple visual inspections are made of the early relay shipments by the purchaser, who might merely count and then store the relays. It is too serious a hardship to both user and vendor to discover only as the relays are being checked out later in completely manufactured units that something is wrong or missing. Losses in both time and money could have been prevented or lessened had adequate attention been paid to the relay shipments as received. If an actual functioning unit in which the relays will be used is not available to inspect at the time the relays start arriving from the vendor, it is well worthwhile for the purchaser to simulate an operating unit accurately for purposes of relay testing. Relay failure found at this stage can be caused by the inability of the furnished product to meet the specification, or it might have resulted from an inadvertent omission in the specification or even a failure on the part of the purchaser to recognize a necessary requirement.

Relay characteristics   When it comes time to make at least a tentative choice, or choices, of relay(s) for a particular job, and the circuit and contact needs have been established, it is the individual relay type characteristics that dictate what kind of relay is best for an application. So that individual relay types can be properly evaluated, they are examined in the following section and their particular strengths and weaknesses are assessed. There is no particular significance to the order in which they have been chosen for purposes of discussion. It is mostly chronological, with the oldest being examined first.

The general-purpose relay   A generalized kind of relay has become designated by its manufacturers and most users as general purpose. This is not in accordance with the

TABLE 4.6 Checklist of Possible Relay Requirements for an Airborne Application[1]

I. Relay function

II. Description of equipment in which relay is used

III. Class of application

A. Military

B. Commercial

C. Industrial

D. Electronic

E. Communications

F. Commercial airborne

G. Other

IV. Applicable documents

A. Military specifications

1. MIL-R-5757E

2. MIL-R-6106F

3. MIL-STD-202C

4. Other

B. Underwriters' Laboratories (UL)

C. Canadian Standards Association (CSA)

D. National Electrical Manufacturers Association (NEMA)

E. Electronics Industries Association (EIA/RETMA)

F. Quality assurance specifications

G. Reliability specifications

V. Environmental tests

A. Nonoperative

1. Thermal shock

2. Sealing

3. Salt spray

4. Humidity

B. Operational

1. RF noise

2. Vibration

3. Altitude

4. Shock

5. Temperature range

(a) −55 to +85°C

(b) −65 to +125°C

(c) Other

6. High- and low-temperature operation

7. Temperature cycling

8. Acceleration

9. Random drop

VI. Contact specifications

A. Form designation

B. Loads (specify each pole separately if loads are different)

1. Current

2. Voltage

3. AC or dc

4. Frequency

5. Resistive

XL Mounting methods

XH. Termination

A. Terminal type

1. Solder

2. Screw

3. Wedge

4. Solderless wrap

5. Pin type (printed-circuit or plug-in)

B. Method of connection

1. Welding

2. Soldering

C. Terminal strength

XIII. Marking

A. Type designation

B. Part number

6. Inductive

  (a) Power factor

  (b) L/R ratio

7. Motor

  (a) Starting-current transient

  (b) Locked-rotor current

8. Lamp

  (a) Inrush current

  (b) Time to reach steady-state current

C. Transient conditions (provide calibrated CRO photograph)

D. Circuit diagram

E. Rate of operation

F. Overload

TABLE 4.6 Checklist of Possible Relay Requirements for an Airborne Application[1] *(Continued)*

VII. Coil specifications	A. Pickup values
A. Resistance	B. Dropout values
B. Impedance	C. Operate time
C. Ac or dc	D. Release time
D. Frequency	E. Contact bounce
E. Voltage	F. Contact chatter
1. Nominal	G. Instrumentation
2. Minimum	H. Temperature
3. Maximum	X. Enclosures
F. Current	A. Open
1. Nominal	B. Dust cover
2. Minimum	C. Hermetically sealed
3. Maximum	D. Size limitations
G. Duty cycle	C. Date code
1. On-off ratio	D. Manufacturer's code
2. Magnitude of on time	XIV. Life expectancy
(a) Minimum	A. Mechanical
(b) Maximum	B. Electrical
H. Repetition rate	XV. Failure criteria
I. Circuit diagram	A. Minor
VIII. Electrical characteristics specifications	B. Major
A. Contact resistance	C. Catastrophic
B. Insulation resistance	XVL Qualification tests
C. Dielectric strength	XVII. Acceptance tests
1. Sea level	XVHI. Procurement factors
2. High altitude	A. Quantity required
IX. Operational specifications	B. Delivery schedule
	C. Cost limitations

NARM definition, which says that any relay not a special-purpose or a definite-purpose relay is a general-purpose relay. Thus, by NARM's definition almost any type of relay could qualify as general-purpose, regardless of shape, size, or construction. This is logical, but usage and habit decree otherwise. The commonly recognized general-purpose (GP) relay usually has a clapper-type armature, leaf springs, button contacts, and an L- or U-shaped heel-piece, with the coil pulling directly on the clapper-type armature, and the movable contacts attached to the armature. General-purpose relays are available in roughly three (or more) duty ranges, light (up to 2 A), medium (2 to 5 and 10 A, as in Figure 4.42), and heavy, or power-type general-purpose (contacts rated is A or more, as in Figure 4.43). It needs to be restated that a contact that is too large for the job will often fail because the actual current is insufficient to break down surface insulation buildup. Therefore, a relay of this type is not the solution for all problems—even though the contact loads to be handled do not exceed the ratings.

Some advantages for the general-purpose relay are relatively low first cost and wide availability. It is usually a shelf item from stocking distributors. Disadvantages are that a

TABLE 4.7 Examples of a Detailed Relay Specification Resulting From Use of Table 6[1].

Item	Checklist Reference
Relay is required to switch audio-frequency circuitry in radio receiver	I
Model 9999 airborne equipment	II
This equipment will be used on commercial airlines, and the following IIIF and IVA documents are applicable: MIL-R-5757E, MIL-STD-202C, and MIL-R 6106F. Only those paragraphs specifically mentioned in this detail specification apply. In case of any discrepancy, the detail specification shall govern	
The following environmental specifications apply:	
1. Thermal shock per Test Condition B of Method 107 of MIL-STD-202C	VA 1
2. Sealing Test II per Paragraph 4.8.4.2 of MIL-R-5757E	VA 2
3. 100-h salt spray test per Paragraph 4.8.13 of MIL-R-5757E	VA 3
4. Humidity per Moisture Resistance Test Method 106A of MIL-STD-202C, except eliminate Paragraph 2.4.2	VA4
5. Vibration Test I of Paragraph 4.8.11.1 of MIL-R-5757E	VB 2
6. Shock Test of 30 g per MIL-R-5757E Shock Type 4, Paragraph 4.8.16.1	VB 4
7. High-altitude performance at 70,000 ft.	VB 3
8. Relay shall operate over ambient temperature range of –65 to +125°C	VB 5b
9. High- and low-temperature test per MIL-R-5757E, Paragraph 4.8.9 shall apply	VB 6
This relay shall have a contact form C (SPDT)	VIA
Both A and B portions of the pole shall handle similar loads of audio-frequency levels of 30 mA min to 1 A max at voltages of 100 mV fo 8 V. Load will be basically resistive with power factor exceeding 0.8	
Normal rate of operation in equipment will be 4 c/min with equal on and off times	VIE
The relay coil resistance shall be 250 Q min at 25°C	VIIA
The nominal dc coil voltage shall be 28 V dc with a range of 24-32 V dc	VIIC and VIIE 1, 2, 3
Coil shall be capable of continuous duty over temperature range of –65 to +125°C	VIIG
Contact resistance shall not exceed 0.02 Q initially when checked by voltmeter-ammeter method with an open-circuit voltage of 1 V dc and a closed-circuit current of 100 mA	VIIIA
Relay contacts shall be closed before applying test-circuit voltage	
Dielectric strength at sea level shall be required by Paragraph 4.8.5.1 of MIL-R-5757E	VIIC 1
Dielectric strength at 70,000 ft shall be in accordance with Paragraph 4.8.5.2 of MIL-R-5757E	VIIIC 2
Insulation resistance of 1 mQ determined per Paragraph 4.8.6 of MIL-R-5757E	VIIIB
Relay shall pick up at 20 V dc max over the temperature range of –65 to +125°C, and shall drop out at 1 to 10 V dc over the temperature range of –65 to +125°C	IXA and IXB
Relay shall be hermetically sealed	XC
Size, mounting, and solder terminals to be per drawing 9999-1 (to be included as part of the specification)	XD, XI, and XIIA 1
Relay shall be marked per Paragraph 3.39 of MIL-R-5757E, items b, e, f, and g only	XIII
Electrically loaded life expectancy shall be per Paragraph 4.8.34 of MIL-R 5757E	XIVB
Failure criteria are categorized as follows:	Xv

1. Minor
   (a) Marking dimensions in error
   (b) 0.1-V deviation beyond allowable limits of pickup and dropout

TABLE 4.7 Examples of a Detailed Relay Specification Resulting From Use of Table 6[1]. *(Continued)*

Item	Checklist Reference
2. Major	
(a) Contact resistance exceeds 0.02 n but is less than 0.5 n	
3. Catastrophic	
(a) Failure of normally open contacts to make contact when coil is energized at 28 V	
(b) Open coil circuit	
5,000 relays required, with delivery to begin 60 days after receipt of order at a rate of 100 relays per week	XVIII

TABLE 4.8 NARM Standard Specification: Electromechanical Relays for Industrial and/or Commercial Applications

1.1.1	Scope
11.1.1	This standard covers general-purpose electromechanical relays for industrial and/or commercial applications
11.1.2	It is not intended to cover contractors, specific-purpose types of relays, circuit breakers, choppers, timers, or smaller allied devices
11.2	Reference Documents
11.2.1	"Engineers' Relay Handbook" sponsored by NARM, dated 1969
11.3	Environmental Section
11.3.1	Altitude
11.3.1.1	Altitude will not exceed 10,000 ft above sea level
11.3.2	Ambient
11.3.2.1	Operating-temperature range—enclosed relays: −20 to +40°C (refer to Par. 12.4.1.1) Open relays: −20 to +40°C
11.3.2.2	Storage-temperature range—enclosed and open relays: −55 to +70°C
11.3.3	Humidity
11.3.3.1	Relative humidity up to 50% will be considered standard
11.3.4	Shock and Vibration
11.3.4.1	Only normal shock and vibration conditions encountered in handling and shipping are considered applicable
11.3.5	Unusual Service Conditions
11.3.5.1	The use of relays at altitudes, ambients, humidity, shock or vibration other than that specified in Par. 12.3.1, 12.3.2, 12.3.3, and 12.3.4 shall be considered as a special application. Other unusual service conditions where they exist will be called out to the manufacturer, such as: (a) Excessive dust (b) Excessive fumes (c) Excessive sprays (d) Excessive corrosion (e) Excessive oil and oil vapor (f) Excessive dampness

TABLE 4.8 NARM Standard Specification: Electromechanical Relays for Industrial and/or Commercial Applications *(Continued)*

11.4	Mechanical and Physical Requirements Section
11.4.1	Physical
11.4.1.1	Enclosure: The relay enclosure refers to a protective enclosure which is fastened to the relay as an integral part at the place of manufacture and not an enclosure into which the complete relay is mounted with or without other components. The following enclosures shall be considered standard: (a) Dust cover (b) Gasket sealed (c) Hermetically sealed
11.4.1.2	Terminals: The following terminals shall be considered standard: (a) Screw type (b) Threaded stud with nut and hardware (c) Solder lug (d) Plug-in for socket mounting (e) Printed circuit board mounting (f) Quick connect (disconnect) (g) Clamp- or crimp-type terminals (h) Solderless wrap
11.4.1.3	Coils: Construction: Coil winding may be untreated, molded, vacuum-impregnated, varnish dipped or brushed
11.5	Electrical Requirements
11.5.1	Voltages: The following nominal voltage ratings shall be considered standard:
11.5.1.1	AC voltage: 6 V; 12 V; 24 V; 48 V; 120 V; 208 V; 240 V; 480 V; 600 V
1.5.1.2	DC voltage: 6 V; 12 V; 24 V; 48 V; 120 V; 240 V
11.5.2	Coils:
11.5.2.1	Range of operation: Relays are to operate satisfactorily over a range of voltage from 85 to 110% of rated nominal voltage on ac coils, and 80% of rated nominal voltage of 110NO of rated voltage on dc coils. Relays will be required to pickup and seal at the minimum voltage with the coil at ultimate operating temperature due to the nominal coil voltage. The coil shall be able to withstand 10% above rated nominal voltage without injury. The above tests to be conducted at 25°C.
11.5.2.2	Duty cycle: Continuous-duty coils will be considered standard except pulse-operated coils as used in latching, stepping, etc., may be considered standard when operated within their specified duty cycle
11.5.2.3	Temperature rise: The temperature rise of the coil or coils shall be limited to the allowable rise for the insulation used.An optional method to determine coil-temperature rise is that specified in UL Standard 508 11.5.2.4 Winding Tolerance: If relay coil resistance is rated or specified, the standard winding tolerance at 25°C shall be +10%
11.5.3	Contacts:
11.5.3.1	Unless specified otherwise, contacts are rated on basis that: (1) Each pole is capable of controlling the rated load (2) All circuits controlled by a given pole are of the same polarity
11.5.3.2	Contacts may be rated in these terms for the following types of loads as specified by the user: (1) Resistive—in terms of continuous current and nominal voltage (2) Motor load—

TABLE 4.8 NARM Standard Specification: Electromechanical Relays for Industrial and/or Commercial Applications *(Continued)*

	(a) Horsepower—in terms of horsepower and voltage
	(b) Continuous current and in-rush current at nominal voltage
	(3) Lamp—in terms of type, watts, and volts
	(4) Inductive—
	(a) DC—DC inductive rating shall call out the amount of inductance by specifying either maximum number of henrys or a maximum L/R ratio
	(b) AC—AC inductive rating is specified by indicating minimum power factor
11.5.3.3	Contact must be able to control rated load (service rating). An optional method to determine performance is that specified in UL Standard 508
11.5.3.4	Life—Ratings must be determined by application requirement. Relay life varies with the application and is not directly related to ratings. When life is specified, the following levels are preferred or should be used as a guideline for actual application requirements: No. of operations (electrical)

(1)	10,000
(2)	25,000
(3)	50,000
(4)	100,000
(5)	250,000
(6)	500,000
(7)	1,000,000
(8)	3,000,000
(9)	5,000,000
(10)	10,000,000
(11)	100,000,000 and over

11.6	Test Section: The performance for relays meeting this standard shall be tests performed on units in accordance with the requirements listed below. All tests shall be performed at room temperature with the relay in its normal mounting position. Performance tests shall include:
11.6.1	Visual inspection to ensure compliance with outline drawings and standards of good workmanship
11.6.2	Pickup voltage—All relays are to pick up at 85% of rated nominal ac voltage and at 80% of rated nominal dc voltage.
11.6.3	Contact and coil continuity—Check coil continuity and/or resistance.Check contact continuity with the relay energized and deenergized
11.6.4	Dielectric strength—Test dielectric strength to values as specified by the manufacturer
11.6.5	Life testing—All relays shall meet manufacturer's standard and/or special ratings when applicable and tested as specified by the manufacturer
1.1.7	Marking
11.7.1	Relays manufactured to satisfy this standard will include as visible minimum marking: (a) Manufacturer's name or trademark (b) Manufacturer's part number (c) Coil rating (voltage and frequency when applicable)
11.7.1.1	The following would be considered optional marking requirements: (a) Contact rating (voltage and current) (b) Customers designation (c) Circuit-connected diagram

general-purpose relay might not fit any particular job well because the coil is a generalization and not readily tailored for marginal current and/or specific timing, it is frequently position-sensitive, the life is less than for many other designs, and its very size, shape, and arrangement invite tampering. It is usually not shock or vibration-resistant, and it frequently presents mounting problems.

FIGURE 4.42 A typical general-purpose relay. *(Potter & Brumfield, Div. of AMF, Inc.)*

The contact rating for relays of this type is usually stated in the vendor's catalog as 2, 5, or 10 A at 120 Vac, 0.8 power factor, or 28 Vdc. If it is capable of handling motor and such kinds of loads, the horsepower rating is given. Underwriters' recognition will be indicated if applicable, as is the case for the relay of Figure 4.42.

Contact resistance is the Ohm's-law resistance measured at closed contacts and is usually in the order of 50 to 100 mΩ for new, clean contacts. Insulation resistance is typically 1000 mΩ, minimum, at 500 Vdc.

Life expectancy for the contacts can be specified at some value such as 100,000 operations at a specifically indicated electrical load. The mechanical life of the mechanism is usually indicated as 10 million operations. For lighter-than-rated loads, the contacts could presumably last as long as the mechanism.

General-purpose relays are used principally in the fields of air-conditioning and heating equipment, household electrical appliances, coin-operated machines, control of low-wattage motors, some lighting controls, and some kinds of elevator controls. A special design of this type of relay is used on automobiles and trucks.

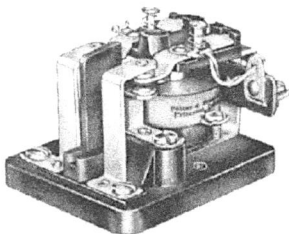

FIGURE 4.43 A typical Underwriters' listed power relay. *(Potter & Brumfield, Div. of AMF, Inc.)*

The power-type relay The appearance of most power-type relays is much like the general-purpose relay, only larger or more rugged (see Figure 4.43). Usually, the insulation is thicker or of superior material, the terminals larger, usually screw type, and in general, favorably looked upon by the underwriters; hence, it might be UL listed. The contacts are adequate for quite heavy current and highly inductive loads, with large armature strokes and contact gaps. Thus, the sensitivity is not great. Contact-current rating is usually 20 to 25 A or more. They are frequently position sensitive and usually not resistant to shock or vibration. There is not much latitude in mounting method or location.

The advantages of relays of this kind are they can best handle heavy contact loads, heavy fixture wire is easily attached to the coil and contacts, repairs can be made by relatively inexpert maintenance personnel, without sophisticated tools, and visual inspection of the contacts to determine probable remaining life is relatively easy. Also, in power-handling and switching situations, the rugged appearance builds confidence in the user's mind that the best relay for the job has been used.

Power-type relays usually are used in these fields: commercial aircraft (when of a specially designed type), air-conditioning and heating equipment, household electrical appliances, electric power control machine-tool control, and some military applications (when of a specially designed type). The specific job is usually electric motor control.

The telephone-type relay Telephone-type relays were developed and perfected during decades of application to the switching and signaling needs of wired communications, where the contacts of the same relay are required to successfully carry wide ranges of power extending from "dry" voltage voice circuits to the medium-power levels used in the actual switching of relays and other electromagnets. This versatility, with respect to handling a variety of power levels, made relays of this type valuable commercially in other fields with similar requirements.

The original telephone-type relays were approximately 4 inches long, 1½ inches high, and varied in width from approximately 1¾ to 2¼ inches, depending on the number of contact springs in the pileup(s). Figure 4.13 is a generalized concept of such a relay.

Size and weight reduction were problems that eventually faced the designers of telephone-type relays, resulting in an intermediate-sized relay, as demonstrated by Figure 4.2. This relay has roughly one-half the volume of the full-sized telephone-type relay, with the reduction being principally one of length.

Still later, designs incorporating the permissive make technique (described earlier), and demonstrated by Figures 4.14 and 4.41, evolved. These relays sacrifice something in the way of controlled timing, contact versatility, and flexibility, but lend themselves particularly well to the modern mounting methods and circuit-design concepts.

A comparison of timing delays obtainable from full-sized telephone-type relays and reduced size, but similar relays is shown by Table 4.9. The miniature, permissive-make type of relays use coils too small for the addition of sleeves or slugs; hence, any timing desired beyond what is inherent requires the addition of capacitors, diodes, or other extraneous circuit devices to accomplish what is wanted.

The advantages for telephone-type relays in general are high mounting density, a practically unlimited kind and quantity of contact forms and materials, good sensitivity, high order of contact reliability, good control of and practical variation in timing, fairly insensitive to mounting position, capable of withstanding moderate shock and vibration both while operating or at rest, and moderately light and small. Telephone-type relays are not readily adaptable to heavy-duty contact loads, usually are not capable of operating under heavy shock or vibration conditions, are not readily adaptable to underwriters' requirements for insulation resistance and voltage breakdown, and are difficult for inexperienced personnel to service on the job (contact springs are small and closely spaced). In an attempt to alleviate the latter difficulty, a great variety of enclosures and mountings are available.

Telephone-type relays have appeared in practically every field of application, sometimes by dint of heavy modification. In general, they are readily applicable to business ma-

TABLE 4.9 Timing Limits of Telephone-Type Relays

Slug or sleeve position	Large size		Reduced size	
	Operate time, ms	Release time, ms	Operate time, ms	Release time, ms
Armature end	100–150 max, 25–40 min	300–750 max, 75–200 min	25–40 max, 10–15 min	40–120 max, 15–25 min
Heel end	4–25	300–750 max, 75–200 min	7–10	40–120 max, 15–25 min
Sleeve (full length of coil core)	25–75	300–750 max, 75–200 min	10–25	40–120 max, 15–25 min

SOURCE: Industrial Products Division, GTE Automatic Electric Company, Inc.

chines, coin-operated machines, telephone and all other wired communications systems, radio and microwave systems, computer input-output devices, electronic data processing, laboratory test instruments, lighting controls, machine-tool control logic, production test equipment, street traffic control, and military ground-defense systems. When properly modified and/or protected from environment shock and vibration, telephone-type relays have been used on commercial aircraft and military aircraft, in aerospace, and on naval shipboard.

The dry-reed (contact) relay Another telephone-industry design innovation has captured a large share of the relay market. In this case, the relay electromagnet (coil) generates a flux that acts directly on the contacts without using any linkage in the form of an armature. Two normally separated, electrically conducting, and magnetic-flux-conducting elements, in a sealed glass envelope, and provide a portion of the main flux path of the coil so that when the coil is energized, these elements are attracted to each other to form a closed contact. They can also be permanent magnetic-flux-biased so that they are normally closed, but are open when the coil is energized sufficiently to neutralize the permanent-magnet flux. This kind of a relay, in its simplest form, is pictured in Figure 4.15.

Thus far, this section has covered relays having armatures to actuate the contacts. Such relays have long been referred to as *electromagnetic* because the armature was attracted by the flux from an electromagnet, as opposed to thermal relays, operating from heat, electrostatic relays, operating from opposite electrical charges, etc. Many people would now like to reserve the designation of *electromechanical* for relays with armatures, and *electromagnetic* for the reed relays. Currently, such a practice is more confusing than clarifying, but perhaps in the future this kind of a distinction might be embraced. If or until then, this class of relays will be most often recognized by the term *reed relay* with further identification as to type, such as dry reed, mercury-wetted reed, and Fereed (principally used in telephony, and is not ordinarily available as a commercial item). These all differ from the resonated relay, which is a frequency-sensitive mechanically resonant device, to be covered later.

The reed switches used in reed relays come in a wide variety of forms and sizes. It has been mentioned that the basic switch is a normally open, or Form A. When biased with a permanent magnet, it becomes a Form B, or normally closed. Originally all Form-C relays were a pair of these wired together. Now there are "true Form C" contacts, in which the movable contact member is caused to move from its normally closed (biased) position to the other position when the operating coil is energized. Sizes of contacts are difficult to categorize because there is little standardization among manufacturers. Roughly, they can be described as:

- The "regular" Bell Laboratories original design, which was approximately 3¼ inches in length (2 inches glass length) and ⁷⁄₃₂ inches in diameter
- The so-called miniature, which is about 1⅝ inches long with a diameter of approximately ⅒ inches
- Microminiature, having a length of 1⅜ inches (or less) and a diameter of less than ⅒ inches.

Other applications are micro-micro, pica, etc. To determine what is meant by these terms, a detailed examination of the manufacturer's literature is required because no standardized terms yet exist regarding size.

There is a tendency to size similarity that makes some comparisons possible. Most relays are grouped roughly into the previous three categories for contact size, and the contact ratings and capabilities result in something similar to Table 4.7 for dry reeds with gold-plated or rhodium (special) plated contacts. Other ratings are achieved by wetting the contacts with mercury, differently shaping the glass envelopes, potting, encasing in epoxy, etc.

There is no clearly defined limit to the number of reed capsules that can be put into a single relay for operation from a common coil. Many manufacturers offer custom packaging, if the job size warrants it. All vendors have multicapsule standard offerings with all makes (Form A), combinations of makes and breaks (Forms A and B), or transfers. As an example of the variety offered, see Figure 4.44.

FIGURE 4.44   An example of possible variations in number of dry-reed switches available in a relay. *(C. P. Clare & Co.)*

Dry-reed relay structures can be categorized as open assemblies, enclosed assemblies, potted or molded assemblies, and hermetically sealed assemblies.

Among printed-board relays, the open assembly is frequently used. The basic structure, typically molded of an electrical-grade plastic, functions as a coil bobbin and provides a means of supporting the reed switches and relay terminals. In many cases, the switch terminals are formed so that they will insert directly into the printed board.

This same structure is often placed in a metal or plastic box, which is then filled with a potting material. In some instances, the box provides terminal support, further simplifying the internal structure or permitting the use of a self-supporting coil. These assemblies can offer improved resistance to environmental or handling stresses.

Molded relays are similar to the potted relays, but differ in the materials used. In these assemblies, the molding material provides the primary mechanical support for the switches and coil and also produces the finished external surfaces. Assemblies similar to the potted relays can be hermetically sealed, using a metal cover and a base with terminals mounted in glass-to-metal seals.

Plug-in and wire-in relay assemblies follow the general patterns described for printed-board relays. Plug-in relays usually use a potting material to support the coil-switch structure within the enclosure carrying the plug. Wire-in relays follow both the open and potted patterns. The DIP has become very much in demand (Figure 4.3).

In general, dry-reed devices can be characterized as quite susceptible to the influences of external magnetic fields. For this reason, and to improve magnetic coupling of the coil to the switches, many relays incorporate some form of magnetic shielding. Metal cases serve this function, as do internal wraps or plates affixed to the coil. In some instances, the magnetic shield is connected to a terminal, which can be grounded to provide electrostatic shielding, but in most cases, the electrostatic shield is nonferrous and separate from the magnetic shield.

Special dry-reed contact relays (high voltage)   Typical dry-reed switches are rather limited in their ability to withstand high voltages across their open contacts (the standard switch rated at 500 V rms). For special applications, switches that can withstand voltages as high as 10,000 V can be incorporated in assemblies similar to those described earlier. Terminal spacing is modified as required to withstand the voltages.

Although modifications for higher voltage typically incorporate special versions of the standard dry-reed capsule, similar versions of the miniature switches are also available. They are limited to about 2000 V.

Power   Reed switches capable of handling power in excess of the typical 15-VA rating of the standard switch capsule are available. Relays incorporating these switches, rated at 50 to 350 VA, are defined as power relays.

High insulation resistance   Reed switches manufactured under carefully controlled processes provide an insulation resistance between contacts in excess of 10^{12} Ω. Although most standard relay assemblies provide shunting paths which appreciably lower insulation resistance, special structures using appropriate materials and processes can preserve the basic high-insulation resistance capability of the switch.

Low thermal voltage   Relays typically produce a voltage between contact terminals as a result of differing temperatures between the junctions of the materials in the assembly. Changing ambient temperatures or heat produced by the relay coil cause temperature gradients within the relay. Dry-reed relays incorporating materials and assembly techniques, which minimize these effects are available.

Low noise   The cantilever reed members in a switch continue to move for a few milliseconds following contact closure. This motion can produce a variation in contact resistance, and it does cause a voltage to be generated between switch terminals. Relays using reeds and structural techniques that minimize the latter effect are called *low-noise relays*.

Low capacitance   Because the contact-overlap area of most reed switches is small, capacitance between contacts is small. When the switch is installed in a coil, this capacitance is paralleled by the comparatively large capacitance of individual reed blades to the coil. The resulting increased capacitance across contacts and the capacitive coupling from coil to reeds can be objectionable in some applications.

By interposing an electrostatic shield between the reed switch and the coil, the paralleling capacitances can be greatly reduced, with capacitance across contacts approaching basic switch capacitance. In multipole relays, the electrostatic shield can be interposed between the switch group and the coil, or can also be interposed between individual switches.

Cross point   Relays used in matrix applications are called *cross-point relays*. Reed relays adapt readily to the various schemes of "no response to one input—response to two inputs" and have been used extensively in matrices.

Logic devices   Reed relays readily adapt to the performance of logic functions through the addition or subtraction of magnetic fluxes produced by multiple coils.

Electrical characteristics   Contact ratings are shown on a generalized basis in Table 4.10. Specific figures from individual manufacturers vary somewhat from these values. Other data, as presented here for such things as timing, sensitivity, contact bounce, and capacitance, are also at variance with the claims of any specific manufacturer or product.

Sensitivity   Power input required to operate dry-reed relays is determined by the sensitivity of the particular reed switch used, by the number of reeds operated by the coil, by the permanent magnet biasing used, and by the efficiency of the coil and the effectiveness of its coupling to the reeds. The minimum input required to effect closure ranges from the very low milliwatt level for a single-capsule "sensitive" unit to several watts for a standard multipole relay.

Operate time   Coil time constant, overdrive, and the characteristics of the reed switch determine operate time. With maximum overdrive, standard reed switches will operate in just under 1 ms; miniature reeds in 500 μs, and microminiature reeds in less than 200 μs. At nominal drive levels, operate times will be two to three times these values.

The other end of the operate time spectrum is less definable because the coil time constant and drive level are the primary determinants. However, with the low inductance typical of reed relay coils, the operate times of even the standard reeds rarely exceed 10 ms.

TABLE 4.10 Dry-Reed-Relay Contact Capabilities[1]

Reed size	Ratings
Standard	Load: 15 VA, 1 A max, 250 V ac max, 3.0 A (carry) Withstanding voltage: 500 rms, 60 Hz Insulation resistance: 10^{11} Ω min Initial contact resistance: 40–100 mΩ Life: 20 million operations at rated load
Miniature	Load: 10 VA, 0.75 A max, 200 V dc max, 1.0 A (carry) Withstanding voltage: 250 rms, 60 Hz Insulation resistance: 10^{10} Ω min Initial contact resistance: 100–200 mΩ (regular), 100 mΩ (with special plating)
Microminiature	Load: 10 VA, 0.50 A max, 100 V dc max Withstanding voltage: 200 rms, 60 Hz Insulation resistance: 10^{10} Ω min Initial contact resistance: 100–250 mΩ (regular), 100 mΩ (with special plating)

Figure 4.45 shows the operate time of one manufacturer's standard reed relay with a single Form A switch.

Release time With the relay coil, dry-reed switch contacts release in a fraction of a millisecond. Miniature and microminiature Form A contacts open in as little as 10 to 20 µs. Standard switches open in 100 µs. Magnetically biased Form B contacts and normally closed contacts of Form C switches reclose in from 100 µs to 1 ms.

If the relay coil is suppressed, release times are increased. Resistor-capacitor suppression usually has the least effect. Zener-diode suppression stretches release time somewhat more. Diode suppression can delay release for several milliseconds, depending on coil characteristic drive level, and reed-release characteristics.

Figure 4.45 also shows a value for expected release time with an unsuppressed coil. Faster and more positive release can be obtained by using two exactly equal, but opposed windings (Figure 4.46).

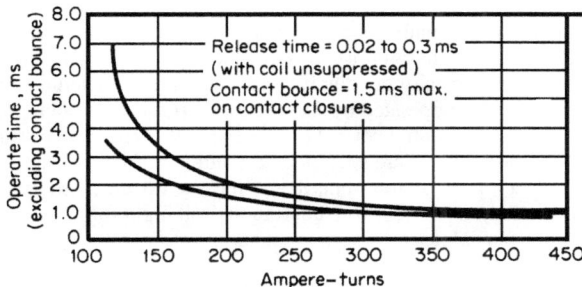

FIGURE 4.45 Typical relationship of operate time to operating coil power for a standard-size switch in a dry-reed relay. *(C. P. Clare & Co.)*

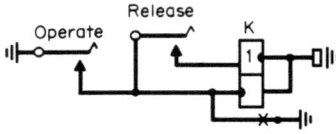

FIGURE 4.46   Forced release of a dry-reed relay

Bounce   As with other hard-contact switches, dry-reed contacts bounce on closure. The duration of bounce is typically quite short, and is in part dependent on drive level. In some of the faster devices, the sum of operate time and bounce is relatively constant as drive is increased, operate time decreasing, and bounce increasing.

Although normally closed contacts—those that are mechanically biased—bounce more than normally open contacts, magnetically biased Form-B contacts exhibit essentially the same bounce as Form As.

Typical bounce times (mechanically biased normally closed contacts) in milliseconds are:

Standard
Normally open   0.50
Normally closed   2.5
Miniature
Normally open   0.25
Normally closed   2.0
Microminiature
Normally open   0.25

Capacitance   Reed capsules typically have low terminal-to-terminal capacitance. However, in the usual relay structure, where the switch is surrounded by a coil, capacitances from each reed to the coil act to increase basic capacitance many times. If the increased capacitance is objectionable, it can be reduced by placing a grounded electrostatic shield between the switch and coil.

Typical capacitance values, in picofarads, for relays using Form-A switches are:

Relay Type	Unshielded		Shielded	
	Across contacts	Closed switch to coil	Across contacts	Closed switch to shield
Standard	1.0	4.0	0.2	5.0
Miniature	0.7	3.0	0.1	3.5
Microminiature	0.4	2.0	0.08	2.5

Where the capacitance from contact to shield is objectionable, greater spacing or unique methods of coupling the coil to the contacts can be used.

Thermal EMF   Because thermally generated voltages result from thermal gradients within the relay assembly, relays built to minimize this effect often use sensitive switches to reduce required coil power and thermally conductive materials to reduce temperature gradients. Latching relays, which can be operated by short-duration pulses, are often used if the operational rate is such that the potential benefit of reduced duty cycle can be realized.

Measurements of thermal emf are specified in a number of ways, each suited to a particular application. One of the more standard, and the one documented in MIL-R-5757E, is measurement at maximum ambient with continuous coil input for a time sufficient to ensure temperature stability. Measured in this manner, relays can be supplied to specifications with limits as low as 10 μV.

Noise   In reed relays, *noise* is defined as a voltage appearing between terminals of a switch for a few milliseconds following closure. It occurs because the reeds are moving in a magnetic field and because voltages are produced within them by magnetostrictive effects. From an application standpoint, noise is important if the signal switched by the reed is to be used in the few milliseconds immediately following closure, if the level of noise compares unfavorably with the signal level, or if the frequencies constituting the noise cannot be filtered conveniently.

When noise is critical in an application, a peak-to-peak limit is established, with measurement made a specified number of milliseconds following application of coil power. Measurement techniques, including the filters that are to be used, are also specified. MIL-R-5757E, for example, sets a peak-to-peak limit of 50 µV at 10 ms with frequencies below 600 Hz and above 100 Hz attenuated.

Vibration   Except at resonant points, reed switches do well when subjected to vibration. With vibratory inputs reasonably separated from the resonant frequency, the relay will withstand relatively high inputs, 20 g's or more. At the resonance of the reeds, the typical device will fail at very low level inputs.

Typical resonant frequencies in hertz are:

Standard	800
Miniature	2500
Microminiature	5000

Shock   Dry-reed relays withstand relatively high levels of shock. Form-A contacts are usually rated as able to pass 30 to 50 g's, 11 ms, half-sine-wave shock, without false contact operation. Switches exposed to a magnetic field tending to close them, such as in the biased latching form, demonstrate somewhat lower resistance to shock. Normally closed contacts of mechanically biased Form-C switches might also fail at somewhat lower levels.

Radiation   The basic reed switch is quite resistant to radiation and has been used to perform control functions in hot environments. Coils and supporting structures utilizing appropriate materials permit the construction of reed relays resistant to radiation.

Dry-reed relays are quite adaptable to electronic applications and hence are found in electronic equipment in many fields, such as logic circuits. Some fields to which they are especially suited are business machines, telephone and other wired communications systems, radio and microwave systems, computer input-output devices, electronic data processing, laboratory test instruments, machine-tool control (logic circuits, particularly), production test equipment, and military ground-defense systems.

Advantages existing for dry-reed relays are that they are small, fast-acting, provide good isolation of input to output, are readily used with solid-state circuitry, mount easily on PC cards, and require relatively small amounts of power. Their chief weakness is a tendency toward contact sticking and unwanted welding. At times, it is difficult to locate the source for this kind of a problem. Such an unsuspected thing as accumulated cable capacitance might be the cause of this kind of failure. Lamps are also a frequent source of such trouble.

Resonant reed relays   This type of relay is designed to respond to a given frequency of coil input current. It consists of an electromagnetic coil that, when energized, drives a vibrating reed with a contact at its end. When the coil input frequency corresponds to the resonant frequency of the reed, the reed will vibrate and cause its contact to touch a stationary contact and thereby close a circuit once each electrical cycle. At other frequencies, the reed does not respond. Sometimes the reed is surrounded for a portion of its length with a permanent-magnet field to provide a constant magnetic bias. Because the vibrating reed closes its contact only for a portion of each cycle, it is often necessary to provide an output circuit that will store these pulses long enough to operate a conventional relay for control purposes.

Resonant-reed relays can be built with a number of reeds having frequency responses in discrete steps, thus providing a device that will produce signals on either side of a desired frequency for control purposes. Resonant-reed relays are also used for a variety of

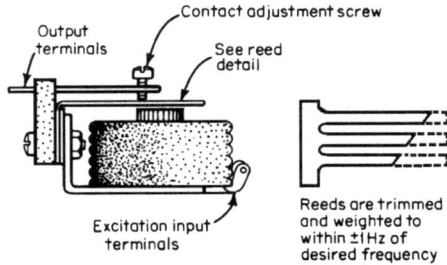

FIGURE 4.47   The physical construction of a typical resonant-reed relay.[3]

applications, where response to frequency only is desired, such as communications, selective signaling, data transmission, and telemetry.

The following material is from Ref. 3.

The physical construction of a typical resonant-reed relay is shown in Figure 4.47. Notice how closely it resembles a conventional relay, having a coil, a relay armature (resonant reed), and a fixed contact. In this case, the fixed contact is adjustable to obtain maximum dwell time from the reed. When the reed is gold plated, it resists environmental contaminants and provides an excellent conductive path. The upper portion of Figure 4.49 shows how the reeds are "tuned."

These reeds respond to almost any frequency in the audio range. For practical operation, the range of 100 Hz to 10 kHz can be considered useful for exciting resonant-reed relays. Harmonic rejection is usually quite good within this range, and simultaneous tone sensing is possible under most circumstances. Thus, if a resonant-reed relay has 8 or 10 tuned reeds, it is quite possible that two or three could be excited at the same time by a similar number of input tones. For practical purposes, the designer would probably select reed frequencies that nullify harmonic interaction if concurrent excitation is required.

The circuit configuration of these devices is shown in Figure 4.48. Obviously, the resonant-reed relay is a very elementary device whose theory of operation is within the understanding of almost anyone. However, its simplicity can be misleading to those who do not understand its versatility and usefulness. A resistance/capacitance filter, such as the one shown in Figure 4.48, can be used to reduce the dc ripple in the output. However, buffer amplifiers working from these devices must be biased below the ripple because it is not practical to eliminate all of it. The common connection for the reeds and excitation coil is optional and is undesirable in some cases.

The functional aspects of resonant-reed relays are their biggest selling point. Regardless of the number of relays used the input consists of a tone generator and a single-pair transmission line. Theoretically, the tone generator will produce programmed discrete frequencies. The programming can consist of presetting potentiometers, or more elaborate means for frequency control.

Figure 4.49 is a simplified diagram illustrating the data-link hookup between the tone generator and the resonant-reed relay network. Noise spikes that could be dangerous to semiconductor circuitry will have little effect on these devices. Therefore, elaborate filtering precautions are not necessary. The resonant frequencies of the reeds can be intermixed or repeated in almost any fashion in the relay stations. There is little or no interaction—even when a reed frequency is repeated on the same relay.

The power output requirement for the programmed discrete-frequency tone generator naturally depends on the number of relays it is driving. A wide range of relay coil imped-

$T\mu F$

100

RRR Contact

Typical contact — filter circuit

Output
A

Channel
1

B

2

C

3

Excitation
voltage

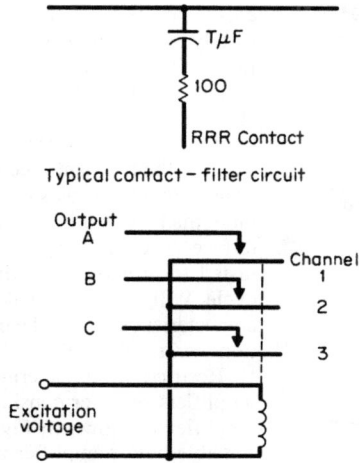

FIGURE 4.48 Circuit configuration and typical contact-filter circuit of a resonant-reed relay.[3] (Programmed discrete frequency.) Although only three channels are shown, as many as 10 to 12 resonant reeds can be excited by a single coil.

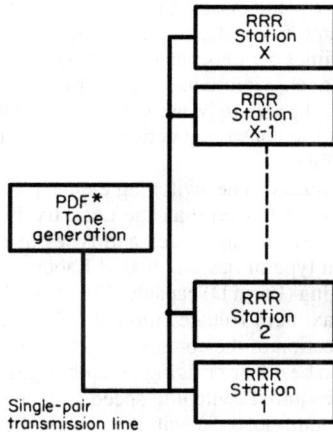

RRR
Station
X

RRR
Station
X-1

PDF*
Tone
generation

RRR
Station
2

RRR
Station
1

Single-pair
transmission line

*Programmed discrete frequency

FIGURE 4.49 Simplified diagram of data-link hookup between tone generator and resonant-reed relay.[3]

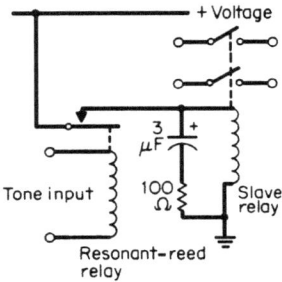

FIGURE 4.50 Method of driving a slave relay from contacts of resonant-reed relay.[3]

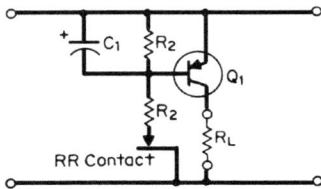

FIGURE 4.51 Method of using a buffer amplifier between a resonant-reed relay and the load.[3]

ances can be used, but an impedance of approximately 5 kΩ should be used for general calculations. In practice, impedances can vary from near 100 Ω to more than 10 kΩ. The relays are tolerant of considerable tone-signal distortion, although it is helpful to keep the signal as pure as practical.

If the resonant-reed relay is to drive a conventional slave relay, as shown in Figure 4.50, nothing more than a simple filter circuit is needed for the interface. The pulsating dc output signal can be adjusted to provide approximately a 50-percent duty cycle. With the additional filtering and the response characteristics of the slave relay, this direct connection works quite well.

Restrictive output criteria might call for buffer amplifiers or other considerations. One such buffer amplifier is shown in Figure 4.51. The number of possible buffer-amplifier configurations is no doubt apparent to the designer.

Because resonant-reed relays are, in effect, gates for tone-modulated signals, they can also be used as logic elements in some applications. To be sure, these are unusual logic elements, but it is not unusual to find sequential tone coding, which could be considered as a form of logic. With this, as with other applications, the designer would do well to consider all of the tradeoff factors before ruling out the use of resonant-reed relays.

Resonant-reed relays have same disadvantages. The contacts do not close with a firm positive closure; hence, many kinds of equipment cannot be easily controlled from the contacts. These relays sometimes demonstrate undesired frequency drift, caused by temperature extremes, tampering, shock, vibration, or other environmental extremes. Advantages, besides those covered in detail previously, are simplicity and low cost.

Because of the nature of the contacting action, loads other than those of the type described above are not advisable.

Mercury-wetted contact relays The switching element in mercury-wetted contact relays is considerably more sophisticated than the deceptively simple-appearing dry-reed switch. There are essentially two design sizes and types of mercury-wetted capsules in common usage. The original type of design, of Bell Laboratories parentage, was a maximum 5-A load-current bridging (Form D) capsule (Figure 4.52). High-pressure gas, sealed in the capsule, provides a maximum voltage rating of 500 and a load capability of 250 VA, but a resistor-capacitor protection of the contacts is essential.

A smaller capsule that can be either bridging or nonbridging and rated at approximately 2 A (or less), depending on required switching speed is demonstrated by Figure 4.52 (see also Figure 4.53). The maximum load capacity for the smaller contact is 100 VA, 500 V, and resistor-capacitor contact protection is required.

In both contact types, the electrical contacting is mercury to mercury, with the contacting faces renewed by capillary action drawing a film of mercury over the surfaces of the contact switching members as the movable contact member is moved from one transfer position to the other. The mercury film is drawn up from a pool at the bottom of the capsule, between the stationary members to provide bridging (Form D) contacting (Figure 4.54). Contact bounce and chatter are eliminated by the dampening of the mercury films. No solid

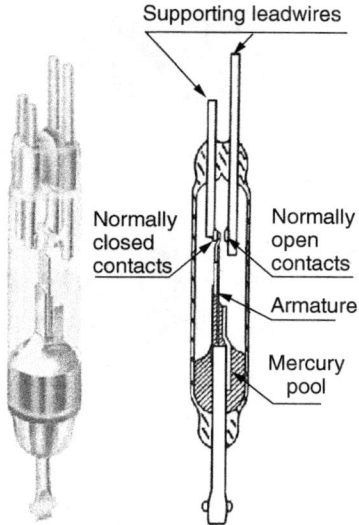

FIGURE 4.52 The basic design of mercury-wetted contact relay. (Contact Form D.) *(C. P. Clare & Co.)*

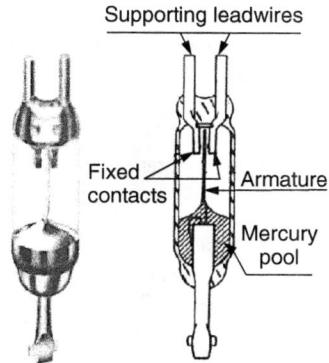

FIGURE 4.53 A smaller and faster mercury-wetted contact relay (can be either Form C or Form D contact). *(C. P. Clare & Co.)*

metal-to-metal contacting occurs; so the contacts are actually renewed on each operation. As a result, wide ranges of signal and power levels can be reliably switched without having the nature of the load affect either contact life or performance.

The action of the mercury is similar for Form-C (nonbridging) contacting, except that spacing is such that a droplet of mercury falls out before bridging can occur (Figure 4.55).

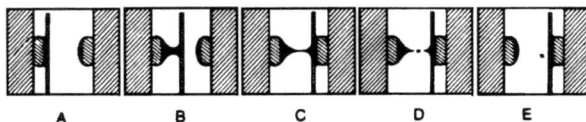

FIGURE 4.54 Form D action (SPDT make before break) in a mercury-wetted contact. *(C. P. Clare & Co.)*

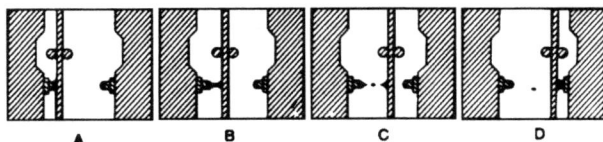

FIGURE 4.55 Form C action (break before make) in a mercury-wetted contact. *(C.P. Clare & Co.)*

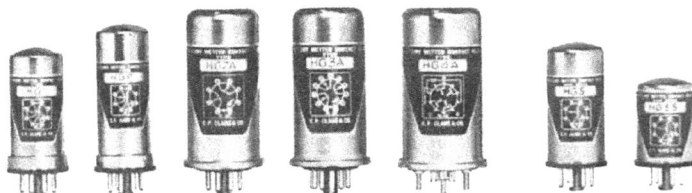

FIGURE 4.56 Octal-base types of mercury-wetted contact relays. *(C. P. Clare & Co.)*

FIGURE 4.57 A small mercury-wetted contact relay for PC board mounting. *(C.P. Clare & Co.)*

For one form of relay, from one to four capsules are operated by a single coil and housed in an octal-base electron-tube type of metal can. Examples of this kind of relay are provided by Figure 4.56. More compact forms, suitable for mounting directly on printed-circuit boards, are also available. With regard to mounting, it must be remembered that these relays are quite position-sensitive and must be used with the capsule right side up and with its axis tilted less than 20 to 30 degrees from the vertical, depending on type and manufacturer. After having been inverted, the contacts are flooded from the mercury pool and might not perform properly for some time. Another problem sometimes encountered is failure at low temperatures because mercury becomes solid at −38.8°C (−37.8°F).

Operating speeds for relays using the large capsule are of the order of 6 ms to transfer and 4 ms to restore. For the smaller capsule, these figures are about 1 ms each. An example of the small relay is Figure 4.57. One manufacturer's performance data for the various kinds of this relay are shown in Table 4.11.

Contact protection is a matter to be fully explored with the vendor, whose recommendations should be followed exactly if

TABLE 4.11 Typical Mercury-Wetted Contact-Relay Capsule Characteristics at Optimum Drive Conditions[9]

Capsule	Timing (speed)			Load (max)			Form C transfer open time, µs		Form D bridging time, µs		Sensitivity, mW	Min on and off time, ms	Contact noise at closure, 1-kHz band-width, µV
	Operate, ms*	Release, ms†	Frequency, Hz	VA	V	I	Min	Max	Min	Max			
High-speed	1.0	1.0	400	50	500	1	50	‡	50	‡	100	1.25	30
Intermediate-speed	1.25	1.25	250	100	500	2	50	350	50	800	100	2.5	60
Standard	2.0	2.0	125	100	500	2	50	500	50	800	100	4.0	75
Heavy-duty	4.0	2.5	80	250	500	5			150	900	500	5.0	50

* Measured at twice minimum operating voltage.
† Measured with nominal coil suppression.
‡ Form T contact; see switch action discussion.

good contact performance is to be achieved. A formula in common usage for a choice of capacitor-resistor for this purpose is:

$$C = \frac{I^2}{10 \ \mu F}$$

$$C_{min} = 0.001 \ \mu F$$

$$R = \frac{E}{10 \ I^{(1 + 50/E)}} \ \Omega$$

$$R_{min} = 0.5 \ \Omega$$

in which I = current immediately prior to contact opening

E = source voltage

A list of the advantages for the mercury-wetted contact relay would include these items: good operating sensitivity, fast contact switching, long contact life, great contact reliability, low and stable contact resistance, bounce-free and chatterless contacting freedom from detrimental effect of environment (such as dust, humidity, and changes in atmospheric pressure, good-contacting capability in the low-power region, protection from igniting hazardous atmospheres, and excellent isolation between coil input and contact output. Shortcomings for this relay is it is position-sensitive, cannot work at low temperatures, generates too much thermal emf under some conditions where it would otherwise be ideal, and is costly. It is ideal for pulsing highly inductive electromagnets, such as rotary stepping switches.

Fields of application for this kind of relay can be air conditioning and heating, business machines, wired communications, computer input-output devices, electric power control, electronic data processing, laboratory test instruments, lighting controls, machine-tool controls, production test equipment, and street-traffic control.

Mercury-wetted reed relays These relays are basically of the same type as the dry-reed relays, except that mercury has been added to the reed capsule during manufacture so that contacting occurs from mercury film to mercury film. There is little standardization, and the ratings and characteristics vary so from manufacturer to manufacturer that if this type of relay seems desirable for a job, the first order of business is to consult with the vendor.

The advantages for this kind of relay are obviously many of those which exist for the dry-reed relay, plus the advantages for mercury-to-mercury film contacting. Disadvantages are essentially the same as those listed for mercury-wetted contact relays. Some vendors claim to have solved the position-sensitivity and tilt problem.

The main purpose in adding the mercury to the capsule was to reduce and stabilize contact resistance, and to increase contact load-handling capability and life. The maximum allowable load varies so much from one reed capsule design to another that the vendor's catalog must be relied on entirely to determine what is available and whether or not it will do the job. One word of caution to be observed is that use of the vendor's prescribed contact protection is a must.

Application areas for these relays are essentially the same as listed for mercury-wetted contact relays, with obviously the fields of aviation, automobiles and trucks, and military being largely incompatible.

Heavy-duty power-type mercury contact relays Because contact erosion, with resulting need for maintenance, is the largest single problem for relays handling heavy power levels on their contacts, it was inevitable that the continuous contact-renewal capabilities of mercury would lead to early consideration of this metal for switching power. Rather than making the contacts through a thin mercury film, as was mentioned, the conduction in power-type mercury contacts is through a pool of mercury. There are two principal ways of accomplishing this. In the first method, the tilted mercury tube which causes the terminals to be bridged when the tube containing the mercury is in one position and nonbridged or open in the other position. In the mercury-displacement technique, a plunger is pulled down into the mercury of the pool so that a bridge of conducting mercury extends from one terminal to the other, thus closing a circuit over a dam that otherwise isolates one terminal from the other. In the latter method, when the coil is deenergized, the plunger floats back up, the mercury returns to refill the pool, and the circuit is opened.

Contact ratings usually vary from 10 to 35 A for 115 Vac and approximately half as much current at 230 Vac. Some types are rated up to 100 A and are UL and CSA listed. Dc ratings are considerably reduced from the ac relays because of the sustained arc possibilities. One example of a relay of this kind is shown in Figure 4.58.

Advantages for the heavy-duty mercury contact relays are these: the contacts give some protection against igniting an explosion in hazardous atmospheres (they are sealed); they are protected from corrosive, dirty, or moist ambient conditions; they are not noisy when operating; there is no appreciable maintenance; and usually there is some saving in weight and size. Disadvantages are, of course, those associated with position-sensitive liquids (such as mercury): inability to function at low temperatures, inability to function where there is severe shock or vibration, and a tendency to delay functioning longer than is the case for the more common power-type relays.

Crystal-can relays The size and appearance of relays changed greatly in World War II from what was in common usage previously. Detrimental environmental service conditions dictated that relays be hermetically sealed, shock- and vibration-resistant, and be made as small and lightweight as possible. In the absence of service-tested new designs aimed specifically at these applications, much of the relay use in World War II had to be satisfied by existing designs modified to meet the service requirements. In many cases, this resulted in compromises with regard to space, weight, and shock and vibration that caused extensive design effort in the direction of producing relays specifically aimed at the military's needs. Today's design is usually called a crystal-can relay or some fraction thereof, such as a half-crystal can or one-sixth-size crystal can. The reference is to the size and shape of the relay housing, originally used to house a frequency-control type of quartz crystal. The dimensions of a full-sized crystal can are usually assumed to be 0.4 by 0.8 by 0.97 inch. A picture of one kind of such relay with three different kinds of terminals is shown in Figure 4.59. Various methods of mounting, as well as kinds of terminals, are used. This is a specific kind, size, and shape of relay, but with many variations on its exterior.

FIGURE 4.58 A heavy-duty, power-type mercury contact relay. (*Magnecraft Electric Co.*)

The distinguishing design features of this relay are the dynamically balanced armature, designed-in resistance to shock and vibration, and hermetic sealing. A majority of them have their terminals spaced on a 0.2-inch grid, which makes them readily inserted into printed-circuit boards. Also, most relays of this type offer only two Form C contact combinations. Because the relays are small and lightweight, only relatively small contacts with fairly light pressures can be operated; hence the contact rating must be restricted to relatively light loads.

Relays of this type are furnished both as conventional energized-on and deenergized-off relays, and as bi-stable magnetically latched relays. Although developed for aerospace

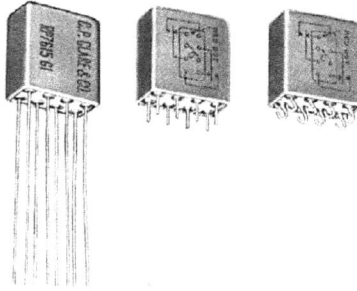

FIGURE 4.59   Typical crystal-can relays.
(C. P. Clare & Co.)

applications, their small size and convenience of mounting have made them a popular choice for many other uses. Later modifications in design in the interest of still smaller size and lighter weight have resulted in the half and one-sixth crystal can sizes. A still newer and smaller design is the TO-5, a small relay in the transistor case indicated by its title. This item will be discussed separately later, since it is of a different size and shape.

Advantages for crystal-can relays have already been enumerated as small, light, vibration and shock-resistant, adaptable to printed-circuit boards and solid-state circuitry, hermetically sealed, plentiful, and moderately inexpensive considering the complexity of design and manufacture. Disadvantages are that the inside mechanisms are inaccessible during use for inspection of remaining life, or even initial correctness and adequacy of adjustment. Because they are rather expensive to build, price is sometimes a disadvantage if a vendor does not have overrun or other kinds of residual from a military-contract job.

TO-5 relays   These relay designs were mentioned previously, but are mentioned separately because they differ from crystal-can relays in size, shape, and some intended applications. Many are essentially hybrid relays, which is covered later. They were aimed directly at logic circuits and relay sensing of low-level signals, originally in aerospace and military applications, but now in such fields as computers and computer-allied equipment.

Besides the one Form-C basic relay, there are these options: with internal diode for coil transient suppression, with internal diodes for coil transient suppression and polarity-reversal protection, and with internal transistor driver and diode-coil suppression.

In addition to the two Form-C basic relay, the options are these: with internal diodes for coil transient suppression and with internal diodes for coil transient suppression and polarity-reversal protection. There is also a magnetic-latching version. Four poles single-throw, might also be provided.

Relays of the TO-5 type have the advantages of being quite compatible with logic circuits and low-level signals, so they are used in computers and in control equipment. Their resistance to shock and vibration suggests their use in military, airborne, and aerospace applications.

Time-delay relay units (TDRs)   An earlier discussion indicated how individual relays had their inherent operate and release times modified by coil modifications and series and shunt circuits connected to the coil windings. In the case of sleeves and slugs, the time-delay features are built into the coil and are barely perceptible externally. The series and shunt circuits are generally introduced into the wiring external to the relay and hence are not really a unit assembly with the relay proper. Recently, as solid-state devices became

FIGURE 4.60 A hybridized plug-in relay with knob-adjustable time delay. *(Potter & Brumfield, Div. of AMF, Inc.)*

common and it became relatively easy to package the delay elements on or in the same enclosure as required for the relay, delay units consisting of hybridized circuits plus the relay were offered commercially. It has been noted that several models of the TO-5 are hybridized, but such hybrids are not especially aimed at producing time-delay features.

The most popular TDRs use a more or less conventional relay plus the required hybrid circuitry (usually on a printed-circuit card or hung on the frame of the relay), plus an enclosure used to combine all these elements into a unit. Adjustable timing, when required, is accomplished by altering the pot settings by means of a knob that can be turned externally, or a slotted shaft for screwdriver setting can be used. Figure 4.60 shows such a unit. A popular form of these is octal-plug mounted, although relays of this kind appear in all kinds of enclosures, with plug-in relay with knob or without the adjustable knob accessible from the outside. The adjustable knob might not be calibrated for identifiable time intervals, depending on need and allowable cost.

Because there is no industry-wide standardization for relays of this type, as to either design parameters or operating characteristics, a search of possible vendor's literature is usually necessary to locate a satisfactory source. All the kinds of timing functions can be handled, such as operate time delay, release time delay, generation of a delay interval with reset, momentary actuation, sequence timing with repetition, pulse generation, and interval timing. Repeat accuracy is a problem.

Because devices of this type can use any kind of relay, the only applications limitations for these timed relays are those that have already been noted for that particular kind of relay, with regard to shock, vibration, temperature, mounting position, etc.

This section has ignored timing relays of the thermal type because they are treated as a specific type later. Also, it is not within the scope of a handbook to examine so diverse a lineup of products as the various kinds and makes of dashpot relays (both pneumatic and oil-retarded), or motor-driven timers with associated relay or relay-contact type of output. A good technical discussion of this subject is presented in Ref. 5. As is pointed out there, some vendor's literature does a good job of stating what can and cannot be expected from their products in the way of accuracy at voltage extremes, repeatable accuracy with various values of elapsed time, and the effect of temperature extremes on initial timing accuracy.

Thermal relays   Not all thermal relays are used to create intentional delay, but there is an unavoidable time lag between current application and contact movement for any relay operating on this principle. Therefore, this relay is suitable for generating time-delay functions. The initial operate time can be well established, but there are two kinds of reoperate times: the operate time when the heater is starting from ambient and the operate time when there is residual heat from a prior operation. Release time is also affected by the length of time the heater was on. For a relay that had reached maximum temperature on operation, called *saturation*, the release time is sometimes several times longer than the operate time. Thermal relays are a heat-integrating device, hence are voltage-sensitive. Thus, if only an operate-time delay of no specific value is required, with adequate time for cooling, the thermal relay can be a satisfactory solution. One obvious use is to take advantage of the thermal time delay to prevent giving an alarm when an infrequently occurring under standard conditions of unlikely frequent repeatability has to be accommodated. The telephone industry, for example, has many such situations. Sometimes this voltage sensitivity is used in supply-voltage sensing and control. Thermal relays operate equally well on alternating or direct current and are free of the too-frequent tendency of other relays to contact chat-

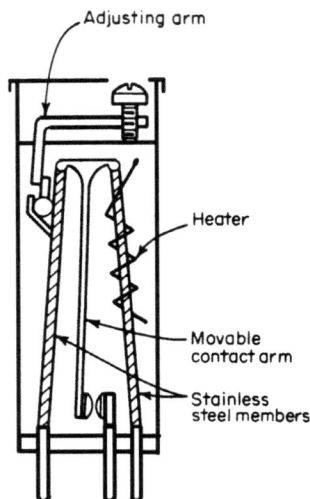

FIGURE 4.61 A schematic diagram of one kind of thermal relay.[1]

ter on alternating current. Thermal relays usually are not position-sensitive or affected by stray magnetic fields. These facts are sometimes used advantageously (Figure 4.61).

Contact capabilities vary from design to design and manufacturer to manufacturer, so the vendors' literature is the best guide in determining whether or not the load at hand can be handled. It is difficult to identify fields of application in which relays of this type are used as against fields in which they are not applicable because their usefulness is oriented more to function than to class of service. If they seem applicable to a particular job for the reasons given, it is time to consult with the vendor to make certain of a complete fit to the job at hand.

Hybrid relays Defined as a combination of mechanical-switch and solid-state circuitry, it is important to note that the mechanical switch might be on either end of the hybrid relay. For example, mechanical contacts are on the output end with movable time delay, but on the input end when gating a triac. As so often happens, usage of a term seems at variance with definition. For example, it has become a custom in some industries, and even by some relay manufacturers, to limit the term *hybrid relay* to relays composed of a reed switch and a solid-state device in combination, for example, a reed relay input and a triac output, or a transistor input to a reed relay. Surely these are hybrid relays, but it was never intended to so limit the term. Any relay and solid-state combination is a hybrid, as originally defined. By definition, then, some forms of the relays covered thus far can be considered hybrids—especially some of the TDRs. However, the combination of relay and one or more solid-state component(s) into a unit deserves considerable attention in a work of this kind, so the matter will be treated in detail. The danger in doing this is that the solid-state art is advancing so rapidly that the sought-after solutions of today might become too cumbersome or uneconomical for tomorrow. An example is the effect of LEDs on solid-state relays. Impractical and almost unknown as a means of isolating the input from the output of such relays at the time much of what follows was first written, they are in use for exactly that purpose today.

In 1967, a NARM committee was commissioned to study hybrid combinations of relays and solid-state devices. They pointed out that because there were advantages for relays over solid-state units, such as transistors and diodes in some situations and the reverse was true for other cases, it appeared logical that a hybrid could in many cases outdo either alone and must certainly be the best engineering for those situations.

The general characteristics of relays vs. solid-state units were examined side by side, and where conclusions could be drawn, they were listed in tabular-comparison form. These tabular details are shown in Table 4.12 and should provide the designer with valuable guidance in making a choice of whether to use conventional relays, hybrids, or solid-state only for any specific cases.

Circuits making use of electromechanical/electromagnetic relays and solid-state components to form a hybrid unit are shown in Figures 4.62 to 4.84. These are obviously generalizations of circuit possibilities, and many relay vendors, as well as users, have their own versions of circuits to do these and other tasks, which (for the most part) they are willing to share. Hybrid circuits seem destined to continue to grow in number and scope as the components change and improve.

TABLE 4.12 Characteristics of Relays and Solid-State Components as Switching Devices[1]

Characteristic	Device	
	Relay	Solid state
	LIFE	
Cyclic	The normal mechanical life of relays (expressed in number of operations) may vary from less than one million operations to hundreds of millions, with some (such as mercury-wetted contact relays) capable of many billions of operations, depending on type and design. The electrical life rating is normally a function of the particular electrical loads being switched	Theoretically, correctly applied semiconductors do not have known wearout modes, since they are essentially mechanically static devices. Cyclic limitations may be encountered which are dependent upon the design, fabrication, construction or application of the semiconductor. (Life is normally expressed in number of hours rather than operations. A good transistor switch can make one million or more operations in 1 s without impairing its useful life.)
Static	The static life of electrically functioning relays may be limited by physical or chemical deterioration of their components. The nature of the conventional relay is such that its contact forms can function in a prescribed manner (blocking or conducting) with or without coil power, depending on type and design. In their blocking condition, relay contacts are inherently immune to transients. Relay contacts dissipate very little power while conducting, because of their normal condition of low contact resistance. In some cases, excessive contact resistance may be encountered owing to contamination, corrosion, or oxidation. Other possible limitations may be coil deterioration (seldom, if ever, encountered in the absence of electrical stresses) and galvanic action between certain dissimilar metals. The design, materials, and manufacturing processes of the relay—along with the application and environmental surroundings—are the ultimate factors that determine static life, and are normally chosen to minimize or effectively eliminate these problems	The static life for electrically functioning semiconductors may be limited by chemical or physical changes affecting the intended function of their junctions. Semiconductors usually require a continuous external driving source, except for latching types, which are internally driven by their output. The maximum junction temperature for semiconductors limits the power dissipated. This internally dissipated power is caused by the forward voltage drop across the device and by the requirements of the device drive. Above-rated voltage transients can destroy or cause a device to go into an unwanted condition. The environmental surroundings, application, design, and fabrication of the semiconductor are the ultimate factors that determine static life

TABLE 4.12 Characteristics of Relays and Solid-State Components as Switching Devices[1] (*Continued*)

Shelf	With proper storage, shelf life is normally not a problem. Hermetically sealed units have a potentially longer shelf life than open units	Generally nonapplicable
Failure modes	Failure modes may be contacts sticking, transferring, or welding; high contact resistance; mechanical failure; coil opening or shorting. Contact sticking and high contact resistance may be intermittent and regarded as misses instead of failures for some applications. Coil failures are usually attributable to excessive voltage, electrolysis or other chemical reactions, or harsh environments. Excessive temperature, especially if prolonged, may deteriorate the insulation, causing the coil to become defective. Most relay failures are fairly easily detected because of visual evidence of failure	Failure modes may be permanent shorts (although opens do occur), inability to block voltage, or leakage current reaching failure proportions. General failure factors related to semiconductors are: exceeding of maximum voltage ratings, e.g., transients; thermomechanical fatigue caused by cyclic temperature surges; chemical reactions, such as channeling; physical changes, such as crystallization of materials; and other associated packaging problems which generally cause greater than intended power dissipation within the device. Most failures are hastened with prolonged temperature increases. Specific failures for semiconductor devices are secondary breakdown found in bipolar transistors, and di/dt and dv/dt found in thyristors. If the commutating dv/dt of a thyristor is exceeded, it will not turn off; and if the static dv/dt of the device is exceeded, it may go into unwanted conduction. Semiconductor-failure detection can become quite involved depending on the knowledge, experience, and equipment required. In many instances there is no visual evidence of failure unless it is heat discoloration

ENVIRONMENT

General	Commercial atmospheres are reasonably well tolerated by most relays in either an unenclosed (open) condition or, if conditions warrant it, in an enclosure. Extreme problems of atmosphere, moisture, particles, etc., may require hermetic sealing. Relays may be designed with radiation-hardened	The types of packaging and small mass of semiconductors make them inherently immune to most environments, particularly shock and vibration. For radiation applications, shielding must be provided

316

Characteristic	Device	
	Relay	Solid state
	ENVIRONMENT (Cont.)	
Temperature	Generally, the ability to withstand heat is ultimately limited by the type of insulating materials employed. Maximum or temperatures above maximum rating, if sustained, will produce a faster deterioration and decomposition of most insulating materials. Above-rated, elevated-ambient temperatures for reasonably short durations can usually be tolerated by most relay designs without causing irreversible changes to the unit. Relay designs are available that can operate in maximum ambients of 125°C, with specials good to 200°C. In general, the contact rating applies over its specified operating-temperature range without derating. Coil resistance varies directly with its temperature, according to the temperature coefficient of the coil-wire material (copper is used almost exclusively)	Essentially, the ability to withstand heat (internal losses plus ambient temperature) is ultimately limited by junction temperature considerations. Above-rated, elevated-ambient temperature surges usually have sufficient inertia to cause irreversible changes in the semiconductor if it is functioning near its maximum capacity. Generally, semiconductors can withstand junction temperature overshoots caused by current surges of several milliseconds. The cumulative effects of the internally dissipated power, combined with the ambient temperature, must not exceed the maximum permissible junction temperature. As the internal power dissipation increases, the maximum allowable ambient temperature decreases. The type of heat sink employed substantially affects semiconductor performance. The proper design for this heat sink may become quite involved, depending upon application and if electrical isolation is required. Prolonged heat exposure hastens chemical and other types of failure. Depending on type, many semiconductors can operate in ambients of 125°C or above if properly applied. Gate sensitivity and gain usually fall off with low temperatures, particularly below −20°C
Contamination	Contamination is of most concern with contacts. Where contact contamination is encountered, the result may vary from slightly increased contact resistance to an electrically open condition. Relay coils, depending on insulation, may be susceptible to certain contaminants which will chemically deteriorate the coil and may result in electrical breakdown and shorting	Contamination is mostly of concern when it is an internal type on a semiconductor element. Where semiconductor pellet contamination is encountered, a decrease in blocking voltage and an increase in leakage current normally results

317

TABLE 4.12 Characteristics of Relays and Solid-State Components as Switching Devices[1] (*Continued*)

External contamination	Unenclosed relays may be affected by undesirable gases and other contaminants, which may require either hermetic or nonhermetic enclosures. Contaminants, such as oxides, on connecting terminals may present difficulties	Semiconductors are essentially immune to external contamination except when encapsulation flaws permit atmospheric impurities to reach the sealed semiconductor junction. Contamination on leads may offer connecting problems
Internal contamination	Internal contamination is generally a result of outgassing of various insulation materials (e.g., in hermetically sealed units). Electrical switching of contacts, at levels producing sparking or arcing, in the presence of outgassing from various organic volatiles, may form contaminants and may promote contact erosion. Internal outgassing and contamination are normally controlled by proper choice of materials, design, and manufacturing methods	Internal contamination consists of entrapment or inclusion of ionizable material inside the sealed package, which may lead to failures. Manufacturing techniques and processes have been developed that provide a high degree of freedom from contamination within the sealed package

RELIABILITY

Rating method	Failure rate is generally expressed in percent per 10,000 operations	Failure rate is usually expressed in percent per 1,000 h
Degree	Relays have demonstrated high component reliability using the above rating method; however, this depends greatly on relay type and use	Under reasonably ideal conditions, extreme reliability can be obtained from semiconductors with the above rating method
Failure rate	The failure rate tends to follow a "bathtub" distribution curve; i.e., it decreases after each consecutive, successful operation, levels off, and does not appreciably increase until mechanical wearout begins	It is generally assumed that the failure rate ranges from constant to slightly decreasing with time, once the infant-mortality period is passed. There is no upturn in failure rate with life
Run-in or burn-in	In some instances where added reliability or stability is desired, relays are given a number of prelife operations (run-in) under predetermined conditions, related to intended use. This tends to minimize early failures. However, extensive run-in will only use up a portion of the useful life	The initial burn-in for semiconductors tends to eliminate devices which would normally fail during the first few hours of operation. It is effective for semiconductor devices having a high initial failure rate followed by a decreasing failure rate. The extent of such testing is usually limited by economic considerations

Characteristic	Relay	Solid state
	RELIABILITY (Cont.)	
System	System reliability is reduced according to the cumulative failures of all the components used. Where the choice of relay or solid-state system is considered, the complexity required for each may be a greater factor in system reliability than is the reliability of the individual component	
Hybrid	The greatest reliability is achieved when the strong points of a certain component offset the weak points of another, as is done in many hybrid devices using both relays and semiconductors	
	ELECTRICAL ISOLATION	
Output/input	Relays have inherently high isolation between output circuits, between output and control (input) circuits, and between control circuits. (Insulation resistance of $\geq$1,000 MΩ and dielectric withstanding voltage of the order of 500 to 1,000 vac are typical.)	Generally, a high degree of electrical isolation between control and output circuits cannot be achieved with junction-type semiconductors. In a limited area of application, a high degree of isolation can be achieved with FET's
High voltage	Isolation is very little affected by voltages which are relatively high compared with nominal system voltages. The loss of dielectric due to momentary exposure to excessively high voltage is usually temporary. The degree of recovery depends on the type of insulating material used	If maximum rated voltage values as given by the manufacturer (usually at 25°C) are exceeded even momentarily, many semiconductor devices will be permanently damaged
Variation	Relay contact isolation normally does not vary substantially with time, temperature, radiation, voltage, etc, unless there is a complete failure under extreme conditions	Semiconductor leakage current is a variable of temperature, time, radiation, and voltage. If device limitations are not exceeded, the variation is reversible, except where due to time (aging)
Electrical noise and magnetic fields	Relays, because of the power and time required for operation, are essentially insensitive to electrical noise. Sensitive relays can be subject to false operation in high magnetic fields unless	In many applications, shielding and signal conditioning are required to prevent false operation caused by electrical noise and electromagnetically induced currents. They do not nor-

TABLE 4.12 Characteristics of Relays and Solid-State Components as Switching Devices[1] (*Continued*)

	shielded. RFI, produced by relay contacts during opening and closing but not while carrying current, is difficult to control. EMI, produced by coils and magnetic circuits, may be suppressed to some extent	mally show sensitivity to static magnetic fields (except Hall devices). Semiconductors generate RFI during turn-on and turn-off while switching ac if they are not turned on at zero current. Thyristors using phase-control techniques may be a serious RFI source because of turning on during each half-cycle (or alternate half-cycles). Various techniques for RFI suppression are possible for some applications
OFF/ON CHARACTERISTICS		
Off/On impedance ratio	The off/on impedance ratio is extremely high	The off/on impedance ratio is moderately high. (FET's may provide significantly higher ratios.)
Power loss	Relays have the ability to handle power with extremely low loss because of low resistance of their closed contact circuitry. The coil power must also be considered to obtain the complete power-consumption picture	Semiconductor ability to handle power is limited by inherent on voltage drop, dependency on heat sinks, and ambient temperature. The base or gate drive power is small compared with the output. However, for some transistors, as output current increases, input current must be disproportionately increased (beta decreases) to obtain desired output saturation
On voltage	The voltage drop across a closed relay contact is the IR product (generally less than 100 mV)	The semiconductor usually has a forward voltage drop from 0.3 to 2.5 V. The voltage drop per junction is approximately 0.3 V for germanium and 0.6 V for silicon, in addition to the voltage drop due to bulk resistivity
On resistance	The on (closed contact) resistance of the relays may vary slightly from cycle to cycle and with life in terms of operations and load. Under adverse conditions, misses may occur because contacts do not close electrically (particles, film, welds on opposite sets of contacts, etc.) or resistance exceeds some predetermined value. The on resistance remains essentially constant with clean contacts. It can increase slightly because of heating of the contact circuitry, and increases with current	Forward voltage drop across a semiconductor is consistent from cycle to cycle. This drop varies with junction temperature. The on resistance generally decreases as current increases, and for some devices such as transistors can be varied with base drive

Characteristic	Device	
	Relay	Solid state

OFF/ON CHARACTERISTICS (Cont.)

Characteristic	Relay	Solid state
Off resistance	The off resistance (open contacts) is affected very little by temperature, voltage, etc.	The off resistance varies with time, temperature, voltage, and radiation. Leakage current increases exponentially with temperature (e.g, may double with every 8 to 10°C increase)

INPUT CONSIDERATIONS

Characteristic	Relay	Solid state
Operating power	Relays are available with operating power from milliwatts to watts, and specials operate on microwatts. Duration of power pulse required for latching relays may vary from less than 1 ms to several milliseconds. Relays having ferrite magnetic circuits can be made to operate from pulses of less than 5 μs duration. Latching relays are normally reset with a power pulse equal to (in some cases less than) the value required to latch. Proper polarity coil voltage must be observed for devices designed to latch magnetically. In special cases, manual-reset features may be provided. Relays do not normally require regulated power supplies	Different semiconductor devices are available with operating power from microwatts to milliwatts. Latching semiconductors, such as thyristors, generally require a 2-μs pulse or greater for latching (conduction turn-on). Latch is lost when conducting current is reduced below holding value. Semiconductors require well-regulated, transient-free, dc power supplies
Operating voltage	Relays operate in response to a wide range of ac or dc coil input voltages as determined by design. Relay coils are generally designed to operate within a ± tolerance of a specified nominal voltage. While insufficient voltage will not permit the relay to operate properly, if at all, greater than maximum specified coil voltage may cause coil deterioration depending on duration and magnitude. Conventional, nonlatching, relays will drop out (return to unenergized condition) when coil voltage is removed or reduced to a value which may be varied widely by design and/or adjustment. Latching relays	Semiconductors easily adapt to a wide range of input voltages via appropriate circuitry. The absolute maximum voltage ratings for the input (and output) of semiconductors were well specified by the manufacturer and are not to be exceeded without risk of permanent damage. The range of operating voltages is predetermined by design

TABLE 4.12 Characteristics of Relays and Solid-State Components as Switching Devices [1] (*Continued*)

	require an input voltage of a specified magnitude and duration for latch and reset. (Proper polarity must also be observed for magnetically latched relays.)	Semiconductors are essentially current-operated devices. Transistors are essentially current amplifiers; i.e., a given input current determines a given output current for a set of conditions. Thyristors require minimum gate-current drive to ensure latching, if load conditions permit. Generally, gate drive can be removed after thyristor turn-on. Semiconductors usually turn off within a matter of microseconds following removal of drive current, whether supplied externally or internally
Operating current	Relays operate over a wide range of predetermined current levels. Even when voltage levels are specified, the electromagnetic relay is essentially a current-operating device whose operation is accomplished when its inherent ampere-turn requirements are met. Again, those comments made under Operating Voltage apply to current-operated relays	
Transients	Relay coils are generally insensitive to transient voltages	High-voltage, short-duration transients can be particularly damaging
Duty cycle	Generally nonapplicable. In some cases, coils may be rated for intermittent duty because of temperature considerations	As the percent duty (conduction time) decreases, the drive-current rating usually increases as long as maximum junction temperature is not exceeded. Large drive currents may be required in power transistors to obtain saturation, and in thyristors to minimize *di/dt* stresses

OUTPUT CONSIDERATIONS

Multiple switching	Relays are available with various types and numbers of contact forms which can be operated by a single input. For most, the choice of circuits switched by each contact form may be of a widely different voltage and frequency from that switched by the others; this choice requires little if any pre-determination in the design of the relay	Solid-state systems feasibly can perform any switching function. However, switching more than one circuit simultaneously using discrete semiconductor components requires proper combination into a workable assembly. The complication, cost, and size of this assembly will increase substantially with the number of poles, magnitude of current, and degree of electrical isolation. (Where the level of switching permits integrated circuits to be used, considerable economic and size advantages may be realized.)

Characteristic	Device	
	Relay	Solid state
	OUTPUT CONSIDERATIONS (Cont.)	
Switching range	Contacts are generally adaptable to a wide current, voltage, and frequency range, since they simply physically connect electrical conductors. However, relay families are usually designed for dry, low, intermediate, or high-level switching. The upper frequency range which they are capable of switching also varies greatly with design	Semiconductors and their assemblies are usually designed for a specific voltage, current, and frequency range. The operating current ratio (output to input) for many semiconductors is extremely high. The frequency range depends greatly on the design of the semiconductor
Voltage	Relay contacts can generally tolerate an exceptionally wide operating range of load voltages without design changes, and they usually recover from short breakdowns or excessive overvoltages. Special designs may be required for high voltages or extreme environmental conditions	Maximum output-voltage ratings are particularly well specified, and it is essential that semiconductors be operated within these ratings to avoid permanent damage
Current	Relays can generally tolerate overload currents of various degrees. As overload current increases, contact sticking or contact welding during break and make will increase. An empirical determination is usually necessary to find out if the relay will function reliably under given overload-current conditions	While most switching-type semiconductors can handle surge currents many times their steady-state ratings, overload currents of relatively sustained duration cannot be tolerated. The junction temperature of the semiconductor is ordinarily the limiting factor in determining the magnitude and duration of its surge-current rating
Transients	The inherent design of relays is such that they are essentially immune to transients	High-voltage, short-duration transients can be particularly damaging to semiconductors. Special protective devices and means are frequently required
Duty cycle	Generally output is independent of conduction duty; however, the maximum permissible cyclic rate of switching is limited by the magnitude of the load being switched	As percent duty (conduction time) decreases, current rating usually increases as long as the maximum junction temperature is not exceeded

323

TABLE 4.12 Characteristics of Relays and Solid-State Components as Switching Devices[1] (*Continued*)

	Relays	Solid-State
Contact bounce	Contact bounce is usually present to some degree during contact make and, in a few designs, during break. (Duration may be from fraction of a millisecond to several milliseconds, and may consist of several closures and openings per contact operation.) Mercury-wetted relays are an exception, since their contacts have essentially no bounce due to the masking effect of the mercury	No contact bounce exists in semiconductors
Amplification	Amplification is possible only in that small signals may control large ones. Extremely high amplification factors are possible using a single relay	Semiconductors amplify the input signal. Some devices are used as switches in that small input signals control large output signals
LOGIC		
Systems	The logic of simple control systems can be used most economically with control relays, particularly since special power supplies and noise-suppression techniques are not required	Solid-state devices lead the field where extensive and complex logic systems are involved or very high-speed operation is required
Speed	Pickup times range from 0.5 to 5.0 ms for reed and other relatively fast operating types to 5.0 to 50 ms for conventional control relays. Dropout times for relays are generally somewhat faster with some reeds having dropout times of 20 μs. Use of overdriving coil voltages to achieve a lower pickup time may increase the severity of contact bounce during make	Operating times for transistors range from nanoseconds for computer types to microseconds for power types. Typical turn-on time for thyristors varies from 0.5 to 5.0 μs while commutation requires from 10 to 50 μs. If thyristor turn-off is achieved when conducted ac goes through zero, turn-off time can be approximately as long as a half-cycle. The above values vary widely with type of semiconductor, circuit, and application
Memory	Memory functions can be easily achieved by latching or stepping relays. In such cases, memory can generally be retained in spite of power loss	Memory is normally accomplished with cores, flip-flops, and integrated circuits. Power loss generally means memory loss, except for core logic or where special auxiliary memory circuits are used
Electrical noise	Electrical noise is normally not a problem because of relay operating speed and power requirements	Signal conditioning and filtering for noise, overshoot, and transients are often required

	Device	
	Relay	Solid state
Characteristic		

LOGIC (Cont.)

Characteristic	Relay	Solid state
Fan-out	Fan-out logic functions per input for conventional relays is generally not a problem. One relay contact can drive many relay coils. Fan-out speeds for relays are usually in milliseconds	Fan-out limits the number of gates which can be driven by one logic element. One gate can fan out to about 10 identical gates. Power gates are available with a fan-out of about 20. Switching times, generally in nanoseconds (both rise and fall time), usually increase with fan-out
Fan-in	Fan-in is the number of inputs to a logic element. Relays are often considered single-input, multiple-output devices. Multiple inputs may be obtained by using diode drivers, and to a limited extent, by separate windings	IC's are often considered to be multiple-input, single-output devices. Rise time is fairly independent of fan-in (fall time increases with fan-in)
Interfacing	Relays can be driven directly by solid-state circuitry. Relays requiring low operating power, such as reed relays, are ideal interfacing devices between solid-state circuits and relays or motor starters	Discrete solid-state power gates can interface directly with relays, solenoids, stepping switches, etc. Integrated microcircuits usually use a transistor to interface with relays and other electromagnetic devices

MAINTENANCE

Characteristic	Relay	Solid state
Installation	While generally not critical in hookup, reasonable care should be exercised in handling of unenclosed relays	Care should be exercised so that the maximum permissible temperature for the semiconductor is not exceeded during solder hookup or potting. Handling of devices is not generally critical except where damage to termination must be considered
Troubleshooting	Technicians are normally able to diagnose failures with reasonable success. (Unenclosed or transparently enclosed relays give visual evidence of operation.)	Specially trained technicians and special test equipment are often needed to analyze problems. Assembly packages consisting of discrete components require module evaluation rather than simple component evaluation

TABLE 4.12 Characteristics of Relays and Solid-State Components as Switching Devices[1] (*Continued*)

	SIZE
Intermediate to high-level switching	In most cases where power or multipole switching are required, the relay is generally smaller and far less complex than the equivalent solid-state unit. However, the life of the equivalent solid-state unit may be many times greater than that of the corresponding relay. The relay does easily adapt to various contact forms and combinations as the number of pole requirements increases. Although the size of the contact and motor assemblies for the relay are usually larger than the required semiconductor devices, this is often offset by the size of the peripheral components (including heat sink) required for the semiconductor devices. The relatively small power dissipated in the relay is usually accommodated by its own radiating surfaces. By contrast, the power dissipated in the semiconductor device, because of its forward voltage drop, generally imposes special heat-sink requirements which will often substantially increase size
Low to intermediate-level switching	Low-level switching applications that permit extreme reductions in heat-sink requirements, especially where extensive logic is performed, favor semiconductors because of savings in weight and size. This is particularly true with microelectronics. For very low-level switching applications, however, difficulties may be encountered with semiconductors where direct device operation is necessary, especially if voltage and current requirements are below that of device operation. Special relays often directly fulfill this application. While semiconductors can also work in this region, appropriate supporting circuitry, power supplies, and peripheral considerations may become increasingly complex and expensive as the switching level becomes extremely low. How critical these considerations are cannot be sharply defined, but applications in the microamperes—millivolts region should be examined to see which method, or combination of methods, of switching is best suited

	COST
Switching levels	Relays are generally economical devices where heavy-load, high-voltage, dry-circuit, or multipole switching is desired	Semiconductors are desirable and economical where low-level, multiple-input switching is used or for applications requiring unusually long life or high speed. Where extensive complex logic switching is required, IC systems are the only practical choice
Application	Worst-case condition studies are only moderately difficult and often done by the relay manufacturer	In general, when solid-state switching circuits are to be mass-produced, particularly from discrete components, comprehensive and thorough worst-case studies usually become quite involved

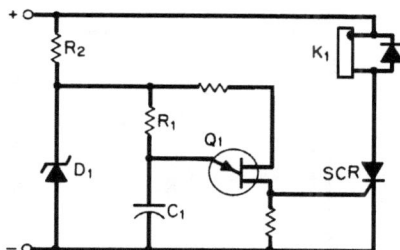

FIGURE 4.62 Hybrid relay to provide time delay on pull-in.[1]

Typical hybrid-relay applications In the following paragraphs, relays are shown in hybrid combinations with solid-state components. Sensors, light-sensitive devices, thermistors, and other devices are illustrated as circuit elements functioning with relays and other solid-state devices.

Time delay (pull-in) Figure 4.62 is a circuit designed to delay a relay on pull-in. Upon application of power, capacitor C1 begins charging through R1 until it reaches the firing point of unijunction transistor Q1. The unijunction transistor triggers the SCR, which energizes relay K1. Resistor R2 and zener-diode D1 form a voltage regulator to allow for a wide range of voltage inputs.

Time delay The circuit of Figure 4.63 simply and accurately delays the dropout of a relay after it is energized. In the quiescent state, no power is applied to the operating components. When S is momentarily closed, transistor Q1 turns on and relay K1 pulls in. The voltage to the circuit is then maintained through the relay contact so that the relay remains energized when S is opened. After a time interval determined by the values of R_1, R_4, and C_1, unijunction transistor Q2 triggers and the discharge of C1 turns off Q1, allowing the relay to drop out. If S is open, the voltage to the circuit is removed and the circuit reverts to its quiescent state.

An output voltage can be obtained from the relay contacts, as shown, or extra contacts on the relay can be used.

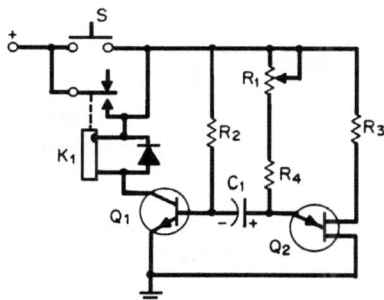

FIGURE 4.63 Hybrid relay to provide time delay on dropout.[1]

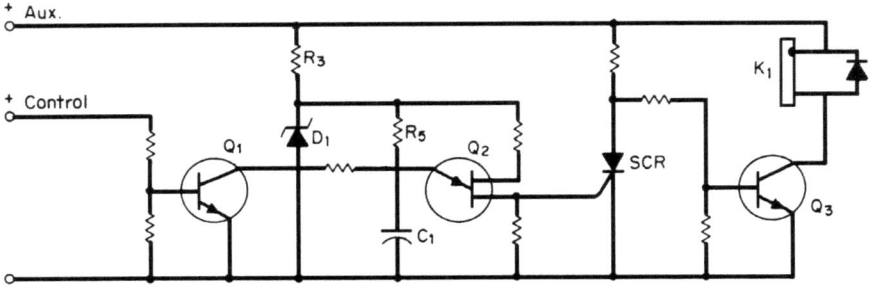

FIGURE 4.64 Hybrid relay with delay on dropout using an auxiliary voltage.[1]

Depending on the supply voltage and the relay used, R2 provides sufficient base current to Q1 to allow the relay to pull in. The size of the capacitor provides sufficient off time for Q1 to allow the relay to drop out. R1 provides the time delay required and the maximum peak-point current of the unijunction transistor. R3 provides the required overall temperature compensation.

Time delay (dropout) The circuit of Figure 4.64 provides a time delay on dropout using an auxiliary voltage. A control voltage is applied to Q1, causing it to horn on and essentially short out C1. The auxiliary voltage causes Q3 to turn on and allows relay K1 to pull in. When the control voltage is removed, Q1 turns off, allowing C1 to charge through R3 and R5. When unijunction transistor Q2 fires, the SCR is triggered, turning off Q3 and deenergizing the relay. R3 and zener-diode D1 allow for a wide range of voltage inputs.

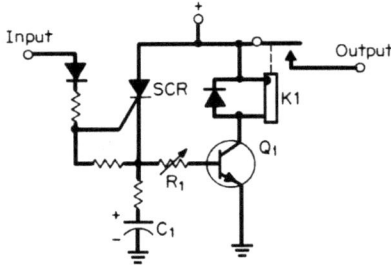

FIGURE 4.65 Hybrid-relay circuit to provide release-time delay without recovery time.[1]

Time delay (no recovery time) The circuit of Figure 4.65 provides a dropout time delay without requiring recovery time. When a positive pulse is applied to the input terminal, the silicon-controlled rectifier (SCR) is turned on and current flows through it to turn on Q1 and charge capacitor C1. When C1 is sufficiently charged, current through the SCR drops below its required holding-current level. The SCR then turns off and C1 begins discharging through R1 into the base of Q1. Meanwhile, Q1, having been turned on, has pulled in relay K1, providing the output connection. As C1 discharges, current in the collector of Q1 and in the relay decreases until it drops below the relay dropout level. Then the relay opens, breaking the output collection. The circuit can be retriggered at any time after SCR shutoff, provided that the trigger amplitude is sufficiently greater than the charge on C1 at that time.

Time delay (pull-in) for two-coil relay A circuit for the time delay of a two-coil relay is shown in Figure 4.66. The two coils of relay K1 are wound so that when power is applied, the resultant magnetomotive force is approximately zero. This does not allow the relay to pull in.

Capacitor C2 is charged through the relay coils and R2. When unijunction transistor Q1 fires, it triggers the SCR into conduction. This essentially shorts out one coil of the relay, allowing a magnetomotive force to develop and pull in the relay.

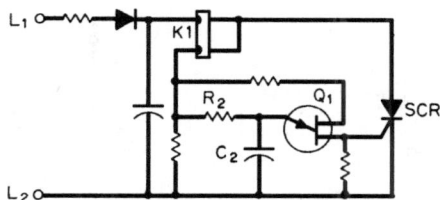

FIGURE 4.66 Hybrid-relay time delay *(pull-in)* using a dual coil.[1]

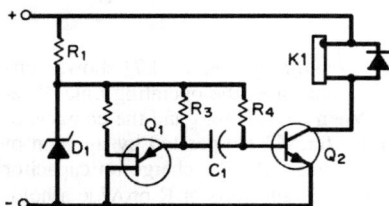

FIGURE 4.67 Hybrid relay as a repeat-cycle timer.[1]

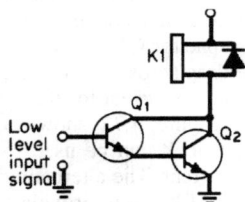

FIGURE 4.68 Hybrid relay functioning from a low-input signal.[1]

Repeat-cycle timer The circuit of Figure 4.67 allows the repeat cycling of a relay at a predetermined frequency. Upon application of power, capacitor C1 charges through R3. R4 provides base current to transistor Q2, turning it on and energizing relay K1. The on time of the relay is determined by R3C1. When the firing point of unijunction transistor Q1 is reached, C1 discharges, causing a negative shift in the voltage on the base of Q2, turning it off and deenergizing the relay. C1 continues to discharge through R4 until the base voltage of Q2 is positive enough to turn on Q2, thus reenergizing the relay. This cycle repeats itself.

Amplifier relay In the circuit of Figure 4.68, the power required of the controlling source can be greatly reduced by combining a relay with an amplifier-driver. This allows use of a low-level input while retaining the inherent ability of a relay to control power circuits of various voltages and currents.

A Darlington amplifier-driver is shown in Figure 4.68. Application of a positive input to the base of Q1 produces an amplified drive to Q2, causing relay K1 to operate.

Triac relay driver The triac relay driver (Figure 4.69) is ideally suited to control ac-operated relays with small input signals. The presence of an input signal causes triac T1 to conduct and, in turn, operate relay K1. The absence of an input signal causes the triac to assume a blocking state, and the relay deenergizes.

Relay driver In the circuit of Figure 4.70 a bistable polar relay is controlled by a one-polarity low-level input to the amplifier-driver circuit. The application of a positive input to Q1 causes it to conduct, lowering the voltage at the R1C1 junction. C1 discharges through Q1, the base-emitter junction of Q3, and R3, turning Q3 on for a period determined by the time constant of the circuit. The short-duration pulse in direction A through K1 causes the relay to assume one of its stable positions. At the termination of the positive input to Q1, Q1 ceases to conduct, the voltage at the R1C1 junction rises, and C1 charges through R1, R2, and the base-emitter junction of Q2. Q2 conducts for a period determined by the circuit; the time constant produces a pulse of current in direction B in K1, causing K1 to assume the other of its stable positions.

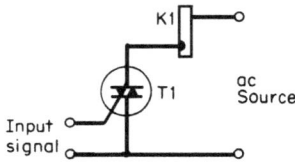

FIGURE 4.69 Hybrid relay functioning from a triac.[1]

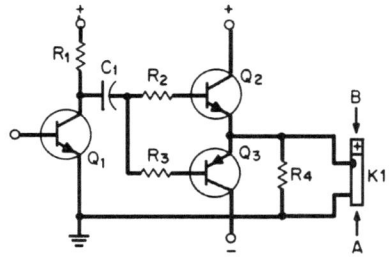

FIGURE 4.70 Bistable polar relay as a hybrid functioning from an amplifier driver.[1]

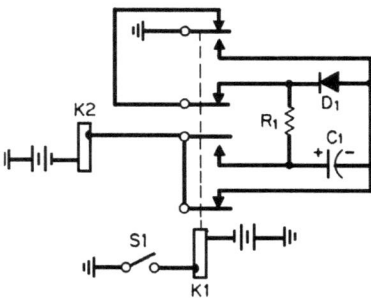

FIGURE 4.71 Hybrid relay for fast operation.[1]

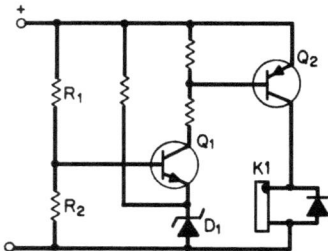

FIGURE 4.72 Hybrid relay as a voltage sensor.[1]

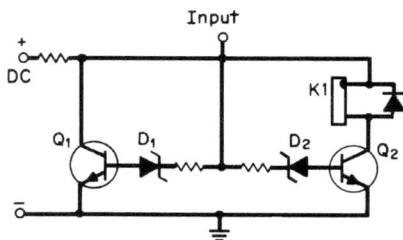

FIGURE 4.73 Hybrid relay as a voltage-limits sensor.[1]

Fast operation Figure 4.71 shows a circuit that decreases the operating time of relays. When S1 is closed and the contacts of K1 trans-fer, K2 is operated by the sum of source voltage and the charge on capacitor C1. Diode D and resistor R provide a holding circuit that con-ducts when the voltage across C1 assumes a polarity opposite to that shown and a magnitude adequate to cause diode turn-on. Momen-tarily, the voltage is doubled. Diode D1 does not conduct until the voltage across capacitor C1 falls to the voltage equal to the drop across resistor R1, resulting from the holding current for K2.

Voltage sensor Figure 4.72 shows a basic dc voltage sensor. It can be used to sense ac voltage or current. The alternating current is rectified and filtered to provide a dc level. For ac sensing, a current trans-former is used, with the output rectified and filtered. Zener diode D1 establishes a refer-ence voltage at the emitter of transistor Q1. Voltage divider R1R2 is set so that when the desired input voltage is reached, Q1 turns on. This allows transistor Q2 to conduct and causes relay K1 to pull in.

Voltage-limits sensor The circuit of Figure 4.73 can be in one of two states: if the input voltage is within the specified limits, the contacts are closed; if it is either higher or lower, the contacts are open. The desired voltage limits are determined by zener diodes D1 and D2.

When the voltage at the input is within the specified limits, Q2 conducts and relay K1 contacts are closed. If the voltage rises, Q1 also begins to conduct, effectively shorting

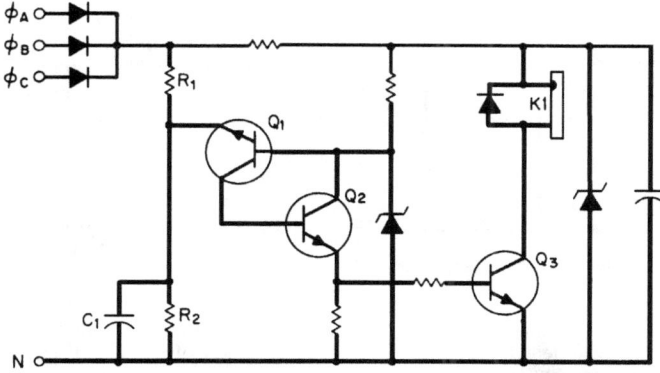

FIGURE 4.74 Hybrid relay as a three-phase, four-wire overvoltage sensor.[1]

out the relay coil. Consequently, the contacts open. If the voltage goes down, both transistors cut off and the contacts again open.

Three-phase, four-wire overvoltage sensor The circuit of Figure 4.74 senses the highest voltage of the three-phase input. The time constant of R1 and C1 is such that a phase voltage after rectification that is higher than the predetermined level, as set up by R1 and R2, causes regenerative transistor switch Q1 and Q2 to turn on. This, in turn, allows transistor Q3 to conduct and to pull in relay K1. When all three phases are below this predetermined point, K1 drops out.

Four-wire undervoltage sensor The detection of voltage below a reference is provided by the circuit of Figure 4.75. Zener diode D4 establishes a voltage reference level to the emitter of transistor Q1. As long as the average dc level of L1, L2, and L3 through diodes D1, D2, and D3 and voltage divider R1 and R3 remains higher than this reference voltage, Q1 remains off. This, in turn, allows transistor Q3 to turn on and pull in relay K1. When the average level of the line voltage drops below the predetermined level, Q1 conducts and turns on Q2, causing Q3 to turn off because the base and emitter now are essentially at the same voltage level. This causes K1 to drop out.

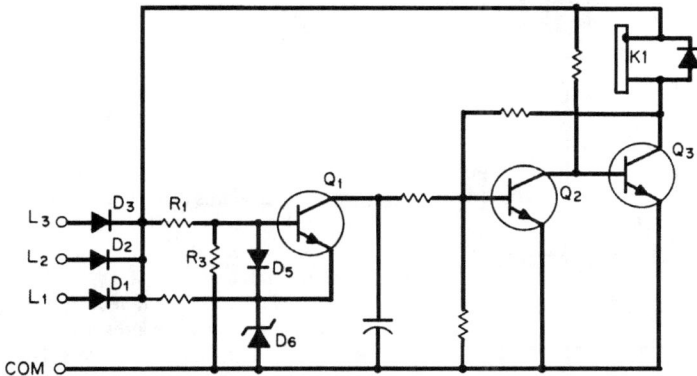

FIGURE 4.75 Hybrid relay as a four-wire undervoltage sensor.[1]

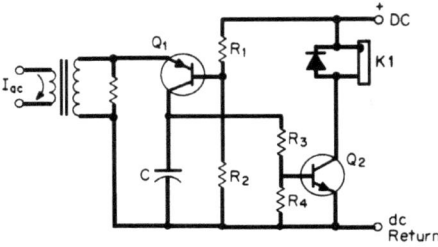

FIGURE 4.76   Hybrid relay as a current sensor.[1]

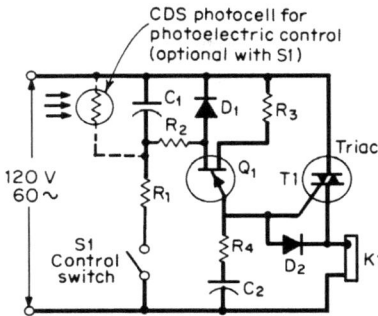

FIGURE 4.77   Hybrid relay as a photoelectric switch.[1]

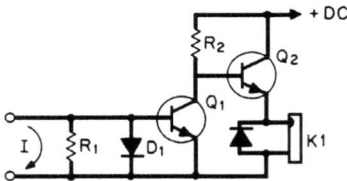

FIGURE 4.78   Hybrid relay as a current sensor.[1]

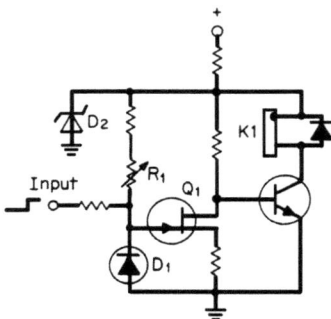

FIGURE 4.79   Hybrid relay as a low-current detector.[1]

Current sensor The circuit of Figure 4.76 monitors the level of an alternating current. The output is a relay, and SPDT contacts can be used to determine overcurrent or undercurrent.

When direct current is applied in the absence of any alternating current, transistor this biased off by R1 and R2. The relative values of R1 and R2 determine this back bias, which is the set point. Q2 is also off and relay K1 remains deenergized.

As the ac level increases, the half-cycle peaks reach a value to forward-bias Q1 and cause conduction. There, pulses charge C1 and cause current flow in R3 and R4. When this current becomes great enough, Q2 is turned on, causing K1 to pull in.

Synchronous photoelectric switch Synchronous switching is turning on only at the instant the ac supply voltage passes through zero and turning off only when current passes through zero. The circuit of Figure 4.77 theoretically provides this function in response to either a mechanical switch or a variable resistance, such as a cadmium sulfide photocell. It should be used with caution because erratic behavior can be caused if the relay has too long an operate time, and coil inductance can cause a current lag, producing too long a time for turn-off.

Current sensor The circuit of Figure 4.78 can detect current level. When a dc voltage is applied, relay K1 pulls in through Q2. Q1 is in the cutoff mode. As long as the current remains below a predetermined value, tunnel diode D1 maintains a very small differential voltage between the base and emitter of Q1. When the current increases above the desired value, the tunnel diode acts to increase this differential voltage, causing Q1 to conduct. This, in turn, effectively shorts out the relay coil, causing it to drop out.

Low-current detection The circuit of Figure 4.79 is a discriminator that detects 1-μA currents with a 1-percent accuracy. This is made possible by driving field-effect transistor Q1 with the output

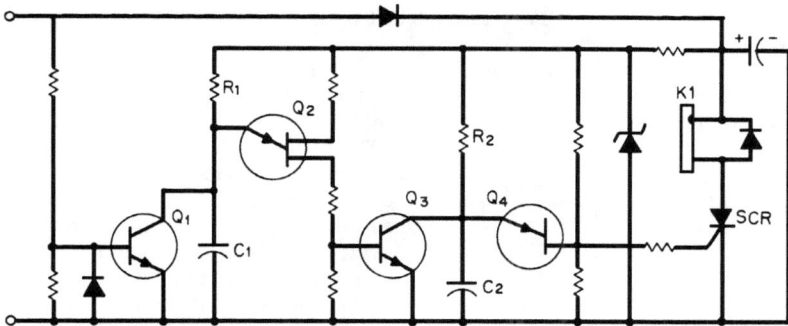

FIGURE 4.80 Hybrid relay as an overfrequency sensor.[1]

from backward diode D1. When the input signal exceeds a preset threshold, the sum of the currents through the backward diode switches the diode to its highest voltage stage. This voltage is then amplified by the FET output stage. The detection threshold can be varied over a broad range by the adjusting resistor.

Overfrequency sensor The circuit of Figure 4.80 detects frequencies in excess of a predetermined frequency. This circuit will cause relay K1 to be operated if the input frequency exceeds a preset value determined by the time constant of R1C1.

On each half-cycle, Q1 is turned on and discharges C1. During alternate half cycles, C1 charges and, at frequencies below the preset limit, it will attain the firing voltage of Q2. Firing of Q2 turns on Q3, which discharges C2.

C2 and R2 establish a time constant long enough that a period exceeding one cycle is required for the charge on C2 to reach the firing voltage of Q4. If the frequency exceeds the preset limit, Q2 and Q3 will not be turned on. C2 can then charge to the firing point of Q4, which will turn on the SCR and K1.

Underfrequency sensor The circuit of Figure 4.81 detects frequencies below a predetermined frequency. This circuit will cause K1 to be energized if the input frequency falls below a preset value determined by time constant R1C1.

When the line frequency is above the set trip point, Q1 discharges C1 on each positive half-cycle of the line voltage before the charge voltage on C1 reaches the intrinsic level of

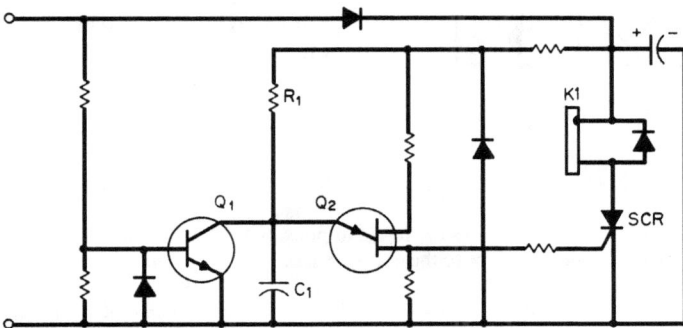

FIGURE 4.81 Hybrid relay as an underfrequency sensor.[1]

the unijunction transistor (UJT). Thus, the UJT never fires when line frequency is above the sense point.

If the line frequency should drop below the set trip point, C1 is allowed to charge to the firing level of Q2. This pulses the SCR, and fault-detecting relay K1 is energized. R1 can be variable to allow for an adjustable trip point.

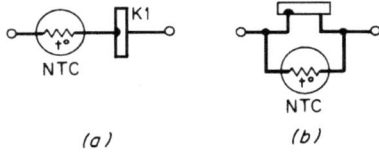

FIGURE 4.82 Hybrid relay used with thermistor.[1]

Thermistors Thermistors can be used with a relay to provide a time delay or present a constant impedance to a source supply. For many designs, the preferred thermistor type provides a resistance that decreases over a wide temperature range (NTC). Simplified circuits are shown in Figure 4.82 and normally are not used in ac relay applications.

Variable-rate driver for stepping switches Spring-driven stepping switches can be driven at controlled rates of less than one step per minute to 20 steps per second by the circuit of Figure 4.83. The coil on time is the minimum necessary, keeping power dissipation at a minimum. The circuit can also be used to drive a relay to produce pulses at a controlled rate.

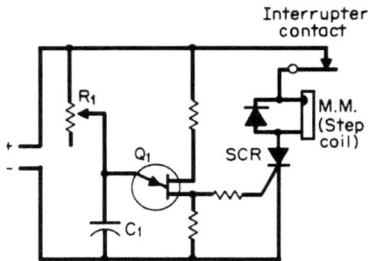

FIGURE 4.83 A variable-rate driver for use with stepping switches. [1]

With voltage applied to the circuit, C1 charges at a rate determined by its value and the setting of R1. When its voltage reaches the level at which Q1 conducts, C1 discharges through Q1 and the gate circuit of the SCR. The SCR turns on and energizes the switch drive coil. As the switch cocks, the interrupter contact opens, turning the SCR off. The switch steps, recloses its interrupter contact, and the solid-state circuit repeats its cycle.

Reset control for a direct-drive stepping switch Two-coil direct-drive (minor) switches can be reliably and economically reset by solid-state control of the rotary magnet (reset) coil (Figure 4.84). Closing S1 applies voltage to the rotary (step) coil, advancing the switch wiper arm one step, and also applies a positive bias to Q1, holding it off. When the switch is advanced to position N, the positive voltage gates the SCR, but Q1 remains off until S1 is opened. With S1 opened, Q1 and the SCR conduct, energizing the release magnet (reset) coil. As the wiper arm reaches the O position, positive bias is applied to Q1, stopping conduction through the release magnet coil.

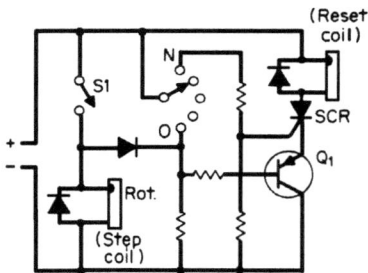

FIGURE 4.84 Reset control for a direct-drive stepping switch.[1]

Solid-state relays (SSR) As the name implies, the solid-state relay (see Figure 4.4) has no moving parts. But it does have the equivalent of a coil (the control) and contacts (the controlled output). In its simplest form, the SSR's output is functionally either a make or a break. More-complicated designs offer multiple contacts, but practical limitations of contact forms usually end with three (called a *three-phase controller*). Thus, the circuit engineer is limited in the circuit logic that can be practically accomplished by the use of solid-state relays. For most present fields of application, this is not much of a handicap in that the logic is obtained in the circuits themselves, to which the relays are attached for implementation.

The solid-state relay has the immediate advantage of appearing to represent the latest in the switching art, thus appealing to both the design engineer and the user. Its use might, in some cases, contribute to the salability of a product because of this appearance of modernity, reliability, and freedom from maintenance. Also, many design engineers feel an empathy for this kind of relay, which might be lacking for conventional relays.

A solid-state relay (SSR) has been identified by one authority as a semiconductor switching device with input terminals isolated from the output switch path. The output switch can be an FET for low-level switching, a bipolar device for medium and power switching, or a triac for ac power switching.

There is as yet little standardization in SSR types, other than that which develops from usage, ratings, and packaging, but there are two commonly recognized broad classifications: ac power switching and ac/dc low-power switching. For a majority of the SSRs in use today, the load is ac, with the triac used as the controlled element. Because a triac always breaks at zero current, its many advantages become immediately apparent in handling heavy current loads and eliminating RFI.

It is frequently important that the input signals be isolated electrically from the output (a feature the conventional relay comes by naturally). To accomplish this in SSRs, it had been a common practice in the past to use a pulse transformer (magnetic coupling). In a more recent technique, shown in Figure 4.85, a gallium arsenide (GaAs) infrared (IR) light-emitting diode (LED) is used to gate the triac. The LED and a photodetector are mounted in close proximity and housed in a six- or eight-lead DIP. When the LED carries current, its emitted light excites the photodetector and couples the input signal to cause triggering of the triac. External interference is eliminated by an opaque covering. No feedback occurs because the coupling is strictly optical.

Advantages for this kind of relay are the following:

- Compatibility with DTL, TTL, and MOS logic
- High sensitivity (typically 8 mW at 4 V)
- High noise immunity and freedom from false actuation
- Elimination of some components in earlier designs, such as coil suppression diodes, and transistor buffers, thus providing some cost reduction

For SSRs in general, there are so many variations in what various manufacturers' relays can do that it is difficult to list any meaningful application data. One manufacturer's data appears as Table 4.13. Ac power-switching relays handle contact loads that vary from a fraction of an ampere up to 40 A, at voltages of 120 and 240, 47 to 63 Hz. The input signal can be either ac or dc, but most often, it is a low-level dc signal in the 3- to 32-V range. Thus SSRs readily accept logic-type input to handle an ac line load at the output. In general, loads that can be handled by SSRs of various types extend from microvolts into high impedance, to 440 V into low impedance. Loads can be resistive, inductive, capacitive,

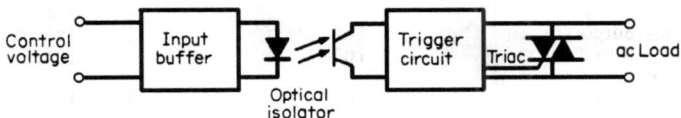

FIGURE 4.85   Schematic of a solid-state relay using an optical isolator. *(Monsanto Commercial Products Co.)*

TABLE 4.13 Typical Solid-State-Relay Specifications as Listed for One Manufacturer's Product

CONTACT-CLOSURE CONTROL

	AC control of ac load, part A	DC control of dc load, part B
Load circuit:		
Load voltage	50–135 vac, 45/60 Hz	6–48 vdc
Load current	6 A with panel mount	
Switch on resistance	0.2 Ω typical	
Switch off resistance	20 kΩ typical	1,600 MΩ typical
Control circuit:		
Control voltage	115 vac typical	Same as load voltage
Control current	0.2 mA typical	100 mA typical
Dielectric strength, device to case	1,500 vac	
Insulation resistance, device to case	3 MΩ at 200 vdc	3 MΩ at 50 vdc
Other characteristics:		
Operating temperature	−55 to +85 °C max, switching rated load, continuous duty	−55 to +100 °C max, switching rated load, continuous duty
Thermal shock	−65 to 100 °C	
Vibration	20g's, 10–2,000 Hz	
Mechanical shock	100g's, 6 ms sawtooth, 1,500g's, 5 ms half-sine	

DC CONTROL OF AC LOAD

	Transformer isolated			LED isolated	
	Part C	Part D	Part E	Part F	Part G
Load circuit:					
Output current	0.25 to 6 A		0.1 to 6 A	6 A	
Load voltage	50 to 140 vac			35 to 140 vac	
Frequency range	45 to 70 Hz				
Transient protection (does not turn on) 0.1 μs rise time	400 V min			300 V min	
Turn-on time (60 Hz)	0.05–0.2 ms			9 ms max	
Turn-off time (60 Hz)	9 ms max				
Off-state leakage	4 ma dc max				
Surge current (1 cycle at 60 Hz)	100 A				
Switch-on voltage drop	0.8–1.6 vac				
Switch-off resistance	10^3 MΩ min				
Control circuit:					
Control voltage	10–30 vdc	4.5–10 vdc	3.5–30 vdc	2.0 vdc min	
Control current	7.0–22 mA	1.0–8.0 mA	1.0–8.0 mA	Approx 10–40 mA	
Turn-off voltage	4.5–6.2 vdc	2.8–3.3 vdc	2.8–3.2 vdc	1.1 vdc min	
Transient input voltage (does not turn on)	14 V peak min (100 μs duration)			Min excitation voltage dependent on turn-on voltage	
Isolation:					
Contacts (output) to case	10^8–10^{13} Ω			10^8–10^{13} Ω	
Coil (input) to case (output)	10^7–10^{10} Ω			10^9 Ω min	
Coil (input) to case	10^7–10^{10} Ω			10^9 Ω min	
Dielectric strength:					
Contacts (output) to case	1,000 vac (rms) min				
Coil (input) to contacts (output)	1,000 vac (rms) min				

SOURCE: Grayhill, Inc.

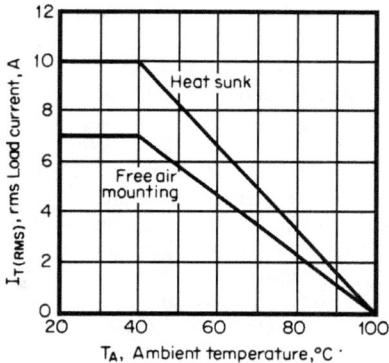

FIGURE 4.86 Typical load current vs. temperature curves for one kind of 10-A triac output, solid-state relay. *(Monsanto Commercial Products Co.)*

and lamp. Allowable junction temperatures must not be exceeded. To make certain of this, the vendor will provide the maximum allowable case and ambient temperatures because these are all that the user can measure. Rated ambient temperatures vary from relay to relay but are usually in the range of −30/−20°C to +80/+100°C, with the allowable percent of rated load current dictating the allowable maximum temperature. The allowable range of storage temperature is somewhat wider. A typical curve sheet for load vs. temperature is shown in Figure 4.86. One surprising aspect of this heat-dissipation problem is that for very heavy load relays, the required heatsinking usually means that the SSR is larger than the conventional power-type relay equivalent.

Early in the decision-making process, the design engineer should find out from the prospective vendor the specific identities of the solid-state components in the prospective SSR relay (such as the kind of transistor, SCR, triac, and FET). Then failures caused by limitations in one or more of these can be avoided and a determination can be made as to what degree of satisfaction and reliability can be expected to result from its use in a particular application.

In general, the advantages for the SSR, besides those noted previously, are practically unlimited life (barring catastrophic failure because of abnormal circumstances), high resistance to shock and vibration, ready mounting on PC boards (for many kinds), extreme operating speed, insignificant operating power, and (in general) a high degree of compatibility with electronic circuitry. For those whose chief interest is the SSR power-type relay application, Ref. 6 will be helpful.

Fields in which SSRs are already being used in considerable quantity are computer and computer-associated equipment, medical electronics, and industrial control. Their advantage in a hazardous atmosphere is readily apparent. Their vulnerability to noise, temperature extremes, and power surges, and their somewhat higher cost are disadvantages that make conventional relays more attractive in some areas. Also large numbers of varied contact combinations are hard to provide on SSRs, leaving the many fields where this need is extensive to conventional relays, at least for the foreseeable future.

Special-purpose relays Relays designed for a special purpose, such as the coaxial relay, and relays that are identifiable as a general class, but have special features added (such as magnetic-latching crystal-can relays) are referred to as *special-purpose relays*. The coaxial relay is not covered in detail elsewhere, but the other special relays have already been examined in detail.

Coaxial relay Figure 4.87 shows one make of this kind of relay. It has at least one set of contacts for switching radio frequency, but it can have additional conventional contacts as well. Because the aim here is to switch high-frequency current with minimum loss, the contact mechanisms are enclosed in a metal chamber with dimensions so that it forms a cavity whose characteristic impedance matches that of the coaxial cable to be attached to it, as nearly as possible. A measure of the effectiveness of the design is its voltage standing-wave ratio (vswr). If the characteristic impedance of the switching mechanism of the relay exactly matches the cable impedance, the ratio is 1:1. This is never realized, but a good ratio that is frequently attained is 1.2:1.

FIGURE 4.87   A typical coaxial relay. *(Magnecraft Electric Co.)*

A relay of the coaxial type is used in microwave switching, as in switching from transmitters to receivers, and also in switching from antenna to antenna and in multiple-antenna systems. Many television installations use coaxial relays for switching purposes. Besides the chambered-contact type of relay in Figure 4.87, there are also dry-reed contact coaxial relays.

Stepping relays   See the section on stepping switches for the distinction between stepping relays and stepping switches.

Latching relays   There are basically two kinds of latching relays, mechanical and magnetic latching. In the first kind, some variety of mechanical catch or toggle action causes the relay to remain in the operated position after the actuating pulse ceases. A second pulse through the same or a different winding restores the relay to its original position. Because most mechanical latches are additions to relays of a type covered earlier, they are not examined in detail again. Mechanical latching is subject to early and rapid wear, maladjustment (usually the result of unauthorized tinkering), and failure from shock or vibration. It has the advantage of simplicity and is easy to understand.

Magnetic latching uses either a permanent magnet to hold the armature in the operated position after the operating pulse ceases or a coil with a permanent-magnet material, where its core takes a magnetic set to hold the armature operated after the actuation occurs. In either case, the armature is restored by having the holding power of the permanent magnetism overcome by a reverse pulse through the same or a different winding. Both these techniques are usually used on relays of a general type, as already examined. Most common magnetic latching relays are of the general-purpose, telephone, or crystal-can (military) types. Relays of this kind are used for memory, overload response, or to aid in resistance to vibration and shock.

Definite-purpose relays   In almost all manufacturers' catalogs, one or more kinds of relays designed to handle a definite task are shown. Some of these indicate or control specific voltages, sequence of operations, or specific machine operations. Because these are not likely to be of general interest, they are not examined further here.

The relay choice   The decision as to what kind of relay to choose can be based on an analysis of the foregoing information and a search of possible vendors' literature. Tables 4.22 and 4.23 at the end of this chapter contain lists of relay vendors and agencies.

In deciding which way to go, the circuit designer has to weigh many things. Tables 4.14 to 4.16 might be of assistance. The EMR is a device with the possibility of a large number of outputs for one or more inputs. The dry-reed relay has the possibility of up to 10 to 12 outputs for a single (or double) input. The SSR is one solid-state element per switching function. The EMR, because of its mechanical coupling, has a definite wear-out disadvantage. Sometimes its exposed contacts are considered a disadvantage. The dry-reed relay, by eliminating mechanical coupling, has some advantage from the reduced-wear standpoint and provides sealed contacts. The SSR has all the advantages listed for such a relay, plus some others (shown in Tables 4.15 and 4.16), but it also presents many problems, most of which are instantly recognized by electronics engineers. The question of what relay is to be selected, therefore, becomes one of weighing advantages against disadvantages and recognizing the necessities of the situation. For example, ask the questions: How necessary is speed of response? Does it outweigh the inherent advantages of the other alternatives? Usually, economics will be the deciding factor.

TABLE 4.14 A Selection Guide and Summary of Electromechanical/Electromagnetic Relay

Kind of relay	Required operating power	Operate time	Release time	Max contact-load current	Max contact arrangement	Size and weight	Remarks
General-purpose (GP) (no particular design shape or arrangement exists that is recognizable as common to all manufacturers)	200 MW to 6 or 8 W, depending upon size of contact load and volume of operating coil. (Sensitive versions of this general relay type are able to operate on approximately 100 mW per pole)	10–25 ms depending upon contact spring load and coil L/R. Operate-time delay up to 50 ms available on some models	5–20 ms depending upon contact spring load and degree of saturation. Release-time delay up to 100 ms available on some models	1, 2, 3, 5, or 10 A, depending upon nature of load and size, shape, and kind of contact. In some cases rating given in horsepower and voltage; e.g., 1/3 hp at 120 V ac	One to eight or nine poles	Less than 1 to 8 in.³, depending upon type and number of contacts, ½–6 oz	This class of relays is so highly varied that it is difficult to establish limits. A thorough search of vendors' literature is necessary to determine the suitability of any specific relay for a particular problem. Advertised life varies from 100,000 operations electrical to 10 million operations mechanical depending upon manufacturer and type of relay
Unless otherwise specified, the insulation-voltage breakdown test is 500 V rms 60 Hz, or more, between all insulated parts. Contact resistance is usually 50 to 100 mΩ, new. Typical insulation resistance, specified, is 1,000 MΩ.							
Power type	2–8 W dc, 3–10 VA, ac	In excess of 10 ms	In excess of 10 ms	10–50 A	One to three poles, single- or double-throw	5–20 in.³, 2–12 oz	This class of relays is highly varied as to size, shape, and style, making it difficult to establish limits. A thorough search of vendors' literature is necessary to determine a relay's suitability for any specific application.
Contact-load ratings are frequently given in specific load capability, such as 1 or 2 hp. Also the maximum figure is sometimes indicated as for a noninductive load. It is significant that maximum current rating is often the same for 115/120 V ac and 24/28 V dc. Insulation-voltage breakdown test is usually 500 V rms 60 Hz between open contacts, 1,000 V rms elsewhere							

TABLE 4.14 A Selection Guide and Summary of Electromechanical/Electromagnetic Relay (*Continued*)

Kind of relay	Required operating power	Operate time	Release time	Max contact-load current	Max contact arrangement	Size and weight	Remarks
Telephone type	200 MW to 12 W, depending upon size of contact load and the copper volume of the coil. In general large relays use 2–8 W, medium 1.5–5 W, and miniature 1–3 W	2 to 15 ms depending upon contact spring load and coil L/R. Operate-time delay of 50 to 100 ms obtainable on larger relays	5 to 20 ms depending upon contact spring load. Release-time delay of 50–100 ms relatively easy to obtain. Release-time delay up to 500 ms available on largest relays with short-lever-ratio armatures	Approximately 1 A for standard contacts. Special contacts are available on some models extending the maximum allowable current up into the light- to medium-duty power-relay ranges	Depending upon the size and design of relay, this is rather easily varied from single-pole single-throw to as many as 13 or more contact springs, single- or double-throw, of two or more pileups	1–20 in.^{3}, depending upon type of delay and number of required contacts	Advertised life varies from 50,000 to 100,000 operations electrical and 3 to 10 million operations mechanical, depending upon manufacturer and type of relay. In general this is a somewhat clearly defined design shape, but the size is quite varied. The contacts are usually either leaf or wire types, stacked in a pileup of considerable height, and fastened to an L-shaped frame. The relay is usually arranged for wiring in a plane on the opposite surface to which the relay mounts

The features which make telephone-type relays attractive are usually the highly varied kinds of contacts and the allowable timing variations. Unless otherwise stated, the insulation-voltage breakdown test is 500 V rms 60 Hz, between all insulated parts and open contacts. Maximum contact resistance is usually less than 50 to 100 mΩ, new.

Type	Power	Operate time	—	Current/Load	Poles & throws	Size & weight	Life / Remarks
Dry-reed and mercury-wetted reed relays	80 MW for lightest load to 6 W for maximum size load. Wattage that coils can dissipate usually varies from two times just-operate wattage to five or more times that value	From 1 ms (max) for single capsule, exclusive of bounce, to 6 ms (max) for large multicapsule units, including bounce	½ ms (max) for single capsule exclusive of bounce to 6 ms (max) for large multicapsule units, including bounce	Varies from 0.10 to 1.0 A depending upon size of capsule and VA of load. The large, or standard, size reed is rated at 1 A switching, 15 VA, 250 V maximum, and 5 A carry	From single-pole single-throw to 10 or more poles, with up to four double-throw combinations	Size varies from a fraction of a cubic inch (single-pole, micro-miniature) to 9 or more cubic inches for the largest multicapsule assemblies. Weight varies from 5 g to 2 oz	Life varies from 5 to 25 million operations at rated load, fully protected, depending upon capsule size and kind. Within allowable and specified conditions the mercury-wetted reeds have longer life and more constant contact resistance than the nonwetted
	Insulation tests vary for various types and sizes. Vendor's information will have to be referred to in each case. Reed relays are attractive because they are compatible with the design concept of discrete components having a low profile when attached to a PC board, and their low coil self-inductance makes them suitable for direct drive from a transistor						
Mercury-wetted contact relay (Although there are only two sizes and shapes of capsules, these may be housed in electron-tube octal-plug housings of from one to four capsules per relay or in	100 MW for small-capsule relays, 500 MW for large-capsule relays	1–2 ms for small capsule and 4 ms for large capsule	1–2 ms for small capsule and 2.5 ms for large capsule	1–2 A for small capsule, 5 A for large capsule	1 Form C or 1 Form D, only, for the small capsule, and 1 Form D only for the large capsule	The size varies from about 4 in³, cylindrical, with a 1.1 in diameter to 0.75 in³ in the form of a rectangular module. The weight varies from about 3 oz for small single-capsule	The popularity of these relays is restricted somewhat by their obvious limitations of weight (mercury is heavy), mounting position (the relay cannot be tilted much), low temperature (mercury becomes a solid in the vicinity of −40°), shock and
	These relays have unlimited life; the contact resistance is nonfluctuating and predictable. They are fast and handle heavy loads quietly, efficiently, and safely						

TABLE 4.14 A Selection Guide and Summary of Electromechanical/Electromagnetic Relay *(Continued)*

Kind of relay	Required operating power	Operate time	Release time	Max contact-load current	Max contact arrangement	Size and weight	Remarks
flat-sided rectangular-module cases with PC pin terminals, a popular arrangement						relay to 10 oz for large 4-capsule relay	vibration (in normal temperature mercury is a liquid), and cost (mercury is expensive)
Crystal-can relays (both the *bathtub* conventional shape and the *top hat*, TO-5, shape)	100 MW, or less, to 750 MW	2 to 5 ms not including bounce. Bounce does not exceed 1.5 to 2 ms for some models, does not exceed 0.250–1.0 ms for others	1.5–4 ms	Up to 5 A, 28 V dc resistive for full-sized can. About 2 A maximum for half size can. Still lower maximum current allowed for the smallest can	2 Form C	¼ crystal-can size to full crystal-can size, and TO-5 size. 0.5 oz or less, up to 0.8 oz for crystal can. TO-5 weighs 0.08 oz or less up to 0.15 oz	Relays of this type were designed for and originally used in military and space-age applications. It was inevitable that because of size and weight reduction other uses would be found, particularly in the sensing of low-level signals. In most industrial applications the inherent resistance to shock and vibration is of little consequence. One of the TO-5's called an op-amp universal relay contains in one TO-5 case, besides the relay, an op-amp, driver, and surge suppressor
		Life is usually not required to exceed 100,000 operations at rated load, for these relays					

Special-purpose relays such as TDR's, latching, coaxial, and thermal are omitted because their operating characteristics are peculiar to their design types and the need they fill. The text discussion for these special relays should be adequate for determining early circuit design and application interest in them. They are not, in most cases, directly competitive with the basic types. For example, thermal relays cannot be compared directly with the basic types because their actuation has a nonspecific inherent time factor and there is no standardization in this class of relay to make such comparison meaningful.

TABLE 4.15 A General Comparison of Electromechanical/Electromagnetic Relays and Solid State Relays[10]

Parameter	EMR	Dry-reed relay	SSR
Life	Dependent upon load and number of operations	Dependent upon load and number of operations	Barring accident, dependent only upon time
Response time	2–5 to 25–50 ms	0.5–2 to 6–20 ms	Fractional micro-seconds
Power gain per $	Excellent	Good	Poor
Contact functions	Multiple	One or more	One
Size-weight	Poor	Good	Excellent

TABLE 4.16 A Comparison of Reed, Solid-State and Hybrid Relay Characteristics[7]

	Reed	All SS	Hybrid
Input specifications:			
Coil voltage	0.6–96 V	3–280 V	2.4–48 V
Coil current	0.66–241 mA	0.5–20 mA	1–33 mA
Coil sensitivity	66–700 mW	1.5–7 mW	60 μW–290 mW
Must pickup voltage (typical 75% of coil V)	0.34–46 V	3.5–90 V	1–36 V
Must dropout voltage (typical 10–20% of coil V)	0.07–15 V	0.4–20 V	0.5–2 V
Pull-in speed	0.5–20 ms	5 μs–1½ cycles	1–25 ms
Dropout speed	60 μs–17 ms	5 μs–1½ cycles	First 0 crossing–½ cycle
Output specifications:			
Max output voltage	500 V	280 V	240 V
Max output current	5 A	40 A	15 A
Max output power	360 VA	9.6 kVA	840 VA
Contact resistance	15–200 mΩ	0.1–2 Ω	0.1 Ω
Contact offset voltage	Virtually 0 V	1 mV–1.6 V	1.0–1.5 V
Off-state leakage current	None	1–5 mA at 120 V	1–5 mA at 120 V
Min on-state current	None	20–100 mA	100–200 mA
Max contact operating frequency	1 kHz	100 kHz	1 kHz
Max lifetime cycles	10^8–10^{10}	$>10^{10}$	5×10^8
Contact bounce	Hg, no; dry, yes	No	Yes and no
Min dV/dt	N/A	20 V/μs	1 V/ms

4.2.7 Cautions

Coil-winding polarity Unless conventional types of relays, as designed for low voltage circuits, are to be used in relatively short-lived equipment, well insulated from ground, it is considered best to connect the negative potential permanently to the outer coil terminal(s) and control the relay by switching the grounded positive potential to the inside terminals of the relay winding(s). This reduces electrolysis to a minimum and, particularly for coil windings of relatively fine wire, adds years of life to relay coils that otherwise might deteriorate from electrolysis in a very short time. The low voltage that was previously referred to is 50 V (or less), as encountered, for example, in wired communications equipment. A notable exception to this rule exists for 24-/28-V equipment intended for the military, where switching of positive ground is frowned upon as uncomfortable for maintenance personnel to work around, although telephone maintenance personnel accept it as a necessary situation and have done so for many years.

Contact-spring polarity The same potential, whichever it is, should be connected to all movable springs. This is an aid to maintenance personnel so that they do not have to consult a blueprint continually for that information, which makes servicing faster and simpler. Also, it lessens the chance of accidental short circuiting, which can destroy relay contacts in an unguarded instant, as can occur when the maintenance man has no recognized standard arrangement to rely upon.

Interaction of windings Relays should not be operated with windings in parallel if at all possible. Occasionally, when relay windings are connected to the same bus on one side and are operated from a common contact closure on the other, they affect each other adversely when their operating circuits are opened, owing to interaction from the back emf generated by their collapsing magnetic fields. This problem can exist for:

- Two or more nonpolarized relays with parallel coils
- A single relay with more than one winding (bifilar or polarized)
- Dc relays using diodes for operation from an ac power source
- Polarized magnetically latching relays

Arcing problems A number of means exist to accelerate the arc extinction when interrupting large voltages and currents. These include:

- Large contact gaps (see Figures 4.88 and 4.89)
- Double-break switch structures
- Magnetic or air-blast arc blowouts for heavy-duty relays
- Arc-suppression circuits

If arcing generates enough plasma to reach from the live circuit to some other circuit or ground that bypasses any current-limiting resistance, an explosively catastrophic flashover

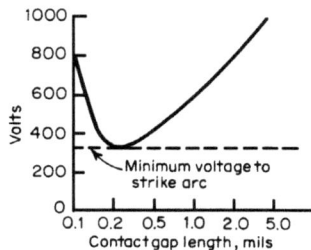

FIGURE 4.88 Breakdown voltages for various air gaps at 1 atm (Paschen effect).[14]

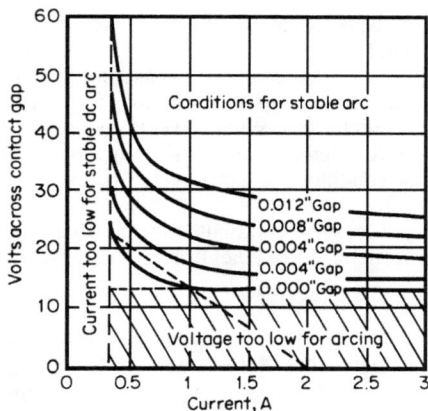

FIGURE 4.89 Voltage-current gap-length relationship for stable and unstable dc arcs.[14]

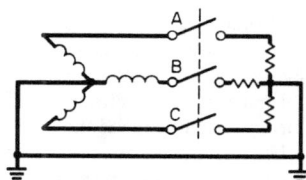

FIGURE 4.90 Conditions for three-phase flashovers in 120-V, Y-connected, four-wire switching circuits. Three-phase flashovers in 120-V Y-connected, four- wire switching circuits can occur with appropriate phase polarities and one contact opening before the other two. In the presence of a conductive plasma, an arc can flash if B is negative and either A or C is positive.[14]

occurs. This hazard particularly affects polyphase power and polarity-reversal circuits. Mandatory for other considerations, these standard design and construction practices often accentuate problems:

- Relays must switch power circuits between the power source and the load, not on the ground side of the load.

- Relays used for switching 115 Vac or higher must have a grounded frame (and can, if enclosed) so that electrical leakage or short circuit in either load or relay does not constitute an operator hazard.

Fault circuits These several examples illustrate fault circuits in common control circuits:

In three-phase Y-connected (or four-wire) circuits (Figure 4.90), a high-voltage (up to 280 V on 120-V systems) low-resistance circuit exists between legs during the arcing, just before the instant of current reversal in any leg. If the conductive plasma cloud reaches from one leg to another, only the capability of line and source limit the current.

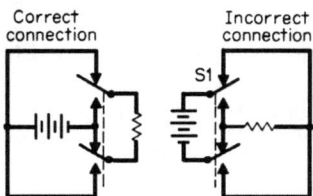

FIGURE 4.91 Problem facing a relay with DPDT contacts when reversing a load. (a) Connected as shown at right, power source can short if one contact sticks (dashed S1 contact). (b) Opposed polarities across fixed switch contacts (especially "correctly connected" left circuit) enhances chance of plasma-arc flashovers.[14]

If contacts are physically near a grounded structure (i.e., a relay frame or can), relay contact chattering can cause a flashover. Flashover can occur if the ac coil power changes slowly, if the relay is improperly adjusted, or as a result of switching or lightning-induced line transients.

The familiar double-pole reversing switch can prove troublesome, even when connected correctly. Unfortunately, this trouble is seldom recognized for what it is. Of course, the switch must be connected so that progressive contact engagement momentarily shorts the load, not the power source (Figure 4.91). Unfortunately, rapidly repeated switch operation (or circuit-induced contact chatter) can generate a power source-shorting plasma cloud. If transient current and voltages exceed the stable arc minimums, the plasma can permit a flashover.

Ironically, safety codes force us to live with potentially catastrophic fault circuits. The problem would not exist if the load-circuit ground leg, rather than its hot leg, could be switched. On the other hand, this might explain why ungrounded relays constitute personnel hazards (short-circuit fault currents greater than 2 to 5 mA)—even though the fault current cannot exceed the load current and a catastrophic fault could never become established. Such a hazard to personnel exists with ungrounded relay cans whenever load voltages and currents exceed the minimum arcing conditions and the design permits the plasma clouds to reach the can.

The causes of these faults and the circuit conditions that contribute to vulnerable situations must be recognized. Generally, the most economical remedies become self-evident:

- Relays must be rated for the voltages they will experience.
- Interleg spacings near contacts in Y-connected 120-V three-phase circuits should be rated at 208 V. There should be adequate space and barriers between contact sets.
- A three-pole relay switching on and off a three-phase 115-V ac load contains about 170 V peak across the contact gap or between any contact and ground. If the same relay is used to switch a load between two unsynchronized three-phase power sources, as much as 340 V peak can appear across a contact gap.
- Provision for the grounded frame and case is mandatory in small hermetically sealed relays in 120-Vac single-phase applications. If the load current and voltage generate enough plasma to reach from the contacts to a grounded part, internal flashover prevention is mandatory.
- When a 28-Vdc inductive load is reversed, a catastrophic fault can occur only during the very brief contact transfer. With stable arc conditions sustained for the duration of contact transfer, the fault path simply jumps whichever fixed contact gap opens first.

These problems can be solved by rearranging the circuit to eliminate uncontrolled fault current, using a relay that tolerates the arc, or using arc suppression to reduce the arcing to an insignificant amount. Above all, a suitable relay should be used for power-supply changeover or, in fault-protection service, a heavy-duty power contactor or circuit breaker.

4.3 SWITCHES

The term switch was defined at the beginning of this chapter so that it could be used in the simplest relay definition. In the rest of this chapter, the word *switch* can have a far more comprehensive meaning. It is applied to all kinds of switching devices, from a simple pair of mating contacts, manually actuated, to electrically operated multipole, multiposition devices, such as stepping switches. Such a range of identities makes this subject difficult to deal with in the space allowable. A recent listing of switch manufacturers consisted of over 160 names, covering a wide range of types of products (see Table 4.24). Consequently, the simpler, well-known, and recognized switches are dealt with sparingly, and the newer and/or more-complicated switching devices are examined in more detail.

Matters such as contact load-handling capabilities, contact protection, and arc and noise suppression are so similar to those of relay contacts of the equivalent type, material, and size that no repetition of ways and means of dealing with such problems is repeated here.

For manually operated switches, not of the toggle or snap-action type, it is possible for an individual to so delay the opening or closing of a pair of contacts or to so reduce the contact pressure by deliberately slow motion that the load-handling capability is seriously reduced. It is not too unusual to see a failed contact set on a manually operated switch that proved inadequate to a load situation, where relay contacts of the same size, shape, material, and mechanical separation and pressure did the job well. For that reason, manually operated switches are frequently used to close the circuit to a relay, which, in turn, handles the actual load. There might well be a wire-size advantage also in doing it that way.

4.3.1 Kinds and Types of Manually Operated Switches

A listing of kinds of switches to be manually operated is quite extensive, and includes the following:

- Pushbutton switches (also called *push keys*)
  - Illuminated and nonilluminated
  - Single and multiple, or ganged
  - Snap action, wiping action, or butt action
  - Locking (push-pull) and nonlocking
  - Alternate (on/off)
- Slide switches
  - Single and ganged
- Lever switches (also called *lever keys*)
  - Locking and nonlocking
  - Toggle
- Turn-button switches (also called *turnkeys*)
- Thumbwheel switches
- Leverwheel switches
- Rocker switches
- Keyboards
- Rotary selector switches
- Matrix-type selector switches

This is obviously a somewhat generalized listing, and many special switches might not fit any of these categories. One vendor even has a listing of his "weird and wonderful switches." Types of mountings, method of attaching circuit connections (such as through the printed-circuit edge of card connectors), open or enclosed bodies (frames), and customized variations of all kinds could lead to further categorization.

Pushbutton switches Push-button switches are available with the contacts remaining operated after the button has been pressed (locking or latching) and with nonmaintained operation after finger removal (nonlocking or nonlatching) . There is also a form with a button that is pulled out to latch and pushed back in to unlatch. In most cases, visual observation determines whether or not latching pushbuttons are in the operated state. An indicator light might be either separate from the button or self-contained. Lighted pushbuttons of two types are shown in Figure 4.92. The lighting can be transmitted through translucent colored buttons or projected as a colored light. Split-screen displays are common.

Contact ratings and life vary to such an extent from one type of switch to another and from manufacturer to manufacturer that it is impossible to present meaningful general data. For example, one vendor will promise a life of 25,000 operations electrical, 50,000 mechanical, at rated load, indicating only limited life, regardless of contact load, but another vendor will quote a life of 10,000,000 operations for a light contact load on a simi-

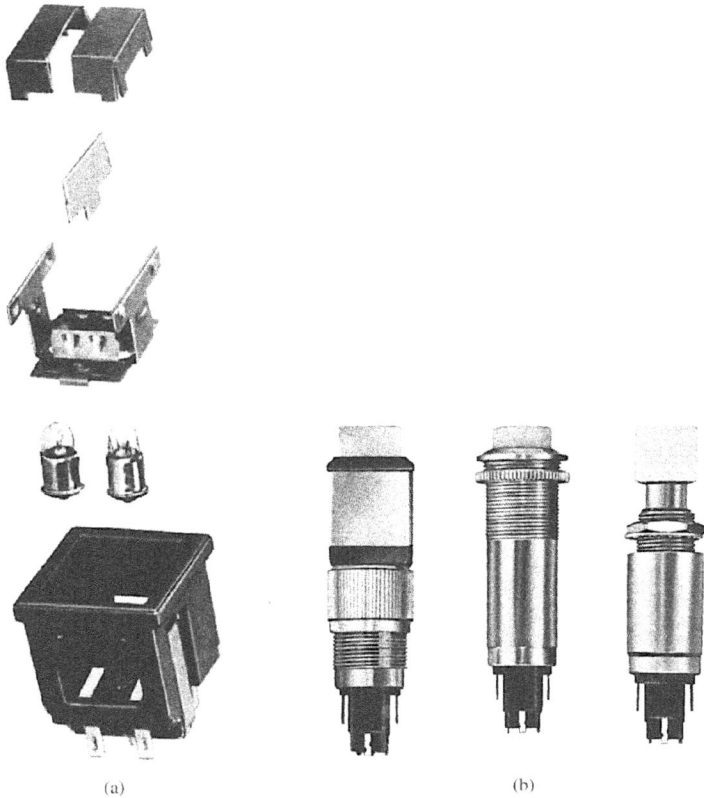

FIGURE 4.92   Lighted push buttons. (a) Dual lamp, exploded view. *(Switch-craft, Inc.)* (b) Round or square button for panel or subpanel mounting. *(Grayhill, Inc.)*

lar switch. Vendor's claims in these matters are to be relied on. This same problem of inability to establish generally meaningful load and life rating applies to all the manually operated switches that follow, also.

Multiple-station pushbutton switches have a common frame with interaction between switches so that they perform interrelated functions that a single switch cannot perform. Also, it is possible to get sequential protection from such a grouping that is impossible to achieve mechanically any other way. Solenoid-operated multistation switches can be actuated from a remote or local position. In one form, this provides an important security feature to prevent aimless or unauthorized pushing of buttons. Table 4.17 is a list of prominent manufacturers of lighted pushbutton switches.

Slide switches   As the name suggests, slide switches take either one of two positions or one of three positions, by means of lateral displacement of the button. They come in single-switch and ganged arrangements. The three-position switches present somewhat of a hazard in correctly identifying the midposition. Contacts can be either toggle-actuated or

TABLE 4.17 A Partial List of Manufacturers of Lighted Push Buttons[13]

Airpax Electronics Exchange
1836 Floradale Ave.
El Monte, CA 91733

Alco Electronic Products
1551 Osgood Street
N. Andover, MA 01845

Arcoelectric Switch Works
PO Box 348
N. Hollywood, CA 91603

Arrow-Hart
103 Hawthorn St.
Hartford, CT 06106

Burgess Switch Co.
777 Warden Ave. N
Scarborough, Ont.
Canada

Carling Electric
505 New Park Ave.
W. Hartford, CT 06110

Chicago Switch
2035 W. Wabansia Ave.
Chicago, IL 60647

Clare-Pendar Div.
PO Box 785
Post Falls, ID 38540

Compu-Lite
17795 "C" Sky Park Cir.
Irvine, CA 92707

Dialight
60 Stewart Ave.
Brooklyn, NY 11237

Drake Manufacturing
4626 North Olcott Ave.
Harwood Heights, IL 60656

Furnas Electric
1000 McKee
Batavia, IL 60510

Gray Hill
561 Hillgrove Ave.
La Grange, IL 60525

Industrial Devices
Edgewater, NJ 07020

Industrial Electronic Engineers
7720 Lemona Ave.
Van Nuys, CA 91405

International Electro
8081 Wallace Rd.
Eden Prairie, MN 55343

Ledex
123 Webster St.
Dayton, OH 45401

Licon Div., Illinois Tool
6616 W. Irving Park Rd.
Chicago, IL 60634

Master Specialities
1640 Monrovia
Costa Mesa, CA 92627

Maxi-Switch
3121 Washington Ave.,
Minneapolis, MN 55411

Micro-Switch, Div. of Honeywell
11 Spring Street
Freeport, IL 61032

Molex
2222 Wellington Ct.
Lisle, IL 60532

Oak Industries, Switch
Crystal Lake Ave.
Crystal Lake, IL 60014

Seacor
598 Broadway
Norwood, NJ 07648

Staco-Switch
1139 Baker St.
Costa Mesa, CA 92626

Switchcraft
555 N. Elston Ave.
Chicago, IL 60630

Symbolic Displays
1762 McGaw
Irvine, CA 92705

UID Electronics
4105 Pembroke Rd.
Hollywood, FL 33021

Unimax Switch
Ives Road
Wallingford, CT 06492

Waldom Electronics
4625 West 53rd St.
Chicago, IL 60632

(a)

(b)

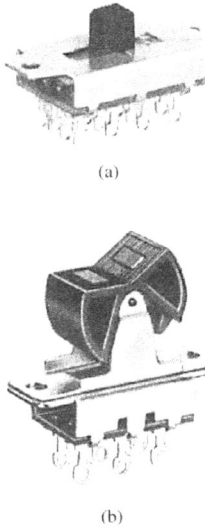

FIGURE 4.93 A typical slide switch (a) and rocker switch (b). *(Switchcraft, Inc.)*

straight deflection-actuated. The contact closure can be momentary, locking, or a combination of both. A slide-switch example is shown in Figure 4.93a.

Lever switches The telephone-industry relative of the lever switch is not available with toggle action and is called a *lever key.* Commercial and industrial versions of these devices can be either toggle-actuated or straight deflection-actuated and either locking or nonlocking. Both two- and three-position switches are available. The three-position switches can be locking to one side and nonlocking to the other side, locking on both sides, or nonlocking on both sides. The telephone-type switch, or "key," because of the nature of its original purpose, has quite a long life (about 10 million operations) when used within the limits of its contact rating.

Turn button switches This is primarily an item made by telephone manufacturers and offered for sale as an industrial item. It operates its contacts when a button is rotated approximately a quarter turn. One of its advantages is that it is easily made either locking or nonlocking and gives ready visual indication of whether it is operated or not.

Thumbwheel and lever switches Rotating a serrated wheel that protrudes above the front surface or operating a lever that sticks out in front activates these essentially similar switches having 8, 10, 12, or 16 discrete dial positions. Modular thumbwheel switches require minimal panel space in setting up multidigit inputs. Integral encoder assemblies convert the thumbwheel position directly to output code. The switch normally contains the circuitry and requires no added wiring (Figure 4.94). Some typical truth tables for the more common codings are shown in Table 4.18. Built-in illumination is available in some models. Many mounting variations and custom assemblies are also offered.

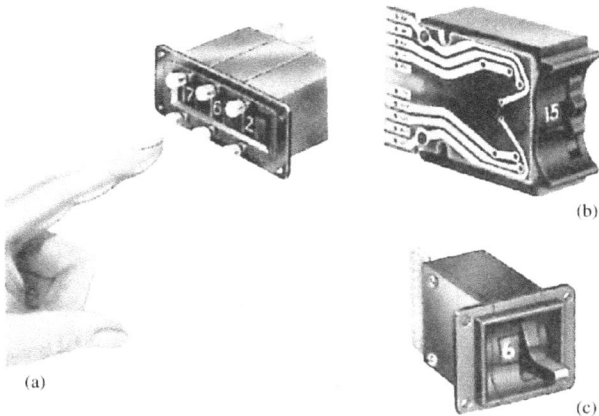

(a)

(b)

(c)

FIGURE 4.94 Typical push button (a) *(Chicago Dynamic Industries, Inc.)*, thumbwheel (b), and lever (c) coded switches. *(Cherry Electrical Products Corporation.)*

TABLE 4.18 Typical Truth Tables for Thumbwheel Switches

Table a — Code: BCD (binary-coded decimal)

Readout symbol	Common C connected to terminals = ●			
	1	2	4	8
0				
1	●			
2		●		
3	●	●		
4			●	
5	●		●	
6		●	●	
7	●	●	●	
8				●
9	●			●

Table b — Code: 10 position decimal

Readout symbol	Common C connected to terminals = ●									
	0	1	2	3	4	5	6	7	8	9
0	●									
1		●								
2			●							
3				●						
4					●					
5						●				
6							●			
7								●		
8									●	
9										●

Table c — Code: binary-coded octal

Readout symbol	Common C connected to terminals = ●		
	1	2	4
0			
1	●		
2		●	
3	●	●	
4			●
5	●		●
6		●	●
7	●	●	●

Table d — Code: BCD complement

Readout symbol	Common C connected to terminals = ●							
	1	2	4	8	1	2	4	8
0					●	●	●	●
1	●					●	●	●
2		●			●		●	●
3	●	●					●	●
4			●		●	●		●
5	●		●			●		●
6		●	●		●			●
7	●	●	●					●
8				●	●	●	●	
9	●			●		●	●	

Table e — Code: BCO complement only

Readout symbol	Com. connected to complement of binary bit output = ●		
	1	2	4
0	●	●	●
1		●	●
2	●		●
3			●
4	●	●	
5		●	
6	●		
7			

Table f — Code: Hexadecimal plus complement plus 2 commons

Common X connected to terminals = ●
Common Y connected to terminals = ○

Readout symbol	Interconnection from common X to binary bit output numbers				Interconnection from common Y to complement of binary bits output number			
	1	2	4	8	$\bar{1}$	$\bar{2}$	$\bar{4}$	$\bar{8}$
0					○	○	○	○
1	●					○	○	○
2		●			○		○	○
3	●	●					○	○
4			●		○	○		○
5	●		●			○		○
6		●	●		○			○
7	●	●	●					○
8				●	○	○	○	
9	●			●		○	○	
10		●		●	○		○	
11	●	●		●			○	
12			●	●	○	○		
13	●		●	●		○		
14		●	●	●	○			
15	●	●	●	●				

SOURCE: Cherry Electrical Products Corporation, Waukegan, Ill.

Pushbutton switches, coding   Pushbutton switches which provide essentially the same functions as thumbwheel switches in that they have the same kind of numerical display, both digital and binary codes, come sealed or not sealed, lighted or unlighted, and are available in bidirectional or unidirectional types. Some fit in the same dimension of panel openings as thumbwheel switches. An example of these devices is shown in Figure 4.94a.

Rocker switches   Rocker switches get their name from appearance and feel. The actuating button is pivoted so that it rocks to either of two operated positions and usually has a neutral or off mid-position. The contact switching elements are usually toggle-activated. The appeal is largely aesthetic because of its good appearance, although its ease of actuation, toggle-switch action, and ease of position identification make it attractive from the good-engineering standpoint. Illumination can be provided by edge lighting of the panel in such a way that light is conducted into the button when tilted. One form is shown in Figure 4.93b.

Keyboards   A wide variety of keysets, keyboards, touch pads, and similarly named components provide a convenient and inexpensive way of providing manual input to electrical and electronic devices. Some are identical in appearance and similar or identical in construction to the touch-calling keynotes of automatic telephony, and others were obviously designed with noncommunication applications in mind. The key arrangement varies from the 12-button (arranged in 4 rows, 3 buttons per row) with button caps labeled 1 through 0 plus * and #, and telephone-type alpha-numeric, to elaborate key arrangements of 20 or more keys following adding-machine and/or typewriter button-cap identity sequence and arrangement. Because it would be impossible to cover all these in this work, only the simplest is examined. Such is probably the telephone-type keyset, without tone generators, although that form, too, is readily available for industrial use, if proper frequency-decoding equipment is at hand for receiving, identifying, and using the received tones.

The keyset in Figure 4.95 has its 12 pushbuttons arranged in the familiar pattern that is standard for all touch-calling telephone instruments. However, it is not equipped with a tone generator. The keyset will find application where serial information is to be encoded in a system and where the cost of tone-decoding equipment cannot be justified. Signaling with the keyset in Figure 4.95 requires a multiple number of leads (the number required depends on the code chosen). Unless considerable distance is involved, dc signaling on a multiple number of leads is quite economical.

Each of the pushbuttons on the keyset controls two normally open contacts of the permissive make design so that contact pressure does not depend on contact over-travel. In addition to the contacts operated by each pushbutton, a common switch is actuated when-

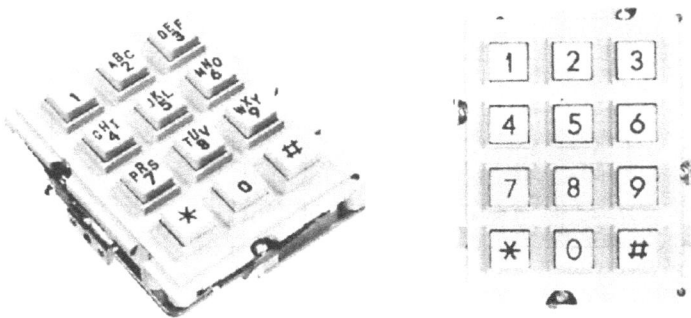

FIGURE 4.95   A typical telephone-type keyset. (*GTE Automatic Electric Co., Inc.*)

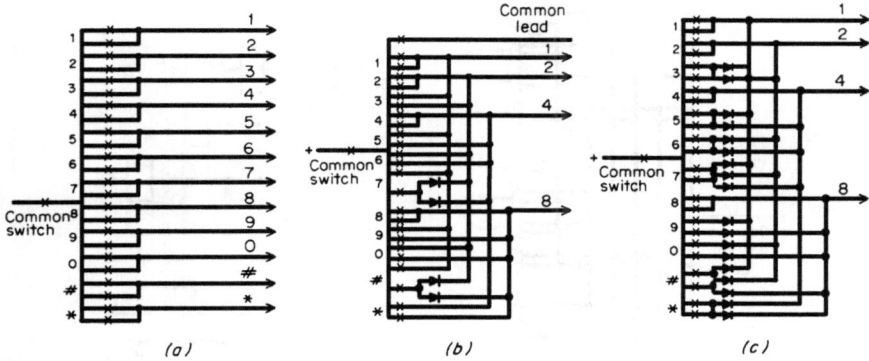

FIGURE 4.96 Examples of connections to the telephone-type keyset for some typical coding. (a) Straight decimal coding. (b) Binary coding. (c) Binary with parallel contacts.

ever any pushbutton is pressed. The contacts on the common switch operate after the pushbutton contacts close and open before the pushbutton contacts reopen. By connecting the common contacts in series with all pushbutton contacts, loads of up to 1 A, 50 VA, can be switched. Figure 4.96a, 4.96b, and 4.96c shows typical applications.

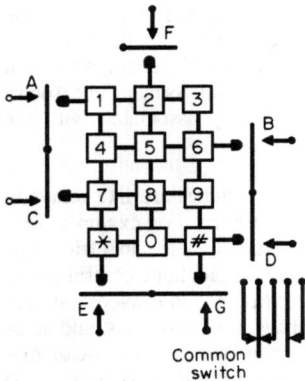

FIGURE 4.97 Schematic drawing showing two-out-of-seven. Code arrangement of telephone-type keyset.

There are seven actuator bars per keyset. Each actuator bar operates one contact form. When any of the 12 buttons is pushed, it moves a unique combination of two actuators to produce a two-out-of-seven contact-code output. Figures 4.97 and 4.98 show how the code format can readily be adapted to most control requirements.

Because they are standard telephone keysets, these keysets have a familiar "feel," which is conducive to fast, accurate operation. Figure 4.97 shows how the two-out-of-seven contact code is derived.

The common switch can be connected in series with the pushbutton contacts to ensure simultaneous switching of those contacts. Used in this manner, when a button is pushed, the two sets of contact springs prepare isolated paths for the two outputs. Immediately thereafter, the common switch closes the circuit through the pushbutton contacts. Upon release, the common switch opens first to break the circuit, then the pushbutton contacts open.

Two methods are shown for converting the two-out-of-seven code generated by the keyset into decimal code in Figure 4.97a and 4.97b.

Rotary selector switches One of the oldest forms of manually operated switches is the rotary switch, so common and so simple it needs little discussion other than to point out what range is available. As can be seen from Figure 4.99, multideck switches are common, almost any number of contacts in each deck are offered, the decks can be enclosed or open, the contacting can be momentary or maintained, and the shaft rotation can be continuous, in either direction, or limited to just 360° of rotation, after which the knob must be re-

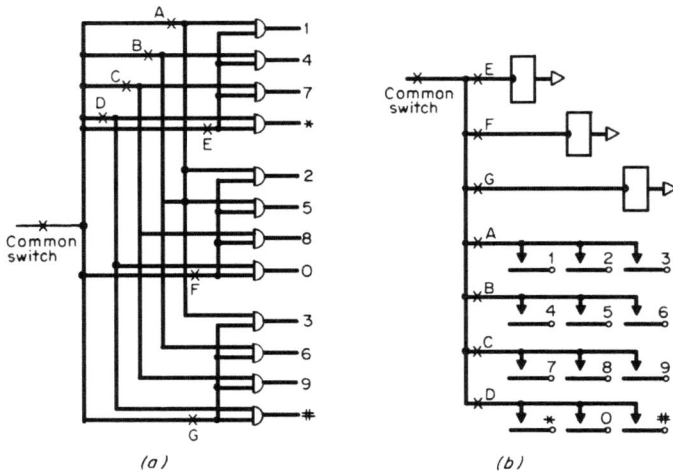

FIGURE 4.98   Two methods of converting two-out-of-seven code to decimal code.

versed. One of the advantages inherent in a rotary switch is that contacting must be in a predetermined sequence. This can, of course, also be a disadvantage that prevents its use. The latter problem is sometimes circumvented by using a special form of the rotary switch, where it is necessary to push the handle or knob in to actuate the connection. Selection of contact layer is achieved in this way on some models. Another special form of the rotary switch is the familiar tap switch, usually associated with the switching of heavy electrical loads.

Grades of manual switches   Manual switches are like relays in that they can be graded by class of service. For example, four common classifications are commercial (as used in office-type equipment), communications (as used in telephone central-office and subscriber equipment), appliance (as used in low-cost devices on vending machines, home appliances, television sets, and automobile dashboards), and industrial (rugged devices for use on machine tools, industrial materials handling and such equipment) where the device has to withstand operator abuse, oils, coolants, chemicals, and dust and dirt. The latter needs are recognized by NEMA in Standard ICS 1-110, with these designations: Type 1—General Purpose, Indoor; Type 3—Dust-tight, Rain-tight, and Sleet (Ice) Resistant, Outdoor; Type 4 and 4X—Water-tight and Dust-tight, Indoor; Type 7—Hazardous Locations, Class I (explosive gases and vapors); Type 9—Hazardous Locations, Class II (explosive dusts); and Type 13—Oil-tight and Dust-tight, Indoor, for protective housing of pilot devices, such as pushbuttons and selector switches.

FIGURE 4.99   A typical multideck rotary switch (manually operated). *(Grayhill, Inc.)*

If equipment is being designed for any of these locations, the ruling NEMA standard should be obtained and the applicable portions included in the purchase specifications.

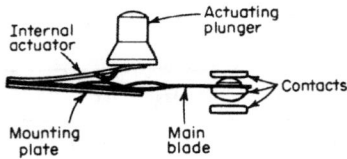

FIGURE 4.100 Schematic diagram of the arrangement of a snap-action switch. Snap-action switches derive their snap action from a springlike main blade. This slotted biposi-tional piece is prestressed by heat treating so that the center member is compressed and the two outside members are under tension. Depressing the actuating plunger forces the internal actuator against the two outside members of the main blade. The relationship between the members is thereby changed, and the blade snaps from the normal position to a second positive position. When the actuating plunger is released, a bias in the blade returns it to the normal position. Advantages: *(1)* High contact pressures result in excellent resistance to vibration. *(2)* Fast transfer time limits arcing and increases load capacity. *(3)* Variety of operating forces obtainable by varying thickness of blade. *(4)* Good re-peatability, because of only one moving part. *(5)* One-piece tumbled blade has no wear points and provides long life.

Sensing switches Switches that are not operated manually or electrically are usually operated from a mechanical stimulus, such as by a float or cam-operated lever. Switches of this type recog-nized by NEMA are cam-operated, control cutout, drum, float, isolating, limit (both control and power circuit), master, pressure, proximity, temperature, and selector. No matter what the in-put, a pair or more of contacts open, close, or transfer to control, indicate, or both. The contacts are usually snap or toggle switches so that small increments of changes will still result in very pos-itive contact action.

Snap-acting switches A snap switch gets its ac-tion from a specially formed and pre-stressed main spring or blade (Figure 4.100). The slotted biposi-tional blade is pre-stressed usually by heat treat-ing) so that the center section is compressed, but the two outside sections are in tension, causing it to reside in an unoperated or normal position (Figure 4.100). Pressing the center section by means of a plunger disturbs the forces within the blade to cause it to assume its other (operated) position with a rapid over-center action. This fast transfer of contacts, or snap action not only aids in extin-guishing arcs because of the speedy contact trans-fer, but also results in good contact pressure, and permits handling relatively heavy loads. In manufacture, the force required for operation and the distance required to deflect the plunger to get snap-over are rather easily controlled within strict tolerances. This permits using the snap switch for relatively precise applications. The repeatability is also rather closely limited because there is only one moving part. These switches are long-lasting because there are no localized extensively deflected areas.

Obviously, the snap switch is adaptable to almost any kind of switching device whether manually, mechanically, or electrically actuated. As a result, it might be found in manually controlled switches, on relays, or mechanically actuated as a limit switch in the control of motion (as on a machine tool), temperature (as in a thermostat), time (as on a motor-driven timer), etc. The big advantage, of course, is that the snap switch can directly control mo-tors, heaters, household appliances, and such from small stimuli without the use of an in-terposing relay. Movement differentials as small as 0.0005 inches permit a snap switch to be rated 25 A and 2 hp at 50 Vac.

The snap switch has several classifications or grades. Precision grades, as the name im-plies, are those whose actuating forces and distance of plunger travel are the most closely controlled. The appliance grade does not require very close tolerance, relatively speaking. There is a size classification of miniature, subminiature, and miniature-subminiature. Some are enclosed in plastic cases, some are open, and some are housed for special envi-ronments. The latter might take the form of a metal housing for attachment to a conduit system; it can be sealed against dust, moisture, or corrosive atmosphere; or it can even be immersion proofed and explosion proofed.

The same precautions apply to the need to pick the correct size and kind of contact to match the load here as were covered, with respect to relay contacts. Ratings given in man-

TABLE 4.19 Typical DC Loads for a 20 A, 250-V, AC General-Purpose Precision Snap-Acting Switch[15]

Contact separation, in	Noninductive			Inductive	
	Direct current, V	Heater load, A	Lamp load, A	At sea level, A	At 50,000-ft altitude, A
0.010	6–8	20.0	3.0	8.0	7.0
	12–14	20.0	3.0	5.0	5.0
	24–30	2.0	2.0	1.0	1.0
	110–115	0.4	0.4	0.03	0.02
	220–230	0.2	0.2	0.02	0.01
0.020	6–8	20.0	3.0	20.0	15.0
	12–14	20.0	3.0	10.0	8.0
	24–30	6.0	3.0	5.0	2.0
	110–115	0.4	0.4	0.05	0.03
	220–230	0.2	0.2	0.03	0.02
0.040	6–8	20.0	3.0	20.0	15.0
	12–14	20.0	3.0	20.0	15.0
	24–30	10.0	3.0	10.0	5.0
	110–115	0.6	0.6	0.1	0.05
	220–230	0.3	0.3	0.05	0.03
0.070	6–8	20.0	3.0	20.0	15.0
	12–14	20.0	3.0	20.0	15.0
	24–30	20.0	3.0	10.0	7.5
	110–115	0.75	0.75	0.4	0.2
	220–230	0.3	0.3	0.2	0.1

ufacturers' catalogs principally apply to alternating current (rapid snap-action contact movement and the zero current on each half-cycle aid these devices in attaining a relatively high load rating for their contact size and distance of separation), but applicable dc ratings are those of Table 4.19. Life expectancy at various ac loads is typically that of Figure 4.101.

Proximity switches The name indicates that switches of this type function without the switch being physically contacted by the stimulus. There are many kinds of proximity switches, some consisting of so many elements and of much complexity that they are really systems. Examples of this are those that operate from a sensing head that recognizes

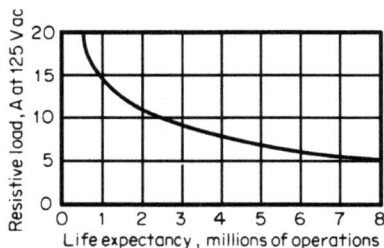

FIGURE 4.101 Life-expectancy curve for a general-purpose precision snap-acting switch.

radio frequency, magnetic bridge, inductive imbalance, photoelectricity (light), ultrasonic beams, or ferromagnetism. This section is confined to the simple proximity switch, principally the dry reed, or the mercury-wetted reed. Anything that can be arranged to move a permanent-magnet past, or close enough, to a reed switch can be the stimulus for operating a reed-type proximity switch. This principle is applied in some designs of keyboards, for example. Others are elevators and escalators, conveyors, machine-tool carriage transverse, etc. Most reed-switch manufacturers have literature suggesting how these switches can be operated from a nearby permanent magnet.

Electrically operated switches Electrically operated switches are available in many forms. As was pointed out, the relay is one example. However, it has been treated separately because of its extremely large number of variations and versatility. The following will deal with some of those remaining, and some, such as motor-driven switches, are so specialized in performance and application that they will not be examined in detail at all.

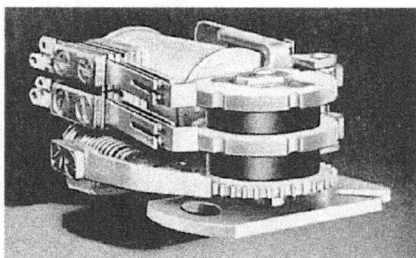

FIGURE 4.102 A typical stepping relay of the indirectly spring driven type. *(GTE Automatic Electric Co., Inc.)*

Stepping switches There has long been confusion in the use of the terms *stepping switch* and *stepping relay*. The line of demarcation is not easily recognized. Because the operating mechanisms can even be identical, or else quite similar, a distinction between stepping relay and stepping switch is hard to make, except by precise definition. Somewhat arbitrarily, therefore, NARM has made a distinction, which seems to be working rather well. NARM defines a stepping relay as a relay with many rotary positions, ratchet-actuated, moving from one step to the next in successive operations, and usually operating its contacts by means of cams. There are two forms: (1) directly driven and (2) indirectly driven, where a spring produces the forward motion on pulse cessation (Figure 4.102). The shorter of two definitions NARM gives for a stepping switch is in the Glossary. The longer one is: A class of electromagnetically operated, multi-position switching devices. Wipers, or groups of wipers, are mounted on a rotatable shaft, which is rotated in steps so that contact is successively made between the wiper tips and contacts that are separated electrically from each other and mounted in a circular arc called a *bank*. The wiper positioning is done electromechanically on successive pulses to actuate the coil. There are two general kinds in common usage, rotary stepping switches and the Strowger two-motion switch:

1. *Direct-acting (two-coil) rotary stepping switch* A directly driven rotary stepping switch is a coil switch in which one electromagnet (called a *rotary magnet, motor magnet*, or *step coil*) and its associated armature provide forward stepping. Immediately on energization, by ratchet action, advancing the wipers; one step for each pulse received, to the desired contacting position. It remains in this position without further coil energization. The rotor is spring-restored, turning in reverse to the route over which it advanced under the control of a second electromagnet (called a *release magnet* or *release coil*) from a single pulse.

2. *Indirect drive, or spring driven, (one-coil) rotary stepping switch* The indirectly driven rotary stepping switch advances the wipers on the return action, or on the release of the armature, following each pulse to the motor magnet (coil). The rotation is unidirectional, one step for each pulse, on pulse cessation. The switch is returned to the nor-

mal or home position, by being stepped forward to the home position either from externally produced pulses or by being self-interrupted.

3. *Bidirectional (two-coil) rotary stepping switch*   A rotary stepping switch of the ratcheting direct-acting type, having two coils and associated stepping mechanisms and capable of rotation either clockwise or counterclockwise.

4. Strowger two-motion switch   A large-capacity switch with 100 discrete positions that is used principally in telephone switching.

In short, the stepping relay differs mainly from the switch in that its contact springs take one of two or three positions as the stepping-relay rotor moves around for a complete revolution. It might do so many times in one complete rotation (Figure 4.102), but the stepping switch wipers sequentially make mechanical contact with different output circuits on each step for a full sweep across the bank (Figure 4.104). In other words, if the movable springs of the contacting elements go only up and down, it is a relay; if they sweep around like the hands of a clock, mechanically contacting different output contacts each time, it is a stepping switch.

The contacts of a stepping relay are usually as large as on any other kind of relay and are capable of handling quite heavy electrical loads. The contacts of a stepping switch are usually limited to relatively light contact loads because they wipe only across their mating contacts, rather than push solidly against them. One bit of added confusion sometimes arises because stepping switches usually have relatively heavy-duty contacts, too, called interrupter contacts, which operate each time the armature operates, plus some other rather heavy-duty contacts, called *off-normal contacts*, which operate after each passage over or sweep of the bank contacts.

Stepping switches are considerably more versatile than relays from the standpoint of the ways in which they can be used for switching. All stepping-switch manufacturers have extensive and informative literature on the ways to use their switches. A good place to start in designing them into a job is to get this helpful information from the manufacturer. Although a stepping switch looks large to an electronics engineer, appears to be cumbersome to mount and wire in to other equipment, and is relatively expensive, one switch can do so much and replace so many other items that its use shrinks all objections to insignificance, in many cases.

The rotary stepping switch   A stepping switch is basically an electromechanical device used to connect one or more input circuits rapidly to an output circuit chosen from a sizable group of such circuits. The switch responds to current pulses supplied by an external source, or operates by interruption of its circuit through interrupter springs on the switch. Stepping switches are widely used to count, sequence, program, select, and control. They are often applied in machine-tool controls, conveyor systems, test equipment, and communication switching.

Rotary stepping switches are available in a variety of sizes and shapes, primarily dependent on the number of contact points in the bank assembly (Figure 4.103). Two basic types are compact switches, which are approximately 4 by 3 by 2 inches, and larger switches, which measure approximately 7 by 6 by 3 inches.

Switch construction   As shown in Figure 4.103, stepping switches consist of a driving mechanism, bank assembly, and wiper assembly.

- *Driving mechanism*   Rotary stepping switches are stepped by a pawl-and-ratchet mechanism, making one step for each current pulse applied to the switch coil. The two types of driving mechanisms are direct and indirect. With direct drive, the armature advances the wiper to the next position when the switch magnet is energized. Thus, the driving force varies with the power supplied to the coil. In the indirect-drive, the wipers are advanced with coil deenergization and resulting release of the armature. The driving force does not vary because it is provided by a spring, in which potential mechanical energy is stored when the armature is attracted to the coil. Indirect-drive switches are most widely used because they offer higher speeds, greater efficiency, and longer life than can ordinarily be expected of direct-drive switches.

- *Bank assembly* The bank assembly, containing individual contact levels built up as a unit, is of a semicircular form, in which the bank contacts are firmly held. The individual levels are composed of contacts insulated from the next level by a phenolic insulation or by molding the contacts in a plastic that has the necessary electrical and physical properties. These individual levels are stacked on top of each other, and are assembled together under pressure to ensure complete tightness. Compact switches generally have 10, 11, or 12 points per level, and from 1 to 12 levels—for a minimum or 10 points to a maximum of 144.

By special wiper arrangements, a larger number of positions per cycle can be provided. For instance, a 10-position switch has its 10 contacts in an arc of 120° and the wiper is triple-ended. As one wiper tip leaves contact number 10, another wiper tip touches contact number 1. The same ratchet mechanism can provide a 30-point switch. Single-ended wipers on three successive levels are staggered 120° apart. Thus, contacts 1 to 10 are on the first physical level, 11 to 20 on the second physical level, and 21 to 30 on the third physical level.

If fewer positions are needed in a particular application, the excess positions can easily be skipped automatically by wiring the switch to self-interrupt past certain contacts.

On larger switches, between 25 and 624 points are available per full-bank capacity. These switches are also available for 50- or 52-point operation by special arrangement of the wiper.

Wiper assembly The wiper assembly is the portion of a rotary stepping switch, which rotates, making electrical contact with the stationary bank contacts. The wiper assembly consists of a shaft-and-hub assembly, the wiper blades, which do the actual contacting; the ratchet-wheel indicating disk, which shows the wiper positioning on the bank; and a cam, which operates the off-normal contact assembly.

Each bank level has its own corresponding wiper level with which it makes contact. Each wiper is made up of two separate phosphor-bronze blades assembled to and properly spaced on the shaft of the wiper assembly. Both ends of each wiper blade are formed into

FIGURE 4.103 The three basic components of a rotary stepping switch.

a wiping tip, and the two blades engage both sides of the bank contacts. When one end is in contact with the bank terminals, the other is off the bank. That is, when one end of the wiper is leaving the last terminal, the other end is approaching the next terminal. Thus, the wipers are engaged with the intended associated bank contacts at all times.

Within a wiper assembly, there can be basically two types of wipers: bridging and nonbridging. A nonbridging wiper leaves one bank contact before engaging the next. Bridging wipers have long, flat tips, which permits the wiper to engage the next bank contact before leaving the preceding one. Bridging wipers are used when the circuit through them must be continuous and unbroken—as in self-interrupted stepping through a bank level. Typical examples of this are absence of ground-searching circuits and homing circuits.

A comparatively new adaptation is the use of normally closed (NC) contacts. Two physical levels are used to make one electrical level of NC contacts by tensioning together the mating contact points. These are opened one at a time by the rotation of a finger on the associated wiper assembly.

In addition to the contacts on the bank-and-wiper assemblies, auxiliary spring assemblies of the off-normal and interrupter type are available. The off-normal spring assembly is a set of contacts actuated by a cam on the wiper assembly. It is used to control an auxiliary circuit or to home the switch; that is, return the wipers to the start position. The interrupter combination is used mainly for self-cycling operation, and sometimes for controlling auxiliary circuits.

Stepping relays   Stepping-switch mechanisms have been adapted for use as cam switches or stepping relays, as shown in Figure 4.104 (lower right) and Figure 4.102. Cams are cut to provide operation or restoration of the associated contact assemblies per the customer's specification.

FIGURE 4.104   Some typical rotary stepping switches and a stepping relay (lower right). *(GTE Automatic Electric Co., Inc.)*

Cam switches with 30, 32, or 36 steps per rotation and with up to 8 cams are available. Cam switches have the advantage that the program is determined by the way the cam is cut so that wiring is usually simplified. Also, cam contacts can switch larger loads (up to 5 A) than can standard stepping-switch contacts.

Electrical characteristics

Operate and release times   The operate and release time ranges for rotary switches depend on mechanical and electrical characteristics associated with the construction of a particular switch. Generally, the cocking time of an indirectly driven switch is approximately 20 to 23 ms, and the armature release time is 8 to 12 ms. Performance timing within these ranges varies between families of switches, and also between switches of the same type.

Operate and release voltages   Rotary stepping switches are designed to operate on a particular nominal voltage, over a limited voltage range. This range must be held to an allowable +5 Vdc at a nominal value of 48 Vdc (or slightly more, depending on ambient conditions and other influencing factors). Any voltage rating under 48 Vdc should have a corresponding smaller voltage range, such as +2½ V at a nominal value of 24 Vdc. Limiting factors in the allowable voltage variation include temperature, shock and vibration, series resistance in control leads, and pulse rate.

Coil resistance   Rotary-stepping-switch coils are designed to be self-protecting for an indefinite amount of time, while being pulsed at the voltage for which they are designed.

If the coil must be energized for long periods, it should be protected by a current-limiting resistor. A 10-W resistor of approximately twice coil resistance is placed in series with the coil. A normally closed interrupter contact in parallel with the resistor shorts the resistor until the switch is almost fully cocked. When the interrupter contacts open, coil current is reduced to a holding value, which can be tolerated continuously. Typical coil resistances are shown in Table 4.20.

Contact bounce   Contact bounce is usually not a problem because nonbridging wipers are usually switched without any power on them—or with only minute current and voltages. Bridging wipers do not bounce appreciably, but if the switch is subjected to extreme vibration, a wiper might bounce beyond its insulated limits and touch the adjacent wiper level. Because of the possibility of the wipers contacting under unfavorable conditions, it is not advisable to place opposite potentials on adjacent wipers.

To prevent contact bounce in the wiper assembly, snubbing washers or barrier insulators between individual wiper levels can be provided in specially engineered assemblies. The snubbing washers tend to damp vibration and the barrier insulator mechanically insulates adjacent levels. Interrupter and off-normal spring combinations are not as likely to respond violently to vibration or shock, but anticipated shock and vibration should be emphasized when ordering a stepping switch.

TABLE 4.20   Typical Rotary-Stepping-Switch Coil Resistance Values[15]

Voltage, V dc	Coil resistance, Ω	
	Regular switches	Oversize switches
24	30	28
48	120	100
120	650	480
6	1.9	1.5

Power consumption varies from 18 to 30 W, depending on resistance and voltage ratings.

TABLE 4.21 Typical Capacitance and Q-Loss Values for Rotary Stepping Switches's[15]

	Compact switches		Large switches	
	Capacitance,* pF	Q loss,† %	Capacitance,* pF	Q loss,† %
Between adjacent bank contacts (same level)	1.0	12	1.7	14
Between adjacent bank contacts (same level, with bank-mounting screw between the two)	1.2	13	1.8	14
Between bank contact and frame	1.2	13	1.3	11
Between bank contact and frame (adjacent to bank-mounting screw)	1.5	16	1.5	12
Between same contact (adjacent levels)	1.2	13	1.7	14
Between wipers (adjacent levels, with wipers sitting on contact)	13.0	48	13.8	46
Between wipers (adjacent levels, with wipers floating between contacts)	11.7	40	11.9	38

* Approximate values at 10 MHz.
† Percentage change from standard Q value of 120 at 10 MHz.

Intercontact capacitance   Typical values of intercontact capacitance and Q losses are shown in Table 4.21 for two-switch families.

Dielectric strength   Minimum dielectric or voltage-breakdown value that can be expected from a standard rotary stepping switch is 500 Vac at 60 Hz, between all mutually insulated points for a period of 1 s. This applies generally to coils and interrupter spring assemblies. Wiper and bank insulation is generally good for 1250 Vac, 60 Hz for a 1-s period.

Contact rating   Standard rotary-switch bank and wiper contacts are commonly rated at 3-A carrying capacity. It is usually good practice to design the circuit so that the wiper is disconnected as it advances from one step to the next. In this way, the contacts are merely carrying the load and are not making or breaking it. However, if no more than 100 mA is interrupted, the full mechanical life of a switch should be reached without detrimental contact effects.

Standard wipers and bank contacts are usually made of phosphor bronze, a material generally corrosion-resistant, except in marine atmospheres. For use in corrosive environments, it is recommended that the switch be hermetically sealed. Phosphor bronze has high electrical conductivity, which is more than adequate when switching circuits at normal contact power levels. Where accuracy and reliability of readings taken through the banks and wipers of a rotary switch are necessary, gold-plated banks and wiping contacts should be used. Gold plating ensures the ability of the switch to provide a constant and low-level resistance path for measurement, indication, and monitoring purposes.

Large switches usually have interrupter contacts composed of tungsten, rated at 450 W at 2 A maximum. Compact switches are rated at 150 W at 3 A maximum. Large switches usually have off-normal spring assemblies with palladium-alloy contacts, rated as 150 W at 3 A noninductive. Compact switch, off-normal, spring assemblies are composed of various materials, such as palladium or platinum alloys, and are rated at 150 W at 3 A.

The primary function of interrupter and off-normal spring-assembly contacts is to make and break the coil circuit, or the coil circuit of an associated switch. This can be accomplished if the combinations are kept in adjustment, and if adequate contact-protection devices are used. This contact protection can be provided by using a resistor-capacitor network, or a varistor or thyristor across the coil.

The varistor is a nonlinear, resistance-varying silicon-carbon device, which draws little current when voltage is applied to the switch coil. However, when the coil circuit is opened, the varistor offers a low-resistance path to short-circuit the high-voltage surge caused by the collapse of the magnetic field.

Physical characteristics

Life expectancy Although rotary stepping switches have been known to operate for long periods of time without attention—even under quite adverse operating conditions—switch life can be extended by proper maintenance. This consists of periodic cleaning and lubrication with the correct lubricant. By exerting proper care, life can be expected to be between 200 and 250 million operations when run self-interruptedly, and about one-half of this when pulsed from an external source.

Temperature limits Most standard rotary stepping switches will function satisfactorily over a temperature range of −10 to 65°C. Switches treated with special lubricants and using special insulating materials can operate over a temperature range of −55 to 85°C. A maximum of 125°C can be tolerated in special cases.

Shock and vibration resistance Normally, rotary switches need not be specially designed for extremes of shock or vibration. Standard rotary stepping switches in an operative condition will withstand 5-g vibration at 5 to 55 Hz, and a 20-g shock load.

Mounting methods Stepping switches can be mounted directly on a panel, or on rubber cushions to dampen sound and limit resonant vibrations. Also, they can be shelf- or base-mounted. Special brackets can be used to meet certain specific mounting requirements. If a switch is to be subjected to extreme shock or vibration, special high-shock mounting frames are available.

Terminal types Standard solder-lug terminals are most commonly used for stepping switches. These terminals are always pre-tinned for soldering convenience. Terminals that mate with taper-tab connectors are also available. These connectors slide over the taper-tab terminal and are crimped into place with a small hand tool.

Enclosure types It is often necessary to seal a rotary stepping switch hermetically. When sealed, the switch cannot be tampered with, corrosive elements cannot reach it, dust and dirt cannot adversely affect it, and the switch is rendered harmless in an explosive atmosphere.

Rotary-switch enclosures vary, primarily depending on the size of the switch and the terminal connections desired. The headers or connectors available are many, but the most common are the solder hook, the plug-in connector, the MS (Military Standard) connector, and the gold-plated ribbon connector.

Things to avoid when using stepping switches

- Do not put opposite potentials on adjacent wipers or bank levels, and preferably not on adjacent bank contacts.
- Do not switch live circuits exceeding the 100-mA noninductive circuit-load rating (or its inductive equivalent) on the wiper-to-bank contacts of nonbridging wipers, unless you are willing to accept some switch life reduction.
- Do not load the bank of a rotary stepping switch to the point where arcing or burning at the wiper tips and bank contacts occurs.
- Do not overpower the driving mechanisms with excessively high voltage.
- Do not locate the power supply for a rotary stepping switch too far away from the switch. Good performance requires that the full-rated voltage is applied to the coil.
- Do not locate the switch too close physically or electrically to sensitive electronic circuitry. The operating coil of the switch is quite inductive and generates all kinds of noise and interference, from which sensitive electronic circuits must be protected.

Solid-state drivers   Programmed control of rotary stepping switches using solid-state components for the control might appeal to electronics engineers. There are many solutions to this problem, but using solid-state circuitry wired directly to the switch itself is of particular interest.

Solid-state control of stepping rate of the rotary stepping switch (drivers)   Solid-state drivers are successfully used, and can be either designed by the user or purchased commercially. One bank of the stepping switch can be used to vary the circuit constants and thus control self-stepping speed from one step to the next, as desired. This permits the creation of an automated-process control with a widely variable preset timing between individual steps, when using only a stepper and appropriate solid-state components.

An example of what can be done in this area is shown by Figure 4.105, which illustrates a somewhat typical, but uncomplicated circuit arrangement, providing the performance indicated in Figure 4.106, with respect to stepping speed of a specific switch. There will be variations in speed from switch to switch. The $R1$ and $C1$ portions of the circuit can be wired through individual bank contacts to give the desired variation in R and C values, and thus control the contact dwell in each wiper position.

Even slower and more varied stepping-time intervals have been achieved by the use of more elegant circuitry, using additional transistors, CSRs, etc. A more elegant circuit is that of Figure 4.83.

Figure 4.107 shows the circuit of a commercially available driver for which less than ±5-percent drift is claimed, with a stepping speed controllable by means of rheostat setting from one step in 5 s to 15 steps per second. By using one set of contacts on the rotary stepping switch to select external timing resistors, the stepping can be programmed to dwell on any particular position for a predetermined time. This permits a program of sequential control that gives the desired dwell on any or all individual contacts.

Exclusive of the control rheostat, the 100- to 130-Vac unit provides the source of dc required by the stepping switch and its control circuitry,

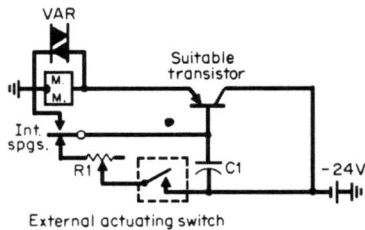

FIGURE 4.105 A simple schematic circuit for a transistor-driven slow-stepping and or variable-stepping-speed rotary stepping switch.[17]

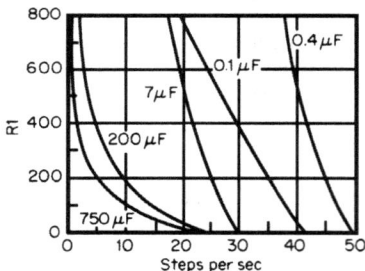

FIGURE 4.106 Typical speed curves for a rotary stepping switch operating in the circuit of Figure 104.[17]

in addition to providing stepping-switch interrupter-contact protection. Similar units are available for operation on 24, 48, and 110 Vdc.

Theory of operation   The basic schematic circuit is shown in Figure 4.107. When 115 Vac is applied to input terminals 1 and 2, capacitor C1 is charged through diode CR1. This voltage appears across the stepping-switch coil and controlled rectifier SCR1 through the normally closed interrupter contacts of the stepping switch. Because SCR1 is off at this time, the coil is not energized.

Simultaneously, capacitor C2 charges through diode CR2 at a rate determined by the (external) control rheostat. When C2 charges to the emitter peak point voltage of unijunction transistor Q1, the latter conducts, gating on SCR1. Current flows through the switch coil, cocking the switch. This opens the interrupter contact, disconnecting the coil and turning off SCR1. The switch then steps, and the action repeats. The control rheostat

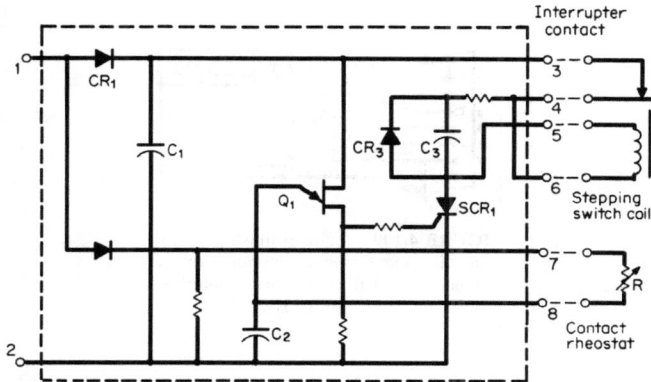

FIGURE 4.107 Basic schematic circuit of a solid-state driver intended for use with a rotary stepping switch. *(Electro Seal Corp., Div. of C. P. Clare & Co.)*

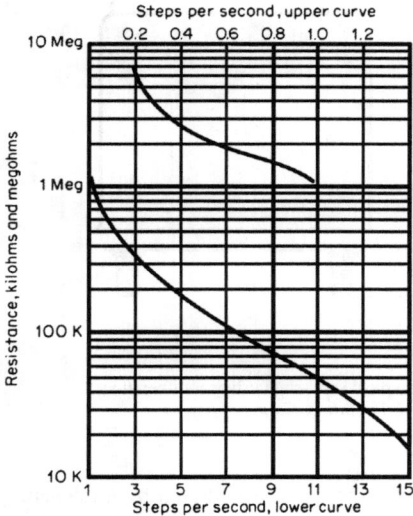

FIGURE 4.108 Typical stepping rate for a stepping switch operated from the driver circuit of Figure 4.101. *(Electro Seal Corp., Div. of C. P. Clare & Co.)*

changes the time required to charge C2 to the firing point of Q1 and changes the stepping rate accordingly. CR3 and CR2 provide arc suppression for the interrupter contacts. The typical stepping rate is shown by Figure 4.108.

Application The stepping-switch driver is designed to operate stepping devices at fixed rates sufficiently slow to permit the bank or wafer contacts to pull in and/or drop out relays and contactors. This can require 10 to 100 ms of contact dwell before stepping to the next position.

The driver offers the further advantage of programming such a device to stay in each position a different length of time (Figure 4.109). Terminal 7 is connected to the wiper of an extra bank of contacts on the stepping switch. As the switch steps, the value of resistance that will give the desired delay for the next step is automatically connected to terminal 8. Delay of as long as 5 s between steps can be obtained. An approximation of the value of resistance required for a given delay can be obtained from the curves in Figure 4.108. Resistors under 26,000 Ω (must be capable of carrying 2 W). Higher values can be rated 1 W up to 0.25 MΩ and ½ W for delay time greater than ¼ s (4 steps per second). A dimensional drawing of the driver in its usual form is shown in Figure 4.110.

Other stepping switches

Rotary solenoid ratcheting One form of nontelephone-type rotary stepping switch consists of a rotary switch, having one or more decks, quite similar in appearance to the

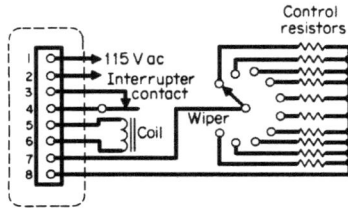

FIGURE 4.109   A circuit illustrating use of a solid-state driver to program a rotary stepping switch for varied and controlled dwell times. *(Electro Seal Corp., Div. of C. P. Clare & Co.)*

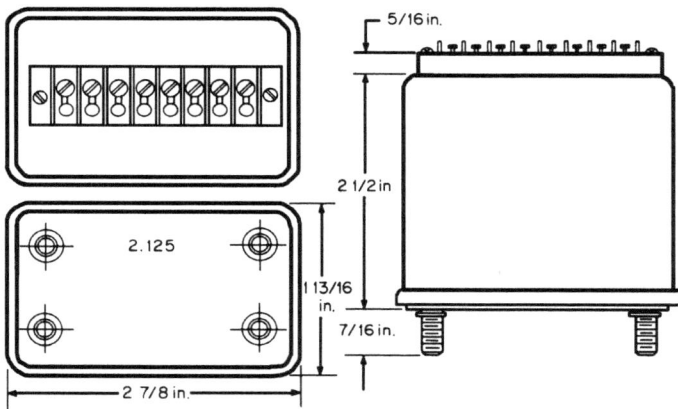

FIGURE 4.110   Dimensional drawing of a typical solid-state driver for operating a stepping switch. *(Electro Seal Corp., Div. of C. P. Clare & Co.)*

manually actuated rotary switch of Figure 4.99, except that it is driven by a ratcheting mechanism powered from an electromagnet. The electromagnet is usually of the rotary solenoid type. Unidirectional stepping switches of this type use only one rotary solenoid, functioning at one end of the rotary shaft, one turning the wipers clockwise and the other turning them counterclockwise. This is particularly useful where an add and subtract requirement has to be met.

100/200 point selection   A telephone-type switch that can select any one of 100 or 200 bank contacts by first moving the wipers to any one of a 10 bank-contact levels and then rotating them to any one of 10 bank positions is provided by the so-called step-by-step, or Strowger (telephone-type), switch. Connection by means of relays that can choose which of two sets of contacts in each rotary position are to be contacted permits a choice of one set of contacts out of a possible 200 sets.

A telephone-type stepping switch that is quite different in appearance with similar bank-choice capabilities is the X-Y switch.

Another of the large telephone-type switches capable of 100-point selection, but operating on a quite different principle, more like a gigantic relay, is the crossbar switch.

A typical 10-by 10-point crossbar switch consists of 100 sets of contacts, arranged in a 10 by 10 grid. At each grid intersection (called a *crosspoint*), the mechanical linkage causes operation of the contact set at that point when a particular pair of electromagnets is actuated. Twenty electromagnets per switch are involved. Ten are arranged along one side of the switch, to control the selection of horizontal row, and a second set of 10 electromagnets are arranged across the top of the switch, to control the vertical columns. When a horizontal row and a vertical column are selected, the contact set at the intersection of the selected row and column is operated. Obviously, this is not a small device. The telephone-type version of a crossbar switch has long been in production by certain telephone-equipment manufacturers, but a commercial version by a nontelephone manufacturer is also available. This commercial version comes in other matrix sizes than 10 by 10, such as 16 by 10 and 21 by 10.

4.4 SUMMARY

This by no means exhausts the subject of kinds, types, and applications of switches, both manually and electronically operated. Vendors' catalogs are a source of additional information, and a rather complete list of relay and switch vendors appears in Tables 4.22 through 4.24. Manufacturers whose names appear in this listing collectively manufacture every type of relay and switch that has been offered commercially. A careful review of their product literature will reveal something for almost any possible need.

TABLE 4.22 A Partial List of Manufacturers of Relays[15]

AMP, Inc.	B/W Controls, Inc.
ASEA, Inc.	Basch-Simpson, Ltd.
Action Electronics Co.	Beckwith Electric Co., Inc.
The Adams & Westlake Co.	California Electronic Mfg. Co., Inc.
Agastat Div., Amerce Esna Corp.	C.P. Clare & Co.
Aircraft Appliances & Equipment, Ltd.	Clark Control Div., A.O. Smith Corp.
Airpax Electronics, Inc.	Cole-Hersee Co.
Alco Electronic Products, Inc.	Compac Engineering, Inc.
Allen-Bradley Co.	Computer Components, Inc.
Allied Control Co., Inc.	Controls Div., Ingraham Industries, Div of
Allis Chalmers	McGraw-Edison Co.
American Design Components	Cook Electric Co.
American Solenoid Co., Inc.	Cornell-Dubilier Electronics
Amtron, Inc.	S.H. Couch Div., ESB, Inc.
Arrow-Heart Inc.	Cox & Co., Inc.
Artisan Electronics Corp.	Cutler-Hammer, Inc.
Auto-Matic Products Co.	Davis Electric Co.
Babcock Electronics Corp.	Delaval, Gems Sensors Div.

TABLE 4.22 A Partial List of Manufacturers of Relays[15] (Continued)

Deltrol Controls/Div. of Deltrol Corp.	Hoagland Instruments
Deutsch Relays, Inc.	Harvey Hubbel, Inc., Industrial Controls Div.
Diversifield Electronics, Inc.	ICS, Inc.
Durakool, Inc.	I-T-E Imperial Corp.
Dynage, Inc.	ITT Jennings Industrial Products
E-T-A Products Co. of America	Imtra Corp.
Eagle Signal, A Gulf & Western Systems Co.	Ingraham Industries-Special Products Div.
Edison Electronics, Div. of McGraw-Edison Co.	Instrument Components Co., Inc.
The Electric Tachometer Corp.	International Ractifier, Semiconductor Div.
Elec-Trol, Inc.	Jaidinger Mfg. Co.
Electronic Applications Co.	Jettron Products, Inc.
Electronic Specialty Div., Daytron System, Inc.	Jewell Electrical Instruments, Inc.
Elemwood Sensors, Inc.	Kilovac Corp.
Essex International, Inc.	Kilovac Corp., Dow-Key Div.
F & B Mfg. Co., Omega Electronics Div.	King Seeley Div., King Seeley Thermos Co.
Federal Pacific Electric Co.	Klockner-Moeller Corp.
Fifth Dimension, Inc.	Kratos
Flight Systems, Inc.	LaMarche Mfg. Co.
Foster & Allen, Inc.	Larson Instrument Co., Inc.
Frost Controls Corp.	Leach International, Inc.
Furnas Electric Co.	Ledex, Inc.
GTE Automatic Electric	Line Electric Co., A Unit of Esterline Corp.
Gemco Electric Co.	Logiteck, Inc.
General Automatic Corp.	MKC Electronics Corp.
General Devices, Inc.	Mack Electric Devices, Inc.
General Electric Co.	Madison Electric Products, Inc.
General Electric Co., General Purpose Control Products Dept.	Madison Laboratories, Inc.
General Electric Co., Semiconductor Products, Electronic Components Div.	Mangnecraft Electric Co.
	Master Electronic Controls
Gordos Corp.	McGraw-Edison Co., Edison Instrument Div.
Grayhill, Inc.	MEKontrol, Inc.
Greentron, Inc.	The Mercoid Corp.
Grigsby Barton, Inc.	Midtex, Inc., Aemco Div.
GTE Automatic Electric Co., Inc.	Micro Switch, A Div. of Honeywell, Inc.
Guardian Electric Mfg. Co.	Monsanto Co., Electronic Special Products Div.
H-B Instrument Co.	N.P.E./Wabash
Hamlin, Inc.	North American Philips Controls Corp.
Hartman Electrical Mfg., Div. of A-T-O, Inc.	North Electric Co., Electronics Div.
Hienemann Electric Co.	Oak Industries, Inc., Switch Div.
Hi-G, Inc.	Omnetics, Inc.,

TABLE 4.22 A Partial List of Manufacturers of Relays[15] *(Continued)*

Payne Engineering Co.	Sterer Engineering & Mfg. Co., Logic Systems Div.
Peco Corp.	Struthers-Dunn, Inc.
Potter & Brumfield, Div. of AMF, Inc.	Sunshine Scientific
Power Control Corp.	Syracuse Electronics Corp.
Prestolite Co., Div. Eltra Corp.	Systems Matrix, Inc.
Princo Instruments, Inc.	T-Bar, Inc., Switching Components Div.
Raytheon Co., Industrial Components operation	Tech Laboratories, Inc.
Regents Controls, Inc.	Teledyne Crystalonics
Relay & Control, Div. of A.W. Sperry Instruments, Inc.	Teledyne Relays
Relay Specialties, Inc.	Tempo Instruments, Inc., Industrial Products Div.
Relays, Inc.	Tenor Co.
Renfew Electric Co., Ltd.	Thermosen, Inc.
Ril Electronics, Inc.	O. Thompson, Inc.
Rhode & Schwarz Sales Corp.	Tokyo Electric Co., Ltd.
Ronk Electrical Industries, Inc. System Analyzer Div.	Universal Relay Corp.
Ross Engineering Corp.	Vapor Corp.
Rowan Controller, Inc., Subs. ITE-Imperial Corp.	Vectrol, Inc.
F.A. Scherma Mfg. Co., Inc.	Wabash Relay and Electronics, Inc.
Schrack Electrical Sales Corp.	Wabco Aerospace Department
Selco Electronics, Inc.	Wapco Mfg. Co.
Shared Technical Services, Inc., An I.E.C. Affiliated Co.	Ward Leonard Electric Co., Inc.
Shigoto Industries, Ltd.	Warner Electric Brake & Clutch Co.
Siemens Corp.	Warren GV Communications
Sigma Instruments (Canada), Ltd.	Watlow Electric Mfg. Co.
Sigma Instruments, Inc.	Western Electric Co., Inc.
Simpson Electric Co.	Westinghouse, Control Products Div.
Slocum Industries, Electronics Div.	Westinghouse Electric Corp.
Smiths Industries, Inc.	Westinghouse Electric Corp., Semiconductor Div.
Solid-State Electronics Corp.	Weston Instruments, Inc.
Sprague Electric Co.	Wilmar Electronics, Inc.
Square D Co.	Zenith Controls, Inc.

TABLE 4.23 Nongovernmental Organizations Having Publications Relating to Relays[1]

AAR	Association of American Railroads 1920 L Street, N.W. Washington, DC 20036	JIC	Joint Industrial Council 2139 Wisconsin Avenue Washington, DC 20007
AIA	Aerospace Industries Association 1725 De Sales Street, N.W. Washington, DC 20036	NARM	National Association of Relay Manufacturers P.O. Box 1649 Scottsdale, AZ 85252
ANSI	American National Standards Institute 1430 Broadway New York, NY 10018	NEMA	National Electric Manufacturers Association 155 East 44th Street New York, NY 10017
EIA	Electronic Industries Association 2001 I Street, N.W. Washington, DC 20006	SAE	Society of Automotive Engineers, Inc. 2 Pennsylvania Plaza New York, NY 10001
IEEE	Institute of Electrical and Electronics Engineers 345 East 47th Street New York, NY 10017		

TABLE 4.24 A Partial List of Switch Manufacturers[13]

Agastat Div., Amerace Esna Corp.	Chicago Switch, Div. F & F Enterprises
Airflyte Electronics	C.P. Clare & Co.
Airpax Electronics, Inc.	Clare-Pendar Co.
Allen-Bradley Co.	Cole Instrument
American Zettler, Inc.	Collectron Corp.
Analog Devices, Inc. Pastoriza Div.	Computer Products Div., Wyle Labs
Ansley Div., Thomas & Betts Corp.	Consolidated Controls Corp.
Arrow-Hart & Hegeman	Control Products, Inc.
Automatic Electric Co.	Controltron Corp.
Automatic Metal Products	Controls Co. of America, Control Switch Div.
Automatic Switch Co.	Corning Glass Works
Beckman Instruments, Inc.	Cunningham Corp.
Bristol Instrument Div., American Chain & Cable	Custom Component Switches
C & K Components, Inc.	Cutler-Hammer, Inc.
CTS Corp.	Dabuen Electronics & Cable
Candy Manufacturing Co.	Daven Div., McGraw Edison
Capitol Machine & Switch	R.B. Denison, Inc.
Carling Electric, Inc.	Dialight Corp.
Carter Mfg. Corp.	Digitran Co.
Centralab/Elexs Div., Globe-Union, Inc.	Disc Instruments, Inc.
Cherry Electrical Products	Dubble A Products Co.
	Dresser Industrial Valve & Instrument Div.

TABLE 4.24 A Partial List of Switch Manufacturers[13] *(Continued)*

Dwyer Instruments, Inc.	Langevin Electromechanical
Electric Regulator Co.	Ledex, Inc.
Electro-Mec Instruments	Leecraft Mfg. Co., Inc.
Electro-Miniatures Corp.	Leeds & Northrup Co.
Electro-Products Labs.	Leviton Mfg. Co.
Electro Switch Corp.	Licon-Ill. Toll Works
Electronic Components for Industry	Linemaster Switch Corp.
Electronic Controls, Inc.	Liquid Level Lectronics
Electronic Engineering Co. Cal.	Litton Industries, Clifton Div.
Electronic Resources, Inc.	Litton Industries, USECO Div.
Euclid Electric & Mfg.	Mack Electric Devices
Farmer Electric Products Co., Inc.	Magnetrol, Inc.
Fifth Dimension, Inc.	Marco-Oak Industries, Div. of Electro/Netics Corp.
Film Microelectronics, Inc.	
Furnas Electric Co.	Mason Electric Co.
Gemco Electric Co.	Master Specialties Co.
The Gems Co., Inc.	McDonnel & Miller, Inc.
General Devices, Inc.	McGill Mfg. Co., Inc.
General Electric	Mead Fluid Dynamics
General Equipment & Mfg. Co.	Mechanical Enterprises, Inc.
General Reed Co.	Metrix Instrument Co.
Gordos Corp.	Micro-Letric, Inc.
Gorn Corp.	Micro Switch, Div. of Honeywell
Grayhill, Inc.	Milliswitch Corp.
Hamlin, Inc.	MINELECO (Miniature Electric Components Corp.)
Haydon, A.W. Co.	
Hatdon Switch & Instrument	Miniature Electronic Components Corp. (MINELCO)
High Vacuum Electronics, Inc.	
Hi-Tak, Corp., Switch Div.	Molex Products Co.
Honeywell Industrial Div.	Donald P. Mossman, Inc.
Humphrey, Inc.	Nanasi Co.
IBM Corp., Industrial Products Div.	Nelson Eletric Div.
ITT Jennings	New England Instrument Co.
Imtra Corp.	New Product Engineering, Inc.
Instrumentation & Control Systems	N.M. Ney Co.
International Electro Exchange of Minneapolis	Northern Precision Labs.
Interswitch	Oak Mfg. Co.
Janco Corp.	Ohmite Mfg. Co.
Jay-El Products, Inc.	Otto Controls
Jordan Controls, Inc.	Philadelphia Scientific Controls, Inc.
	Pollak, J. Corp.

TABLE 4.24 A Partial List of Switch Manufacturers[13] *(Continued)*

Precision Sensors, Inc.	Subminiature Instrument Corp.
Qualitrol Corp.	Switchcraft, Inc.
PBM Controls	Synchro Star Products, Inc.
RCL Electronics, Inc.	Tann Controls Co.
Reed Switch Development Co., Inc.	Tapeswitch Corp. of America
Remvac Components, Inc.	Tech Labs., Inc., Bergen & Edsall
Robertshaw Control, Acro Div.	Tele-Dynamics, Div. of AMBAC Industries,
Ross, Milton Co.,	Inc.
Sage Laboratories, Inc.	Texas Instruments, Controls Products Div.
Saico Controls Div.	Therm-O-Disc
Schrack Electrical Sales Corp.	Torq Engineered Products
Seacor, Inc.	Truco, Inc.
Sealectro Corp.	UMC Electronics Co.
Shallcross Mfg.	Unimax Switch, Div. of Maxson Electronics
Shelly Associates, Inc.	Corp.
Sparton Southwest	United Control Corp., Subs Sundstrand Corp.
Spectrol Electronics Corp.	Wabash Magnetics, Inc.
Stackpole Components Co.	Wanteck Data Communications
Staco, Inc.	Wheelock Signals, Inc.
Standard Controls, Inc.	Zenith Controls, Inc.
Standard Instrument Corp., Div., Automatic Timing and Controls, Inc.	

4.5 CONTINUING DEVELOPMENTS

Technology in this field continues to expand. One special area that should be mentioned is the continuing progress on standardization of solid-state relays, with respect to size, shape, configuration, and operating characteristics. Up-to-date information is available at any given time from the National Association of Relay Manufacturers.

Another recent development that typified continuing developments is an optically encoded keyboard for which all the encoding is retained in the key switch, a single piece of stamped steel, free-form eroding wear, and contact bounce. It is claimed that improved efficiency and simplicity with lower cost result in a keyboard immune to environmental disturbances.

REFERENCES

1. National Association of Relay Manufacturers, "Engineers' Relay Handbook," Hayden Book Co., New York, 1969.
2. Oliver, Frank J: "Practical Relay Circuits," Hayden Book Co., New York, 1971.

3. Kear, Fred W.: Take Another Look at Resonant-Reed Relays, *EDN*, vol. 18, no. 10, June 20, 1973.

4. Kaetsch, Philip W.: Suppressing Relay Transients, *Electro-Technology*, December 1968.

5. Lippke, James A.: Time-Delay Relays, *EEE*, September 1968.

6. Andriev, N.: Power Relays—Solid State vs. Electromechanical, *Control Eng.*, January 1973.

7. Thompson, Stephen A.: Relays: Form Versus Function, *Electron. Eng.*, October 1971.

8. "Correed Handbook and Application Manual," GTE Automatic Electric Co., Northlake, Ill.

9. "Technical Applications Reference for Mercury-wetted Contact Relays, Dry Reed Relays and Mercury-Wetted Reed Relays," C.P. Clare & Co., Chicago, Ill.

10. Deeg, W.L., and R.H. Marks: "Clareed Control Concept and Its Systems Application," C.P. Clare & Co., Chicago, Ill.

11. "Designers' Handbook and Catalog of Reed and Mercury-wetted Contact Relays," Magnecraft Electric Co., Chicago, Ill.

12. "Designer's Handbook and Catalog of Time Delay Relays," Magnecraft Electric Co., Chicago, Ill.

13. 1972/73 Systems Designers' Handbook, *Electromech. Des.*, 1972–1973.

14. 1973/1974 Systems Designers' Handbook, *Electromech. Des.*, August 1973.

15. 1973–1974 Electric Controls Reference Issue, *Mach. Des.*, Apr. 26, 1973.

16. Ashby, J.D.: Stepping Switches, *Mach. Des.*, Apr. 26, 1973.

17. "How to Use Rotary Stepping Switches," GTE Automatic Electric Company, Inc., Northlake, Ill.

CHAPTER 5
BATTERIES

Mirna Urquidi-Macdonald

Engineering Science and Mechanics, The Pennsylvania State University
MVMESM@ENGR.PSU.EDU

5.1 BATTERY HISTORY

The first discoveries in the field of electrical cells were related to the area of primary (non-rechargeable) batteries. The first source of continuous direct current, the primary battery, had a slow development.

In 1678, the Dutch naturalist Jan Swammerdam made a significant observation about the contraction of muscles in living (or freshly killed) organisms in contact with certain metals. This observation was completely disregarded until 1786, when Galvani carried out his experiments and concluded that "animal tissues were in some manner a source of electricity."

In 1792, Alessandro Volta, Professor of Natural Philosophy at the University of Pavia, became interested in Galvani's experiments and came to the conclusion that electricity was the result of the mere contact of dissimilar metals. His work was virtually unknown until 1800 when, in a letter to the Royal Society of London, he claimed to have built an apparatus of perpetual power (Figure 5.1).[1-4]

Several important discoveries and technological developments resulted from the Volta pile:

- Electrolysis of water
- Electrical arc
- Isolation of alkali metals
- Discovery of magnetic effect of currents
- The electrochemical telegraph
- Theory of oxidation of metals
- Faraday's theory of electrochemical process
- Electroplating and electro-metallurgy
- Engraved copper plates by Voltaic action

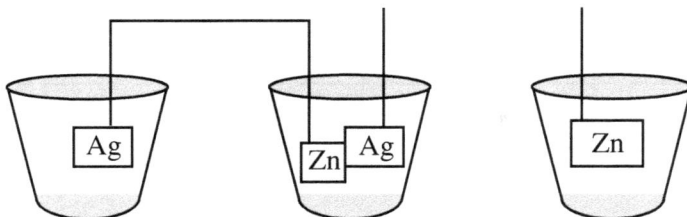

FIGURE 5.1   The Volta pile as *couronne de tasses*.

Also, several improvements to the initial design, which were considered milestones in the field of battery research, lead to an increase in commercial battery use (1830–1880). The following represent important battery improvements:

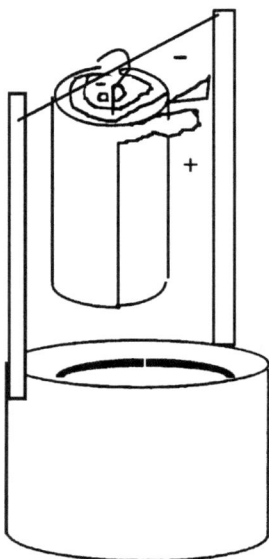

FIGURE 5.2   Helical battery assembly to increase area.

• Mechanical on/off switching mechanism
• Higher power by increasing plate area (Figure 5.2)
• Two fluid batteries (Daniell's cell, Figure 5.3)
• Gravity separation of cell electrolytes (Figure 5.4)
• Leclanche dry cell

Despite the commercial importance of primary batteries, technological progress on batteries after 1880 failed to keep pace with industrial expansion. The introduction of the radio in 1920 marked a period of renewed interest in batteries. It was observed that some electrochemical systems permit current to pass in both directions (i.e, the electrodes pass current in two opposite directions). This process occurs if the applied potential across the electrodes is not too large. The application of these observations gave rise to the design of rechargeable batteries or secondary batteries. The current passed in one direction to charge the battery and in the inverse direction to discharge the battery. Since their initial development, the use of secondary batteries has expanded rapidly, with lead-acid, nickel-iron, and nickel-cadmium being prime examples.

Modern battery technology has come a long way since the 1800s. The principles remain the same, but technological and scientific advances have permitted the fabrication of batteries for different needs. The efficiency of electrochemical converters (batteries) is the highest (90%), followed by heat engine (40%), and photoelectric (solar cells) (12%); compared to thermoionic (8%), and nuclear (1%). But, the cost of electrical energy obtained from electrochemical process is more expensive than that obtained with heat engines. However, because packaged electrical energy is convenient—and it is offered by batteries—users are prepared to pay this higher price for energy in some applications. Modern batteries have a wide range of shapes, output power, and applications. The commercial production of batteries in the United States alone exceeded a market of $18.5 billion dollars per year[5] a decade ago, and is still increasing. The immediate future of primary and secondary batteries seems to be assured, particularly with the advent of portable electronics, load-leveling, electrical vehicles, and stand-by power, to name but a few applications.

FIGURE 5.3   Gravity-type cell.

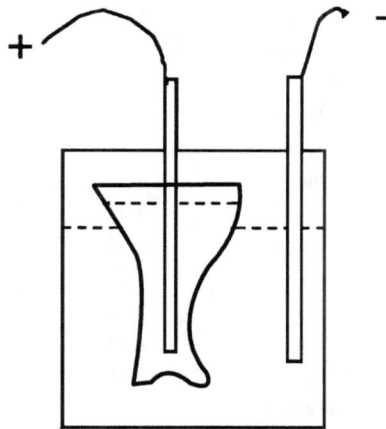

FIGURE 5.4   Original form of Grove's nitric-
acid cell. The battery contained two fluids. The
material of the inside container was later modified by Daniell.

Although both types of batteries might be suitable for a specific application, restrictions imposed by the application generally favor one over the other. The following section
briefly reviews the areas of application for both primary and secondary batteries, their
characteristics and limitations, and the factors that determine the choice for a specific use.

5.2 THEORETICAL CONCEPTS

5.2.1 Work Function at the Metal-Metal Junction

When two dissimilar metals are in contact, there is a tendency for (electrons) to pass from
one metal to the other. Metals have different affinities for electrons. The amount of energy

required to remove an electron from a metal surface, in a vacuum, to an infinite distance is called the *work function*. As the value of the work function increases, the tendency of the metal (electrode) to capture electrons increases and the tendency of the metal to give away electrons in an ionization process decreases.

The energy required to transfer electrons from one metal to another metal is determined by the difference between their work functions (listed in Table 5.1 for some metals). Thus, if Φ_1 is the work function for Metal 1 and Φ_2 is the work function of Metal 2, the energy required to transfer electrons from 1 to 2 is:[6]

$$\Delta E = (\Phi_1 - \Phi_2) \, F \tag{5.1}$$

TABLE 5.1 The Thermionic Work Functions of the Metals

Metal	Thermionic Work Function (V)
Potassium	2.12
Sodium	2.20
Lithium	2.28
Calcium	3.20
Magnesium	3.68
Aluminum	4.10
Zinc	3.57
Lead	3.95
Cadmium	3.68
Iron	4.70
Tin	4.38
Copper	4.16
Silver	4.68
Platinum	6.45

The electrons will move from one metal to another in the direction in which the energy is minimized (from the lower work function to the higher work function). A current that is determined by the internal resistance of the junction and the difference between the work function for the two metals is generated.

The scenario applies strictly to a metal/metal junction and is the basis of the thermocouple effect. However, batteries involve opposing metal/electrolyte junctions that are fundamentally different because of charge-transfer reactions. The science of these junctions lies in the realm of electrochemistry. Hence, it is necessary to briefly address the underlying principles; in this case, from a somewhat historical perspective. In 1832, Michael Faraday recognized that the amount of chemical changes occurring at a metal/electrolyte interface are proportional to the quantity of electricity passing through the cell. The proportionality constant, relating the charge and the number of equivalents (gr. formula weight/electron number, n), is known as *Faraday's constant*, $F = 96485.3$ Coulomb/mol-eq. or Coulomb/faraday. Faraday's law is expressed as:

$$Q = nFN_i \tag{5.2}$$

Q is the amount of electricity (C) coulombs, n is the number of electrons involved per molecule of the reacting species, and N_i is the number of moles of material transformed during the reaction.

When a metal is in contact with a solution, chemical reactions can occur at the metal/solution interface and ion/electron pairs are formed, with the electrons residing in the metal and the ions remaining in the solution: $M ==> M^{n+} + ne^-$. If the solution contains M^{n+}, then the reverse reaction, $M^{n+} + ne^- ==> M$ will also occur. When these rates are equal, the system is said to be in equilibrium and because both reactions are affected by the potential difference across the metal/solution interface, we can define an equilibrium potential. The equilibrium potential of a single metal/electrolyte (i.e., M/M^{n+}) interfaces does depend on the work function, but because only potential differences can be measured, the concept of equilibrium in this system is hypothetical. Instead, refer the equilibrium potential to that of another electrode, which has been chosen to be the Standard Hydrogen Electrode (SHE, itself a hypothetical electrode). However, the equilibrium potential of the M/M^{n+} couple is now a potential difference (right terminal minus left terminal) for the cell SHE/M^{n+}/M, and can be calculated from thermodynamics.

5.2.2 The Electrochemical Cell

It is possible to store the chemical energy from the reactions occurring at the interfaces between two different metals. Because an oxidation-reduction reaction can be resolved in two partial reactions, oxidation and reduction, the electrons produced by one reaction and consumed by the other reaction could be "used" to perform an electrical work if we could separate the two reactions. These can be arranged if the two reactions are physically separated and the oxidation reaction occurs at a place (the anode) and the reduction reaction occurs at another place (the cathode). To "withdraw that energy on command," dissimilar metals (or electrodes) are submerged in an electrolyte, which is ionically conductive.[7] The electrolyte "helps" to transport the ions from one electrode to the other electrode. These processes close the circuit when the electrodes are externally connected. The external flow of electrons constitutes a current, which can be used to drive a motor and to perform some work. The potential difference between the electrodes and the current flowing when the electrical circuit is closed define the power of the system. The system is called an *electrochemical cell*. The potential difference across electrodes is called the *cell potential*.

Accordingly, the simplest electrochemical cell consists of an ionic conductor sandwiched between two electronic conductor electrodes that support charge transfer reactions on their surfaces (Figure 5.5). The voltage difference is composed of two voltage differences from each of the two electrodes and a voltage loss because of the finite internal resistance. Because the internal IR voltage drop represents a voltage loss from the cell, every effort is made to reduce the internal resistance to an absolute minimum.

At the electrode interfaces, as mentioned, two reactions occur:

1. An oxidation reaction on the negative electrode (or anode) on discharge, which produces electrons, (reduced state $==>$ oxidized state + ne^-);
2. A reduction reaction, which accepts electrons (oxidized state + $ne^- ==>$ reduced state) and occurs on the positive electrode (or cathode) during discharge. During discharge, electron current through the external circuit flows from the negative electrode (where oxidation occurs) to the positive electrode (where reduction occurs). On charging (in the case of a secondary battery), the direction of current flow is reversed, so reduction occurs at the "negative" electrode and oxidation occurs at the "positive" electrode.

The open-circuit potential (OCP) is the voltage difference between two terminals or the difference between each of the electrode potentials measured versus the same reference elec-

$$E = E_2 - E_1$$

ELECTROLYTE

P
o
t
e
n
t
i
a
l

$E = E_2 - E_1$

E_1

E_2

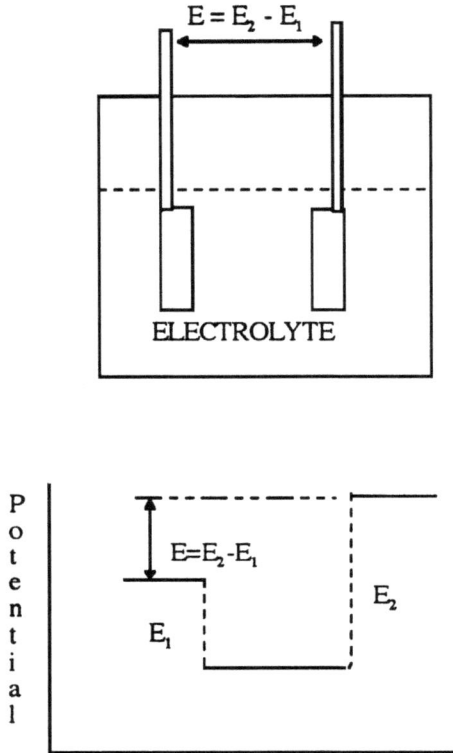

FIGURE 5.5   Scheme of an electrochemical cell.

trode, when the electrical circuit is open and no current is passing through the external circuit. Other names for the OCP are the *cell potential*, and the *cell null* or *rest potential* (because no current flows from the cell).[8] Notice that the OCP generally does not correspond to the equilibrium potential (it is always less than that value) because of the existence of irreversible parasitic reactions that occur within the cell. This issue is reviewed later in this chapter.

The OCP between the electrodes decreases when energy is removed from the cell.

An important parameter used to specify an electrochemical cell is the constancy of the OCP as a function of the extend of the discharge. In some cells (batteries), the OCP voltage decreases only slightly as the cell is discharged; in other batteries, a rather pronounced change of the OCP occurs. In the former, the activities of the reactants remain constant. This frequently occurs when the reactants and products are pure components of some phase (an example is water or a solid). Constancy in the concentrations of species (in the electrolyte) is a common attempt to maintain the potentials of the electrodes as constant as possible, but it generally fails because of the generation of local concentration gradients and because of changes in the resistivity of reaction product phases.

The theoretical equilibrium cell potential can be calculated if the reactions occurring at the electrode interfaces are known, and if the relevant thermodynamic data is available. Consider a cell with the overall reaction (cathode plus anode):

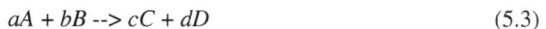

$$aA + bB \, \text{-->} \, cC + dD \tag{5.3}$$

The cell voltage (emf) of a cell depends on the activity (or partial pressure) of each species involved in the reaction, and it is given by the Nernst equation:

$$E = E^o - (RT/nF) \, Ln \, (W) \tag{5.4}$$

Where E is the emf of the cell in volts; E^o is the standard potential of the cell (this parameter depends on the temperature and on the cell material); T is the temperature; n is the equivalent charge transferred from anode to cathode per mole (mole-eq.) of reaction; the number of moles of electrons (faradays) transferred through the external circuit when the cell reaction occurs, F is Faraday's constant, R is the gas constant (1.987 cal/degree-mole or 8.314 joules/degree-mole), and

$$W = (a_C)^c (a_D)^d / \{ (a_A)^a (a_B)^b \} \tag{5.5}$$

The quantity a_i is the activity of the i-th component (A, B, C, and D in Reaction 5.3), electronic components are omitted.

When a chemical reaction is capable of doing work, as in a galvanic cell, there is a decrease in free energy when the reaction occurs. This concept is used to define equilibrium of the cell, a concept that is different from the equilibrium potential of the cell, which is covered later in this section. If current is drawn from the cell until the cell reaction has come to equilibrium, the emf must become zero, and no reactions occur; therefore, the conditions of equilibrium are calculated for $E = 0$. By using Nernst equations (Eq. 5.4) equal to zero, the value of W (the quotient of Reaction 5.3) at equilibrium conditions can be calculated if E^o is given. The values W obtained for this condition gives the rate constant of the reaction occurring in the cell, at chemical equilibrium (i.e., $K = W = (a_C)^c (a_D)^d / \{ (a_A)^a (a_B)^b \}$).

The emf of the whole cell reaction, E, and the work available by the reaction, ΔG, are related. Cell equilibrium potential E, and the standard cell potential, E^o, are related to Gibbs free-energy changes (ΔG) and the standard free-energy changes (ΔG^o) for the reaction as:

$$\Delta G = -nFE \tag{5.6}$$

$$\Delta G^o = -nFE^o \tag{5.7}$$

Table 5.2 lists standard Gibbs energies of formation, $\Delta_{fG} i^o$, (in kJ/mol) for selected neutral and ionic species (i). This data can be used to calculate ΔG^o for a whole reaction (Reaction 5.3) as:

$$\Delta G^o = c\Delta_{fG} C^o + d\Delta_{fG} D^o - a\Delta_{fG} A^o - b\Delta_{fG} B^o \tag{5.8}$$

The letters A, B, C, and D indicate the species involved in Reaction 5.3.

If the cell operates at a temperature and pressure that is different from standard conditions ($T = 25°C$, $P = 1$ atm), the standard energy function must be integrated over the temperature and pressure range considered:

$$\Delta G^o(T,P) = \Delta G^o(T_o,P_o) - \int_{T_o}^{T} \Delta S^o(T)dT + \int_{P_o}^{P} \Delta V^o(P)dP \tag{5.9}$$

where T_o and P_o designate standard conditions (25°C and 1 atm, respectively), ΔS^o is the change in the standard entropy for the cell reaction, and ΔV^o is the change in the standard volume. The change in standard entropy, ΔS^o, is readily calculated from the change in the standard heat capacity:

$$\Delta S^o(T,P) = \Delta S^o(T_o,P_o) + \int_{T_o}^{T} \frac{\Delta C_P^o}{T}dT + \frac{1}{T}\int_{P_o}^{P} \Delta C_P^o(P)dP \tag{5.10}$$

TABLE 5.2 Standard Gibbs Energies $G°/kJ\ mol^{-1}$ for Selected Neutral and Ionic

AgBr(s)	−95.92	Ag⁺(aq)	77.08	Br⁻ (aq)	−102.76	
AgCI(s)	−109.59	Al^{3+}(aq)	−485	Cl⁻ (aq)	−131.02	
Ag_2O(s)	−11.20	Ca^{2+}(aq)	−553.57	CO_3^{2-} (aq)	−527.82	
$CaCO_3$(s)	−1128.79	Cd^{2+}(aq)	−77.65	$Fe(CN)_6^{3-}$ (aq)	729.4	
CO_2(g)	−394.38	Cu+(aq)	50.3	$Fe(CN)_6^{4-}$ (aq)	695.08	
CO_2(aq)	−386.2	Cu^{2+}(aq)	65.7	HCO_3^- (aq)	−586.98	
Hg_2CI_2(s)	−210.33	Fe^{2+}(aq)	78.9	HSO_4^- (aq)	−755.91	
HgO(s)	−58.54	Fe^{3+}(aq)	−4.5	HS⁻ (aq)	12.08	
H2(ℓ)	−237.13	Hg_2^{2+} (aq)	153.57	I⁻ (aq)	−51.65	
H_2O_2(aq)	−134.03	Hg^{2+}(aq)	164.67	I_3^- (aq)	−51.48	
H_2S(aq)	−27.84	In^{3+}(aq)	−97.95	MnO_4^- (aq)	−447.1	
$PbCI_2$(s)	−313.94	Mn^{2+} (aq)	−228.1	MnO_4^{2-} (aq)	−500.6	
PbO_2(s)	−218.96	Ni^{2+} (aq)	−46.4	OH⁻ (aq)	−157.24	
$PbSO_4$(s)	−813.76	Pb^{2+} (aq)	−24.18	S^{2-} (aq)	86.34	
ZnO(s)	−318.30	TI⁺ (aq)	−32.47	SO_4^{2-} (aq)	−744.53	
$Zn(OH)_2$(s)	−555.06	Zn^{2+} (aq)	−147.19	ZnO_2^{2-}(aq)	−384.4	

and ΔV^o can be estimated to sufficient accuracy for most purposes from molar volume data. The methods of calculating the integrals are described extensively in the literature.

Half-cell It is useful to think of a cell as consisting of two half-cells, with each half-cell contributing its share to the emf of the whole cell. A half-cell consists of an electrode and the electrolyte surrounding that electrode.

Also, frequently, in battery evaluation, it is necessary to separately characterize the cathode and the anode. The potential of a single electrode cannot be measured, but the combined potential of the whole cell is measured. Because, absolute potential cannot be measured, a reference electrode is needed to measure the potential of a single electrode. An arbitrary electrode can be used as a reference point to measure the potential of a single electrode, but the standard that has been adopted in electrochemistry is the Standard Hydrogen Electrode (SHE).[9,10]

The convention adopted is that the potential of the SHE is equal to zero. In this way, the potential of a single electrode is expressed as the potential of that electrode measured with respect to the SHE. The sign convention is such that metals more active than hydrogen gas have a negative sign, and all half-cell reactions are written in the reduction sense. Therefore, most tables display potentials with reference to the SHE. If other reference electrodes are used to measure the potential of a working electrode, it is necessary to make a scale conversion between the reference used and the SHE, to compare the measurements to values contained in standard tables. Examples of frequently used reference electrodes are the Saturated Calomel Electrode (SCE Hg_2Cl_2/Hg, $E_{Calomel-ref}$ = 0.244 V vs. SHE at 298.15 K), and the silver/silver-chloride reference electrode (Ag/AgCl $E_{silver-ref}$ = 0.199 V vs. SHE at 298.15 K). In any event, the potential of a given electrode can be determined by inserting a reference electrode into the cell so that the tip of the reference electrode is in electrolytic contact with the battery electrolyte. This commonly requires the use of a microelectrode, or at least a fine Luggin probe coupled to an external reference electrode, because the interplate distance in most batteries is very small to minimize the cell resistance. In many cases, one is not interested in evaluating the thermodynamic properties of the battery electrode, but rather one is interested in examining the kinetic performances (for example, by measuring the impedance). In this case, it might be acceptable to use a pseudo-reference electrode, which might simply consist of a wire (e.g., oxidized tungsten).

When only one electrode is evaluated, the electrode equilibrium potential can be calculated from the Nernst equation (Eq. 5.4), in a manner that is analogous to that for estimating the cell potential, as described. In this case, E^o is the corresponding standard electrode potential (Table 5.3 shows some standard electrode potential for selected reactions). This is the potential of a "single" electrode, measured with respect to a reference electrode (usually SHE), and with all components in their standard states (activities equal to one). For nonstandard conditions, the half-cell potential is given by the Nernst equation, which is written for the half-cell reaction.

$$aA + xH^+ + ne^- \rightleftharpoons bB \qquad (5.11)$$

in the form:

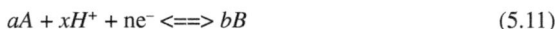

$$E = E^o - (RT/nF) \, Ln \, (a_B{}^b/\{a_A{}^a a_{H+}{}^x\}) \qquad (5.12)$$

where:

$$E^o = -\Delta G^o/nF \qquad (5.13)$$

with ΔG^o corresponding to the change in standard Gibbs energy for the whole reaction (hydrogen reaction: $nH_2 \rightleftharpoons ne^- + 2nH^+$ plus Reaction 5.11),

$$aA + (x - n) \, H^+ + (n/2)e^- \rightleftharpoons bB + (n/2)H_2 \qquad (5.14)$$

that is:

$$\Delta G^o = b \, \Delta_f G_B^o - a\Delta_f G_A^o - (x - n)\Delta_f G_{H+}{}^o + (n/2)\Delta_f G_{H2}{}^o \qquad (5.15)$$

Equation 5.12 is frequently written in an alternate form as an explicit function of pH:

$$E = E^o - (RT/nF) \, Ln \, (a_B{}^b/a_A{}^a) - (2.303RT/F)(x/n)pH \qquad (5.16)$$

where $pH = -\log_{10}(a_{H+})$. This equation form for the equilibrium potential of a half-cell reaction is very convenient because it allows one to detect a possible dependence of the equilibrium potential for the cell on pH (generally, an undesirable characteristic). This issue is covered further in another section, with specific reference to the lead-acid battery.

TABLE 5.3 Standard Oxidation Potentials at
25°C on the Hydrogen Scale

Aqueous acid solutions	
Electrode reaction	$E°$ (V)
$Ag^+ + e^- = Ag$	+0.7991
$Ag^{+2} + e^- = Ag^+$	+1.9800
$AgBr + e^- = Ag + Br^-$	+0.0713
$AgBrO_3 + e^- = Ag + BrO_3^-$	+0.5460
$AgC_2H_3O_2 + e^- = Ag + C_2H_3O_2^-$	+0.6430
$AgCl + e^- = Ag + Cl^-$	+0.2222
$Ag_2CrO_4 + 2e^- = 2\,Ag + CrO_4^{-2}$	+0.4640
$AgI + e^- = Ag + I^-$	−0.1518
$AgIO_3 + e^- = Ag + IO_3^-$	+0.3540
$Ag_2MoO_4 + 2e^- = 2Ag + MoO_4^{-2}$	+0.4860
$AgNO_2 + e^- = Ag + NO_2^-$	+0.5640
$Ag_2SO_4 + 2e^- = 2Ag + SO_4^{-2}$	+0.6540
$Ag(S_2O_3)_2 + e^- = Ag + 2S_2O_3^{-2}$	+0.0170
$Al^{-3} + 3e^- = Al$	−1.6620
$AlF_6^{-3} + 3e^- = Al + 6F$	−2.0690
$Am^{-3} + 3e^- = Am$	−2.3200
$Am^{-4} + e^- = Am^{-3}$	+2.1800
$AmO_2^+ + 4H^+ + 2e^- = Am^{+3} + 2H_2O$	+1.7210
$AmO_2^{+2} + e^- = AmO_2^+$	+1.6390
$AmO_2^+ + 4H^+ + e^- = Am^{+4} + 2H_2O$	+1.2610
$AmO_2^{+2} + 4H^+ + 3e^- = Am^{+3} + 2H_2O$	+1.6940
$As^+ + e^- = As$	−2.9230
$As + 3H^+ + 3e^- = AsH_{3(g)}$	−0.6070
$HAsO_2(aq) + 3H^+ + 3e^- = As + 2H_2O$	+0.2476
$H_3AsO_4 + 2H^+ + 2e^- = HAsO_2 + 2H_2O$	+0.5600
$Au^+ + e^- = Au$	+1.6910
$Au^{+3} + 3e^- = Au$	+1.4980

Source: A. de Bethune and N. Loud. *Standard Aqueous
Electrode Potentials and Temperature Coefficients at
25°C.* C. Hampel, Publisher, Skokie, IL, 1964.

TABLE 5.3 Standard Oxidation Potentials at 25°C on the Hydrogen Scale (2) *(Continued)*

Aqueous acid solutions	
Electrode reaction	$E°$ (V)
$AuBr_2 + e^- = Au + 2Br^-$	+0.9560
$AuBr_4^- + 3e^- = Au + 4Br^-$	+0.87 (60°c)
$Au(CNS)_4^- + 3e^- = Au + 4CNS^-$	+0.6540
$Au(CN)_2 + e^- = Au + 2\ CN^-$	−0.6000
$AuCl_4^- + 3e^- = Au + 4Cl^-$	+1.0000
$Au(OH)_3 + 3H^+ + 3e^- = Au + 3H_2O$	+1.4500
$Ba^{+2} + 2e^- = Ba$	−2.9060
$Be^{+2} + 2e^- = Be$	−1.8470
$BiO^+ + 2H^+ + 3e^- = Bi + H_2O$	+0.3200
$Bi_2O_4 + 4H^+ + 2e^- = 2BiO^+ + 2H_2O$	+1.5930
$BiOCl + 2H^+ + 3e^- = Bi + H_2O + Cl^-$	+0.1600
$Bk^{+4} + e^- = Bk^{+3}$	+1.6000
$H_3BO_3 + 3H^+ + 3e^- = B + 3H_2O$	−0.8690
$HBrO + H^+ + e^- = \frac{1}{2}\ Br2_{(1)} + H_2O$	+1.5950
$Br_{2(1)} + 2e^- = 2Br^-$	+1.06520
$BrO_3^- + 6H^+ + 5e^- = \frac{1}{2}\ Br_{2(1)} + 3H_2O$	+1.5200
$C_{(graphite)} + 4H^+ + 4e^- = CH_{4(g)}$.	+0.1316
$CCl_4 + 4H^+ + 4e^- = 4Cl^- + C + 4H^+$	+1.1800
$CO_2 + 2H^+\ 2e^- = HCOOH_{(aq)}$	−0.1990
$CH_3OH_{(aq)} + 2H^+ + 2e^- = CH_4 + H_2O$	+0.5880
$C_2OH_{2(g)} + 2H^+ + 2e^- = C_2H_{4(g)}$	+0.7310
$C_2H_4 + 2H^+ + 2e^- = C_2H_6$	+0.5200
$(CNS)_2 + 2e^- = 2CNS^-$	+0.7700
$Ca^{+2} + 2e^- = Ca$	−2.8660
$Cd^{+2} + 2e^- = Cd$	−0.4029
$Ce^{+3} + 3e^- = Ce$	−2.4830

Source: A. de Bethune and N. Loud. *Standard Aqueous Electrode Potentials and Temperature Coefficients at 25°C.* C. Hampel, Publisher, Skokie, IL, 1964.

TABLE 5.3 Standard Oxidation Potentials at 25°C on the Hydrogen Scale (3) *(Continued)*

Aqueous acid solutions	
Electrode reaction	E° (V)
$Ce^{+4} + e^- = Ce^{+3}$	+1.6100
$(CH_3)_2SO + 2H^+ + 2e^- = (CH_3)_2SO + H_2O$	+0.2300
$HCHO_{(aq)} + 2H^+ + 2e^- = CH_3OH_{(aq)}$	+0.1900
$CI_2 + 2e^- = 2\ CI^-$	+1.3595
$HClO + H^+ + e^- = \frac{1}{2}\ CI_2 + H_2O$	+1.6300
$HClO_2 + 2H^+ + 2e^- = HClO + H_2O$	+1.6450
$ClO_2 + H^+ + e^- = HClO_2$	+1.2750
$ClO_3 + 3H^+ + 2e^- = HClO_2 + H_2O$	+1.2100
$ClO_4^- + 2H^+ + 2e^- = ClO_3 + H_2O$	+1.1900
$HCNO + H^+ + e^- = \frac{1}{2}\ C_2N_2O$	+0.3300
$CO^{+3} + e^- = CO^{+2}$	+1.8080
$CO^{+2} + 2e^- = CO$	−0.2770
$HCOOH_{(aq)} + 2H^+ + 2e^- = HCHO_{(aq)} + H_2O$	+0.0560
$Cr_2O_7^{-2} + 14H^+ + 6e^- = 2Cr^{+3} + 7H_2O$	+1.3300
$HCrO_7^{-2} + 7H_3O^+ + 3e^- = Cr_3^+ + 11\ H_2O$	+1.1950
$Cr^{+3} + e^- = Cr^{+2}$	−0.4080
$Cr^{+3} + 3e^- = Cr$	−0.7440
$Cu^+ + e^- = Cu$	+0.5210
$Cu^{+2} + e^- = Cu^+$	+0.1530
$Cu^{+2} + 2e^- = Cu$	+0.3370
$Cu^{+2} + Br^- + e^- = CuBr$	+0.6400
$Cu^{+2} + 2CN^- + e^- = Cu(CN)_2^-$	+1.1200
$Cu^{+2} + Ci^- + e^- = CuCl$	+0.5380
$Cu^{+2} + I^- + e^- = CuI$	+0.8600
$CuBr + e^- = Cu + Br^-$	+0.0330
$CuCl + e^- = Cu + Cl^-$	+0.1370

Source: A. de Bethune and N. Loud. *Standard Aqueous Electrode Potentials and Temperature Coefficients at 25°C.* C. Hampel, Publisher, Skokie, IL, 1964.

TABLE 5.3 Standard Oxidation
Potentials at 25°C on the Hydrogen
Scale (4) *(Continued)*

Aqueous acid solutions

Electrode reaction	E° (V)
$CuI + e^- = Cu + I^-$	−0.1852
$Eu^{+3} + e^- = EU^{+2}$	−0.4290
$F_2 + 2e^- = 2F^-$	+2.8700
$F_2 + 2H^+ 2e^- = 2HF_{(aq)}$	+3.0600
$F_2O + 2H^+ + 4e^- = H_2O + 2F^-$	+2.1500
$Fe^{+2} + 2e^- = Fe$	−0.4402
$Fe^{+3} + e^- = Fe^{+2}$	+0.7710
$Fe(CN)_6^{-3} + e^- = Fe(CN)_6^{-4}$	+0.3600
$Fe(CN)_6^{2-} + e^- \rightarrow Fe(CN)_6^{4-}$	+0.356
$FeO_4^{-2} + 8H^+ 3e^- = Fe^{+3} + 4H_2O$	+2.2000
$Ga^{+3} + 3e^- = Ga$	−0.5290
$Gd^{+3} + 3e^- = Gd$	−2.3970
$GeO_2 + 4H^+ + 4e^- = Ge + 2H_2O$	−0.1500
$H^+ + e^- = H_{(g)}$	−2.1065
$\frac{1}{2} H_2 + e^- = H^-$	−2.2500
$2H^+ + 2e^- = H_2$	0
$Hf^{+4} + 4e^- = Hf$	−1.7000
$Hg_2^{+2} + 2e^- = 2Hg$	+0.7880
$2Hg^{+3} + 2e^- = Hg_2^{+2}$	+0.9200
$HgBr_4^{-2} + 2e^- = Hg + 4Br$	+0.2230
$HgI_4^{-2} + 2e^- = Hg + 4I^-$	−0.0380
$HgCl_2 + 2e^- = 2Hg(l) + 2Cl^-(aq)$	+0.2680
$I_2 + 2e^- = 2I$	+0.5355
$I_3 + 2e^- = 3I$	+0.5360
$ICl_2 + e^- = 2Cl^- + \frac{1}{2} I_2$	+1.0560

Source: A. de Bethune and N. Loud. *Standard
Aqueous Electrode Potentials and Temperature
Coefficients at 25°C.* C. Hampel, Publisher,
Skokie, IL, 1964

TABLE 5.3 Standard Oxidation
Potentials at 25°C on the Hydrogen
Scale (5) *(Continued)*

Aqueous acid solutions	
Electrode reaction	$E°$ (V)
$In^{13} + 3e^- = In$	−0.3430
$HIO + H^+ + e^- = \frac{1}{2} I_2 + H_2O$	+1.4500
$IO_3 + +6H^+ + 5e^- = \frac{1}{2} I_2 + 3H_2O$	+1.1950
$H_5IO_6 + H^4 + 2e^- = IO_6^- + 3H_2O$	+1.6010
$IrBr_6^{-3} + e^- = IrBr_6^{-4}$	+0.9900
$IrCl_6^{-2} + e^- = IrCl_6^{-3}$	+1.0170
$IrCl_6^{-3} + 3e^- = Ir + 6Cl^-$	+0.7700
$K^+ + e^- = K$	−2.9250
$La^{+3} + 3e^- = La$	−2.5220
$Li^{+3} + e^- = Li$	−3.0450
$Lu^{+3} + 3e^- = Lu$	−2.2550
$Mg^{+2} + 2e^- = Mg$	−2.3630
$Mn^{+2} + 2e^- = Mn$	−1.1860
$Mn^{+3} + e^- = Mn^{+2}$	+1.5100
$MnO_4 + 4H^+ + 2e^- = Mn^{+2} + 2H_2O$	+1.2300
$MnO_4 + e^- = Mn\ O4^{-2}$	+0.5640
$MnO_4 + 4H^+ + 3e^- = Mn\ O_2 + 2H_2O$	+1.6950
$MnO_4 + 8H^+ + 5e^- = Mn^{+2} + 4H_2O$	+1.5100
$MO^3 + 3e^- = Mo$	−0.2000
$\frac{3}{2} N_2 + H^+ + e^- = HN_{3(aq)}$	−3.0900
$N_2 + 5H^+ + 4e^- = N_2H_5$	−0.2300
$HN_3 + 3H^+ + 2e^- = NH_4 + N_2$	+1.9600
$HN_{3(eq)} + 11H^+ + 8e^- = 3NH_4^+$	+0.6950
$N_2H_5^+ + 3H^+ + 2e^- = 2NH_4^+$	+1.2750
$H_3OH^+ + 2H^+ + 2e^- = NH_4^+\ H_2O$	+1.3500
$2NH_3OH^+ + H^+ + 2e^- = N_2H_5 + 2H_2O$	+1.4200

Source: A. de Bethune and N. Loud. *Standard Aqueous
Electrode Potentials and Temperature Coefficients at
25°C.* C. Hampel, Publisher, Skokie, IL, 1964.

TABLE 5.3 Standard Oxidation Potentials
at 25°C on the Hydrogen Scale (6) *(Continued)*

Aqueous acid solutions	
Electrode reaction	E° (V)
$NO_3^- + 3H^+ + 2e^- = HNO_2 + H_2O$	+0.9400
$NO_3^- + 4H^+ + 4e^- = NO + 2H_2O$	+0.9600
$2NO + 2H^+ + 2e^- = H_2N_2O_2$	+0.7120
$2NO_3^- + 4H^+ + 2e^- = N_2O_{4(g)} + 2H_2O$	+0.8030
$HNO_2 + H^+ + e^- = NO + H_2O$	+1.0000
$2HNO_2 + 4H^+ + 4e^- = H_2N_2O_2 + 2H_2O$	+0.8600
$2HNO_{2(aq)} + 4H^+ + 4e^- = N_2O_g + 3H_2O$	+1.2900
$N_2O_4 + 2H^+ + 2e^- = 2HNO_2$	+1.0700
$N_2O_4 + 4H^+ + 4e^- = 2NO + 2H_2O$	+1.0300
$H_2N_2O_2 + 2H^+ + 2e^- = N_2 + 2H_2O$	+2.6500
$H_2N_2O_2 + 6H^+ + 4e^- = 2NH_3 OH^+$	+0.3870
$Na^+ + e^- = Na$	−2.7140
$Nb^{+3} + 3e^- = Nb$	−1.0990
$Nb_2O_4 + 10H^+ + 10e^- = 2Nb + 5H_2O$	−0.6440
$Nd^{+3} + 3e^- = Nd$	−2.4310
$Ni^{+2} + 2e^- = Ni$	−0.2500
$NiO_2 + 4H^+ + 2e^- = Ni^{+2} + 2H_2O$	+1.6780
$Np^{+3} + 3e^- = Np$	−1.8560
$Np^{+4} + e^- = Np^{+3}$	+0.1470
$NpO_2^{+2} + e^- = NpO_2^+$	+1.1500
$NpO_2 + 4H^+ + e^- = Np^{+4} + 2H_2O$	+0.7500
$O_{(g)} + 2H^+ + 2e^- = H_2O$	+2.4220
$O_2 + H^+ + e^- = HO_2$	−0.1300
$O_2 + 2H^+ + 2e^- = H_2O_2$	+0.6824
$O_{2(g)} + 4H^+ + 4e^- = 2H_2O_{(1)}$	+1.2290
$O_3 + 2H^+ + 2e^- = O_2 + H_2O$	+2.0700
$HO_2 + H^+ + e^- = H_2O_2$	+1.4950

Source: A. de Bethune and N. Loud. *Standard Aqueous Electrode Potentials and Temperature Coefficients at 25°C.* C. Hampel, Publisher, Skokie, IL, 1964.

TABLE 5.3 Standard Oxidation Potentials
at 25°C on the Hydrogen Scale (7) *(Continued)*

Aqueous acid solutions	
Electrode reaction	E° (V)
$H_2O_2 + H^+ + e^- = OH + H_2O$	+0.7100
$H_2O_2 + 2H^+ + 2e^- = H_2O$	+1.7760
$OH + H^+ + e^- = H_2O$	+2.8500
$OsO_{4(c)} + 8H^+ + 8e^- = Os + 4H_2O$	+0.8500
$P_{(white)} + 3H^+ + 3e^- = PH_{3(g)}$	+0.0630
$H_3PO_2 + 2H^+ + e^- = P_{(white)}$	−0.5080
$H_3PO_4 + H^+ + 2e^- = H_3PO_2 + H_2O$	−0.4990
$H_3PO_3 + 2H^+ + 2e^- = H_3PO_3$	−0.2760
$Pb^{+2} + 2e^- = Pb$	−0.1260
$PbBr_2 + 2e^- = Pb + 2Br^-$	−0.2840
$PbCl_2 + 2e^- = Pb + 2Cl^-$	0.2680
$PbI_2 + 2e^- = Pb + 2I^-$	−0.3650
$PbO_2 + 4H^+ + 2e^- = Pb^{+2} + 2H_2O$	+1.4550
$PbO_2 + SO_4^{-2} + 4H^+\ 2e^- = PbSO_4 + 2H_2O$	+1.6820
$PbSO_1 + 2e = Pb + SO_4^{-2}$	−0.3588
$Pd^{+2} + 2e^- = Pd$	+0.9870
$PdBr_4^{-2} + 2e^- = Pd + 4Br^-$	+0.6000
$PdCl_4^{-2} + 2e^- = Pd + 4Cl^-$	+0.6200
$PdCl_6^{-2} + 2e^- = PdCl_4^{-2} + 2Cl^-$	+1.2880
$Pt^{2+} + 2e^- = Pt$	+1.2000
$PtBr_4^{-2} + 2e^- = Pt + 4Br^-$	+0.5810
$PtCl_4^{-2} + 2e^- = Pt + 4Cl^-$	+0.7300
$PtCl_6^{-2} + 2e^- = PtCl_4 + 2Cl^-$	+0.6800
$Pt(OH)_2 + 2H^4 + 2e^- = Pt + 2H_2O$	+0.9870
$PtS + 2H^+ + 2e^- = Pt + H_2S$	0.2970
$Pu^{+3} + 3e^- = Pu$	−2.0310
$Pu^{+4} + e^- = Pu^{+3}$	+0.9700
$PuO_2^+ + 4H^+ + e^- = Pu^{+4} + 2H_2O$	+1.1500

Source: A. de Bethune and N. Loud. *Standard Aqueous Electrode Potentials and Temperature Coefficients at 25°C.* C. Hampel, Publisher, Skokie, IL, 1964.

TABLE 5.3 Standard Oxidation Potentials
at 25°C on the Hydrogen Scale (8) *(Continued)*

Aqueous acid solutions	
Electrode reaction	E° (V)
$PuO_2^+ + e^- = PuO_2$	+0.9300
$PuO_2 + 4H^+ + 2e^- = Pu^{+4} 2H_2O$	+1.0400
$Ra^{+2} + 2e^- = Ra$	−2.9160
$Rb^+ + e^- = Rb$	−2.9250
$ReO_2^+ + 4H^+ + 4e^- = Re + 2H_2O$	+0.2513
$ReO_4^- + 4H^+ + 3e^- = ReO_2 + 2H_2O$	+0.5100
$ReO_4^- + 8H^+ + 7e^- = Re + 4H_2O$	+0.3620
$Rh^{+3} + 3e^- = Rh$	+0.8000
$RhCl_6^{-3} + 3e^- = Rh + 6Cl^-$	+0.4310
$RuCl_5^{-2} + 3e^- = Rh + 4Cl^-$	+0.6010
$S_{(rhombic)} + 2e^- = H_2S_{(aq)}$	+0.1420
$H_2SO_3 + 4H^+ 4e = S + 3H_2O$	+0.4500
$2H_2SO_3 + H^+ + 2e^- = HS_2O_4 + 2H_2O$	−0.0820
$2H_2SO_3 + 2H^+ + 4e^- = S_2O_3^{-2} + 3H_2O$	+0.4000
$4H_2SO_3 + 4H^+ + 6e^- = S_4O_6^{-2} + 6H_2O$	+0.5100
$S_2Cl_2 + 2e^- = 2S + 2Cl^-$	+1.2300
$SO_4^{-2} + 4H^+ + 2e^- = H_2SO_3 + H_2O$	+0.1720
$2SO_4^{-2} + 4H^+ + 2e^- = S_2O_6 + 2H_2O$	−0.220
$S_2O_8^{-2} + 2e^- = 2SO^{4-2}$	+2.0100
$Sb + 3H^+ + 3e^- = Sb_{3(g)}$	−0.5100
$Sb_2O_3 + 6H^+ + 6e^- = 2Sb + 3H_2O$	+0.1520
$Sb_2O_3 + 2H^+ + 2e^- = Sb_2O_4 + H_2O$	+0.4790
$Sb_2O_3 + 6H^+ + 4e^- = 2SbO^+ + 3H_2O$	+0.5810
$Sc^{+3} + 3e^- = Sc$	−2.0770
$Se + 2H^+ 2e^- = H_2Se_{(aq)}$	−0.3990
$H_2SeO_3 + 4H^+ + 4e^- = Sc_{(grey)} + 3H_2O$	+0.7400
$SeO_4^{-2} + 4H^+ + 2e^- = H_2SeO_3{:}H_2O$	+1.1500
$S_2O_6^{-2} + 4H^+ + 2e^- = 2H_2SO$	+0.5700

Source: A. de Bethune and N. Loud. *Standard Aqueous Electrode Potentials and Temperature Coefficients at 25°C.* C. Hampel, Publisher, Skokie, IL, 1964.

TABLE 5.3 Standard Oxidation Potentials at 25°C on the Hydrogen Scale (9) *(Continued)*

Aqueous acid solutions	
Electrode reaction	E° (V)
$Si + 4H^+ \ 4e^- = SiH_{4(g)}$	+0.1020
$SiF_6^{-2} + 4e^- = Si + 6F$	−1.2400
$SiO_{2(quartz)} + 4H^+ + 4e^- = Si + 2H_2O$	−0.8570
$Sm^{+3} + 3e^- = Sm$	−2.4140
$Sn^{+2} + 2e^- = Sn_{(white)}$	−0.1360
$Sn^{+4} + 2e^- = Sn^{+2}$	+0.1500
$SnF_6^{-2} + 4e^- = Sn + 6F^-$	−0.2500
$Sr^{+2} + 2e^- = Sr$	−2.8880
$Ta_2O_5 + 10H^+ + 10e^- = 2Ta + 5H_2O$	−0.8120
$TeO_{2(c)} + 4H^+ + 4e^- = Te + 2H_2O$	+0.5290
$H_6Te_{6(c)} + 2H^+ + 2e^- = TeO_2 + 4H_2O$	+1.0200
$TeOOH^+ + 3H^+ + 4e^- = Te + 2H_2O$	+0.5590
$Te + 2H^+ + 2e^- = H_2Te_{(g)}$	−0.7180
$Th^{+4} + 4e^- = Th$	−1.8990
$Ti^{+2} + 2e^- = Ti$	−1.6280
$Ti^{+3} + e^- = Ti^{+2}$	−0.3690
$TiF_6^{-2} + 4e^- = Ti + 6F$	−1.1910
$TiO^{+2} + 2H^+ + e^- = Ti^{+3} + H_2O$	+0.0990
$TiO^{+2} + 2H^+ + 4e^- = Ti + H_2O$	+0.8820
$Ti^+ + e^- = Tl$	−0.3363
$Ti^{+3} + 2e^- = Tl^+$	+1.2500
$TlBr + 2e^- = Tl + Br^-$	−0.6580
$TlCl + e^- = Tl + Cl^-$	−0.5568
$TlI + e^- = Tl + I^-$	−0.7520
$U^{+3} + 3e^- = U$	−1.7890
$U^{+4} + e^- = U^{+3}$	−0.6090
$UO_2^{+2} + e^- = UO_2^+$	+0.0500
$UO_2^{+2} + 4H^+ + 2e^- = U^{+4} + 2H_2O$	−0.3300

Source: A. de Bethune and N. Loud. *Standard Aqueous Electrode Potentials and Temperature Coefficients at 25°C.* C. Hampel, Publisher, Skokie, IL, 1964.

TABLE 5.3 Standard Oxidation Potentials
at 25°C on the Hydrogen Scale (10) *(Continued)*

Aqueous acid solutions

Electrode reaction	$E°$ (V)
$UO2^+ + 4H^+ + e^- = U^{+4} + 2H_2O$	+0.6200
$V^{+2} + 2e^- = V$	−1.1860
$V^{+3} + e^- = V^{12}$	−0.2560
$VO^{+2} + 2H^+ + e^- = V^{+3} + H_2O$	+0.3590
$V(OH)_4^+ + 2H^+ + e^- = VO^{+2} + 3H_2O$	+1.0000
$V(OH)_4^+ + 4H^+ + 5e^- = V + 4H_2O$	−0.2540
$WO_{3(c)} + 6H^+ + 6e^- = W + 3H_2O$	−0.0900
$Y^{+3} + 3e^- = Y$	−2.3720
$Zn^{+2} + 2e^- = Zn$	−0.7628
$ZR^{+4} + 4e^- = Zr$	−1.5290

Aqueous basic solutions

Electrode reaction	$E°$ (V)
$AgCN + e^- = Ag + CN^-$	−0.0170
$AgCN_2^- + e^- = Ag + 2CN^-$	−0.3100
$AgCO_3 + 2e^- = 2Ag + CO_3^{-2}$	+0.4700
$Ag(NH_3)^{2+} + e^- = Ag + 2NH_3$	+0.3730
$Ag_2O + H_2O + 2e^- = 2Ag + 2OH^-$	+0.3450
$Ag_2O_3 + H_2O\ 2e^- = 2AgO + 2OH^-$	+0.7390
$Ag_2S + 2e^- = 2Ag + S^{-2}$	−0.6600
$Ag(SO_3)_2^3 + e^- = Ag + 2SO_3^{-2}$	+0.2950
$H_2AIO_3^- + H_2O + 3e^- = AI + 4OH^-$	−2.3300
$AsO_2^- + 2H_2O + 3e^- = As + 4OH^-$	−0.6750
$AsO_4^{-3} + 2H_2O + 2e^- = AsO_2^- + 4OH^-$	−0.6800
$H_2BO_3^- + H_2O + 3e^- = B + 4OH^-$	−1.7900
$Ba(OH)_2 + 8H_2O + 2e^- = Ba + 8H_2O + 2OH^-$	−2.9900

Source: A. de Bethune and N. Loud. *Standard Aqueous Electrode Potentials and Temperature Coefficients at 25°C*. C. Hampel, Publisher, Skokie, IL, 1964.

TABLE 5.3 Standard Oxidation Potentials
at 25°C on the Hydrogen Scale (11) *(Continued)*

Aqueous basic solutions	
Electrode reaction	$E°$ (V)
$Be_2O_3^{-2} + 3H_2O + 4e^- = 2Be + 6OH^-$	−2.6300
$Be_2O_3 + 3H_2O + 6e^- = 2Bi + 6OH^-$	−0.4600
$BrO^- + H_2O + 2e^- = Br^- + 2OH^-$	+0.7610
$BrO_3^- + 3H_2O + 6e^- = Br^- + 6OH^-$	+0.6070
$Ca(OH)_2 + 2e^- = Ca + 2OH^-$	−3.0200
$Cd(CN)_4^{-2} + 2e^- = Cd + 4CN^-$	−1.0280
$CdCO_3 + 2e^- = Cd + CO_3^{-2}$	−0.7400
$Cd(NH_3)_4^{+2} + 2e^- = Cd + 4N_3$	−0.6130
$Cd(OH)2 + 2e^- = Cd + 2OH^-$	−0.8090
$CdS + 2e^- = Cd + S$	−1.1750
$ClO^- + H_2O + 2e^- = Cl + 2OH^-$	+0.8900
$ClO_{2(g)}^- + 2e^- = ClO_2^-$	+1.1600
$ClO_2^- + H_2O + 2e^- = ClO^- + 2OH^-$	+0.6600
$ClO_3^- + H_2O + 2e^- = ClO_2^- + 2OH^-$	+0.3300
$ClO_4^- + H_2O + 2e^- = ClO_3^- + 2OH^-$	+0.3600
$CNO^- + H_2O + 2e^- = CN^- + 2OH^-$	−0.9700
$CoCO_3 + 2e^- = Co + CO_3^{-2}$	−0.6400
$Co(OH)_2 + 2e^- = Co + 2OH^-$	−0.7300
$Co(OH)_3 + e^- = Co(OH)_2 + OH^-$	+0.1700
$Co(NH_3)_6^{+3} = e^- = Co(NH_3)_6^{+2}$	+0.1080
$CrO_2^- + 2H_2O + 3e^- = Cr + 4OH^-$	−1.2700
$CrO_4^{-2} + 4H_2O + 3e^- = Cr(OH)_3 + 5OH^-$	−0.1300
$Cr(OH)_3 + 3e^- = Cr + 3OH^-$	−1.3400
$Cu(CN)_2^- + e^- = Cu + 2CN^-$	−0.4290
$Cu(CNS) + e^- = Cu + CNS^-$	−0.2700
$Cu(NH_3)_2^+ + e^- = Cu + 2NH_3$	−0.1200
$Cu_2O + H_2O + 2e^- = 2Cu + 2OH^-$	−0.3580
$2Cu(OH)_2 + 2e^- = Cu_2O + 2OH^- + H_2O$	−0.0800
$Cu_2S + 2e^- = 2Cu + S^{-2}$	−0.8900

Source: A. de Bethune and N. Loud. *Standard Aqueous
Electrode Potentials and Temperature Coefficients at
25°C.* C. Hampel, Publisher, Skokie, IL, 1964.

TABLE 5.3 Standard Oxidation Potentials
at 25°C on the Hydrogen Scale (12) *(Continued)*

<table>
<tr><th colspan="2">Aqueous basic solutions</th></tr>
<tr><th>Electrode reaction</th><th>$E°$ (V)</th></tr>
<tr><td>$FeCO_3 + 2e^- = Fe + CO_3^{-2}$</td><td>−0.7560</td></tr>
<tr><td>$FeO_4^{-2} + 2H_2O + 3e^- = FeO_2^- + 4OH^-$</td><td>−0.9000</td></tr>
<tr><td>$FeO_4^{-2} + 4H_2O + 3e^- = Fe(OH)_3^- + 5OH^-$</td><td>+0.7200</td></tr>
<tr><td>$Fe(OH)_2 + 2e^- = Fe + 2OH^-$</td><td>−0.8700</td></tr>
<tr><td>$Fe(OH)_3 + e^- = Fe(OH)_2 + OH^-$</td><td>−0.5600</td></tr>
<tr><td>$FeS + 2e^- = Fe + S^{-2}$</td><td>−0.9500</td></tr>
<tr><td>$Fe_2S_3\ 2e^- = 2FeS + S^{-2}$</td><td>−0.7150</td></tr>
<tr><td>$H_2GaO_3^- + H_2O + 3e^- = Ga + 4OH^-$</td><td>−1.2190</td></tr>
<tr><td>$HGeO_3^- + 2H_2O + 4e^- = Ge + 5OH^-$</td><td>−1.0300</td></tr>
<tr><td>$HfO(OH)_2 + H_2O + 4e^- = Hf + 4OH^-$</td><td>−2.5000</td></tr>
<tr><td>$2H_2O(l) + 2e^- = H_2(g) + 2OH^-(aq)$</td><td>−0.8280</td></tr>
<tr><td>$Hg(CN)_4^{-2} + 2e^- = Hg^- + 4CN^-$</td><td>−0.3700</td></tr>
<tr><td>$HgO_{(r)} + H_2O + 2e^- = Hg + 2OH^-$</td><td>+0.0980</td></tr>
<tr><td>$HgS_{(black)} + 2e^- = Hg + S^{-2}$</td><td>−0.6900</td></tr>
<tr><td>$H_2O + e^- = H_{(g)} + OH^-$</td><td>−2.9345</td></tr>
<tr><td>$2H_2O + 2e^- = H_2 + 2OH^-$</td><td>−0.8280</td></tr>
<tr><td>$In(OH)_3 + 3e^- = In + 3OH^-$</td><td>−1.0000</td></tr>
<tr><td>$IO^- + H^2O + 2e^- = I^- + 2OH^-$</td><td>+0.4850</td></tr>
<tr><td>$IO_3 + 3H_2O + 6e^- = I^- + 6OH^-$</td><td>+0.2600</td></tr>
<tr><td>$H_3\ IO_6^{-2} + 2\ e^- = IO_3^- + 3OH^-$</td><td>+0.7000</td></tr>
<tr><td>$Ir_2O_3 + 3H_2O + 6e^- = 2Ir + 6OH^-$</td><td>+0.0980</td></tr>
<tr><td>$La(OH)_3 + 3e^- = La + 3OH^-$</td><td>−2.9000</td></tr>
<tr><td>$Lu(OH)_3 + 3e = Lu + 3OH^-$</td><td>−2.7200</td></tr>
<tr><td>$Mg(OH)_2^- + 2e^- = Mg + 2OH^-$</td><td>−2.6900</td></tr>
<tr><td>$MnCO_{3(ppl)} + 2e^- = Mn + CO_3^{-2}$</td><td>−1.4800</td></tr>
<tr><td>$MnO_2 + 2H_2O + 2e^- = Mn(OH)_2 + 2OH^-$</td><td>−0.0500</td></tr>
<tr><td>$MnO_4^{-2} + 2H_2O + 2e^- = MnO_2 + 4OH^-$</td><td>+0.6000</td></tr>
<tr><td>$Mn(OH)_3 + e^- = Mn(OH)_2 + OH^-$</td><td>+0.1500</td></tr>
<tr><td>$Mn(OH)_2 + 2e^- = Mn + 2OH^-$</td><td>−1.5500</td></tr>
</table>

Source: A. de Bethune and N. Loud. *Standard Aqueous
Electrode Potentials and Temperature Coefficients at
25°C.* C. Hampel, Publisher, Skokie, IL, 1964.

TABLE 5.3 Standard Oxidation Potentials
at 25°C on the Hydrogen Scale (13) *(Continued)*

Aqueous basic solutions	
Electrode reaction	$E°$ (V)
$MoO_4^{-2} + 4H_2O + 6e^- = Mo + 8OH^-$	−1.0500
$N_2H_4 + 4H_2O + 2e^- = 2NH_4OH + 2OH^-$	+0.1100
$NO_3 + H_2O + 2e^- = NO_2 + 2OH^-$	+0.0100
$2NH_2OH + 2e^- = N_2H_4 + 2OH^-$	+0.7300
$Na_2UO_4 + 4H_2O + 2e^- = U(OH)_4 + 2Na^+ + 4OH^-$	−1.6180
$NiCO_3 + 2e^- = Ni + COH_3^{-2}$	−0.4500
$Ni(NH_3)_6^{+2} + 2e^- = Ni + 6NH_{3(aq)}$	−0.4760
$NiO_2 + 2H_2O + 2e^- = Ni(OH)_2 + 2OH^-$	+0.4900
$Ni(OH)_2 + 2e^- = Ni + 2OH^-$	−0.7200
$NiS + 2e^- = Ni + S^{-2}$	−0.8300
$O_2 + e^- = O_2^-$	−0.5630
$HO_2^- + H_2O + e^- = OH_{(aq)} + 2OH^-$	−0.2450
$HO_2^- + H_2O + 2e^- = 3OH^-$	+0.8780
$O_2^- + H_2O + e^- = OH^- + HO_2^-$	+0.4130
$O_2 + 2H_2O + 4e^- = 4OH^-$	+0.4010
$O_2 + H_2O + 2e^- = HO_2^- + OH^-$	−0.7600
$O_{3(g)} + H_2 + 2e^- = O_2 + 2OH^-$	+1.2400
$OH_{(g)} + e^- = OH^-$	+2.0200
$HOsO_5^- + 4H_2O + 8e^- = Os + 9OH^-$	−0.0150
$P_{(white)} + 3H_2O + 3e^- = PH_3 + 3OH^-$	−0.8900
$PbCO_3 + 2e^- = Pb + CO_3^{-2}$	−0.5090
$HPbO_2^- + H_2O + 2\ e^- = Pb + 3OH^-$	−0.5400
$PbO_2 + H_2O + 2e^- = PbO_{(r)} + 2OH^-$	+0.2470
$PbS + 2e^- = Pb + S^{-2}$	−0.9300
$Pd(OH)_2 + 2e^- = Pd + 2OH^-$	+0.0700
$H_2PO_2^- + e^- = P + 2OH^-$	−2.0500
$HPO_3^{-2} + 2H_2O + 2e^- = H_2PO_2^- + 3OH^-$	−1.5650
$Pt(OH)_2 + 2e^- = Pt + 2OH^-$	+0.1500
$PuO_2(OH)_2 + e^- = PuO_2OH + OH^-$	−0.2340

Source: A. de Bethune and N. Loud. *Standard Aqueous
Electrode Potentials and Temperature Coefficients at
25°C.* C. Hampel, Publisher, Skokie, IL, 1964.

TABLE 5.3 Standard Oxidation Potentials
at 25°C on the Hydrogen Scale (14) *(Continued)*

Aqueous basic solutions

Electrode reaction	$E°$ (V)
$PuO_4^{-2} + 2H_2O + 2e^- = HPO_3^{-2} + 3OH^-$	−1.1200
$Pu(OH)_3 + 3e^- = Pu + 3OH^-$	−2.4200
$Pu(OH)_4 + e^- = Pu(OH)_3 + OH^-$	−0.9630
$ReO_2 + H_2O + 4e^- = Re + 4OH^-$	−0.5770
$ReO_4^- + 2H_2O + 3e^- = ReO_2 + 4OH^-$	−0.5940
$ReO_4^- + 4H_2O + 7e^- = Re + 8OH^-$	−0.5840
$Rb_2O_3 + 3H_2O + 6e^- = 2Rh + 6OH^-$	+0.0400
$RuO_4^- + e^- = RuO_4^{-2}$	+0.6000
$S + 2e^- = S^{-2}$	−0.4470
$2SO_3^{-2} + 3H_2O + 4e^- = S_2O_2^{-2}$	−0.5710
$SO_4^{-2} + H_2O + 2e^- = SO_3^{-2} + 2OH^-$	−0.9300
$S_4O_6^{-2} + 2e^- = 2S_2O_3^{-2}$	+0.0800
$2SO_3^{-2} + 2H_2O + 2e^- = S_2O_4^{-2} + 4HO^-$	−1.1200
$SbO_2^- + 2H_2O + 3e^- = Sb + 4OH^-$	−0.6600
$Sc(OH)_3 + 3e^- = Sc + 3OH^-$	−2.6100
$Se + 2e^- = Se^{-2}$	−0.9200
$SeO_3^{-2} + 3H_2O + 4e^- = Sc + 6OH^-$	−0.3660
$SeO_4^{-2} + H_2O + 2e^- = ScO_3 + 2OH^-$	+0.0500
$SiO_3^{-2} + 3H_2O + 4e^- = Si + 6OH^-$	−1.6970
$Sn(OH)_6^{-2} + 2e^- = HSnO_2^- + H_2O + 3OH^-$	−0.9300
$SnS + 2e^- = Sn + S^{-2}$	−0.8700
$HsnO_2^- + H_2O + 2e^- = Sn + 3OH^-$	−0.9090
$Sr(OH)_2 + 2e^- = Sr + 2OH^-$	−2.8800
$Sr(OH)_2 + 8H_2O + 2e^- = Sr + 2OH^- + 8H_2O$	−2.9900
$Te + 2e^- = Te^{-2}$	−1.1430
$TeO_3 + 3H_2O + 4e^- = Te + 6OH^-$	−0.5700
$TeO_4^{-2} + H_2O + 2e^- = TeO_3^{-2} + 2OH^-$	+0.4000
$Th(OH)_4 + 4e^- = Th + 2OH^-$	−2.4800
$Tl(OH) + e^- = Tl + OH^-$	−0.3430

Source: A. de Bethune and N. Loud. *Standard Aqueous
Electrode Potentials and Temperature Coefficients at
25°C.* C. Hampel, Publisher, Skokie, IL, 1964.

TABLE 5.3 Standard Oxidation Potentials at 25°C on the Hydrogen Scale (15) *(Continued)*

Aqueous basic solutions	
Electrode reaction	E° (V)
$Tl(OH)_3 + 2e^- = TlOH + 2OH^-$	−0.0500
$Tl_2S + 2e^- = 2Tl + S^{-2}$	−0.9000
$UO_2 + 2H_2O + 4e^- = U + 4OH^-$	−2.3900
$U(OH)_3 + 3e^- = U + 3OH^-$	−2.1700
$U(OH)_4 + e^- = U(OH)_3 + OH^-$	−2.2000
$HV_6O_{17}^{-3} + 16H_2O + 30e^- = 6V + 33OH^-$	−1.1540
$WO_4^{-2} + 4H_2O + 6e^- = W + 8OH^-$	−1.0500
$ZnCO_3 + 2e^- = Zn + CO_3^{-2}$	−1.0600
$Zn(CN)_4^{-2} + 2e^- = Zn + 4CN^-$	−1.2600
$Zn(NH_3)_4^{+2} + 2e^- = Zn + 4NH_{3(aq)}$	−1.0400
$ZnO_2^{-2} + 2H_2O + 2e^- = Zn + 4OH^-$	−1.2150
$Zn(OH)_2 + 2e^- = Zn + 2OH^-$	−1.2450
$ZnS_{(wurtzite)} + 2e^- = Zn + S^{-2}$	−1.4050
$H_2ZrO_3 + H_2O + 4e^- = Zr + 4OH^-$	−2.3600

Source: A. de Bethune and N. Loud. *Standard Aqueous Electrode Potentials and Temperature Coefficients at 25°C.* C. Hampel, Publisher, Skokie, IL, 1964.

Example: Calculation of cell equilibrium potential To illustrate the calculation of a cell equilibrium potential, E, a battery has an anode of Zn metal and a cathode of carbon felt in contact with Br_2/Br^-. The electrolyte is composed of a solution of Zn^{2+} and Br^-. The half-cell reactions are:

$$Zn \longrightarrow Zn^{2+} + 2e^- \text{ (anode on discharge)} \qquad (5.17)$$

and

$$Br^2 + 2e^- \longrightarrow 2Br^- \text{ (cathode on discharge)} \qquad (5.18)$$

$$Zn + Br_2 ==> ZnBr_2(aq) \text{ (cell reaction)} \qquad (5.19)$$

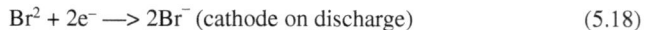

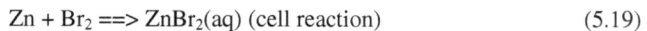

The half-cell potentials are obtained directly from the Nernst equation as:

$$E_{ox(Zn|Zn+2)} = E^o_{(Zn|Zn+2)} - (RT/2F) \, Ln(a_{Zn+2}/a_{Zn}) \qquad (5.20)$$

and:

$$E_{red(Br_2|Br-)} = E^o_{(Br_2|Br-)} - (RT/2F) \, Ln(a_{Br-2}/a_{Br_2}) \qquad (5.21)$$

where the activity is defined as $a_L = \gamma_L \, m_L$, with γ_L and mi corresponding to the activity coefficient and the molal concentration, respectively. The activity of $ZnBr_2$ (aq) is written as:

$$a_{ZnBr2(aq)} = (a_{Zn^{+2}})(a_{Br^-})^2 = m_{Zn^{+2}}m_{Br^-}{}^2(\gamma_2\gamma_1{}^2) \tag{5.22}$$

where γ_1 and γ_2 are the single ion activity coefficients for Br^- and Zn^{2+} respectively. Denoting the molal concentration of $ZnBr_2(aq)$ as m (mol/kg H_2O), we have $m_{Zn^{+2}} = m$, and $m_{Br^-} = 2m$, so $ZnBr_2(aq) = (\gamma m)^3$, and:

$$\gamma_m = (4m^3)^{1/3}\,(\gamma_2\gamma_1{}^2)^{1/3} = m\,(4\gamma_2\gamma_1{}^2)^{1/3} \tag{5.23}$$

Thus, the relationship between the activity coefficient of the salt (γ) and those for the individuals ions is:

$$\gamma = (4\gamma_2\gamma_1{}^2)^{1/3} \tag{5.24}$$

The whole cell potential is calculated from the Nernst equation from the two half cells in the battery $Br_2/Br^-|Zn^{2+}/Zn$ as:

$$E_{cell} = E_{RHS} - E_{LHS} = E^\circ(Zn|Zn^{+2}) - E^\circ(Br_2|Br^-) - (2.303\ RT/2F)\log(a_{Br_2}/\{a_{Zn^+}+2a_{Br^-2}\}) \tag{5.25}$$

and hence:

$$E_{cell} = E^\circ{}_{cell} + (2.303\ RT/2F)\log(a_{Zn^+}+2a_{Br^-2}) - (2.303\ RT/2F)\log(a_{Br_2})$$
$$= E^\circ{}_{cell} + (2.303\ RT/2F)\log(4m^3\gamma_2\gamma_1) - (2.303\ RT/2F)\log(a_{Br_2}) \tag{5.26}$$

where a_{Br_2} is the activity of bromine. The half-cell standard potentials in aqueous solution and room temperature, E°, are usually given in tables (Table 5.3 lists the standard oxidation potentials at 25°C on the hydrogen scale; for example, it can be read: $Zn^{2+} + 2e^- = Zn$, $E^\circ = -0.7628$ V, or $Zn = Zn^{2+} + 2e^-$, $E^\circ = +0.7628$ V).

The theoretical storage capability of the battery can be calculated when the number of electrons participating in the reaction, the standard equilibrium potential, and the molecular weights of components participating in the reaction are known. An example applied to the lead-acid battery is shown in Figures. 5.6 and 5.7.

From the analysis in Figures 5.6 and 5.7, it is clear that two important parameters exist that maximize the gravimetric power of a battery:

- Gibbs free energy of the reaction must be large (high cell standard equilibrium voltage (E°), and
- The components involved in the reaction must be of a low weight.

Multireactions occurring at the same electrode surface Two or more reactions can occur simultaneously on an electrode. An important situation arises when one of the reactions occurs as an oxidation process and the other as a reduction reaction on the same electrode. Then, it is possible for the two reactions to occur in the same electrode surface without any net current flow from the electrode into an external circuit. An important example is provided by the corrosion of iron:

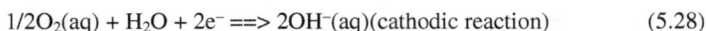

$$Fe(s) \Longrightarrow Fe^{2+} + 2e^- \text{ (anodic reaction)} \tag{5.27}$$

$$1/2O_2(aq) + H_2O + 2e^- \Longrightarrow 2OH^-(aq)\text{(cathodic reaction)} \tag{5.28}$$

| positive electrode: | $PbO_2 + 3H^+ + HSO_4^- + 2e^-$ <====> $PbSO_4 + 2H_2O$ |
| negative electrode: | $Pb + HSO_4^-$ <====> $PbSO_4 + H^+ + 2e^-$ |

| cell reaction: | $Pb + PbO_2 + 2H^+ + 2HSO_4^-$ <====> $2PbSO_4 + 2H_2O$ |

free enthalpy of formation
(standard values) ΔG(kJ)
$T = 25°C$:   0   -219.2   0   2(-753.5)   2(-811.7)   2(-237.4)

free enthalpy of reaction: $\Delta G = -372.2$ kJ (standard)

standard equilibrium voltage: $E^o = -\dfrac{\Delta G}{nF} = \dfrac{372.2}{2 \cdot 96,500} = 1.928$ Volt

$(\alpha_{H^+} ; \alpha_{HSO_4^-} = 1 \, mol/l)$

equilibrium voltage: $E = E^o + \dfrac{RT}{F} \ln \dfrac{\alpha_{H^+} \cdot \alpha_{HSO_4^-}}{\alpha_{H_2O}}$

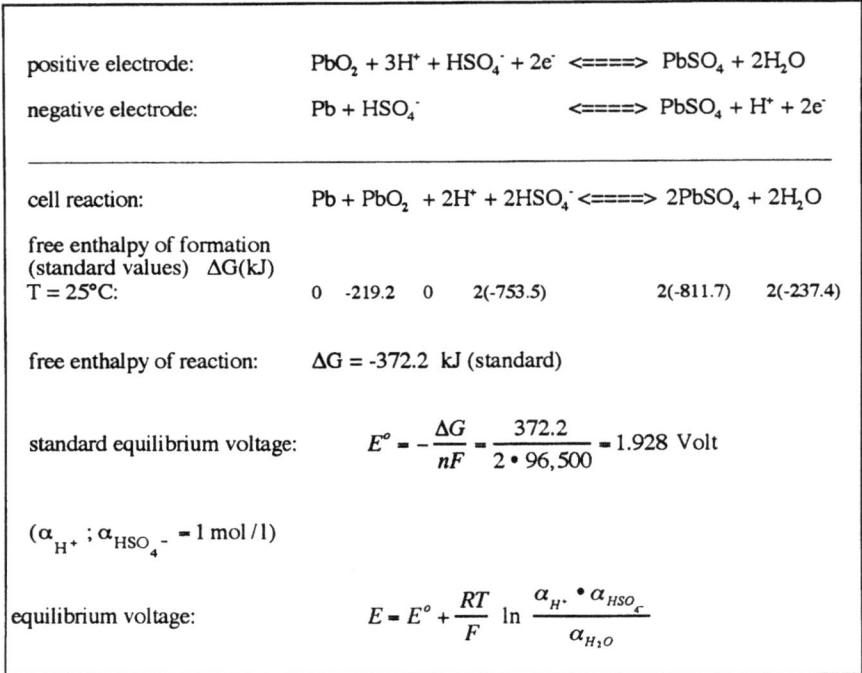

FIGURE 5.6 Lead-acid battery; reaction equation and equilibrium cell voltage.

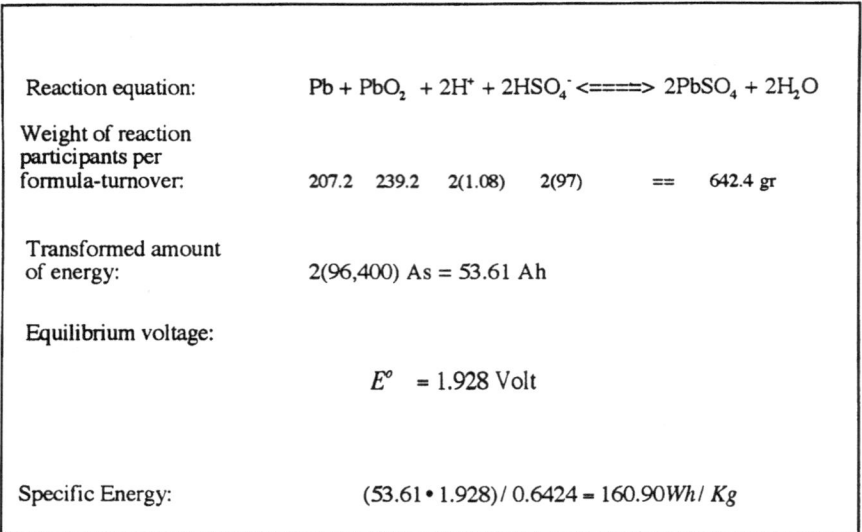

| Reaction equation: | $Pb + PbO_2 + 2H^+ + 2HSO_4^-$ <====> $2PbSO_4 + 2H_2O$ |

Weight of reaction
participants per
formula-turnover: 207.2   239.2   2(1.08)   2(97)   ==   642.4 gr

Transformed amount
of energy: 2(96,400) As = 53.61 Ah

Equilibrium voltage:

E^o = 1.928 Volt

Specific Energy: $(53.61 \cdot 1.928)/0.6424 = 160.90 Wh/Kg$

FIGURE 5.7 Lead-acid battery; theoretical storage ability per unit of weight.

The sample of iron adopts a potential called the *mixed potential* so that the currents caused by these reactions are equal and opposite, resulting in zero net current flow, and the potential falls somewhere between the equilibrium potential for the individual reactions. In the event that more than one reaction occurs on the same electrode, and that one reaction dominates the others (the current produced or consumed by the other reactions in negligible), then the mixed potential approaches the equilibrium potential of the dominant reaction. However, in many instances, the mixed potential lies far from the desired equilibrium potential, with the consequence that energy density of the battery is significantly reduced. Such a case is the aluminum/air battery, in which parasitic hydrogen evolution on the aluminum anode reduces the potential from the equilibrium values -2.4 V_{SHE} to about -1.6 V_{SHE} when measured in 4 M KOH at 25°C.

Activity The term of *activity* must be well understood because many thermodynamic equations, such as the Nernst equation, are strictly applicable only to ideal systems.

When a solid or a liquid is placed in an enclosed space, it evaporates to some extent. As the partial pressure of the vapor increases, the number of molecules that return to the surface of the solid or liquid also increases until finally, the rate of molecules evaporating and condensing on the surface reach an equilibrium. The vapor pressure at equilibrium is the vapor pressure of that substance. Each component of a solid or liquid has a certain vapor pressure, which is less that the vapor pressure of the pure substance at the same temperature. The other substances in the system also occupy an area of the surface, thus the rate of molecules passing to vapor phase is smaller, and the corresponding vapor pressure of each substance is smaller than if the system was pure.

When dealing with mixtures, we are interested in knowing the weight, the volume, the number of moles of each component of the mixture, and the total volume and total weight of the mixture. *Mole fraction* is defined as the ratio of mole of a substance to the ratio of mole of all components of the mixture. Francois Marie Raoult (1886) found that the partial pressure of each component (A), p_A, of a mixture is nearly equal to the mole fraction, X_A, of that component times the vapor pressure of the pure component (p_A^o), (i.e., $p_A = X_A P_A^o$). In fact, Raoult's law is a good approximation if the different molecules forming the mixture have similar size and polarities.

Solutions that closely follows Raoult's law are called *ideal solutions*. A given substance in solution can exhibit a positive or a negative deviation from Raoult's law, depending if the forces between molecules is weaker or stronger than for those forces acting between similar molecules. A way to measure the deviation from Raoult's law uses the concept of activity, *a*. Activity is expressed as the ratio between the partial pressure of a substance, *A*, in a mixture, p_A, to its partial pressure of the same pure substance, p_A^o (i.e., $a_A = p_A/p_A^o$). It follows, that for ideal solutions ("ideal" being that which follows Raoult's law), the activity is equal to the mole fraction, X_A. Obviously, all substances belonging to a mixture for which size and polarity are similar (independent of the molar fraction) will follow closely Raoult's law; but it also can be expected that all substances, independent of size and polarity, belonging to a mixture in very small proportions (dilute solutions) will follow closely Raoult's law. This statement was made by William Henry in 1804, and it is known as *Henry's law*: in a dilute solution, the vapor pressure of the solute is proportional to its mole fraction, $p_A = k_A X_A$; k_A is a proportionality constant and varies with temperature. Furthermore, in a dilute solution the molarity and the molality, m_A, of a substance are proportional to its mole fraction ($m_A = \sigma X_A$, σ is a constant that depends on the identity of the solvent). Consequently, the vapor pressure of the solute in a dilute solution, p_A, is also proportional to its molarity or molality (i.e., $p_A = (k_A/\sigma)m_A$). Accordingly only, a_A is equal to $[k_A/(\sigma p_A^o)]m_A$, where $[k_A/(\sigma p_A^o)]$ is a constant depending on the temperature and the natures of substance A. If the solution is diluted $[k_A/(\sigma p_A^o)] = 1/\sigma \sim 1$, and only for dilute solutions can concentrations be substituted for activities.

Also, activity can be defined as a parameter that represents the "thermodynamic" concentration that is required for the species to obey an ideal thermodynamic relation, as is also defined as:

$$a = \gamma(m/m^o) = [k_A \sigma^o/(\sigma k_A^o)] \, (m/m^o) \tag{5.29}$$

where m is the molal concentration, m^o is the molal concentration of the species in the standard state, and γ is the activity coefficient. The standard state for a pure substance is defined as the pure substance in its most stable state at the temperature of interest.

Activity coefficient   From a thermodynamically point of view, the activity coefficient, γ, is defined as the quantity by which the ratio m/mo must be multiplied by for the chemical potential of the species of interest (any specie) to obey the ideal law:

$$\mu = \mu^o + RT \, Ln[\gamma m/m^o] = \mu^o + RT \, Ln(a) \tag{5.30}$$

where a $(= \gamma m/m^o)$ is the activity, μ is the chemical potential, μ^o is the chemical potential in the standard state—i.e., it is the value that m must be multiplied by so that μ varies linearly with $Ln(\gamma m)$.

The activity coefficient of a single ion cannot be measured because we cannot measure the chemical potential of the same species. However, using various electrostatic theories, most notable the Debye-Huckel theory, it is possible to estimate single ion activity coefficients.

The activity of a dissolved gas can be measured by determining the partial pressure of the gas above the solution, then using the deviation from the ideal Henry's law to determine the activity coefficient. Likewise, the activity of the solution can be measured by determining the vapor pressure of the solvent and asserting the deviation from the ideal of Raoult's law. The determination of the activity coefficient of a salt can be affected with a variety of techniques, including isopiestic measurements, and electrochemical cells, to name but two. In all cases, one seeks to detect and ascertain the extent of deviation from the system from the ideal.

Fugacity   The concept of *fugacity*, *f*, is sometimes used as a means of understanding the concept of activity. The activity in a given phase is defined as the ratio of the fugacity of the given component to it standard fugacity:

$$a = f/f^o \tag{5.31}$$

For a gaseous mixture, the ratio of fugacities can be replaced by the ratio of partial pressures to a very good approximation. Similarly, for a liquid or solid, the fugacity can be approximated as the vapor pressure ratio, with the standard state being defined as the pure substance.

Unless the total pressure is greater than 100 atm, the activity of the standard state of a pure liquid or a pure solid phase can be considered equal to unity. For a solvent in a solution, the fugacity of the standard state, f^c, at a given temperature is usually defined as the fugacity of a pure solvent at that particular temperature and at a pressure of 1 atm.

The origin of the concept of fugacity can be traced to deviations of real gaseous systems from ideal thermodynamic laws:

$$\mu = \mu^o + RT \, Ln(f) \tag{5.32}$$

$f = \gamma p/p^o$, where γ is the fugacity coefficient, p is the partial pressure and p^o is the total pressure on the gaseous mixture. The quantity μ is the chemical potential. It is defined as the partial derivative of the Gibbs energy, G, with respect to the number of moles (n_i) with all the other variables held constant. The quantity μ^o is the standard chemical potential, and it is a constant independent of p^o, n_i, and p, but depends on temperature.

Thus, in a manner that is analogous with that used in defining activity and the activity coefficient, the fugacity coefficient, γ, is the quantity by which the partial pressure must be multiplied so that μ varies linearly with $Ln(\gamma p/p^o)$ at constant temperature. As with the activity coefficient, the fugacity coefficient can be calculated from a knowledge of intermolecular interactions as embodied in an equation of state.

Electrochemical series Another historically important concept in electrochemistry is the electrochemical series. When an element is put in contact with a solution, a potential (not necessarily the equilibrium potential) can be measured using a reference electrode. The elements are then arranged in order of decreasing potential, with the most positive at the top of the list and the most negative at the bottom, corresponding to the most noble and the most active, respectively. The farther apart elements appear on the chart, the greater the energy available in the reaction involving the two couples. The utility of the electrochemical series is limited because the results are a function of the pH of the solutions. Mixed potentials often control the measured potential because many elements can establish redox reactions with more than one species. In this case, the measured potential cannot be easily related to the thermodynamics of the system. In any event, the electrochemical series is a useful concept in classifying the electrochemical properties of real systems.

5.2.3 Cells Containing Two Electrolytes Separated by a Membrane

An electrolyte is a substance that, when dissolved in water or some other liquid, produces ions that enhance electrical conductivity. The principle of electroneutrality applies in a solution; when a compound is dissolved, the same number of positive and negative charges are formed. If the electrolyte dissolves completely in the solution, it is said that the electrolyte is "strong." If the electrolyte does not dissolve completely, it is termed a "weak" electrolyte. When electrolytes decompose, they achieve a state of equilibrium. The equilibrium constant is the ratio of product activities divided by the reactant activities (e.g., by using Reaction 5.3, W is defined in Eq. 5.4). The equilibrium constant can be calculated from the standard Gibbs energy change for the reaction.

The overall electrochemical reaction in a cell (reactions occurring at the anode and cathode) requires the transport of ions from one side of the cell to the other. However, in many cases, it is undesirable to have the anolyte (the solution in contact with the anode) being in contact with the catholyte (i.e., the solution in contact with the cathode). This normally occurs when the anolyte contains a specie that might react directly with the cathode or the catholyte might contain a specie that might react with the anode. In these cases, the two electrolytes are separated by a membrane that might simply provide a physical barrier (a separator) or might allow the passage for only specific ions (a semi-permeable membrane). That membrane must be selected so that it will permit the desired ions to pass while preventing the passage of the undesired ion. The presence of a permeable barrier (a separator or a membrane) will greatly increase the electric resistance in the ionic conductor. The resulting undesirable barrier resistance gives rise to an IR potential drop, as well as to a junction potential.

To understand the cause of this latter potential loss, consider two aqueous electrolytes separated by a membrane. Assume that a membrane separates two aqueous solutions containing different concentrations of LiBr. Ions pass through the barrier according to their concentration gradients, not their size. If there is a difference between the ion concentrations on the two sides of the barrier, ions will move from the higher to the lower concentration side. However, not all ions move with the same speed; in this example, Br⁻ ions move about twice as fast as Li⁺ ions. As a result, the low-concentration side of the solution will develop a negative charge, relative to the high concentration side, resulting in potential drop. The difference in potential will depend on the concentrations at the membrane interface and on the mobility of the ions.

Metal ions transported across artificial membranes can be separated into two general classes: passive or active transport. In passive transport, the ions are transported by diffusion because of concentration gradients. In active transport, the cations are transferred by coupling with the movement of mobile protons and electrons. Proton-driven transport represent the most familiar system. It is based on a large change in cation-complexing ability of synthetic organic compound that exhibit selectivity to a given ion over others. The overall action can be seen as a coupled movement of the metal ion and a proton ($[H^+]$) across the membrane in opposite directions. Thus, a pH (= $-\log_{10}([H^+])$) gradient can be utilized to transport metal ions against its concentration gradient.

5.3 ELECTROLYTES

Several types of electrolytes can be used in batteries. The capacity and the discharge current of the battery will depend on the type of electrolyte chosen. Figure 5.8 shows the characteristic range of batteries as a function electrolyte type.

The electrolyte is a very important component of the cell. It is the medium in which the reactions occur. Ions are transported from one electrode to the other, and occasionally participate in the electrochemical reaction. A very important electrolyte characteristic is the specific conductivity. The ohmic resistance of the cell is closely related to the ohmic resistance of the electrolyte, which is related to the electrolyte conductivity.

The electrolyte can be a liquid, gas, or solid, or exist as a two-phase system as organic or inorganic salts (see Figure 5.9).

5.3.1 Solid Electrolytes

Demands for miniaturization of a system yield to new battery concepts and designs. Extremely thin batteries have been developed to meet industrial applications and requirements. Thin or paper (~50 Mm) batteries use solid polymer electrolytes. The development of thin-battery technology depends on research involving suitable polymer materials and understanding interface problems.

The projected values for this technology exceed specific energies of 200 Wh/Kg, and possess a cycle life above 500 cycles. Operating temperatures between –40°C and 250°C seem possible for these batteries. The typical conductivity of the electrolyte is about 10^{-3} to 10^{-7} (ohm-cm)$^{-1}$. The higher possible values of conductivity's (above 10^{-3}) are desirable for producing high-energy systems. Because high conductivity is an essential quality in this type of electrolytes, several strategies have been developed to improve the conductivity of these systems at low temperatures.

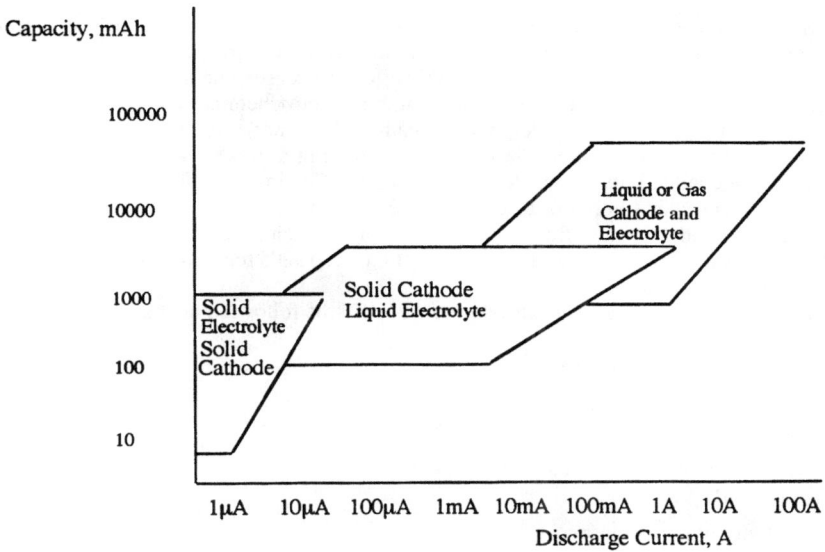

FIGURE 5.8 Capacity versus discharge current for several types of cathodes.

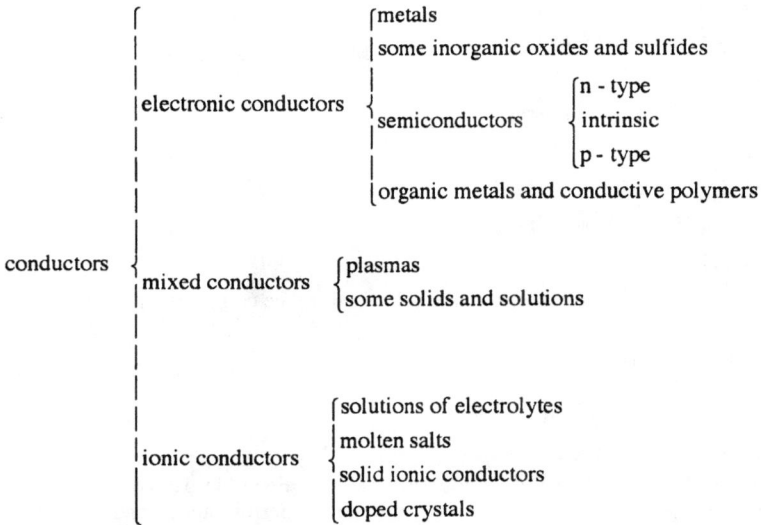

FIGURE 5.9 Generic type of electrolytes.

Strategies that attempt to improve conductivity in a solid electrolyte include plasticizing the polymer by the use of a specific salt or solvent. However, solvents used to increase the conductivity of the electrolyte often decreases the life of the system. Increasing conductivity often yields unfavorable electrochemical and photochemical reactions at the interfaces (anode/polymer, cathode/polymer) and detrimentally affects the overall battery life. Researchers have been putting a great deal of effort into increasing the conductivity of the electrolyte without affecting the other properties of the system. As Owens[11] summarized, great progress has been done in improving the solid electrolyte conductivities from 10^{-5} S/cm for a typical cell (Ag/V_2O_5) in the 1950s/1960s to conductivities of 10 to 0.5 S/cm in the 1980s for the electrolyte $RbAg_4I_5$ in a typical cell Ag/Me_4NI_5.

In general, an ideal solid electrolyte should fulfill the following characteristics:

- High ionic conductivity (10^{-1}–>10^{-2} S/cm)
- Negligible electronic conductivity (~0)
- Stability across electrodes
- High decomposition potential
- Thermal stability
- Selectivity of ionic transport
- Easy fabrication
- Mechanical suitability
- Low cost

The solid electrolyte technology offers advantages such as:

- Solid-state construction
- Stability over a wide range of temperatures
- Safety
- Easy manufacturing processes
- Flexible shape factor
- Rugged construction
- Leak-proof hermetic-sealed systems
- Fast kinetics
- The electrolyte is not consumed in the cell reaction; the composition remains constant

The disadvantages are:

- Shape change with cycling
- Electrode/electrolyte energy losses
- Aging of the electrolyte/electrode interfaces (i.e., electrode passivation, or dendrite growth on the polymer, reflecting on the reliability, performances, and life of the cell)
- Contamination of the cell as a result of impurities during fabrication

Among the most promising secondary batteries (which use solid electrolyte) are those that use lithium anodes. Table 5.4 summarize some lithium-based battery properties.

5.3.2 Fluid Electrolytes

Fluid (liquid and gas) electrolytes have the following advantages:

- High specific energy density
- High kinetic activity
- No shape change problems
- Electrodes can be easily regenerated (in secondary batteries)

Among the main disadvantages:

- Self discharge
- The active specie is usually highly corrosive
- It may require an ion selection membrane
- Electronic or ionic insulators might form on the collector surfaces
- Container leakage

5.4 ELECTRODES

The cathode and anode that "sandwich" the electrolyte are called *electrodes* (anode and cathode). The properties of good electrodes are:

- Low equivalent weight
- Low equivalent volume
- High electrochemical character (see electrochemical scale)
- Good stability against the electrolyte
- High electronic conductivity
- Some ionic conductivity at the interface oxide film

In advanced batteries, specific energies values lie over a wide domain, as illustrated in Figures 5.10 and 5.11. Among the batteries that promise higher energies and power are the lithium-based batteries.

5.4.1 Anode

A good choice in anode material would be an electrode that presents a very negative standard potential, with respect to the cathode. The theoretical energy output is limited by the

TABLE 5.4 Ionic Conductivity (σ) and Activation Energy (E_a) for a Number of Li$^+$-Based Solid Electrolytes

Solid Electrolyte	Form[a]	σ at T (Ω^{-1} cm^{-1})	E[b](eV)	T-range[c] (°C)	References[d]
Li$_3$N	sc	1.2 × 10^{-3} at 27° C	0.26	20–200	(1–3)
	sc	1 × 10^{-5} at 27°C	0.45	20–250	(1–3)
	pc	1.5 × 10^{-3} at 27°C	0.26	15–200	(4)
	pc	3.7 × 10^{-8} at 25°C	0.56	25–520	(5)
	pc	2.0 × 10^{-4} at 25°C	0.16	25–180	(5)
	pc	—	0.53	—	(6)
	pc	—	0.55	—	(7)
Li$_3$N-LiI-LiOH	pc	1 × 10^{-3} at 25°C	0.21	–50–300	(8)
Li-β-Al$_2$O$_3$	sc	1.3 × 10^{-4} at 25°C	0.16	–100–180	(9)
		2.3 × 10^{-3} at 200°C	0.31	180–800	(9)
(Li Na)β-Al$_2$O$_3$	sc	1 × 10^{-3} at 25°C	—	—	(10)
LiI	sc	3.1 × 10^{-3} at 350°C	1.05	340–420	(11)
	pc	1 × 10^{-7} at 27°C	0.36	25–150	(12)
		~9 × 10^{-4} at 285°C	0.92	250–350	(12)
	pc	5.5 × 10^{-7} at 27°C	0.43	—	(13)
	pc	1.3 × 10^{-6} at 60°C	0.44	25–200	(14)
	pc	~1 × 10^{-7} at 50°C	—	—	(15)
LiI·H$_2$O	pc	1.6 × 10^{-5} at 27°C	0.60	25–120	(20)
LiI·2H$_2$O	pc	2 × 10^{-6} at 50°C	3.67	40–70	(15)
LiI·3H$_2$O	pc	2.8 × 10^{-6} at 50°C	1.07	20–60	(15)
LiBr	sc	1.4 × 10^{-3} at 450°C	1.29	440–450	(11)
LiBr + •55%Mg^{++}	sc	3.8 × 10^{-3} at 300°C	0.39	270–540	(11)
LiBr	pc	1.6 × 10^{-9} at 30°C 10^{-5} at 350°C	0.56	30–300 350–500	(12) (12)
LiCl	sc	6.58 × 10^{-4} at 500°C	1.47	480–570	(11)
	pc	~7 × 10^{-4} at 500°C	1.42	420–550	(12)

Material	Method	Conductivity		Temp. range	Ref.
LiI·Al$_2$O$_3$(~40%)	pc	~7.4 × 10^{-9} at 72°C	0.59	30–350	(12)
	pc	1.7 × 10^{-7} at 182°C	0.42	182–375	(16)
		2.3 × 10^{-5} at 394°C	1.31	375–520	(16)
	DSES	1.2 × 10^{-5} at 27°C	0.44	—	(17)
	DSES	1 × 10^{-4} at 27°C	0.40	–50–300	(18)
		1 × 10^{-1} at 300°C	—	—	(18)
	DSES	5 × 10^{-6} at 27°C	0.44	25–200	(19)
		2 × 10^{-4} at 127°C	—	—	(19)
LiI·H$_2$O-Al$_2$O$_3$ (35%)	DSES	1 × 10^{-5} at 27°C	0.59	25–125	(14)
LiI·H$_2$O-Al$_2$O$_3$ (60%)	DSES	1 × 10^{-4} at 25°C	—	—	(20)
LiI·H$_2$O-SiO$_2$ (60%)	DSES	8.7 × 10^{-5} at 25°C	—	—	(20)
LiCl·Al$_2$O$_3$ (25%)	DSES	2.5 × 10^{-6} at 182°C	0.39	125–300	(16)
		6.9 × 10^{-4} at 394°C	0.51	300–440	(16)
LiCl-SiO$_2$ (25%)	DSES	2.3 × 10^{-5} at 182°C	0.32	170–330	(16)
Li$_2$SO$_4$(β)	pc	2.5 × 10^{-6} at 300°C	1.09	280–470	(21)
		1.5 × 10^{-3} at 500°C	1.98	480–560	(21)
Li$_2$SO$_4$(α)	pc	1.03 at 600°C	0.36	575–860	(22)
Li$_2$WO$_4$(α)	pc	1.43 at 700°C	0.36	684–738	(23)
Li$_{14}$Zn(GeO$_4$)$_4$	pc	0.2 at 400°C	0.19	250–400	(24)
(LISICON)		0.13 at 300°C	—	—	(24)
		1.25 × 10^{-2} at 300°C	—	—	(25)
Li$_4$SiO$_4$-Li$_3$PO4(40%)	pc	2 × 10^{-6} at 50°C	0.52	50–300	(25)
	pc	~1 × 10^{-4} at 100°C	0.50	310–510	(26)
LiAlCl$_4$	pc	1.2 × 10^{-6} at 25°C	0.44	25–146	(27,28)
Li$_5$AlO$_4$	pc	2.3 × 10^{-7} at 200°C	1.20	120–380	(29)
Li$_5$GaO$_4$	pc	7.5 × 10^{-7} at 200°C	0.85	120–380	(29)
Li$_5$ZnO$_4$	pc	1.2 × 10^{-6} at 200°C	0.71	120–380	(29)
Li$_4$B$_7$O$_{12}$Cl$_{.68}$Br$_{.32}$	sc	6.5 × 10^{-4} at 200°C	0.49	50–230	(30)
Li$_8$Zr$_{1.8}$Ta$_2$P$_3$O$_{12}$	pc	1.5 × 10^{-3} at 200°C	0.40	85–230	(30)

TABLE 5.4 Ionic Conductivity (σ) and Activation Energy (E_a) for a Number of Li$^+$-Based Solid Electrolytes *(Continued)*

Solid Electrolyte	Form[a]	σ at T (Ω^{-1} cm^{-1})	E_a[b](eV)	T-range[c] (°C)	References[d]
Li$_{2.25}$C$_{.75}$B$_{.25}$O$_3$	pc	~1 × 10^{-3} at 200°C	0.54	70–230	(30)
LiHf$_2$(PO$_4$)$_3$	pc	1.1 × 10^{-3} at 200°C	0.39	85–230	(30)
Li$_{3.75}$Si$_{.75}$P$_{.25}$O$_4$	pc	~1 × 10^{-3} at 200°C	0.57	30–230	(30)
Li$_{3.4}$Si$_{.7}$S$_3$O$_4$	pc	~1 × 10^{-3} at 200°C	0.53	70–230	(30)
LiI-Li$_2$S(37%)-P$_2$S$_5$(18%)	glass	~10^{-3} at 25°C	0.31	25–120	(31)
Li$_5$AlO$_4$	glass	~2 × 10^{-3} at 227°C	0.50	25–300	(32)
Li$_5$GaO$_4$	glass	~9 × 10^{-4} at 227°C	0.52	25–260	(32)
Li$_5$BiO$_4$	glass	3.8 × 10^{-4} at 227°C	0.48	25–230	(32)
LiTaO$_3$	glass	1.3 × 10^{-3} at 227°C	0.38	25–300	(33)
LiNbO$_3$	glass	8.1 × 10^{-4} at 227°C	0.40	25–300	(33)
LiNbO$_3$ + 10%WO$_3$	glass	1.9 × 10^{-3} at 227°C	0.39	25–300	(33)
Li$_2$GeS$_3$	glass	4 × 10^{-5} at 25°C	0.51	—	(34)
Li$_2$O-LiF-B$_2$O$_3$-Li$_2$SO$_4$-Li$_2$SO$_3$ (composition (mole%))					
8·8 8·6 30·6 17·0 15·0		3.5 × 10^{-4} at 200°C	0.57	27–170	(35)
8·8 13·6 30·6 17·0 15·0		~6 × 10^{-4} at 200°C	0.55	25–170	(35)
8·8 18·6 30·6 17·0 15·0		8.5 × 10^{-4} at 200°C	0.54	≥25	(35)
4·8 25·5 31·7 15·7 12·3		1.3 × 10^{-3} at 200°C	0.54	≥25	(35)
Poly (ethylene oxide)-LiSCN PEO-LiSCN (%Li~5)	pmer	~2 × 10^{-6} at 60° C	Non Arrhenius	40-150	(36)
Poly (propylene oxide)-LiCF$_3$So$_3$	pmer	~2 × 10^{-5} at 60°C	Non Arrhenius	40-150	(36)

a sc, pc, DSES and pmer refers respectively to single Crystal, polycristal, dispersed phase solid electrolyte, and polymer.

b Activation energy E_a quoted above is derived from log σ vs $1/T$ plot. Literature values obtained from log σ vs. T plots E'_a using the relation $E_a = E'_a - kt$ (Shahi, 1977) where k is the Boltzman constant and T the average temperature in Kelvin.

c (1) VonAlpen et al. (1977); (2) VonAlpen et al. (1979); (3) Rabenau (1978); (4) VonAlpen et al. (1979); (5) Boukamp and Huggins (1976); (6) Masdupuy (1957); Bishop et al (1966); (8) Obayashi et al. (1981 a,b). (9) Whittingham and Huggins (1972). (10) Farrington and Roth (1977); (11) Haven (1950); (12) Ginnings and Phipps (1930); (13) Schlaikjer and Liang (1971); (14) Pack et al. (1979); (15) Rudo et al. (1980); (16) Li-Chuan et al. (1980); (17) Liang (1973 b); Liang et al. (1978); (19) VonAlpen and Bell (1979); (20) Skarstad et al. (1980); (21) Shahi (1883); (22) Kvist and Lunden (1965); (23) Kvist and Lunden (1966); (24) Hong (1978); (25) VonAlpen et al. (1978); (26) Hu et al. (1976); (27) Weppner and Huggins (1977); (28) Weppner and Huggins (1976); (29) Huggins (1977); (30) Shannon et al. (1977); (31) Robert et al. (1980); (32) Glass and Nassau (1980); Glass et al. (1978); (34) Souguet et al. (1980); (35) Boehm and Angell (1979); (36) Armand et al., (1978); (37) Armand et al (1979).

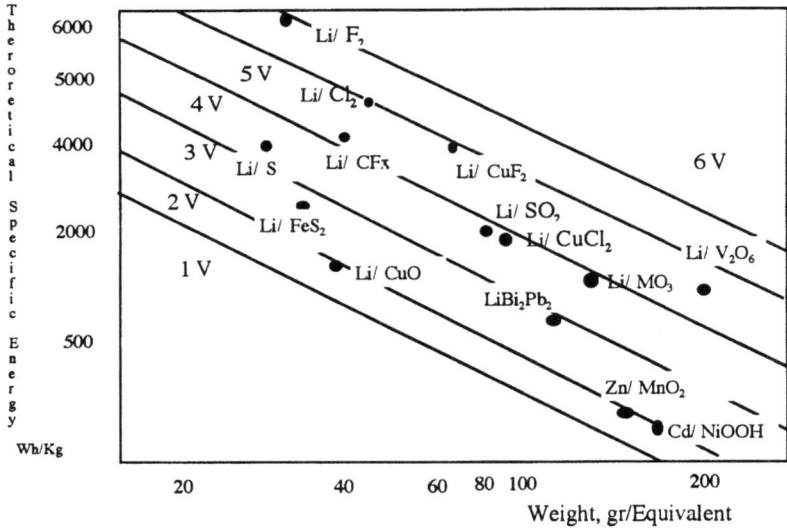

FIGURE 5.10   Theoretical specific density versus weight of several commercial systems.

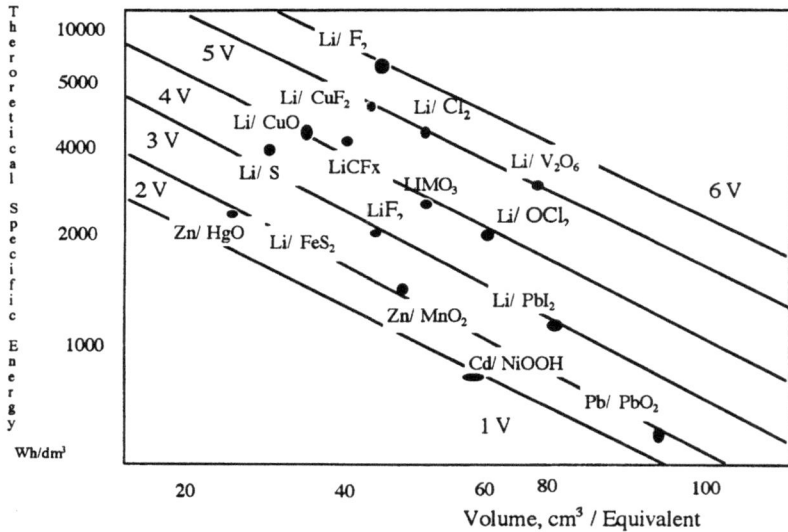

FIGURE 5.11   Theoretical specific energy versus volume of commercial systems.

potential difference between anode and cathode. Suitable properties of an anode/electrolyte system involve a one-step electron reaction, with a nontoxic, low-cost material, and low weight. High specific energy (energy by unit of weight or volume) depends upon a large natural potential difference between the cathode and anode and the weight or volume of the system. Lithium possesses the most negative reduction potential of all existing metals; accordingly, it constitutes the anode material that loses electrons most easily to form positive ions and yields the greatest energy per unit of weight (theoretical gravimetric energy ~11,300 W-hr/kg).

5.4.2 Cathode

The most electropositive materials are the halogens, and all oxidizers can be, theoretically, used as cathode material because of their strong affinity for electrons. In many cases, halogens result in an ineffective cathode material because of their high toxicity and high reactivity with other materials used in the battery system (corrosion of containers). Metals are a more common choice. Metals more commonly used as the negative plate are Fe, Cd, and Zn. In general, Zn is widely used because of its low weight and price, compared to Cd. Oxygen is known as an excellent oxidizer; accordingly, it is an excellent cathode material. The so-called "air-cathode" is a design cathode that uses oxygen from the air as a cathode.

5.5 LOSSES IN BATTERIES

A general concern when using batteries is the efficiency of the system. Efficiency is closely related to the losses (internal and external) occurring in the battery system. The real performance of a given battery is separated from the theoretical performance by the losses.

The experimental performance of a battery can be very different to the theoretical performance because of internal and external losses.

- Internal losses refer to the losses due to heat, parasitic reactions, internal impedance of the cell, and losses occurring at the electrode/electrolyte interfaces.
- External losses refer to losses occurring outside the electrodes/electrolyte system.

The Nernst equation provides a means to calculate the theoretical equilibrium potential of an electrode or of a cell (two electrodes) when one or several reactions are occurring at the electrodes surfaces. We also reviewed the means to calculate the specific energy (Figure 5.7) from the calculation of the equilibrium potential. However, the results do not account for losses, which can occur at several places in the cell. The departure of the theoretical cell terminal can be caused by any of the following internal loses:

- I-R losses in the bulk of the electrolyte and in the leads
- Losses associated with interfaces
- Heat losses

The performance of the cell will depend on how energetically the reactions are occurring at the electrode interfaces, how fast those reactions occur, and how well the ions transport through the electrolyte. The slowest process will dictate the overall cell behavior. Therefore, it is important to identify the process that control the cell behavior to prioritize the importance of losses that affects the overall behavior of the cell.

5.5.1 I-R Losses

The calculation of I-R drop is simple if the current is distributed uniformly. For example, if a cathode is separated from an anode by an electrolyte through distance x, if the exposed area of the electrodes is A, and the specific conductance of the electrolyte is σ, the I-R drop of current I passing through the cell is:

$$\eta_{ir} = I\,x/(\sigma A) \tag{5.33}$$

Problems arise in porous cathodes, (example MnO_2 and C) because high E_{ir} drops can occur within the pores, or for very complicated geometries, where a clear definition of x and A is difficult to obtain. In these cases, estimation of the I-R loss is frequently only approximate.

5.5.2 Losses Associated with Interfaces

In battery research, establishing the type of overpotential that is present at a given electrode can be worthwhile. This classification can be used as a guide to optimize the performance of the battery. Knowing the limiting problems of a battery makes it possible to address those difficulties, in a rational, scientific manner.

Overpotential losses   The losses associated to electrodes interfaces can be related to the overpotential (or overvoltage, or electrode polarization), η. *Overpotential* is defined as the difference between the measured electrode potential, E, and the open-circuit potential for that electrode (which is defined as the potential at which the sum of all partial currents, including metal dissolution and hydrogen evolution, for example, is zero). The factors contributing to the overpotential are:

Concentration overpotential   Concentration overpotential, η_c is the type of overpotential that is associated with the increase in concentration of the products, and the decrease in concentration of reactants, at an electrode surface because of the limited rate of transport.

A potential barrier caused by the concentration gradient can be formed at the electrode surface, yielding the overpotential. Concentration overpotential can occur in the cathode or in the anode, or in both electrodes. Typical problems resulting in concentration overpotential are:

- Restricted mass transport because of the existence of a stagnated electrolyte layer at the electrode surface
- Restricted transport through a reaction product film on the electrode surface
- The presence of a gelled electrolyte

The problem, which is not well understood, can only be minimized by increasing the rate of mass transport of the species in question to or from the surface, or by increasing mass transport by convection, or by reducing the thickness of electrolyte. This might require imposition of forced convection or decreasing the viscosity of the electrolyte.

Activation overpotential   Activation overpotential is denoted by η_A. This type of overpotential loss occurs when the rates of the reactions at the electrodes are limited by charge-transfer processes so that a large voltage across the interface is required to cause the reaction to proceed at the desired rate. Overpotential arising from the irreversibility of a charge transfer step is one of the most common types of activation overpotential. In general, this problem will be more likely to occur if the reaction at the electrode/electrolyte interface involves more that one-electron transfer. Another loss of this type occurs when two competitive reactions occur in the same interface simultaneously (in a uniform manner or on separate zones or patches of the electrode surface) so that the electrode exhibits a mixed potential, rather than the thermodynamic reversible potential.

For example, corrosion losses of this type occur in Al/Mg systems; the OCP of aluminum, and magnesium in acid solutions are remote to both the reversible metal dissolution and the hydrogen-electrode reaction potentials. Accordingly, it is not likely that the reaction at the anode can be approximated by the I/V curve for metal dissolution, centered on the reversible potential, perhaps because an oxide is formed on the surface of the metal or because the suppression of the hydrogen evolution reaction is difficult, as a result of the ubiquitous presence of water.

In the case of active anodes (e.g., Al, Mg, Zn), hydrogen evolution causes the mixed potential to be displaced significantly from the reversible M/M^{n+} potentials, thereby causing significant losses. The hydrogen reaction is generally less reversible on less active metals.

Sometimes it is beneficial to amalgamate the anode to suppress the hydrogen evolution at the OCP, and hence to reduce self-discharge and to displace the OCP of the anode in the negative direction. This is (or was) a common practice with Zinc anodes. Some organic and inorganic additives to the electrolyte have similar effects; however, they have to be selected judiciously to prevent interference with the metal dissolution reaction.

For example, if activation overpotential is the dominant loss, we would seek to improve catalytic properties of the electrode surface to speed up the reactions. Cell performance can also be improved by eliminating impurities (particularly if they inhibit the electrode reactions), or by increasing the contact area between electrolyte and electrodes.

5.5.3 Ohmic Overpotential

Ohmic overpotential is designed by η_o. Ohmic overpotentials are caused by the I-R drop through the electrolyte between the anode and the cathode, and sometimes through poorly conducting film at the interfaces. These are also referred to as *interfacial I-R drop*, to differentiate them from the IR losses that can occur in the bulk of the electrolyte. Electrode/electrolyte systems that develop a thick insulating film are generally poor energy sources. For example, the valve metals (Ta, Ti, W, Zr), although possessing high energy densities, are poor anodes because of their very low discharge rates and because of the IR losses across the oxide films that normally exist on their surfaces.

5.5.4 Heat losses

Heat loss refers to the heat generated in an electrochemical cell. The heat generated during a single reaction can be calculated using the following equation:

$$Q = \Delta H/n + FE \tag{5.34}$$

Q is the heat liberated (negative) or consumed (positive), E is the actual cell potential, ΔH is the enthalpy of the reaction, n equals equivalents of charge transferred from anode to cathode per mole of reaction, and F is the Faraday constant.

The enthalpy of the reaction during an electrochemical reaction is frequently measured using calorimeter techniques. Enthalpy can be easily calculated if the temperature coefficient of resistance [$= dE/dT$ (Volt/K)], the temperature at which the reaction takes place (T), and the *OCP* (open circuit potential) are known; it is calculated using:

$$\Delta H = -nF \left[OCP - T \times (dE/dT) \right] \tag{5.35}$$

n and F have been defined.

Heat losses can be caused by the heat delivered during the chemical reaction, the heat from internal resistance of the cell, and the losses caused by inefficiencies in the system resulting from the withdrawal of electrical energy generated by the cell.

The heating effect caused by the internal resistance of the battery is different from the heat produced during the electrochemical reactions (exothermic or endothermic). This heat represents a loss that adds up to the overall losses of the battery. This heat loss can be calculated by ($I_T^2 R_T t$)/4.18 (in cal/t), where t refers to time in seconds, I_T to the total current arrangement in amperes (one or more batteries can be considered). R_T is the total resistance arrangement in ohms. When an array of several batteries is used, the heating effect will become independent of the battery array (batteries in parallel or in series); however, it will be a function of the ventilation arrangements intrinsic to the layout design. The design will depend on the current and voltage requirements.

5.6 MEANS TO IDENTIFY TYPES OF LOSSES

In battery research, establishing the type of losses and quantifying those losses can be of great importance in optimizing the battery performances. Several techniques are used to study electrode process and losses (Table 5.5). Electrode study techniques aim to indirectly or directly measure the rate constants and the kinetic parameters of the reactions occurring at the interfaces. The rate constants provide a clear idea of the determining reaction

TABLE 5.5 Electrical Methods for the Study of Electrode Processes

General class	Specific method	Maximum values[a] standard rate constant (k_s) (cm/sec)
Steady state[b]	Direct potentiostatic	10^{-2}
	Direct galvanostatic	10^{-2}
	Indirect galvanostatic	10^{-2}
Non steady state	Voltage-step function	1
	Single current-step function	1
	Double current-step function	10
	Charge step function	1
	Linear sweep voltammetry and polarographic	10^{-1}
	a-c impedance measurements	some 10^{-1} – some 10^2
	Faradaic rectification	10 - 100

a Estimates indicate highest values of the standard rate constant k_s which can be determined for a reaction of the form A $\rightarrow$ B + e$^-$ with approximately equal concentrations of reactant (A) and product (B).

b Estimates of k_s for steady-state methods are applicable for forced convection of a conventional type.

step that dominates the cell behavior. By estimating their values, you can be able to determine the process that dominates the overall behavior of the battery (the slowest one).

Neglecting time and budget considerations, a research technique can be selected based on the precision and accuracy to measure the rate constants occurring at the electrodes. Table 5.7 lists the measurable rate constant ranges available with different techniques. The techniques can be classified as steady-state techniques and nonsteady-state techniques. The steady-state techniques are more commonly used because of their simplicity and the cost efficiency of the equipment needed to perform the study. Among nonsteady-state techniques, ac impedance is the more popular available.[12–14]

Additional techniques can be used to evaluate changes and degradation that occur in a reacting cell. Among those techniques are:

5.6.1 Electron Microscopy

Information on composition, morphology, concentration profiles of elements on electrodes can be obtained from electron microscopy. Two main types of electromicroscopes are:

- Scanning electron microscopy
- Transmission electron microscopy

5.6.2 X-Ray Diffraction and Neutron Diffraction

X-ray diffraction and neutron diffraction are two methods used for substance and phase identification. Diffraction patterning to find crystallographic information is also used. Extended X-ray absorption is able to explore fine structure (EXXAFS).

5.6.3 Electron Spectroscopy

Quantitative information on covalent and ionic compounds formed on the electrode surface can be obtained by using this technique. Two main types are:

- Secondary ion mass spectroscopy (SIMS)
- Infrared Raman Spectroscopy

5.6.4 Thermogravimetric Analysis (TGA) and Differential Thermal Analysis (DTA)

Thermogravimetric analysis (TGA) and differential thermal analysis (DTA) are used to study structural phase changes (for example, from solid to liquid, then solid again).

5.7 BATTERY TYPES

Cells are currently available with a voltage range about 1.5 to 3.9 V; with capacity ranging approximately between 1 mAh to 20,000 A; miniature cells to considerable-volume cells; life spans from few days to several years; and with operational and storage temperatures ranging from –70°C to +400°C. Electrolytes, ranging from gas to solid and from inorganic to organic, are used in batteries, and with electrodes of all matter states.

High-performances batteries (MW range) are obtained with liquid or gas cathodes (dissolved cathodes) and organic and inorganic liquid or gas electrolytes. Low to medium energies (W range) are obtained with solid metallic cathodes, and organic liquid electrolytes. Low-energy application cells (MW range) are obtained using solid cathodes and solid electrolytes (see Figure 5.8). Among all available systems, the lithium-systems are the most salient, with respect to the gravimetric power and energy that they produce.

In general, batteries are not classified by the type of cathode or electrolyte they are made of, but by whether or not the electrode is consumed. This is equivalent to stating that the battery is classified by its need to be replaced with a new system or not. Those systems for which the electrodes are consumed are called *primary cells* or *primary batteries.* Systems for which the electrodes are not replaced and for which the reactions occurring at the electrodes can be switched back and forth (a cycle) during charge and discharge are called *secondary cells* or *batteries.*

5.7.1 Primary Batteries

Nonrechargeable cells are characterized by a simple design and generally they will not include any electrical moving part. The ranges of power, size, and type of applications of most typical primary batteries are shown in Table 5.6. Primary batteries are generally high-energy batteries. The term *high-energy batteries* applies to primary batteries with a minimum energy density of 100 W-hr/lb.

The main research area in battery design is devoted to increase their power and energy density (gravimetric or volumetric) and to control the losses. The design of batteries of this type constitutes an art that involves technology, engineering, science, and a great deal of good luck expressed though imagination, and knowledge. Much of this research work is experimental and must be performed in the laboratory under a well-controlled environment. It is highly recommended that the researchers involved in primary battery research have a solid knowledge of electrochemistry.

5.7.2 Secondary Batteries

Secondary batteries are cells for which the direction of the electrode reactions can be reversed by supplying electrical energy to the cell. This is known as *recharging.* An external dc source is used to drive the electrons from the cathode to the anode (the external source can be another battery). Such reversal behavior in the cell will happen when the voltage supplied to the cell exceeds the open-circuit value. Thus, secondary batteries are rechargeable. In general, they represent a more sophisticated type of design, compared to primary batteries.

Secondary batteries have a difference in behavior between charge and discharge, caused by the change in concentration of the species in the electrolyte or the surface modification in the electrodes. The difference is expressed as current efficiency. It is calculated as the ratio of the current produced during discharge to the current produced during charge. Current efficiency is also expressed as the energy efficiency, which is measured by the ratio of the discharge voltage-times-the discharge current to the charge voltage-times-the charge current.

TABLE 5.6 Summary of Some Commercial Primary Batteries and their Characteristics

System	Load Requirements	OCP V	Specific Capacity, Ah/dm³	Energy Density, Wh/dm³	Energy Density, Wh/Kg	Operating Temp. Range, °C	Life	Applications
Carbon-Zinc	a) 1.2 V aa) 6V, 0.3A-60 pulses/min ab) 12V, 0.3A; ac) 12V, 0.6A; ad) 9V, 0.025A; ae) 18V, 0.025; af) 6V, 0.2A;	a) 1.5 b) 1.7		a) 120-152 b) 120-250	a) 55-77 b) 60-110	a) -7 => +54 b) -40+54	-------- aa)1200h ab)8h after 12 months Storage; ac) 3h; ad) 200h@4h/day ae)200h@4h/day; af)20days@0.5h/day b) Self discharge 10%/year	------------ a)Roadside hazard lamp b)Intruder alarm Syst. c)Emergency fluorescent lamp d)Frequency meter e)Wide-range-oscillator f)Tape recorder
Mercury-Zinc	1.24 V a) 13.4V,0.02A b) 13.4V,10uA c) 1.34V,0.5A d) 13.4V,15mA 2h/day e) 1.25V,0.016-28 Ah	1.35		300-600	99-123	-40 => +60 e) -40=> +60	a) 100h b) 18months c) 1pulse/18months d) 50h e) 2 years	a)Laser beam Detector b)Solid state relay c)Detonator d)Radio microphon e)TV, walkie-talkies, watches, instruments, hearing aids, radios, etc.
Mercury-Cadmium	1.2V a) 15mA pulse	1.5		120-152	55-77	-40 => +70	------- a) 48h	a)Sealed metering
Mercury-Cadmium-Bismuth	1.17 V	-------		201	77	-20 => +90		
Zinc-Air	1.2 V	1.4		180-900	>220	-29 => +52		a)Radios, flash lights, shavers, toys,calculators, high-current drain, etc.
Mercury dioxide	a) 1.25V,0.1-23Ah	1.5		122-263	66-99	-40 => +54.4	a) 2 years	
Silver-Zinc	1.5 V a) 1.6V, 0.035-210Ah	1.6		400-550	110-126	-40 => +54	----	a)Hearing-aids, watches, instruments
Li Sulfur dioxide	2.75 V	2.9		420	260-330	-40 => +60	a) 2 years; 5% Self discharge	
Li₂S	---	2.27	1290	660	2930	-54 => 60		
LiV₂O₃	2.4	3.4			264			

TABLE 5.6 Summary of Some Commercial Primary Batteries and their Characteristics (*Continued*)

System	Load Requirements	OCP V	Specific Capacity, Ah/dm³	Energy Density, Wh/dm³	Energy Density, Wh/Kg	Operating Temp. Range, °C	Life	Applications
Lithium-Thionyl chloride	3.2-3.4 V	3.66		1080-3000	500-1500	-54 => +50	<1%/year Self discharge	
LiF		6.05	1050	6340				
LiCl		4.0	750	3000				
LiBr		3.5	690	2430				
LiTe		1.76	1150	2030				
Li₂O		3.0	1340	4020	800-1000		Efficiency ~30%	Mechanically rechargeable
LiMnO₄		3.5	890	3100				
LiTiS₂		2.14	560	1200				
LiO₂+Ag		2.24	920	2700				
Li₂S+Fe		a) 1.84 b) 1.75	a) 1410	a) 2590 b) 400	b) 150	-60=>+120		
Li-Sulphur-Dioxide		2.95 -3.1	400	235-524	275	-55=>+70	1%/year Self discharge	Cost per KWh $300.
Li-ThionylChloride	Maximum Continuous rate 1000 mA; nominal capacity rating 7Ah	3.7	900	432		-40=>+150	Very low self discharge per month (0.03%).	Cost per KWh $490.
Li-Sulfuryl-Cloride		3.9	900	400		-30=>+150	Very low self discharge per month (0.02%).	Cost per KWh $360.
Li-Manganese-Dioxide	>200	3.0-3.6	450-750	150-364		-20 =>+70	Self discharge 1%/year	Cost per KWh $360
Li-Copper-Oxide		1.5	550	240		-40=>+70	Very low self discharge per month (0.01%).	Cost per KWh $360.
Li-Water	Discharge Life 20 hours	1.25		4,000		0=>+30		Submarine applications
Zn-Alkaline-Manganese		1.5		150-270	9-20	-30=>+55	Self discharge <3%/year	Lantern
Al-Air		1.9		<500	<500	-40=>+60	<20% efficiency	Telecommunications
Li-Vanadium pentoxide		3.4		670	264	-54=>+54	Self discharge<1%	

Note: This table reflects a compilation of data reported by several authors; in some cases the range reported for similar system by several authors is listed. Sometimes the reported information seen incomplete.

The electrodes constituting a reversible cell (charge and discharge reactions are possible) are called *reversible electrodes*. The three main types are:

- A material in contact with an electrolyte of the same chemical composition. Electrodes of this type are reversible, with respect to the ions of the electrode material.
- A metal and an insoluble salt of the same metal, in contact with an electrolyte containing the ions that form the insoluble salt with the metal of interest:

$$M(\text{solid}) + X^- <=> MX(\text{solid}) + e^-$$

- An inactive metal in contact with an electrolyte containing both oxidizing and reducing agents.

Energy and power are important properties of secondary batteries. The term *high-energy batteries* applies to secondary batteries with a minimum energy density of 50 W-hr/lb.[17–19] Table 5.7 illustrates typical ranges of power, sizes, and principal characteristics of some commonly used secondary batteries.

The overall problem in secondary battery, in some cases, is the cycle life, which is reduced by irreversible changes that occur at the electrodes when the current is reversed. In other cases, it is the lack of reproducibility that can be obtained under identical fabrication conditions, which severely limits the reliability of secondary batteries. The main stream of research focuses on understanding the process and parameters that control the cycle life and the reproducibility of secondary batteries. This research includes both experimental and theoretical work, which is used to improve the understanding of interfacial problems that play key roles in determining battery performances and cycle life.

5.8 DEFINITION OF ENGINEERING VARIABLES COMMONLY USED

5.8.1 Basic Variables

Inefficiencies   An engineering word for defining losses.

Capacity   The capacity (or load capacity, K of a battery) is the quantity of electricity that a charged battery can deliver before it becomes exhausted. It is usually expressed in Coulomb-hours.

Batteries are classified by their ability to supply a certain amount of current in a certain period of time until the cut-off voltage is reached. This value is also called the *capacity* of the battery and it is given in mAh or Ah. The capacity of the battery is not a constant value and it depends on the load, temperature, design characteristics, and electrochemical characteristics of the system. Higher discharge currents result in smaller capacities. Given a typical application, a nominal voltage range, a load range and type of operation, the capacity can be determined. The type of application will dictate the system requirement and the battery-capacity needs.

Capacitance   The amount of possible stored charge in a cell or battery is proportional to the current and the time (as the integral of current × time) needed to accumulate charge Q. The amount of charge is proportional to the potential difference between electrodes. The constant of proportionality relating charge and potential difference is called the *capacitance* (sometimes also improperly called the *capacity*):

$$Q = -C\Delta V \tag{5.36}$$

TABLE 5.7 Summary of Some Commercial Secondary Batteries and their Characteristics

System	Life Cycle	Charge time,hr	Maximum Discharge Rate, C	OCP V	Energy Density, Wh/dm³	Energy Density, Wh/Kg	Operating Temp. Range, °C	Charge and Retention	Efficiency %	Applications and Remarks
Nickel-Cadmium, sealed	a) deep: 300-2000 shallow: 3000-5000; >1500@80%DOD b) 2500@100%DOD c)>500; 4-8 years	a) 1h@1.2V b) 10-1500Ah c) 0.1-10Ah	a) 15	a) 1.35 b)1.28 c)1.3	a) 37-91.5 b) 25-50 c) 50-120	a) theoretical 240 Actual 26-52 b) 15-35 c) 13-35	a)-40 =>+70 b)- 60=>+60 c)- 40=>+70	a) 60days 50%Capacity 27oC b) <5%/year Self discharge c) 10-25%/month Self discharge	a) 60-70	a) Initial cost high, reliable. High loss (12%)/100 cycles. High self discharge (100%l in 15 days) Expensive ~$1,000-2000.00/KWh
Pb-acid, sealed	a) deep: 100-300 shallow: 300-1000 b) 500 (50%DOD)	a) 14h, nominal 10h 1.90V@2h rate	a) 20	a) 1.3-1.35 b) 2.1	a) 43-85 b) 70	a) 18-33 b) 35	a) -40=>+60 b) 0=>+40	a) 1.5 years b)5 years; self discharge 2%/month	a) 75-80	a) Initial cost low, reliable. Low cost ~$250/Kwh. b) portable
Alkaline MnO₂, sealed	deep: 50 shallow: 50-100	nominal 14h;			36.6	theoretical: 336 Actual: 13-50	-20=>+70		60	
Ni-Zn	100-200	1h@1.6V	10	1.71	67-134	theoretical 374 Actual 33-80	-44 => +82	1 year	65	Low and high rate capabilities over wide temperature range. High loss (10%)/100 cycles. Expensive ~$5,000.00/KWh
Lead-acid, unsealed	a) 10-600; 500@80%DOD b) 200	a) 1h@1.8V	a) 20	a) 2.10 b) 2.1	a) 31-122 b) 80	a) 15-26 b) 30	a) -54 =>+54 b)-18=>+70	a) 90 days b)3-5 years; self discharge 203%/month	65	a) Inexpensive ~$200.00/KWh; reliable
Ni-Hydrogen	>1000@80%DOD		20	1.4	55	49-80	-20 =>+30	15%@10°C/72 hr; 1-3%/year (wet cell); 5-15% (dry cell)	65	Long life, reliable over years. Cost estimated per Kwh $50,000.
Silver-Hydrogen				1.5	60	90-100	+4 =>+27			High energy density over 0.5-1 year life
Zn-Chlorine	>1000@80%DOD				130	80-100			50-70	Very high density applications

422

Type	Cycle Life	Charge Rate	Self-Discharge	Voltage	Power Density	Energy Density	Temperature Range	Shelf Life	Efficiency	Comments
Nickel-Cadmium vented	1500 cycles@70%-80%DOD; 10,000 cycles @20% DOD		50	1.35	90-	40-62	+51=>+60		74	Cost per Kwh is estimated to be $1000-2000. Low self discharge.
Nickel-Metal Hydride	>1500 30,000 at 40% DOD 13,000 at 60% DOD		4	1.35	120-	60-80	-10 =>+40		80	Self discharge completely in 15 days. Cost per Kwh $3,500.00. Capacity loss ~4%/100 cycles
Nickel-Iron, vented ZnO_2; Zinc/Air Aluminum/Air			50	1.35 1.4 1.4	120 120 100	53 110 150	-10 =>+60 0 =>+45 +2= =>+60			Cost per KWh $500.00
Silver-Zinc, sealed	10-200	1.5h@1.5V	30-50	1.86	55-610	theoretical 440 Actual 37-220	-42 =>+73	1 year	70-75	Where maximum energy density and voltage is required 0.5-1 year. High loss (50%)/100 cycles.Cost ~$475-2,500.00/KWh
Ag-Cd, sealed	deep: 700 shallow: 3-5000	1h@1.1V	10-50	1.4	24-165	theoretical 310 Actual 37-250	-44 =>+71	2 years		High energy density, ability for voltage regulation 0.5-3 years. High loss (18%)/100 cycles. Cost.~650-3,300.00/KWh
Sodium/Sulfur			30	2.07	300	190	+350			Cost per Kwh $2,000.
Li-MoS$_2$	a) >50-200@100% DOD; 200 Ah rate of discharge b) 200 cycles, 10 years			a) 1.7-1.9 b) 1.1-2.4	b) 171	a) 214-421 b) 61	a) -20=>+55 b) -54=>+55		High efficiency	b) 5%/year Self discharge
Li-NbSe$_3$ Lithium-Thionyl chloride	>150-200			1.7-1.8		214-384				High rate energy density; long shelf safety; excellent storage capabilities

TABLE 5.7 Summary of Some Commercial Secondary Batteries and their Characteristics (*Continued*)

System	Life Cycle	Charge time, hr	Maximum Discharge Rate, C	OCP V	Energy Density, Wh/dm³	Energy Density, Wh/Kg	Operating Temp. Range, °C	Charge and Retention	Efficiency %	Applications and Remarks
LiAl-C										Lowe capacity than primaries. Excellent life cycle
LiAl-FeS₂	>1000		(100-400 Ah)	1.35	200-	105-150	+400=>+50 0	0.3-3.0%/month self discharge		
Li-Sulfur Dioxide	>30			3.1	290-	235-524				Very low self discharge per month (0.01%). 40% capacity loss/100 cycles
Li-Copper-Choloride	a) >140			a) 3.2-3.4 b) 2.4	a) 240- b) 630	a) 130-665 b) 300	b) -40=>+55	b) 1%/year self discharge		Very low self discharge per month (0.01%). 10% capacity loss/100 cycles. Limited sizes available.
Li-Manganese Dioxide	>200			3.0-3.2	220-	100-364				Low self discharge per month (1%). 10% capacity loss/100 cycles. Medical applications
Lithium ions				4.1	250	115	-20=>+65			Medium self discharge/month(12 %); good capacity loss/100 cycles (1%)
Lithium Polymer	10 years, 1000 cycles at 80% DOD			3.5		>200	25			

System	Cycle Life / DOD	Voltage		Energy	Temp. Range	Self Discharge / Notes
Li₂.₅Cd-TiS₂	>100-300@100% DOD; 1000 cycles @50% discharge	1.94		322		
Li-TiS₂		2.1		29-417		
Li-CoO₂	>50@100% DOD	4.0		465		
Li-SOCl₂	4.5-10.5Ah Capacity@1.5V;	3.66			-40=>+280	
Li-V₆O₁₃	a) >50@100% DOD b) 300-300@100°C; 50-100@20°C	a) 2.2 b) 3.2	b) 250	a) 361 b) 200	b) -40=>+60	1% Self discharge
Li-NbSe₃	>200@100% DOD	1.8		330-430		
Li-CFx		3.3-3.5	1139	450	-45=>+85	Self discharge 1%/year
NaS		1.75	597	235		Low drain, Long life; miniature cells

Note: This table reflects a compilation of data reported by several authors; in some cases the range reported for similar system by several authors is listed. Sometimes the reported information seen incomplete.

425

Capacitance is measured in coulomb/volt, or Farad. Therefore, a battery system can be represented as a capacitor that stores energy. The energy stored, U, between two plates separated by a gap is equal to $Q^2/2C$. If we assume that the battery system can be represented by a parallel plate capacitor, the capacitance of the system can be calculated by:

$$-Q/\Delta V = C = \varepsilon\, A/d \tag{5.37}$$

ε represents the permittivity of the gap between plates, and is expressed in Farad/m, A is the area of the plates, and d is the distance between plates (width of the gap). It is obvious from Eq. 5-37 that a large capacity will require a very narrow gap.

Storage energy   An interesting way of storing energy utilizes the concept of double-layer chemical capacitors. The double-layer capacitor utilizes large surface-area electrodes and a liquid electrolyte to form a charge storage layer. This charged layer is of the order of magnitude of few angstroms. The capacitance, C, stored energy, U, and energy density, U_d, of a cell or battery can be calculated by using:

$$U = 1/2\ C\Delta V^2 \tag{5.38}$$

$$U_d = \epsilon\Phi/2 \tag{5.39}$$

ϵ ($= K\epsilon_0$, K is the relative dielectric constant of the medium, ϵ_0 is the permittivity of free space) is the dielectric constant of the dielectric between plates; ϕ is the electrical field between the plates.

One method used to maximize the storage capabilities is by increasing ϵ or increasing the ratio A/d in Eq. 5.37. The use of very small distances between electrodes and small double layer size dramatically increases the specific power of the cell.

Time constant   If a battery or cell is kept at a constant applied voltage difference, the current will decay with time. The current decays exponentially to zero. When the current decays to 0.37 I_o (or I_o/e, where e is the natural logarithm constant, and I_o is the initial current), the time at which $I = 0.37I_o$ is the time constant of the system. The rate of current decay is determined by $RC = t_{constant}$, where R and C represent the impedance of the cell or battery as an RC parallel circuit equivalent.

Off-load voltage   Off-load voltage, V_o, is the same thing as OCP. It is dependent upon the system design and temperature. $V_o = f$ (system, T), V_o is expressed in volts.

On-load voltage   On-load voltage, V_a, is the potential drop between battery terminals, when a load closes the circuit. It can be expressed as a function of discharge current, I_D, temperature, T, and applied voltage as $V_E(I_D, T, V)$. It has units expressed in Volts.

Cut-off voltage   Cut-off voltage is the minimum charging voltage in a secondary battery. It is usually expressed by V_S, and is a function of the design requirements. The units are generally volts.

Discharge voltage   The discharge voltage, V_D, is the discharge voltage for a given load. The discharge voltage depends on the current, temperature, and system's design and properties. $V_D = f$(system, T, I_D), and its units are volts. As on-load voltage but refers only to the discharge.

Charging potential   Charging potential, V_L, for secondary cells, is the maximum voltage at which the system is being charged. The charging potential depends on the charging current, temperature, and system design. $V_L = f$ (system, T, I_C), and its units are in volts.

Discharge current   Discharge current, I_D, for a secondary battery is the maximum discharge current that a system undergoes for a given load. It is a function of temperature and applied voltage, $I_D(T, V)$, and it has units of amperes.

Volume of cell   The volume of a cell or battery, v, generally has units of dm.[3] It is important to notice that some authors report their findings with respect to a partial and not a total volume (for example, anode volume).

Gravimetric weight   Gravimetric weight is the weight of a cell or battery, G, generally in units of Kg. It is important to notice that some authors report their findings with respect to a partial and not a total weight (for example, anode weight).

Maximum power point   Maximum power point is the power obtained from a battery when only the I-R drop is significant, and other losses are not considered (for example, aging effects).

Nominal power   Nominal power is the power consumption for a direct current battery device. It can be defined as: $P_N = V_D \times I_D$. V_D is the operating voltage, or discharge voltage, or average discharged voltage, and I_D is the discharge current of the battery. If the efficiency of the cell is η, the nominal power is $P_N = V_D \times I_D \times \eta$. The calculated nominal power represents an average value.

Nominal voltage   Nominal voltage is the maximum voltage during the discharge of a fully charged battery. The voltage drops as time passes, from the nominal to the cut-off voltage. When an appliance is composed of several cells, the nominal voltage of the appliance corresponds to the average nominal voltage of the cells. It is a value that characterizes the system. It is usually a function of the system, $V_n = f$ (system), and is expressed in volts.

5.8.2 Dependent Variables

Some of the terms already defined are defined as independent variables, and others dependent of the independent variables. Some of those derivated variables that are frequently used in batteries are listed.

- Power: $P = I_D \times V_D$
- Power density: $P_G = P/G$
- Energy content: $U(I_D, V_S, T, \text{design}) = V_B^2 \times C/2$
- Energy density: $U_G = U/G$
- Storage life: $L = \Delta U/U \times \text{time}$
- Dc resistance: $R_i = \Delta V_B /\Delta I_D$
- Energy cost: *money/U*

In all the variables defined, the most important characteristics to be considered in battery design are:

- High energy and power density
- Stable discharge voltage
- Wide temperature range for use and storage
- Size and weight according to standards
- Ease in manufacturing and construction
- Low cost
- Shock resistant
- No leakage
- No hazardous (fire, explosions, toxic, etc.).

5.9 *GENERAL GUIDELINES FOR BATTERY SELECTION*

The selection of a battery varies according to its uses. The type of battery used depends on whether the battery is used in commercial, in research, or in engineering application. Batteries perform differently in different conditions, and they are an integrated part of an overall electrical design. Considerations such as weight, volume, service life, operational temperature range, power, energy, cost, safety, demand profiles, mechanical resistance and characteristics, thermal characteristics, and the type of discharge (continuous, intermittent, or pulse) are only a few of the variables engineers must understand to select a battery. Engineering battery selections often result in a tradeoff between system needs and available battery performance.

In selecting a battery, the following basic criteria should be followed:

1. Determine the battery requirements, according to the following analysis areas: physical, electrical, environmental, and special considerations. It is important to establish relative importance of requirements; to compare the characteristics of each battery against the user requirements; and to determine the best compromise. Among the questions to make to select a good system are the following: is the required battery system:

   • Automatically activated?

   • Manually activated?

   • Nonrechargeable or rechargeable?

2. Additional information will be needed for automatically activated systems; the following information will help to narrow down the choice of automatically activated batteries:

   • Method of activation: mechanical or electrical

   • Activation time

   • Stand time after activation

   • Orientation during activation

   • Temperature during storage

3. For a rechargeable secondary system will need, in addition, to know:

   • Cycle life

   • Wet shelf life

   • Charge retention

   • Method of charging

   • Maintenance

   • Self discharge

4. In general, an important element is cost effectiveness, and the following information must be known:

   • Payload

   • Operating life

   • Performances in wide temperature range

   • Maintenance cost

   • Storage

   • Replacement cost

Battery selection is best achieved by setting out a list of minimum requirements, limitations imposed by the system and minimum performance of the selected battery systems.

5.10 ACKNOWLEDGMENTS

This summary is possible because of the long hours and intensive effort of many researchers and engineers that contributed to the field of battery research. I would like to offer my deepest acknowledgment to all of them. I would also like to thank my husband, D.D. Macdonald, for his contributions to this review. Finally, I would like to acknowledge the two programs/organizations that sponsored my time and made this summary possible: the Summer NASA-ASEE faculty fellowship program; and The Office of Research and Development in Washington, Research Contract No. 95F 144700*000, which actively supported my research efforts in the field of lithium batteries.

5.11 REFERENCES

1. *The Primary Batteries*, Vols. 1 and 2, Edited by G. W. Wiley and N. C. Cahoon, Wiley, New York, 1971 – 1976.
2. Alfred Niaudet, *Elementary Treatise on Electric Batteries*, from the French translated by L. M. Fishback, 2nd Edition, John Wiley & Sons, New York, 1882.
3. S.R. Bottone, *Galvanic Batteries: Their Theory, Construction, and Use*, Whittaker & Co., London, 1902.
4. W.R. Coper, *Primary Batteries: Their Theory, Construction and Use*, D. Van Nostrand Co., London, 1902.
5. *Battery Reference Book*, Edited by T. R. Crompton, Butterworths International, 1990.
6. L. Pauling, *General Chemistry*, W. H. Freeman, 1953.
7. F. Brescia, J. Arents, H. Meislish, and A. Turk, *Fundamentals of Chemistry—A Modern Introduction*, Academy Press, 1966.
8. H.B. Oldham and J.C. Myland, *Fundamentals of Electrochemical Science*, Academic Press, 1994.
9. *Battery Technology Handbook*, Edited by H.A. Kiehne, Marcel Dekker, New York and Basel, 1989.
10. *Lithium Batteries*, Edited by Jean-Paul Gabano, SAFT, Academic Press, Potiers, France, 1983.
11. B.B. Owens and P.M. Skarstand, "Ambient Temperature Solid State Batteries," *Proceedings of the 8th International Conference on Solid State Ionic*, Canada, 1991.
12. "Impedance Spectroscopy," *Proceedings of the Symposium on High Temperature Electrode Materials and Characterization*, Edited by D.D. Macdonald, and A.C. Khandhar, Electrochemical Society, 1991.
13. D.D. Macdonald and M.L. Challingsworth, "The Impedance Characteristics of Nickel-Cadmium Batteries." *Internal Report CAM*. The Pennsylvania State University, 1992.
14. *Emphasizing Solid Materials Systems*, Edited by J.R. Macdonald, Wiley, New York, 1982.
15. *High-Energy Batteries*, Edited by Raymond Jasinski, Plenum Press, New York, 1967.
16. *Modern Battery Technology*, Edited by C.D.S. Tuck, Ellis Horwood, 1991.
17. *Materials for Advanced Batteries*. Edited by D.W. Murphy, J. Broadhead, and B.C.H. Steele, Plenum Press, New York, 1980.
18. *Proceedings of the NASA Aerospace Battery Workshop*, Edited by NASA, 1991.
19. G. Halpert, et al. "Status of Development of Rechargeable Lithium Cells," *Journal of Power Sources*, 47, 287 – 294, 1994.

.

CHAPTER 6
OVERCURRENT PROTECTIVE COMPONENTS

Carl E. Lindquist
Vice President, New Product Development
San-O Industrial Corporation

6.1 INTRODUCTION

Circuits within electronic equipment need to be protected against overvoltage and over-current fault conditions. A suitable fault-protection system will prevent fault conditions from harming the local circuitry or presenting a safety hazard to operating personnel, and will isolate the fault condition, keeping it from propagating to adjacent circuitry.

Much has been written about fault protection concerning power plants and other heavy-current applications, but comparatively little about low-power equipment protection. Consequently, low-power equipment circuit protectors are too-often chosen with inadequate knowledge of how they work and their real-world limitations. This chapter covers the basic principles of circuit protection and will describe protectors in enough detail so that design and component engineers will be able to select the optimum protection component, or combination of components.

A comprehensive fault-protection system will consist of overvoltage, overcurrent and thermal-protective devices, or some combination of the three. The protective actions of these devices complement each other. Overvoltage devices are not specifically covered in this chapter of the *Handbook*. Further information can be obtained from voltage-protector manufacturers, such as CP Clare Corporation or Joslyn Electronic Systems Corporation. Thermal protectors are covered briefly in Section 6.2.5.

This chapter describes "supplementary" overcurrent protectors with heavy emphasis on fuses. A *supplementary protector* (for instance, a fuse) is a device that will be protected against overload by another current-interrupting component (a secondary or a primary protector) closer to the power source. Protectors included within free-standing electronic equipment will usually be "supplementary" in nature. "Branch" (or "secondary") circuit protection must always be provided, between the equipment and the power source, in addition to the "supplementary" protector.

Overcurrent protection is placed at a location, which protects "downstream" circuitry, normally at a point where there is an effective reduction in size of the current-carrying medium. For example, a relatively short length of 12-AWG powercord wire can be rated at 25 A (25 amperes), and is protected by an appropriate circuit breaker. Several 18-AWG powercord wires, rated at 10 A, can branch from the main cord. An overcurrent of (for example, 20 A or 200%), which might harm the 18-AWG powercord carrying the overcurrent or the load circuitry, would not be noticed by the main powercord fuse. For this reason, each 18-AWG powercord also requires a fuse or other protector. Figure 6.1 illustrates the isolation aspect of proper overcurrent protection. Loads 1 and 4 are protected by fuses. Load 2 is protected by a thermistor. Load 3 is protected by a thermal switch. A fault in lines 1 or 4 will open that line's fuse, and isolate the fault, preventing it from affecting other lines. Similarly, the thermistor in line 2 will go into a high-resistance state, or the switch in line 3 will open, thus protecting and isolating those lines.

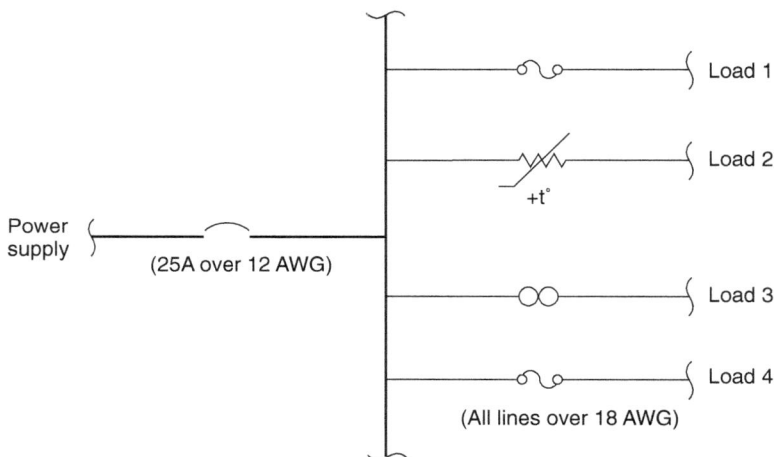

FIGURE 6.1   Overcurrent protection at points of reduction in current-carrying capacity.

6.2 PROTECTOR CATEGORIES

Overcurrent protectors can be grouped into categories, according to their mode of operation.

6.2.1 Fuses

A fuse is a device that protects a circuit by fusing (melting) open its conductive element when an excessive ("overcurrent" or "short circuit") current flows through it.

Some types of fuses are designed for high-voltage and high-current applications. Such fuses are of interest mainly to those engineers who must control large amounts of power. For information about such fuses, refer to specialized texts in that field—especially the excellent book by Wright and Newbery.[3]

The electronics engineer is more likely to be interested in "miniature," and "micro" fuses, of the kind usually chosen to protect electronic equipment. These fuses usually

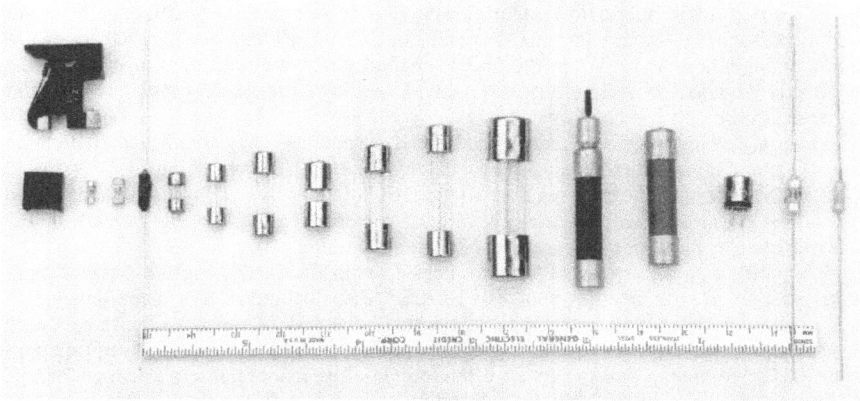

FIGURE 6.2   Typical miniature fuses. *(Courtesy San-O Industrial Corp.)*

(but not invariably) have cylindrical bodies with ferrule-type end caps, and are rated in a range from several milliamperes to a few tens of amperes, and maximum rated voltages of about 250 V or less. There are a few "miniature" fuses that are rated as high as 600 V. But, these are the exception. Figure 6.2 shows a selection of typical miniature fuses.

Technical standards for this type of fuse fall into one of two "camps." There is what will be called in this chapter the "UL 198G/CSA C22.2 #59" standards grouping, and the "IEC 127 standards grouping." UL 198G *Fuses for Supplementary Overcurrent Protection* and CSA Standard C22.2 #59 *Fuses* are, respectively, the U.S.A. and Canadian standards, which have governed fuses of this type for decades. UL 198G and CSA C22.2 #59 are very similar, so fuses built to this standards grouping are dimensionally and functionally similar. It should be noted that these two standards referenced are being replaced with one new combined standard: UL 248/CSA-C22.2 No. 248. Protectors tested in the past to UL 198G/CSA C22.2 #59 are accepted as is with no further testing. All new products submitted for approvals will be tested to UL 248/CSA-C22.2 No. 248. Purely because of habit, I will continue to reference the older standards unless a specific deviation is being identified. Several other Pacific Rim nations have standards similar to UL 198G/CSA C22.2 #59. Most of the rest of the world bases its fuse standards on IEC 127 *Cartridge Fuse-Links for Miniature Fuses*. IEC 127 requirements (for other than microfuses) differ radically from UL 198G/CSA C22.2 #59 requirements, to the extent that the two are mutually exclusive, and no fuse can simultaneously meet both standards groupings. More information about fuse standards is provided later in this chapter.

There is no universally accepted definition of the word "fuse." IEC defines a fuse to comprise all the parts of the fuse assembly, including the element itself, and the supporting hardware. In IEC parlance, the "fuselink" is that part of the fuse assembly which must be replaced when the fuse blows. In common North American usage, the "fuse" is understood to be the "fuselink." The discussion in this chapter will follow the North American convention.

A fuse consists of a conductive metallic link housed in a suitable package. The assembly includes the fusible element itself and contacts at each end of the link that permit attachment to mating connection points in the circuit. The link is a positive temperature-coefficient device; the resistance of the link increases with increasing temperature, and temperature increases with increasing current. During normal, nonovercurrent conditions, a fuse will simply pass current through its conductive element. The fuse element will reach and maintain a stable temperature. When the fuse is challenged by a circuit current increase, the conductive-element temperature rises quickly. Eventually, the element melts, resulting in an open circuit.

Fuses are available in both leaded and unleaded versions. They are also available with any of several delay characteristics, and with any of several voltage ratings. UL 198G and CSA C22.2 #59 provide approval for two "speeds" of fuse: *time delay*, sometimes called *slow blow* or *surge proof*, and *non-time delay*," sometimes called *normal blow*. IEC 127 has two categories of fuse speed: *quick acting* and *time lag*.

The writing of fuse standards has not kept up with fuse technology, so several other fuse speeds not explicitly mentioned in UL/CSA/IEC documents have been developed. These fuses are often chosen for applications that have unusual current-time-temperature fuse requirements. Safety agencies might not approve use of these fuses in specific applications. The following five speeds are often found in fuse manufacturers' catalogs for IEC 127 fuses and sometimes UL/CSA-qualified fuses:

TABLE 6.1   IEC 127 Fuse Speed Codes

Speed terminology	Letter code
super quick-acting	FF
quick-acting (normal blow)	F
medium time lag	M
time-lag	T
super time-lag (time delay)	TT

For purposes of fuse approval categorization, super quick-acting, quick-acting, and medium time-lag fuses are classified as *quick-acting* fuses under IEC 127. Time-lag, and super time-lag fuses are classified in the *time-lag* category. Similarly for UL-198G fuses, a fuse faster than normal blow will still be classified for UL/CSA approval purposes as *normal blow*, and a fuse slower than time delay will be classified as *time delay*. The requirements for classification as a *time-delay* fuse under UL-198G are specific. A fuse can have substantially delayed action, yet still not be classified as *time delay* unless it meets the requirements of *time-delay* fuses.

Because the FF, M, and TT speeds are not yet specified by IEC 127, there is no standard set of requirements. Manufacturers may differ in their interpretations of the requirements for FF, M, and TT fuses. Care should be taken by design and component engineers to specify exactly which current-time-temperature characteristics are expected. Also note that the time delay of a UL/CSA fuse is not the same as that for the IEC time-lag fuse. The specific differences are covered later in this chapter.

Fuses without intentional time-delayed action   Single-element fuses are the most common type of nontime-delay fuse, and are usually the type intended when the word "fuse" is used. The element is most often a single wire of uniform cross-section, typically made of silver, copper and its alloys, nickel, or nickel chromium. With increasing current, the escalating temperature of the element eventually causes the fuse to melt "open." If the overcurrent is sufficiently large, heat generation will be so rapid that heat dissipation is minimal, and the element vaporizes, rather than melts. With any fuse element, the temperature distribution along the element length can be viewed as a "Normal" (Gaussian) distribution—hottest at the center and cooler toward either end cap. When the temperature at the hottest point reaches the melting point of the element material, the fuse begins to open. The time required for melting to occur is called the *melting time*. Current does not actually stop flowing until the element melts open and arcing ceases. The period of arcing is called the *arcing time*. Variables, such as open-circuit voltage and the amount of short-circuit current available from the source can cause the arcing time to exceed the melting time. The total time required for the current to cease flowing through the circuit is called the *clearing time*. Therefore, *clearing time = melting time + arcing time*.

Twisted-pair element fuses are made by using wires composed of two elements of dissimilar materials having different temperature coefficients and melting points. These two wires are twisted tightly together along their entire length, becoming a "compound" element. When current flows through the compound wire, the percentage of current flowing through each element is a function of its resistivity at the temperature at any instant. The element components begin at room temperature with one having higher resistance (or voltage drop per unit length) than the other. At some current level (temperature), the resistance of both wires is equal. At this point, both conductors carry equal current. As current increases beyond this point, the opposite wire becomes the high-resistance conductor.

The shifting of current from one wire to the other because of changes in temperature alters the behavior of the compound element under short-duration surges. This type of fuse element is more surge tolerant (i.e., slower responding to an overload) than is a single-element type. Twisted-pair element configurations are generally used in medium time-delay fuse designs.

Solid-matrix fuses are a relatively new type of small fuse that maintains stable operation in spite of today's harsh manufacturing environments. The stability is achieved by enclosing the link in a relatively small amount of ceramic filler. The filler and element assembly is placed in the fuse package. Welding, rather than soldering, is used to attach the element to contact posts, which further enhances the stability of the fuse. The fuse is totally sealed to prevent contamination of the fusing chamber by cleaning solvents or other foreign matter. It is claimed that this type of fuse has much greater *mean time between failure (MTBF)* under stable circuit operating conditions than standard soldered-element fuses.

Use of the ceramic filler permits very thin fuse links, and therefore very low I^2t. The lack of solder in this fuse design eliminates changes associated with solder reflow in the fuse during bonding to the final circuit assembly. For their physical size, these fuses have relatively high ac and dc voltage ratings and interrupting-current (short circuit) ratings. Some have gold elements that are welded or ultrasonically bonded to the contact terminals.

Solid matrix construction is associated with newer, high-precision, high-reliability fuses, and more specifically, those fuses meant for printed wiring board (PWB) and surface-mount applications. In these applications, high wave-solder temperatures and cleaning-solution contamination can be a major concern. The term *solid matrix* is associated with Bussmann Fuses, and one example is Bussmann's 1608FF.

Because they are very fast-acting devices, the 1608FF will usually be selected with a slightly higher current rating than would be chosen for a standard fuselink. The fuselink is encased within ceramic, making it very resistant to mechanical fatigue. This type of fuse is able to withstand surges having I^2t ratings that are very near the melting I^2t of the fuse.

Although these fuses have relatively high-voltage and short-circuit ratings for their physical size, remember that they are quite small. They will often be UL recognized, rather than UL listed, and will have short-circuit ratings typically much lower than standard 3AG or 5- × 20-mm fuses.

Fuses with intentional time-delayed action   The primary purpose of the time delay fuse is to provide surge withstand capability while limiting low-level, long-term overloads, and heavy short-circuit currents. Under North American (UL and CSA) standards, a time-delay fuse must meet all requirements of a normal (nontime delay) fuse for all parameters. The time-delay fuse must also be capable of withstanding 200% rated current for a minimum time of five seconds for fuses rated up to 3.0 A (or 12 seconds for fuses rated from 3.1 A to 30 A) without opening or experiencing any noticeable degradation.

Consider that time-delay fuses must delay 5 seconds minimum before opening. One fuse consistently requires 7 seconds to open, and another nominally "equivalent" fuse requires 15 seconds. Both meet the UL/CSA requirements and are referred to as *time-delay* fuses. But it can hardly be said that they are "equivalent" for circuit protection. Thus, it is not enough that two fuses from different manufacturers are similarly classed as "time delay." They might have different constructions, leading to different current-time-temperature characteristics. The data sheets of both fuses must be read carefully to determine if they are truly equivalent.

The terms *time lag* and *slow blow* are frequently used as synonyms for *time delay*. I prefer to avoid the use of terms other than *time delay* for UL 198G/CSA C22.2 #59 fuses, and *time lag* for IEC 127 fuses, because of the risk of misinterpretation when other terms are used. Slo-Blo®, for example, is a proprietary trademark used both for time-delay and medium time-delay fuses. The terms *slow blow* and *Slo-Blo®* are phonetically similar, so many component and design engineers assume they are functionally identically, which might not be true. The issue of correct fuse time response is treated in more detail in section 6.4.6. Figure 6.3 shows a selection of constructions for deliberate time-delayed fuse action.

Dual-element fuses are constructed with two separate fuse elements in either series or parallel combination.

In the series type of construction, the fusible link consists of a series combination of a "high-temperature" wire and a coil spring, which are soldered together using a low-melting-

FIGURE 6.3   Time-delay/time-lag fuse constructions. (*Courtesy San-O Industrial Corp.*)

point alloy of tin, lead, or bismuth. Under conditions of high-current overload, the high-temperature element, sometimes called the *heater wire*, usually composed of a nickel or copper alloy, ruptures (vaporizes) before the low-temperature alloy melts. The spring quickly contracts back into the cap so that a large gap is created to minimize arcing. The high-temperature wire is designed to be somewhat slower-acting than a normal fuse. Its position in the fuse enclosure (immediately adjacent to one fuse cap) provides a heatsink effect, further slowing the fuse operation. At lower overcurrents, the soldered joint melts first, and again the spring contracts back into the cap, thus preventing partial opening of the fuse. The fuse designer will normally ensure that the low-temperature alloy used is eutectic (or have phase-change characteristics close to that of a eutectic) to minimize chances of partial melting during temporary low-level overloads.

The spring is made of a hard drawn copper alloy. Care is taken by the fuse designer to ensure that the spring does not anneal before the heater wire ruptures or the low-temperature alloy melts, which would compromise the spring action. To prevent overheating and annealing of the spring, or when high total fuse resistance cannot be tolerated, fuses often have a flexible bypass wire in parallel with the spring to carry the current.

The series dual-element fuse construction can serve as a thermal limiter. The melting point of the low-temperature element can be specified, and is often around 160°C. The fuse will open when the ambient temperature reaches the melting point when the current is zero. However, do not rely on this feature, unless it is specified by written agreement with the fuse manufacturer. Without the agreement, the manufacturer can arbitrarily change the design of the fuse so that the melting temperature changes.

Parallel construction of a dual-element fuse provides surge-withstand capability in a different manner. Because fuse-clearing time is a function of the element reaching its melting point, a parallel dual-element fuse requires much more energy to open than does a single-element fuse, all else being equal, because the physical midpoints of two elements must be raised to the melting temperature, rather than just one element. When one of the two parallel elements opens, the second opens very quickly. The extra energy required to reach melting results in a time delay. This method is frequently used in very low current time-delay fuses, in which the element is also spring loaded.

Some time-delay fuse designs combine series and parallel techniques. Such fuses will have a short pair of parallel elements, usually tied to a spring, which are joined by a low-melting-temperature bonding material.

The dual-element design will generally provide the greatest amount of delay, compared to other time-delay fuse types. However, this extra delay is not without penalty. For series dual-element design, variability in the low-temperature element mass and spring-load force result in a relatively larger standard deviation for the I-t curves. Similarly, with parallel dual-element designs, the two "arms" of the fuse are generally not identical and the current split between the two is seldom balanced. In nearly every such device, one of the two elements is carrying more current than the other. This also results in relatively larger standard deviation for the I-t curves.

Spring-force variation, current imbalance, and the complex construction inherent in dual-element fuse design often make this type of time-delay fuse less repeatable than other types. Many manufacturers today are adopting the newer spiral-wound element, where possible, to avoid the variabilities associated with dual-element fuses.

M effect construction, also called *loaded-link construction*, is sometimes used by fuse manufacturers. This consists of a globule (a so-called "diffusion pill") of low-melting temperature metal, such as tin, surrounding a silver- or copper-wire fuse element. Overcurrents of moderate magnitude cause the tin to melt and the two materials to diffuse into each other. The resulting alloy has increased resistance, causing the element to rupture at the location of the diffusion pill. An alternate element construction has a composite wire, consisting of a tin core surrounded by a sheath of silver. The silver or copper element allows low fuse resistance and low operating fuse surface temperatures at rated current. The M effect allows the fuse element to melt at much lower temperatures than pure silver or copper would otherwise require. The increased mass of the element causes somewhat slower fusing, thus providing the short delay that is required in the medium time-lag category. These fuses will withstand approximately 10× rated current for 10-ms duration, provided that the interval between successive pulses is greater than 30 seconds, which allows the element to cool sufficiently between pulses. By choosing suitable materials and dimensions, medium time-lag characteristics can be obtained that would not be possible in fuselinks containing elements of just one material. This construction has been popular with the North American fuse manufacturers because of the resulting unique characteristics. However, there might be some difficulty in maintaining consistency in fuse parameters for this type of construction. Alloying can be initiated with a temporary overload and cease when the overload is removed. The resulting "new" alloy element will not have the same characteristics as the initial fuse design. The M effect is covered in detail by Wright and Newbery in *Electric Fuses*, and is described in a 1950 publication of the same title by H.W. Baxter[3].

Single spiral-wound element fuses are constructed by winding a single element of an appropriate conductor, such as silver or copper alloy, tightly around a heatsinking core. This core is often fabricated from a rigid high-alumina ceramic, or a similar flexible ceramic yarn. High-alumina ceramic is a material that provides excellent element heatsinking while being electrically nonconductive. Because the length of the element is longer than that of a straight element, the spiral-wound element must be thicker for a given resistance. The resulting higher mass of the spiral-wound element, combined with the heatsinking effect of the core, causes the rate of temperature rise of the spiral-wound element to be slower than that of a straight element. Thus the fuse is relatively slow to open. The pitch of the element turns is maintained to close tolerance to ensure consistent fuse parameters. This type of construction results in a fuse that is more repeatable than a dual-element fuse. However, the maximum length of time delay is usually less, and dc resistance is usually higher than with dual-element construction.

A note of caution regarding any spiral-wound fuse: This construction guarantees that the fuse will have measurable inductive reactance at higher frequencies. Although the fuse element impedance is negligible in typical fuse applications, such as dc or power frequency ac circuits, it can be significant in other types of circuits. Specifically, telecommunications lines or computer circuits that carry data must be evaluated carefully. Although the data signal frequency can be relatively low, retention of sharp corners on a square wave relies on the presence of high-frequency components. If these high-frequency components are attenuated by the inductance of the fuse, the corners will become rounded and data integrity can be jeopardized.

Nonalarm vs. alarm fuses *Nonalarm fuses* are the most common type. The term *fuse* used without adjective is understood to mean a nonalarm fuse. Nonalarm fuses do not contain a provision for calling attention to their operated state, although in most cases, the operated state of an open-link fuse or a fuse with a clear (transparent) enclosure can be determined by close visual inspection of the fuse element. Fuses, in which the element is not clearly visible, must be checked using some type of continuity tester.

Alarm-type fuses are available in a variety of physical packages. These types of fuses fall into two categories: spring-loaded and nonspring loaded indicators. Several examples are shown in Figure 6.4. The primary reason for using alarm fuses is the ease with which the fuse status can be verified either locally, or via a sensing element, at a remote location.

Spring-loaded indicator fuses constitute the vast majority of alarm fuses. Circuit protection is accomplished as with standard nonindicating fuses. The element is overloaded, melts, and opens the circuit between the power source and the load. In most designs, the element is tied to a spring-loaded "flag" that becomes visible upon fuse operation.

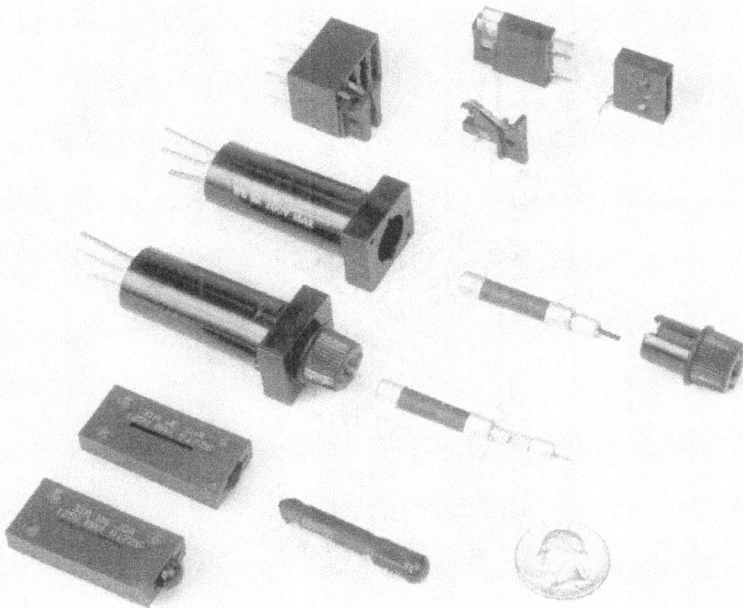

FIGURE 6.4   Typical alarm indicating fuses. *(Courtesy San-O Industrial Corp.)*

These fuse elements are designed to be under constant tensile stress because of the spring loading and must therefore be constructed of materials with superior tensile strength. The "flag" or "indicator pin" is an example of a local visual alarm. If a third contact point is added to the fuse holder (it generally is), a remote alarm signal can also be generated. Operation of this type of fuse is the same as for the local alarm device, except that a portion of the fuse below the "flag" is conductive and designed to strike this third (alarm) position, connecting the battery with the alarm circuit. The majority of this type of fuse are used in telecommunications circuits.

Figure 6.5[1] shows one popular spring-loaded alarm fuse and a multiple position holder, variations of which are manufactured by San-O, Bussmann, Reliance Fuse, and Littlefuse. These assemblies are available in single or multiple position units. This particular fuse type is available with safety features, such as transparent snap-on shields. In the San-O AX-1 version, the combination of the snap-on shield and flange around the back portion of the fuse body completely seals the fusing chamber. For applications where loss of the shield (thereby allowing accidental contact with the alarm spring) is of concern, the shield can be permanently bonded to the fuse. Otherwise, these shields are normally reusable.

Where multiple position fuse blocks are used, care must be taken that the maximum current rating for each position, as well as the overall maximum current rating for the block is known. If, for instance, a rating for a UL-recognized 10-position block is 10 A per position and 30 A maximum for the block, the maximum rating for each fuse would be 3.0 A. In contrast, if the UL-recognized 10-position block is rated at 10 A per position, and 100 A per block, then the maximum rating for each fuse would be 10 A.

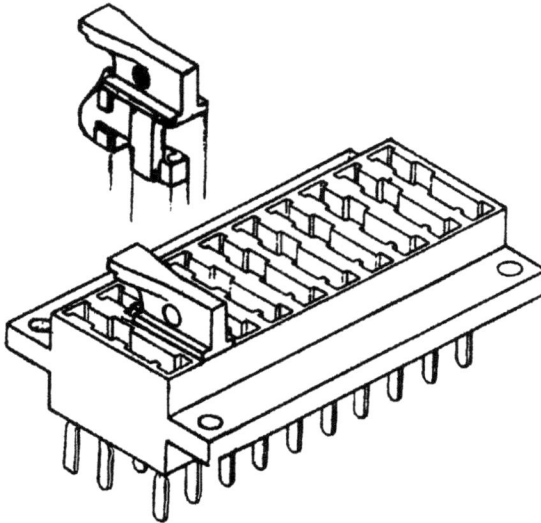

FIGURE 6.5   AX-1 alarm indicating fuse assembly.*(Courtesy San-O Industrial Corp.)*

This specific type of fuse assembly requires a close tolerance fit between fuse and holder. Because different manufacturers can use different materials, and can have slightly different tolerances, intermixing of fuses and holders from different manufacturers should be checked carefully for dimensional fit and plating compatibility.

Other commonly encountered spring-loaded indicator fuses include the AT&T Type 70, AT&T Type 35, and Northern Telecom QFF.

Nonspring-loaded indicator fuses are similar to spring-loaded fuses, but they do not normally use mechanical switching from one fuse status to the other. There are several methods of accomplishing the alarm status. One method is to place an LED with a series resistor across the element or in series with a resistor, between a separate power source and the circuit load. In the first case, the resistance value is chosen to limit the current through the LED, based on the open circuit voltage. In the second, a separate alarm circuit voltage is provided, making the open-circuit voltage irrelevant to the LED operation as long as the circuit voltage is always higher than that of the alarm circuit. With either design, the LED illuminates when the fuse element opens. AT&T Types 77 and 78 LED indicator fuses (Figures 6.4 and 6.6) are manufactured by San-O Industrial and have been popular in the telecommunications industry.

This type of fuse has two important differences when compared to the spring-loaded alarm-type fuse. Because the element is not placed under tensile stress by an alarm spring, a wider range of element materials can be used. For example, a pure silver element having very low dc resistance could be used in the LED fuse. Because pure silver is very ductile, it could not be successfully used in a spring-loaded fuse. (Note: The low dc resistance of silver does not guarantee low voltage drop in operation. Devices should be compared using voltage drop at the normal operating current.) Secondly, the LED fuse requires a current to flow through the LED and the load—even when the element has opened. The LED current is provided by a separate low-voltage supply and should be very low when the fuse

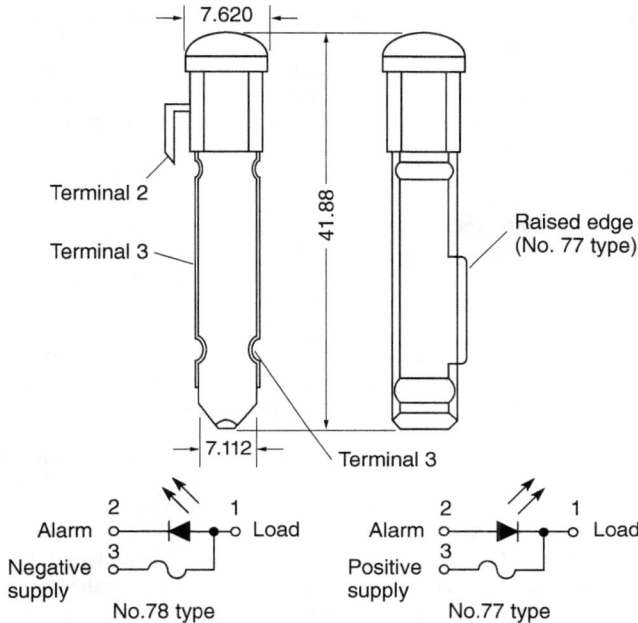

FIGURE 6.6   AT&T #77/78 LED alarm indicating fuses. *(Courtesy AT&T)*

element opens, but it is present and the operator can be exposed to some voltage at the load. Lastly, if the circuit (or load) is damaged extensively during the fuse operation, resulting in a significant rise in total circuit impedance, the LED might not have sufficient current flowing to light, and the "open" fuse state might not be visually indicated.

The LED fuse is a feasible method of circuit protection in many applications. But, as with any other current-limiting device covered in this chapter, unique characteristics of this protector must be recognized and carefully reviewed to ensure compatibility with the intended application.

Another way to accomplish the alarm function is to use an indicating fuseholder. In this case, the fuse itself can be an ordinary nonindicating fuse, but the fuseholder assembly contains an LED, neon, or incandescent lamp electrically in parallel with the fuse. When the fuse blows, the lamp lights. Normally the indicating lamp uses the same voltage supply as the fuse. So, unlike the LED fuses, it is possible that the operator can be exposed to source voltage at the load. For this reason, indicating fuseholders are scrutinized carefully by safety approval agencies (such as UL) prior to acceptance, to ensure that the continuing flow of current after fuse operation cannot result in injury to personnel. These fuseholders are generally UL recognized, rather than UL listed.

Lightning surge-withstand fuses   *Lightning surge-withstand fuses* are relatively recent innovations (10 to 15 years) in the telecommunications industry. They can cause confusion for engineers and buyers. These fuses are listed under UL 198G and certified under CSA C22.2 #59 as a standard fuse. However, industry has additional requirements. These include minimum and maximum surge withstand, and 600-Vac to 1000-Vac line-cross interruption capabilities. Certain safety criteria set in UL 1459 and UL 497 make these types of fuses necessary in some applications.

Lightning surge-withstand fuses are typically constructed of a single spiral-wound element or twisted-pair element. They are often classified in the UL time-delay category. However, these fuses are sometimes classified in the medium-time delay category, which is considered a normal (nontime delay) operating fuse by UL. Lightning surge-withstand fuses will tolerate very short-duration, relatively high-current surges without degradation. Their primary purpose is to terminate potentially damaging overload currents, while permitting transient surges of very short duration that will not damage the circuit (such as low-level lightning) to pass. The use of surge-withstand fuses ensures desired circuit protection under normal operating conditions, but reduces "nuisance" operations. Fuses in this category have special construction requirements and are, therefore, generally more costly than other fuses.

The surges are defined by a peak current and two points of time measured in microseconds. A typical waveshape will be listed as "25A, 10×560 (μs)," read as "Twenty-five amperes, ten by five-sixty microseconds." This defines a waveshape rising from 10% to 90% of the peak ("$0.90\times25 = 22.5A$") in less than 10 μs, then falling to 50% of the peak (12.5A) in 560 μs. (Note: units of time are assumed to be μs if no other unit is specifically stated). Other popular waveshapes in the industry include "25A, 10×1000" ("25 amperes peak, ten by one thousand microseconds"); "30 A, 10×560;" "50 A, 10×1000;" and "70 A, 10×560."

One manufacturer might have a 25-A, 10-x-560 device listed under UL 198G as a 450-mA fuse, and another manufacturer might have an electrically equivalent device listed under UL 198G as a 375-mA fuse. Although the two devices are essentially identical, the UL 198G standard rating will prevent the design or components engineer from grouping them for purposes of alternate sourcing. The fuse manufacturer decides whether the fuse rating will be 450-mA or 375-mA when the device is submitted for approval to UL. Thereafter, the manufacturer must advertise only the current rating for which UL recognition was granted, 375 mA or 450 mA, whether or not the other current rating could be met.

Another confounding factor for the engineer seeking "equivalent second-sources" is that these fuses can have identical electrical descriptions, but different physical sizes (e.g., 1.5 AG, 2 AG, 3 AG, 5 × 20 mm, etc.).

Because of the precision associated with this type of protector and its location (typically in a telecommunications line), it is recommended that the fuse always be soldered into the circuit. Use of clips or other mechanical contacts will decrease reliability and increase the probability of line noise over time.

As surge-withstand protection becomes better understood, and users insist on compatibility, industry standards for surge-withstand fuses will evolve.

6.2.2 Fusible Elements

Fusible links are sometimes used in place of fuses. The term simply refers to a replaceable "link" or element. The body, or casing, is reusable and the link can be replaced when the protector operates. The enclosure used can be an insulative tubular type, or it can be the interior of a cabinet. Usually, fusible links are mechanically larger and are associated with primary and secondary protection. They are typically approved by safety agencies for primary or secondary applications.

Fusible resistors are another form of protector. They behave like conventional resistors under normal conditions, but act like fuses under overcurrent conditions.

One version of fusible resistor is constructed like an ordinary resistor, except that it is designed to "open" during an overcurrent condition. Modern forms of this version of fusible resistor are generally of carbon, film, or wirewound construction. They are available in resistance values ranging from ohms to tens of kilohms, in tolerances ranging from 1% to 10% and in sizes ranging from 250 mW to 3.0 W. Fusing-to-operating current ratios are roughly 10:1, depending on type. These fusible resistors are usually flameproof, in accordance with the latest revision of EIA-325 *Flammability Tests for Electronic Components*, but do not usually carry safety agency approvals. Flame-proofing is necessary because these resistors reach relatively high temperatures before opening. For this reason, such fusible resistors should be elevated above the surface of printed circuit boards. Fusing time depends on the design of the fusible resistor, but in general these are not fast-clearing devices. Typically, the clearing time is 1-5 seconds at 20× operating current. Lower resistance values tend to clear slower. Fusible resistors do have predictable fusing times, but the time depends on many variables, and should be established by test for every application. This type of fusible resistor is generally made by resistor, as distinct from fuse, manufacturers. Examples are the BW series of fusible resistors by RCD, and the FA8 series by IRC.

Another version of fusible resistor is a 125-mW resistor incorporated in the casing of a standard alarm fuse (e.g., several versions of the AT&T Type 70 and the San-O AX70). When the resistor is overloaded sufficiently to cause it to open, the two halves are pulled apart by an alarm spring. The new position for the alarm spring ensures that the connection between the battery and load is open, and the connection between the battery and alarm contact is made. A third version is very similar, but the resistor is connected to a spring under tension on the surface of a printed wiring board. Neither the spring nor resistor is covered and there is a tendency upon fusing of the resistor for the spring assembly to disconnect entirely from the board and land elsewhere on the board. This type of protector is often referred to as a "snapper" and usage was generally limited to telecommunications boards. Surge-type fuses have replaced snappers today because of safety concerns with the snapper.

A larger spring-loaded fusible resistor uses either a 500-mW or 1.0-watt resistor encased in a high-temperature plastic sleeve. The two halves of the sleeve are anchored to the two ends of the resistor and a large steel spring is compressed between the two halves of the sleeve. With nominal ambient temperature and normal circuit current applied to the protector, the resistor holds the assembly together and current continues to flow. If the temperature or current becomes too great, the resistor burns in half and the spring forces the plastic sleeves to partially separate, opening the circuit. Current ratings associated with this type of fusible resistor are quite low, less than 250 mA. These, also, are typically found in telecommunications circuits.

None of the fusible resistors described is recommended for use in ambient temperatures above 60°C, without comprehensive evaluation of the effect of high temperature on the clearing current and failure mode. In all cases, the use of fusible resistors should be studied carefully to determine the variables involved and to ensure the safety of the protector. Agency approvals do exist for some types of fusible resistor, and proof of certification should be sought from the manufacturer of the protection device, along with a listing of test procedures used for the approval.

6.2.3 Circuit Breakers

Circuit breakers are another type of overcurrent protector. Breakers are a viable alternative to fuses, provided that care is taken to thoroughly understand the functioning of the device. The three basic types of breakers are: thermal, thermal-magnetic, and magnetic.

Thermal breakers work by temperature-sensing alone. In a controlled-temperature environment, the tripping temperature is a function of current carried by the breaker. When the current increase is sufficient to cause the trip-point temperature to be reached, a bimetal strip disengages the contacts, and the breaker opens. It can be mechanically reset after the actuating components cool. This type of current protector is quite slow to respond to overcurrent conditions, and ambient temperature plays a large role in the trip point. Because of the slow response characteristic, it can be used where there are heavy current surges at turn-on. Nuisance tripping is minimized. A small thermal breaker is normally pushbutton reset, but it can be available with a rocker-arm reset.

This breaker will typically be the least-costly type. Thermal breakers are considered to be physically more robust than magnetic breakers, and are used in applications where there are high levels of shock and vibration. Ambient temperatures should be stable and within the acceptable range of the manufacturer's ratings. Speed and high precision should not be critical factors.

Thermal-magnetic breakers are probably the most common type of circuit breaker in use. Most breakers used in service entrance and distribution panels are of this type. This is basically the same as a thermal breaker with a magnetic function that speeds up the action at heavy overloads. As with the thermal breaker, heat build-up caused by heavy continuous current flow, or high ambient temperatures, will typically reduce the current trip point of the thermal-magnetic breaker.

Magnetic breakers are typically the most costly type and are considered more-precise devices. Opening of the contacts is based on the strength of a current-generated magnetic field overcoming a precision spring load. If the force of the magnetic field does not reach the tripping limit, the circuit continues uninterrupted. However, if the magnetic field force exceeds the spring force, the breaker contacts open, interrupting the current flow. Because the operating speed of this type of breaker is relatively high, many manufacturers have included damping mechanisms to ensure that the breaker is not tripped during a normal inrush or periodic surge. The damping mechanisms slow the breaker response. Section 6.5 covers circuit breaker technology in more detail.

6.2.4 Current-Limiting Devices

Current-limiting (the glossary explains the two definitions for current limiting) *devices* are another class of protectors. Unlike fuses or breakers, which interrupt the current, these mechanisms only limit the current flow to some low, but finite, value. They do not actually open the circuit.

Positive temperature coefficient (PTC) thermistors are widely used for overcurrent protection. These devices undergo a large, abrupt increase in resistance when an overcurrent or high ambient temperature heats them above a specific point. This section focuses on PTC thermistors optimized for overcurrent/overtemperature protection.

PTC thermistors are usually made of doped barium titanate. Interaction of semiconductance and ferroelectricity cause the resistance to change with temperature. Below a transition temperature, or Curie Point, the device is in a low-resistance state (in the order of 1 Ω to 100 Ω) and has a small negative-temperature coefficient. Above the Curie Point, the dielectric constant of the material drops, increasing the barrier potential between individual crystals. Because free electron flow is restricted, the material resistivity rises. Further temperature increases cause total device resistance to increase exponentially by several orders of magnitude, and produce a switch-like response.

When used for overcurrent operation, a PTC thermistor is placed in series with the load. The current rating of the PTC thermistor is chosen so that during normal operation, I^2R heating is insufficient to raise the temperature of the device above the Curie Point. When an overcurrent occurs, I^2R heating raises the temperature of the PTC thermistor above the Curie Point, switching the thermistor into the high-resistance state. The load current decreases to several milliamperes, and most of the source voltage is dropped across the PTC thermistor. In a correctly selected PTC thermistor, the heat generated by the residual current is sufficient to keep the PTC thermistor in the high-resistance state. The PTC thermistor will reset, and return to its low-resistance state if it is allowed to cool below its transition temperature, such as by turning off the power source or clearing the fault. Keystone Carbon, and Therm-O-Disc are two companies that offer this type of PTC thermistor "resettable fuse." Manufacturers' data sheets and application notes contain formulas to calculate operating and switching currents as functions of device parameters and operating temperature. The ceramic PTC thermistor resistance-vs.-temperature characteristic is normally plotted as a curve, similar to the shape in Figure 6.7.

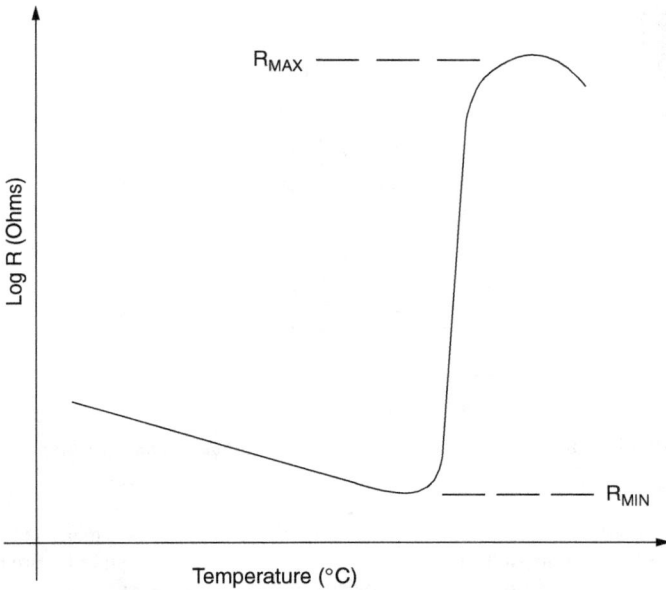

FIGURE 6.7   Ceramic PTC thermistor R-T curve. (*Courtesy Raychem Corp.*)

Polymeric PTC devices are a recent improvement of the PTC current-limiting idea. Polymer PTCs are made of conductive particles, such as carbon, scattered in a polymer matrix. Resistivity of the material is about an order of magnitude lower than that of ceramic PTC thermistors, which allows operating currents much higher than possible with ceramic thermistors. Under normal operating conditions, little I^2R heat is generated, and the conductive particles remain in close contact. When overcurrent occurs, the polymer heats above its transition temperature. The polymer matrix expands, causing the conductive particles to lose contact, which sharply increases device resistance. As with the ceramic PTC thermistor, I^2R heating is normally enough to keep the polymer PTC latched in the high-resistance state.

Polymer PTC devices have several advantages over the older ceramic PTC thermistors. Polymers generally have lower resistance than ceramics (considerably less than 1 Ω), thus I^2R heating is smaller, so polymer PTC devices can be rated to higher currents. Also, the difference between low-resistance and high-resistance values is often much greater for polymer PTC devices than for ceramic PTC thermistors. Polymer PTC resistances can change by six or seven orders of magnitude. Ceramic PTC thermistors can change as little

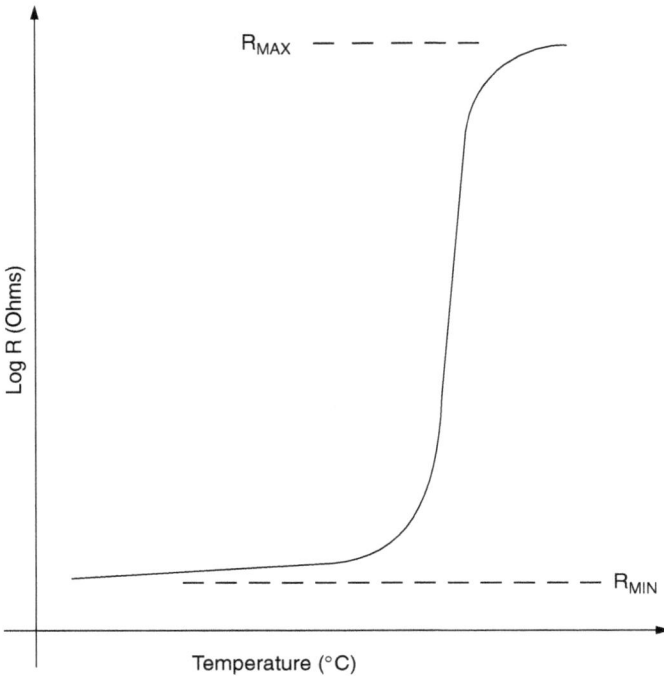

FIGURE 6.8   Polymeric PTC thermistor R-T curve. *(Courtesy Raychem Corp.)*

as one or two orders of magnitude. Thus, the current limiting of polymer PTC devices is more drastic. The polymeric PTC resistance vs. temperature is shown in Figure 6.8.

One disadvantage of polymer PTC devices in some applications is that the polymer PTC can exhibit "post trip resistance," whereas the ceramic PTC thermistor does not. In other words, when switched back from the high-resistance state, the polymer PTC can have a somewhat higher steady-state resistance than it had originally.

Both ceramic PTC thermistors and polymeric PTC devices are relatively slow-opening, $\approx$ 10 ms to 10 seconds for polymeric PTC devices, and 100 ms to 100 seconds for ceramic PTC thermistors. The actual switching time depends on the magnitude of the overload current. For the polymeric PTC, it can be as little as 10 μs for severe overload conditions, or as much as several hundred seconds for very light overloads. These devices are also quite sensitive to ambient temperatures, with respect to their current-carrying capability. Thus, PTC devices will require temperature derating for well-defined and stable high- or low-temperature applications. They might not be suitable for circuits that will experience relatively large excursions of ambient temperature in normal use. Although temperature derating is necessary to some degree for all overcurrent protectors, PTC devices are more sensitive to ambient temperature changes. It should be remembered that PTC devices will reset to the low-resistance state if current-induced device heating falls below a threshold. The PTC will reset if $V^2/4R_L < P_d$, where R_L is load resistance, V is circuit voltage, and P_d is the power dissipated in the tripped state. Even though this is highly unlikely outside the laboratory setting, there are circumstances that can cause the PTC to recycle continually.

Raychem's PolySwitch®, Therm-O-Disc's PolyFuse®, and Bourns' MultiFuse® are examples of polymer PTC resettable fuses. These manufacturers provide application notes and engineering assistance.

PTC "resettable fuses" have significant advantages over other protection devices in some applications. They do not require manual resetting or replacement. Unlike fuses, circuit protection cannot be defeated by incorrect substitution. PTC devices are typically more rugged than fuses. Some PTC protector devices are UL recognized. However, parameters (such as interrupting current, maximum voltage, etc.) should be reviewed carefully for the intended application.

Heat coils are resistance wire devices that are placed in series as current limiters. Although they are considered current-limiting, they can open the circuit upon actuation. Line fluctuations or surges are temporarily current-limited by a rapid rise in resistance of the wire coil. Ultimately, if the heat generated by the current passing through the spring-loaded resistance wire is heavy enough, or of sufficient duration to cause soldered connections to melt, current-carrying contacts are opened and current flow ceases. The primary application of heat coils is in telecommunications equipment lines. These devices are extremely slow-acting and not very precise. Heat coils can simply open the circuit, or have an additional function of shunting excess current to ground by using a spring-loaded contact. An electrical and/or visual alarm feature is often included.

6.2.5 Thermal Protectors

Thermal protectors are current-interrupting devices that are triggered by temperature. They are available in both nonresettable and resettable versions.

Nonresettable thermal protectors usually consist of two silver-plated contacts held closed by a pellet of temperature-sensitive material, such as an organic chemical material. The contacts have forces applied by a spring under compression that will separate the contacts if the pellet is removed. When a predetermined temperature is reached, the pellet melts, permitting the contacts to open. This type of protector is fairly precise for limiting circuit operation to a predetermined upper temperature limit. It will have adequate maximum voltage and short-circuit current limits for many applications. When such a device fails (opens), it must be replaced or the entire assembly must be discarded. These thermal protectors are often found in low-cost consumer applications, such as hair dryers and coffee makers, for which repair is not cost-effective and therefore not expected by the manufacturer.

A cross section of a typical nonresettable thermal protector is shown in Figure 6.9. Aid in assessing the actual circuit parameters for a nonresettable thermal device is available in the form of a "dummy" protector that has all of the components of the actual protector, except the thermal pellet has been replaced with a nontemperature-sensing pellet and a thermocouple is inserted. This will permit the design engineer to determine actual temperature ranges that will be encountered where the protector is to be placed.

In the design shown in Figure 6.9, it must be remembered that the metal case is actually one of the electrical contacts. Care must be taken that this case does not contact other live portions of the circuit. Further, the orientation, with respect to circuit ground, will determine whether the case remains "live" after opening. Attempts to electrically insulate this case will affect its thermal conductivity and the response time of the protector must be considered during application engineering. Appropriate procedures must be used when installing this type of thermal protector into the circuit. Because it is a thermally activated device, heat reaching the protector (e.g., from solder or welding) must be minimized to ensure that it does not operate or become damaged during installation.

Resettable thermal protectors, also called *thermal switches*, are typically bimetal devices. As the temperature increases, the bimetal disk moves or "snaps" to a point where

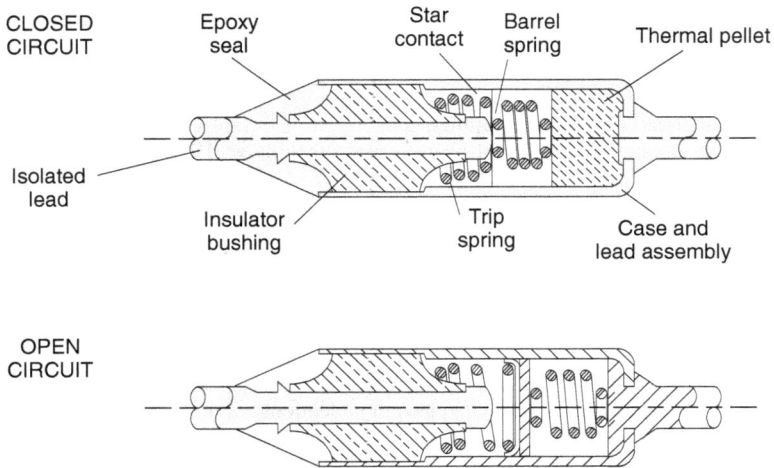

FIGURE 6.9   Nonresettable thermal protector. *(Courtesy Therm-O-Disc)*

current-carrying contacts are forced to open. After sufficient cooling, the protector can be designed to reset automatically or with a manual reset button. In well-designed bimetal thermal switches, both the opening and closing temperatures, are controlled. As with the nonresettable protector, it is preferable that temperature rise caused by contact heat dissipation be minimized in the protector. Also, opening and closing the contacts in this type of protector should be "snap-action" to minimize arcing. If the points open or close slowly, they will quickly degenerate because of the long arcing times.

Agency approvals (e.g., UL, CSA, VDE, MITI, etc.) are available for nonresettable and resettable protectors. The maximum voltage, current and interrupting ratings should be reviewed carefully and, as with the types of protectors covered earlier, the device to be used should be tested in the actual application during product development to determine its performance.

Technical and engineering support can be obtained from the associated protector manufacturer. Design engineers using such protectors should take full advantage of the protector manufacturer's services.

6.3 OVERCURRENT PROTECTOR SELECTION CRITERIA

6.3.1 Protector Selection Issues

The proper selection of a protector type is a function of the application and design engineer's preferences. High speed and precision for low-current, low-voltage applications would generally best be met by a fuse. Where a resettable protector is desired and ambient temperature is not an issue, a PTC can best meet the designer's needs. Applications where temperature sensitivity is a problem can be better-suited to magnetic circuit breakers.[1]

In some situations, the overcurrent condition that would damage the circuit is only slightly larger than the normal current. It may not be possible to find a fuse current rating that reliably passes the normal current over an extended period, but opens with the overcurrent. In this case, a thermal switch might be the best protector. For instance, small transformers, when subjected to low-level fault conditions, often will overheat without increasing the current sufficiently to cause a fuse to open. In such cases, a thermal switch operating on temperature might be the best protector option. Safety agencies will frequently require that both a fuse and thermal switch be used to achieve adequate protection.

The equipment's intended market can determine the choice between fuse and circuit breaker. If the equipment is to be marketed worldwide, a fuse is most often a better choice than a circuit breaker. The lower cost of fuses is certainly a factor, but more important is the ease with which fuse-protected products can be modified to accommodate the various input voltages worldwide—*Electric Current Abroad*, (latest edition)[2].

For example, a 6.0-A fuse (protecting equipment) connected to a North American 120-Vac power source, might be replaced by a 2.5-A fuse for 220- to 250-Vac overseas power sources. When used with an appropriate fuseholder, the correct fuse can be easily and quickly added to the equipment in the final stages of manufacture to match the intended destination of the equipment. A circuit breaker is normally difficult to replace. Different circuit-breaker models have to be stocked for North American and overseas markets. Once installed in equipment, circuit breakers require more time and labor to replace than do fuses.

Once the protector type has been chosen, the following checklist of questions will help determine the correct selection of protector parameters. Most of these questions apply to all protectors, but a few will only be meaningful for one category of protective device.

- Is the application ac or dc? This helps determine which safety agency approvals are required.
- If ac, what is the frequency? There will seldom be any problem in use in a 50- or 60-Hz application. However, if the application is in the 400-Hz or 35-kHz frequency range, additional information will be needed to determine adequacy of the selection.
- What is the open-circuit voltage? Does the agency approval for the suggested protector meet these minimum requirements?
- What is the typical steady-state current flowing through the circuit? This figure will be used to determine the current rating of the most appropriate protector.
- Which agency approvals will be necessary? The use of one protector with all necessary approvals could be advantageous when equipment is intended for marketing in several countries.
- What is the maximum short-circuit current available that the protector must have the ability to interrupt, and at what open-circuit voltage?
- Will the series resistance (or shunt leakage) of the protector be a factor in circuit function? This is an important parameter in very low voltage (or very high voltage) applications.
- Will a surge current be present during turn-on or at other times? Use of a time-delay device might be necessary to avoid nuisance blowing/tripping.
- What is the peak value and duration of any surge current?
- What is the power factor for the given circuit? If it is less than 0.7, have appropriate allowances been made for a fuse rating?
- What is the maximum acceptable voltage for the circuit? The selected overvoltage limiter should have a breakdown voltage below this figure.

- What is the maximum anticipated current during an overvoltage condition? The protector must be able to withstand this current, yet still function.
- What is the ambient temperature in the operating environment? Derating curves should be used to ensure proper operation of the protector.
- Where in the equipment enclosure will the protector be mounted, and how much room is available?
- Are other protectors being used in the same circuit? If so, how will they affect each other?
- How are the devices to be connected in the circuit? (e.g., hand soldered to a PWB, wave soldered to a PWB, wire wrapped, clipped, mounted in a holder, surface mounted, etc.)
- The component-insertion technique determines fuse packaging. Should the fuse be bought in tape and reel? bulk? blister pack?
- How will the manufacturing environment affect the component? Can the component be modified to withstand the anticipated stresses? Will a fuse have to undergo a potentially damaging wave-solder process?
- If mechanical connections are chosen, what platings are on the mating connector and the protector? If dissimilar platings are used, will they provide a satisfactory junction over the product's expected lifetime?
- Is visual (local and/or remote) indication of an open circuit necessary?
- Is it preferable that the protector be easily replaced by the end user, or should access to the protector be restricted? The answer to this question can disqualify some types of protective devices.
- Are solvents or other cleaning solutions or methods used? Some cleaning solutions disqualify certain circuit protectors from consideration, or require that additional precautions be taken to prevent contamination.
- Will the end product be subjected to severe conditions? (e.g., temperature, humidity, corrosive atmosphere, shock and vibration, etc.)

A current- or thermal-limiting device must always have a voltage rating equal to, or higher than the open circuit voltage. The short-circuit current rating for any protector must always be higher than or equal to the highest anticipated current load at the rated voltage. UL 198G/CSA C22.2 #59 fuses should be selected to carry between 50% and 75% of the rated current, unless constraints dictate otherwise. IEC 127 fuses, however, should be selected so that the fuse's rated current is within $\pm 10\%$ of the expected circuit current.

6.3.2 Second-Sourcing

If the application is not dependent on some critical and precise parameter, the protector can be defined using only a few constraints. Manufacturers' part numbers can be chosen, based on their published data sheets and cross-reference tables. However, where high precision and/or unique function are important, the device should be described in a customer's "source control drawing" using parameters given in a supplier's "fuse data sheet," as shown in Figure 6.16, with all pertinent vital parameters listed. For applications requiring exact definition of performance requirements, cross-reference lists are generally not acceptable and can lead to wrong component selection. When a "source control drawing" has been generated, it must be made part of any purchase agreement to ensure ongoing adherence to the requirements. Naturally, the date or revision number for this document must be included.

For example, one common complaint about a time-delay fuse is that one manufacturer's device works perfectly while another's frequently opens without apparent cause. Consider a UL 198G time-delay fuse. A cross-reference list claims both manufacturers' fuses are equivalent. From UL's point of view, both meet the UL 198G minimum time of 5 seconds at 200% current criterion. Looking at the two fuses more closely, the one that appears to work well is a dual-element device and is capable of withstanding the 200% overload for more than 12 seconds. The other (designated the "problem fuse") has a spiral wound element that will open at 200% in 6 or 7 seconds. Are they equivalent enough to include in a cross-reference table? That depends on the application.

What if the situation were reversed and the circuit designer's application could only withstand 200% for 8 seconds maximum? In this case, the spiral-wound element device would provide adequate protection. However, substitution of a dual-element device from the cross-reference table can result in serious damage.

Another example relates to certain 250-Vac fuses approved to UL 198G. To meet the standard, the fuse must interrupt a minimum of 500 A (for example) at the rated voltage. If one manufacturer's fuse can interrupt 10,000-A at 250 Vac and another manufacturer's fuse can only interrupt 500 A at 250 Vac, both will meet this particular UL requirement. Both will be UL listed. Are the two fuses equivalent? Not if the circuit designer needs more than 500 A of short-circuit (interrupting) capability at 250 Vac.

Similar equivalency questions affect circuit breakers, overvoltage protectors and thermal protectors of different types. Second sources of protector components are not necessarily equivalent! Cross-reference tables should be regarded with prudent skepticism. Reputable protector manufacturers maintain engineering staffs to help the design, component or compliance engineer choose the right protector for the application.

Selection of a fuse for protection of a power supply will require that the peak in-rush current, duration of the in-rush, open-circuit voltage, short-circuit current, typical steady-state current, and ambient temperature all be known. If, for instance, the typical current will be 2 A at 125 Vac and we are designing for a UL or CSA application, a "normal" 4.0-A/125-Vac protector fuse would be selected. If, however, the in-rush current reaches 50 A for 100-ms duration, the correct selection would probably be a "time-delay" 4.0-A/125-Vac fuse.

Choosing the proper fuse for a very low-voltage application can be problematic. Consider a typical 1-A, 5-V power supply. A typical dc resistance for a 2.0-A fuse could be 200 mΩ. However, voltage drop across the fuse at 1 A can be nearly 1 V. Can this circuit function reliably on the remaining 4 V?

6.3.3 Agency Approvals

Safety Agency Approvals for protectors are usually required. Most circuit protectors are approved (e.g., listed, recognized, certified, etc.) by agencies such as Underwriters Laboratories (UL of USA), Canadian Standards Association (CSA of Canada), and Ministry of International Trade and Industry (MITI of Japan). Elsewhere, national agencies, such as SEMKO for Sweden, and VDE for Germany, approve protectors, according to standards promulgated by the International Electrotechnical Commission (IEC).

Agency approvals for fuses and other protectors can be extremely difficult to interpret. Approval of a given component by one or more of the referenced agencies must be substantiated in writing by the component manufacturer to ensure that the certification exists and is sufficient to meet the customer's minimum criteria.

Equipment manufactured and sold in the U.S.A. normally must use protectors approved by UL or another recognized agency. The UL *listing* of a protector means that it successfully met all requirements of the referenced UL standard. UL *recognition* of a protector means that some part of the standard was not met, or not evaluated, or that the protector is not meant to be sold as a self-contained unit without further UL approval. The assembly incorporating UL-recognized protectors must be tested by UL with the protector in place to ensure that it will meet the end-use requirements. If a fuse is UL recognized, rather than listed, the fuse manufacturer should be asked what reason led to that classification. Was it a failure or deviation from the standard that led to this more limited status? It should not be interpreted that a recognized component is necessarily not as good as a listed component. It is very important that the designer understand the difference between recognized and listed status and be aware of the impact that this can have on the end product.

Equipment manufactured and sold in Canada must use protectors approved by CSA. CSA's certification serves the same purpose as UL's listing. CSA has recently added their equivalent to the UL recognition program. USA and Canadian standards are usually very similar. The recently adopted UL 248/CSA-22.2 No. 248 does make either agency's test results acceptable to the other's, in most instances.

A discussion of approval agencies would not be complete without a clarification concerning "approvals" and "equivalency," with regard to component substitution. Nominally similar components from several different manufacturers can be approved by an agency as "equivalent." This should never be taken as assurance that all devices will function identically, or even satisfactorily in a given application. The reason is that agency criteria for component approval must be liberal enough to ensure that all potential manufacturers can pass. That is, the approval agencies can dictate minimum requirements for components, such as fuses, but beyond the minimum requirements, manufacturers' fuses can differ. The agency cannot know actual end requirements on the device. Therefore, the approval status of a device cannot by itself guarantee correct fuse operation in applications where more stringent requirements than the approval agency tests for are necessary. This is especially true of fuses for precision applications.

For these reasons, fuses from different sources cannot be assumed to be "equivalent" without careful investigation by the fuse specifier. Furthermore, comparisons between different protector types, such as between fuses and circuit breakers, to determine "equivalency" can be confusing. Consider a comparison of a circuit breaker that has been UL listed as a 14,000-A interrupting rating, 125-V device, with a fuse that is UL listed as a 10,000-A interrupting rating, 125-V device. You could conclude that the protector capable of interrupting 14,000 A is more robust than that able to handle only 10,000 A, but this conclusion might not necessarily be valid.

It is important to understand that interrupting ratings for protectors assume certain test conditions. Fault currents for fuse tests are defined as minimums available from the source, assuming that the feed wires have minimal resistance. A circuit breaker is tested for short-circuit capability under a UL standard that was specifically written for circuit-breaker evaluation. The procedure required by UL for circuit breakers limits the short-circuit current by specifying significantly smaller cross-section feed wires than those used in fuse testing. This adds considerably to the circuit impedance, reducing the actual short-circuit current, challenging the protection device to less than 10,000 A. The voltage drop caused by heavy currents over the feed lines reduces voltage "seen" at the protector.

This comparison is not meant to disparage circuit breakers. It is intended to convey the importance of knowing that different protector types use different test conditions to determine ratings. Tests to evaluate fuses and circuit breakers are not directly correlatable, so rating numbers are not directly comparable. Unless devices are covered under the same standard and the same category within that standard, testing should be carried out by the circuit designer or component engineer to determine protector suitability for the application. Protector manufacturers are usually eager to give engineering assistance.

Safety agencies usually require that their logo be marked on approved protectors. Figure 6.31 shows some of the most common logos.

6.4 FUSE TECHNOLOGY

6.4.1 Introduction

The most common fuse size for supplementary fuses in North America is the 0.25" × 1.25"3AG. *AG* represents *Automobile Glass*, which reveals its origin in the automotive industry. Nonglass fuses of the same dimensions were given an *AB* suffix, such as 3AB, indicating the outer tube was Bakelite, ceramic, or some other substance. The largest fuse size shown in Table 6.2 is 5AG, called *the midget* because, even at 0.406" × 1.50", it is smaller than the smallest, 0.563" × 2.00", fuse recognized by the National Electrical Code. The 5.2-mm × 20-mm fuse size, (often shortened to "5×20mm") was first used in Europe, but is now a standard fuse size everywhere, including North America and some Pacific Rim nations. Figure 6.10 shows the comparative sizes of small-dimension fuses.

TABLE 6.2 Common Cylindrical Fuse Sizes
(courtesy San-O Industrial Corp.)

Size	Diameter	Length
1AG	0.250"/6.4mm	0.625"/15.9mm
1.5AG	0.157"/4.0mm	0.354"/9.0mm
2AG	0.177"/4.5mm	0.588"/14.9mm
3AG	0.250"/6.4mm	1.250"31.8mm
4AG	0.381"/7.1mm	1.250"/31.8mm
5AG	0.406"/10.3mm	1.500"/38.1mm
6AG	0.250"/6.4mm	0.875"/22.2mm
8AG	0.260"/6.4mm	1.000"/25.4mm
5 × 20mm	5.2mm	20mm
Pico	0.087"/2.2mm	0.295"/7.5mm
Pico	0.161"/4.1mm	0.394"/10.0mm
Nano™	0.080"/2.0mm	0.225"/5.7mm

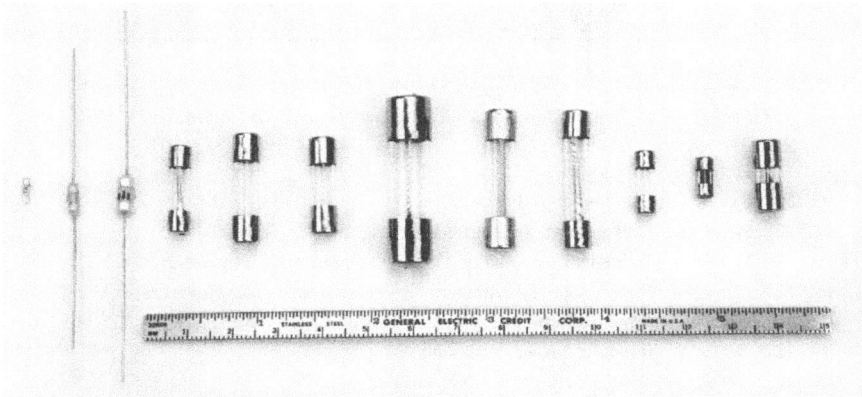

FIGURE 6.10 Small dimension fuses. *(Courtesy San-O Industrial Corp.)*

6.4.2 Fuse Standards

Fuse standards define ratings and other performance criteria. It is important to understand these standards so that fuses can be properly selected and specified.

ANSI/UL 198G, *Fuses for Supplementary Overcurrent Protection* governs fuses manufactured for use in the USA. The terms *small-dimension fuse*, *miniature fuse*, and *micro-fuse* are often used to refer to supplementary fuses covered by this standard.

UL 275 (SAE J544), *Automotive Glass Tube Fuses* governs certain low-voltage (32 Vdc) fuses used in automobile applications. Interrupting ratings are not required. This type of fuse should never be substituted for a UL 198G listed or recognized fuse.

CSA Standard C22.2 #59, *Fuses* governs fuses manufactured for use in Canada. CSA ampere-rating and interrupting-current requirements are similar to UL 198G. CSA has different temperature-rise requirements and can differ on power factor requirements. Therefore, fuses with certain current ratings might not have both UL and CSA approval.

UL 248/CSA-C22.2 No. 248, as noted earlier, has replaced the previously referenced UL and CSA standards for new fuses being submitted for listing or recognition. There have been some changes with this new standard and it should be reviewed carefully by the component engineer to ensure that devices continue to meet safety- and design-related requirements.

IEC 127, *Cartridge Fuse-Links for Miniature Fuses* (and supplements) governs fuses manufactured in most countries outside North America and certain Pacific Rim countries. The IEC organization writes standards, but does not give approval for fuses. Approval is given by national agencies, such as SEMKO (Sweden) and BSI (Great Britain), which base their Standards on IEC 127. VDE (Germany) will provide approval for fuse holders, but fuses are usually not necessarily VDE approved. Approval by any agency of a CENELEC member country is usually accepted throughout Europe. SEMKO (Sweden) is the only major test agency that requires that miniature fuses actually be tested to IEC 127. Other agencies, such as VDE, merely require that the fuse manufacturer declare conformance to IEC 127 or a comparable national equivalent. All SEMKO-approved fuses carry a stylized "S" mark, which is embossed on one of the fuse caps. SEMKO approval is a guarantee that the fuse meets IEC 127. Thus, SEMKO approval should be specified on all fuses for which IEC 127 requirements are essential.

IEC 127 is in the process of undergoing major revision. Based on preliminary information, this International Standard will be subdivided into several parts, as shown in Table 6.3:

TABLE 6.3 Standard IEC 127 Sections

IEC 127:	Miniature fuses (general title)
IEC 127, Part 1:	Definitions for Miniature Fuses and General Requirements for Miniature Fuselinks
IEC 127, Part 2:	Cartridge Fuse-Links
IEC 127, Part 3:	Sub-Miniature Fuse-Links
IEC 127, Part 4:	Universal Modular Fuse-Links (UMF)
IEC 127, Part 5:	Guidelines for Quality Assessment of Miniature Fuse-Links
IEC 127, Part 6:	Fuse-holders for Miniature Cartridge Fuse-Links (Previously IEC 257)
IEC 127, Part 7:	(Free for further documents)
IEC 127, Part 8:	(Free for further documents)
IEC 127, Part 9:	Test-holders and Test-circuits
IEC 127, Part 10:	User Guide

In Japan, the MITI standard governs fuse approvals in accord with the Electrical Appliance and Material Control Law (EAMCL). Two main types of fuses fall under this standard (Class A and Class B), with the Class A being faster acting. The Class-A fuse is similar in operating characteristics to the North American "Normal" acting fuses covered by UL and CSA. In fact, it is possible for a given fuse to carry approval markings for UL, CSA, and MITI. The Class-A or Class-B fuse in Japan must be marked with the appropriate letter enclosed in a circle. The choice of class to be marked is left to the discretion of the fuse manufacturer and is based on the fuse operating characteristics. Although the Class-A fuse, rated up to 60 A, must open in 60 minutes or less (melting time) at 135% of rated current, the Class-B fuse must perform within the same time limits at 160% rated current. Additionally, both Class-A and Class-B fuses must open at 200% rated current in 2 minutes maximum (melting time) for 30 A and lower-rated fuses, and 4 minutes maximum for those fuses rated above 30 A, but less than or equal to 60 A.

A third group of fuse types includes what are described as *motor fuses*. Motor fuses, rated up to and including 60 A, must open (melting time) at 135% rated current in 120 minutes or less, and in 4 minutes or less at 200% rated current. Further, these fuses must open in not less than 3 seconds (melting time), but not exceed 45 seconds, at 500% rated current.

A possible fourth group of fuses are those defined as *special fuses*, including "microfuses." This group is very similar to the "microfuses," defined by international (IEC) and North American agencies.

The United Kingdom is the only country that requires a fused power plug. The reason has to do with the ring wiring system used in British buildings. The fuse used in British power plugs must conform to BS 1362, *General Purpose Fuse Links for Domestic and Similar Purposes*, and is a unique size: 1.0" long (25.4 mm).[3]

Selection of a fuse must be performed with care because the safety of the equipment operator is at stake. Therefore, it is necessary to know which characteristics of a fuse are important. As stated earlier, approval agencies differ in their philosophies and techniques for rating fuses.

6.4.3 Fuse Parameters

The conductive element of a fuse is a metal wire or ribbon with a positive temperature coefficient of resistance. In the normal operating state, the element has a finite, relatively small resistance. Resistance generally increases as fuse current rating decreases. For example, a common 5- × 20-mm 15-A fuse has a typical dc resistance of 4.3 mΩ, but a 3AG 62-mA fuse can have a typical dc resistance of 152 Ω. Current through the element results in power generation, according to the formula $P = I^2R$. The power generation results in heating the wire to a temperature at which heat loss through radiation, conduction, and convection equals heat generation. As temperature rises because of increasing current, the resistance of the element also rises, which increases the I^2R power generation in the element, until temperature equilibrium is reached. At some critical current, whose value is determined by the construction of the fuse, the temperature does not stabilize, but instead increases continuously until the element melts.

In summary: Below the critical current, temperature stabilizes at some equilibrium value below the melting point of the fuse element material. Above the critical current, temperature would increase past the melting point of the fuse element, but it is prevented from doing so by the melting of the element.

When the fuse element melts, it opens a gap, and an arc is formed. Electrical power in the form $E \times I$ is dissipated in the arc. This power heats the remaining element material at the ends of the gap. The material melts back, and the distance of the gap continually grows. Eventually, the gap becomes sufficiently wide that the voltage cannot sustain the arc, and the arc is extinguished. Arc physics is highly complex, and is not addressed further here. For more information, see the books by McCleer, and Wright and Newbery.[3]

This sequence of events occurs if the overcurrent condition is relatively moderate. If overcurrent is severe, such that no appreciable heat dissipation occurs from the fuse element to its surroundings, the element vaporizes, rather than melts.

The *current rating* is the most obvious fuse-selection parameter. Under UL 198G and CSA 22.2 #59, a 3AG normal fuse rated 1.0 A must be capable of carrying 110% rated current continuously at 25°C. The word *continuously* implies forever, but as a practical matter, UL and CSA have defined continuously to mean that the current continues to flow for a minimum of four hours. To expedite production testing, this was modified so that 110% rated current is continuously carried and the fuse temperature is stabilized for three consecutive 10-minute intervals. More recently, the new UL 248 / CSA-C22.2 No. 248 standard has reduced this testing to four consecutive 5-minute intervals. The essence of the definition of current rating for a fuse is the maximum current for which the fuse-element temperature is stable with time, and is below the melting point of the fuse material.

In practice, a UL-rated 1.0-A fuse should not normally be used in a steady-state circuit exceeding 750 mA. This 75%-of-rated-current rule is considered too high by conservative designers; they would limit the steady-state current to not more than 60% of rated current. This is a matter of individual choice and, in many cases, reflects a difference between designing for a planned product life of 2 years versus a planned product life of 20 years. In either case, continued use above 60% or 75% of the rated level will significantly shorten the life expectancy of the fuse.

This UL 198G-qualified (nontime delay) 1.0-A fuse must be capable of opening within 60 minutes at 135% of rated current, and within 2 minutes at 200% rated current. Actually, UL 198G states that at 135%, two of three fuses must open within the 60-minute time limit. The third fuse, if it has not opened, must open within 5 minutes at 150% of rated current. A typical 1.0-A fuse will open in approximately 2.0 seconds at 135% rated current and in approximately 0.34 second at 200%. The actual clearing time for a fuse at either of these points will normally be significantly shorter than the allowable UL limits. This is emphasized here to show that the 60-minute and 2-minute requirements respectively for UL are absolute maximums for purposes of ensuring safety. Real fuses can open at substantially different times. Designers too often use these maximums, rather than obtaining and using typical values available from fuse manufacturers, and then wonder why the fuse does not behave as expected.

A component or compliance engineer can generate a specification that is tighter than the agency requires. If the customer sets limits that are too tight, the specification can be impossible to meet at a reasonable cost. However, safety agency limits are loose enough that there is much room to set tolerances well within maximums allowed for safety approval. Customer component specifications can fix the requirements of the fuse well enough that true parameter equivalence between fuses purchased from different manufacturers is guaranteed. Even if a customer decides to accept the manufacturer's normal limits, the limits should always be specified in writing to avoid dispute. The manufacturer should certify in writing that his product will meet the customer's requirements. Evidence offered by the fuse manufacturer of "typical" values does not constitute certification. A well-written "source control drawing" included with the purchase order, and accepted by the manufacturer, will normally suffice.

IEC 127 has a different philosophy for current rating of fuses. Table 6.4 shows some of the differences between UL/CSA and IEC 127 fuses. An IEC 127 fuse rated at 1.0 A will comfortably carry 1.0 A continuously. In that sense, IEC 127 fuses are more robust than UL 198G fuses of equal nominal current rating. IEC 127 requires fuses to withstand 120%

TABLE 6.4A IEC 127 differentiates between classes of fuses as follows:

IEC 127 sheet number	Clearing time	Symbol	Breaking capacity
I	Quick	F	High
II	Quick	F	Low
III	Time-Lag	T	Low
V	Time-Lag	T	High

(Reprinted from Panel Components Corporation's Export Designer's Reference & Catalog #8)

of rated current for a minimum of 100 cycles. It defines each cycle as composed of a period of 60 minutes operation at 120% of rated current, followed by a period of 15 minutes of zero current, followed by a period of 60 minutes at 150% of rated current.

IEC 127 and UL 198G/CSA 22.2 #59 standards differ enough that a fuse that meets one generally cannot meet the other. As Table 6.4B shows, these standards have radically different clearing-time requirements. For instance, a UL 198G/CSA C22.2 #59 fuse must open within one hour at 135% of rated current. In contrast, an IEC 127 fuse must be able to carry 150% of rated current for at least one hour. These requirements are mutually exclusive.

For these reasons, a fuse cannot carry both a UL listing to UL 198G and approval to IEC 127. Some fuses do, however, carry both UL recognition to IEC 127 and approval by some other agency (such as SEMKO, to IEC 127). UL tests these fuses to confirm their compliance with the IEC 127 standard, but will not normally allow the use of such fuses in UL-listed electrical equipment without further testing in that equipment. Without specific documentation from the approval agency and/or fuse manufacturer, the end user cannot be certain of the conditions used for testing under which UL recognition was granted.

Ambient temperature can significantly influence fuse operation. Most fuses are designed to meet rated parameters at just above room temperature: 25°C for UL 198G/CSA C22.2 #59 fuses, and 23°C for IEC 127 fuses. Circuit designers generally need not be concerned with deviations of less than ±20°C from this nominal. Although this small temperature deviation will result in changes in fuse rated current and voltage drop, it will not normally exceed 10% of the nominal. This is well within the typical time/current manufacturing variation from one fuse to another.

However, when ambient temperature deviation exceeds ±20°C, correction factors should be incorporated to optimize circuit protection and fuse life. A fuse's current-carrying capacity decreases as the ambient temperature rises. Correction factors are available from the fuse manufacturer in the form of curves similar to Figure 6.11. At elevated ambient temperatures, a fuse with a higher current rating must be used to provide the desired overload protection. Similarly, a lower temperature environment will necessitate use of a fuse having a lower current rating to provide the desired overload protection. Remember that the opening time of fuses is also affected. Fuse opening time decreases with increasing temperature. If ambient temperature excursions much greater than ±20°C regularly occur, or if temperature excursions exceed the ±20° limit in an unpredictable manner, a fuse manufacturer should be consulted to help choose the correct fuse.

Interrupting rating, also known as *short-circuit rating* or *breaking capacity*, defines a fuse's ability to clear a fault condition without being destroyed in the process. Short-circuit current is the maximum current that the fuse is guaranteed to interrupt safely in the circuit

TABLE 6.4B Time-Current Characteristics . . . A Comparison of North American and International Types

| Percent of fuse rating | Ampere range | North American U.L. 198G and CSA C22.2, No. 59 | | | | International IEC 127 (International Electrotechnical Commission, Publication 127) | | | | | | | |
		Fast-acting Min.	Fast-acting Max.	Time delay Min.	Time delay Max.	Sheet I Quick, high Min.	Sheet I Quick, high Max.	Sheet II Quick, low Min.	Sheet II Quick, low Max.	Sheet IV Time-lag, low Min.	Sheet IV Time-lag, low Max.	Sheet V Time-lag, high Min.	Sheet V Time-lag, high Max.
110%	0–30A	Cont.	—	Cont.	—								
135%	0–30A	—	1 hr.	—	1 hr.								
150%	32mA–5.3A	—	—	—	—	1 hr.	—	1 hr.	—	1 hr.	—	1 hr.	—
200%	0–3.0A	—	2 min.	5 sec.	2 min.								
	3.1–30A	—	2 min.	12 sec.	2 min.								
210%	32mA–6.3A					—	30 min.	—	30 min.	—	2 min.	—	30 min.
275%	32mA–3.9A					.01 sec.	2 sec.						
	4A–6.3A					.01 sec.	3 sec.						
	32–100mA							.01 sec.	.5 sec.	.2 sec.	10 sec.		
	125mA–6.3A							.05 sec.	2 sec.	.5 sec.	10 sec.		
	1A–3.15A											1 sec.	80 sec.
	3.15A–6.3A											1 sec.	80 sec.
400%	32–100mA					.003 sec.	.3 sec.	.003 sec.	.1 sec.	.04 sec.	3 sec.		
	125mA–6.3A					.003 sec.	.3 sec.	.01 sec.	3 sec.	.15 sec.	3 sec.		
	1A–3.15A											95ms.	5 sec.
	3.15A–6.3A											150 ms.	5 sec.
1000%	32–100mA					—	.02 sec.	—	.02 sec.	.01 sec.	.3 sec.		
	125mA–6.3A					—	.02 sec.	—	.02 sec.	.02 sec.	3 sec.		
	1A–3.15A											10 ms.	100 ms.
	3.15A–6.3A											20 ms.	100 ms.

(Reprinted from Panel Components Corporation's Export Designer's Reference & Catalog #8)

TABLE 6.4C Breaking Capacity of North American and International Fuse Types

Standard	Voltage & rating	Amperage rating	Breaking capacity
UL 198G & CSA C22.2, No. 59	125 VAC	All	10000A
	250 VAC	0–1A	35A
	250 VAC	1.1–3.5A	100A
	250 VAC	3.6–10A	200A
	250 VAC	10.1–15A	750A
	250 VAC	15.1–30A	1500A
IEC 127			
Sheet I	250 VAC	All	1500A
Sheet II, III	250 VAC	0–3.5A	35A
	250 VAC	3.6–6.3A	10 times rated current (36–63 Amps)
Sheet V	250 VAC	All	1500A

(Reprinted from Panel Components Corporation's Export Designer's Reference & Catalog #8)

at the rated voltage—normally 125 V rms with a power factor of 0.7 to 0.8. A UL-listed fuse will have a short-circuit current rating of 10,000 A at 125 V rms. A lower capability (e.g., 5000 A) at 125 V rms can be approved by UL as a recognized component. Lower interrupting ratings apply for 250-V fuses (Table 6.4c).

IEC 127 short-circuit current for high breaking capacity (HBC) fuses is 1500 A, and for low breaking capacity (LBC) fuses is 35 A or 10 times rated current, whichever is greater. This test is always carried out at 250 V rms for IEC 127 fuses. A recent revision to the IEC 127-2 specification has added an enhanced breaking capacity (EBC) fuse. This fuse has a

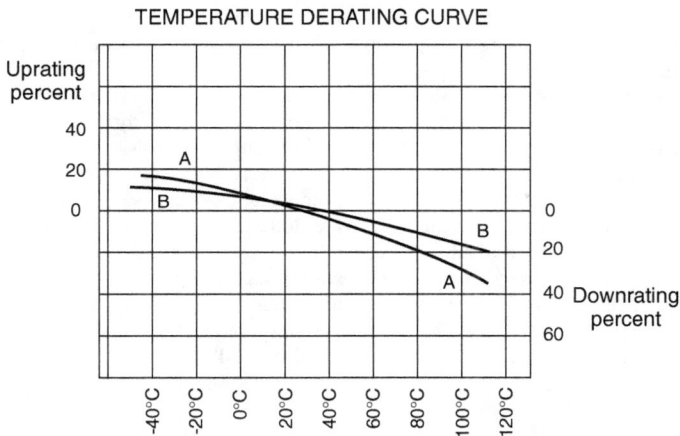

Curve A - Time delay fuses

Curve B - Normal (quick) acting and medium lag fuses

FIGURE 6.11 Effects of ambient temperature on current-carrying capacity. *(Courtesy San-O Industrial Corp.)*

breaking capacity of 750 A at 250 V rms. Further, there are efforts to replace the low breaking capacity fuse with one having a 150-A rating at 250 V rms.

The fuse chosen for an application must have an interrupting rating equal to or greater than the maximum current available from the supply into a short circuit. This maximum current is called the *short-circuit prospective current*, or the *prospective fault current*.

Fuses qualified to IEC 127 as *low breaking capacity devices* have a much lower interrupting rating than do fuses qualified to UL 198G. How is it that IEC 127 fuses are used successfully? The answer is found in examination of the prospective short-circuit current available from power sources and typical load impedances. Assume that the main supply is an ideal 220-Vac voltage source, and the load is connected through a 2-meter 18-AWG powercord. The powercord will have a total wire resistance of about 100 mΩ. Adding the plug and other termination contact resistance, a 10-meter 14-AWG feed wire from the service entrance panel at approximately 170 mΩ, and at least 12 mΩ for the fuse itself, will total a minimum resistance of 500 mΩ. Adding in the inevitable source impedances and inductive reactances, will mean the actual prospective fault current is normally within the maximum rating of 63 A, which is 10 times the largest IEC 127 fuse current rating of 6.3 A.

Many applications will require a higher breaking capacity than is available with the low breaking capacity device. Where protective short-circuits can be higher, high breaking capacity fuses must be used at this time. Their 1500-A interrupting rating at 250 Vac is at least as high as that required for UL/CSA and is more than adequate in most cases. As noted, the new 750-A "medium breaking capacity" fuse and change from the present "low breaking capacity" limits to the new 150-A level should provide more cost-effective options.

If all circuits were connected to the power source with 18-AWG wire through standard wall outlets, and loads of a few ohms were always present, short-circuit levels could be relaxed to those used by the IEC. However, the same fuse can be used in an industrial application where the feed is a 10-AWG cable or even a bus bar. The typical UL listed fuse would meet all of these stresses, but the IEC device cannot! UL and CSA conservatively choose to require high interrupting ratings for miniature fuses because of the possibility that the fuse might be used in applications with large prospective currents.

Power factor is the ratio of real power to apparent power, or $VI \cdot \cos \theta$, where θ is the phase angle between voltage and current. Breaking capacity (or short-circuit current) is measured at a specified power factor. A power factor less than 1 indicates a circuit with energy-storing components (e.g., capacitor or inductor) are present. This stored energy is additive to the line capability. Thus, it presents additional total energy that the fuse is required to interrupt, essentially increasing the apparent $I^2 t$ ("let-through current") value. This increased stress requires additional energy interrupting capability for the fuse.

A fuse can be required to withstand the short-circuit limits at a power factor of 0.7 to 0.8 because operation at low power factors is the worst case. The same fuse would be able to safely interrupt more energy if the test were done at a power factor of 1.0. This is another parameter that would result in a fuse being recognized rather than listed by UL per 198G. Failure of the power-factor criteria, even though all other requirements were met, would restrict the fuse to UL recognition, rather than listing. UL 198G required a power factor of 0.7 to 0.8. CSA certification had always required testing at only a power factor of 1.0 for small fuses. The new UL 248/CSA-C22.2 No. 248 standard has been harmonized to require that all short circuit testing for fuses with a voltage rating of 125 Vac or greater to be done at a 0.7 to 0.8 power factor.

Rated voltage, which is usually marked on the fuse body, indicates that the fuse will safely interrupt its rated short-circuit current in circuits where the source voltage is equal to or less than its rated voltage. The IEC 127 rated voltage is always 250 V rms. UL 198G/CSA 22.2 #59 fuses are usually rated at 125 V rms. Fuses are available at higher voltage ratings (generally 250 V rms), but at the cost of a substantially reduced interrupting rating. The interrupting rating current of some manufacturers' fuses is as high as

10,000 A at 250 V rms. However, UL requires only that the fuse safely interrupt between 35 A and 1500 A, as a function of rated current, at 250 V, to qualify for listing (Table 6.4c). These are the minimum requirements.

Information about a fuse's ability to meet more than the minimum requirements should be available from the fuse manufacturer. If these enhanced ratings are necessary for the fuse's intended application, the circuit designer or components engineer should request that product certification be provided by the fuse manufacturer in writing, as well as evidence of agency approval.

Notice that UL-recognized components can have rated voltages other than 125 V or 250 V. One such example is 300 Vdc, with a short-circuit current of 1000 A, for some telecommunications applications. Although 60 Vdc is generally associated with telecommunications, the 300-Vdc level is still required in older teletype systems (±130 Vdc) and microwave offices (+130 Vdc, –48 Vdc). Many of these facilities use vacuum tubes with a 250-Vdc plate voltage and have 130-Vdc battery strings on site.

A 60-Vdc rating is satisfactory for U.S. telecommunications (e.g., 48 Vdc nominal plus 5 Vdc float and some safety margin), but European telecommunications will often require an 80-Vdc rating (e.g., 60-Vdc nominal battery voltage, plus the 5- to 10-Vdc float and some safety margin).

Fuses are most often rated at power frequencies, 50 or 60 Hz. There is little difference in behavior at these two frequencies. A fuse whose parameters are calibrated at 60 Hz will work satisfactorily at 50 Hz. Lower frequency and dc voltages can cause problems. Total fuse operating times at lower frequencies are longer, which means higher energies are released in the fuse. For this reason, low-frequency and dc applications can require voltage derating of the fuse. In dc applications, the voltage rating can be only half the ac rating, and high values of circuit inductance can necessitate further voltage derating. High-frequency power applications, such as 50-kHz induction heating applications can also require voltage derating of fuses because of skin effect and other considerations. Fuse manufacturers can give recommendations for voltage derating in unusual fuse applications.

I-t *curves* provide a method of showing the relationship between current and clearing time for a given fuse. Figure 6.12 shows an I-t graph form often used by North American manufacturers for UL 198G fuses. The curve is plotted on log-log scale, with current on the ordinate and time on the abscissa. The curve represents average values. The total clearing time (T_c) shown on the curve is equal to the melting time (T_m) plus the arcing time (T_a). At low current levels (e.g., 200%), the total time is almost exclusively melting time. As the overload current increases, the arcing time increases to the point that arcing time accounts for the majority of the total clearing time. Note that the standard deviation of clearing time at the low-current levels is very large. Measurement of the low current-level clearing time can be difficult because of variables present that can significantly affect the ambient environment (e.g., air passing over the fuse under test providing a cooling effect, etc.). These inconsistencies will result in a very wide standard deviation (variance) for clearing time in this area of the curve. In contrast, as illustrated in Figure 6.13, the standard deviation at high current levels is quite small and relatively unaffected by external factors.

We will now review the I-t curve shown in Figure 6.13 in greater detail.

Each point on the I-t curve is a "typical" value of clearing time t_c for a given "I." This "typical" value is arrived at by testing several samples under controlled overload conditions. The normal distributions shown in Figure 6.13 represent the distribution resulting from N samples at each point. Keep in mind that these distributions are not associated with the actual coordinates used in this graph. They are simply being shown to help understand the reason for variation about the I-t curve.

For North American agency-approved fuses (e.g., UL and CSA), "t" for a given "I" from 75% to 110% of rated current is quite unpredictable. The fuse can be permanently degraded without opening as a result of irreversible metallurgical changes from thermal

SAN-O INDUSTRIAL CORPORATION

SOC FUSE TYPE TSCR - 3 A

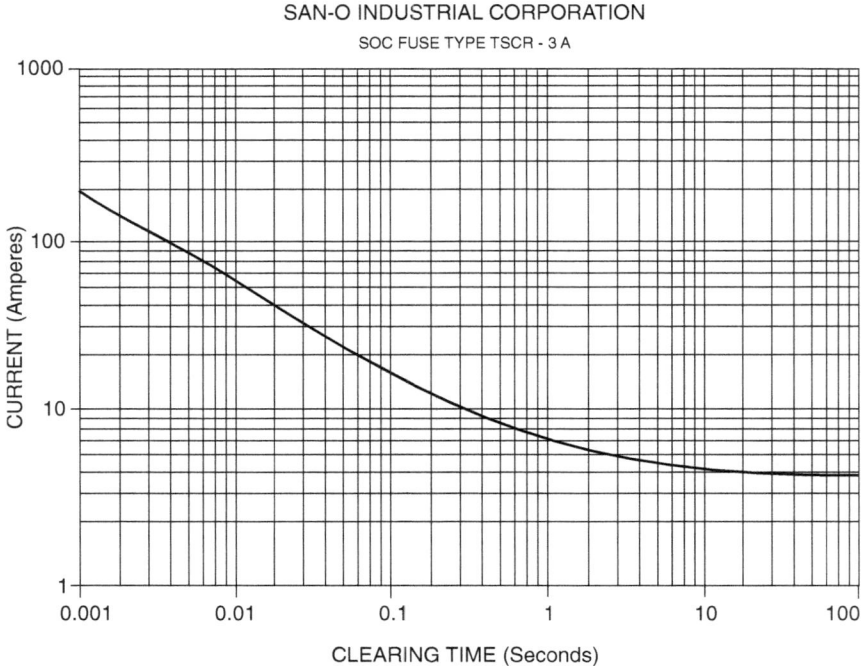

FIGURE 6.12   Typical UL 198G I-t fuse clearing time curve. *(Courtesy San-O Industrial Corp.)*

stresses in the element. At 75% rated current or less, for instance, the fuse will normally provide a full life. Aging will be minimal. Continuous current (I) above 75% is likely to seriously shorten the life of the fuse. Actual life to be expected will vary, depending on element materials selected and fuse construction. But for all fuses, there will be an overcurrent point in this region for which the fuse element characteristics will be permanently changed.

Even though the current is subsequently reduced to an acceptable level (e.g., less than 75% rated current), the fuse is no longer the same as it was prior to being stressed. At a continuous current ranging from 110% to 135%, the fuse will ultimately open. However, the time for opening will vary greatly from fuse to fuse of a given type in this region of the I-t curve. Inconsistencies in fuse operation in this region (e.g., 75% to 135% of rated current) are primarily caused by subtle differences in fuse construction and ambient condition variables. The amount of energy being applied to the fuse element can be affected significantly by variations in air flow across the fuse or the mass of the connecting clips. Ambient temperature will greatly affect the actual time (t) for a given current within this range.

As the current increases above the 135% point, the variables causing the large deviations in time of operation become less significant. It will be evident that the standard deviation for a statistically significant sample size will decrease as the overcurrent increases. This is true until very large overloads are approached (e.g., approaching the interrupting rating), where the unpredictability associated with significant fuse arcing will again cause the standard deviation to increase. These heavy overload currents are generally well above those represented in the I-t curve region.

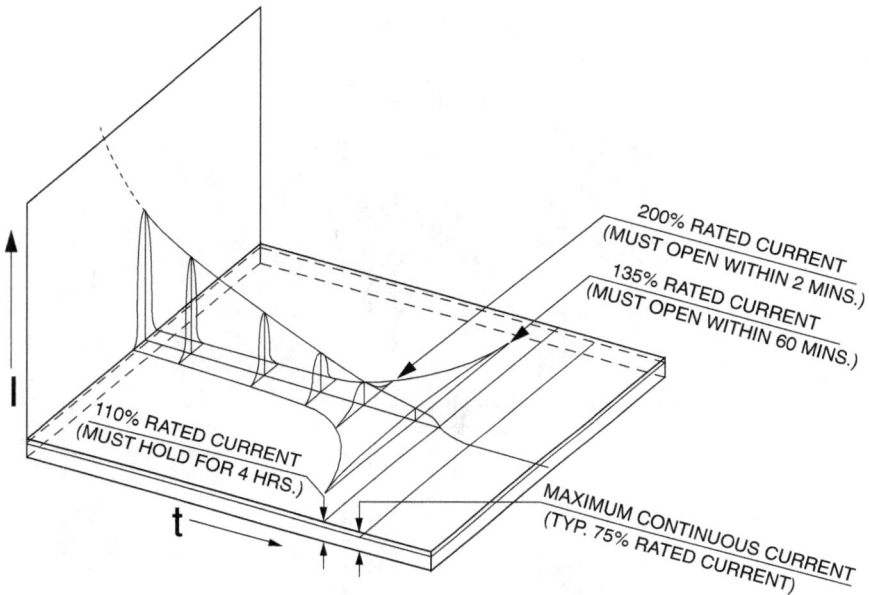

FIGURE 6.13   Typical fuse clearing time standard deviation.

Similar conditions apply to fuses approved by other agencies (e.g., SEMKO, DEMKO, etc.), but at slightly higher overloads. The 75% rated current for the North American agencies is approximately equivalent to the 100% rated current range for the European agencies testing to IEC 127. Likewise, the 135% rating moves to approximately 150% when comparing.

Summarizing, an I-t curve can be viewed as having a large standard deviation for time (t) for a given overload current (I) at low overload levels. The standard deviation for t will narrow as the overload current (I) continues to increase. This results in creating a "band" around the nominal I-t curve providing upper and lower limits. Again, the deviation from the nominal seen at this point is the result of variations encountered in:

1. Fuse components
2. Fuse-assembly operations
3. Subsequent circuit-assembly practices
4. Circuit applications

The first two items must be controlled by the fuse manufacturer and are ultimately reflected in the price of the component. The latter two items are controlled by the equipment manufacturer.

As noted, clearing time at high current levels is very repeatable. However, the temptation to extrapolate the I-t curve beyond the maximum current plotted should be tempered with the understanding that any data on the curve is only an approximation or statistical average, so extrapolations are dubious.

Although the I-t curve shown in Figure 6.12 is one of the typical curves used in the industry, Figure 6.14[1] shows another format used just as often. This format places "time" (t) on the ordinate, rather than the abscissa—a mathematically more-correct arrangement, but

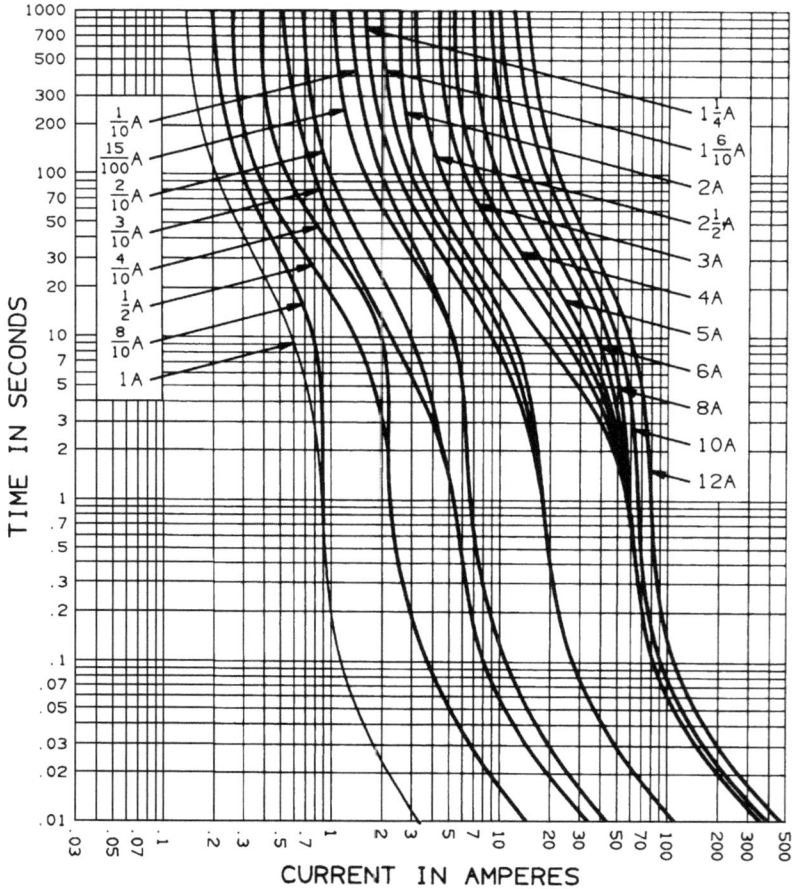

FIGURE 6.14 Typical UL 198G I-t fuse melting time curves. *(Courtesy Gould Scawmut)*

not necessarily as intuitive as the first example. The use of either curve is acceptable. Fuse manufacturers appear to be split evenly or which representational format more effectively conveys the concept of clearing time as a function of current. It might be of interest that circuit-breaker manufacturers, as well as some primary fuse and secondary fuse manufacturers, appear to use the format of Figure 6.14 exclusively.

A subtle distinction that makes the two curves difficult to compare is shown in Figures 6.12 and 6.14. Figure 6.12 is a plot of current vs. clearing time, and Figure 6.14 shows current vs. melting time. This difference is significant—especially at high overcurrents, where arcing time dominates the clearing time.

Finally, the curve in Figure 6.12 is descriptive of only one current rating. Many manufacturers will combine several current ratings on a single curve (Figure 6.14). A single curve for each fuse makes it easier to update curves for the fuse manufacturer and lessens the chances for error in readings. Multiple curves, however, do make selection of an optimal current rating easier because all values are simultaneously in view during the selection.

Figure 6.15 shows a third type of chart, often used by manufacturers of IEC 127 fuses. Here, time (t) is the ordinate, and "multiples of rated current" is the abscissa. The chart is an interval graph showing a minimum melting time curve, and a maximum melting time curve for any current rating or set of current ratings. The curves are defined by IEC 127 minimum and maximum limit points, called *gates* or *checkpoints*, which are often shown on the graphs as directional triangles. The fuse manufacturer guarantees that his fuse's melting (not clearing) characteristics will fall between the two curves, under test conditions specified in IEC 127, but will not guarantee exactly where between the curves, except by special arrangement with the customer. Because application conditions can differ from IEC 127's standardized test conditions, the fuse's time-current characteristics can differ somewhat from the curve area.

Dc resistance is frequently referred to as the *cold resistance* of the fuse. It is generally measured using a Wheatstone bridge or ohmmeter, at no more than 10% of the fuse's nominal rated current. Fuses with very low current ratings can have relatively high dc resistances.

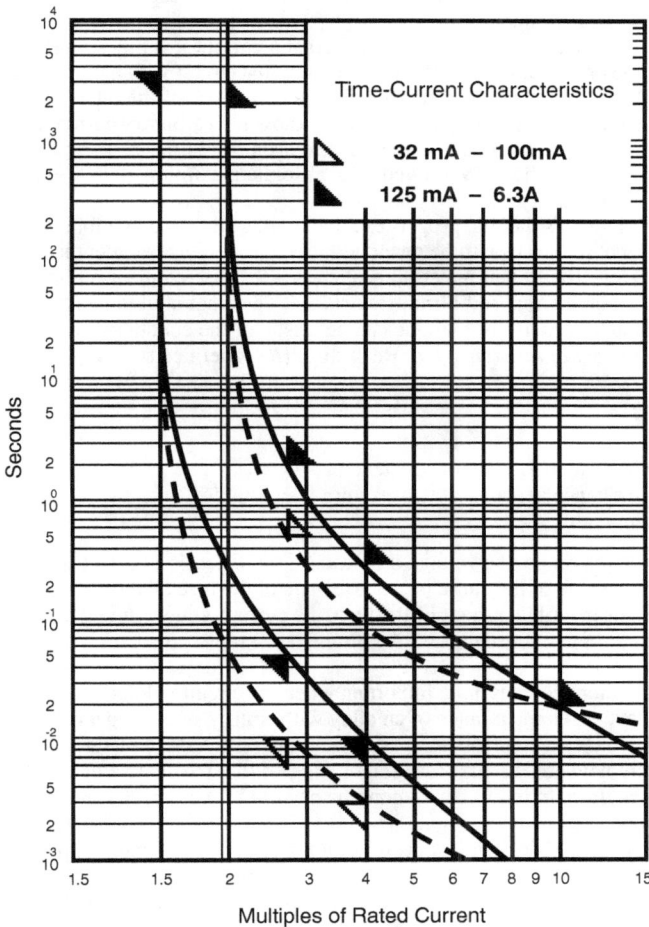

FIGURE 6.15 Typical IEC 127 I-t fuse melting time curves.

A 50-mA fuse can have a dc resistance around 150 Ω, which can be problematic in low-voltage circuits. Fuse elements with very low current ratings must necessarily be very thin, and therefore relatively fragile. For this reason, materials with higher tensile strengths are often used for fuses with low current ratings. These materials often have higher resistivities. Higher-resistivity materials permit thicker, more-robust fuse elements.

Although dc resistance is an interesting characteristic of a fuse (and good for a quick check to determine that a fuse has the correct current rating), voltage drop is a much more valuable measurement.

Voltage drop is a dc measurement of the voltage across the fuse element at a designated current. It can be thought of as a measure of "hot resistance" or "loaded resistance." It is important that this measurement be limited to the fuselink itself. Contacts, such as clips, will have a voltage drop from the clip to the fuse end cap. In some circumstances, this voltage drop will be significant. Special care should be taken to ensure that any clip-to-fuse-cap voltage drop is not included in the measurement. Using voltage drop data and Ohm's Law, the equivalent resistance can be determined. Voltage drop and dc resistance are two distinctly different parameters.

Usually, the voltage drop will be measured at a current level meaningful to the circuit designer. If the protector is a 10-A fuse to be used in a circuit with a normal operating current of 3.0 A, the voltage drop would be calculated at 30% of the rated current. If, on the other hand, the normal operating current is not known, an arbitrary current level of 50% to 75% is typically selected. Because the voltage drop test is meant to be a nondestructive test, the currents used should be limited to an absolute maximum of 75% of the fuse's rated current for UL 198G fuses.

The fuse element's thermal properties affect voltage drop calculations. Current through the fuse element causes heat to be generated. Because the fuse element is normally a positive temperature coefficient material, heat causes the fuse-element resistance to increase to a level at which thermal equilibrium is reached. The amount of increase in resistance per degree of temperature rise is a unique characteristic of the conductive material, called the *thermal coefficient of resistance (α)*. Resistance (R) is defined by:

$$R = R_0(1 + \alpha t) \ \Omega$$

where,

α = Thermal coefficient of resistance $\Omega/°C$
R_o = Resistance at standard temperature (usually 25°C)
t = temperature delta, °C

Within a given range, the voltage drop caused by increasing current is a linear function of current. The slope of the curve is characteristic of the fuse element material. When current approaches the rated current of the fuse, the current-temperature curve becomes nonlinear (α is no longer constant) and the temperature rapidly reaches the melting point of the element.

This parameter becomes quite important when the circuit voltage is small (e.g., 5 Vdc). A fuse can have an element made of an alloy with a unity slope, and a nominally "equivalent" fuse can have an element (such as constantan) with a slope of zero. A lower dc resistance measurement for the first fuse might suggest that the first sample would have less circuit voltage drop. In reality, the second fuse could have a lower voltage drop at the operating current level.

Voltage drop specification is a useful tool for ensuring that a given fuse meets necessary constraints. However, specification of this parameter can limit the fuse manufacturer to using only certain element materials for a given fuse current rating. This can result in adding unnecessary cost to the fuse. So, voltage drop should be added to a fuse specification only after consultation with prospective fuse manufacturers.

I^2t is a measure of the thermal energy (heating effect) of a fault current in ampere-squared-seconds. The heating effect of a current is I^2Rt (energy), or more generally, the integral $\int i^2Rdt$. In a series circuit, in which all elements see the same current, the heating effect in each element is proportional to I^2t. I^2t is given as a "fusing" parameter for some fuses and such devices as diodes and thyristors. Each electronic component's I^2t rating is the time integral of current that will destroy the device. A fuse is said to "let through" an amount of time integral of current (energy) before the fuse blows. So, if a fuse's I^2t rating is smaller than the I^2t rating of the electronic devices it protects, the fuse will open before the other devices can be damaged. The I^2t parameter is defined in applicable UL and IEC standards for certain large primary and secondary current-limiting fuses. The I is rms amperes. I^2t rating is a function of the voltage, fault current, frequency and closing angle—the closing angle being the point along the sinusoidal curve at which the switch is actually closed. All of these parameters must be clearly defined to provide data useful to the circuit design engineer. I^2t tests are not defined in standards covering miniature and subminiature fuses (UL 198G, CSA C22.2 #59, UL 248/CSA-C22.2 No. 248 or IEC 127), so comparisons between fuses from different manufacturers should use comparable test methods. I^2t values are usually only meaningful for clearing times of less than a half cycle—approximately 8 ms for 60-Hz applications.[10]

Manufacturers of some components, such as diodes, publish I^2t ratings for their products. In these cases, a fuse protecting such items should have a lower I^2t rating than that of the device being protected. Specialized "super fast (semiconductor) fuses" are manufactured to protect semiconductors.[4] Actual values of I^2t are most easily obtained by contacting the technical sales or engineering departments of the fuse manufacturers.

Specification and qualification (and correct second-sourcing of fuses) requires specific agreement between customer and fuse manufacturer on the expected characteristics of the fuse. It is not sufficient merely to require that a fuse have safety-agency approvals because two fuses might have identical agency approval, yet have different current-time-temperature characteristics. A fuse manufacturer will typically supply a one-page complete description of the fuse characteristics. Figure 6.16 shows a typical data sheet from San-O Industrial Corporation.

A good data sheet will give measured points so that an I-t curve can be drawn, if necessary. Clearing time data are usually given as "typical" values, unless the customer specifies that "minimum" and "maximum" values must also be supplied. Of course, the manufacturer's engineering staff and the customer's engineering staff must agree on the upper and lower bounds of parameters. This is especially true of the "low overcurrent" parameter.

The customer should have his own specification drawing, often called a *source control drawing*, which defines all necessary aspects of the fuse behavior, packaging, and safety agency approvals.

Mechanical dimensions must also be evaluated when designing a fuse into the circuit. Adequate room should be left for the fuse. Typically, the fuse is the last component to be selected and the circuit layout is fixed beforehand. This can force the selection of a fuse with less than the best possible characteristics. A fuse's ability to interrupt high currents at the rated voltage is a function of the physical design of the device. All other parameters being equal, the larger the fuse body, the greater the interrupting capability. This relationship can be altered through use of various fillers and/or high-strength glass.

Arc-quenching filler materials surround the fuse element in fuses designed for high breaking capacity. These materials work by falling into the space previously occupied by the element after it melts. The filler material is resistive, so the effective length of the arc is increased, and the arc extinguishes sooner. The most common filler is granular, normally quartz sand. Grain size, consistency of grain size, and packing density are important variables that are carefully monitored by fuse manufacturers. Consistent packing density is achieved by filling, packing, and refilling several times during processing. Adherence to

SAN-O INDUSTRIAL CORPORATION

ELECTRICAL CHARACTERISTICS

PART NO: __TSCR__   AMPS _3.0_   VOLTS _125_

BODY: _GLASS TUBE - RADIAL PIGTAIL TYPE: NOMINAL 20 x 5mm_

SAFETY AGENCY APPROVALS

UL FILE E39265:	LISTED	_XXX_
	RECOGNIZED	___
	PENDING	___
CSA FILE LR 34647:	CERTIFIED	_XXX_
	PENDING	___
MITI:	APPROVED	_XXX_

IEC AGENCY (_____):   CERTIFIED   ___   LBC ___   HBC ___

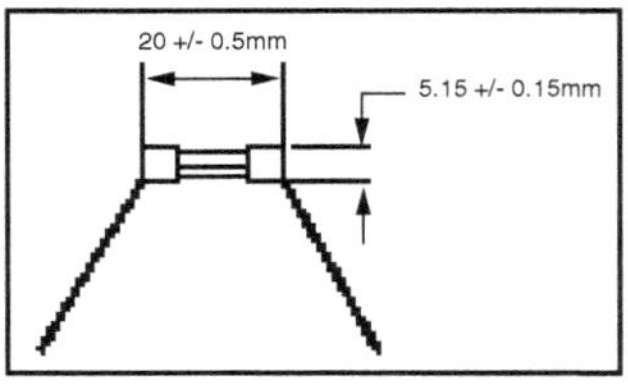

20 +/- 0.5mm

5.15 +/- 0.15mm

COLD RESISTANCE (_mΩ_):   MAXIMUM _43_   MINIMUM _40_   AVERAGE _41_

VOLTAGE DROP: _____   (@ ___ % Rated Current)

TESTS PER   __UL 198G__ - MINIATURE FUSES

CURRENT CARRYING CAPACITY: _110%_   Rated Current Continuously

SHORT CIRCUIT RATING: _10,000A_   125Vac   _N/A_   250Vac

TEMPERATURE TEST: Less than 70°C Rise @ _110%_

TYPICAL CLEARING TIME TESTS (125 Volts):

	135%-4.35A	2x-6.0A	3x-9 .0A	5x-15A	10x-30A	20x-60A
Maximum	4.4s	0.78s	0.29s	100ms	29ms	84ms
Minimum	3.5s	0.72s	0.28s	82ms	27ms	65ms
Average	3.9s	0.76s	0.28s	95ms	28ms	78ms

REVISION _A_   02/05/91

FIGURE 6.16   Typical manufacturer's fuse data sheet. (*Courtesy San-O Industrial Corp.*)

this procedure is crucial because of the tendency of loose filler to pack further during normal shipping and handling. If inadequate final packing allows any portion of the fuse element to become exposed, fuse operating characteristics might be compromised. Further, if the filler packing density is not constant from fuse to fuse, consistency of fuse operation can be jeopardized.

Filler material acts as a fuse-element heatsink to some degree. Therefore, variation in either grain density, size, or thermal conductivity will affect fuse precision and repeatability. Also, during heavy short-circuit overloads, the melting and/or vaporizing conductive element material in a filled fuse will combine with nearby quartz particles to form a normally nonconductive fulgurite. The vaporized metal plating out on adjacent filler residue can, on rare occasions, create a high-resistance current path through the filler, rather than the open circuit desired.

A vacuum can also be considered to be "filler" material. Depending on the degree of evacuation, the short circuit and/or voltage rating of a fuse can be significantly improved because an arc cannot be sustained in a vacuum. However, use of a vacuum is not practical in the types of fuses covered in this chapter. Reliability of such a fuse depends on maintenance of a hermetic seal for the fuse cavity—an extremely difficult and expensive task. Any leak will destroy the fuse's ability to interrupt at the rated current and voltage. Such a leak would not normally be noticed and the results could be catastrophic.

Fillers do enhance the power-handling capability (e.g., higher voltage and/or current) for supplementary fuses of smaller physical size. But, they will likely do so at the expense of fuse parameter consistency. Use of a filled fuse should, therefore, be limited to applications where voltage and current demands outweigh these risks. In the IEC 127 5- $\times$ 20-mm fuse, unfilled low-breaking-capacity fuse parameters are more repeatable and consistent than the parameters of filled high-breaking-capacity fuses.

The use of a physically larger, unfilled, fuse is usually preferred over use of smaller, filled fuses, which have parameter-value uncertainties. In North America, the common, unfilled 3AG fuse size generally provides the most precision and highest interrupting capability at the lowest price.

High-strength glass tubing is often used to increase interrupting current ratings. If the fuse uses high-strength glass, the amount of short-circuit current and/or overload voltage is increased significantly without use of fillers. It is not possible to know the interrupting rating of the fuse merely by appearances. Regardless of what the construction of a fuse body appears to be, a statement of the precise makeup of the material and its capabilities should be requested from the fuse manufacturer if doubt exists.

The customer can be assured of consistent safety standards because approval agencies retest fuses on an ongoing basis to guarantee their adherence to critical safety requirements.

Microfuses are a special class of fuse. Very small fuses are advertised regularly as *UL listed* and *CSA certified* at 125 Vac and even 250 Vac. How can they be so tiny and yet continue to meet the minimum requirements for listing?

UL and CSA have designated this unique classification as *microfuses* (IEC 127-3 uses the term *subminiature*). A microfuse is any fuse that can fit within a one milliliter of volume, (one cubic centimeter), not including leads. This class consists of fuses identified in manufacturers' catalogs as "microfuses" and "picofuses," or any other term (e.g., *1.5AG*) so long as the size constraints apply. Examples of microfuses are shown in Figure 6.17. Such fuses need only meet a 50-A interrupting rating at 125 V, rather than the 10,000-A interrupting rating at 125 V, required for UL listing of a "miniature" fuse. This is a significant difference!

Unlike "miniature fuses" per UL 198G, which must carry 110% rated current continuously and open within two minutes at 200% rated current, "microfuses" must be capable of carrying 100% rated current continuously, open at 150% rated current within 10 minutes, and open at 200% of rated current within 60 seconds to obtain a UL listing. Many microfuses have failed to meet the 150% rated current opening requirement per this UL standard, but will open at 200% rated current and were, therefore, UL recognized, rather than listed. The new UL 248/CSA-C22.2 No. 248 standard has changed this requirement to treat the microfuses the same as the larger fuses. IEC 127-3 has slightly different opening time requirements.

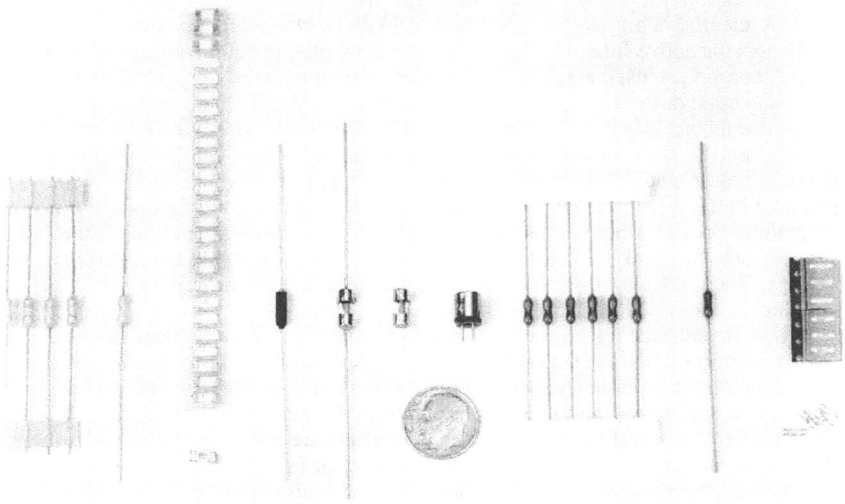

FIGURE 6.17   Typical microfuses. *(Courtesy San-O Industrial Corp.)*

6.4.4 Fuse Construction

Fuse dimensions and construction vary widely. Many different designs are available, including the standard tubular (cartridge) fuse, exposed element fuses for screw terminal mounting, and unique fuse and fuse holder sets. The basic cartridge fuse is the most popular type.

Cartridge fuse construction is accomplished by bonding the element ends with caps at each end. These end caps are mounted firmly on the ends of an insulative tube body. For softer material bodies, such as impregnated phenolic, and vulcanized fiber, the end caps can be staked by forming protrusions from the end caps into the body material. The staking operation must be carried out in at least two places, diametrically opposed, for each metal cap. More than two stakes are usually used and they are oriented symmetrically about the cap/body circumference. End cap edges that are formed into the relatively soft body material around the entire end cap circumference (360°) at both ends, represent the most desirable staking method. Any end-cap motion after staking can significantly reduce the life of the fuse, so it should be avoided.

Hard body tubing, such as glass or ceramic, must be glued to the end caps. Glue is applied in such a manner as to ensure that the seal is restricted to 60% to 70% of the circumference. The remaining circumference provides small openings at each end of the fuse for venting gases during short-circuit operation. The glue must be a good adhesive for both materials being joined, and must be capable of retaining strength at significantly elevated temperatures, such as can be encountered by an opening fuse. A device that qualifies as a microfuse, or one in which the short circuit current is significantly below the standard 10,000 A at 125 V, can be totally sealed. Other small-dimension fuses can also be totally sealed if the agency certification that identifies them as a sealed-type fuse is available and/or actual parameters used in testing are given in writing by the manufacturer.

End caps are live contacts and should never be permitted to directly contact other parts of the circuit. It is quite common to find several leaded fuses laying next to one another without consideration that circuit protection has been compromised if the caps contact the adjacent fuse caps. An often-necessary corrective action is to insulate the entire body of the fuse to ensure isolation from adjacent fuses. It would be preferable and less expensive to first implement proper placement and spacing.

Specialized fuses can have plastic bodies. It is essential that information on the plastic's flammability, deflection temperature, resistance to arc tracking and resistance to chemicals be obtained from the manufacturer. Altering the chemical characteristics (such as using a colorant) in a plastic can have detrimental affects on one or more of these parameters.

Another type of fuse can consist of a totally exposed element suspended between two plated copper pads on a flame-retardant insulative sheet. This is usually limited to low-voltage, low short-circuit applications. The element material in this type fuse can have either a high or low melting temperature.

End caps must be fabricated from conductive metal, usually copper, copper alloy, or brass, of appropriate and consistent thickness. The metal selected will determine the rigidity of the cap. It is important to note that if the metal chosen is too soft, it will deform when placed in high-compression-quality clips. Caps fabricated of brass need a barrier plating to prevent zinc migration through the final plating layer to the cap surface.

Element materials, configurations, and manufacturing consistency are key factors in guaranteeing fuse reliability and parameter precision. Manufacturers of fuses consider information regarding the details of the materials used in the fuses to be proprietary. Still, the component, compliance and circuit design engineer can learn a lot by detailed inspection of fuse element configurations (single wire, dual wire, twisted pair, spiral wound, resistor, spring loaded, etc.). Other factors, such as the type and amount of solder at each junction, type of solder flux used, bonding methods (such as soldering or welding), termination point location for the element from fuse to fuse and element tension are not so easy to inspect visually. These are but a few of the variables involved in good fuse design and manufacture.

The final assembly of a cartridge fuse can be carried out in several ways. The element can be drawn diagonally through the tubular body and wrapped over at each end. End caps with a predetermined amount of solder and bonding cement applied are then placed over the ends of the tube and element. The element is cut just prior to fully sliding the end caps to their resting point and the end caps are heated to cause the solder to reflow. This method is used for cartridge fuses and leaded fuses with butt-welded leads. The soldering technique is often referred to as a "blind" solder.

Another method of construction consists of threading the element through a hole in the center of one end cap, the core of the body tube, and the center of the opposite end cap. Bonding cement is placed on the end caps beforehand, and the unit is slid together until the caps butt against the ends of the insulative tube. The element is then soldered to the outside of a slightly indented surface on each end cap. A second coat of solder is added to the indentation to build it up to a flush or slightly convex surface. If a leaded fuse is desired, the lead can be added during the second solder operation. Leads can be oriented radially or axially, with the latter using the center hole in the cap as a centering guide.

All cements used must be able to withstand temperatures associated with soldering operations and still provide a sound mechanical connection. As covered in the following section, solder connections are preferred over welds.

6.4.5 Fuse Connection Methods

Fuses can be mounted in many different ways. This section covers clips, blocks, holders, leaded fuses, and surface mounts. This section is limited to "supplementary protectors." Comments here do not necessarily apply to larger secondary or primary devices.

Fuse clips provide a convenient means of incorporating a fuse into a circuit design. The fuse can be replaced by qualified personnel with minimal effort. Clips can be made of many materials, but the most common are (spring) brass, phosphor bronze, and beryllium copper. In addition, each type of clip can have a variety of methods to connect to the load and power source—spade lug, solder terminals, wire wrap, rivet mount and through hole, to name just a few. The selection of the correct material and means of connection for an application is as critical to circuit reliability as proper selection of the fuse.

UL and other agencies do not have formal test programs for fuse clips. Current ratings are provided by the manufacturer, but each application must be evaluated to determine if the selected clip is suitable. It is good practice to put a label next to the fuse clip, stating the correct replacement fuse type and rating. Figure 6.18 shows typical clips and fuse blocks.

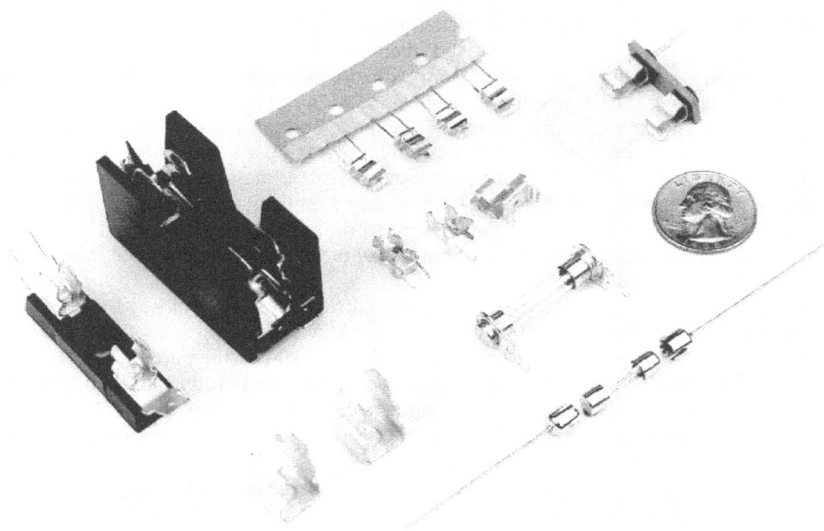

FIGURE 6.18   Fuse hardware. *(Courtesy San-O Industrial Corp.)*

Brass clips are generally acceptable for steady-state currents of up to 5.0 A and in applications in which high reliability is not paramount. Some manufacturers of brass clips have recommended maximum currents as high as 30 A, but I would not recommend the use of brass clips at such high currents. Reputable manufacturers will make brass clips out of "spring brass" and this should be noted in the clip description.

Brass material is used for the least-expensive clips. A major component of brass, zinc, makes this alloy less desirable than others, unless appropriate steps are taken to restrict migration of the zinc to the contact surface. If not protected carefully with an appropriate barrier layer, the zinc migrate to the surface of some platings, where it will oxidize. Zinc oxide is a tenacious, electrically insulating, coating that will cause temperature rise at the junction because of the increasing resistance. This is a very slow degradation, often taking several years to reach a point where the effects are noticed.

Phosphor bronze is a better material for clips than brass, but this author would still limit its use to circuits having a steady-state current that does not exceed 5.0 A. Reliability is reputed to be slightly better than the brass clips. As with the brass clips, some manufacturers recommended maximum currents are as high as 30 A. However, it is my opinion that where safety and reliability are important, currents above 5.0 A should be avoided when using phosphor bronze fuse clips. These clips will be somewhat more costly than brass clips, but the extra cost can generally be justified in critical applications, where clips must be used.

Beryllium copper is probably the best material available for fuse clips. Although beryllium copper is difficult to work with and rather expensive, these difficulties can be overlooked in higher-current applications. Beryllium copper clips can be used with confidence in circuits with steady-state currents as high as 15 A.

Beryllium copper is the only material of the three mentioned that requires heat treating after blanking and forming. Dimensional stability during heat treating can be a problem. However, if the process is carried out correctly, this material is capable of meeting tight dimensional and mechanical tolerances. Care should be exercised in handling any alloys containing beryllium because this metal and its salts are toxic.

Copper-clad steel clips provide excellent contact pressure and spring characteristics. However, their applications must be investigated thoroughly to ensure that the amount of conductive material is sufficient for the maximum anticipated current. The steel cross-sectional area is much less conductive than a copper-based alloy, and cannot be relied on to carry significant current.

Mounting techniques for fuse clips are another consideration. The clips must provide a satisfactory electrical and mechanical connection with the circuit. A basic rule to follow when optimizing reliability is to eliminate mechanical connections in the electrical path wherever possible. With this in mind, soldering a clip directly to a circuit contact or printed wiring board (PWB) after riveting is preferable to riveting alone. A rivet can be used to mechanically secure the clip to the PWB, but all electrical connections must be soldered. A rivet should never be used as a current conductor, unless it was specifically designed for that purpose and confirming test data is made available.

Spade lug (or NEMA Quick Connect) wire connections to a clip are acceptable at lower current ratings (3.0 A or less) and where reliability is not an overriding issue. Higher currents can be safely managed using mechanical spade lug connections, but careful evaluation must be made to confirm safe operation over time. This is an example of a solderless connection being used for electrical contact.

The spade lug is often designed to be used on a clip as either a quick connect terminal or a solder terminal. Solder is definitely the connection method of choice.

Fuse blocks, shown in Figures 6.18 and 6.19, are comprised of fuse clips, mechanically mounted to an insulative base. Blocks are approved by UL and CSA, but international

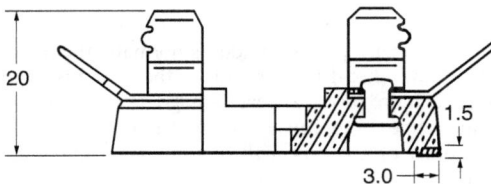

FIGURE 6.19 Fuse block. *(Courtesy San-O Industrial Corp.)*

agencies generally do not require approval. Fuse blocks are capable of handling fuses with current ratings as large as 30 A. Even in a block, however, a beryllium copper clip should be chosen for fuses above 15 A.

The same connection method constraints that apply to fuse clips also apply to fuse blocks. The sketch in Figure 6.19 reveals a potential for mechanical connection in the current path. The spade lug and clip are shown as two distinct components held together by a rivet. In this type of component, it is essential that the spade lug and clip be soldered together to ensure a satisfactory electrical connection has been made.

Fuse blocks are usually less expensive than post-style (bayonet) fuseholders, and are used inside equipment where the fuse is expected to be rarely or never changed.

Fuse holders are available in a wide variety of forms, with the most common being the bayonet type. This type of holder will have a spring either in the cap or in the base. Both types are shown in Figures 6.20 and 6.21. Figure 6.20 contains "bayonet"-type fuse holders, and Figure 6.21 shows several types of "in-line" fuse holders. In the holders shown in Figure 6.20, fuses can be held in a "fuse carrier" that has been incorporated into the fuse holder cap. Each carrier size can hold a different-sized fuse. In this way, one fuseholder can be used to house either a 3AG or 5- × 20-mm fuse, depending on the fuse carrier provided for the cap. Where fuseholder total length is important, a fuseholder that accepts only 5- × 20-mm fuses requires less space behind the mounting panel because the 5- × 20-mm fuse is shorter than the 3AG fuse (20 mm vs. 32 mm).

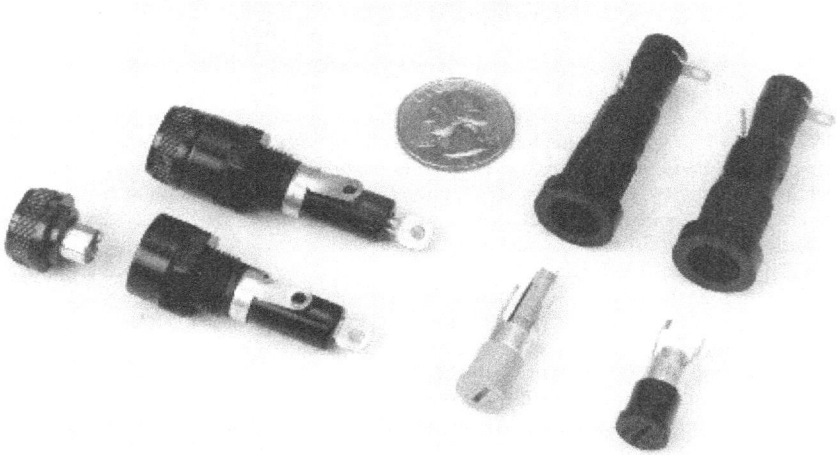

FIGURE 6.20   Typical bayonet fuseholders. *(Courtesy San-O Industrial Corp.)*

The cap portion of a bayonet-type fuse holder is normally the hottest operating part of the entire assembly. It follows that if the spring is in the cap, it is exposed to higher temperatures and, therefore, is more prone to heat damage. Fortunately, caps can be replaced when the spring becomes annealed because of elevated operating temperatures over time. A well-designed fuse holder of this type will minimize current flow through the spring and/or maximize spring material conductivity.

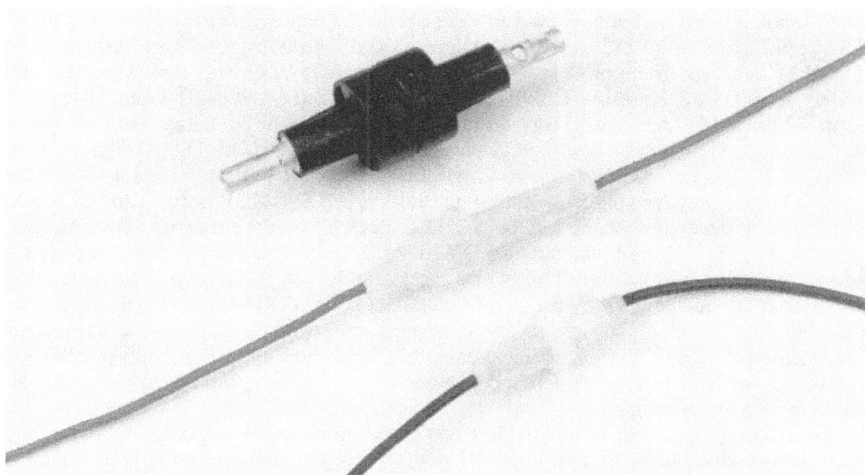

FIGURE 6.21 Typical in-line fuseholders. *(Courtesy San-O Industrial Corp.)*

If the spring is located in the base of the holder, it should not be used as a current conductor. This also applies to the in-line holders. Unfortunately, a spring in the base is not as easily replaced as in the cap. A very popular bayonet-type holder with the spring in the base uses a conductive plunger that is spring-loaded to achieve an excellent connection without requiring the spring to carry current. With this design, the spring can be steel, making it much less susceptible to annealing. Care must be taken with this type of holder to avoid restricting the motion of the rear contact, the plunger. Field applications sometimes use a wire as heavy as 10 AWG to minimize voltage drop along the feed path. Whether the feed wire is soldered to the plunger while a fuse is in place or with the fuse cavity empty, plunger movement will be restricted, leading to serious consequences. Such a heavy wire will not permit the plunger to move freely. Free plunger movement is necessary to ensure intimate contact with the fuse. Contrary to popular belief, fuses do vary considerably in overall length! Where plunger compliance movement has been compromised, the result is either poor contact between the plunger and the fuse, or a total inability to install the fuse.

An excellent test when evaluating a bayonet-type holder is to press on the cap until it seats against the body, while the holder is operating under power. Preferred designs maintain continuity during this test. Inadequate designs will result in an undesirable open circuit at some point of the cap's travel as it is pressed home.

Fuseholders designed to comply only with UL and CSA requirements often have knurled caps that can easily be removed to change a fuse without any tools. Various international standards, such as IEC 65, *Safety Requirements for Mains Operated Electronic and Related Apparatus for Household and Similar General Use* and IEC 257, *Fuse-holders for Miniature Cartridge Fuse Links* require that fuseholders be shock-safe, or "touchproof" in IEC terminology. The fuseholders are designed to require the use of a tool, usually a slotted screwdriver, and incorporate additional insulation and insulation barriers to eliminate the presence of live conducting surfaces during fuse-change operations.

Fuseholders chosen for equipment marketed solely in North America should have UL recognition and CSA certification. UL and CSA logos, and statements of maximum current and voltage must be printed on the body of the fuseholder.

Fuseholders that accept 5- × 20-mm fuses should be required to have UL recognition, CSA certification, and VDE and SEMKO approvals. SEMKO test results are generally accepted by agencies in the other Nordic countries, whereas VDE tests on fuseholders are widely accepted by the rest of CENELEC agencies. SEV (Switzerland) requires separate approval for fuseholders sold in Switzerland, but does not require separate approval when the fuseholder is sold in Switzerland as part of electrical equipment. The specifier should note that various test agencies do not all rate fuseholders for service at the same current. SEMKO limits post-type panel-mounting fuseholders to 6.3 A, VDE to 10 A, and UL/CSA to 12 to 30 A, depending on the service. SEMKO and VDE limits are derived from current limits on IEC 5- × 20-mm (and similar) DIN fuses. UL and CSA limits are based on temperature tests that attempt to simulate worst-case scenarios. UL, CSA, and SEMKO require that their approval marks be visible on the fuseholder body. VDE and SEV do not allow the fuseholder manufacturer to have an approval mark appear on the body of the fuseholder unless, in the case of VDE, the manufacturer holds a "Gutachten mit Fetigungsuberwachung" (expert report with factory surveillance). None of the major agencies require approval marks on the fuse carrier.

Leaded fuses are meant to be used in PWB applications, screw terminals or, less often, with compression box terminals. Leads of any type are frequently referred to as *pigtails*. The leads can be of the permanent or push-on type. The leads can be of an axial (Figure 6.22), radial (Figure 6.23) or "hairpin" (Figure 6.24) configuration. Choice of lead configuration is a function of the final application and insertion method that will be used. For example, axial or hairpin leads would normally be required where leaded fuses will be auto-inserted from tape. The hairpin configuration is especially useful where the available footprint on a PWB is limited and there is sufficient height to accommodate the device. Radial-lead devices are more common for hand insertion. Two types of radial-lead devices are shown in Figure 6.23, externally soldered radial-lead and formed axial-lead fuses.

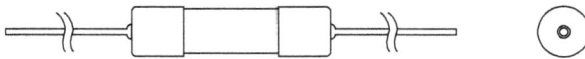

FIGURE 6.22   Axial-leaded fuse. *(Courtesy San-O Industrial Corp.)*

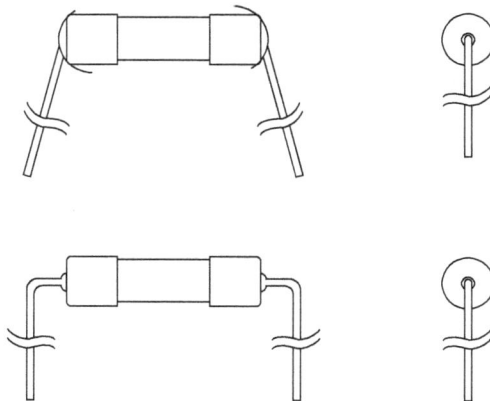

FIGURE 6.23   Radial-leaded fuse. *(Courtesy San-O Industrial Corp.)*

Permanent leads are either butt welded or soldered to the end cap. Permanent leads are a far more reliable method for connecting a fuse to a circuit compared to alternatives with mechanical junctions, such as fuse clips or push-on caps. Also, if a fuse is hard-wired to a circuit, it is less likely that untrained individuals will either deliberately or accidentally replace the device with a nonequivalent fuse.

Common sense should be sufficient to alert the end user that a frequently blowing fuse is likely a sign of a circuit problem. Unfortunately, common sense is not so common. Improper corrective actions are frequent. End users repeatedly replace fuses without failure analysis, substitute fuses with different time characteristics than intended for the application (such as replacing a quick-acting fuse with a time-lag fuse), or simply increase the fuse rated-current value. Even more pernicious is replacement of a blown fuse with a fuse having the same current rating, but lower voltage and short-circuit ratings. This is frequently done for economic reasons by end users that do not understand the significance of these parameters.

Butt-welded leads are coaxial with the barrel of the fuse, with sufficient precision to make them preferable for tape and reel applications. The use of a welded connection minimizes deviation in the overall length of the fuse body. Also, there is less chance of the weld being weakened or changed when the component is soldered into position on a board. However, because the lead is butt-welded to the end cap prior to fuse assembly, the actual solder connections of the element to the end cap cannot be visually inspected without destroying the component. Although the solder holding the element in place in this type of fuse is not visible, it can be reflowed during board-level assembly operations. Effective process control, aided by readily available and well-understood nondestructive test methods, can overcome this deficiency.

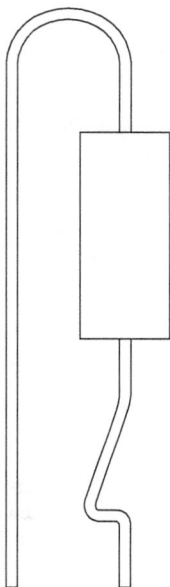

FIGURE 6.24 "Hairpin" leaded fuse. *(Courtesy San-O Industrial Corp.)*

Butt-welded leads are usually axial leads, but the leads can be bent at right angles to form a radial-lead device. Care must be taken when bending a butt-welded lead so as to not weaken the joint by bending too close to the end cap. A minimum distance along the lead of from two- to three-lead diameters should be left between the end of the weld deformation and the inside radius of the bend.

Soldered leads can be oriented either coaxially with the barrel of the fuse or radially to the fuse axis. In either configuration, the advantage of soldered leads over welded leads is that the element connection to the end cap is made externally, providing visual verification of a good solder bond. On the other hand, soldering is less precise in centering the lead of axial-leaded fuses on the end cap so that soldered-lead fuses are less suitable for tape and reeling. Also, both the axial and radial version of soldered-lead fuses are more prone to visible heat damage because of soldering-iron or wave-solder heat. Also note that the use of external solder tends to increase the variation of overall body length because the solder beads, or domes, at each end.

Axial-leaded fuses, especially the smaller sizes (such as pico-fuses) are available taped and reeled per the latest revisions of EIA standard EIA-296, *Lead Taping of Components in Axial Lead Configuration for Automatic Insertion*, or the IEC equivalent IEC-286-1, *Tape Packaging of Components with Axial Leads on Continuous Tape*. Tape and reeling facilitates auto-insertion of fuses onto circuit boards. Tape and reeling of radial, and hairpin-leaded fuses is also possible, but much less common.

Push-on end caps are another method of obtaining a leaded fuse. Push-on end caps permit the flexibility of producing or purchasing only nonleaded cartridge fuses and making up leaded fuses, as needed. Fuses with soldered leads are not covered by IEC 127 at the time of this writing, and so are not permitted by the countries that use IEC 127. Push-on caps are frequently used where leaded fuses are desired. Figure 6.18 shows examples of a push-on lead.

Push-on end caps do add slightly to the length and diameter of the fuse being used. They must be considered to have the same weaknesses as any other mechanical junction in the current path. Some care must be taken when using this type of connector in production lines, where the board might be subjected to harsh environments. Boards must be adequately rinsed to ensure that all potentially corrosive materials (e.g., flux, detergents, etc.) have been removed from the area between the fuse cap and push-on lead cap. If these materials are left behind, the combination of heat, time, and electric current accelerate degeneration of the fuse assembly. When push-on end caps are used, special attention should be given to tightness of fit between the fuse and cap. Some applications require that a minimum pull strength of 10 or 20 newtons be maintained; other applications need only 5 newtons. These engineering parameters can only be determined by the designer. If the pull strength is too low, and the fuse-cap assemblies are auto-inserted from tape and reel, boards can come off the production line with end caps soldered firmly to the printed wiring board, but with no fuse between them (i.e., the fuse separated from the cap).

Permanently leaded fuses are tested by UL and CSA to ensure that the leads are of sufficient cross section and high conductivity to safely handle steady-state and overload currents. The relatively small cross-section of the leads generally limit permanently leaded fuses to 15 A or less. The difference in thermal resistance between a fuse cap held tightly in a clip and a fuse having thin leads can result in different I-t operating characteristics. The difference can be significant enough that UL and CSA will only agree to qualification of the leaded version, or that the unleaded fuse is given recognition by UL, rather than listing. In some borderline cases, the addition of leads changes I-t characteristics enough to push nonleaded time-delay fuses into the normal classification because of the loss of clips, which act as heatsinks. As stated, IEC 127 does not (at the time of this writing) cover leaded fuses, so some agencies (such as SEMKO) do not usually consider leaded fuses eligible for approval.

Because push-on leads can be installed on any fuse with the correct end-cap diameter, the compliance and component engineer must restrict fuses used with end caps to those having current ratings within the constraints of the push-on leads, which can be 8 to 10 A or so. Parameters that must be checked for the assembled fuse and clip include the voltage drop at normal operating current, short-circuit current, and temperature rise.

Finally, push-on leads are usually meant to be used with a specific manufacturer's fuse. Tolerances for given fuse diameters and push-on lead cap diameters are generally far too large to ensure a tight fit with other manufacturers' fuses. For this reason, the use of push-on leads as a means to permit fuse replacement in a circuit is not an acceptable practice. The push-on lead was not designed to be used as a fuse clip in situations where fuses can be expected to be removed and replaced. The entire assembly should be discarded when fuse replacement is required.

Surface-mount fuses are available, and their use is just now becoming widespread. Issues such as physical size standardization are being addressed by organizations such as the EIA. Field replacement difficulties and product stability during processing are among those issues still to be addressed. Although these fuses are mainly used in low-voltage circuits (e.g., 12 V or less), care must be taken that the fuse is safe with higher voltages that can exist on the board because of the possibility that this voltage can be temporarily applied across the surface-mount fuse. The selected fuse should be able to protect the circuit under these conditions. Figure 6.25 shows two views each of four typical surface-mount fuse configurations. The vertical and horizontal scales are approximately representative of 1-mm intervals.

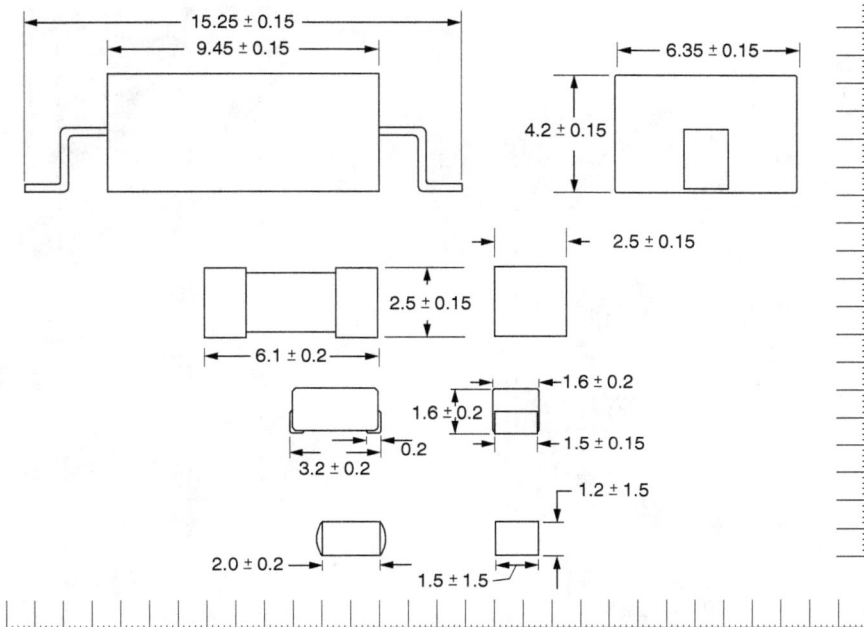

FIGURE 6.25 Relative size surface-mount fuses. *(Courtesy San-O Industrial Corp.)*

The tendency is to include surface-mount fuses in the UL category of microfuses because there is no other applicable category. This means they must be able to withstand a short-circuit current of 50 A at 125 Vac to become a UL-listed fuse. Maintaining this requirement while keeping the package small is a real challenge. However, some surface-mount fuses are as small as 1.5 × 2 mm with UL and CSA ratings of up to 3 A at 125 Vac and 70 Vdc. The largest fuse shown in Figure 6.25, for example, is available with a 250-Vac rating. This would be able to meet IEC 127 requirements.

Ideally, a surface-mount device should be able to be submerged in a wave solder bath for several seconds without undergoing electrical changes. However, such a requirement causes fuse manufacturers additional design challenges. Most surface-mount fuses are capable of withstanding a "reflow" solder process. This is where solder paste is applied to the junction between the printed wiring board and fuse terminals and reflowed with the addition of heat. They are not necessarily capable of withstanding "dip," "wave," or "drag" processes, where the fuse is actually submerged in the solder bath. Specific capabilities should be discussed with the fuse manufacturer before proceeding with a design requiring other than the "reflow" process.

Because of the manner in which these devices are installed on a PWB, standardization in physical size would be of tremendous benefit in establishing alternate sourcing. This is just beginning to happen. EIA PDP-100, *Mechanical Outlines* for several surface-mount and other miniature fuses now exist. More will be added as the demand for standardization grows. Examples of some of the typical surface-mount devices are shown in Figure 6.26 to provide some perspective of their actual size, when compared to a U.S. postage stamp.

An option for standardization for the electronics manufacturing industry has been to provide a tiered contact pad on the PWB, permitting perhaps two different lengths of fuses

FIGURE 6.26   Typical surface-mount fuses. *(Courtesy San-O Industrial Corp.)*

to be used. This presumes that the two fuses have been screened and found similar enough to meet circuit requirements. The use of physically equivalent packages is preferred. As with other types of fuses, just because they appear to be mechanically equivalent does not ensure electrical equivalence. Surface-mount fuses using wire elements are often grouped with those using thin and thick films as the fuse link. Although they appear to be similar devices, it is imperative that the end user evaluate the critical electrical and mechanical characteristics to ensure interchangeabil:ty. In many instances, there are increases in the minimum fusing current with the thin- and thick-film fuses, when compared with element-type fuses. Although the element-type fuse is still guaranteed to open at 200% rated current per UL 248, the thin- and thick-film devices can require as much as 300% rated current overload to open.

Costs of surface-mount fuses have been relatively high, compared to through-hole devices. The costs are coming down as usage increases. Also, fuse replacement has been complicated by the different-size devices and special tools required. Most boards using surface-mount fuses at this time are designed so that if the fuse opens, it is likely that some other component on the board has failed. It is intended that the entire board be replaced and the old one discarded. One manufacturer provides a surface-mount holder that can be soldered to a printed wiring board with a fuse in place. However, this introduces a mechanical connection into the circuit and can reduce reliability. As with other mechanical-type connections, plating compatibility must be reviewed carefully by the end user when a holder is used.

As noted, both the EIA and IEC are presently reviewing surface-mount standards aggressively. When these standards take effect, surface-mount fuses will increase greatly in popularity.

Platings for fuse clips, end caps and leads must be selected with care. Typical plating materials include tin, tin-lead (solder), nickel alloys, copper alloys, silver, and gold. An acceptable barrier layer should be used under any final plating material, where deemed necessary. Fuses used in clips or holders must be evaluated differently than those that will be soldered directly into the circuit. Platings to be used in all cases must be solderable. However, those to be used on electrical contacts, such as end caps and clips, must retain a low contact resistance over time at elevated temperatures.

Tin or *tin-lead solder* is the most common and usually preferred plating for high-reliability, low-cost applications. Tin is seldom used in its pure form because of solderability constraints. However, tin with at least 3% lead remains solderable using electronic-grade nonactivated or mildly activated fluxes, such as LR or LRMA rosin, specified in the latest revision of EIA standard EIA-402, *Liquid Rosin Fluxes*. Although 3% is usually considered acceptable, 10% will generally provide optimum wetting in critical applications. Tin-lead alloys also provide excellent electrical contact over extended periods. These materials can be slightly more expensive and/or difficult to obtain because of the environmental aspects associated with lead. But, the benefits of low-contact resistance over time and excellent solderability more than make up for the added cost. The plating is normally called out as "bright tin" and a minimum of 3% to 10% lead is specified.

The sliding friction of tin-lead on tin-lead is high. Excellent lubricants are available, however, that can significantly reduce the friction without adversely affecting other aspects of the design. Stabilant-22[5] is one such lubricant. Stabilant-22 is a liquid that becomes conductive in the presence of an electric field. Although Stabilant-22 is not primarily sold as a lubricant, it does have lubricating qualities.

These platings will provide excellent long-term contacts in controlled atmospheres with relatively clean air and wide variation in humidity. But, care must be exercised where these contacts can be placed in areas with high levels of pollutants found in some industrial locations, or areas where exposure to salt spray can be a factor.

Nickel and copper alloys are also very popular and they meet the needs for both solderability and good long-term contact resistance. Contact resistance is somewhat higher than a tin-lead plating, but costs are lower and the slight increase in contact resistance will not normally be a factor in the end application. These types of platings containing zinc should be avoided for the same reasons as covered in the brass clips section.

Silver is an excellent plating material for maintaining low contact resistance, but silver will contaminate a wave solder bath. Use of silver plate on one part of the contact (the fuse cap, for instance) and some other plating on the other (the clip), defeats the purpose of the silver plate. The contact resistance is only as good as the worst of the two platings.

Gold has good contact resistance characteristics and it is not prone to degradation as other metals are. Gold, like silver, can contaminate a wave solder bath. However, the rate of contamination is much lower for gold than for silver and usually found acceptable. A plus for gold is its self-lubricating nature. Where tin-lead contacts without a lubricant will present unacceptably-high sliding friction, gold will provide a consistently low insertion and withdrawal force without a lubricant. Unwanted intermetallic compounds are formed where gold is soldered. In normal applications, gold can be used successfully with solder if the gold does not exceed 4 μm in thickness.

Regardless of which plating is selected, the plating on the fuse should be the same as that on the fuse clip (or other mechanical contact) where they are used. Intermixing of plating materials in a mechanical contact can result in contact degradation over time and serious overheating problems. Platings containing zinc should be avoided in mechanical contacts, but are perfectly acceptable for solder connections.

Body materials for fuses must be electrically nonconductive, physically strong, have acceptable arc-tracking characteristics, good resistance to moisture, and be nonflammable. Such materials include glass, ceramics, impregnated paper tubing, and some plastics.

Normally, high interrupting current and/or high-voltage applications will necessitate the use of glass or ceramic. A ceramic fuse is often thought to be "stronger" than a glass fuse. However, the strength of a material can be a very misleading parameter. Ceramic materials are considered very hard and brittle. This does not imply that they are stronger than glass. Hardness and embrittlement can be signs of structural weakness.

Materials (such as glass and ceramic) are constantly being enhanced. Agency approvals will always ensure that the materials used were expected to meet the intended applications. Without any other information, a ceramic body fuse should be selected for heavy-duty applications over glass. However, some forms of glass can achieve superior strength at lower costs than ceramic materials.

Environmental conditions   Ambient conditions can affect fuse operation. Temperature and humidity are the most common of these. The use of fuses at significantly elevated temperatures in circuits without current derating will result in shortened fuse life. Surge or inrush currents can cause fuses to appear to be opening prematurely. Use of fuses at significantly low temperatures in circuits without current-rating correction, can result in damage to the circuit components as a result of uncontrolled overcurrent.

High humidity can be a problem when fuse clips or pressure contacts are used. Under these conditions, the design or components engineer should ensure that current-carrying contacts do not degrade, a condition that can possibly result in higher contact resistance. Similar concerns arise where solder flux used in fuse manufacture has not been adequately cleaned prior to delivery.

Another potential problem can exist if highly saturated air becomes trapped in the fuse enclosure. This is possible because fuses are usually not hermetically sealed. As the fuse is cooled, the trapped water vapor will condense on the interior surface of the fuse cavity. If a heavy short-circuit current is applied to the fuse, the condensate will immediately change from a liquid to a gas and the fuse is likely to rupture.

Hostile environments include all conditions in which the fuse life will be shortened or the fuse function can be jeopardized. Both extremely elevated temperatures and high humidity are considered to be hostile environmental factors. The types of environmental challenges that can be considered dangerously hostile include salt water spray, heavily polluted atmosphere, severe acceleration force, intense atmospheric pressure changes, or combinations of these.

Salt water spray, for instance, is associated with shipboard equipment or coastal installations. These applications require special attention during the design stage to ensure that adequate isolation is provided and all metal platings are appropriate. A heavily polluted atmosphere is not only found in manufacturing environments. Sufficient pollutants exist in any metropolitan area to raise concern over the reliability of electrical contacts.

Severe acceleration force, shock, and vibration are common on commercial jet aircraft. Applications for fuses have included electronic packages designed to be fired from a howitzer! Relatively fast atmospheric pressure changes occur often in commercial jet aircraft.

Selection of a fuse for equipment designed to be lowered to the floor of the deepest part of the ocean is problematic.

Sufficiently hostile environments require a special engineering effort. Where equipment will be exposed to heavily polluted areas, fuse manufacturers have begun to recommend that gold plating be used to ensure stable contact resistance. If the fuse is to be soldered directly to the circuit, gold plating can and should be avoided.

Cleanliness of fuses must be guaranteed from the time the fuse is being manufactured right through to equipment delivery and operation. Specifiers of fuses should require that good handling practices be followed. The latest revision of EIA 383, *Preparation for the Delivery of Electrical and Electronic Components* should be called out in fuse-specification documents.

Fuses (and other components) can be machine-inserted ("auto-inserted") into a PWB. The stuffed board is then sent through a preheat, flux and solder wave, and finally the board is cleaned. Because fuses must be vented to permit safe short-circuit operation, they are susceptible to taking on, or being contaminated by, cleaning agents. In the past, the cleaning was usually accomplished with volatile solvents, such as Freon. Volatile solvents evaporate before the boards are ever handled by human hands. With fresh volatile solvents, there is no visible trace that the fuse cavity had ever contained cleaning solution.

If the volatile cleaning solvent has been allowed to become laden with solids, a residue that is invisible to the unaided eye can be left on the inside and outside surfaces of the fuse body during cleaning. This residue is usually organic (made up of flux and oils), and it can carbonize during fuse operation. A carbon track that bridges between the two end caps can act as a conductor and the fuse function can be compromised. This problem can be caused by severely contaminated volatile solvent. Generally, no precautions need be taken when subjecting a standard vented fuse to a solvent-type bath, where the cleaning solution will quickly evaporate if it enters the fuse chamber. For obvious reasons, contamination levels of the cleaning solutions must be maintained within acceptable limits.

Contamination problems occur with aqueous detergent cleaning (water wash) operations. The aqueous detergent cleaning solutions used are not volatile. Usually, any solution that gets into the fuse chamber remains there. If even a small amount of condensate appears in the fuse chamber, safe fuse operation can be compromised. A high short-circuit current with any moisture remaining inside the fuse body cavity can cause the liquid to immediately vaporize, with the combination of forces because of the severe electrical, thermal, and vapor pressure shock. This results in rupture of the fuse body. Solutions used in these types of cleaning baths are also usually very corrosive.

Care must be taken to prevent an aqueous detergent cleaning solution from entering the enclosed, but vented, fuse chamber. This is a very challenging task. A description of what happens inside a fuse chamber during the soldering operation is essential to understanding the problem: As the fuse is heated during flux application, preheating and wave soldering, the air in the fuse chamber expands. This expansion increases fuse internal pressure temporarily, but the venting designed into the fuse permits the excess air volume at the elevated temperature to escape. The fuse is then moved into the cleaning region of the production line, which is normally cooler. As the fuse cools, a slight vacuum is created in the fusing chamber. The cleaning solution is sprayed onto the outside surface under high pressure. Some of this solution is pulled into the fuse cavity through the vents.

One method of preventing the solution from entering a fuse chamber during "water wash" cleaning is to completely seal the fuse. A UL-listed fuse is typically vented by gluing the end caps to the body using just enough glue to seal from 60% to 70% of the circumference of the cap. This leaves enough of a gap between the body and end caps to effectively vent the fuse during short-circuit operation. If the manufacturer completely seals the fuse, the short-circuit withstand capability will possibly be dropped to the point where the device no longer meets minimum agency requirements.

Another method of restricting the unwanted solution is to cover the fuse body and end caps with suitable heat-shrink tubing. Care should be taken to ensure that the tubing meets all UL flammability requirements. The tubing will usually prevent cleaning solution from entering the fuse cavity, yet it will expand slightly during short-circuit operation to permit the fuse to continue to meet agency requirements.

Still another method to keep cleaning solution from entering the fuse body is to carefully control cleaning-process temperatures. If the body temperature continues to increase, and never decreases during the cleaning process, a vacuum is never formed until the fuse is well beyond the cleaning environment and fluid is not drawn into the cavity.

Regardless of what method is used to seal the fuse cavity, adequate production trials should be run to ensure that the corrective action taken is sufficient to meet the actual conditions encountered. Each production line must be evaluated to confirm that the method of isolating the fusing cavity is effective.

6.4.6 Fuse-Rating Selection

To properly select a fuse, the following device parameters must be considered:

1. Voltage rating
2. Interrupting rating
3. Current rating
4. Fuse time response and I-t characteristic
5. I^2t rating

Selection of fuse-voltage rating   The fuse voltage rating is related to the fuse-interrupting rating, in that a fuse is rated to interrupt a certain maximum current at voltages up to the rated voltage. The fuse voltage rating is only critical when the fuse is trying to open. Most manufacturers recommend that the fuse-voltage rating should be greater than or equal to the maximum possible circuit voltage impressed on the fuse when the fuse element opens.

Some manufacturers modify this recommendation by noting that the prospective current in most electronic secondary circuits is much lower than the interrupting rating of the fuse. The "dead-short" fault condition produces only a low energy, nondestructive arc. One manufacturer[6] suggests that the fuse rating can be exceeded when the prospective fault current is no more than 10 times the current rating of the fuse, so it is common practice to specify 125-V or 250-V fuses for secondary circuit protection or higher. In practice, many 3AG fuses sold today are rated at 250 V, but this is usually at a greatly reduced interruption rating (e.g., 1500 A or less). A few 3AG fuses are available with 10,000-A/250-V ratings, but these are exceptional.

If a customer elects to exceed approval agency published limits, (justified merely by the customer's assumptions concerning expected normal vs. worst-case conditions), the customer must accept increased liability exposure. A fuse design application that works most of the time can be more dangerous from a safety and legal point of view than a fuse that is obviously inappropriate. If special application requirements dictate operation of a fuse beyond its rated limits, the fuse can usually be resubmitted to approval agencies to obtain their blessing for use of the fuse under special circumstances. For instance, a fuse to be used in a 300-Vdc application with a potential 1000-A short circuit can be had by simply discussing the factors with the approval agency and providing appropriate retesting— even though the selected fuse might not already be thus rated.

As covered in Section 6.4.3, fuse voltage ratings are given for 50-Hz or 60-Hz operation. Use of the fuse at very low frequencies or dc, or at high frequencies (such as 50 kHz), can require significant voltage derating. Fuse manufacturers can offer advice on appropriate derating.

Selection of the fuse-interrupting rating The interrupting rating of a fuse must be chosen to exceed the worst-case prospective fault current of the application.

The interrupting rating of all UL/CSA-qualified miniature fuses is 10,000 A at 125 V, which (in most electronic applications) is far more than the prospective fault current. The minimum required interrupting rating of a UL/CSA-qualified fuse at 250 V is reduced per Table 6.4.

IEC 127 fuses have one of three interrupting ratings. "High breaking capacity" (HBC) fuses have a rating at 250 Vac of 1500 A at a power factor between 0.7 and 0.8. They are normally quartz-filled and, in most cases, the fuse body is ceramic, and thus opaque. The new enhanced breaking capacity (EBC) fuses have a rating at 250 Vac of 750 A at a power factor of between 0.7 and 0.8. "Low breaking capacity" (LBC) fuses have an interrupting rating at 250 Vac of 35 A, or 10 times the rated current, whichever is greater, at a power factor of 1.0. The tube for both low and enhanced breaking capacity fuses is made of a transparent material, usually glass. LBC fuses are cheaper than EBC or HBC fuses, and are adequate for many secondary electronic fuse applications. In general and perhaps oversimplified terms, IEC 127 is based on the premise that many fuse applications can tolerate far lower interruption current ratings than those required by UL 198G/CSA C22.2 #59. Although this is most often the case, the exceptions could prove to be catastrophic, so careful evaluation of the circuit prospective-fault current should be performed. Members of the IEC 127 working groups are considering the possible upgrade of the LBC fuse from its present interrupting level to 150 A in the near future.

The interrupting ratings for both UL/CSA and IEC 127 fuses are based on ac voltages. IEC 127 fuses are always evaluated at 250 Vac. Interrupting ratings at dc voltages are often, but not always, lower than interrupting ratings at ac voltages. Manufacturers can be asked to provide dc interrupting ratings for their fuses.

Selection of fuse current rating The criteria for fuse current rating are as follows: First, the fuse current rating must be sufficiently greater than the normal operating current of the load. Second, the fuse operating current must be less than that current that causes unacceptable heating in the load.

Electronic loads are nonlinear, and typically have start-up power transients. Because fuses are thermal devices, they respond to rms (rather than peak values) under low-level overloads. Investigation of normal operating loads should be done with rms-measuring instruments. Multiplier oscilloscopes, such as the analog Philips 3265 or some digital oscilloscopes, are useful in analyzing situations with low power factor.[7] Investigation should be made of the start-up power transients, and an appropriate time-delay fuse should be chosen if these transients are significant.

Caution should be exercised in the use of a digital scope for determining clearing time of a fuse under heavy overloads (very fast response times). The electronics industry is constantly pressing for faster-acting fuses. Some fuses, such as microfuses, can open within a few µs or less. Although analog oscilloscopes are preferred to record such events, digital oscilloscopes can be used provided the sampling rate for the digital scope is adequate to provide a clear description of what is occurring during the fuse-opening process. This would be at least 10 samples during the fuse-operating time. Further note that the effective sampling rate for a digital scope is specified differently for continuous sampling versus single-shot sampling. The effective sampling rate for a single shot event (such as a fuse-clearing time measurement) for a 125-MHz digital scope can be only 3 MHz.

An often-quoted rule is to select a fuse current rating 33% to 100% higher (for UL/CSA fuses) than the worst-case normal operating current. IEC 127 fuses, however, are supposed to be chosen at the normal worst-case operating current, with little or no nominal current derating. IEC 127 current ratings represent the currents that are intended to typically flow through the circuit. UL 198G/CSA 22.2 #59 fuses should never see continuous circuit currents exceeding 75% of rated current. Both types of fuses are supposed to be further derated, or rerated according to ambient temperature rerating charts (such as Figure 6.11).

This rule results in an adequate choice of fuse current rating in many cases. A more exact analysis is based on consideration of the overcurrent condition, which will cause unacceptable heating in the load.[8] The operating time of the fuse can be matched closely, but will always be less, at any current level, than the time period for which the circuit can withstand the overcurrent condition. Faults that cause unacceptable overheating are not necessarily short-circuits. In fact, short circuits will cause any reasonably rated fuse to open quickly, so they are not relevant when calculating unacceptable overcurrents. The faults that cause unacceptable overheating are those that result from low, but finite, resistances. The fuse specifier should identify the critical component(s), whose failure will result in an unacceptable overcurrent. The fuse current rating should be chosen so that the fuse blows when that critical component fails. For instance, consider the snubber network in a switching power supply, consisting of a capacitor and resistor. If the capacitor shorts, the power supply will probably continue to operate, but the resistor will be forced to dissipate excessive power, which can cause excessive heating of nearby materials. The current rating of the fuse should be chosen so that the fuse blows if the snubber capacitor fails. Similar what-if analysis can be performed on any components within electronic equipment.

Occasionally, it is found that the difference between the normal operating current and the maximum desirable fault current needed to protect circuit components is smaller than the increment of available fuse ratings. In such cases, the next higher-rated fuse can be selected and a thermal switch added to further protect against low-level current overloads. The thermal switch will protect against long-term, low-level current overload and the fuse will protect against short-duration, higher-current-level faults. If a resolution to this type of problem (such as using the thermal protector) is not implemented, the circuit designer can be forced to increase the power-dissipation capacity of all affected circuit components. This will usually increase their physical size and cost.

Components are often changed on bills of materials in a product's manufacturing life cycle. The compliance or components engineer should be vigilant that these changes do not compromise the overcurrent protection system. An anecdote illustrates this point: A crowbar circuit works by deliberately creating an overcurrent situation. A fuse is expected to blow when the crowbar operates. In one design, the crowbar originally worked as expected. Then some apparently unrelated engineering change added a connector in the fuse-crowbar circuit, and reduced the connecting wire size. As a result, when on an occasion the crowbar SCR fired because of some circuit malfunction, the fuse did not blow; too much resistance had been added to the SCR circuit with the connector and the smaller wire! The crowbar caught fire instead. The fuse should have been changed to accommodate the new wire and connector, or perhaps the entire overcurrent-protection system should have been modified.

UL 198G/CSA C22.2 #59, and IEC 127 fuse standards differ enough that a fuse that meets one generally cannot meet the other. For instance, a UL 198G/CSA C22.2 #59 fuse must open within one hour at 135% of rated current. In contrast, an IEC 127 fuse must be able to carry 150% of rated current for at least one hour. These are mutually exclusive requirements. For equal nominal-current ratings, the IEC 127 fuse will have a significantly higher current-carrying capacity than a UL 198G/CSA C22.2 #59 fuse.

Selection of fuse time response   Selection of fuses should take into account transient changes in operating current. Time-delay fuses ("time-lag" in IEC 127 terminology) are preferred for circuits with transient power-on current surges. Such surges are caused by motors, transformers, incandescent lamps, capacitive loads, etc. If a "normal-blow" fuse were used in these circuits, it might be necessary to choose a current rating in excess of 300% of the circuit's normal operating current. Even then, nuisance openings might occur. Time-delay fuses allow sizing of the fuse to safely limit normal long-term operating currents, while allowing transient currents to pass.

Nontime-delay ("quick-acting" in IEC 127 terminology) fuses have no intentional built-in time delay, and are best suited for circuits without transient in-rush currents, such as constant resistance loads. Time-current or current-time charts (such as Figures 6.12, 6.14, and 6.15) are provided by manufacturers. They are average curves, provided as a design aid, but are not generally considered as part of the fuse specification. In other words, the fuse is guaranteed to comply to safety agency specifications and to other specifications the fuse manufacturer puts in writing, but not to "average" values shown in graphs. If a fuse specifier wishes to guarantee certain parameter values, it must be included in the part specification to which both the specifier and the fuse supplier agree. There is nothing wrong with requiring extra parameter specification; fuses made by reputable manufacturers usually do exceed UL/CSA/IEC minimum requirements by a comfortable margin; however, extra parameter verifications can raise the cost of the fuse and reduce opportunities for second sourcing.

As mentioned, UL 198G has only two categories of time-current characteristics: time delay and nontime delay (normal). The only difference between time-delay and nontime-delay (Table 6.4) is that the time-delay fuse has a maximum and minimum time rating at 200% of rated current, whereas the nontime-delay has only a maximum time rating at 200% of current rating. Both fuses are required to open within 2 minutes at 200% current, but only the time-delay fuse is required to wait at least 5 seconds (0 to 3.0 A) or 12 second (3.1 to 30 A) before blowing. The CSA C22.2 #29 requirements are similar. Both the UL and CSA are safety standards, and as such, establish minimum and maximum requirements only, leaving a great deal of room for variations. Manufacturers have taken advantage of this by tailoring time-delay fuses for different applications. Some of these fuses are UL listed, others are UL recognized. It should be noted however, that the "looseness" of the UL and CSA requirements mean that nominally equivalent fuses from different manufacturers might not, in fact, behave the same way. They should be tested for equivalence.

Most time-delay fuses are either dual-element or spiral-wound. Dual-element fuses have typically longer delays than spiral-wound fuses. Spiral-wound fuses are typically more precise. Voltage drop for circuits of 5.0 V or less must be evaluated carefully when using the spiral-wound element. This is because the spiral-wound element's significantly increased length and, therefore, increased resistance and voltage drop. In low-voltage, low-energy circuits, spiral-wound fuses of 1 A or less tend to absorb, without opening, all available energy. Dual-element designs are less likely to have this limitation.

IEC 127 "time-lag" fuses serve the same purpose as UL/CSA time-delay fuses. However, as Table 6.4 shows, the minimum and maximum limits are very different. Table 6.4 shows that there is no IEC 127 class of time-lag fuse, which produces "several seconds" of protection at two to three times the rated current. A nonofficial IEC 127 style time-lag fuse has become popular to fill this gap. The "super time-lag" fuse is approved by several agencies, such as SEMKO, but is not part of IEC 127. The super time-lag fuse gives 3-s (63 mA to 100 mA) and 5-s (125 mA to 4.0 A) minimum delay, in a low breaking capacity fuse. The maximum melt time is 100 s (63 mA to 100 mA) and 200 s (125 mA to 4.0 A).

I^2t rating and fast fuse protection Some fuse applications have the opposite problem from applications requiring time delay. Some fuses need to be blown extra fast. UL 198G and CSA C22.2 #59 do not have provision for very fast-acting fuses, but fuse manufacturers have developed specialized fuses for this purpose, some of which are UL recognized. These fuses are characterized by small I^2t ratings, and are often called *semiconductor fuses*. A very fast-acting fuse will generally be *current-limiting*, which means that the fuse blows before the full prospective fault current begins to flow. The current limiting is accomplished by limiting the time the fuse continues to conduct current. Three diagrams, Figures 6.27 through 6.29, can assist in clarifying this concept.

Figure 6.27[9] shows the available current to the circuit through the fuse. Operation of a fuse in a fraction of the time required for a half sine wave (8.33 ms for 60 Hz, and 10 ms for 50 Hz) so that peak let-through current is less than the peak available current, is defined as current limiting. A well-designed very fast-acting fuse will not only ensure that the fault will be cleared in a very short time, but will also limit the value of the prospective current.

Figure 6.28 shows the relative effect of current squared (αI^2). This is a measure of the mechanical (electromagnetic) force caused by the peak current. Figure 6.28 also shows that current limiting can dramatically decrease the mechanical force imposed on the protected circuit.

Figure 6.29, by adding the dimension of time, in the form of I^2t, shows how current-limiting dramatically decreases the thermal energy released in the protected circuit.

IEC 127 does not specify very fast-acting fuses at present. SEMKO and a few other agencies do approve an IEC 127-like fuse, called a "super quick-acting" fuse with high breaking capacity, for rated currents from 1.6 A to 10 A, and fusing integrals in the 0.1- to 8.1-A^2s range.

Elements in very fast-acting fuses have little mass, enabling them to operate more quickly. Low-cost very fast-clearing fuses often have a single strand of extremely thin

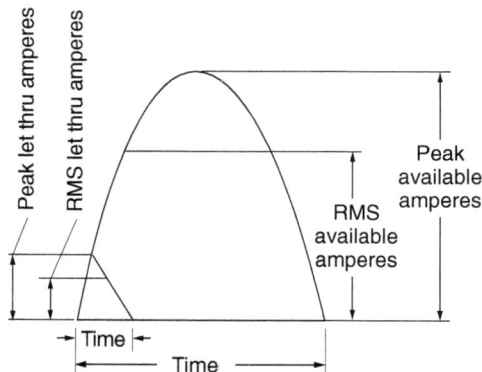

FIGURE 6.27 Let-through current (I). (*Courtesy Gould Shawmut*)

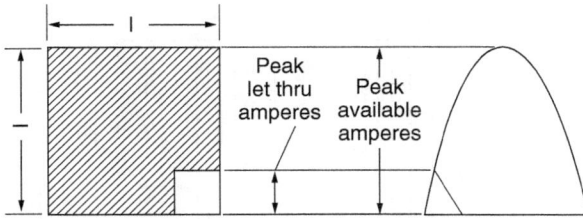

FIGURE 6.28   Mechanical force (I^2). *(Courtesy Gould Shawnut)*

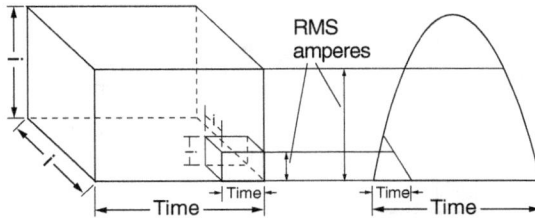

FIGURE 6.29   Thermal energy (I^2t). *(Courtesy Gould Shawnut)*

wire, and can have a region of narrowed cross-section, which helps decrease I^2t let-through energy. This wire is subject to mechanical stress and vibration, and will deteriorate over time—even at currents below the rated value. Manufacturers often recommend that the I^2t of periodic inrush currents do not exceed 20% to 30% of the fuse's I^2t rating to minimize the aging of the fuse. Typical expected operating lives of 1000 hours at the rated current are often quoted for this type of fuse.

The more expensive types of very fast-acting fuses have a fuse element surrounded by a filler material, such as quartz. The filler has a thermal conductivity high enough to carry away heat from the fuse element during normal operating currents, thus providing a much longer fuse service life. However, the thermal conductivity is low enough that under overcurrent conditions, there is very little heat transfer. The fuse element reaches melting or vaporization temperature quickly. Quartz-filled fuses have the added advantage that the filler gives mechanical support to the fuse element, thus protecting it from vibration.

Reliance on I^2t ratings must be tempered with a sound understanding of the source of data for both the fuse and the device to be protected. I can respond to the source for the fuse I^2t. In Figure 6.30, points for I^2 for a given t on the curve are generated directly from the fuse I-t curve. Remember, this is a typical I and there will be some deviation from nominal values plotted. The amount of deviation is a function of the fuse quality and the value of t. As you recall from 6.4.3, the standard deviation for N samples narrows as the percent overload increases. Maintaining a low standard deviation will minimize the I^2t derating that must be used. A common t for which the I^2t value of a fuse is requested is 8.3 ms—the time for one half cycle at 60 Hz. This is considered the worst-case t. As a result, it is most frequently given by the solid-state device manufacturers in their literature. A good-quality fuse at this overload should be able to protect the load successfully with a 25% derating. A superior-quality fuse might only need a 5% to 10% derating. A second line above the nominal

SCC SSFC 2A

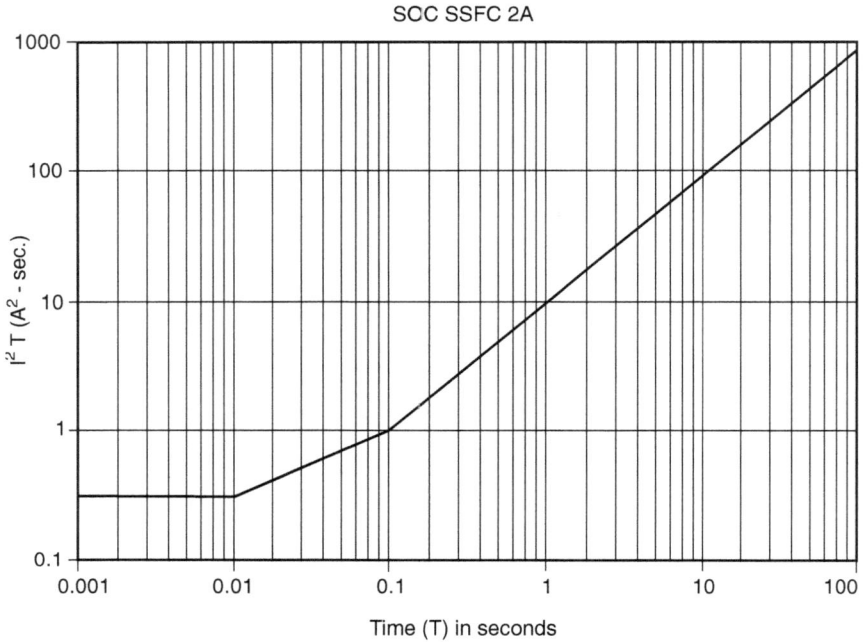

FIGURE 6.30 I^2t vs. t curve. *(Courtesy San-O Industrial Corp.)*

I^2t, representing the derated I^2t, is often shown. I would prefer to have the application engineer establish this second line, based on the actual circuit and components involved.

Before we proceed further with this subject, it must be emphasized that I^2t is a measure of the thermal energy of a fault current. Therefore, the t used should be representative of the entire period of current flow. This is the clearing time (t_C), not the melting time (t_M). The difference between the two, arcing time (t_A), adds as much as 10 to 15% to the time required to open at the typically requested 8.3 ms. The deviation increases as the t decreases, unless an arc-quenching filler has been used in the fuse construction. The melting time is a much more consistent value, but is not as meaningful in this evaluation.

Something often overlooked when selecting a fuse for energy limiting at a particular t is the probability of that type of fault actually occurring. Great care is taken to match the fuse and the solid-state load I^2t at a particular value of t.

What if the actual t encountered is of some other value? Is the protection still certain?

A specific fuse will have a specific I^2t curve. Likewise, a specific solid-state device will have its own I^2t curve. Because of significant differences between fuses and solid-state devices in construction and materials, it is extremely unlikely that these curves will match. We have ensured that the two curves will cross at the point for which I^2t was calculated. Unless the curves are identical, this is likely to be the only point that they will have in common. I have personally tried to obtain an I^2t curve for a solid-state device being protected by a given fuse and have not been able to locate one. Only a single point is provided for the solid-state devices investigated. Although severe short circuits will generally be noticed within the first half cycle, other forms of overload might not.

When evaluating the effectiveness of a fuse I^2t for a given task, there is no better method than bench testing in the actual circuit. Regardless of what the literature and curves would indicate, actual test results will be the only way to know how well the design will perform. For this reason, I consider the I^2t rating for a given fuse (and component to be protected) only as a useful tool to select the initial fuse and component values to be tested. Further, if the circuit design is considered important enough to require I^2t protection, it is essential that the components' characteristics be frozen through the use of a formal component specification. This type of stringent control will likely limit the sources as well as increase the cost. These factors must be weighed against the overall costs and/or risks associated with not providing I^2t protection.

In general, a very fast-acting fuse should be chosen that has a smaller I^2t rating than the device it is supposed to protect.[10] However, there are many variables in the application of very fast-acting fuses, and the advice of fuse manufacturers should be sought before committing to any type or rating. It might be found more cost-effective to use semiconductors with higher current rating, which withstand the higher current let-through by ordinary, lower-cost nontime-delay fuses, than to invest in very fast-acting fuses.

The fuse manufacturer can recommend a periodic fuse replacement policy for some of the more-fragile very fast-acting fuses. In such cases, the end-user of electronic equipment should be notified to schedule fuse replacement.

6.4.7 Agency Approvals for Fuses

Safety agencies publish standards, which set minimum requirements that a fuse must comply with to receive some form of approval. In North America, UL 198G and CSA C22.2 #59 are de facto standards, which all fuse manufacturers have met. As noted early in this chapter, this has now changed to UL 248/CSA-C22.2 No. 248. In the USA, UL publishes the fuse standard, and conducts the testing. The U.S. Federal Government, through the Occupational Safety and Health Administration (OSHA), allows other so-called Nationally Recognized Testing Laboratories (NRTL), to conduct fuse testing. Theoretically, they could test to the UL requirements, or even set their own fuse standards. But, only UL approves fuses. In Canada, CSA both sets the standard and tests the fuses.

UL 198G and C22.2 #59, are very similar, though there are differences in detail that sometimes have meant that certain fuses were eligible for UL or CSA approval, but not vice versa. CSA C22.2 #59 differed from UL 198G in certain temperature and power factor requirements, so not all fuse current ratings had both UL and CSA approval. The recent UL 248/CSA-C22.2 No. 248 standard is improving the problems previously associated with UL/CSA equivalence.

As explained earlier, UL recognition is a limited form of UL approval. A fuse recognized by UL can have passed most, but not all, of the requirements for UL listing. The UL listing of a fuse means that it successfully met all requirements of the referenced UL standard. UL recognition of a fuse means that some part of the standard was not met, or not evaluated. The assembly incorporating UL recognized fuses must be tested by UL with the fuse in place to ensure that it will meet the end-use requirements. Another use for UL's Recognized Components Program is to test a fuse type for which no UL standard exists, such as for new and advanced types of fuses. In such cases, the fuse manufacturer and UL agree beforehand what the fuse characteristics should be. If the fuse is safe, and works in accordance with the manufacturer's declarations, the fuse gets UL recognition. UL recognition is often used to test 5- $\times$ 20-mm fuses made in North America to the IEC 127 standard. These fuses are not the same as 5- × 20-mm fuses made to UL 198G. The latter 5- × 20-mm fuse can be UL listed or recognized. A frightening fact is that these two very different fuses from an electrical standpoint need carry no distinguishing markings. They simply have the markings required for a recognized component. In practice, however, a fuse recognized to IEC 127 will usually also carry an approval logo from another agency, such as SEMKO, which a 198G-recognized fuse will not.

In many instances, devices that are UL listed and CSA certified can also receive (Japanese) MITI approval, even though MITI criteria are defined somewhat differently.

Non-North American agencies (such as SEMKO, DEMKO, VDE, and BSI) generally base their national fuse standards around IEC 127. Some countries have specific standards for special fuses, such as British standard BS 1362 for power-cord plug fuses. IEC 127 fuses are usually 5- × 20-mm, but there is provision in IEC 127 for a 0.25" × 1.25" fuse. This is meant for replacement of UL 198G fuses in equipment that was manufactured in North America. The IEC 0.25" × 1.25" fuse does not have the electrical characteristics of a UL 198G fuse, and will not normally be used in new equipment.

UL 198G/CSA C22.2 #59 on the one hand, and IEC 127 on the other, are mutually exclusive, so no fuse can be qualified to both sets of standards. It appears that this situation will continue for some time to come. However, work is beginning on harmonization of some device standards. A great deal of negotiation will be needed to harmonize UL 198G and IEC 127. It appears unlikely that the two "camps" of fuse standards will ever be fully reconciled.

Safety is always the primary concern of both the approval agency and the equipment manufacturer. After the safety of the equipment operation has been ensured by the selection of a fuse that meets the appropriate agency approvals, protection of the equipment becomes the next major hurdle. Twenty years ago, the typical purpose of a fuse was to provide "gross" circuit protection—keep a device or equipment from burning! A standard fuse was available from several sources to meet the needs. If a known temporary overload existed, such as starting current for a motor, a time-delay fuse would solve the problem. Today, requirements for circuit protection can be more exacting.

Work is proceeding with agencies, such as the EIA, to generate fuse-design standards that will provide more precise information for the end user in fuse selection. These standards will still rely on safety agency testing and certification to control the broad limits needed to ensure safe fuse operation. However, they will build on this base to provide a procedure for equipment manufacturers to specify truly "equivalent" fuses for second sourcing and equipment maintenance. Figure 6.31 shows the safety agency approval logos most likely to be found on fuses.

6.4.8 Fuse Marking and Packaging

Marking and packaging for fuses reflects approval status and ratings. Fuses designed to UL 198G/CSA 22.2 #59, will bear the manufacturer's logo, the UL listing or recognition logo, and/or CSA logos, a current rating, and a voltage rating. Exceptions to the marking requirement can be made for very small fuses. In that case, approval status must be marked on the next "higher" packaging, such as on the reel or box of fuses, as packaged by the fuse manufacturer. Current rating can be fractions of an ampere (e.g., ¼A), decimals of an ampere (e.g., 0.250 A), or milliamperes (e.g., 250 mA).

IEC 127 fuses will normally be marked with the speed characteristic (TT, T, M, F, or FF), the breaking characteristic (H or L), a current rating in amperes or milliamperes, a voltage rating, the manufacturer's logo, and agency approvals. The most common agency approval is SEMKO. VDE and BSI approvals are also used, especially on fuses that are

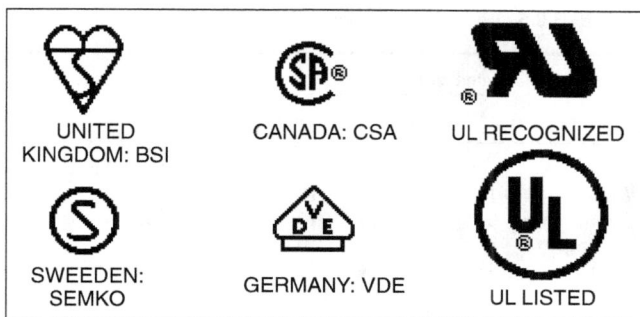

| UNITED KINGDOM: BSI | CANADA: CSA | UL RECOGNIZED |
| SWEEDEN: SEMKO | GERMANY: VDE | UL LISTED |

FIGURE 6.31 Common safety-agency approval logos.

similar to IEC 127, but (for some reason) are not eligible for SEMKO approval. Such fuses can have current ratings higher than IEC 127 allows, or can have TT, M, or FF speed characteristics, which are not yet addressed by IEC 127.

Color codes are sometimes used on IEC 127 fuses to indicate the current rating and fuse speed characteristic. Color coding is not required by any safety agency, but has become widely (though informally) used by fuse manufacturers. IEC has been reluctant in the past to support color coding of fuses for several reasons: Colors can change over time and colors can rub off in vibratory bowl feeders for auto insertion machines. Also, a significant percentage of the population is color-blind. However, Appendix A of IEC 127 does now list an optional color-banding system to indicate both speed and current rating. Four bands are used. The first two bands indicate rated current. The third band indicates the multiplier and the fourth band indicates the speed characteristic. The current-rating digit code is identical to the familiar code used for resistor values. Nominal rating is expressed in mA by the first two digits (band 1 and 2) and a corresponding multiplier (band 3). The speed characteristic is indicated by a fourth double-width color band. For example, a violet/green/brown/red banded fuse would indicate a 750-mA, quick-acting fuse.

North American fuse manufacturers do not ordinarily use this color scheme for miniature fuses. Although some North American manufacturers do color-code microfuses, and, on request, their miniature fuses, the code does not necessarily match the IEC 127 scheme. Table 6.5 details the IEC 127/A color-code scheme.

Nonleaded fuses are usually packed by the manufacturer in bulk lots or in blister packs. Leaded fuses are available in axial tape and reel, per the latest revision of EIA 296 *Lead Taping of Components in Axial Configuration for Automatic Handling* (or equivalently, IEC 286-1), "hairpin" tape and reel per EIA-468 *Lead Taping of Components in the Radial Configuration for Automatic Handling* (or equivalently, IEC 286-2) or with radial lead forming. Surface-mount fuses can be bought in bulk, or on reel per EIA 481 *Taping of Surface Mount Components for Automatic Handling* (or equivalently, IEC 286-3).

Customers can specify the latest revision of EIA 383 *Preparation for the Delivery of Electrical and Electronic Components* on their source control drawings to ensure proper packaging and handling of fuses from manufacturer to customer.

6.4.9 Fuse Quality, Reliability, and Failure Analysis

Manufacturers maintain tight control on all operating characteristics to ensure consistent, accurate performance of fuse products. The procedures adopted by the various manufacturers

TABLE 6.5   IEC 127 Fuse Color Code

	Rated current		Multiplier	Speed characteristic
Color	Band 1	Band 2	Band 3	Band 4
Black	—	0	10^0	FF = Very Quick-Acting
Brown	1	1	10^1	
Red	2	2	10^2	F = Quick-Acting
Orange	3	3	10^3	
Yellow	4	4	10^4	M = Medium Time-Lag
Green	5	5	10^5	
Blue	6	6	10^6	T = Time-Lag
Violet	7	7	10^7	
Grey	8	8	10^8	TT = Super Time Lag
White	9	9	10^9	

vary in detail, but always the manufacturer will use in-process controls and a comprehensive quality-assurance program. In practice, Acceptable Quality Levels (AQL) for fuses, including "failures" because of mismarking, or shipment of a wrong fuse value, are better than 50 ppm.[8] In some markets, it is, or will shortly be, necessary for fuse manufacturers to certify compliance with the ISO 9000 series of quality-assurance standards. The U.S. equivalents to ISO 9000 are known as the *ANSI/ASQC Q90 series*. Canada's standard CSA Q420-87 incorporates some portions of the ISO 9000 series.

Manufacturers generally avoid discussion of reliability of a fuse in terms of MTBF (Mean Time Between Failure) or FITs (Failures in 10^9 hours). Numbers are sometimes quoted, but for the following reasons, such figures might not be relevant to the user's particular application. For example, Bell Communications Research, "Bellcore," a company that develops standards for the USA telecommunications industry, lists a steady-state failure rate of 15 FITs ($\therefore MTBF = 67 \times 10^6$ hours), for a generic fuse in a benign environment.[11] Bellcore lists the average failure rate during the first year of the fuse's operation (sometimes called the *infant mortality period*) as 60 FITs. Bellcore considers these numbers to be consistent with observed field-failure reports for telecommunications equipment, but there is no justification for assuming that these numbers apply for electronics equipment in general. The most valuable interpretation of the Bellcore listed failure rates can be as a comparison with other electronic components. Bellcore considers a fuse, in a fuseholder, to be roughly comparable in reliability to a silicon diode or low-power transistor.

As-manufactured vs. as-used reliability   Most manufacturers do not provide reliability information in fuse data sheets, for the reason that it is difficult to gather field failure rate data, and even more difficult to accredit statistical significance to gathered data. In most instances of fuse opening, the fuse did not "fail;" it simply did its job—opening when a hostile-current condition arose. There is often no way to distinguish between a "bad" fuse and anomalous circuit conditions.

The only certain way to know if a specific sample fuse will work is to blow it under controlled conditions. Naturally, once the fuse is blown to "prove" its quality, it is too late to use the device! Quality and reliability can only be measured on a statistical basis, with data from blown samples being extrapolated to draw conclusions about the quality and reliability of the entire batch. However, as with field-derived MTBF figures, MTBF numbers derived from laboratory tests are suspect. Tightly controlled laboratory conditions can produce MTBF values grossly different from real-life conditions. In real applications, fuse MTBF is strongly dependent on such variables as: the nature of the fuseholder, current derating, current and voltage waveform, application temperature, and vibration conditions, etc. It is possible for a fuse manufacturer and a fuse user to agree on a standardized set of test conditions, but such testing is usually very expensive. Large numbers of good fuses would have to be destroyed in the testing program to guarantee stipulated quality and reliability levels, and there would still be residual uncertainty about how such fuses would behave in uncontrolled field conditions.

The fuse, as manufactured, can have excellent quality and reliability, but these can be compromised by an incorrect application. If a fuse is driven beyond its design limits (e.g., a UL 198G fuse driven beyond 75% of rated current value and 100% of rated voltage in a high ambient-temperature environment with excessive shock or vibration, etc.) in the application, reliability will be reduced. To guarantee adequate reliability, it is essential that sufficient time and effort be invested in proper fuse selection and usage. Fuse manufacturers have staff that can help the design, components, or compliance engineer choose the correct fuse for the intended application.

A common error is to use a fuse size at the limits of its capability. In general, the larger fuses (3AG) have more repeatable parameters, are more robust, and tend to be less expensive. If available space is a problem, a 5- × 20-mm fuse can be substituted. Even though the ratings on the devices appear to be equivalent (e.g., a minimum of 10,000-A interrupting rating at 125 V), the actual capability of the larger device will usually be superior. Thus, even if, for instance, a 2AG fuse can appear to have the same ratings as a 3AG or 5- × 20-mm fuse, the reliability of the larger fuses will probably be higher. Use of a larger fuse where space is limited can be facilitated by use of a "hairpin" configuration (Figure 6.24), which will allow the fuse to be stood on end.

Fuse-failure analysis The term *fuse failure* has to be defined carefully. Fuse failure is used too often in reference to a fuse that has actually performed correctly. A fuse that has truly failed is one that opened without an overcurrent condition present or one that did not open when an overcurrent condition was present.

Visual inspection of a blown fuse can provide a great deal of information regarding the circumstances existing at the time of fuse operation. Failed devices should be saved for further evaluation if the failure is suspicious.

If, in a single-element cartridge fuse, only a very small portion of the center of the element is missing, it indicates that a low-energy overload condition had occurred, such as a mild overcurrent.

If the interior of the glass tube is coated with a metallic and/or black plating, it can be concluded with some certainty that the overload condition resulting in the open fuse was a heavy short. The plating occurs because a very high fault current rapidly melts and vaporizes the fuse element. High magnetic fields caused by the high currents propel the melted and vaporized metal to the walls of the fuse, thus imparting a mirror-like finish.

An open at one end of the element usually indicates that the fuse opened as a result of a severe mechanical shock or that the element was nicked or damaged during manufacture, causing an electrically and mechanically weak point in the link.

Another "failure" mode often overlooked is an open fuse with the element appearing to be intact. In many instances, further examination will show that the fuse opened because of reflowed solder. The solder will frequently be visible on the exterior of the fuse at the boundary between the glass tube and end cap. This would indicate that the fuse had overheated due to poor clip-to-fuse contact or an ambient temperature far exceeding normal operating conditions. It is likely that any fuse clips involved have been weakened (or annealed) because of the elevated temperature over time and they should be changed prior to replacement of the fuse. Naturally, the circuit performance should be reviewed to ensure that current levels have returned to normal.

Of all of the possible "failures" listed, only the nicked element would qualify as a true failure. The other open fuses have provided needed protection during circuit stress.

6.4.10 Fuse Strategy for Worldwide Equipment Marketing

Electronics manufacturing companies would like to be able to distribute their finished products throughout the world. A major constraint is power source differences. Outside North America, most nations provide 250 Vac as their standard outlet voltage. Some are at 50 Hz, others at 60 Hz. This means a single fuse cannot be chosen to serve in both markets.

A circuit-protection strategy based on the use of fuses is complicated by the existence of the two commonly used sizes: the 0.250"-×-1.25" (6.3 mm × 32 mm) 3AG standard in North America, and the smaller 5- × 20-mm fuse standard in the rest of the world. The 3AG fuse is generally available in North America and in certain Pacific Rim nations. Elsewhere, and especially in Europe, the 3AG size is available only for fuse replacement in North American-made equipment. 5- × 20-mm fuses qualified to UL and CSA blowing characteristics are rapidly becoming more popular in North America, but the differences between UL 198G/CSA C22.2 #59 and IEC 127 preclude a fuse being rated identically under both sets of standards.

Companies that market their products worldwide should not choose the 3AG-size fuse. Overseas customers would have the problem of trying to source replacement fuses locally, which might not be easy. Furthermore, most commonly used fuseholders that accept 3AG-size fuses (only) are not approved by international agencies, such as BSI, SEMKO, and VDE. All things considered, the specification of a 3AG fuse is not an acceptable alternative.

A second strategy is to specify 5- × 20-mm fuses for use in every market. This alternative offers the simplicity of a single fuse size. Also, because the 5- × 20-mm fuse is shorter than the 3AG fuse (20 mm vs. 32 mm), its fuseholder uses less space behind the equipment panel. Fuseholders that accept only 5- × 20-mm fuses are available with UL recognition, CSA certification, and approval from international test agencies (such as BSI, VDE, and SEMKO). In addition, some power-entry modules, which combine a fuseholder with an ac power outlet, switch or voltage selector, and an EMI filter accept only the 5- × 20-mm fuse.

Although the fuse size remains the same worldwide for this strategy, it is important to recognize that two different 5- × 20-mm fuses will probably be required: a UL/CSA approved fuse for North America, and an IEC 127 approved fuse for use internationally. In one respect, this is only a minor inconvenience; different current ratings imposed by higher voltage ratings will dictate use of different fuses, anyway. However, the different fuses will be virtually identical in appearance, and it is easy to confuse them. The only consistent differences are the markings embossed on the fuse cap, and they can be difficult to read.

Fuse specifiers need to remember that there are two "styles" of 5- × 20-mm fuses made in North America. One style is UL listed/CSA certified to North American standards, and can be used in equipment destined for North American markets. The other style of 5- × 20 mm is UL recognized/CSA certified to IEC 127, and may not be used in equipment sold in North America. This second style of 5- × 20-mm fuse can only be used in equipment manufactured for sale outside North America.

A third strategy is to specify the 3AG fuses for use in North America, and 5- × 20-mm fuses elsewhere. This can be accomplished with one fuseholder, provided that the fuseholder is designed to accept both sizes of fuse, via special fuse carriers. Such fuseholders do exist, and they are approved by both UL/CSA and international agencies. This strategy allows the equipment manufacturer to ensure that the product is user-friendly, in that the end-user will have a fuse that can easily be replaced in his local area.

The third strategy is the one most-often followed today. As the popularity in North America of 5- × 20-mm fuses grows, it is expected that the second strategy will predominate.

One method successfully used to circumvent the 5- × 20-mm vs. 3AG problem has been to fuse in the secondary of the circuit, rather than the primary. Every safety agency permits such protection, but there are very specific restrictions associated with secondary protection. It is likely that if the point is reached where all agencies have been satisfied, there will be only one fuse vendor available for the unique fuse ultimately selected. Most equipment manufacturers do not like being single-sourced.

6.5 CIRCUIT BREAKER TECHNOLOGY

Most electronic equipment is protected against overcurrent by fuses. Occasionally, a mechanical circuit breaker is found to be more suitable. Three types of circuit breakers were described earlier: thermal, thermal-magnetic, and magnetic.

The thermal breaker is used primarily where crude (imprecise) overcurrent protection is required and severe ambient temperatures are not an issue. The most common thermal breaker utilizes a bimetallic or three-layer sandwiched metallic element, which changes shape by differential expansion of the metals as it is heated by current passing through. At a predetermined overcurrent, the bimetallic element triggers a tripping mechanism, which is often a spring-loaded switch in higher-quality thermal breakers. A high-quality thermal breaker will open a 10,000 A fault at 250 V in 40 or 50 ms. The speed of a thermal breaker varies directly with temperature, and with the square of the current. Thermal breakers are often used to protect wires because the reaction of the thermal element to overcurrent-induced heat is similar to the behavior of the protected wire.

Unless compensated, thermal breakers tend to be sensitive to ambient temperature, so they should be used only at stable operating temperatures. Use at very low or very high temperatures requires recalibration. The relatively high thermal mass of most thermal breakers makes them natural time-delay overcurrent protectors. The thermal mass "integrates" heat energy, so thermal breakers are not generally frequency sensitive or prone to nuisance tripping because of fast current transients. Thermal breakers are sensitive to shock and vibration, however, when carrying full-rated current. Correct selection and application of thermal breakers requires detailed consideration of the surrounding thermal environment. Ambient operating temperatures, effective heatsinking, and allowable voltage drop must all be considered. Figure 6.32 shows the outline of a typical thermal circuit breaker manufactured by Square D. Mechanical Products is another leading manufacturer of thermal circuit breakers.

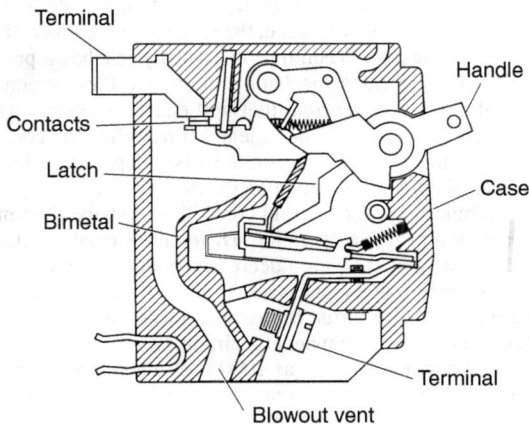

FIGURE 6.32 Outline of thermal circuit breaker. *(Courtesy Square D Co.)*

One specialized type of thermal breaker, the "hot-wire" breaker, functions by the expansion and contraction of a hot wire element, which operates a tripping mechanism. Because the temperature of the wire at time of trip is 425°C to 475°C, ambient temperature has very little effect on breaker operation. Because the hot wire has less thermal mass than bimetallic elements, the hot-wire breaker operates somewhat faster, but the voltage drop across the element is larger.

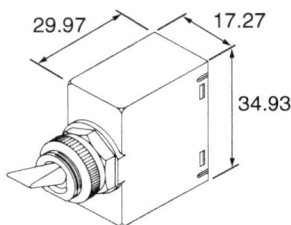

FIGURE 6.33 Small magnetic circuit breaker. *(Courtesy Aipax)*

The magnetic breaker far surpasses the other types in terms of precision and repeatability. Two prominent manufacturers are Airpax and Heineman Electric. The magnetic breaker can be made to operate dependably at as low as 10 mA. The fastest magnetic breakers can trip within three or four ms, although tens or hundreds of ms is more usual. Figure 6.33 shows an Airpax SNAPAK, which is a magnetic circuit breaker, designed for limited space requirements. Current ratings range from 100 mA to 20 A, with a short-circuit capacity of up to 1000 A. This small breaker can be used for circuit protection as well as an on/off switch.

The magnetic breaker operates via a coil that creates a magnetic field, which triggers a tripping mechanism at some predetermined value of overcurrent. Thermal energy is not involved, so the magnetic breaker is better able to operate in extreme temperature environments. The trip point is relatively independent of ambient temperature.

Without a specific means to delay the opening of a magnetic breaker, it will respond at low overloads (e.g., 200%) very quickly. Such magnetic breakers are known as *instantaneous trip*. The speed of tripping varies depending on circuit conditions and the design of the magnetic breaker. They are sensitive to inrush current, vibration, and shock, and should not be used where these factors exist.

As with fuses, overly quick or overly sensitive action in a magnetic circuit breaker is often undesirable. One technique is to slow down the response by use of a hydraulic dashpot. An iron-core plunger is placed in a hermetically sealed, nonmagnetic tube filled with a viscous liquid, often an oil or silicone. The core is spring-loaded and held in position outside the breaker windings. Under a moderate overcurrent condition, the iron core is moved through the liquid by current-generated magnetic flux into the center of the breaker windings. As the core moves toward the pole piece, the magnetic reluctance of the magnetic circuit containing the coil decreases. When the flux density reaches a predetermined level, the armature is attracted, the breaker mechanism trips, and the contacts open. The "ultimate trip current" is the minimum current that will ensure a reliable trip of the breaker. This trip point occurs after a predetermined (delay) time, when the core has made its full travel in the tube. The ultimate trip current is relatively independent of temperature, and it depends on the number of ampere-turns and on the design of the dashpot.

Viscosity of the damping liquid is closely controlled to shape the time-response characteristic and to maintain repeatable operation. Different viscosities cause different delay characteristics. The viscosity of the liquid decreases with temperature. Correspondingly, the response time of the magnetic breaker (but not the ultimate trip current) decreases as temperature increases, which is a virtue in many applications. Figure 6.34 illustrates the operating principles of a dashpot in a magnetic circuit breaker.

Under high overload conditions, such as 1000% of rated current, the breaker trips instantaneously. The "instantaneous trip current" is the value of current required to trip the circuit breaker without causing the core to move into the tube. This is possible because the leakage flux caused by high overcurrents attracts the armature directly and trips the breaker.

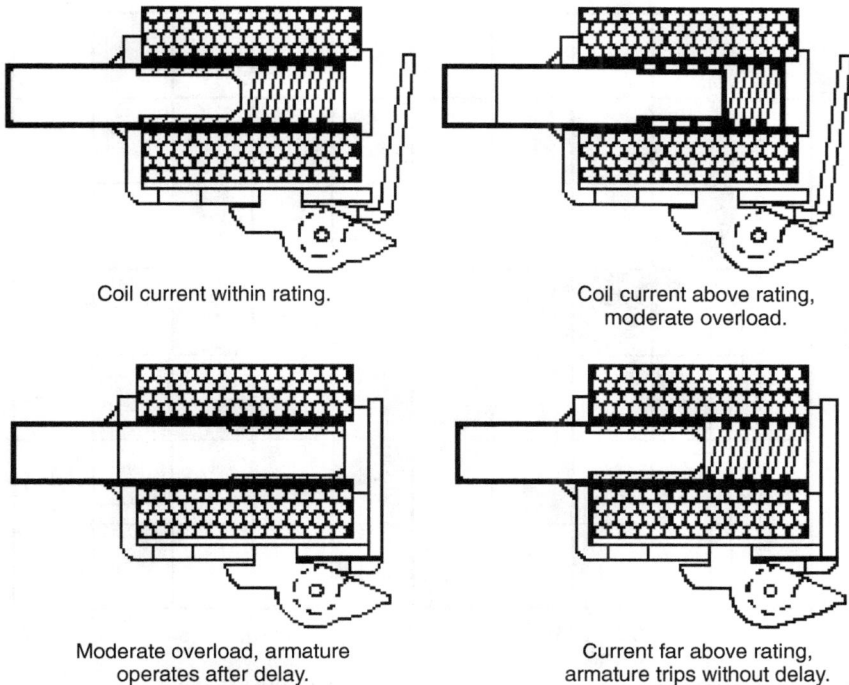

Coil current within rating.

Coil current above rating, moderate overload.

Moderate overload, armature operates after delay.

Current far above rating, armature trips without delay.

FIGURE 6.34   Operation of a magnetic circuit breaker. *(Courtesy Airpax)*

The temperature-independence of trip current illustrates an important virtue of magnetic circuit breakers not shared by thermal breakers or fuses. The magnetic breaker will trip reliably at a given overcurrent, for example 125% of rated current, at all reasonable ambient temperatures. However, the trip time of the dashpot-damped magnetic circuit is temperature-dependent. Delay time decreases with increasing temperature, which is often desirable. It is characteristic of dashpot-damped circuit breakers that they are position-sensitive.

If the breaker is designed to operate so that the axis of the dashpot tube is horizontal, then the trip current and delay time of the breaker will be different if the breaker is held vertically or at an angle. If gravity opposes the travel of the core, the breaker will require more current to trip, and the trip point will increase. If gravity aids the travel of the core, the breaker will trip at lower-than-rated hold current, and the trip point will be reduced. Dashpot-damped circuit breakers will have normal mounting orientation, often marked on the breaker body. The circuit breaker manufacturer should be consulted to determine how nonnormal mounting orientation affects trip parameters.

Figure 6.35 illustrates the time-current relationship of a typical magnetic circuit breaker. The two curves divide the time-current plane into three areas. The top curve defines the "upper-trip time limit." The bottom curve defines the "lower-trip time limit." Below the curves, the breaker "must trip." Above the curves, the breaker "must carry." In the band between the curves, the breaker will trip, but the time required will vary between the upper and lower limits. This describes the components of a breaker I-t curve. These curves, as those for other overcurrent protectors, are estimates. A more practical description

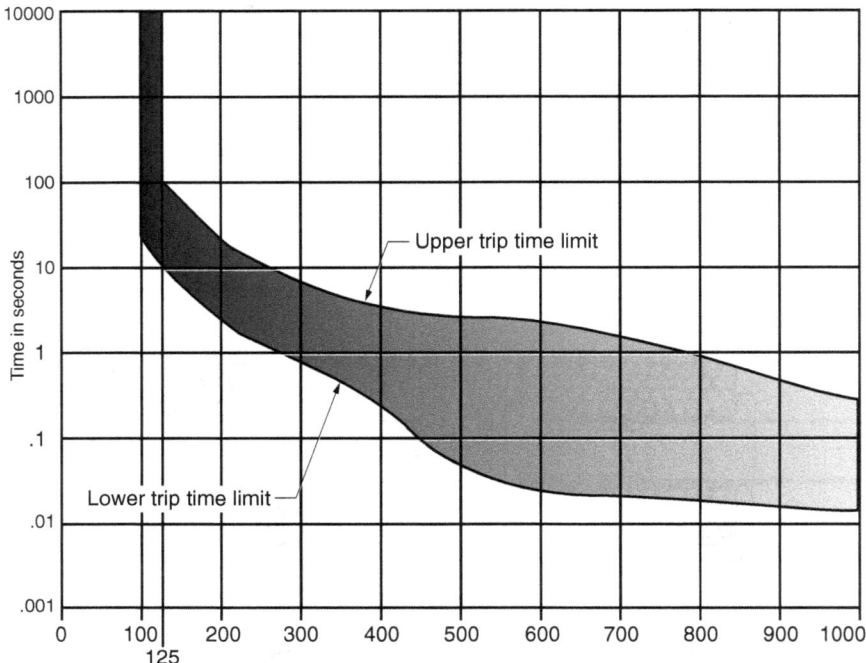

FIGURE 6.35 Magnetic circuit breaker time-current relationship. *(Courtesy Airpax)*

would be that the magnetic circuit breaker is designed to carry 100% of the rated load when mounted in the vertical position. The circuit breaker must trip when the current level reaches the minimum trip level (typically 125% of rated current). It must be remembered that the mounting orientation is critical to the proper operation of a magnetic breaker.

Figure 6.36 illustrates the operating mechanism of a typical dashpot-damped magnetic circuit breaker. This somewhat larger breaker combines overcurrent protection with an off/on switch. The magnetic circuit consists of the frame 1, armature 2, delay core 3, and pole piece 12. The electrical circuit consists of terminal 4, coil 5, contact bar 6, contact 7, contact 8, and terminal 9. As long as the current remains at or below 100% of rated current, the mechanism will not trip, and the contacts will remain closed (Figure 6.36a). Under these conditions, the electrical circuit can be open and closed by moving the toggle handle 10, on and off.

As current is increased, the magnetic flux generated in coil 5 moves the delay core 3 against the spring 11. At every level of overcurrent between the rated current and the ultimate trip current, there is an equilibrium position for the delay core 3 in the delay tube. Under conditions of moderate overcurrent, such as 125% of rated current, the magnetic flux generated in coil 5 is sufficient to move the delay core against spring 11, to a position where it rests against pole piece 12 (Figure 6.36b). The movement of the core against the pole piece increases the flux in the magnetic circuit enough to cause the armature 2 to move from its normal position (Figure 6.36a) to a position against the pole piece 12 (Figure 6.36b). As the armature moves, it trips sear pin 13, which triggers the collapsing link of the mechanism, thus opening the contacts. A high current overload will cause sufficient leakage flux to attract the armature, regardless of the position of the delay core.

Thermal-magnetic breakers are hybrids of thermal and magnetic breakers. The main tripping mechanism is thermally operated, but a magnetic feature gives instantaneous tripping under conditions of high-current overload. The magnetic circuit usually consists of a few turns of a large cross-section conductor in series with the thermal element. The magnetic assist does not affect normal thermal trip-response time, but on high overload, the current level generates sufficient magnetic force to trip the breaker without waiting for the bi-metal to deflect.

Thermal-magnetic and magnetic breakers are frequency sensitive. There will be a considerable difference in operating current, depending on whether the application is dc, 50 Hz, 60 Hz, or 400 Hz. Correction factors to calibrate the breaker for any application can be recommended by the circuit breaker manufacturer's support engineering staff.

One important category of circuit breakers is the "high-inrush" type. A standard breaker will be capable of typically withstanding up to 600% rated current for several seconds to accommodate surge resulting from a motor startup. High-inrush breakers, however, can withstand short surges of up to 3000% that can be encountered in high-capacitance circuits or ferroresonant power supplies.

An extremely important characteristic in selecting circuit breakers is that they should normally be "trip-free." In other words, closing the circuit by resetting the lever or pushing the reset button should result in repeated safe opening, as long as a current exceeding the rating of the device is present. A trip-free breaker cannot be kept closed by manual restraint.

All circuit breakers pass current through contacts. These contacts pit and degrade with use. The pitting can be minimized by incorporating arc-quenching mechanisms. One form of mechanism, an arcing grid, also called *arcing baffle*, consists of a series of metallic plates under the influence of a magnetic field. The grid surrounds the corridor through which the separating contacts pass. The arc is thereby split into several parts and quickly quenched as the contacts open.

It is important to keep circuit-breaker contact resistance as low as possible. This can be achieved though designing the contacts to have a wiping action as they are opened and closed. The "wiping action" of breaker contact points in high-quality breakers is very important. But the only time this operation is normally accomplished is during reset. Because of the importance of low contact resistance, a schedule should be established to ensure periodic cycling of each breaker so that they are properly maintained and operating.

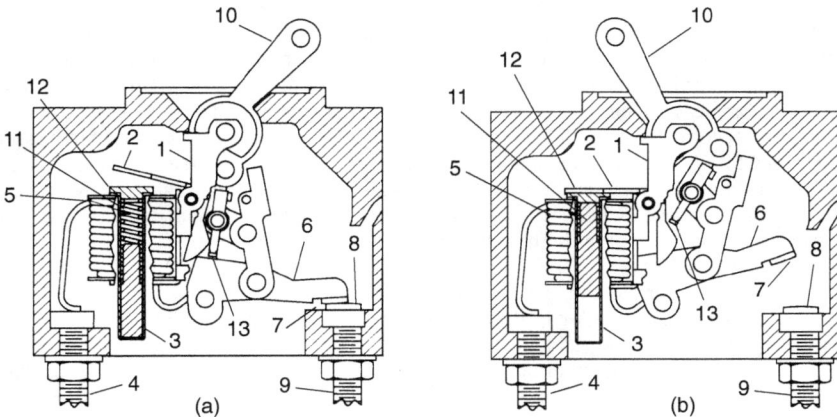

FIGURE 6.36 Operation of a magnetic circuit breaker. *(Courtesy Airpax)*

Safety-agency approvals are normally required for circuit breakers. Larger breakers are listed by UL to UL 489 *Molded Case Circuit Breakers and Circuit Breaker Enclosures*. CSA has similar certification procedures for service entrance or branch circuits published as sections of CSA 22.2. Smaller breakers are recognized under UL 1077 *Supplementary Protectors for Use in Electrical Equipment*. Strictly speaking, UL restricts the phrase "circuit breaker" to those devices that are *Listed* to UL489. Other components are *Recognized* as "circuit protectors," even though they operate on the same principles as circuit breakers. Some breakers are covered by UL 1416 *Overcurrent and Overtemperature Protectors for Radio and Television-type Appliances*.[12]

Outside North America, breaker standards of the type of interest to electronics engineers conform to IEC 934 *Circuit Breakers For Equipment, IEC 380 Safety of Electrically Energized Office Machines*, and IEC 435 *Safety of Data Processing Equipment*. Both IEC 380 and IEC 435 have been superseded by IEC 950 *Safety of Information Technology Equipment Including Business Equipment*. In turn, IEC and national standards in the European Economic Community will shortly be superseded by EC-wide standards.

Regulations on both sides of the Atlantic for circuit breakers are changing too rapidly to be captured in handbook form. Compliance engineers, and the engineering department of circuit-breaker manufacturers can provide up-to-date information.

It must be remembered that safety-agency approvals for circuit breakers are based on different specifications than those for fuses. The appropriate agency specification(s) should be reviewed to determine that the device meets the intended application's minimum requirements. Specifiers should insist on appropriate safety-agency approval for all overcurrent-protection devices.

As is true with other types of overcurrent protectors, it should be accepted that most breakers will eventually require replacement. Although breaker MTBF can be significantly improved by proper breaker selection, and by periodic on/off cycling of the breakers throughout their life, many breakers will still require replacement. The use of an agency-approved breaker will ensure that the mode of failure is safe. Breaker "failure" can be subtle: Breakers can fail by not being resettable. A breaker's demise is often discovered as a result of turning the breaker off (or having it trip off because of an overcurrent condition) and not being able to reset the breaker back on.

Circuit designers should anticipate a method for breaker replacement, should that become necessary. If the breaker is used as the power switch in a design, the feed side is "hot," unless the entire branch is shut down. Inclusion of a separate switch that can be used to open the breaker feed line, or incorporation of a disconnect method, should be considered.

Circuit breakers from different manufacturers are often not mechanically interchangeable. Both panel-mounting and bus-mounting dimensions must be reviewed carefully.

If the breaker is of the rocker-switch reset type, it is not meant to be used as a switch, unless the approval agency has tested it as a switch. The minimum number of switching cycles and specified test conditions that must be met for an agency-approved switch (such as UL or CSA) are far more demanding than those for a breaker. A breaker can be thought of as a switch that has a high short-circuit-withstand capability, but usually not the trip cycle capacity of a true switch. Bellcore assigns a failure rate for circuit breakers used for power on/off that is 10 times higher than circuit breakers meant for constant on-protection only (2550 vs. 255 FITs).

Much of the information regarding magnetic circuit breakers in this section was taken from the *Airpax Guide To Circuit Protection*, Chapter A. This document is an excellent resource when selecting and sizing circuit breakers.

Further information about circuit-breaker technology, including accessible mathematical derivations, can be found in *The Theory and Practice of Overcurrent Protection* by Patrick J. McCleer.

6.6 GLOSSARY OF
CIRCUIT-PROTECTOR TERMINOLOGY

alarm indicator   Any means by which protector status can be determined.

arcing time   Period of current flow after circuit-protector activation and subsequent to element melting or contact opening.

circuit breaker, trip-free   Breaker whose pole(s) cannot be manually maintained closed when carrying current that would otherwise automatically trip breaker to the open position.

clearing time   Entire period of current flow after circuit protector activation. In a fuse, this period is the sum of the melting (or vaporization) time and arcing time.

current limiting   The term *current limiting* has two distinct meanings. *Current limiting* describes a class of overcurrent-protection device that operates by limiting current to a low and constant value. *Current limiting* is alternately used to describe a fuse that blows (opens) before full short-circuit current can be delivered to a load by a source.

electrical (remote) alarm   Any means by which protector status can be determined at a distance from the protector itself.

fault (current)   Abnormal (high) flow of current caused by defect in circuit. Commonly called a *short circuit*. A fault current is generally considered to be any overcurrent greater than 800% of normal operating current.

fusing factor   The ratio of the minimum fusing current and the fuse-rated current, usually in the range 1.2 to 2.0 to ensure adequate protection against long term overcurrent.

fuse block   An assembly of fuse clip pairs, appropriately spaced on an insulative base, used for providing electrical and mechanical contact between a fuse and an associated circuit.

fuse clip   An electrically conductive component mounted to an insulative base (e.g., printed wiring board) that is designed to provide electrical and mechanical contact between a fuse and an associated circuit.

fuse holder   Any of a variety of devices used to house a fuse and provide electrical and mechanical connection to a circuit. These can be referred to as *bayonet, panel-mount* or *board-mount* holders.

I-t curve   Plot of current vs. clearing time or melting time used to select appropriate protector-current rating for a given function.

lightning-surge withstand   Maximum current surge that a current protector can endure repeatedly without opening. Waveshape used (e.g., 10 × 560, 10 × 1000, etc.), pulse duration, period between pulses, and ambient temperature are critical to repeatable test results.

limiter   Overcurrent devices that are used for heavy-overload, short-circuit protection only. Limiters cannot be designed to open under mild overcurrent conditions.

line cross   Test performed on current protectors to determine low-level-overload failure mode. This test is structured to simulate a power line falling on a communications line, creating a high-impedance path to ground. Under this test, current is started low and gradually increased (e.g., 15-minute intervals) in predetermined increments (e.g., 5% of rated current) until the protector opens or some other event occurs on the circuit under test, which opens the circuit.

melting time   Period of current flow after circuit protector activation required to melt a fuse element.

normal fuse   Accepted designation for standard fuse with no time delay. Often referred to as *nontime-delay* or *quick-acting fuse*.

power factor   Ratio of real power to apparent power.

rated voltage   Maximum voltage at which the protector will continue to operate safely.

short-circuit current   Maximum current that a protector can interrupt safely at the rated voltage. Frequently referred to as *fault current* or *interrupting rating*.

time-delay fuse   Accepted designation for fuse having any predefined delay in operation between time of fault and circuit interruption. Often referred to as *slow blow* or *time lag*.

visual (local) alarm   Any means by which protector status can be determined located on the protector itself.

voltage drop   The potential measured across the circuit protector under load conditions. This measurement is limited to the device only and does not include voltages across connecting devices, such as fuse clips. Unless otherwise noted, voltage is measured at 50% of device current rating.

voltage rating   Maximum voltage the device can withstand and still safely interrupt overload current under short-circuit conditions.

6.7 BIBLIOGRAPHY

The Choice of Protection, *Airpax, Cambridge, MD. Contains information concerning various types of circuit breakers.*

Circuit Breaker Handbook, *Heineman Electric Company, Lawrenceville, NJ.*

Component Technology and Standardization, *Genium Publishing, Schenectady, NY, 1988, vol. II, chapter 16:* Protective Devices. *This chapter describes fuses and circuit breakers primarily for military and aerospace applications.*

Electric Fuses, *H.W. Baxter, The University Press, Edward Arnold & Co., London, 1950.*

Electric Fuses, *A. Wright & P.G. Newbery, Peter Peregrinus (on behalf of IEE), 1984. A comprehensive treatment of fuse technology.*

Export Designer's Reference & Catalog, *Panel Components Corporation, Oskaloosa, IA. Contains detailed information about designing electronic equipment to meet power-supply conditions worldwide.*

Electronic Designer's Protection Handbook, *Bulletin EDPH, Bussmann, Cooper Industries, St. Louis, MO.*

Quick Guide to Overcurrent Protection, *Heineman Electric Company, Lawrenceville, NJ.*

Overcurrent Protection Handbook, *Reliance Fuse, Des Plaines, IL. Contains discussion of large fuses.*

The Theory and Practice of Overcurrent Protection, *Patrick J. McCleer, Mechanical Products, Inc., 1987, Jackson, MI. A comprehensive treatment of overcurrent protection issues, with emphasis on circuit breakers.*

6.8 REFERENCES

1. Patrick M. Craney, *Protecting Semiconductor Devices: Circuit Breakers vs. Fuses, Electronics*, January 5, 1978, pp 163–167.
2. *Electric Current Abroad*, (latest edition). U.S. Department of Commerce, Bureau of Industrial Economics, Washington, D.C. Contains a detailed listing of power supply voltage conditions country by country.
3. Further discussion is found in *Electric Fuses*, A. Wright, P.G. Newbery, Peter Peregrinus, London, 1984, section 6.2 *Domestic Plug Fuses*; and *Electric Fuses*, H.W. Baxter, Edward Arnold & Co., London, 1950.
4. *Semiconductor Fuse Applications Handbook*, available from International Rectifier Corp, El Segundo, CA, 1972 gives a detailed discussion on the technology and selection criteria of semiconductor fuses. Much of the information in the *IR Handbook* is also available in *Electric Fuses*, by Wright and Newbery.
5. DW Electrochemicals, Ltd, 9005 Leslie Street, Unit 106, Richmond Hill, Ontario L4B 1G7, Canada, (416) 889-1522.
6. See for instance, *Littlefuse Designer's Guide*, EC-101, Littlefuse, Inc., Des Plaines, IL 60016.
7. (Four Quadrant) multiplying oscilloscopes are discussed in Rein van Erk, *Oscilloscopes: Functional Operation and Measuring Examples*, McGraw-Hill, New York, NY, 1978.
8. Rich Nute, *Selecting a Fuse Value*, Product Safety Newsletter, IEEE EMC Society, Vol. 3, #3, May/June, 1990.
9. *Gould Shawmut Book of Electrical Information*, The Newburyport Press (1988), Gould Shawmut, Newburyport, MA.
10. *A Fuse-Thyristor Coordination Primer*, Motorola Application Note AN-568, 1972.
11. *Reliability Prediction Procedure for Electronic Equipment*, Bellcore Document TR-NWT-000332, Issue 3, September 1990.
12. See Chapter 7, *The Theory and Practice of Overcurrent Protection*, ibid, for discussion on the differences between UL 489 and UL 1077.

CHAPTER 7
FILTERS

Michael M. Driscoll

Advisory Engineer, Northrup Grumman Corporation

Christopher R. Vale

Advisory Engineer, Northrup Grumman Corporation

7.1 INTRODUCTION

Electronic filters are used to modify and enhance the spectrum of electrical signals such as radio and television broadcasts, covert communications, radar transmissions and returns, and other information carrying signals. Here we describe the function, workings, and styles of these filters.

Three elementary components form the simple signal-delivery system in Figure 7.1: the generator is the source of electrical signals, with source resistance R_S; a perfectly lossless transmission line is used to deliver all the signal available from the generator to the load; and the load R_L. The delivery of all the available signal to the load, without the use of transformers ($R_S = Z_O = R_L$) is called the *impedance-matched case* (not to be confused with the signal-matched, or North-sense, case). If there is spectral energy in the signal that must not be delivered to the load, then a filter can be inserted somewhere in the system. Figure 7.2 shows the filter in the middle of the transmission line with all the components having the same impedance, R.

The filter can prevent the delivery of the unwanted spectral energy by absorbing it and converting it into heat or it can reflect it back toward its source. This chapter treats only the second type. Thus, in some applications, undesirable nonlinear effects arise from the arrival of reflected energy back at the generator. Special techniques to combat these problems are sometimes required.

Some varieties of filter functions are illustrated in Figure 7.3, using "brick-wall" response shapes, which, of course, are not realizable. The lowpass function (Figure 7.3a) is intuitive; it can be used to reject unwanted interfering high-frequency signals or to remove

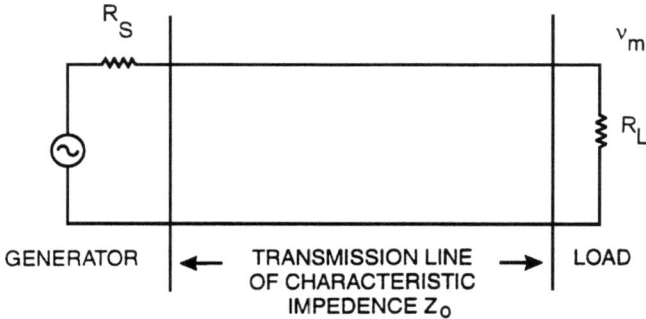

FIGURE 7.1   Elementary signal-delivery system.

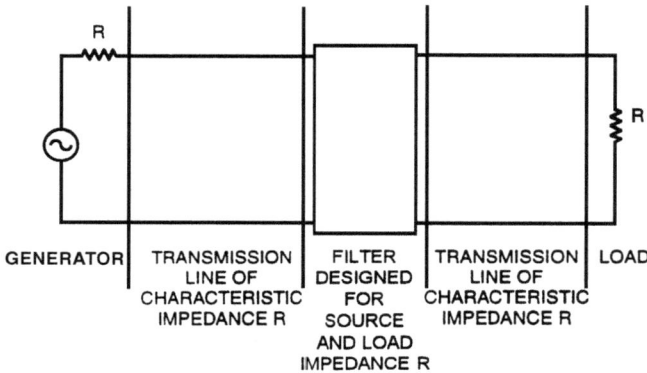

FIGURE 7.2   Filter in a signal-delivery system.

alternating current components from a direct-current power supply. The most well-known use of the lowpass filter to act as a prototype for all other styles: all catalog-like tabulations, such as in Zverev,[1] are lowpasses and transformation procedures exist for making their characteristics appear in the gain response of most classes of bandpass, highpass, and band-reject filters. The highpass (Figure 7.3b) is complementary in function to the lowpass. Whereas the latter passes everything up to a certain frequency (called the *cut-off frequency*), the highpass rejects everything up to the cut-off frequency and passes all signals beyond it. For instance, if the lowest frequencies of the spectrum of Figure 7.3 were impulsive noise, the highpass filter illustrated would remove it.

Bandpass filters are probably the most useful and common signal-modifying filters. Roughly, they can be categorized according to relative bandwidth with somewhat ill-defined limits. Very wideband filters (Figure 7.3c) are those with fractional bandwidth greater than about 25 percent, where fractional bandwidth is given by:

$$Fractional\ Bandwidth = \frac{F_2 - F_1}{F_0} = \frac{BW_{3dB}}{F_o} \tag{7.1}$$

and $BW_{3dB} = F_2 - F_1$ is the passband width at the 3-decibel down points with respect to the maximum output in the passband; and F_0 is the center frequency given by:

$$F_0 = \sqrt{F_1 F_2} \tag{7.2}$$

F_1 is the lower 3-dB frequency and F_2 is the upper 3-dB frequency. The inverse of fractional bandwidth is called the *loaded Q* and is a very important design parameter that is covered later.

$$Q_L = \frac{F_0}{BW_{3dB}} \tag{7.3}$$

AMPLITUDE

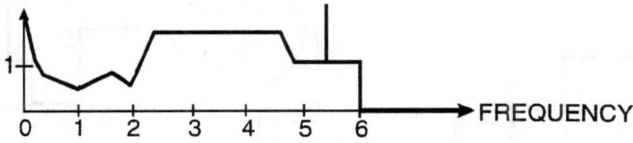

INPUT SPECTRUM

FILTER RESPONSE

FILTER OUTPUT

GAIN

AMPLITUDE

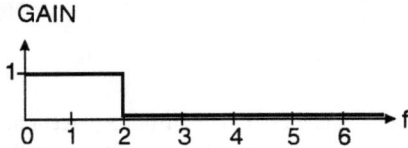
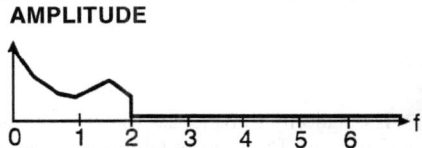

(a) LOW PASS

GAIN

AMPLITUDE

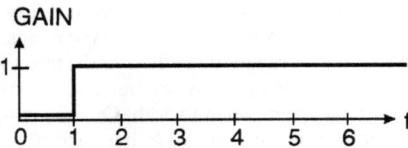
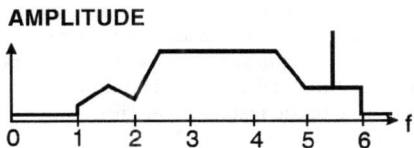

(b) HIGH PASS

GAIN

AMPLITUDE

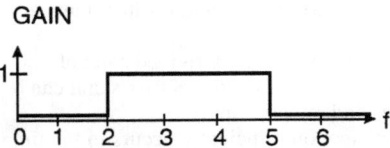
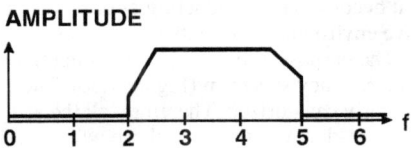

(c) WIDE BAND PASS

FIGURE 7.3   Various idealized filters and their effect on a signal spectrum.

FILTER RESPONSE   FILTER OUTPUT

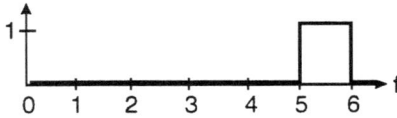

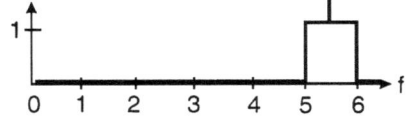

(d) NARROW BAND PASS

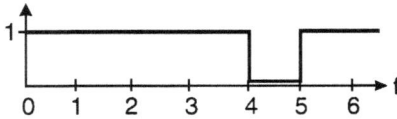

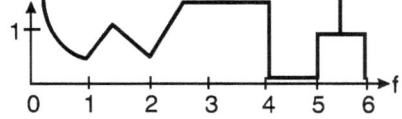

(e) BAND REJECT

FIGURE 7.3   Various idealized filters and their effect on a signal spectrum. *(Continued)*

The majority of bandpass filters have fractional bandwidth ranging from 3 percent up to 25 percent. They can be fabricated with lumped elements (capacitors and coils) or with distributed resonators and microwave cavities. For narrowband filters, the quality factor, called Q, of the individual elements or resonators is critical: it must be substantially larger than the Q_L of the filter, if good performance (like the transformed lowpass shape) and low insertion loss are to be achieved.

Narrowband filters (Figure 7.3d) have fractional bandwidth less than 3 percent. Large values of element Q are achieved with special elements like polished microwave cavities, acoustic resonators and cryogenic resonators. Resonator Qs range in the thousands, tens to hundred thousands, and in the millions, respectively. Simple-coupled cavities and piezo-electric acoustic resonators are necessary to achieve narrow passband width. Quartz crystal resonators can have Q values up to one million; but achieving "wide bandwidth" is difficult. Wider filter bandwidth can be achieved using lithium tantalate and ceramic resonators, rather than quartz. Very high Q can be achieved with cryogenic electromagnetic resonators; but the principle modern thrust is in filters fabricated using special materials that become superconducting at or below 90 Kelvin. Narrow-bandwidth filters must also have environmentally stable resonators.

The purpose of band reject, or notch filters, is to remove a predetermined band of a signal frequency spectrum (Figure 7.3e). They are the inverse of bandpass filters and can be similarly categorized. Therefore, all the same rules and remarks apply.

Time-domain filters are those whose design or function principally occurs in the time domain; that is, either the impulse response or the step response or the phase response must meet prescribed objectives. The concept of filter performance in the time domain has been developed by Blinchikoff and Zverev in *Filtering in the Time and Frequency Domain*.[2]

7.2 FILTER RESPONSES

This section covers the various ideal filter-selectivity shapes in a more practical sense. Design procedures for passive realizations using only reactances, v.i.z. coils (or inductors) and capacitors, will be indicated. Because all shapes are derivable from lowpass networks, more emphasis will be placed on them.

7.2.1 Passive Lowpass Filters: The Frequency-Domain Ideal

The performance of elementary forms of filters is intuitive. For example, the circuit shown in Figure 7.4a is a coil of inductance L inserted between a generator and a load: direct current (dc) is delivered to the load, but very high-frequency signals are not. This is a one-pole lowpass filter. A fundamental relation describing the performance of this elementary filter for all frequencies is called the *frequency response* (the *gain*). It is the magnitude of the ratio of two voltages: the voltage delivered to the load when the filter is "inserted," v_0, versus the voltage that would be delivered to the load by the generator alone, v_m (Figure 7.1). For the case shown, the magnitude response is:

$$\left|\frac{v_0}{v_m}\right| = \frac{1}{\sqrt{1 + \left(\frac{\omega}{\omega_c}\right)^2}} \tag{7.4}$$

where $\omega = 2\pi f$ and f is frequency in Hertz, where $\omega_c = (R_S + R_L)/L$ is the cut-off frequency in radians per second. At this frequency, the voltage delivered to the load is the maximum passband voltage divided by the square root of two. The maximum occurs at 0 radians per second (dc). In the vast majority of cases, the cut-off frequency or the bandwidth is defined at the frequency where the transfer function is 3 dB smaller than the maximum passband output because:

$$A = 10\log\frac{1}{2} = -3 \text{ dB}$$

Thus, it is also called the *half-power point*.

Figure 7.4b shows the dual of the series inductor; it is a shunt capacitor inserted between the same generator and load resistance. In this case, the cut-off frequency is:

$$\omega_c = \frac{1}{\left(\dfrac{R_S R_L}{R_S + R_L}\right)C} \tag{7.5}$$

Using this value for cut-off, the magnitude response function is identical to Eq. 7.4. The difference between these two identically performing lowpass filters is that the first creates its stop band by presenting a high impedance to the generator at frequencies beyond the cut-off and the second does it by presenting a low impedance. This subtlety can be important, depending on the application.

The gain, Eq. 7.4, is plotted in Figure 7.5. As shown, it is a relatively crude approximation, compared to the "brick-wall" response (Figure 7.3a). It can be converted to an additive quantity (called *attenuation*). Thus, attenuation in decibels:

$$A = 20\log\sqrt{1 + \left(\frac{\omega}{\omega_c}\right)^2} \tag{7.6}$$

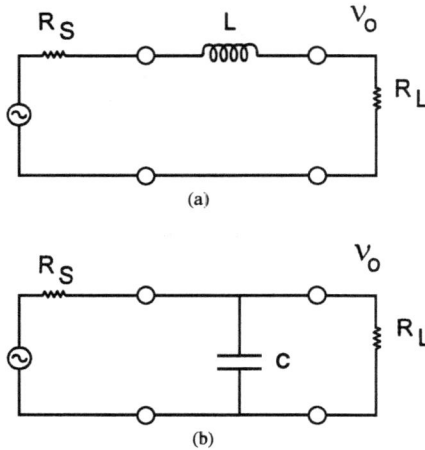

(a)

(b)

FIGURE 7.4   Elementary lowpass filters.

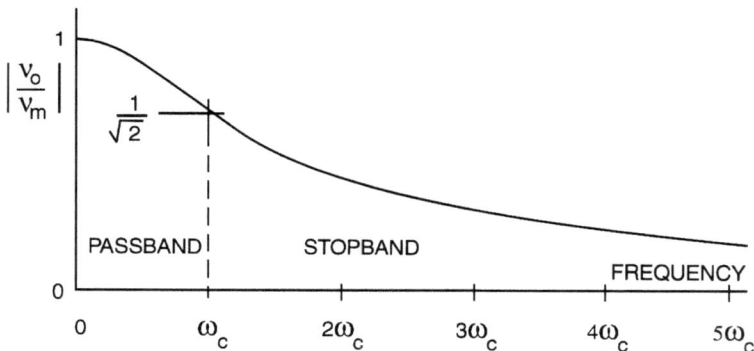

FIGURE 7.5   Gain, or magnitude of the transfer function for an elementary lowpass filter.

This is the universal descriptive ordinate for filter performance and it is plotted in Figure 7.6. It serves to illustrate the salient features of a realistic lowpass filter and the simple transfer function is an easily analyzed foundation for higher-order networks.

The well-known Butterworth or maximally flat-gain filter performs with this attenuation function:

$$A_B = 20 \log \sqrt{1 + \left(\frac{\omega}{\omega_c}\right)^{2N}} \tag{7.7}$$

where N is the number of poles or the order of the network; it is the same as the number of reactances in the lowpass or highpass structure. It is also equal to the number of inductors in an N-pole bandpass, either of the "full-bandpass" type or of the coupled-resonator type. Plots of Eq. 7.7 are shown in Figure 7.7 for up to $N = 9$ and with the cut-off frequency normalized to one radian per second. It is clear that as N approaches infinity the "brick-wall"

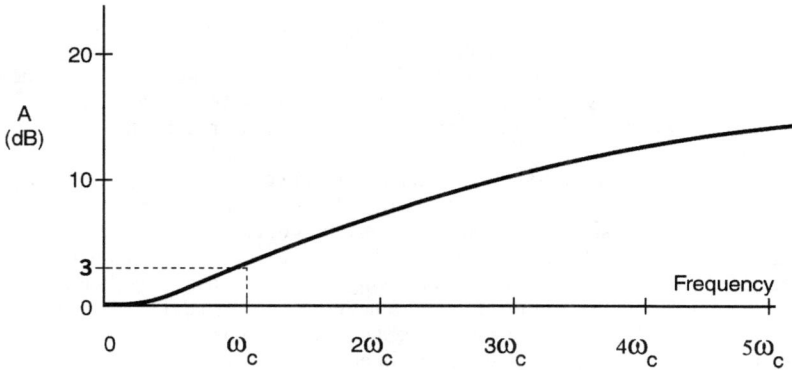

FIGURE 7.6 Attenuation of elementary lowpass filter.

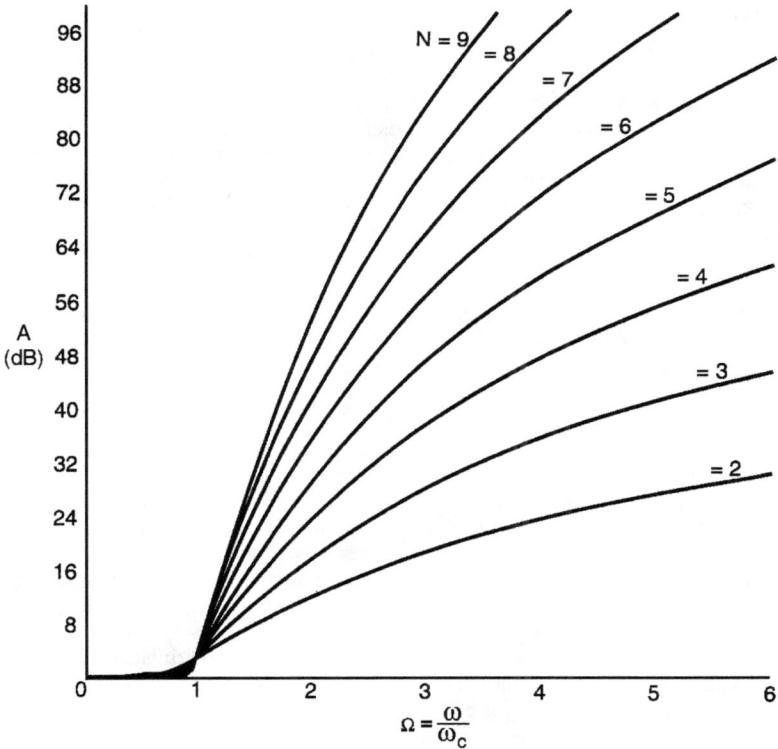

FIGURE 7.7 Attenuation of Butterworth filter for an order of 2 to 9.

performance is approached; but, with real-circuit components, as N approaches infinity, the midband insertion loss also approaches infinity along with the group delay and the number of elements in the network and the size, weight, and expense. The task of the filter users is to specify only the minimum selectivity that will serve their needs; and the task of the designer is to meet those needs with the minimum number of components; both efforts lead to economy.

The hard work (the actual synthesis of the networks) has been accomplished, yielding the normalized tabulation shown in Table 7.1 (also from Zverev); the remainder of the filter design task is quite simple. The table lists the values of capacitance and inductance of 1-Ω filters with the cut-off frequency equal to 1 radian per second and having the performance of Figure 7.7. There is a choice of having the first element being a capacitor to ground or an inductance in series; or, more functionally expressed, the filter can be designed to exhibit either high or low input impedance in the stopband. These normalized filters serve as prototypes for an infinite variety of all types of filters. This section covers the procedure for obtaining various lowpass filter responses. As an example, having determined which order of Butterworth filter (i.e., number of poles) serves the need using Figure 7.7, the normalized element values are found from Table 7.1. To change impedance level R (Figure 7.2), simply multiply every inductance value by R and divide all capacitance values by R. Finally, to move the cut-off frequency from 1 radian per second to f_c, divide every inductance and every capacitance by:

$$\omega_c = 2\pi f_c \tag{7.8}$$

This completes the design of an N-pole Butterworth lowpass filter, which, when inserted between a generator of internal impedance R and a load having impedance R, will have a cut-off frequency of f_c and will deliver selectivity performance shown in Figure 7.7.

Although the ninth-order Butterworth selectivity or shape factor is obviously more rectangular, and closer to Figure 7.3a than Figure 7.6, it is not the best that can be done. Even a 30-pole Butterworth has a ratio of only 1.26 for the 60-dB bandwidth divided by the 3-dB bandwidth. This is called the *60-dB selectivity*. Other functions for the attenuation response exist that for the same number of elements are superior to the Butterworth in terms of selectivity. One is called the Chebyshev filter because the gain function uses Chebyshev polynomials. Thus:

$$\left|\frac{v_0}{v_m}\right| = \frac{1}{\sqrt{1 + \varepsilon^2 C_N^2(\omega)}} \tag{7.9}$$

where ε is a real number less than 1 (which determines ripple amplitude), C_N is the Nth order Chebyshev polynomial as a function of radian frequency, ω. N is the order of the network in the same sense as it is for the Butterworth filter. When normalized to pass through 3 dB at one radian per second, all Chebyshev responses have higher selectivity than the same order of Butterworth filter. Humpherys[3] has shown that in the limit, as Chebyshev ripple amplitude goes to zero, the Chebyshev filter becomes a Butterworth filter; a result that is intuitive, but rigorously proven in his book.

The performance of Chebyshev filters having 0.1-dB passband ripple and 3-dB attenuation at 1 radian per second is shown in Figure 7.8. The increased attenuation at a given multiple of the cut-off frequency is evident. Even higher selectivity is available if higher passband ripple can be tolerated, but the user must be cautious because increased gain ripple introduces increased phase distortion, as well as the obvious effect of the ripple itself on the desired output spectrum. Furthermore, the time response exhibits more ringing, meaning a higher level of impulse-response overshoot, as the shape in the frequency do-

TABLE 7.1 Normalized Element Values for Butterworth Lowpass Filters

LOW PASS ELEMENT VALUES

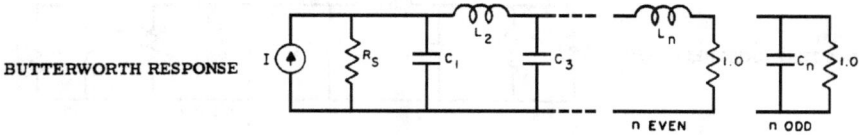

BUTTERWORTH RESPONSE

n	R_s	C_1	L_2	C_3	L_4
2	1.0000	1.4142	1.4142		
	1.1111	1.0353	1.8352		
	1.2500	0.8485	2.1213		
	1.4286	0.6971	2.4387		
	1.6667	0.5657	2.8284		
	2.0000	0.4483	3.3461		
	2.5000	0.3419	4.0951		
	3.3333	0.2447	5.3126		
	5.0000	0.1557	7.7067		
	10.0000	0.0743	14.8138		
	INF.	1.4142	0.7071		
3	1.0000	1.0000	2.0000	1.0000	
	0.9000	0.8082	1.6332	1.5994	
	0.8000	0.8442	1.3840	1.9259	
	0.7000	0.9152	1.1652	2.2774	
	0.6000	1.0225	0.9650	2.7024	
	0.5000	1.1811	0.7789	3.2612	
	0.4000	1.4254	0.6042	4.0642	
	0.3000	1.8380	0.4396	5.3634	
	0.2000	2.6687	0.2842	7.9102	
	0.1000	5.1672	0.1377	15.4554	
	INF.	1.5000	1.3333	0.5000	
4	1.0000	0.7654	1.8478	1.8478	0.7654
	1.1111	0.4657	1.5924	1.7439	1.4690
	1.2500	0.3882	1.6946	1.5110	1.8109
	1.4286	0.3251	1.8618	1.2913	2.1752
	1.6667	0.2690	2.1029	1.0824	2.6131
	2.0000	0.2175	2.4524	0.8826	3.1868
	2.5000	0.1692	2.9858	0.6911	4.0094
	3.3333	0.1237	3.8826	0.5072	5.3381
	5.0000	0.0904	5.6835	0.3307	7.9397
	10.0000	0.0392	11.0942	0.1616	15.6421
	INF.	1.5307	1.5772	1.0824	0.3827
n	$1/R_s$	L_1	C_2	L_3	C_4

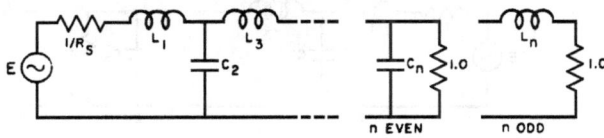

TABLE 7.1 Normalized Element Values for Butterworth Lowpass Filters *(Continued)*

LOW PASS ELEMENT VALUES

BUTTERWORTH RESPONSE

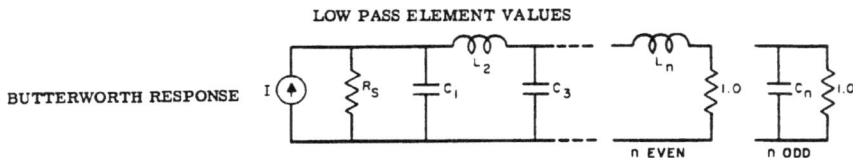

n	R_s	C_1	L_2	C_3	L_4	C_5	L_6	C_7
	1.0000	0.6180	1.6180	2.0000	1.6180	0.6180		
	0.9000	0.4416	1.0265	1.9095	1.7562	1.3887		
	0.8000	0.4698	0.8660	2.0605	1.5443	1.7380		
	0.7000	0.5173	0.7313	2.2849	1.3326	2.1083		
	0.6000	0.5860	0.6094	2.5998	1.1255	2.5524		
5	0.5000	0.6857	0.4955	3.0510	0.9237	3.1331		
	0.4000	0.8378	0.3877	3.7357	0.7274	3.9648		
	0.3000	1.0937	0.2848	4.8835	0.5367	5.3073		
	0.2000	1.6077	0.1861	7.1849	0.3518	7.9345		
	0.1000	3.1522	0.0912	14.0945	0.1727	15.7103		
	INF.	1.5451	1.6944	1.3820	0.8944	0.3090		
	1.0000	0.5176	1.4142	1.9319	1.9319	1.4142	0.5176	
	1.1111	0.2890	1.0403	1.3217	2.0539	1.7443	1.3347	
	1.2500	0.2445	1.1163	1.1257	2.2389	1.5498	1.6881	
	1.4286	0.2072	1.2363	0.9567	2.4991	1.3464	2.0618	
	1.6667	0.1732	1.4071	0.8011	2.8580	1.1431	2.5092	
6	2.0000	0.1412	1.6531	0.6542	3.3687	0.9423	3.0938	
	2.5000	0.1109	2.0275	0.5139	4.1408	0.7450	3.9305	
	3.3333	0.0816	2.6559	0.3788	5.4325	0.5517	5.2804	
	5.0000	0.0535	3.9170	0.2484	8.0201	0.3628	7.9216	
	10.0000	0.0263	7.7053	0.1222	15.7855	0.1788	15.7375	
	INF.	1.5529	1.7593	1.5529	1.2016	0.7579	0.2588	
	1.0000	0.4450	1.2470	1.8019	2.0000	1.8019	1.2470	0.4450
	0.9000	0.2985	0.7111	1.4043	1.4891	2.1249	1.7268	1.2961
	0.8000	0.3215	0.6057	1.5174	1.2777	2.3338	1.5461	1.6520
	0.7000	0.3571	0.5154	1.6883	1.0910	2.6177	1.3498	2.0277
	0.6000	0.4075	0.4322	1.9284	0.9170	3.0050	1.1503	2.4771
7	0.5000	0.4799	0.3536	2.2726	0.7512	3.5532	0.9513	3.0640
	0.4000	0.5899	0.2782	2.7950	0.5917	4.3799	0.7542	3.9037
	0.3000	0.7745	0.2055	3.6706	0.4373	5.7612	0.5600	5.2583
	0.2000	1.1449	0.1350	5.4267	0.2874	8.5263	0.3692	7.9079
	0.1000	2.2571	0.0665	10.7004	0.1417	16.8222	0.1823	15.7480
	INF.	1.5576	1.7988	1.6588	1.3972	1.0550	0.6560	0.2225
n	$1/R_s$	L_1	C_2	L_3	C_4	L_5	C_6	L_7

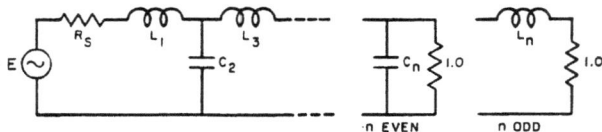

TABLE 7.1 Normalized Element Values for Butterworth Lowpass Filters *(Continued)*

LOW PASS ELEMENT VALUES

BUTTERWORTH RESPONSE

n	R_s	C_1	L_2	C_3	L_4	C_5	L_6	C_7	L_8	C_9	L_{10}
	1.0000	0.3902	1.1111	1.6629	1.9616	1.9616	1.6629	1.1111	0.3902		
	1.1111	0.2075	0.7575	0.9925	1.6362	1.5900	2.1612	1.7092	1.2671		
	1.2500	0.1774	0.8199	0.8499	1.7779	1.3721	2.3874	1.5393	1.6246		
	1.4286	0.1513	0.9138	0.7257	1.9852	1.1760	2.6879	1.3490	2.0017		
8	1.6667	0.1272	1.0455	0.6102	2.2740	0.9912	3.0945	1.1530	2.4524		
	2.0000	0.1042	1.2341	0.5003	2.6863	0.8139	3.6678	0.9558	3.0408		
	2.5000	0.0822	1.5201	0.3945	3.3106	0.6424	4.5308	0.7594	3.8825		
	3.3333	0.0608	1.9995	0.2919	4.3563	0.4757	5.9714	0.5650	5.2400		
	5.0000	0.0400	2.9608	0.1921	6.4523	0.3133	8.8538	0.3732	7.8952		
	10.0000	0.0198	5.8479	0.0949	12.7455	0.1547	17.4999	0.1846	15.7510		
	INF.	1.5607	1.8246	1.7287	1.5283	1.2588	0.9371	0.5776	0.1951		
	1.0000	0.3473	1.0000	1.5321	1.8794	2.0000	1.8794	1.5321	1.0000	0.3473	
	0.9000	0.2242	0.5388	1.0835	1.1859	1.7905	1.6538	2.1796	1.6930	1.2447	
	0.8000	0.2434	0.4623	1.1777	1.0200	1.9542	1.4336	2.4189	1.5318	1.6033	
	0.7000	0.2719	0.3954	1.3162	0.8734	2.1885	1.2323	2.7314	1.3464	1.9812	
9	0.6000	0.3117	0.3330	1.5092	0.7361	2.5124	1.0410	3.1516	1.1533	2.4328	
	0.5000	0.3685	0.2735	1.7846	0.6046	2.9734	0.8565	3.7426	0.9579	3.0223	
	0.4000	0.4545	0.2159	2.2019	0.4775	3.6706	0.6771	4.6310	0.7624	3.8654	
	0.3000	0.5987	0.1600	2.9006	0.3539	4.8373	0.5022	6.1128	0.5680	5.2249	
	0.2000	0.8878	0.1054	4.3014	0.2333	7.1750	0.3312	9.0766	0.3757	7.8838	
	0.1000	1.7559	0.0521	8.5074	0.1153	14.1930	0.1638	17.9654	0.1862	15.7504	
	INF.	1.5628	1.8424	1.7772	1.6202	1.4037	1.1408	0.8814	0.5155	0.1736	
	1.0000	0.3129	0.9080	1.4142	1.7820	1.9754	1.9754	1.7820	1.4142	0.9080	0.3129
	1.1111	0.1614	0.5924	0.7853	1.3202	1.3230	1.8968	1.6956	2.1883	1.6785	1.2267
	1.2500	0.1388	0.6452	0.6762	1.4400	1.1420	2.0779	1.4754	2.4377	1.5245	1.5861
	1.4286	0.1190	0.7222	0.5797	1.6130	0.9802	2.3324	1.2712	2.7592	1.3431	1.9646
10	1.6667	0.1004	0.8292	0.4891	1.8528	0.8275	2.6825	1.0758	3.1895	1.1526	2.4169
	2.0000	0.0825	0.9818	0.4021	2.1943	0.6808	3.1795	0.8864	3.7934	0.9588	3.0072
	2.5000	0.0652	1.2127	0.3179	2.7108	0.5384	3.9302	0.7018	4.7002	0.7641	3.8512
	3.3333	0.0484	1.5992	0.2358	3.5754	0.3995	5.1858	0.5211	6.2118	0.5700	5.2122
	5.0000	0.0319	2.3740	0.1556	5.3082	0.2636	7.7010	0.3440	9.2343	0.3775	7.8738
	10.0000	0.0158	4.7005	0.0770	10.5104	0.1305	15.2505	0.1704	18.2981	0.1872	15.7481
	INF.	1.5643	1.8552	1.8121	1.6869	1.5100	1.2921	1.0406	0.7626	0.4654	0.1564
n	$1/R_s$	L_1	C_2	L_3	C_4	L_5	C_6	L_7	C_8	L_9	C_{10}

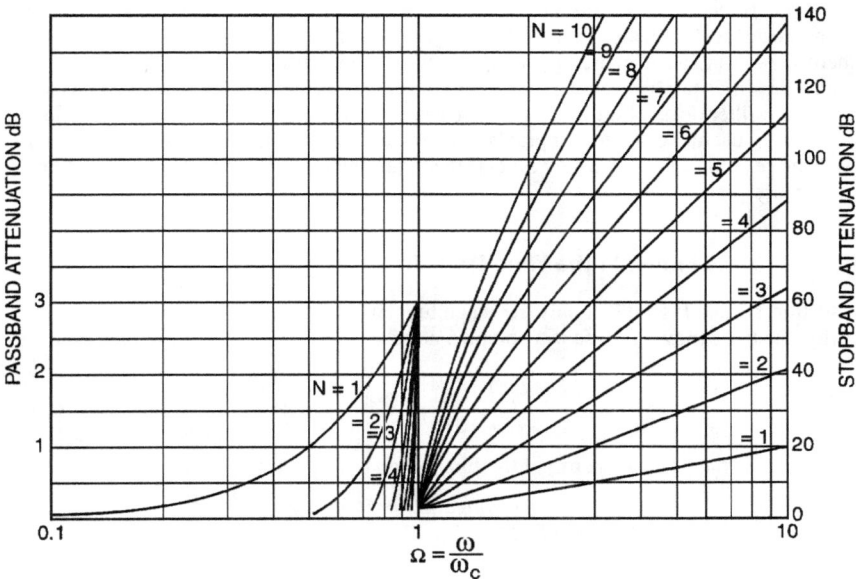

FIGURE 7.8 Attenuation of 0.1-dB ripple Chebyshev filters for an order of 2 to 10.

main tends toward the "brick wall." A high-order, high-ripple Chebyshev filter can have impulse overshoots even larger than the theoretically "ideal" rectangular filter. The latter version of the brick wall filter is presumed to have linear phase; therefore, the impulse response, $h(t)$ is:

$$h(t) = \frac{\sin(at)}{at} \tag{7.10}$$

where t is time and a is a constant related to the filter cut-off frequency. The Chebyshev impulse response is worse because its phase is not at all linear. Surface Acoustic Wave (SAW) filters can closely approximate the ideal rectangular filter, but only in the bandpass form. A monotonic response shape that is sharper than the Chebyshev in the vicinity of the cut-off is based on Legendre polynomials. Although its performance and element values are listed[1], it is seldom actually used. The rare use of this filter type is likely because of the passband amplitude rolloff near the cut-off frequency and the severe phase distortion in the passband.

If selectivity is the only consideration in choosing a filter-response shape, then still one more step that can be taken toward the rectangular ideal: to design a filter that introduces gain zeros, (or notches), in the stopband located near the passband cut-off frequency. Until the mid-1930s, they could only be designed by the image parameter method; they were called *m-derived filters*. That procedure was computationally simple, but was imprecise in performance. Wilhelm Cauer then introduced a synthesis technique that was the reverse-computationally involved, but precise in performance. In modern parlance, these are called *elliptic* or *Cauer-parameter filters*. The application of post-war electronic computing to the synthesis of these desirable filters resulted in normalized design tables introduced by R. Saal and E. Ulbrich in 1958.[4] A more complete set of element-value tables for response shapes from 4th to 9th order was published by Telefunken in 1968.[5] Some of these were also published in Zverev. The benefit, and the somewhat unlikely drawback of the elliptic filters is shown in Figure 7.9. Up to a finite stopband attenuation level, chosen by the designer, a given-order elliptic filter is sharper than the same-order, same-ripple Chebyshev. That is, it achieves the specified attenuation closer to the passband. The Chebyshev filter stopband attenuation does not stop and return tangent to a given level like the elliptic filter does, but it keeps on increasing (theoretically). However, the elliptic filters are very useful because most practical filters don't exhibit more than 80-dB maximum stopband rejection. The price they exact is the requirement for more elements and, in many cases, more-precise elements.

7.2.2 Lowpass Filters in the Time Domain

In section 7.1 and Figure 7.3, only the magnitude of the frequency response is covered. In reality, the frequency response is a complex number, and it is completely expressed by:

$$\frac{v_0}{v_m} = \left|\frac{v_0}{v_m}\right| e^{j\Phi(\omega)} \tag{7.11}$$

where $\phi(\omega)$ is the phase response of the filter given by:

$$\phi(\omega) = \tan^{-1}\left(\frac{imaginary\ part\ of\ v_o/v_m}{real\ part\ of\ v_o/v_m}\right) \tag{7.12}$$

The magnitude is an even function of ω; therefore, it is symmetrical around $\omega = 0$. The phase response is an odd function of ω; thus, it is anti-symmetrical around $\omega = 0$.

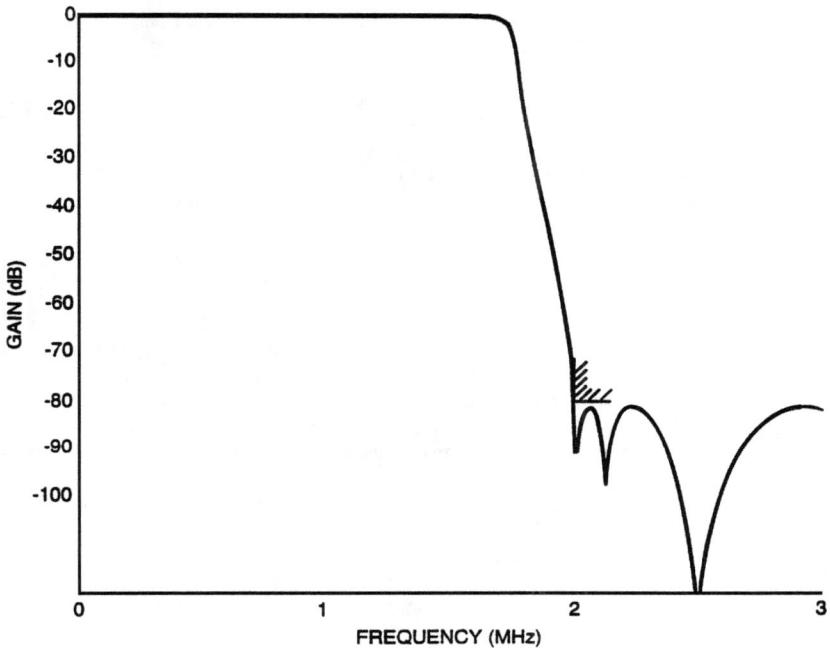

(a)

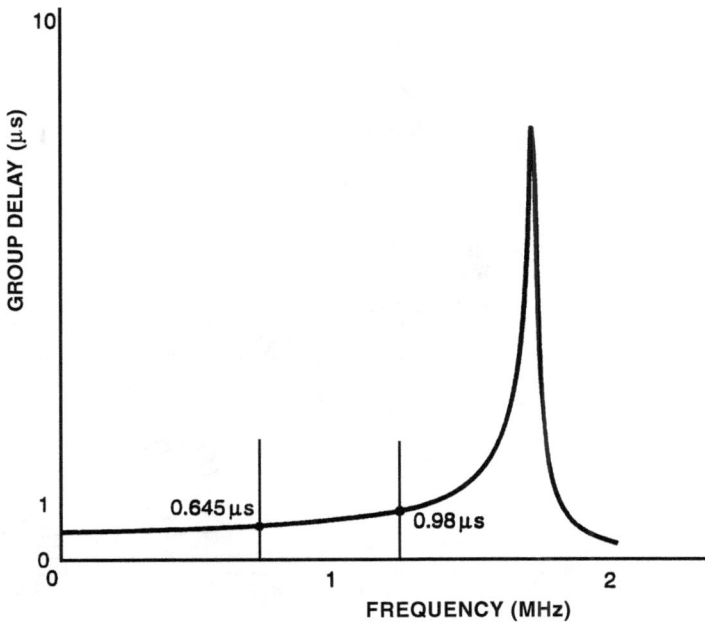

(b)

FIGURE 7.9   Measured elliptic, or cauer parameter, lowpass filter response: (a) attenuation, (b) group delay.

The impulse response, $h(t)$, of a filter is determined by taking the Fourier transform of its frequency response function. That is:

$$h(t) = \frac{1}{2\pi} \int_{-\infty}^{\infty} \left| \frac{v_0}{v_m} \right| \varepsilon^{j\Phi(\omega)} \varepsilon^{j\omega t} d\omega \qquad (7.13)$$

The result of this mathematical manipulation is that the shape of $h(t)$ is principally determined by the shape of the magnitude response function and the position of the centroid of $h(t)$ in time is principally given by:

$$\tau(o) = -\frac{d\phi}{d\omega}\bigg|_{\omega = 0} \qquad (7.14)$$

The parameter function $\tau(\omega)$ is called the *group delay* and it is, in general, not a constant as a function of ω. If it were constant, an input pulse would have its spectrum shaped by the filter gain response and the resultant time shape would be delayed by a time value calculated from Eq. 7.14, simply the negative of the phase slope. Without all-pass phase correction, all filters having attenuation responses tending toward the frequency-domain-ideal exhibit significant phase distortions; and therefore, nonconstant group delay in the passband. The group delay for the Butterworth filters and for Chebyshev filters with 0.1-dB ripple is shown in Figures 7.10 and 7.11, respectively. In the attenuation figures and group delay figures, it is almost intuitive that the normalized abscissa, called Ω in Blinchikoff and Zverev[2], changes to (ω/ω_c) for the general lowpass filter having a 3-dB cut-off frequency ω_c and evaluated at any chosen frequency ω. Succinctly:

$$\Omega = \frac{\omega}{\omega_c} \qquad (7.15)$$

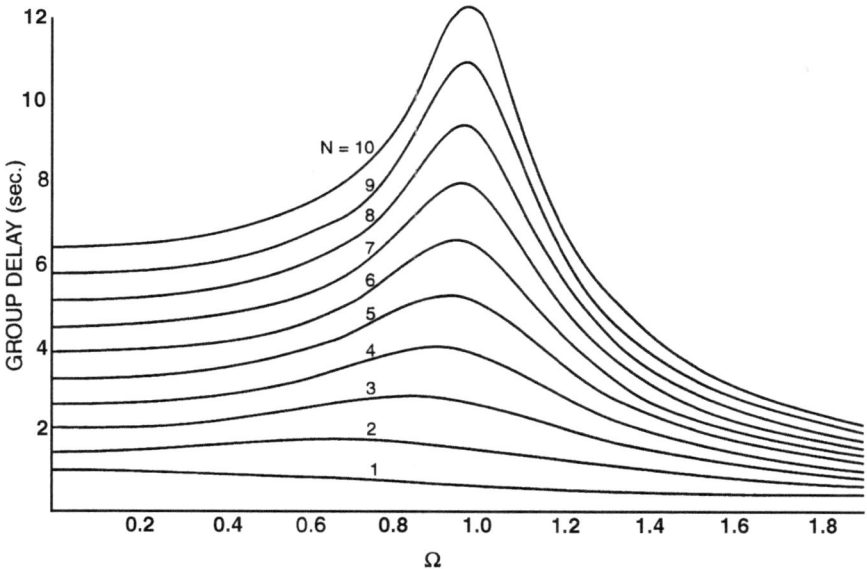

FIGURE 7.10   Group delay of Butterworth filters.

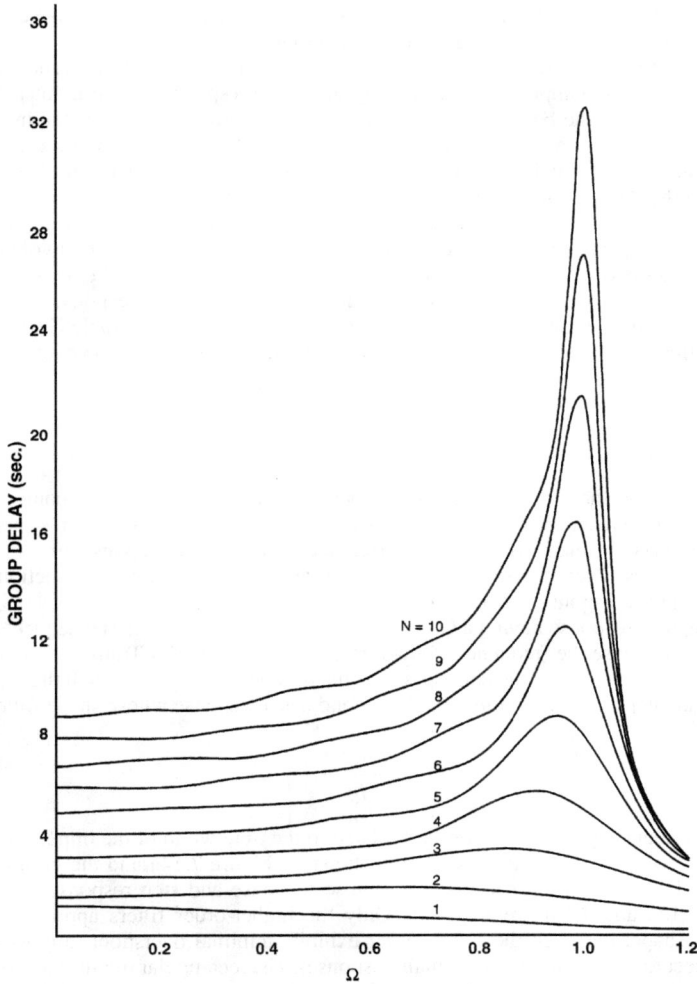

FIGURE 7.11 Group delay of 0.1-dB ripple Chebyshev filters.

It is not so clear what the ordinate numbers mean for the same lowpass filter. Blinchikoff and Zverev have shown that for *any* style filter the transformed group delay, $D_T(\omega)$, is given by:

$$D_T(\omega) = D_L(\omega)\Big|_{\omega = \Omega} \times \frac{d\Omega}{d\omega} \qquad (7.16)$$

The first term on the right is the ordinate number from the normalized curves 7.10 and 7.11. The second term transforms that number into the true denormalized group delay, which is measurable on a network analyzer. The lowpass is particularly simple because:

$$\frac{d\Omega}{d\omega} = \frac{1}{\omega_c} \qquad (7.17)$$

Thus, the ordinate from the normalized curves is simply divided by the radian cut-off frequency to obtain the correct value of denormalized group delay.

If time-response overshoot is a concern then the rectangular, brick wall, filter response is not correct for the application. Correcting the phase response until it is linear will not solve the problem; the best that can be done using this procedure makes the impulse response function approach $(\sin x)/x$ or -13 dB overshoot, relative to the maximum pulse output. The time domain ideal is that, if a unit impulse (or reasonable facsimile thereof) is applied to the input, then the output time response is "compact." *Compact* means that the impulse response of the filter consists of a single main pulse with no leading or trailing auxiliary pulses, or overshoots. In the lowpass case, it means that the output voltage never crosses the zero volts axis. The impulse response of a Butterworth filter evaluated from Eq. 7.13 is shown in Figure 7.12a and its time integral, the step response, is shown in 7.12b. Similarly, the same two parameters are shown for the 0.1-dB ripple Chebyshev filter in Figures 7.13a and 7.13b. To denormalize the time response for a lowpass having cutoff frequency ω_c, simply divide the abscissa number by ω_c. That is:

$$t = \frac{t_N}{\omega_c} \tag{7.18}$$

Notice that as the order of these filters increases and the selectivity becomes sharper, then the impulse response overshoots get larger; and in a pulse-transmission system, there would be unacceptable inter-symbol interference. The impulse response crosses the 0-V axis many times. There are several classes of filters whose lowpass gain function is such that the impulse response is compact.

An impulse response having a Gaussian pulse shape or rectangular pulse meets this requirement. Because the frequency-domain response is the Fourier Transform of the time-domain impulse response, the corresponding magnitude responses are such that the first is a Gaussian-shaped lowpass filter and the second has a response whose shape follows:

$$\frac{\sin(bf)}{bf} \tag{7.19}$$

where f is frequency and b is a constant related to the time width of the impulse response. A Gaussian-shaped frequency response is shown in Figure 7.14a and the group delay is shown in Figure 7.14b. The Gaussian impulse response and step response is shown in Figures 7.15a and 7.15b, respectively. Only the higher-order filters approach the true Gaussian shape; but even the fourth-order exhibits minimal overshoot (almost -60 dB, with respect to the maximum of the main response). The rectangular impulse response cannot normally be realized with a lowpass filter; but surface acoustic wave (SAW) technology allows fabrication of filters with a near-rectangular impulse response in a bandpass filter. To obtain compact impulse response in these two cases, the frequency response selectivity has been compromised (the lowpass response does not rise to high attenuation quickly outside of the passband). In fact, it is generally true that the most desirable time response is normally obtained from filters with a poor shape factor (slowly increasing attenuation response). The synchronously tuned type shown in Zverev also falls in this class. It guarantees no overshoot in the time domain, but exhibits the poorest shape factor of any conventional filter.

In practice, the users of filters would like high selectivity to reduce noise and interference, *and* no impulse response overshoots so that information carrying pulse trains will have no inter-symbol interference. In fact, designers of filters have, for years, been searching for frequency-response shapes that are both selective in the frequency domain and have small impulse-response overshoots. Zverev[1] has published several ideas along these lines.

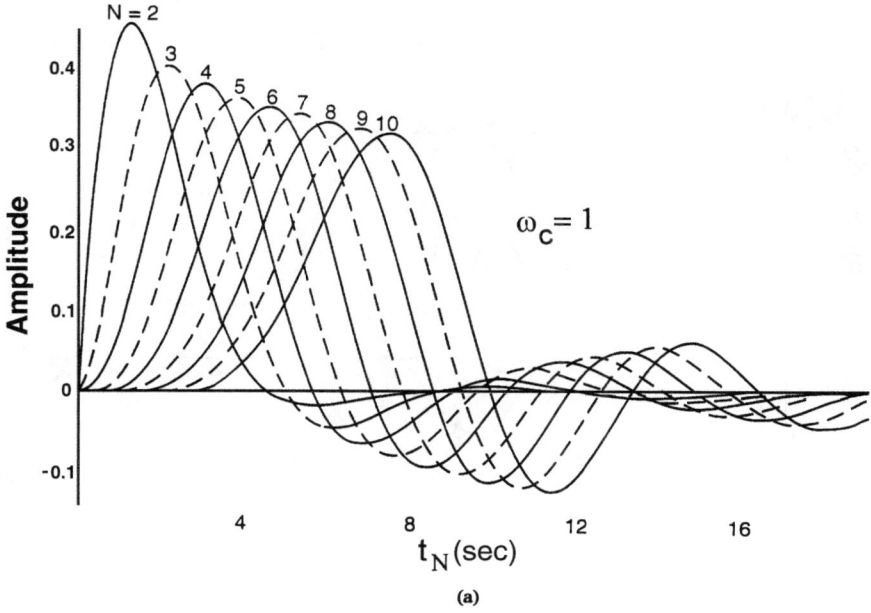

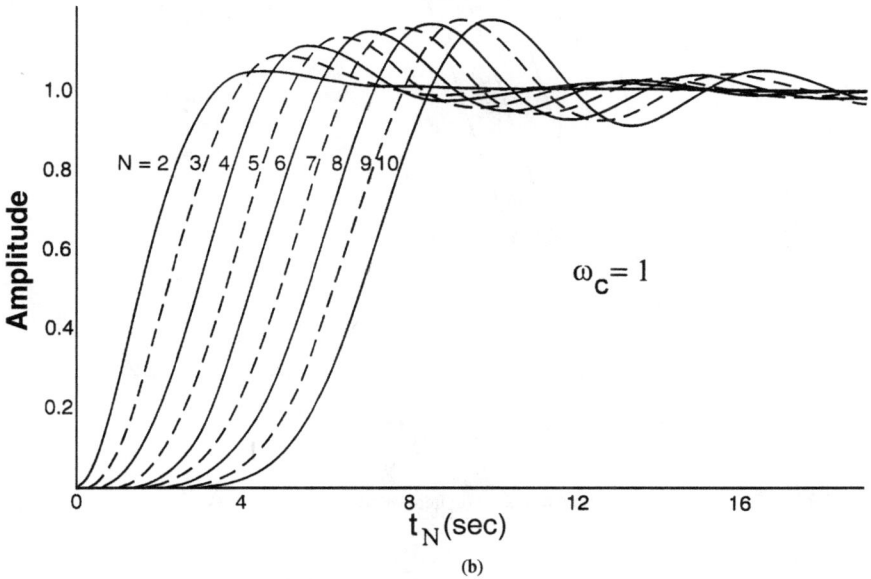

FIGURE 7.12 Butterworth-filter time response: (a) impulse response, (b) step response.

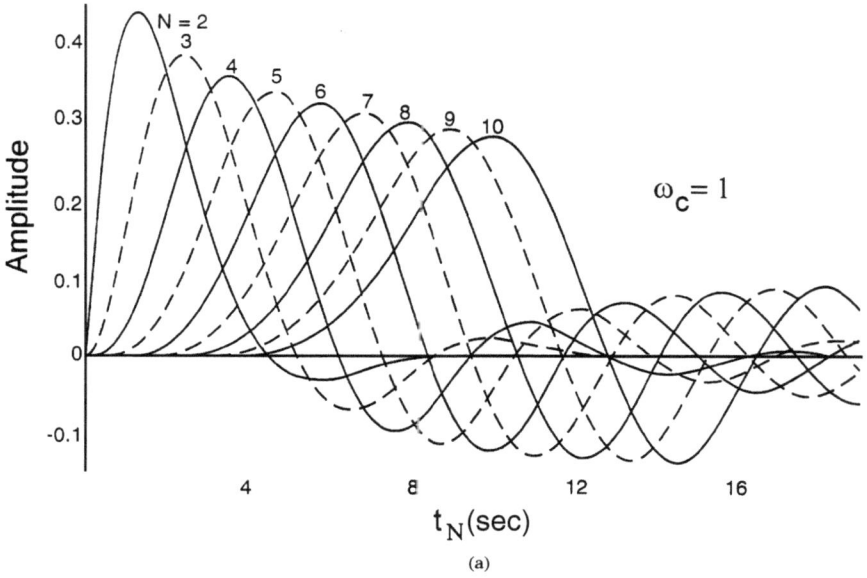

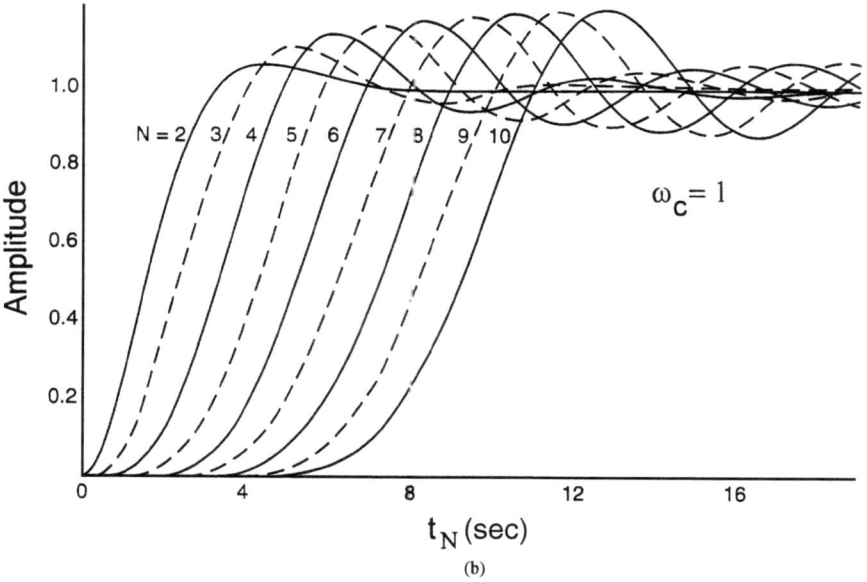

FIGURE 7.13 Chebyshev-filter time response: (a) impulse response, (b) step response.

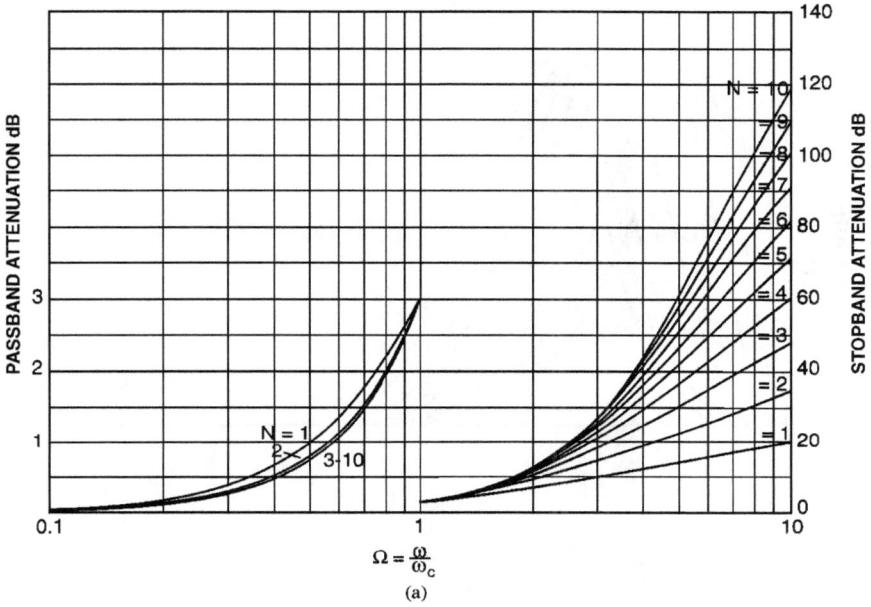

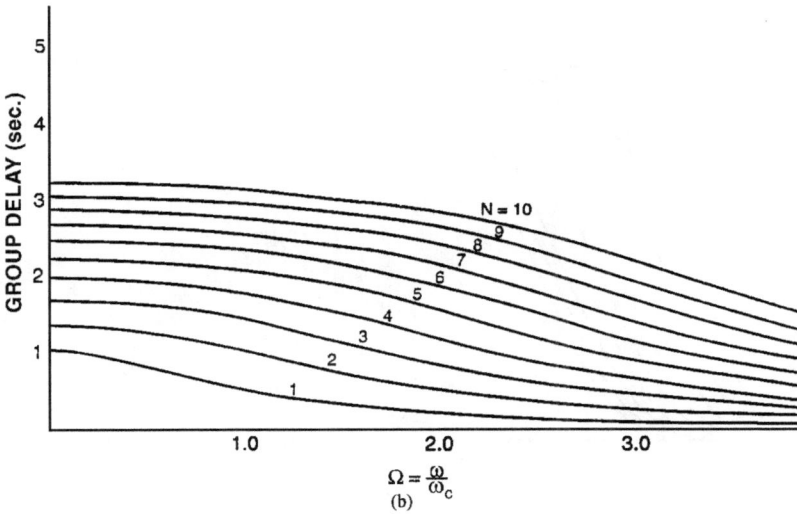

FIGURE 7.14 Gaussian-filter frequency response: (a) attenuation, (b) group delay.

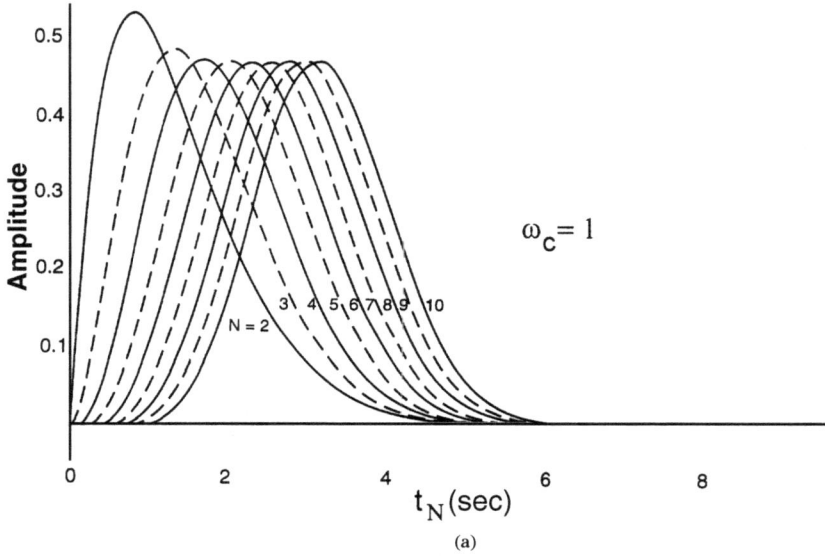

(a)

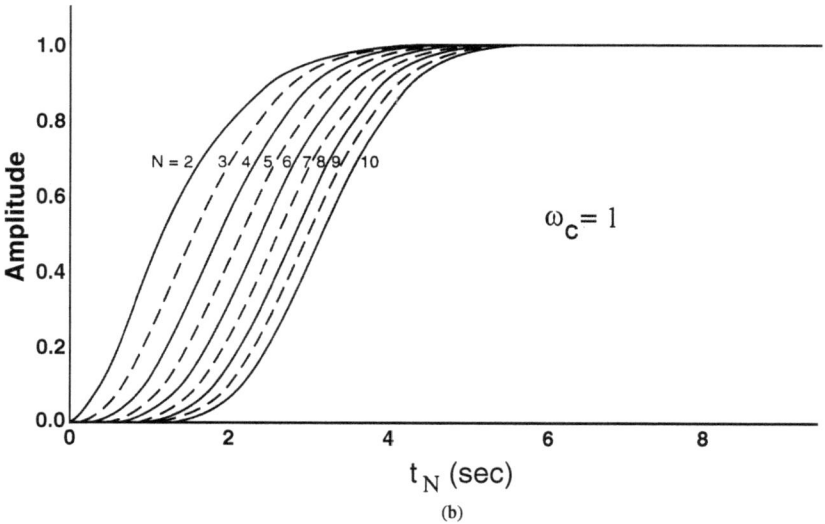

(b)

FIGURE 7.15 Gaussian-filter time response: (a) impulse response, (b) step response.

The filter shapes derived from Bessel functions, for instance, offer maximally flat group-delay response and slightly greater selectivity, compared to Gaussian filters. However, the impulse-response overshoots are slightly larger. A different approach given by Humpherys,[6] is to make the passband group delay equiripple like the passband amplitude response in a Chebyshev filter (Chebyshev group delay). This approach yields more selectivity, but at the expense of increased impulse-response overshoot. Another notion of Humpherys is for filters that exhibit a Gaussian shape up to X dB in the stopband and then break away with a Chebyshev attenuation skirt. Tables for these designs are available[1] with X equal to 6 dB and 12 dB. The Gaussian to 6 dB has the best selectivity at the expense of impulse-response overshoot and the Gaussian to 12 dB has better impulse-response overshoot at the expense of selectivity. All of these frequency responses can be made more selective by introducing gain zeros in the stopband, in the elliptic fashion. Indeed, the effect on the impulse response is minimal. In some cases, there is minor improvement in the relative size of the overshoot.

The necessity of the detailed information concerning lowpass filters is that most characteristics of a lowpass transform into characteristics of bandpass filters. Table 7.2 lists the frequency-domain filters from the lower selectivity filters at the top to the more selective filters at the bottom. Table 7.3 lists the time-domain filters—from those having the biggest impulse overshoots at the top to those having the smallest impulse overshoots at the bottom. For some practical orders of filters, there might be very little difference in performance for those near the middle, and some might even interchange positions.

TABLE 7.2 Lowpass Filters with Increasing Selectivity for a Given Order

Butterworth
Low Ripple Chebyshev
High Ripple Chebyshev
Low Ripple Elliptic or Cauer
High Ripple Elliptic or Cauer

TABLE 7.3 Lowpass Filters with Decreasing Impulse Overshoot for a Given Order

Gaussian to 6 dB
Gaussian to 12 dB
Equiripple Group Delay, 0.5° Phase Ripple
Equiripple Group Delay, 0.05° Phase Ripple
Bessel, or Maximally Flat Group Delay
Gaussian
Synchronously Tuned

Whether requirements are totally and simply selectivity or totally and simply time-domain response, remember that a designer and manufacturer of filters will test units on a precision network analyzer with accurate source and load resistance and short lead connections to the input and the output. Under these conditions, many filters will exhibit almost ideal performance in the frequency and time domain. The user of the units must put forth the effort to provide an accurate load and source resistance. Furthermore, if a 70-dB stopband is required, then the circuitry and layout without the filter must provide at least this much isolation before the filter is installed.

7.2.3 Highpass Filters

Highpass filters yield a gain-vs.-frequency response that is complementary to the lowpass. Its normalized element values are easily derived from the normalized low-pass values, such as those given in Table 7.1 and all the element-value tables in Ref. 1. The highpass transformation used to derive element values from the lowpass tables is

$$j\omega \rightarrow \frac{1}{j\omega}$$

The result of applying it to normalized lowpass elements is:

- Each inductor is replaced with a capacitor.
- Each capacitor is replaced with an inductor.
- The numerical value of the new element is the reciprocal of that of the old element.

The procedure is illustrated in Figures 7.16a and 7.16b. Once the desired normalized highpass is obtained, it is denormalized to the desired cut-off frequency and impedance using the procedure outlined in the paragraph containing Eq. 7.8. The attenuation performance of highpass filters can be found from the normalized lowpass plots (Figures 7.7 and 7.8), but the abscissa becomes:

$$\Omega = -\frac{\omega_c}{\omega} \tag{7.20}$$

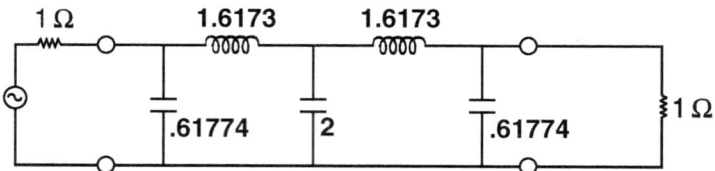

(a)

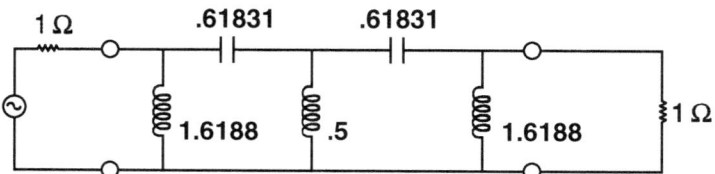

(b)

FIGURE 7.16   Prototype five-pole Butterworth filters normalized to $\omega_C = 1$ radian/sec and $R_S = R_L = 1\ \Omega$: (a) lowpass, (b) highpass.

Even though Ω is a negative number, the use of the curves is still valid because gain (and attenuation) is a symmetrical function, with respect to $\omega = 0$, so simply use the magnitude of Ω to enter the curves. However, unlike the narrow bandpass case to be treated later, the group-delay curves and time-domain responses are not preserved. In fact, the second term on the right of Eq. 7.16 becomes:

$$\frac{d\Omega}{d\omega} = \frac{\omega_c}{\omega^2} = \frac{\Omega^2}{\omega_c} \qquad (7.21)$$

Thus, the ordinate from the normalized lowpass group-delay plots (Figures 7.10 and 7.11) is multiplied by Ω^2/ω_c. The highpass group delay is distorted by this term; but, as Ref. 2 points out, a filter with theoretically infinite bandwidth cannot be expected to maintain the finite lowpass characteristics.

As for the transient, or time response, it is also not elegantly related to the lowpass time performance. The time responses of individual denormalized highpass designs must be determined from the inverse Fourier transform of the frequency-domain performance. Section 4.3.2 of Ref. 2 shows some highpass normalized transient responses and the procedure for denormalizing them.

Finally, in measuring performance of highpass filters no caveats are necessary if a modern network analyzer is used because the sweeping signals are very pure with very little harmonic content. However, with less-elegant, less-expensive equipment, it is still possible to accurately measure highpass filter responses. A simple evaluation system would contain a signal generator, a high-frequency voltmeter or powermeter, and a frequency counter. One problem with this arrangement is that the signal generator output is likely to have finite harmonic content, which (of course) will fall in the passband of a highpass filter. Thus, use of a broadband voltmeter or powermeter will give false information when you attempt to measure the stopband response. To guard against this, temporary lowpass filters have to be made to reduce the harmonics for the stopband test.

Also recognize here that infinite passband width, such as promised by an ideal highpass filter, is not possible. Therefore, stray capacity, distributed capacity, lead inductance, etc., are going to limit the upper reaches of the passband. These spurious elements must be intelligently minimized to maximize passband width.

7.2.4 Bandpass Filters

Ideally, bandpass filters are simply a translation of the lowpass function, which is symmetrical about $\Omega = 0$, to some desired center frequency (F_0) given by Eq. 7.2. F_1 is the lower 3-dB band edge and F_2 is the upper 3-dB band edge. The bandpass transformation used to derive the bandpass element values from lowpass element values is:

$$j\omega \rightarrow Q_L\left(j\frac{\omega}{\omega_0} + \frac{\omega_0}{j\omega}\right) \qquad (7.22)$$

where Q_L is given by Eq. 7.3. The detailed steps resulting from this transformation are:

- The normalized lowpass elements are changed to move the cut-off frequency to the desired bandpass bandwidth using Eq. 7.8 with $f_c = BW_{3dB}$; the impedance level is also changed to that required using the procedure introduced just before Eq. 7.8. This results in the new values L_{LP} and C_{LP}.

- A capacitor is inserted in series with every inductor having value that resonates with it at F_0; thus:

$$C_{BP} = \frac{1}{4\pi^2 F_0^2 L_{LP}} \qquad (7.23)$$

and an inductor is inserted in parallel with every capacitor having value that resonates with it at F_0.

$$L_{BP} = \frac{1}{4\pi^2 F_0^2 C_{LP}} \tag{7.24}$$

This procedure, illustrated in Figure 7.17, yields a bandpass filter having center frequency F_0, bandwidth BW_{3dB}, and a geometrically symmetrical shape about F_0 that is determined by the selected lowpass prototype response.

The frequency response of the bandpass is found using the abscissa of the normalized lowpass curves with:

$$\Omega = Q_L \left(\frac{\omega}{\omega_0} - \frac{\omega_0}{\omega} \right) \tag{7.25}$$

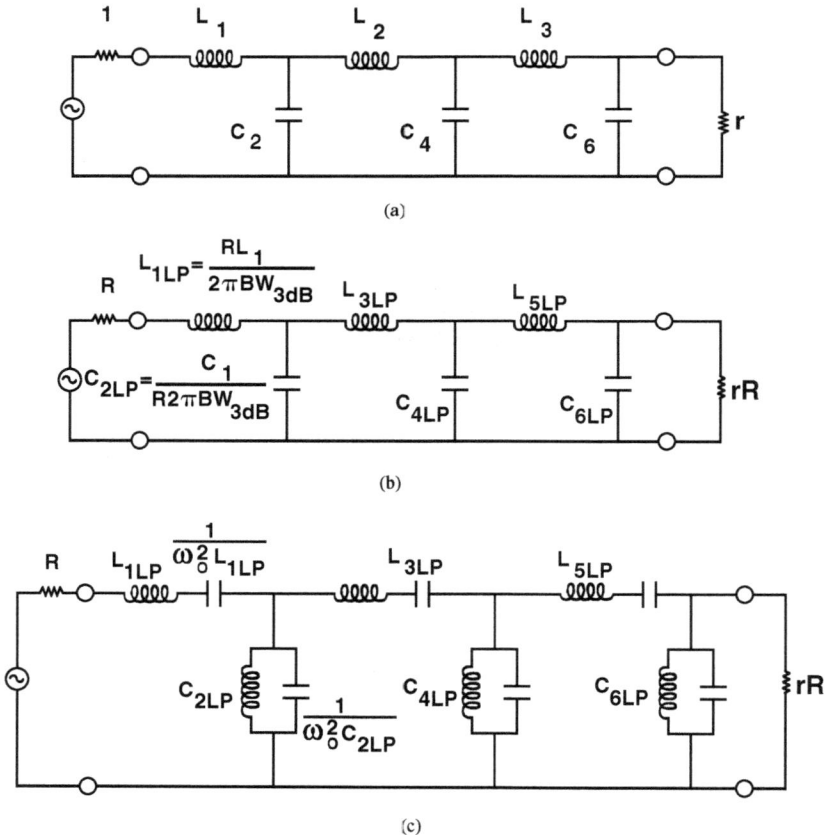

FIGURE 7.17 Filter denormalization procedure: (a) prototype lowpass filter, (b) denormalization to impedance level $= R$ and cut-off frequency $= BW_{3dB}$, (c) transformation to bandpass centered at ω_0.

As in the highpass case, the *magnitude* of Ω is used to enter the curves. Also, a given level of attenuation and its corresponding value of Ω in the normalized curve becomes a ratio of bandwidths in the bandpass filter. For example, from Figure 7.8, a 6-pole Chebyshev with 0.1-dB ripple achieves 60-dB at 2.3 times the 3-dB cut-off frequency. Our bandpass filter derived from it will have a ratio of:

$$\frac{BW_{60\,dB}}{BW_{3\,dB}} = 2.3$$

with the 60-dB locations *not necessarily* arithmetically symmetrical about F_0. For narrow-bandwidth filters, the geometric symmetry approaches arithmetic symmetry thus simplifying the extrapolation of the bandpass performance. Wider-bandwidth filters do not have simple arithmetic symmetry. Their performance is covered in reference 2, section 4.4.

The normalized lowpass group-delay performance is distorted by the second factor in Eq. 7.16.

$$\frac{d\Omega}{d\omega} = \frac{1}{2\pi(F_2 - F_1)} \left[1 + \left(\frac{\omega_0}{\omega} \right)^2 \right] \tag{7.26}$$

For narrowband filters, passband frequency does not depart far from ω_0. Thus, the passband group delay for this case is practically undistorted from the normalized curves shown, for instance, in Figures 7.10 and 7.11. For medium-bandwidth filters, the effect of Eq. 7.26 must be taken into account; for bandwidths wider than 25 percent, severe distortion of the normalized group delay occurs. However, the value of bandpass group delay can still be calculated by multiplying the normalized delay corresponding to the abscissa from Eq. 7.25 by 7.26.

A practical problem with the "full" bandpass schematic is that the ratio of the series inductance to the shunt inductance is

$$\frac{L\,series}{L\,shunt} = L_1 C_2 Q_L^2 \tag{7.27}$$

Even for bandwidth as wide as 10 percent ($Q_L = 10$), the series inductance value is about 100 times bigger than the shunt value. For a narrower bandwidth, the value of one or the other inductance will be impractical. A possible approximate solution to this dilemma is to let the series element be an acoustic resonator (quartz, ceramic, or lithium tantalate) because they typically have equivalent series-inductance orders of magnitude bigger than ordinary wire-wound inductances on powdered iron or ferrite cores. The approximation is completed by semi-precise resonating the acoustic resonator's parallel capacitance.

For bandwidths beyond 10 percent ($Q_L < 10$), the time response of the bandpass filter is not simply derived from the normalized lowpass time response. Such filters must be completely designed, analyzed, and the Fourier transform extracted from the steady-state analysis.

For bandpass filters with less than 10-percent bandwidth ($Q_L > 10$) coupled resonator-ladder configurations are the most practical. Figure 7.18 shows two forms of coupled resonator filters. The first is a low-impedance network, where series-resonant circuits in series are coupled with capacitors connected to ground. The second is a high-impedance network where anti-resonant circuits to ground are coupled with capacitors in series. This design procedure is very flexible: although capacitors are shown for coupling elements, the resonators can also be coupled with inductors (it is useful to do so in some cases), or the resonators can be coupled with a mix of capacitors and inductors. Ernest Green[7] first introduced the idea of normalized coupling coefficients (k's) and normalized Qs (q's) for the design of coupled resonator filters. Tables of k's and q's have been published in references

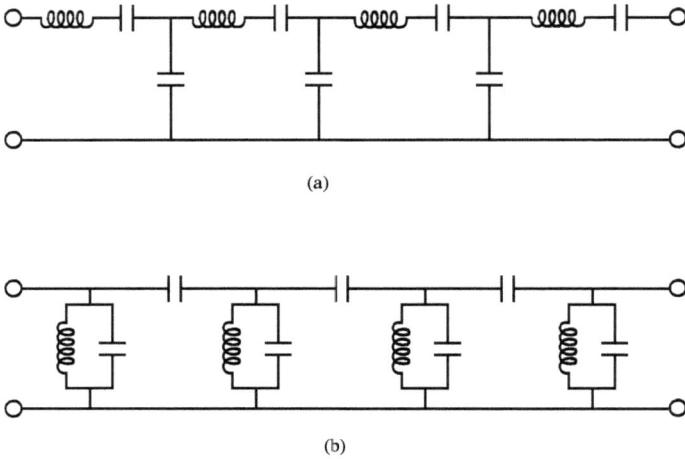

FIGURE 7.18   Bandpass filters: (a) low impedance, (b) high impedance.

(1) and (7) and *Reference Data for Radio Engineers*.[8] Section 4.5.8[2] gives detailed instructions for the design of coupled resonator filters, using the many tables of normalized coupling coefficients and Qs. However, this very popular design technique is not the only method for creating coupled resonator filters.

Humpherys[3] shows in his Section 7.2 that the k and q method can be derived from application of an elementary circuit artifice called the constant reactance impedance inverter. Indeed, the intermediate concept of k's and q's can be dispensed with and the constant reactance inverters of Figure 7.19 can be applied directly to normalized lowpass filters (Table 7.1) to directly obtain the coupled resonator filter. The constant reactance inverter is almost self-explanatory in the figure. Surmising the existence of positive or negative reactances that are constant in value versus frequency, and if these reactances are connected properly (Figure 7.19) with two of the same sign coupled by one of the opposite sign, then

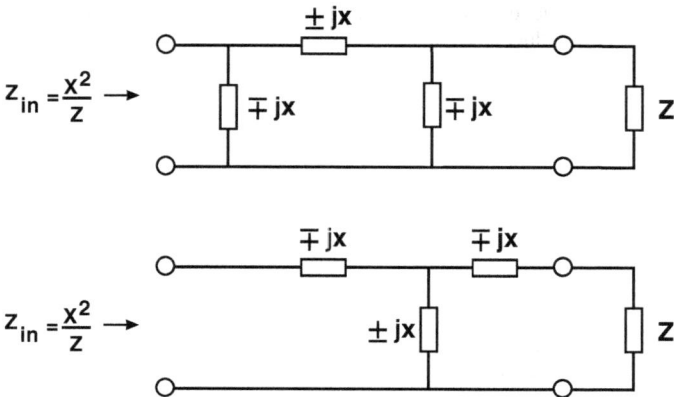

FIGURE 7.19   Constant reactance impedance inverters.

a lumped approximation to a quarter-wavelength transmission line has been achieved. Thus, any impedance connected to the output is inverted by the constant-reactance inverter when viewed at the input, and the numerical value of the inversion is scaled by x^2 where x is the element reactance. Over narrow bandwidths (such as those where coupled resonator filters are valid, for example, *relative bandwidth* < 10 percent and Q_L > 10), capacitors and inductors can be used to approximate constant-reactance elements. Thus, $-jx$ can be approximated at F_0 by a capacitor of value:

$$C = \frac{1}{2\pi F_0 x}$$

and $+jx$ by an inductor of value:

$$L = \frac{x}{2\pi F_0}$$

The k and q method and the constant reactance inverter technique, lead to great circuit flexibility in the bandwidth range of applicability. For example:

- The circuit can be designed to use only one value of inductance and can often be selected for high unloaded Q and/or reasonable resonating capacitance value.
- The termination impedances can be chosen independent of the inductance without using transformers.
- The circuit can be configured to accommodate expected stray capacity to ground at every node.
- Chebyshev filters of even order can have equal terminating impedances.
- Acoustic resonators and electromagnetic resonators can be accommodated in the final network realization.

For narrow-bandwidth filters, whether coupled resonator or "full bandpass," the geometric symmetry closely approaches arithmetic symmetry; thus:

$$F_0 = \frac{F_1 + F_2}{2} \tag{7.28}$$

and the attenuation at frequency F on the skirts of a narrow-bandwidth filter can be found from the normalized lowpass curves. The abscissa is entered at the *magnitude* of:

$$\Omega = \frac{F - F_0}{\left(\dfrac{F_2 - F_1}{2}\right)} \tag{7.29}$$

and the attenuation is acquired using the appropriate curve. Thus, the new abscissa simply means "units of half-bandwidths." This same value for Ω is used to enter the abscissa of the group-delay curves and find $D_L(\Omega)$ and the narrow bandpass group delay is found from:

$$D_{NB} = \left[\frac{1}{\pi(F_2 - F_1)}\right] D_L(\Omega) \tag{7.30}$$

For all practical purposes, the narrow-bandwidth filter translates the steady-state characteristics of a lowpass filter up to a center frequency and scales the group delay by the reciprocal of π times the BW_{3dB}. It also scales the time response.

The impulse response and step-response plots of normalized lowpass filters, such as Figure 7.12 and the fine collection in Ref. 1 can be used to predict the impulse-response *envelope* of narrow-bandwidth bandpass filters. Within the envelope is a sine wave of frequency F_0. Reference 2 shows a plot illustrating this in Figure 4.22; and Section 4.5.7 in the same reference gives detailed derivation and examples. The normalized time, t_N, which is the abscissa of the curves, is related to real time, t, and the bandwidth by:

$$t_N = \pi(F_2 - F_1)t \tag{7.31}$$

It is in the context of the narrow-bandwidth filters, whether coupled resonator or full bandpass, that the loaded Q of Eq. 7.3 assumes importance. All reactances depart from the ideal of simply storing charge (capacitors) or storing current (inductors) by also dissipating energy in element loss (resistance). The series-resonant circuits of Figure 7.18a have associated series resistance that should ideally be zero; and the anti-resonant circuits of Figure 7.18b have associated parallel resistance that should ideally be infinite. That parameter used to account for element loss is expressed by the unloaded Q of the resonator or its elements. Q is defined in any resonators as:

$$Q = \frac{Energy\ Stored\ per\ cycle}{Energy\ Dissipated\ per\ cycle}$$

For inductors, the loss is usually approximated as a resistance in series with an ideal inductor and in capacitors as resistance in parallel with an ideal capacitor. In the context of bandpass-filter performance, it is desirable that the ratio of resonator Q to Q_L be a large number, ideally greater than 10. Although filters can be made to roughly perform with this ratio as low as 3, the midband insertion loss will be high and the expected characteristic ideal performance will not be achieved. Reference 2, Chapter 6 examines this problem in detail and shows the performance of Butterworth filters with low-Q resonators. It should not be inferred that narrow-bandwidth filters are difficult to design or that theoretical selectivity cannot be achieved. For narrow-bandwidth filters, the unloaded Q of the individual elements, both the inductors and the capacitors must be factored into the design. One method for dealing with finite element design is to use a predistorted design. Theoretically, such a design proceeds from the synthesis of complex poles located so that the expected resonator losses shift them into perfect Butterworth, Chebyshev, or whatever alignment. Practically, such a design has very poor input and output impedance, very high insertion loss, and a burden of extra precision required on the element values. It is sometimes better to design the filter for wider bandwidth and tolerate the resulting loss in selectivity.

7.2.5 Bandreject Filters

As the bandpass is a translation of the lowpass function, so the band-reject filter is a translation of the highpass function. Its elements are derived from the lowpass prototype with the transformation:

$$j\omega \rightarrow \frac{1}{Q_L} \left[\frac{1}{j\dfrac{\omega}{\omega_0} + \dfrac{\omega_0}{j\omega}} \right] \tag{7.32}$$

This is a double transformation performing two functions already covered:

- It converts the prototype lowpass to a highpass filter like Eq. 7.20 with the band-reject width given by Eq. 7.8, where $f_c = BW_{3dB}$.
- It moves the highpass function out to center frequency $\omega_0 = 2\pi F_0$.

Figure 7.20 illustrates the procedure. It also shows implied intermediate steps; and it highlights the problem introduced with Eq. 7.27. In the bandreject case, the ratio of the capacitors, series to shunt, is given by Eq. 7.27. The practical result is the same: it is impossible to make both resonating inductors become practical values. An approximation is possible, as in the bandpass case, by making the shunt elements acoustic resonators and then resonating out the unwanted static capacity. But this, of course, is a flagrant approximation. It results in putting the bandreject portion of the filter frequency response in the middle of a broad bandpass response. Still, it can be an acceptable technique because most systems operate with a relatively narrow signal spectrum of interest. For example, the analog approach to pulse doppler radar uses bandreject filters positioned inside a finite-bandwidth bandpass response. In wider-bandwidth filters, the inductance ratio problem is alleviated.

The corresponding abscissa value for entering the normalized steady-state curves is given by:

$$\Omega = -\frac{1}{Q_L} \left[\frac{1}{\dfrac{\omega}{\omega_0} - \dfrac{\omega_0}{\omega}} \right] \qquad (7.33)$$

Use the *magnitude* of Ω to enter the curves; the remarks in the corresponding section on bandpass filters near Eq. 7.25 also hold for band-reject filters. Reference 2, Section 4.7 treats this in detail.

As for the group delay, the second term on the right from Eq. 7.16 becomes:

$$\frac{d\Omega}{d\omega} = \frac{\Omega^2}{\omega_2 - \omega_1} \left(1 + \frac{\omega_0^2}{\omega^2} \right) \qquad (7.34)$$

Thus, the normalized prototype lowpass group delay is severely distorted by this term. This does not imply that the group delay cannot be accurately determined. The shape of the group delay is not a translation of the lowpass function—even for narrow bandwidths.

Approximations not requiring acoustic resonators are possible for those narrower bandwidths, where the constant-reactance inverters are valid. In Ref. 3 at Section 7.3, the use of constant-reactance inverters is shown to yield bandreject filters enclosed in either lowpass or highpass structures. Figure 7.21 shows two possible configurations for four-pole band-reject filters. The prototype, its conversion to highpass and the result of applying two different constant-reactance inverters is shown. The first realization is a bandreject characteristic in a lowpass filter; the second is the same characteristics in a highpass filter. The features are:

- The inverters are chosen so that all the resonators utilize the same value inductance and capacitance.

- The choice of filter terminating impedance, R, can scale the inductance to be favorable at the band-reject frequency; subsequently broadband transformers can be used to match the impedance to the actual external circuit impedance values.

- Acoustic resonators, such as quartz crystals, can be accommodated in both schematics if the ratio of parallel-to-series capacitance is large enough (> 250 for quartz; > 10 for lithium tantalate).

In the narrowband reject case where the inverter technique is valid, some simplifications occur. For instance:

$$\Omega = -\frac{\dfrac{F_2 - F_1}{2}}{F - F_0} \qquad (7.35)$$

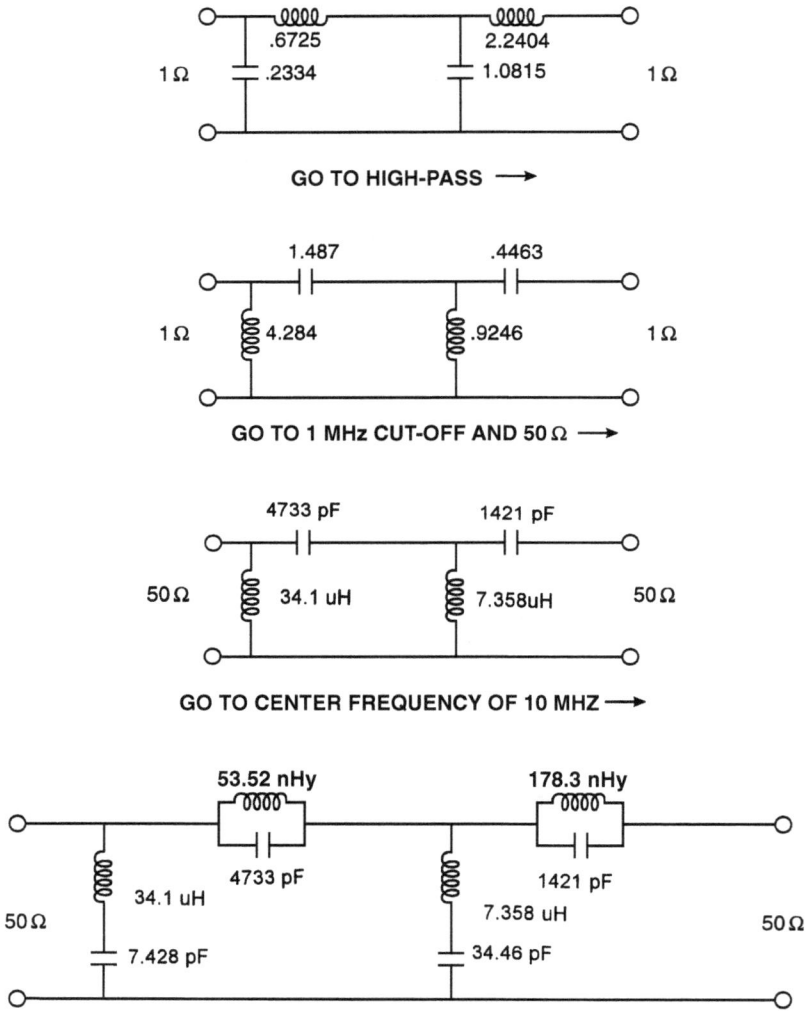

FIGURE 7.20 Lowpass- to band-reject transformation process for maximally flat group-delay filter (Bessel).

Then the meaning of the abscissa for band-reject filters is "reciprocal of half-bandwidths." The narrowband simplification of Eq. 7.34 is:

$$\frac{d\Omega}{d\omega} = \frac{2\Omega^2}{\omega_2 - \omega_1} \tag{7.36}$$

which does not preserve the characteristics of the lowpass prototype.

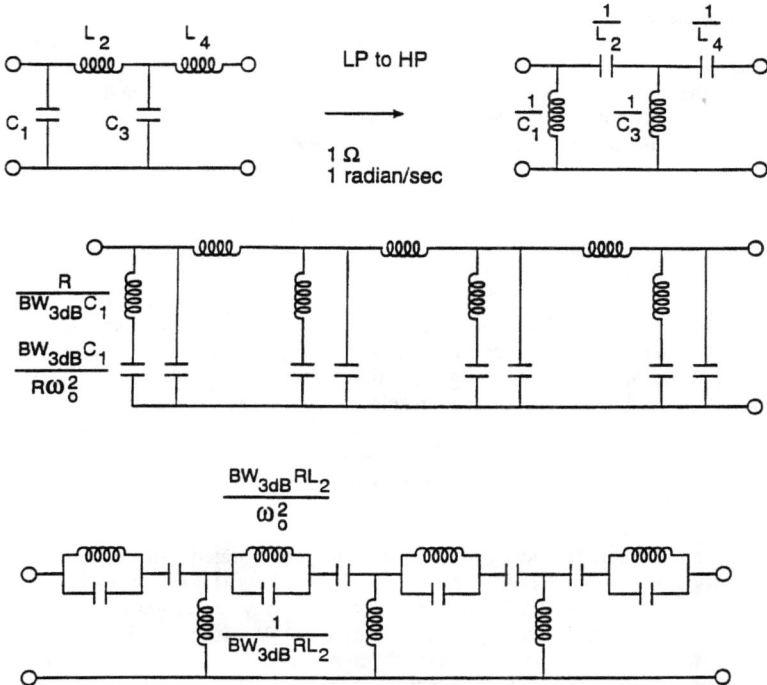

FIGURE 7.21 Normalized prototypes and two versions of narrowband reject filter resulting from the application of constant-reactance inverters.

As for the time response, an analogy to the narrow-bandpass case holds; there the envelope of the bandpass impulse response is the same shape as the impulse response of the lowpass from which it was derived. So too, the envelope of the narrow band-reject impulse response is the same shape as that of the highpass from which it was derived. The normalized time, t_N, is related to real time, t, and the bandwidth by Eq. 7.31. Similarly, the remarks about the Q of the reactances in the bandpass section apply here also.

Again, the bandreject filter is as awkward to measure as a highpass and for the same reason: the harmonics of many elementary signal generators will appear in the passband of a bandreject filter and might give false readings on a broadband voltmeter or powermeter when measuring the stopband attenuation. Use of a modern network analyzer eliminates this measurement problem.

7.3 APPLICATIONS OF FILTERS

Filters are important components in almost all electronic hardware; from simple commercial devices (such as hand-held pagers) to complex radar and electronic warfare systems. The predominant applications of filters in electronic hardware are described in this section.

7.3.1 Receivers

Communication and radar system receivers use filters primarily as a means of allowing passage of the desired signal through the various receiver gain and demodulation circuits while rejecting unwanted or interfering signals that can desensitize the receiver. These interfering signals include those originating outside the receiver that are picked up at the receiver antenna, or they might be generated in the receiver itself in the form of unwanted intermodulation products generated in the receiver frequency conversion and other nonlinear circuitry. Filters can be, and often are, used to separate incoming signals occurring in different frequency bands to route them to individual receivers for processing and information extraction.

Preselector filters   Figure 7.22 shows a block diagram of a typical superheterodyne receiver. As shown in the block diagram, preselector filter(s) are used at the receiver input or "front-end" to separate the desired signal, which might be very small in amplitude, from unwanted signals and broadband noise. Preselector filters are required to have very low insertion loss so that the receiver sensitivity or noise figure not be degraded. Wideband receivers might use preselector filter banks or tunable filters. Preselector filters often have lumped element, distributed element, cavity, or waveguide technologies, depending on the signal frequency.

IF filters   Intermediate frequency or (IF) filters are incorporated in the receiver after the signal downconverter(s). Most receivers use local oscillator (L.O.) signals to reduce the frequency of the received signals into a band where narrowband filtering and signal demodulation are more easily accomplished (Figure 7.22). Because the IF filter is normally the narrowest in the receiver, its design determines overall receiver signal-discrimination characteristics, frequency and time response, and noise content. IF filter insertion loss is usually not critical because of the use of preceding gain stages in the receiver front end. IF filters commonly have lumped-element, distributed-element, bulk-acoustic (i.e., quartz crystal), or surface acoustic-wave technologies, depending on the filter operating frequency, selectivity, and time-response requirements.

Matched filters   Matched filters are used for the detection of chirp signals or other expected special pulse-signal spectra and are widely used in radar for target detection. The meaning of the word *matched* here is that the filter is matched to a specific expected pulse; when that pulse is the input to the filter, there exists no other pulse of equal energy that yields a higher output signal-to-noise ratio. This is the "North-sense" meaning of matched.[17] In a receiver, the proper use of a matched filter directly increases the detectability of signals; it is equivalent to increasing transmitter power, but is far less expensive. A chirp signal is a pulse that contains a carrier signal whose instantaneous frequency changes linearly within the pulse. In the receiver, the matched filter to the chirp signal provides the mechanism for compressing (correlating), the received pulse energy in time and reducing the time sidelobes of the compressed pulse, thus improving signal detectability. Surface acoustic-wave technology is often used and is especially suited to provide the amplitude and phase response required in the matched filter.

7.3.2 Signal Generators

Modern communication and radar transmitters have signal generators that are capable of providing frequency agility, both in the generation of transmitted signals and their reception (by providing frequency agility in the receiver local oscillator signals). In addition,

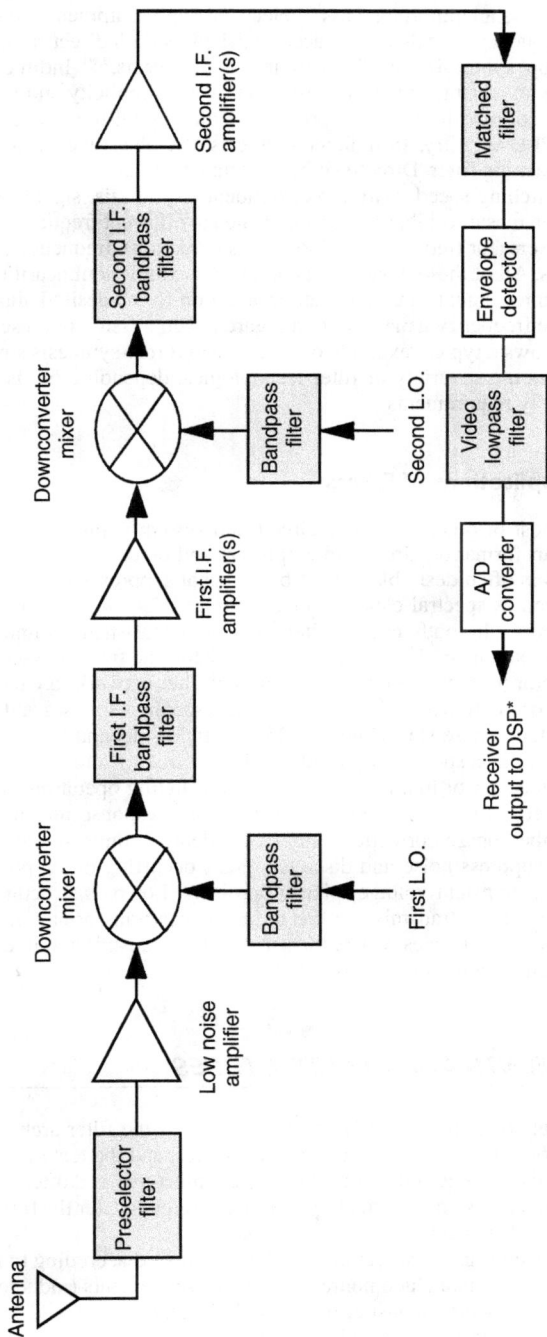

FIGURE 7.22 Superheterodyne receiver block diagram showing filter locations (shaded).

signal generators comprise an important class of electronic test equipment. Coherent generation of multiple frequency signals can be accomplished using indirect frequency synthesis (phase-lock loop synthesis) or direct-frequency synthesis.[9,10] Indirect-synthesis signal generators have an advantage with regard to hardware simplicity and programmability, largely in connection with the use of programmable, digital logic-level frequency dividers used in the PLL circuitry. In indirect synthesis, the phase-locked loop acts, in essence, as a carrier-tracking filter. Direct-synthesis signal generators offer superior spectral purity and fast switching speed. With direct frequency synthesis, signals are usually derived from a common master oscillator, and are created at different frequencies by using frequency algebra: mixers (for frequency addition and subtraction), frequency multipliers, and frequency dividers. All of these devices depend on and exhibit nonlinearities. As a result, spurious or unwanted signals are generated in addition to the desired signal. Filters play a major role in the frequency generator with regard to suppression of these unwanted signals. Figure 7.23 shows a typical example of filter use in direct-synthesis signal generation. Signal generators use a variety of filter technologies, depending on the operating frequency and selectivity requirements.

7.3.3 Additional Applications of Filters

Filters are used in a host of other electronic circuit and system applications. Filters are used as impedance-transformation circuits in amplifiers and other devices for maximum power transfer and, as is often desirable, out-of-band signal suppression. Extremely narrowband filters are used as spectral cleanup filters (Figure 7.24) or as an oscillator frequency-control element.[11] In both cases, filter environmental-stress immunity is an important performance parameter. In the case of an oscillator, the frequency-control filter is often a single resonator (one-pole filter), and additional filter networks are used to eliminate the possibility of oscillation at undesired resonator resonant responses and to provide the proper amount of loop phase shift (Figure 7.25).[11] Single-sideband filters provide an intentionally disymmetrical response to provide a high degree of attenuation of an unwanted sideband in a receiver or in a frequency-generator mixing operation. All-pass filters ideally provide zero attenuation and prescribed phase response and are used for introduction of prescribed phase correction and/or time delay. Lumped-element lowpass structures are used to suppress noise and discrete signals occurring on dc power-supply lines and can be used as transient voltage spike suppressors. Lowpass structures are also used to approximate lengths of transmission line in impedance transformation and delay-line applications at lower frequencies, where the use of actual transmission line is impractical because of large line length (signal wavelength).

7.4 FILTER ELEMENTS AND ARCHITECTURES

The types of individual components used in filters, as well as the filter architecture, depend on the filter response type (i.e., lowpass, bandpass, etc.) and the technology utilized (L-C, acoustic, transmission line, etc.). In addition, architectures are often selected or modified to perform impedance transformations within the filter, to alter the filter response symmetry, or to satisfy other specific performance goals.

Individual filter elements and architectures can be categorized according to filter technology type. Filter components include nonresonant reactive elements (inductors and capacitors) and resonant elements (acoustic, distributed element, cavity, and waveguide). The loss associated with filter components is considered a primary design parameter, and

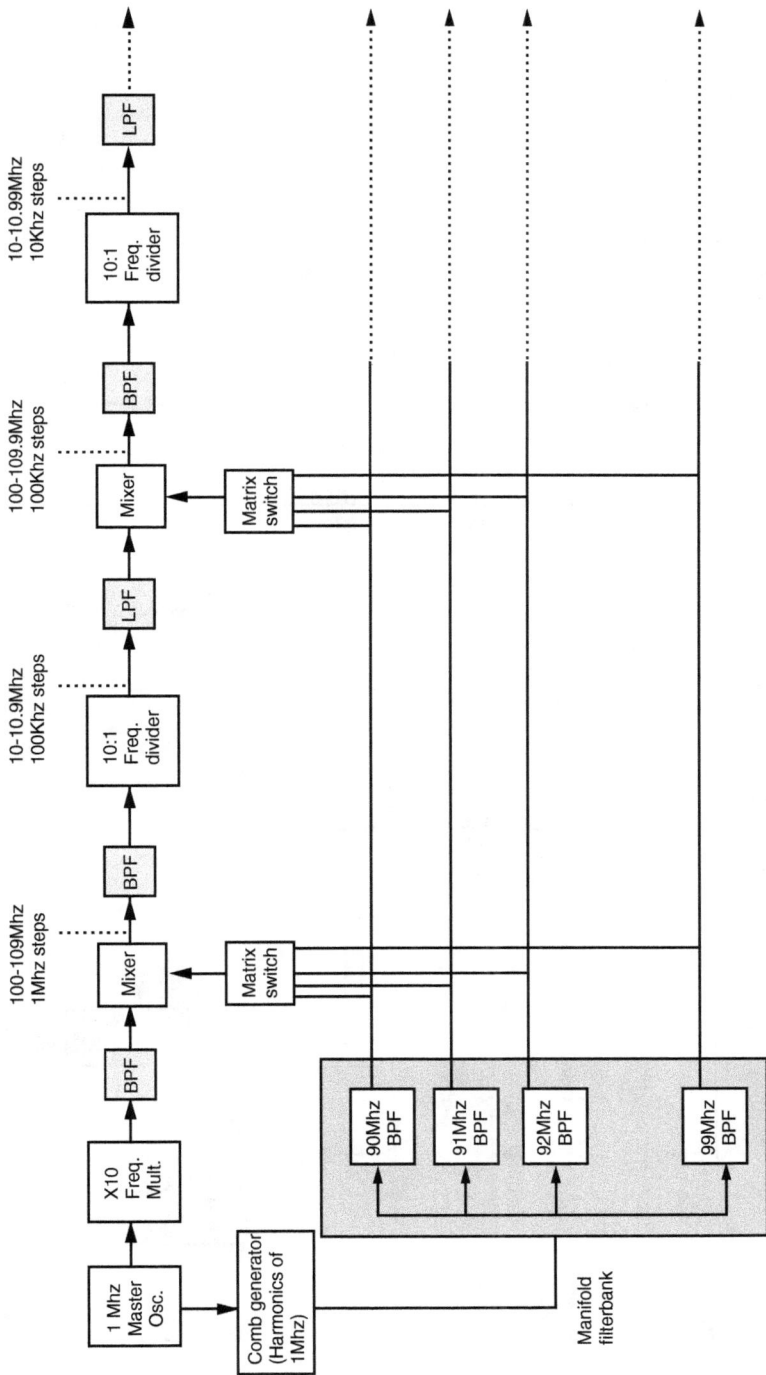

FIGURE 7.23 Iterative mix-and-divide synthesizer showing filter locations (shaded).

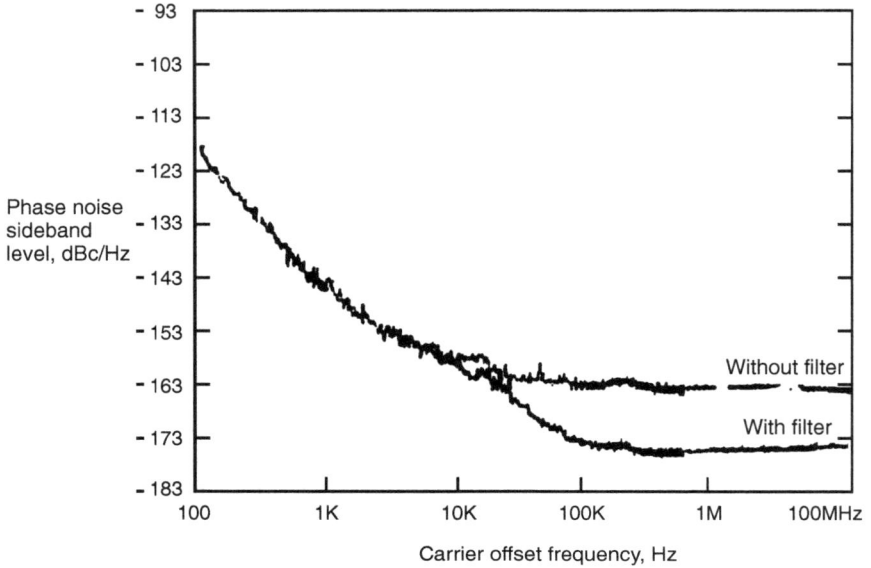

FIGURE 7.24   Carrier signal phase noise-reduction using a SAWR bandpass filter *(©1988, IEEE.)*.

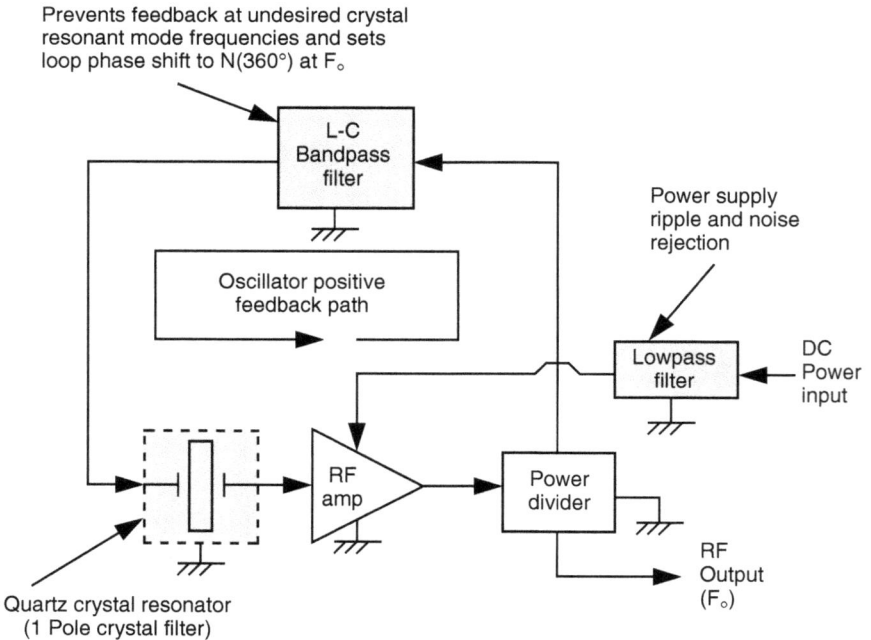

FIGURE 7.25   Crystal-oscillator block-diagram showing filter location (shaded).

components are normally evaluated, according to their unloaded quality factor (Q). Filter component and/or resonator Q is particularly important in bandpass and bandstop filters because it constitutes a measure of the attainable filter selectivity and insertion loss performance. Attainable Q in lumped elements (inductors and capacitors) is associated with wire resistivity and core losses, and in metalization and dielectric material loss, respectively. Distributed element (transmission line), cavity, and waveguide Q is also associated with metalization and dielectric material loss. In addition, Q degradation can occur in microwave resonators because of signal energy lost via excitation of undesired, coupled modes. In the case of piezoelectric or acoustic resonators, Q limitations result from electrical (metalization) and acoustic (piezoelectric material) losses, and also as a result of excitation of coupled modes. Additional insertion loss can also occur as a result of electrical impedance mismatch or inefficient conversion of the external electrical signal energy into acoustic energy and vice-versa. Table 7.4 shows a comparison of filter element Q as a function of technology type and the operating frequency range for commonly used filters. Capacitors have not been included because in almost all LC filters, inductor Q is lower than that of the filter capacitors. Refer to Chapters 2 and 3 of this handbook for additional information with regard to capacitive and inductive components.

TABLE 7.4 Filter Element Frequency Range and Unloaded Q

Component	Type	Frequency range (fo)	Unloaded Q factor
Inductor	Laminated core	10 Hz–10 kHz	5–50
	Ferrite core	5 kHz–5 MHz	50–500
	Powdered iron	100 kHz–100 MHz	10–200
	Air core	10 MHz–500 MHz	10–100
Piezoelectric	Quartz BAW resonator	10 kHz–200 MHz	$(Q)(fo)=10^{13}$
	Quartz SAW resonator	200 MHz–2 GHz	$(Q)(fo)=10^{13}$
Helical	Resonator	30 MHz–1 GHz	200–1000
Transmission line	Resonator	1 GHz–20 GHz	50–5000
Dielectric ring or cylinder	Resonator	1 GHz–20 GHz	1000–20,000
Cavity/waveguide	Resonator	2 GHz–100 GHz	1000–20,000

7.4.1 Ladder Architectures

Inductors and capacitors constitute the basic building blocks for lumped element filters. The type of inductor and capacitor (core material, dielectric, etc.) used in the filter depend on the absolute inductance and capacitance values required. In general, the value of filter element inductance and capacitance varies as the inverse of the filter operating frequency. In addition to Q, other important inductor and capacitor performance parameters are size, cost, manufacturing tolerance, adjustability, reliability, and signal power-handling capability. Figure 7.26 shows the basic architectures for lowpass, highpass, bandpass, and band-reject filters, covered in detail in Section 7.2. The highpass filter topology is obtained by reversing the positions of the inductors and capacitors in the lowpass design so that the reactance-value magnitudes of the series and shunt elements are identical at the filter cutoff frequency. Similarly, the bandpass architecture is obtained from the basic lowpass

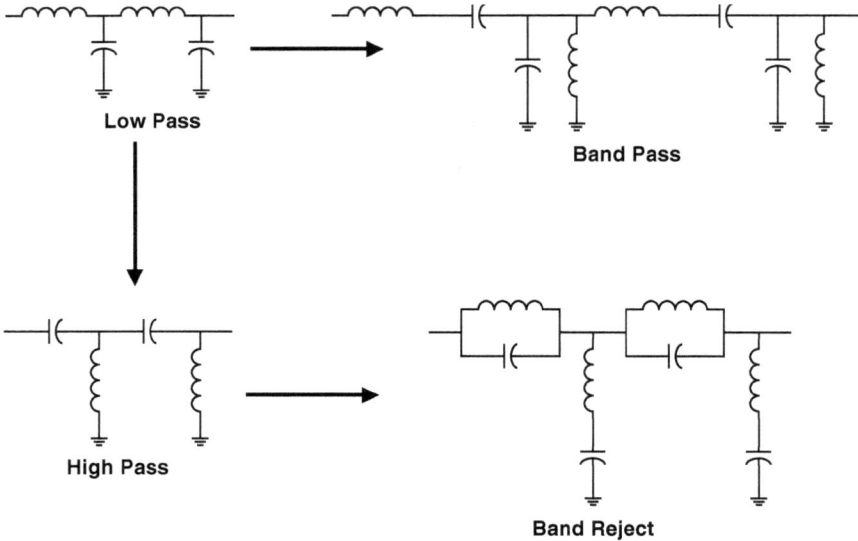

FIGURE 7.26   Basic ladder-filter topologies.

structure by "resonating" each series inductor with a series capacitor and each shunt capacitor with a parallel inductor, at the filter center frequency. The band-reject topology is obtained from the highpass structure by resonating each series capacitor with a parallel inductor and each shunt inductor with a series capacitor. The actual design of each of the Figure 7.26 filter topologies are usually derived using the lowpass structure as a design starting point.

Figure 7.27 shows a second lowpass topology also described in section 7.2 that is the basis for filters having transmission zeros (or attenuation peaks) in the stopband region. These classes of filters normally provide sharper "skirt" selectivity in the region between the filter passband and stopband in exchange for a finite level of stopband rejection. Highpass, bandpass, and band-reject topologies are formed in the same manner from the lowpass prototype structure for the Figure 7.27 topology as well.

The analysis and design of distributed-element filters can be accomplished using ladder-type lumped-element, equivalent electrical circuits.

7.4.2 Lattice Architectures

A lattice structure is also the basis for filter topologies. The primary use of a lattice topology is in the design of relatively wide bandwidth (i.e., several percent) quartz-crystal bandpass filters.[1,2,12] Figure 7.28 shows the basic lattice structure and its equivalent circuit applied to use with quartz-crystal resonators. As shown in the figure, the resulting filter topology includes a balanced transformer as part of the filter structure. In the stopband region of the filter, the motional-impedance portion of the quartz-crystal-equivalent electrical circuit (L_{S1}, C_S, R_s) is extremely large, and the crystal electrical impedance is essentially that of its inter-electrode or static capacitance (C_o). Under this condition, each arm of the transformer contains identical series capacitive impedances, and reasonably high stopband at-

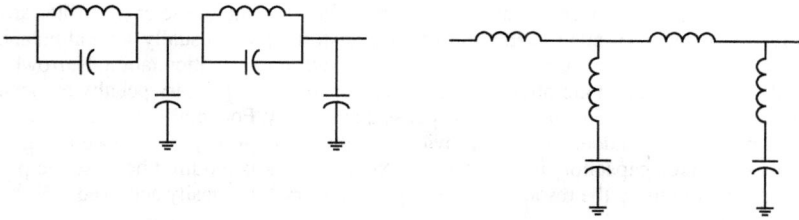

FIGURE 7.27 Lowpass elliptic-filter topology.

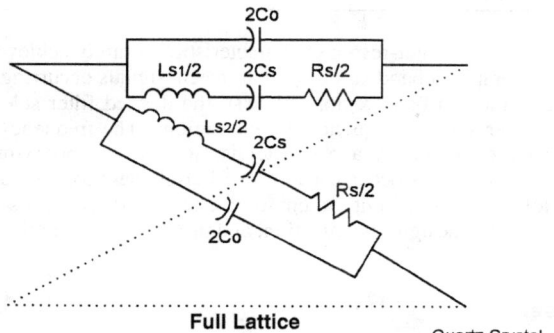

Full Lattice

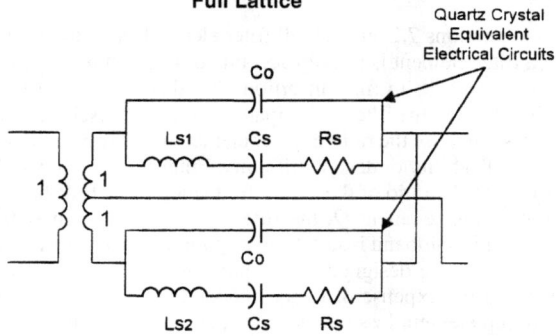

Balanced Transformer Realization

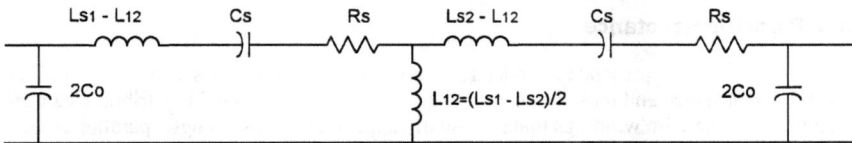

Equivalent Ladder Network

FIGURE 7.28 Lattice filter topology applied to quartz-crystal filters.

tenuation is maintained over a relatively large bandwidth. A multipole crystal-filter architecture can include crystals in each arm of the transformer, and usually several balanced transformer structures are cascaded to achieve increased stopband attenuation. Narrowband crystal filters can be and are often realized in ladder form, the primary penalty associated with ladder realization being stopband-response disymmetry. For small relative bandwidths (< 0.05 percent), the ladder structure provides exceptionally good performance using only crystals with fixed capacitors in the network. No alignment is required because the precision required in setting the resonant frequency of each crystal is easily achieved.

7.5 REAL VS. IDEAL FILTERS

The so-called "ideal filter-response characteristic" is rarely achieved. It is not possible to design, for example, a bandpass filter that rejects signals occurring at all frequencies, except those in the passband region. At best, the desired filter selectivity is usually only maintained over several frequency-decade regions. The frequency and/or time-response characteristics exhibited by a real filter is, at best, an approximation of that desired. Performance limitations occur because of filter-element loss, parasitic reactance, manufacturing tolerances, sensitivity to environmental stress, etc. These effects must be taken into account in the design, specification, and use of the filter in the overall system.

7.5.1 Loss

As covered in Sections 7.2 and 7.4, all filter elements are "lossy," and the degree of resistive loss in a filter element is usually accounted for as unloaded quality factor or unloaded Q. Resistive loss results in filter insertion loss; that is, some of the desired energy in the signal passing through the filter is dissipated in the filter itself. As covered in Section 7.2.4, filter designers often use the ratio of filter element Q to overall filter loaded Q_L as one measure of filter realizability or design difficulty. Filter Q_L is usually defined for bandpass and bandstop filters as the ratio of the center frequency to 3 dB of bandwidth.

For a given filter element Q, the filter response will depart from the theoretical response, resulting in midband insertion loss, poorer selectivity, and a change in bandwidth. However, filters can be designed by the predistorted method[1] to exhibit the lossless-element response at the expense of larger midband insertion loss. Table 7.5 shows the relationship between element loss (Q), filter Q_L, and midband insertion loss for three types of 5-pole filters using a "predistorted" design. The individual element values for these filters differs from those used in the lossless case. Figure 7.29 shows the attenuation of a 5-pole Chebyshev (0.1-dB ripple) filter for various values of q_o, after renormalizing each response to have a 3-dB cut-off frequency of unity.[13]

7.5.2 Parasitic Reactance

In addition to loss, the presence of undesired ("parasitic") reactance has a detrimental effect on filter performance and must be taken into account. Inductors contain distributed capacitance between the wire windings that are usually approximated as a single, parallel capacitor. The frequency at which the self-inductance of the inductor resonates with the parallel capacitance is termed the *self-resonant frequency (SRF)*. Similarly, capacitors contain lead inductance and exhibit a SRF where the capacitance resonates with the lead inductance. At frequencies approaching the SRF, the apparent inductance and capacitance of inductors and

TABLE 7.5 Predistorted Design, Multipole Filter Insertion Loss

5 Pole Butterworth		5 Pole Chebyshev (0.1 dB ripple)		5 Pole Bessel	
q_0	I.L.	q_0	I.L.	q_0	I.L.
32.4	1.04	68.1	0.935	10.4	2.04
16.2	2.26	34.1	2.01	5.22	4.14
10.8	3.66	22.7	3.24	3.48	6.29
8.09	5.26	17.0	4.67	2.61	8.49
6.47	7.15	13.6	6.36	2.09	10.7
5.39	9.42	11.3	8.42		
		9.73	11.1		

q_0 = ratio of filter element unloaded Q to the overall filter Q_L
I.L. = filter insertion loss in dB

capacitors deviate from the prescribed value and the apparent Q decreases. Above the SRF, inductors exhibit capacitive, rather than inductive, reactance and capacitors exhibit inductive, rather than capacitive, reactance. The result is usually a pronounced loss of filter selectivity (i.e., stopband attenuation, in the case of a lowpass or bandpass filter) at high frequencies. Additional parasitic reactance can be encountered in connection with printed wiring board (PWB) metallization (i.e., track inductance and bonding-pad capacitance). For lumped-element filters operating in the range above 30 MHz and using PWB interconnects, great care must be taken with regard to the layout of the printed wiring board to minimize parasitic reactance. The dielectric constant and the loss tangent of the PWB material can introduce deleterious effects on the filter performance. Generally speaking, it is best to use materials with low dielectric constant and very small loss tangent. Several methods are available to the designer for dealing with these effects. In some cases, the filter inductors and capacitors can be implemented as a combination of two or more elements (series inductors or parallel capacitors) each chosen to have relatively higher and non-coincident self-resonant frequencies. An alternative is to allow the filter high-frequency response to deteriorate, but to cascde a second filter in series to provide the additional lost selectivity. Figures 7.30 and 7.31 show examples of how this technique can be applied.

7.5.3 Harmonic Resonances

The transmission-line segments of distributed-element filters exhibit a resonant-like behavior when the line length is a multiple number of quarter or half wavelengths. Cavity filters also exhibit a variety of resonant modes at different frequencies. The effects of these undesired resonances can be minimized as previously described by cascading several filter elements or distinct filters so that the undesired resonances do not overlap. In the case of cavity filters, coupling into, out of, and between resonators can also be adjusted to favor certain resonant modes.

7.5.4 Piezoelectric Filter Harmonic and Inharmonic Response

Piezoelectric resonators (such as bulk-wave quartz crystal) and the filters they are used in are designed to operate using a specific mode of vibration or acoustic wave propagation in

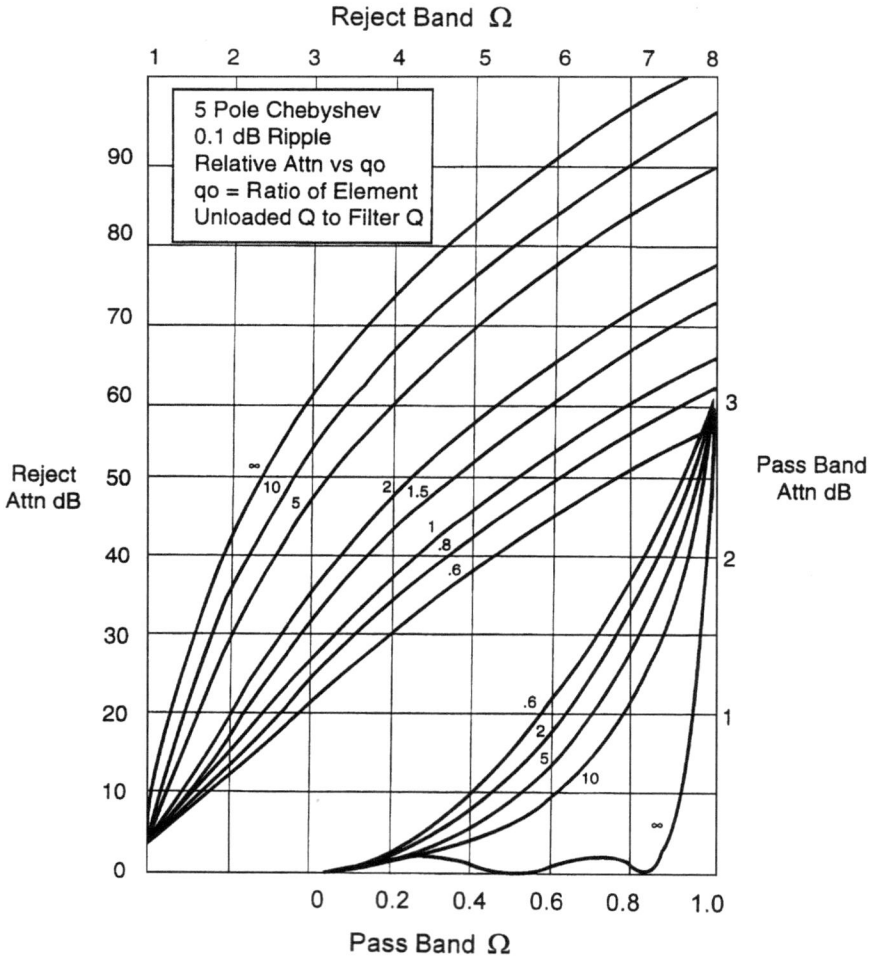

FIGURE 7.29 Lossless design lowpass filter selectivity using lossy components.

the resonator. Piezoelectric resonator excitation in the desired mode occurs at frequencies in the filter passband. When/if undesired modes are excited at filter out-of-band frequencies, the presence of these modes can result in unintended signal transmission through the filter. One way to minimize this problem is to design the filter so that the individual resonator-desired resonant frequencies (that occur in the passband region) are slightly offset from one another so that the undesired-mode resonator responses do not overlap. Figure 7.32[12] shows how the effects of undesired resonant modes can be suppressed using this technique. Furthermore, cascaded sections of half-lattices or crystal-ladder filters automatically suppress the effects of individual crystal spurious responses that might be frequency coincident. The worst-case circuit topology, in terms of spurious effects, is the single half-lattice filter.

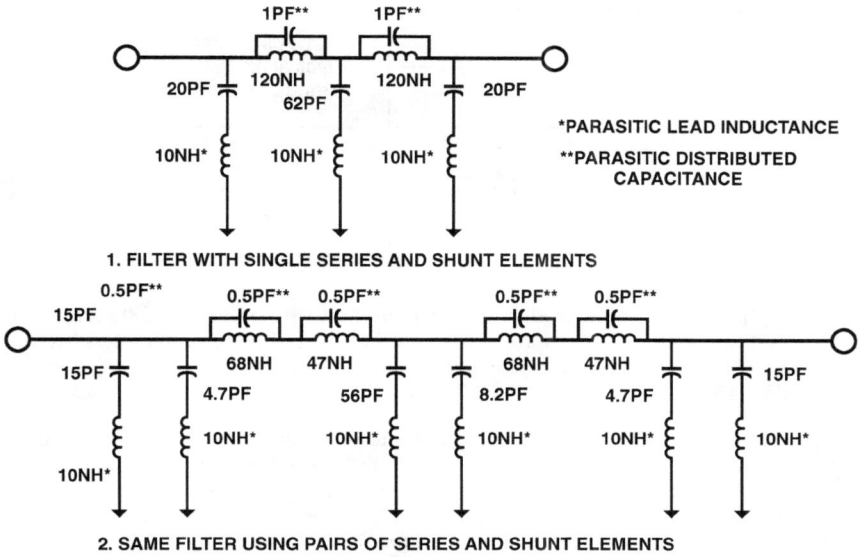

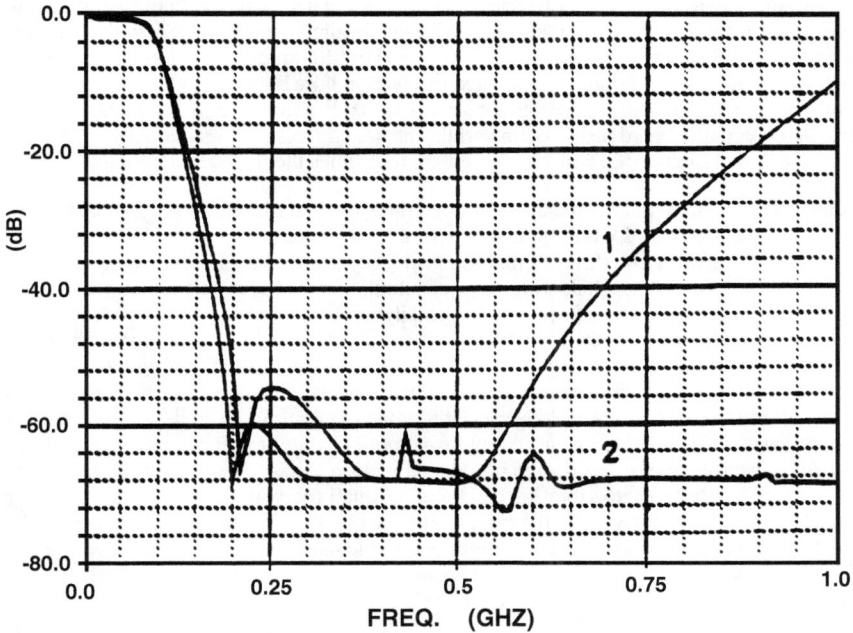

FIGURE 7.30 Use of multiple components to reduce effects of parasitic reactance: (a) schematics (b) frequency responses.

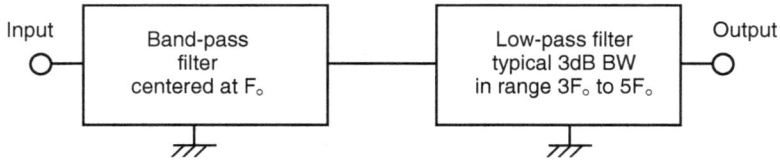

FIGURE 7.31 Use of cascaded filters to reduce effects of parasitic reactance.

7.5.5 Environmental Effects

Filters usually need to be designed for minimal sensitivity to environmental stress, such as temperature variation and vibration. This is especially true of narrowband filters because of the increased degree of phase and amplitude error, resulting from a given, stress-induced filter frequency shift.

Temperature All of the filter elements' characteristics, including the enclosure itself, change by varying amounts with temperature. In many cases, the effect on filter performance can be pronounced. As an example, in a bandpass filter having one percent bandwidth, but using inductors or capacitors with a 300-ppm/°C temperature coefficient, a 33°C temperature change would result in the entire filter-passband response moving out of band. Filter-response sensitivity to temperature results from individual-element and/or resonator variation with temperature. In the case of transmission line or cavity filters, the temperature coefficient (T.C.) of concern is that associated with the dielectric material, transmission-line conductor length, and metal-enclosure dimensions. Secondary effects must also be taken into account. For example, a VHF filter might use 10-pF shunt capacitors having low T.C. (30 ppm/°C) mounted on PWB, whose bonding pad to ground capacitance (occurring across the 10-pF capacitor) is 2 pF and has a T.C. of 600 ppm/°C associated with the board dielectric material. The net effect is an effective 12 pF capacitor with a T.C. of 125 ppm/°C, which is more than four times the intended amount.

Vibration Vibration must also be a source of concern to the filter designer and user. Table 7.6 shows an example of vibration levels typically encountered in various hardware environments.[14] The primary effect of vibration is induced phase and amplitude modulation of the signal(s) passing through the filter. You can think of the filter frequency response (both amplitude and phase) as being sensitive to, and dynamically changing as a result of vibration. The frequency sensitivity to vibration can result from mechanical motion in the components themselves (i.e., inductor wires), motion in the printed wiring board, filter enclosure walls, etc., or in the case of piezoelectric resonators, stress on the resonator itself. Adjustable components can be especially prone to vibration. In addition, at certain vibration rates, the filter components or enclosure might exhibit mechanical resonances that can significantly increase the degree of induced motion. Although the effective filter center frequency sensitivity to vibration can be quite small (parts in 10^8 per g), the narrower the filter bandwidth, the greater the group delay or phase-vs.-frequency slope, hence, the greater the induced amount of phase change across the band. As an example, the vibration-induced phase modulation occurring in a narrowband quartz crystal or SAW filter used in a tactical jet fighter aircraft radar system can significantly degrade system performance—even though the sensitivity of the resonators to vibration is quite small (on the order of several parts in 10^9 to 10^8 per g.).[14]

Contaminants The immediate or gradual introduction of contaminants into the filter can degrade performance. Humidity can adversely effect the response of filters having ex-

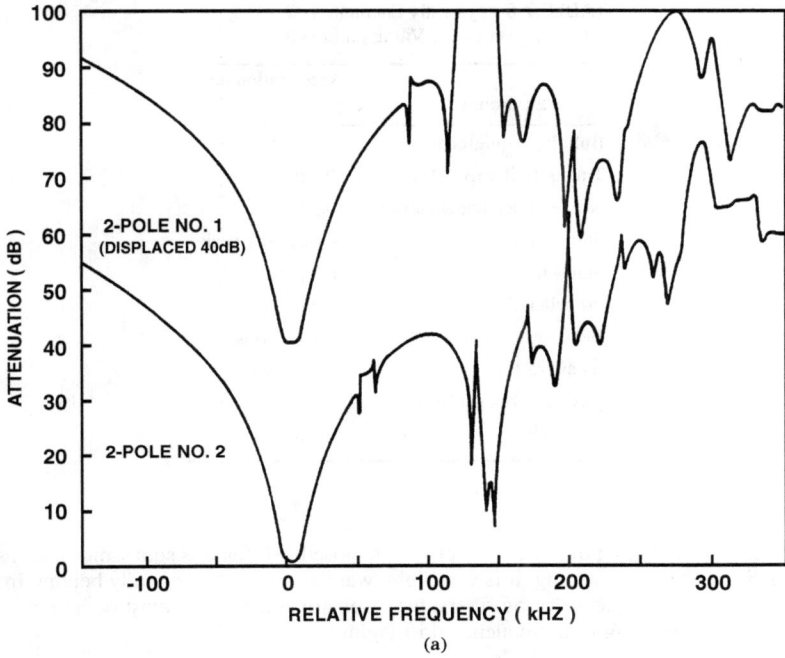

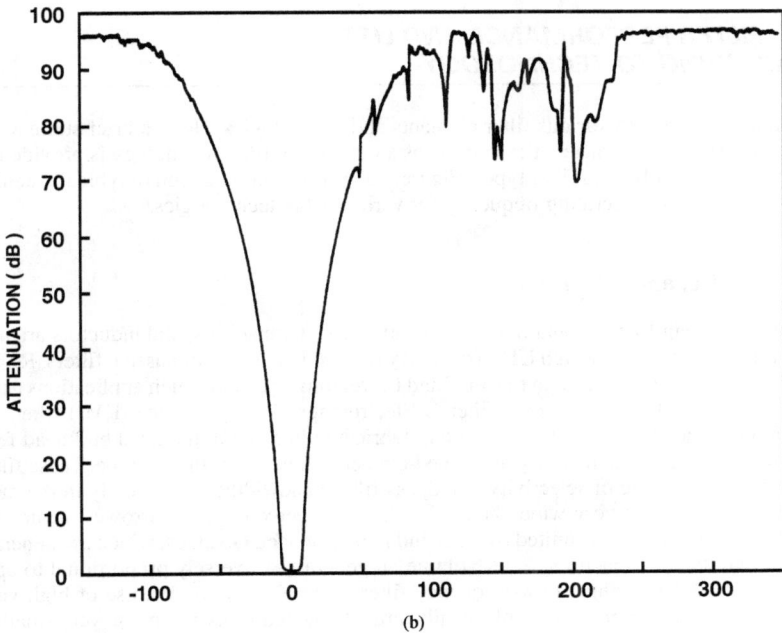

FIGURE 7.32   Suppression of undesired resonant modes in a quartz crystal filter: (a) attenuation in one, two-pole section, (b) attenuation of two cascaded sections. (©*1985 by Academic Press, Inc.*)

TABLE 7.6 Typically Encountered
Electronic Hardware Vibration Levels

Environment	Acceleration level (in g's)
Buildings, quiescent	0.02 rms
Tractor trailer (3–80 Hz)	0.3 peak
Armored personnel carrier	0.5 to 3 rms
Ship - calm seas	0.02 to 0.1 peak
Ship - rough seas	0.8 peak
Propeller aircraft	0.3 to 5 rms
Helicopter	0.1 to 7 rms
Jet aircraft	0.02 to 2 rms
Missiles - boost phase	15 peak
Railroads	0.1 to 1 peak

posed elements. Potting the filter elements with closed-cell foam is sometimes used as an alternative to hermetic sealing. It is very light, waterproof, and electrically benign. In extremely narrowband piezoelectric filters, the resonators themselves must be hermetically sealed to decrease long-term frequency drift (aging).

7.6 FILTER PERFORMANCE AND USE ACCORDING TO TECHNOLOGY

Section 7.4 described various filter elements and technologies. Here, a brief summary of filter performance characteristics and use as a function of filter technology is provided for the more commonly used filter types. Figure 7.33 shows a comparison of typically achievable bandwidth vs. operating frequency for various filter technologies.

7.6.1 RC, RL, and LC Filters

Lumped element filters, containing individual resistors capacitors, and inductors are used across the entire VLF through UHF frequency ranges. In the case of passive filters, RL and RC filters are normally used in the VLF and LF regions, mostly in such applications as dc-supply ripple reduction (lowpass filters). Electromagnetic Interference (EMI) filters are commonly available as LC lowpass filters fabricated in a small, threaded bulkhead feed-through connector. The majority of lumped-element filters are of the LC type. These filters provide a wide range of selectivity: bandpass-filter bandwidths are typically in the range of several percent to very wide bandwidth (i.e., 100 percent). The narrowest achievable bandwidths are normally limited by filter-inductor unloaded Q values, which are generally lower than that of capacitors. Filter volume, in general, is inversely proportional to operating the frequency range. Low-frequency filters normally require the use of high-value inductors and capacitors that are physically large; high-frequency filters use very small capacitor and inductor values that can be obtained, for example, as tiny chip components.

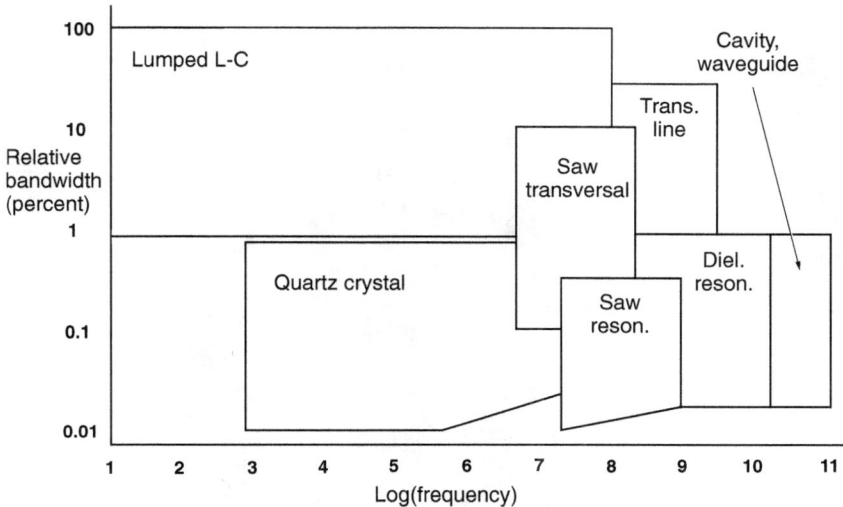

FIGURE 7.33   Bandwidth vs. operating frequency for various filter technologies.

Wide-bandwidth filters having noncritical frequency-accuracy specifications can usually be implemented using "nearest standard value" inductors and capacitors. For filters requiring more exact response characteristics, adjustable components or the use of selectable, fixed-value components can be used to "tune-up" the filter. In the case of high-frequency filters, this tuning procedure might consist of physically spreading or closing the distance between turns of a coiled-wire inductor or snipping the length of a pair of twisted wires, whose inter-wire capacitance is one of the filter elements. In lower-frequency filters, in which the inductors contain many turns of wire wrapped in or around an iron core, the fine inductance setting is achievable by adjusting the number of turns. In the frequency band 5 kHz to 500 kHz, pot-core inductors are offered with tuning or adjustment capability. Because of the wide frequency range and wide choice of filter-response characteristics obtainable using lumped-element filter technology, these types of filters are used in virtually every variety of commercial and military communication and radar systems. A photograph of a typical LC filter utilizing ferrite pot-core inductors is shown in Figure 7.34.

7.6.2 Acoustic Wave Filters

Bulk acoustic-wave (BAW) and surface acoustic-wave (SAW) filters use piezoelectric materials to transform the electrical signal to an acoustic wave propagating through or along the surface of the crystal or other propagating medium. Quartz is the most commonly used piezoelectric material because of its stability and low acoustic loss, along with ease of precise machining, polishing, etching, and metallization qualities. Other commonly used piezoelectric filter materials include PZT (lead zirconate titanate), lithium niobate, and lithium tantalate. In general, acoustic-wave filters are used in applications that require filter bandwidths too narrow to be achieved using LC technology. SAW transversal filters provide simultaneous frequency and time responses not ordinarily achievable using other technologies, and are often used for the generation and reception of coded signals.[12]

FIGURE 7.34   Typical LC filter utilizing ferrite pot-core inductors.

BAW filters   Bulk acoustic-wave filters include quartz-crystal and ceramic piezoelectric filters, and electromechanical filters. Also included are thin-film resonators (TFRs), also referred to as *film bulk acoustic resonators (FBARS)*, which are covered separately in Section 7.7. Electromechanical filters can be fabricated to operate from several kHz to several hundred kHz. Crystal filters span the frequency range from units of kHz to hundreds of MHz. Ceramic filters typically span a frequency range from tens of kHz to tens of MHz. The so-called AT-cut quartz-crystal resonator is probably the most widely used bulk wave-filter resonator, primarily because of a combination of high Q and excellent frequency stability. AT-cut resonator filters cover the frequency range from 1 to several hundred MHz, and filter bandwidths are in the 0.001-percent to 1-percent range. For filter bandwidths greater than 0.1 percent, balanced transformers are normally used to neutralize the effects of resonator parallel capacitance over a large frequency range. Multipole BAW filters can be designed to use multiple, individually fabricated and packaged resonators reactively coupled using external inductors and/or capacitors, or they can be designed on a single quartz blank utilizing acoustic coupling. The latter form of quartz crystal filter is commonly referred to as a *monolithic filter*. Figure 7.35 shows a sketch and associated schematic diagram for a typical quartz crystal filter.

SAW filters   SAW filters can also be implemented in several ways; as either a transversal filter or a resonator filter. The transversal filter is fabricated using thin-film metal transducers deposited on the surface of the piezoelectric substrate. The transducers consist of

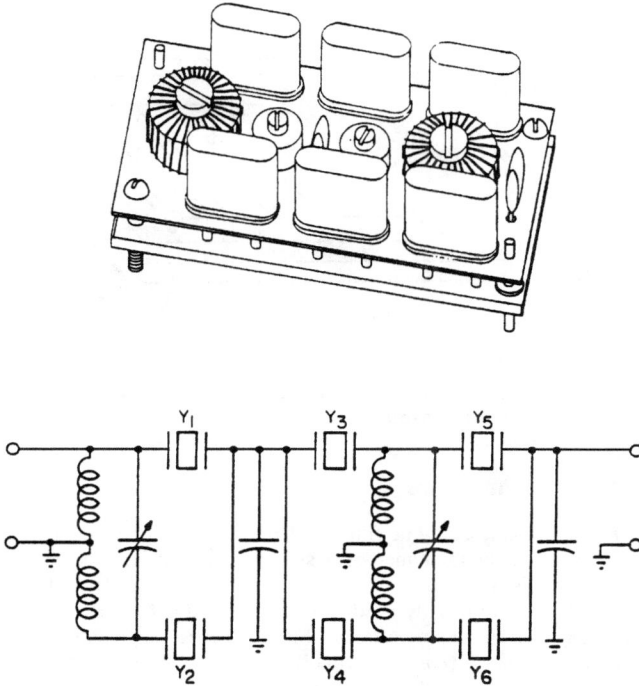

FIGURE 7.35 Sketch and schematic for a six-pole crystal bandpass filter. (*1985 by Academic Press, Inc.*)

parallel, interdigital electrodes or fingers spaced at ¼ wavelength and are electrically connected for alternating polarity. The transducer finger lengths are weighted by design to achieve the desired filter frequency and/or time response, such as those required for receiver-matched filters. These responses are often difficult to achieve with other technologies. Figure 7.36 shows a sketch of a SAW filter substrate. The interdigital metal fingers have nonuniform overlap; the finger-overlap pattern, called *apodization*, is designed to mirror the desired filter impulse response in time—the frequency response is the inverse Fourier Transform, thereof. In fact, any finite time length signal can be synthesized using the finger overlap representation as long as:

- A piezoelectric substrate having enough physical length to represent all of the time signal is available.
- The center frequency can be represented on the substrate with a finger-pair.

The first caveat gives an upper limit of about 100 μs to the pulse length. The second gives an upper limit in frequency of about 2 GHz. Because the surface acoustic-wave velocity along the substrate (from the "sending" transducer to the "receiving" transducer) is orders of magnitude slower than the speed of light, the acoustic wavelength (and resulting filter size) is dimensionally small.

A SAW resonator filter consists of reactively or acoustically coupled, individual resonators. In general, individual SAW resonator filters are fabricated as low-order types (two to three poles each) and exhibit lower insertion loss, compared to transversal filters.

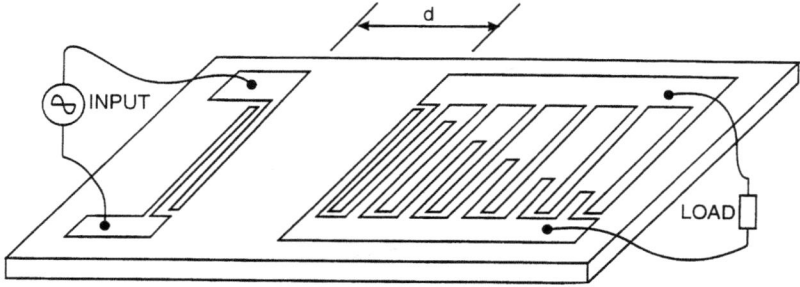

FIGURE 7.36   SAW filter apodization pattern.

Figure 7.37 shows photographs for several typical types of SAW filters, together with a typical 2-pole SAW-resonator transmission-response measurement.

7.6.3 Transmission Line Filters

Distributed-element, transmission-line filters utilize end- or edge-coupled lengths of transmission line. The lines can be implemented in stripline, microstrip, or coaxial form. In most cases, the line length is a quarter or a half wavelength at the filter center frequency. Transmission-line filters are usually used at frequencies above 1 GHz. Large-percentage bandwidths are possible. The degree of attainable small-percentage bandwidth is related to the transmission-line unloaded Q and is normally in the range of several percent. Depositing the lines onto or between high dielectric-constant substrates, or surrounding them by high dielectric-constant material lowers the wavelength and hence, the filter volume. Microstrip transmission-line filters are very popular because they can be fabricated as a metallization pattern on the same printed board or ceramic substrate as the other electronic components as part of the normal plating process. Figure 7.38 shows examples of various transmission-line filter topologies.[15] Distributed-element filters are used where their small size and modest cost are important: namely in the microwave portions of communications and radar systems.

The ceramic block filter is a transmission-line filter consisting of a molded or machined block of dielectric material containing through-holes. Metal is plated inside the holes, forming quarter-wave "posts," and the plating on the outside of the block forms the outer enclosure metallization. This technology is being used extensively in the wireless market because of its small size and low cost. A ceramic block filter is shown in Figure 7.39.

The helical resonator is a special case in which the transmission-line center conductor is coiled (usually by winding it on a threaded form) within a metal enclosure, which forms the outer conductor (Figure 7.40). Coiling the line inner conductor reduces the resonator volume, and modest-sized helical-resonator filters can be fabricated at frequencies over the entire VHF range (30 to 300 MHz).

7.6.4 Dielectric Resonator Filters

The term *dielectric resonator filter* normally refers to a filter whose individual resonators consist of cylinder rods or tubes of dielectric material—either mounted onto a substrate or "post" mounted inside a metal enclosure (Figure 7.41). The electromagnetic field of a given resonant mode is largely confined in and near the ceramic resonator. The field strength outside the resonator falls off to small values at distances much shorter than the free-space wavelength. Therefore, the primary resonator loss mechanism is usually the loss tangent ex-

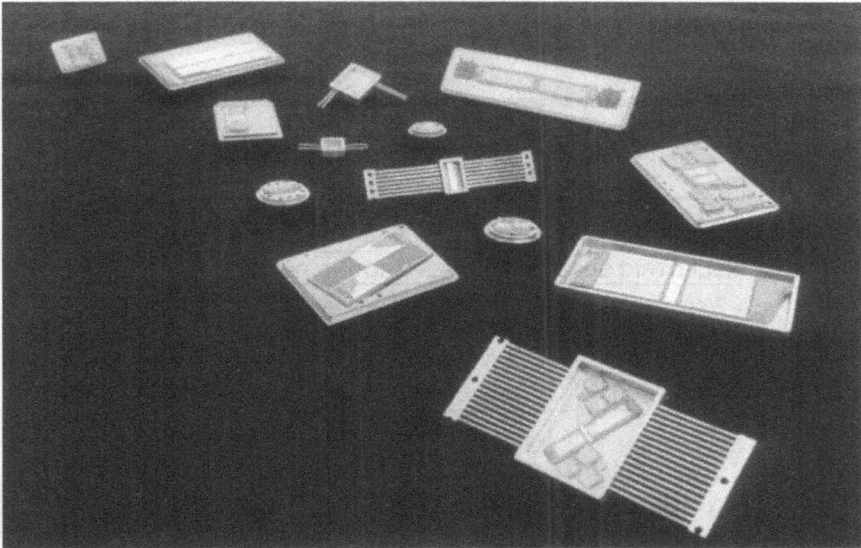

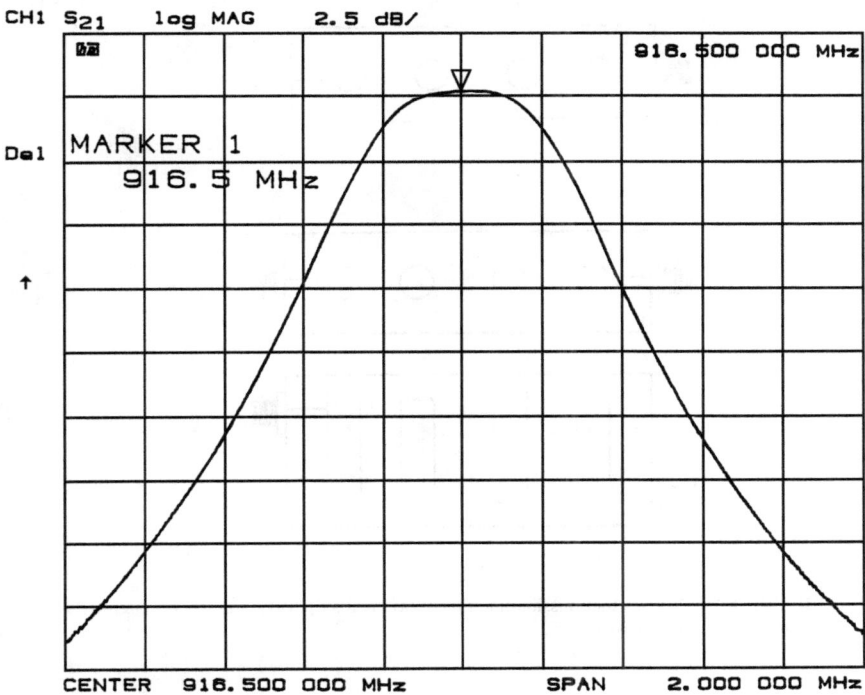

(b)

FIGURE 7.37 SAW filters: (a) photographs, (b) SAW resonator filter transmission response. (*Courtesy SAWTEK, Inc.*)

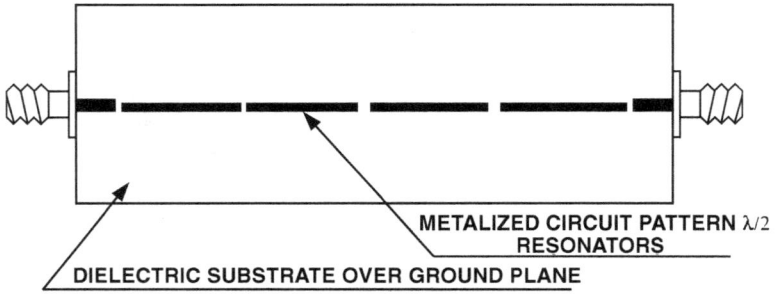

(a)

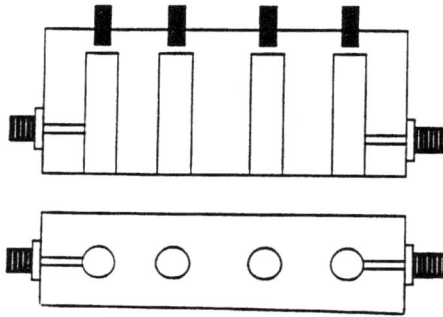

(b)

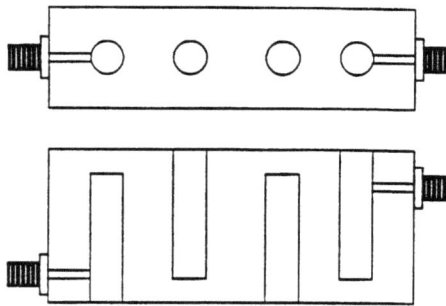

(c)

FIGURE 7.38   Various transmission-line filters: (a) microstrip, (b) combline, (c) interdigital. (*Courtesy of Dr. K. A. Zaki.*)

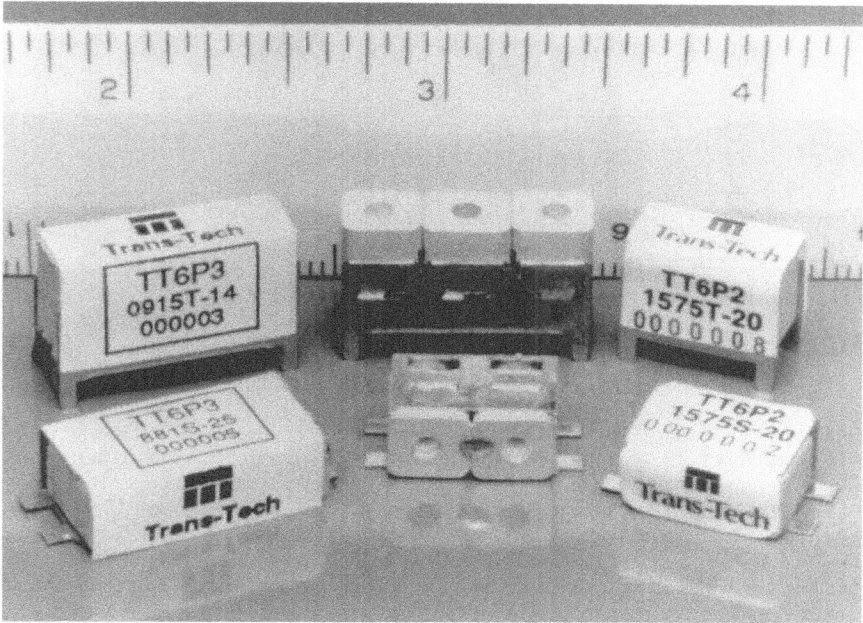

FIGURE 7.39 Photographs of ceramic block filters. (*Courtesy Trans-Tek, Inc.*)

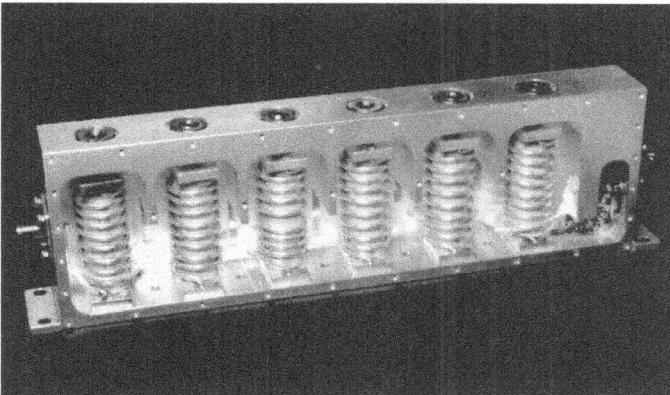

FIGURE 7.40 Photograph of a helical resonator filter. (*Courtesy Salisbury Engineering, Inc.*)

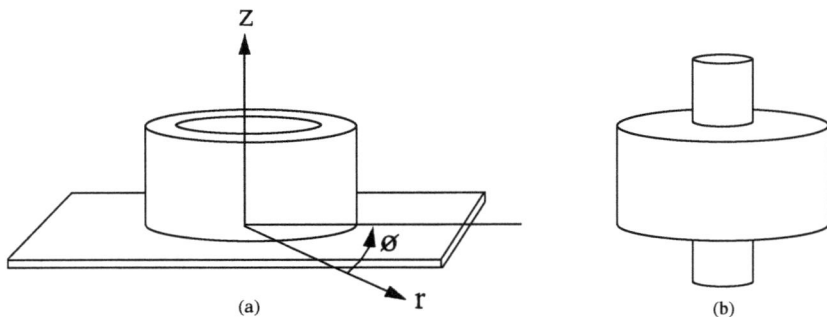

FIGURE 7.41   Sketch of a dielectric resonator: (a) board-mounted, (b) post-mounted.

hibited by the dielectric material. Ceramic materials are available that exhibit low temperature coefficient while supporting resonator unloaded Q values in the thousands at resonant frequencies in the range of 2 to 20 GHz. Angularly symmetric modes having no ϕ variation that can exist in dielectric (and cavity and waveguide) filters are termed *transverse electric (TE-)* and *transverse magnetic (TM-)*, according to the absence of electric or magnetic fields along the z-axis. These modes are further defined by two subscripts that define the number of half-period field variations in the radial and axial directions.[16] Other modes that can exist in dielectric resonators, which have variation along the ϕ-direction are hybrid (or HE) modes. Hybrid modes have both electric and magnetic fields along the Z-direction. They are identified by subscripts and the order of the resonant frequency.

7.6.5 Cavity and Waveguide Filters

Cavity and waveguide filters consist of adjacent hollow metal enclosures propagating similarly termed modes, as described for dielectric resonators. Electrical coupling between resonators is usually via small openings or irises in the enclosure walls. Filling the cavity or waveguide with dielectric material is a method for reducing the effective signal wavelength and filter volume. Cavity and waveguide filters span a typical operating frequency band from several GHz to 50 GHz with bandwidths in the 0.1- to 5-percent range. Figure 7.42 shows typical dielectric resonator and waveguide filter construction.[15]

7.7 EMERGING FILTER TECHNOLOGIES

Several promising filter technologies are currently under development. One of the technologies involves the use of ultra-lowloss, high-temperature superconductors. Another technology involves the use of tiny, piezoelectric filters included in and compatible with the processing of monolithic microwave integrated circuit (MMIC) chips.

7.7.1 High-Temperature Superconducting (HTS) Filters

HTS filters use high-temperature superconducting films instead of traditional metal conductors. This class of filter is usually implemented in stripline or microstrip form (Figure 7.43).

BAND PASS

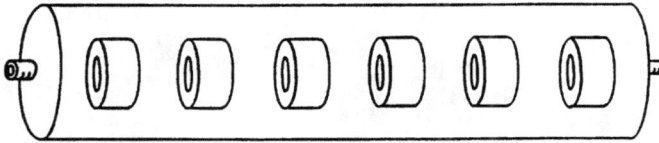

BAND STOP

(a)

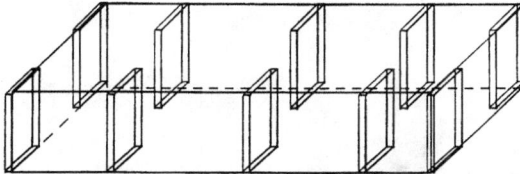

INDUCTIVE WINDOWS (MODERATE BANDWIDTHS)

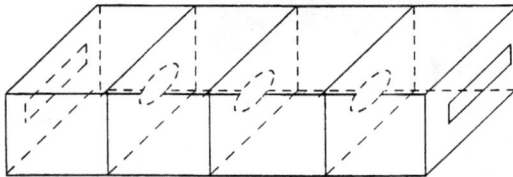

DIRECT COUPLED USING IRIS (NARROW BANSWIDTHS)

(b)

FIGURE 7.42 Microwave filter construction: (a) dielectric resonator, (b) waveguide. (*Courtesy Dr. K. A. Zaki.*)

The ultra-lowloss nature of the films (high resonator unloaded Q) allows distributed-element filters to be fabricated with much lower passband loss, for a given degree of selectivity, compared to metal transmission-line filters. Generally, HTS filters find use in receiver front ends. They must be operated at temperatures on the order of 80 K (below the superconducting material transition temperature), and require the use of liquid-nitrogen based cryocoolers.

7.7.2 Film Bulk Acoustic-Resonator (FBAR) Filters

FBAR filters (sometimes referred to as *thin-film resonator* or *TFR filters*) use very small, thin films of piezoelectric material that are sputter-deposited onto semiconductor sub-

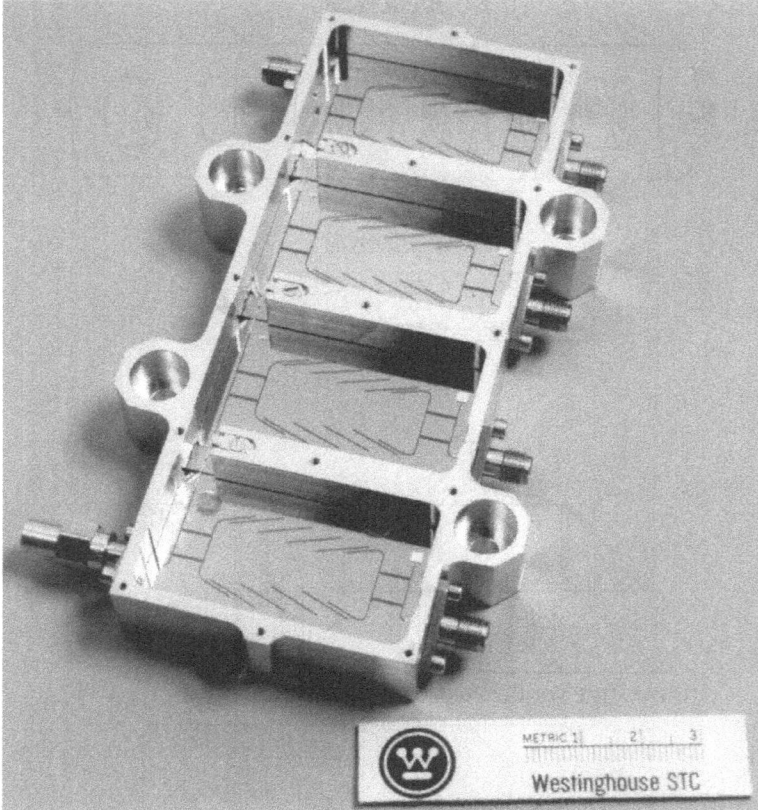

FIGURE 7.43 High-temperature superconducting filter. (*Courtesy Westinghouse Science and Technology Center.*)

strates. Substrate material is then chemically or ion-etched under the deposited film so that the piezoelectric membrane is free to vibrate (Figure 7.44). In some cases, a thin layer of doped semiconductor is used as an etch stop and can also be used to temperature-compensate the resonant frequency of the FBAR. Aluminum-nitride and zinc-oxide films have been extensively used as the piezoelectric. The FBAR is primarily a UHF (300 MHz to 3000 MHz) technology. With unloaded resonator Q values on the order of 500 to 1000, the FBAR exhibits the highest ratio of Q to volume of any known resonator. An obvious advantage of this type of microelectronic filter is that can be processed along with active devices as a fully compatible, integrated-circuit element. A primary stumbling block with regard to bringing this technology to fruition involves the development of a high-volume, low-cost process for accurate control of film uniformity and resonant frequency.

Initial FBAR filters were designed as both single layer as well as stacked devices. Recently, another version of the FBAR has been developed. The solidly mounted resonator (SMR), based on a concept by Newell and developed by TFR Technologies, Inc., has the piezoelectric layer deposited onto a set of quarter-wave layers of alternating high and low acoustic-impedance films that confines all acoustic energy within the transducer

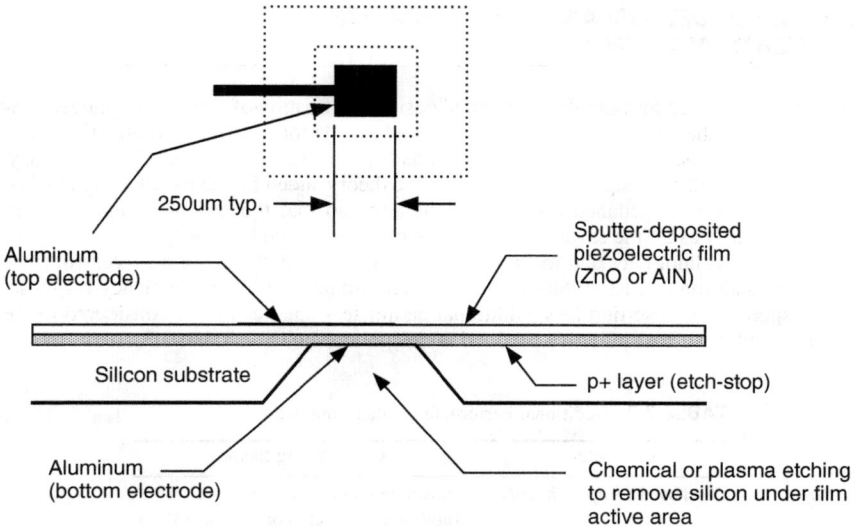

FIGURE 7.44   Cross-sectional view of an FBAR resonator.

region. Figure 7.45 shows a SMR ladder filter composed of three series resonators and two shunt resonators. The advantage of SMR filters is that they do not require a membrane support and hence, do not require backside etching. A third thin-film approach being developed by TFR Technologies uses planar dielectric-resonator technology. This is an electromagnetic resonator approach in which a coplanar waveguide line is loaded with interdigitated capacitors to decrease the phase velocity and hence reduce the size of the resonator. By using high dielectric-constant film, the resonator size can be further reduced.

FIGURE 7.45   FBAR bandpass filter. (*Courtesy TFR Technologies, Inc.*)

7.8 FILTER DESIGN AND PERFORMANCE
SIMULATION METHODS

The process utilized for both the design and design verification of filters has changed dramatically with the advent of circuit synthesis and simulation software. To effectively use these modern tools, however, the designer needs to be aware of the "basics." That is to say, one should be able to design the filter using basic theory, aided by the use of design tables, selectivity charts, impedance transformations, etc., so that the use of design software serves primarily as a time saver and not a substitute for design knowledge. Knowledge of various filter technologies and technology limitations is mandatory for a successful filter design. In addition to basic filter-performance requirements, such as frequency response, time response, and insertion loss, additional parameters that need to be considered in the development of an cost-effective production design are shown in Table 7.7.

TABLE 7.7 Additional Performance Filter Parameters

Parameter	Contributing factor
Purchased component cost	Allowable component tolerance Number of different component values Allowable temperature coefficient Number of component vendors
In-house cost	Manufacturing tolerances Assembly time Alignment time Test equipment cost Test time
Performance	Reliability (derating, etc.) Environmental stress (temperature, vibration, etc.)

7.8.1 Filter Design and Design Aids

Usually the first step in the design of a filter is to choose the optimum (and sometimes only) technology that can be used to satisfy the frequency and/or time-response requirements. This choice is normally determined by the operating frequency and bandwidth requirements of the filter. For example, a 20-MHz bandpass filter having 0.1-percent bandwidth would likely use quartz-crystal technology; a 10-GHz bandpass filter having 5-percent bandwidth would likely use transmission-line resonators.

Once the filter technology is determined, the complexity or "order" of the filter (number of resonators or poles) and filter response shape is selected, based on the required frequency selectivity and/or time-response requirements. As described in section 7.2, the often-used response types include Butterworth, Chebyshev, Bessel, Gaussian, Elliptic, etc. Charts showing lowpass filter frequency, phase, delay, and time responses (usually normalized to 1 radian/sec bandwidth) are used to select the filter-response type and order.[1,3,4] In addition, design charts showing the impact of finite element Q on filter-response characteristics and insertion loss (such as those shown in Figure 7.29 and Table 7.5) are used to determine whether or not to use a predistorted design.[13]

Next, the actual design of the filter is accomplished using appropriate design tables and/or formulas. For lumped-element filters, the tables usually consist of tabulated low-pass element values normalized to 1-radian/sec bandwidth, or alternatively tabulated values for end-resonator loaded Q and inter-resonator coupling coefficients.[1] The resulting filter schematic (especially lumped-element, bandpass, and band-reject types) might include unrealistic element values, and often an additional design step consists of performing impedance transformations within the filter to allow use of readily obtainable components. Figure 7.46 shows a step-by-step example of the design process for a 30-MHz, 5-MHz bandwidth, 3-pole, Butterworth bandpass filter.

7.8.2 Filter Performance Simulation

The most important use of the computer in facilitating the design of filters is not so much in connection with determination of the design itself, but in the calculation of the filter frequency and time responses, and also with regard to performing accurate statistical analyses of the effects of manufacturing tolerances on yield. In this regard, a host of software packages are available that perform these analyses/calculations for both lumped- and distributed-element filters. The filter response shown in Figure 7.30 was easily and quickly obtained using circuit simulation software. In addition to statistical-analysis capability, the software packages usually include optimization routines that allow "customizing" the filter to modify the response characteristics to other than those exhibited by conventional filter types. It is particularly important in any filter design (especially an optimized design) to check overall filter-response sensitivity to element-value variation.

7.9 FILTER-PERFORMANCE VERIFICATION

Verification of actual filter performance is usually accomplished via measurement of the transmission response of the individual filter itself. Alternatively, measurements can be made with the filter installed in the "next higher assembly" hardware, where it will actually be used. More often than not, the filter is specified to operate between industry-standard, 50-ohm generator, and load impedances. The automatic network analyzer (ANA) is currently the most-often used tool for measuring of filter and other electronic circuit small signal (i.e., linear) response characteristics. ANAs are available that can accurately measure filter input and output impedance or VSWR, transmission response (phase and amplitude), and group delay; some can also operate over the entire low- to microwave-frequency bands with large dynamic-range capability. Alternatively, a variable-frequency signal generator, together with a power meter or RF voltmeter can be used (Figure 7.47). Most RF test equipment is designed to operate in a 50-Ω system to allow filter connection to the test equipment using standard (50-Ω characteristic impedance) coaxial cable. The effects of coaxial connecting cable can be "calibrated out" of the measurement, and the use of cable allows, for example, the filter to measured in a temperature chamber located many electrical wavelengths away from the test equipment. For this reason, non-50-Ω filters are often installed in a specially designed test fixtures that transform the filter impedances to 50 Ω. Filter time (i.e., pulse response) characteristics can be measured using an RF gating switch and RF detector (Figure 7.48).

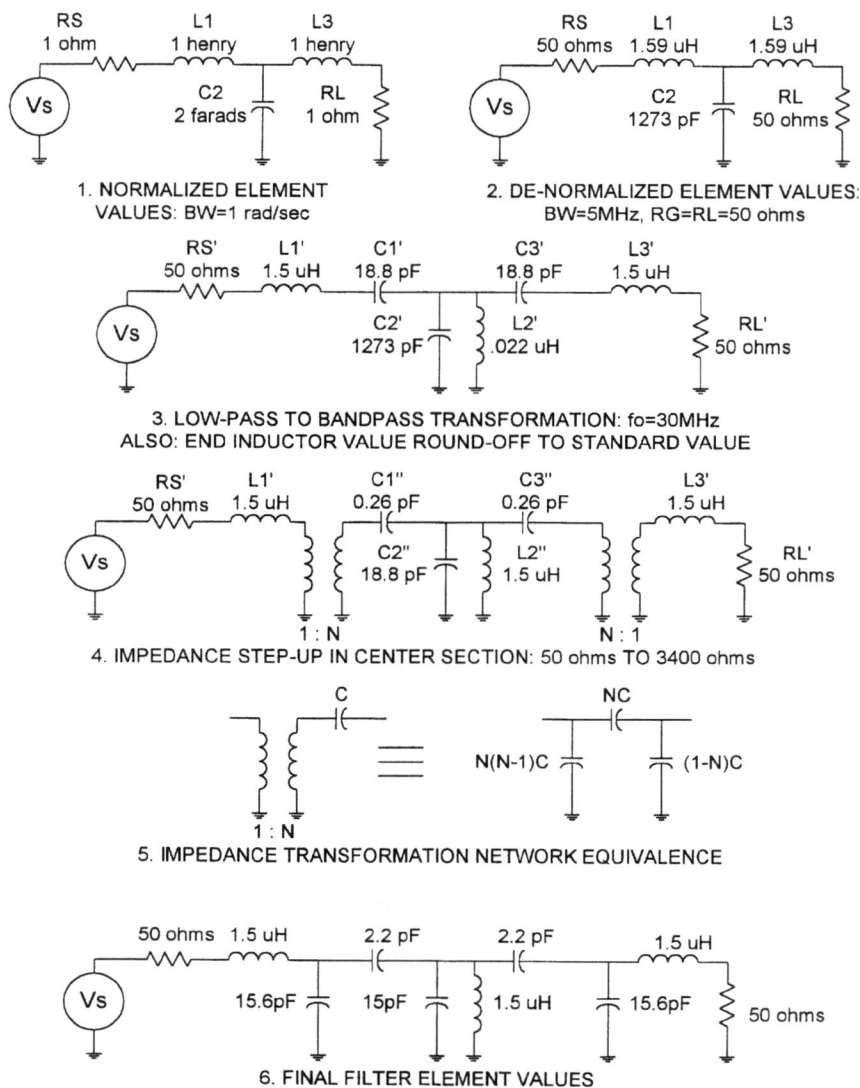

FIGURE 7.46 Step-by-step design of a three-pole bandpass filter.

7.10 REFERENCES

1. Zverev, A.I., *The Handbook of Filter Synthesis*, John Wiley, New York, 1967.
2. Blinchikoff, Herman J. and Zverev, A.I., *Filtering in the Time and Frequency Domains*, John Wiley, New York, 1976.

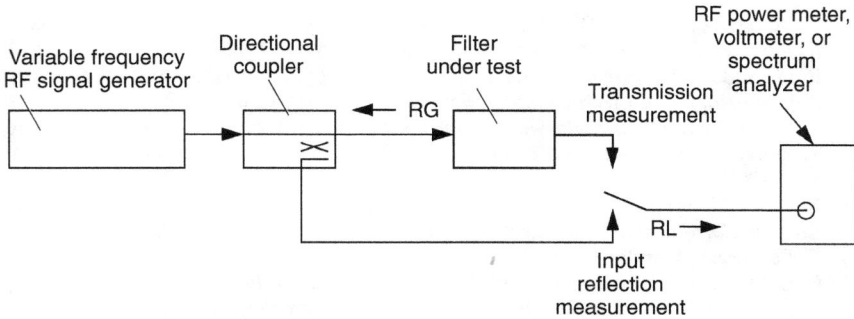

FIGURE 7.47   Filter frequency-response measurement test set-up.

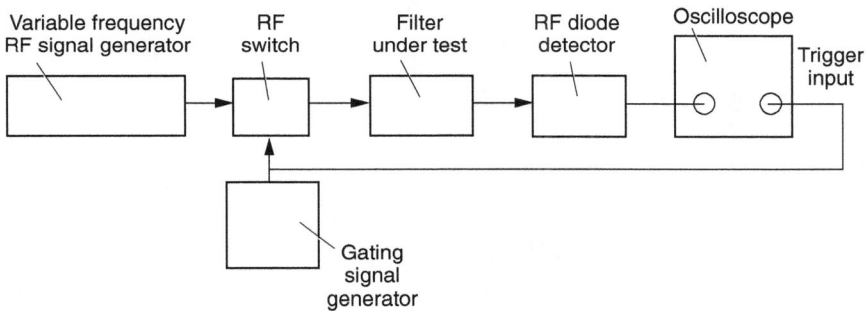

FIGURE 7.48   Bandpass-filter time (step) response measurement test set-up.

3. Humpherys, DeVerl S., *The Analysis, Design, and Synthesis of Electrical Filters*, Prentice Hall, Englewood Cliffs, N.J., 1970.

4. Saal, Rudolf, and Ulbrich, E., *On the Design of Filters by Synthesis*, IRE Transactions CT-5, 1958, pp. 284–327.

5. Saal, Rudolf, *Der Entwurf von Filtern mit Hilfe des Kataloges normierter Tiefpasse*, AEG-Telefunken, Backnang, West Germany, 1968.

6. Saal, Rudolf, and Entenmann, Walter, *Handbook of Filter Design*, Dr. Alfred Hüthig, Heidelberg, 1988.

7. Green, Ernest, *Amplitude—Frequency Characteristics of Ladder Networks*, Marconi's Wireless Telegraph Company, Chelmsford, Essex, 1954.

8. Various authors, *Reference Data for Radio Engineers*, Federal Telecommunication Laboratories.

9. V. Manasseewitsch, *Frequency Synthesizers Theory and Design*, John Wiley and Sons, New York, 1976.

10. V. F. Kroupa, *Frequency Synthesis Theory Design and Applications*, John Wiley and Sons, New York, 1973.

11. M. M. Driscoll, *Low Noise Microwave Signal Generation Using Bulk and Surface Acoustic Wave Resonators*, IEEE Trans. on UFFC, Vol. 35, No. 3, May 1988, pp. 426–434.

12. E. A. Gerber and A. Ballato, *Precision Frequency Control, Volume 1, Acoustic Resonators and Filters*, Academic Press, Inc., New York, 1985.

13. R. G. Anderson, *Design Data for Filters with Low-Q Elements*, Report No. SP6808-301, Westinghouse Electric Corporation, Baltimore, Maryland, 1968.

14. J. R. Vig, *Quartz Crystal Resonators and Oscillators: A Tutorial*, US Army Electronic Technology and Devices Laboratory Tech. Report No. SLCET-TR-88-1 (Rev. 3.0), Jan., 1990.

15. K. A. Zaki, *Microwave Filter Design Course Notes: Prepared for Westinghouse Electric Corporation*, Department of Electrical Engineering, University of Maryland, College Park, Maryland, Aug., 1995

16. G. L. Matthaei, L. Young, and E. M. T. Jones, *Microwave Filters, Impedance-Matching Networks, and Coupling Structures*, McGraw-Hill, New York, 1964.

17. D.O. North, *Analysis of the Factors Which Determine Signal/Noise Discrimination in Radar*, Report PTR 6-C, RCA Laboratories, June 1943.

CHAPTER 8
CONNECTOR AND INTERCONNECTION TECHNOLOGY

Robert S. Mroczkowski
AMP Incorporated, Harrisburg, PA

8.1 CONNECTOR OVERVIEW

This chapter provides an introduction to electrical/electronic connectors. Both the function and structure of connectors are covered. Because of space limitations, the coverage is limited. References are provided for those interested in more detailed discussions of individual topics.

8.1.1 Connector Function

A functional definition of a connector is an electromechanical device which provides a separable interface between two electronic subsystems without an unacceptable effect on signal integrity or loss in power.

The key elements in this definition are "electromechanical," "separable," and "unacceptable." A connector is an electromechanical device because it uses mechanical means, spring deflections, to create an electrical interface. These separate functions also tie into the other two key words. Providing "separability" is basically a mechanical function, while ensuring that the connector does not introduce "unacceptable" electrical effects on the system is an electrical function, which depends on the mechanical stability of the connector contact interfaces.

Separability is the major reason for using a connector, and it may be required for a number of reasons, among them being ease in manufacturing of subassemblies, repair or upgrading or, increasingly, portability of electronic equipment. Separability requirements are generally stated in terms of the mating force and number of mating cycles a connector must support without degradation. Mating forces become increasingly important as the number of positions in the connector, or pin count, increases. Connectors are now available with over a thousand positions. Depending on the function of the connector, the required number of mating cycles can vary from a few to several thousand.

In this chapter, the major electrical parameter is connector resistance. The allowed change in resistance over the application life of the connector is used as the measure of "unacceptable." The meaning of "unacceptable" is strongly application dependent. In particular, the allowed resistance change depends strongly on whether the connector is used in a signal or power application. Signal applications are, in general, more tolerant of increases in resistance than power applications.

8.1.2 Connector Applications: Levels of Interconnection

Levels of interconnection is a concept that allows for the description of connectors in terms of their application. Although a number of levels of interconnection approaches exist, this chapter follows that of Granitz.[1] In this approach, the levels of interconnection are defined by the two points within the electronic system which are being connected and not the connector type or function. The six levels of interconnection are illustrated and described in Figure 8.1.

The levels of interconnection concept highlights several different connector issues and requirements. In general, as the level of interconnection increases the number of mating cycles a connector will experience increases. The exposure of the connector to untrained users increases in the same manner. The standards which a connector must meet also vary with the level of interconnection. Currently standards requirements are more detailed at levels 2 and 5 and 6 than at the other levels. Standards on levels 3 and 4 are, however, increasing. Levels 3 through 6 are the levels which include more "traditional" connectors. Pin count requirements peak at level 3 and, in general, decrease from levels 4 through 6. It should also be noted that some connector types, in particular cable connectors, see applications at more than one level appearing at levels 4 and 5. The discussion in this chapter is limited to levels 3 through 6, the levels in which traditional connectors are used. The interconnection devices

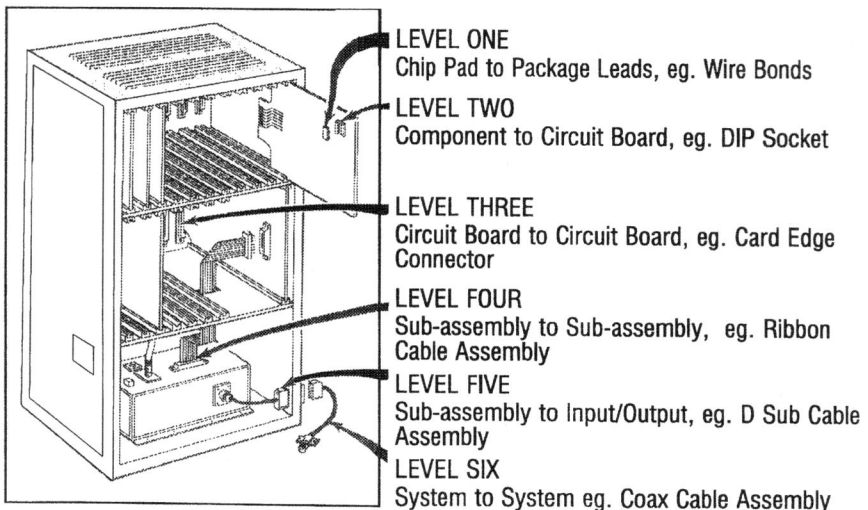

LEVEL ONE
Chip Pad to Package Leads, eg. Wire Bonds

LEVEL TWO
Component to Circuit Board, eg. DIP Socket

LEVEL THREE
Circuit Board to Circuit Board, eg. Card Edge Connector

LEVEL FOUR
Sub-assembly to Sub-assembly, eg. Ribbon Cable Assembly

LEVEL FIVE
Sub-assembly to Input/Output, eg. D Sub Cable Assembly

LEVEL SIX
System to System eg. Coax Cable Assembly

FIGURE 8.1 Schematic illustration of the six levels of packaging. (*Courtesy AMP Incorporated.*)

in level 2 are more properly described as sockets. Socket requirements, however, are becoming increasingly similar to those of connectors as socket pin counts increase.

8.1.3 Connector Types

One method of characterizing connector types follows the levels of interconnection approach in that the connector type is defined by the permanent connection. According to this scheme there are three major connector types:

- board-to-board, levels 3 and 4
- wire-to-board, levels 4 and 5
- wire-to-wire, levels 5 and 6

In this discussion, the term *wire* also includes cables of various constructions. A brief description of each type follows, with additional discussion to follow in Section 8.9.

Board-to-board connectors Board-to-board connectors have experienced the greatest impact of the advances in microprocessor technology. Tremendous increases in chip functionality and, therefore, input/output (I/O) requirements require similar increases in pin count in level-3 connectors. Level-3 connectors have also experienced more demanding requirements in impedance control because they are closest to the chip and face the greatest demands on maintaining signal integrity. These two requirements, especially pin count, have led to a transition from card-edge connectors to multi-row, two-piece connectors to satisfy high-density connector requirements.

Wire-to-board connectors Wire-to-board connectors, particularly at level 4, can also face controlled-impedance requirements. These connectors are increasingly cable assemblies, rather than discrete wire to more effectively address the wire-handling requirements for high-pin-count connectors.

Wire-to-wire connectors Wire-to-wire connectors are generally farther away from printed wiring boards and chips and, with the exception of coaxial connectors, face less demanding impedance requirements. Wire-to-wire connectors are often external to equipment and, therefore, ruggedness and grounding/shielding requirements could become important considerations.

8.1.4 Connector Applications: Signal and Power

The two basic connector applications are signal and power. Signal applications are characterized by low current and voltage requirements. Power contacts/connectors, in contrast, generally address higher current and, often, higher voltage applications.

Signal applications Signal applications span the range of currents from microamps to hundreds of milliamps, with this range expanding in both directions as electronic applications increase in complexity. Driving voltages for signal applications are generally a few volts.

An additional requirement facing signal connectors is impedance control, as mentioned in the previous section. Connector design for impedance control is a major area of development in connector technology. Approaches using ground pins, open pin field connectors, and ground planes are being developed (Section 8.8).

Power applications Power contacts/connectors face an additional requirement: thermal management. Managing the Joule, or I^2R, heating which accompanies current flow becomes an important design/selection criterion.

Two approaches to power distribution are dedicated power contacts and using multiple "signal" contacts in parallel. The application considerations for these two approaches differ significantly and are covered in Section 8.8.

This completes the overview of connector function. Attention turns to connector structure—how the design and materials of manufacture of a connector affect connector performance.

8.1.5 Connector Structure

A connector can be described as consisting of four components as outlined by Mroczkowski[2] and schematically illustrated in Figure 8.2. At this point, each component is considered in an overview fashion with a more detailed discussion to follow.

Contact interface There are two types of contact interfaces that must be considered: separable and permanent. The separable interface, created when the plug and receptacle halves of the connector are mated, has already been mentioned as the primary reason for using a connector. Permanent interfaces must also be created between the connector and the functional elements that are being connected. Both metallurgical (soldered and brazed/welded) and mechanical (crimped, insulation displacement and compliant pin) per-

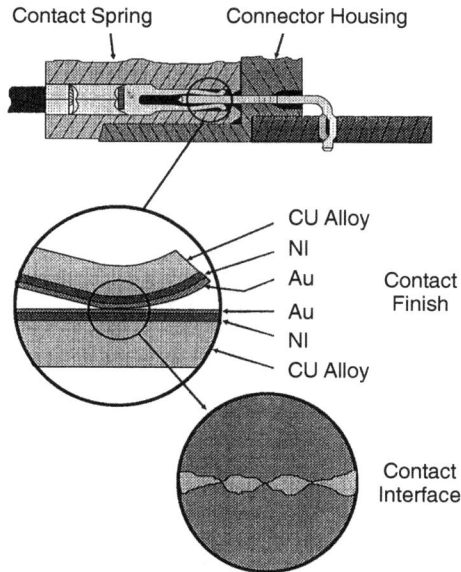

FIGURE 8.2 Schematic cross section of a connector. Four components of the connector are illustrated and described in the text. (*Courtesy AMP Incorporated.*)

manent connections are widely used. Figure 8.2 illustrates a crimped connection and a compliant press-in connection. Separable connections are covered in Section 8.6 and permanent connections in Section 8.7. A stable, low-resistance contact interface requires that a metallic contact interface be established. Control and elimination of surface films is a major factor in creating such an interface. It is the contact finish which has the strongest impact on the formation and disruption of surface films.

Contact finishes   The contact finish has two major functions, to provide corrosion protection for the contact spring and "optimizing" film management to allow the creation of a metallic interface. The two basic classes of contact finishes—noble (gold and other precious metals) and non-noble (tin, silver and nickel)—differ significantly in the design considerations for, and mechanisms of, film management. Contact finishes are covered in Section 8.3.

Contact spring   The contact spring also has two functions. Electrically, the contact spring provides continuity between the permanent and separable connections. To achieve acceptable performance the contact spring must be metallic. Mechanically, the contact spring must be capable of creating the contact forces which establish and maintain the separable and permanent contact interfaces. The contact spring must also be capable of forming the desired permanent connection, mechanical or metallurgical. Mechanical permanent connections might require severe forming, such as in crimped connections or spring characteristics, such as in IDC connections. Metallurgical connections, in particular soldered connections, usually rely on a plating, generally tin-lead, on the spring to maintain solderability.

Copper alloys are the dominant contact spring materials because of a good combination of mechanical and electrical characteristics as are covered in Section 8.4.

Connector housing   The connector housing also performs electrical and mechanical functions. Electrically the housing insulates the contacts from one another. Mechanically, the housing captures and supports the contact springs and maintains the contact spacings of both the separable and permanent connections. Dimensional control of the separable interface must be maintained to ensure appropriate mating of the connector. Dimensional control of the permanent interface is necessary to ensure proper registration of the permanent connections with the wire/cable or printed wiring board (PWB). A variety of thermoplastic polymers with acceptable electrical and mechanical properties are used in connectors. Assembly requirements—in particular, surface mounting technology—strongly influence polymer selection for connector housings. The application temperature is also an important consideration in material selection. Connector housings are covered in Section 8.5.

Connector design   The performance and reliability of connectors depends on the material selection and design of these four connector components. For separable connections, the contact finish, the normal force and mating interface geometries determine the connector contact resistance as well as the mating force and mating durability of the connector. For permanent connections minimizing the magnitude of connection resistance and maximizing its stability are major considerations. Ensuring formation of a metallic interface and a suitable residual force distribution through controlled deformation of the conductors and connectors are dominant considerations for mechanical permanent connections. The connector housing must maintain its insulating characteristics and dimensional integrity throughout the manufacturing/assembly process and over the desired operating life of the connector. Selection of an appropriate connector for a given application requires consideration of the operating environment, mechanical, chemical and thermal, as well as the electrical requirements the connector must meet.

8.2 THE CONTACT INTERFACE

It can be argued that the structure of the contact interfaces determine the electrical and mechanical performance of a connector. In fact, a not completely facetious description of a connector has been given as "contact interfaces held together by supporting structures" [Whitley, private communication]. An understanding of the basic structure of contact interfaces is necessary to understand their importance in connector performance. The following sections are expressed in terms of separable interfaces, but similar considerations apply to the interfaces of permanent mechanical connections.

8.2.1 Contact Interface Morphology and Contact Resistance

As mentioned previously, the contact interface is created when the plug and receptacle contacts are mated. Williamson[3] provides an informative analogy for the process of creating a contact interface in terms of bringing two mountainous regions in contact, e.g., Vermont on top of New Hampshire. This analogy illustrates the importance of surface roughness on contact interface formation. Figure 8.3 schematically illustrates the creation of a contact interface. Only the high spots on the surface, called asperities or a-spots, actually come in contact. Because of their small size and radii the asperities deform plastically even at low applied loads.[4] The asperities deform to create a contact area sufficient to support the applied load. For typical connector interfaces only a small portion of the surfaces, a percent or less, are in contact. The distribution of the contact spots is determined by the geometries of the mating surfaces. For example, as illustrated in Figure 8.3, spherical surfaces in contact will result in a circular a-spot distribution.

This interface morphology gives rise to an electrical resistance, termed constriction resistance, at the contact interface. The source of constriction resistance, as described by Holm[5] is a fundamental effect and can be illustrated by consideration of Figure 8.4. The asperity microstructure of contact interfaces causes current flow to be "constricted" to

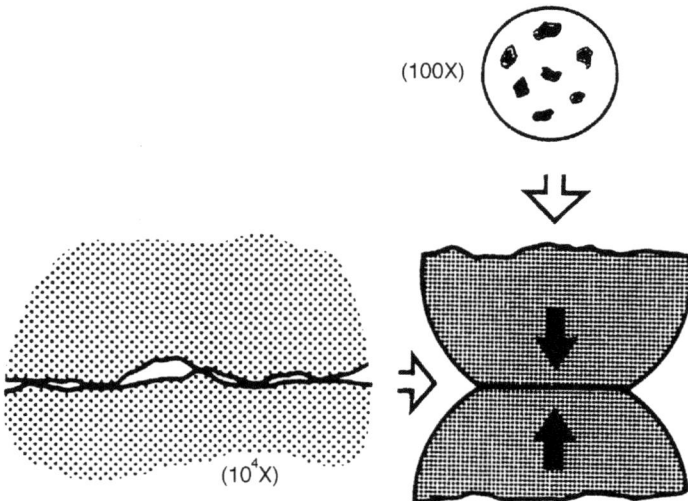

(100X)

(10^4X)

FIGURE 8.3 Schematic illustration of the morphology of a typical contact interface. (*Courtesy AMP Incorporated.*)

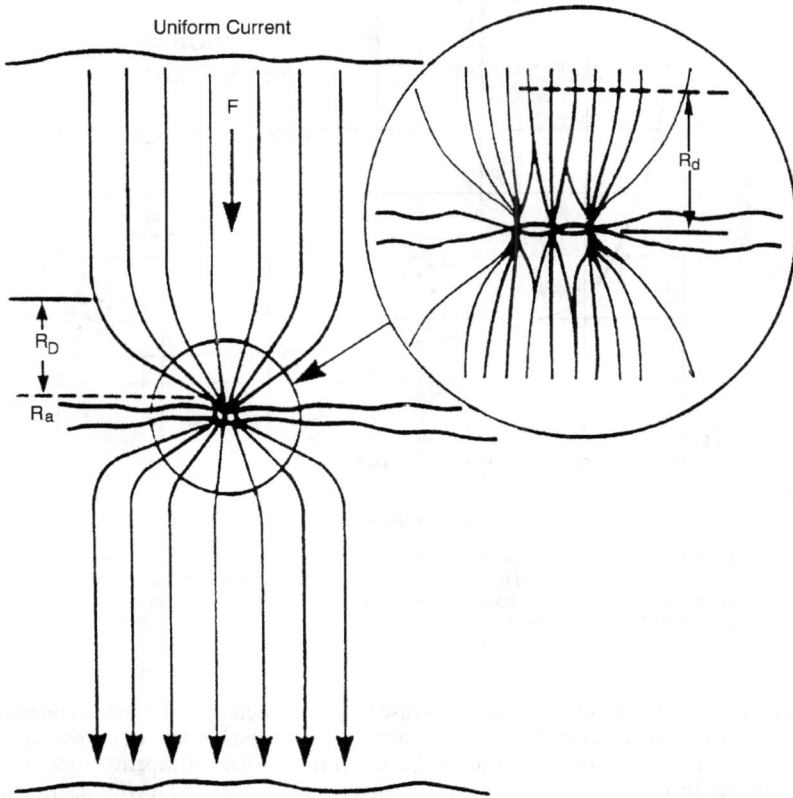

FIGURE 8.4 Schematic illustration of the current distribution at a contact interface which results in constriction resistance.

flow only through the asperity contact spots, hence the terminology "constriction resistance." According to Holm[5] for a single spot the constriction resistance is given by:

$$R_c = \frac{\rho}{2a} \tag{8.1}$$

where ρ is the resistivity of the material and a the radius of the asperity contact spot. For the purposes of this discussion, it can be stated that the asperity contact distribution affects contact resistance as illustrated in Figure 8.5. The equations inset in Figure 8.5 indicate the constriction resistance for a single asperity and for multiple asperities distributed over a geometric area depending on the contact geometry. The multispot equation:

$$R_c = \frac{\rho}{2na} + \frac{\rho}{D} \tag{8.2}$$

where n is the number of asperity contacts and D the diameter of the area over which the contacts are distributed is more relevant to typical contact interfaces. The first term indicates the effect of multiple asperity contact resistances in parallel and is indicated as R_a in

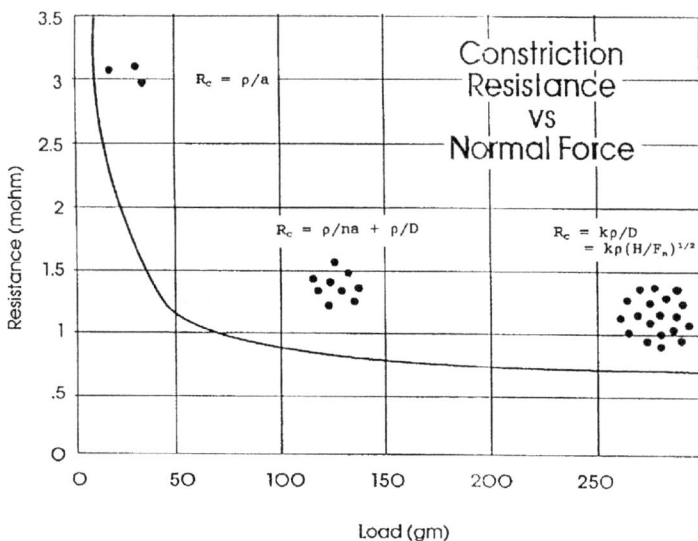

FIGURE 8.5 Typical relationship between contact resistance, R_c and contact force, F_n resulting from the variation in asperity number and distribution as contact force increases. Equations relating contact resistance to interface parameters are shown. (*Courtesy AMP Incorporated.*)

Figure 8.4. The current constriction in this case occurs very close to the contact interface. The second term indicates the effect of the current constriction because of the overall distributed contact area and is indicated as R_D. When the number of asperity contacts becomes large, several tens, the second term, which depends on the asperity distribution, dominates. In such cases the contact interface behaves as though the entire distributed area of a-spots is conducting. Under these conditions, the distributional area, and hence its diameter, assuming a circular asperity distribution, can be calculated from the contact geometry and hardness of the materials in contact according to:

$$R_c = k\rho \left(\frac{H}{F_n} \right)^{1/2} \tag{8.3}$$

where k is a coefficient including the effects of surface roughness, contact geometry and elastic/plastic deformation, H is the hardness, and F_n the contact normal force. This simple equation, however, has a complex interpretation in that, in addition to the interactions in k, in real cases the appropriate hardness is the composite hardness of the finish and the contact spring. Despite these limitations, the basic form of the equation is useful in understanding material/design effects on constriction resistance, in particular, that the contact normal force is a major factor in determining contact resistance.

It should also be noted that the calculation assumes metallic contact; that is, any surface films are completely displaced because of the deformation of the surfaces. Surface films, whether present initially or occurring during connector application, are a major degradation mechanism for contacts interfaces. The contact finish is a major factor in film management because it determines the types of films which will form and how readily they can be disrupted. This topic is covered in more detail in Section 8.3.

If surface films are not displaced, they result in a resistance in series with the constriction resistance. Film resistances can be very high and are highly variable. The variability arises from a combination of several factors. Film resistance depends on the thickness, composition, and structure of film. Each of these factors, in turn, depends on the conditions under which the film was formed. The composition of the environment, temperature and humidity are particularly influential in determining film characteristics. For these reasons, avoiding film effects by disruption and displacement is an important aspect of connector design. The contact normal force and geometry are two design parameters that strongly affect the effectiveness of film displacement.

8.2.2 Contact Interface Morphology and Mechanical Performance

The same asperity contact distribution responsible for constriction resistance determines the mechanical characteristics of friction and wear at the contact interface. Friction, in turn, influences the connector mating force and the mechanical stability of the contact interface. Wear processes impact on the number of mating cycles a connector can experience before showing any effect on performance. Both friction and wear processes depend on asperity contact interfaces as illustrated in Figure 8.6.

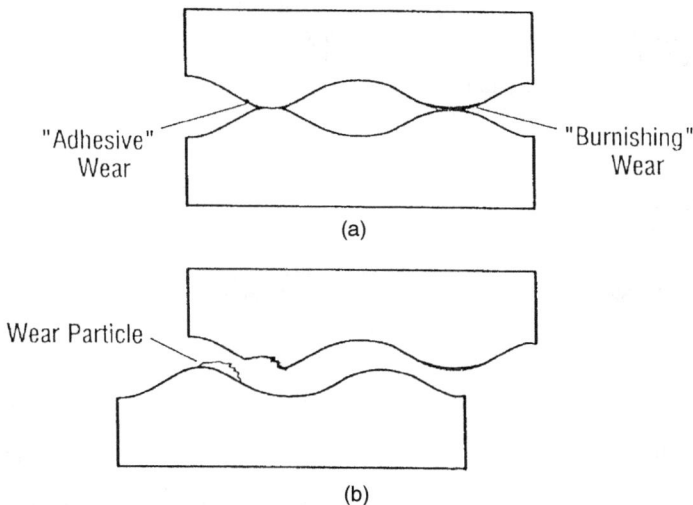

FIGURE 8.6 Schematic illustration of the asperity structure of contact interfaces as it relates to friction and wear mechanics. (*Courtesy AMP Incorporated.*)

As covered in Bowden and Tabor,[4] friction and wear processes depend on the location at which the interfaces separate during relative motion. Plastic deformation of asperity contacts can lead to cold welding at the asperity interface. In fact, the strength of the cold-welded interface might be higher than the cohesive strength of the base metal because of the work hardening that occurs during deformation. Two junctions are illustrated in Figure 8.6. Junction a is "stronger" than junction b having experienced a larger amount of plastic deformation, therefore, cold welding. Under a force tending to shear the contact interface, junction a might break away from the original interface resulting in a wear particle and metal transfer, as shown in the lower part of Figure 8.6. Junction b, with lower deformation, and, therefore, less work hardening and cold welding, might separate at or near the original interface so that little wear or metal transfer results.

The wear process at junction a is referred to as adhesive or galling wear, while that at junction b is burnishing wear. Wear tracks in the two cases are rough or smooth, respectively. If the transferred wear particle from junction a breaks loose, it might act as an abrasive at the interface and lead to a third-wear mechanism, abrasive wear as covered in Antler.[6] The same process, separation of cold-welded junctions under a shear stress, is the source of friction. The friction force, in fact, is the force necessary to shear the multiple cold welds, which occur at contact interfaces. Friction forces at the separable interface are a major factor in determining connector mating force, as is covered in Section 8.6.

8.2.3 Summary

The asperity model of contact interfaces provides insight into both the electrical and mechanical characteristics of contact interfaces. Contact interface morphology depends, simply speaking, on the surface roughness, the force on the interface and the geometry of the surfaces in contact. Surface roughness strongly influences the number of asperity contacts created. The force on the interface, the contact normal force, and the hardness of the surface determine the total contact area. As mentioned, it is a composite hardness of the contact finish and the spring material that determines the effective hardness. The geometries of the contact springs determines the area over which the asperities are distributed. This explains why the contact finish, contact normal force, and geometry are major design parameters, and each is covered in more detail—contact finishes in Section 8.3 and contact force and geometry in Section 8.6.

8.3 THE CONTACT FINISH

As mentioned earlier, the two major reasons for using a contact finish are: corrosion protection for the base metal of the contact spring and optimization of the properties of the contact surface with respect to establishing and maintaining the contact interface, in particular, a metallic contact interface.

Section 8.2 covered the structure of the contact interface and its relationship to constriction resistance and mechanical durability. Surface films and film resistance were briefly mentioned. But film management, during mating as well as during use, is arguably the most critical factor in connector performance. The selection of the contact finish is one of the basic means of controlling film effects.

In the great majority of connectors the contact finish is applied by electroplating. The electroplating process is not covered here. Basic references relevant to connector electroplating include Durney[7] and Reid and Goldie.[8] Clad or inlay finishes see limited use in connectors. An overview of cladding processes relevant to connectors is available in Harlan.[9]

8.3.1 Contact Finish and Corrosion Protection

In most cases, connector contact springs are made from copper alloys that are subject to corrosion in typical connector operating environments—through oxidation or sulfidation, for example. Application of a contact finish, in effect, seals off the contact spring from the environment and prevents corrosion. For this function, the contact finish itself must be corrosion resistant, either intrinsically (noble metal finishes) or by forming passivating surface films (tin and nickel).

8.3.2 Contact Finish and Interface Optimization

Optimization of the contact interface in this context is film management. There are two approaches to film management, first, avoiding film formation and, second, disruption of the films, which might exist during the mating of the connector. These two approaches define the difference between noble, or precious metal, and non-noble finishes.

Noble finishes—gold, palladium and some alloys of these metals—are intrinsically free, or relatively free, of surface films. Under such conditions, metallic contact is straightforward, at least initially. The potential for film formation at other points in the connector system, in particular exposed contact spring surfaces, must be given consideration, however. Factors that must be considered include maintaining the nobility of the contact surface against extrinsic factors such as contamination, base metal diffusion, and contact wear. Nickel underplates are important in providing such protection.

Non-noble finishes, in particular tin, are covered with a surface film, usually an oxide. The utility of tin contact finishes derives from the fact that the oxide film is easily disrupted on mating and metallic contact areas are readily established. The application concern with tin finishes is reoxidation of the tin, in the form of fretting corrosion, the major degradation mechanism for tin contact finishes.

Silver is best considered as non-noble because it is subject to sulfide and chloride corrosion. Nickel, too, is non-noble because it readily forms a surface oxide.

The difference between noble and non-noble finishes is reflected in a number of connector design considerations.

8.3.3 Noble Finish Overview

Noble metal contact finishes are systems in which each component performs multiple functions. A noble metal contact finish system consists of a noble metal surface, usually gold, over an underplate, usually nickel, over the contact spring material, generally a copper alloy. Typical thicknesses of the platings are 0.4 to 0.8 μm for the gold, or in some cases gold over palladium alloy, and 1.25 to 2.5 μm for the nickel underplate.

The function of the noble metal surface is to provide a film free metallic contact surface. But the presence of a precious metal surface does not in itself guarantee a film free surface. The finish must be continuous. Discontinuities in the finish, such as porosity and scratches, can result in corrosion at sites where base metal is exposed. Wear-through of the noble surface metal also serves to expose base metal. Wear-through of the finish can occur through mating of the connector or mechanically or thermally induced relative motions of the contacts during application. Such small-scale motions, of the order of a few to a few tens of micrometers, are referred to as fretting. The importance of fretting action is that it can lead to contact resistance degradation under a variety of application conditions.

Diffusion of the base metal consituents of the contact spring to the surface can also result in surface films. Each of these potential degradation mechanisms is moderated by the nickel underplate.

Noble finishes and the nickel underplate For a detailed discussion of the functions of a nickel underplate, see Mroczkowski.[10] At this point, a summary of the benefits of such an underplate is sufficient.

1. Nickel, through the formation of a passive oxide surface, seals off the base of pore sites and surface scratches reducing the potential for corrosion of the underlying copper alloy contact spring.
2. Nickel provides an effective barrier against the diffusion of base metal consititents to the contact surface where they could result in films.[11]
3. Nickel provides a hard supporting layer under the noble surface that improves contact durability.[6]
4. Nickel provides a barrier against the migration of base metal corrosion products reducing the potential of their contaminating the contact interface.[10]

The first three benefits allow for equivalent or improved performance at reduced noble metal thicknesses. The effects of discontinuities have been moderated, the noble metal is not used as a diffusion barrier and the durability is improved. The fourth benefit allows a reduction in the size of the contact area which must be covered with the noble metal finish. All these functions serve to maintain the nobility of the contact surface finish at a reduced cost.

The most common noble metal contact finishes are gold, palladium and alloys of these metals. These finishes vary in their degree of "nobility" and they are considered separately.

Gold Gold provides an ideal contact finish because of excellent electrical and thermal characteristics and corrosion resistance in virtually all environments. Pure gold, however, is relatively soft and gold contact platings are usually alloyed, with cobalt being the most common, at levels of a few tenths of a percent to increase the hardness and wear resistance. A typical hardness for soft gold is in the range of 90 $Knoop_{25}$, and cobalt-hardened gold contact finishes show hardnesses in the range of 130 to 200 $Knoop_{25}$. Because of this combination of characteristics, gold is the dominant contact finish for connectors which must provide high reliability in demanding applications.

Cost reduction objectives have led to the development of selective plating practices (to reduce the required area which must be plated) and reductions in thickness (permitted by the use of nickel underplates) to reduce gold finish costs. Further cost reductions can be realized by the use of alternative noble metals, most commonly gold flashed palladium or palladium alloy.

Palladium Palladium is also a noble metal, but does not equal gold in corrosion resistance or electrical/thermal conductivity. It is, however, significantly harder than gold (200 to 400 $Knoop_{25}$), even without alloying, which improves durability performance. Palladium can catalyze the polymerization of organic deposits and result in contact resistance degradation under fretting motions. Palladium is, therefore, not as noble as gold although the effect of these factors on connector performance depends on the operating environment, as covered by Antler and Sproles.[12] Whitley, Wei, and Krumbein[13] state that, in most applications, palladium is used with a gold flash (of the order of 0.1 μm) to provide a gold contact interface.

Noble metal alloys The performance of gold alloys in connector applications has been mixed. The major problems have occurred because of the loss of corrosion resistance, which often accompanies alloying using base metals. When this factor is accounted for, or avoided by selection of appropriate alloying agents, satisfactory performance can be realized in some connector applications. The major gold alloy used in connectors is WE1, a gold/silver/platinum (69/25/6) alloy. WE1 is available only in inlay form and sees limited use.

There are two palladium alloys in use, palladium(80)-nickel(20) and palladium(60)-silver(40) as reviewed by Whitelaw[14] and Antler, Drozdovich, and Haque,[15] respectively. The palladium-nickel alloy is electroplated and the palladium-silver finish is primarily an inlay. In general these finishes include a gold flash to counter the lower corrosion resistance of these alloys compared to gold.

8.3.4 Non-Noble Finishes

Non-noble contact finishes differ from noble finishes in that they always have a surface film. The utility of non-noble finishes depends on the ease with which these films can be disrupted during connector mating and the potential for recurrence of the films during the application lifetime of the connector.

Three non-noble contact finishes, tin (in this discussion "tin" includes tin-lead alloys, most commonly 93/7 tin/lead for mechanical connections and 60/40 tin/lead for soldered connections), silver and nickel are used in connectors. Tin is the most commonly used non-noble finish; silver offers advantages for high-current contacts, and nickel is used in high-temperature applications.

Tin The utility of tin as a contact finish is a result of the ease which the tin oxide surface is disrupted and displaced. The mechanism of disruption is schematically illustrated in Figure 8.7.

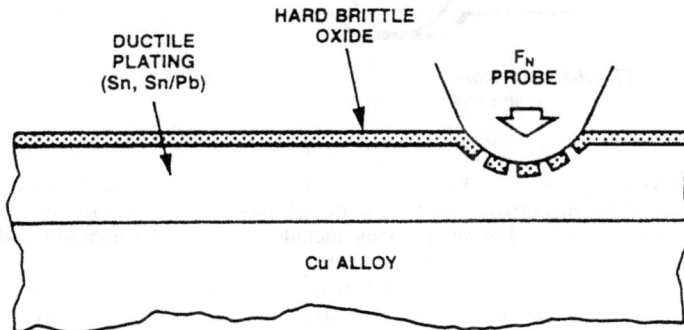

FIGURE 8.7 Schematic illustration of the mechanics of tin oxide disruption under an applied force. (*Courtesy AMP Incorporated.*)

The oxide disruption is facilitated by the large disparity in mechanical properties between the tin and tin oxide. The tin oxide is thin (a few tens of nanometers), hard and brittle. In contrast the underlying tin, generally in the range of 2 to 5 μm in thickness is soft and ductile. This combination of properties results in cracks in the oxide under an applied load. The load transfers to the soft ductile tin which flows easily, opening the cracks in the oxide, and tin then extrudes through the cracks to form the desired metallic contact regions.

Generation of a tin contact interface for tin finishes is, therefore, relatively simple. Normal force alone might be sufficient, but the wiping action that occurs on mating of connectors virtually ensures oxide disruption and creation of a metallic interface. The potential problem with tin, however, is the tendency for reoxidation of the tin at the contact interface if it is disturbed. This process is referred to as *fretting corrosion*, as covered by Bock and Whitley.[16] As mentioned previously, *fretting* refers to repetitive small-scale motions, of the order of a few to a few tens of micrometers. The "corrosion" aspect of fretting corrosion is the reoxidation of the surface as the tin surface is repeatedly exposed, as schematically illustrated in Figure 8.8. The end result is a buildup of oxide debris at the contact interface leading to an increase in contact resistance.

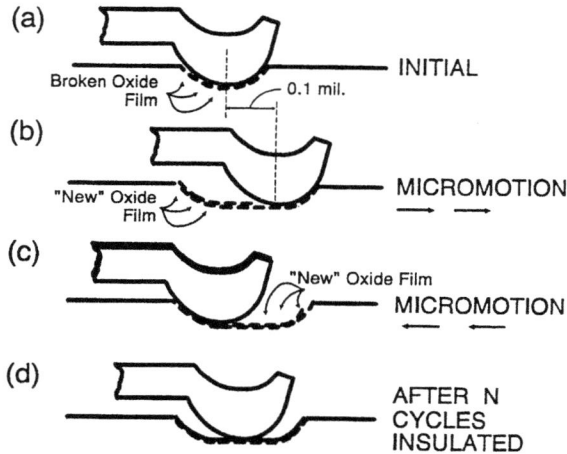

FIGURE 8.8   Schematic illustration of the mechanism of fretting corrosion. (*Courtesy AMP Incorporated.*)

Under severe conditions, fretting corrosion can lead to open circuits in a few hundred fretting cycles. Fretting corrosion is the major degradation mechanism for tin contact finishes. Driving forces for fretting motions include mechanical (vibration and disturbances/shock) and thermal expansion mismatch stresses.

Two approaches to mitigating fretting corrosion in tin and tin alloy finishes are high normal force (to reduce the potential for motion) and contact lubricants (to prevent oxidation). Each has been used successfully, and each has its limitations. High normal forces limit the durability capability of the finish, which is already low because of tin being very soft, and result in increased mating forces, which limit the realizable pin count in tin-finished connectors. Contact lubricants also have limitations. They require secondary operations for application, either during connector manufacture or in assembly, have limited temperature capability, and might also result in dust retention.

Tin contact finishes, however, can provide acceptable electrical performance at both low (millivolts and milliamps) and high (volts and amps) circuit conditions if fretting corrosion can be avoided. Unfortunately, the susceptibility of a connector to fretting is application dependent and often difficult to assess.

Silver   Silver is considered a non-noble contact finish because of its reactivity with sulfur and chlorine. Silver films differ in their effects on connector performance. Silver sulfide films tend to be soft and readily disrupted. In addition, silver sulfide does not result in fretting corrosion. Silver chloride films, however, are harder, more adherent, and more likely to have detrimental effects on contact performance. In addition, silver is susceptible to electromigration, which can be a problem in some applications. The end result is that silver sees limited use in connectors. Despite these limitations, silver is a candidate for high-current contacts because of its high electrical and thermal conductivity and resistance to welding. Silver finish thicknesses are generally in the range of 2 to 4 μm.

Nickel In the discussion of nickel underplates, it was noted that nickel forms a passivating oxide film that reduces its susceptibility to further corrosion. This passive film also has a significant effect on the contact resistance of nickel. In this case, the favorable disparity in properties that gives tin its utility is not present. Both the nickel oxide and the base nickel are hard. Under these conditions, cracking and separation of the oxide and extrusion of the nickel to the surface are more difficult to realize. For this reason, nickel finishes require higher contact normal forces to ensure film disruption. The self-limiting oxide on nickel, however, makes it a candidate for high-temperature applications. It should be noted that nickel contact finishes are susceptible to fretting corrosion by the same mechanism as tin finishes. This potential should be taken into account in connectors using thin noble metal finishes. Wear-through of the finish because of mating cycles or fretting action, which exposes the nickel underplate, can result in fretting corrosion.

8.3.5 Selection of a Contact Finish

Selection of an appropriate contact finish depends on the application requirements the connector must meet. Of particular concern are the required number of mating cycles, the operating environment, and the electrical requirements.

Mating cycles The number of mating cycles a connector must support varies from a few to several thousand. Level-3 connectors generally require a low number of mating cycles, and level-6 requirements can run into the thousands.

The mating durability of a connector depends on the contact finish, the normal force, and the contact geometry. With respect to contact finishes, the dominant factor is the hardness of the finish. From this viewpoint, the relative durability performance of typical finishes is summarized in Table 8.1. It should be noted that the durability performance of a contact finish depends very strongly on the contact geometry, normal force, surface roughness, and state of lubrication of the interface. The statements in Table 8.1 are based on results from flat test coupons traversed by a spherical rider. Gold-flashed palladium or

TABLE 8.1 Selected Mechanical Characteristics of Typical Contact Finishes

	Hardness (Knoop@25g)	Durability	Coefficient of friction
Gold (soft)	90	Fair	0.4/0.7
(hard)	130/200	Good	0.3/0.5
Palladium	200/300	Very Good*	0.3/0.5
Palladium (80)–Nickel (20)	300/450	Very Good*	0.2/0.5
Tin	9/20	Poor	0.5/1.0
Silver	80/120	Fair	0.5/0.8
Nickel	300/500	Very Good	0.3/0.6

*with gold flash

palladium alloy finishes provide the highest durability capability. Silver and tin finishes are severely limited in the number of mating cycles they can support. Contact lubricants are often used to improve the mating cycle capability of noble finishes. An appropriate contact lubricant can improve durability capability by an order of magnitude under favorable conditions.

In addition, durability depends on the finish thickness. The relationship between mating cycles and thickness is roughly linear for a given contact force/geometry configuration.

The effect of normal force on mating life has two aspects. In general, wear rates increase with mating force. In addition, the wear mechanism might change from burnishing to adhesive wear as normal force increases as covered in Antler.[17,18]

The effect of contact geometry on mating cycles results from localization of the contact areas. High curvatures result in narrow contact wear tracks and increased wear for a given value of normal force.

Operating environment   Operating temperature and corrosion severity will obviously impact finish selection. Corrosion of contact surfaces, in particular for noble metal contact systems, has received a great deal of attention in recent years in terms of both the composition of the environments and the corrosion mechanisms in those environments. For example, see Bader, Sharma and Feder,[19] Abbott,[20,21] Mroczkowski[22] and Geckle, and Mroczkowski.[23] These workers have shown that chlorine and sulfur are the major contributors to corrosion mechanisms in noble metal plated connectors. Under test conditions intended to simulate typical connector operating environments,[21] the corrosion resistance of noble metal finishes decreases in the following order: gold, palladium, and palladium(80)-nickel(20) alloy.[22] The importance of these differences will depend on the severity of the operating environment. As mentioned, the nickel underplate used in noble contact finishes reduces the corrosion susceptibility of the connector. In addition, shielding of the contact interface by the connector housing reduces corrosion by limiting access of the environment to the contact interface.

Somewhat surprisingly, tin finishes show good stability with respect to corrosion. This is because of the passivating surface oxide, mentioned previously, which protects the surface from further corrosion in most operating evironments. Fretting corrosion, however, must always be taken into consideration for tin finishes.

Temperature limitations for noble metal finishes show a similar pattern to corrosion resistance. In both cases, the order is determined by the presence of the non-noble constituent in the palladium alloys, because it is the alloying agents that are susceptible to corrosion. Hard gold finishes, even though the alloy content is of the order of tenths of a percent, are also subject to oxidation as operating temperatures increase. In general, soft, or pure, gold finishes are recommended for temperatures above 125°C.

Tin has a temperature limitation because of an increasing rate of intermetallic compound formation, a reduction in the already low mechanical strength and an enhanced oxidation rate. The interaction of these factors results in a recommendation that tin not be used above 100°C in conventional connector designs.

Thermal cycling is another important environmental consideration and is arguably the major driving force for fretting corrosion of tin systems. In addition, thermal cycling accelerates the effects of humidity on connector degradation.

Electrical requirements   In general operating voltages for electronic connectors are decreasing, but the opposite is true for currents. Although metallic contact interface performance is not dependent on circuit voltage or current, corrosion/film effects on contact resistance do show such a dependence. Unfortunately current/voltage effects on interfaces containing films are highly unstable. The voltage necessary for film breakdown is variable, being dependent on the film structure and thickness, as covered by Wagar.[24] For this reason relying on voltage, especially low voltages, to effect film breakdown is suspect. In addition, the resistance realized on dielectric breakdown might depend on the current flowing in the circuit at the time. Such variability is not acceptable in electronic connectors, so mechanical disruption of surface films and avoiding corrosion during applications are critical to connector performance.

In low-current applications, non-noble contact systems are more susceptible to noise because of interface films and variations in contact resistance as shown by Abbott and Schrieber.[25] In high-current, high-power applications where low interface resistance must be maintained, film resistance susceptibility must be minimized to avoid thermal runaway because of Joule heating as resistance increases.

In principle, because both noble and non-noble interfaces can be metallic, both systems are capable of equivalent performance. In practice, however, the greater susceptibility of the non-noble systems to film formation result in a reliability risk factor.

8.3.6 Summary

Selection of a contact finish depends on a wide range of application requirements, primarily durability and resistance stability. Precious metal finishes provide greater inherent stability, but tin finishes might be acceptable if the conditions necessary to reduce fretting susceptibility can be satisfied. Table 8.1 provides a brief compilation of contact finish characteristics of relevance to connector performance. Antler[26] provides a recent summary of contact finish practices and trends.

8.4 CONTACT SPRINGS

The contact spring performs two separate functions in a connector:

1. As a mechanical element, it provides the contact force that produces and maintains the separable contact interface and allows for the creation of the permanent connection interface.
2. As an electrical element, it carries the signal, or power, from the separable interface to the permanent interface and then to the circuit element to which it is connected.

This section is directed primarily to the first function, but a few words on the second are in order.

8.4.1 Contact Spring and Electrical Requirements

The resistance of a connector consists of three components (Figure 8.9).

R_{CONN}, is the resistance introduced at the permanent connection, whether it is soldered or mechanical. The bulk resistance, R_B, is that introduced by the contact spring and depends on the spring material resistivity and the geometry of the spring. The third component, the resistance of the separable interface, is commonly referred to as the contact resistance, R_C. These three components are present in both halves of the connector, the separable interface resistance being shared in common. For a typical connector the values of these resistances are of the order of tens/hundreds of microhms, several milliohms and a milliohm, respectively.

The effect of the contact spring is largest in the bulk resistance, the largest component of connector resistance. The resistivity of the contact spring material, therefore, is an im-

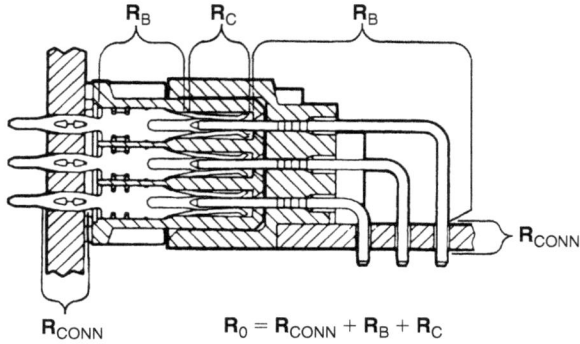

FIGURE 8.9   Schematic connector cross section illustrating the sources of connector resistance. Permanent connection, R_{CONN}, bulk, R_B and contact, R_C, resistance contributions are indicated. (*Courtesy AMP Incorporated.*)

portant property. This is particularly relevant for power applications. In such cases, millivolt drop budgets in power distribution can place severe limitations on bulk resistance. In addition, Joule heating, which depends primarily on bulk resistance, and the associated temperature rise can limit the current rating and application temperature of connectors. In signal applications the bulk resistance is of secondary importance in most cases. In such applications stability of the resistance is more important than the initial value.

The range of electrical resistivities of the copper alloys most commonly used in connectors is in the range of four to five times the resistivity of copper. For power contacts higher conductivity materials such as C19500, C19500, and C17410 are often selected to minimize bulk resistance contributions. Table 8.2 contains a compilation of electrical and mechanical properties of copper alloys relevant to connectors.

8.4.2 Contact Springs and Mechanical Requirements

In contrast to bulk resistance, which is variable only with respect to material selection for a given spring geometry, mechanical design considerations are much more complicated and involve a number of trade-offs. The important factors and trade-offs are somewhat different for separable and permanent interfaces.

TABLE 8.2   Selected Properties of Typical Contact Spring Materials

	Young's modulus, (E) (10^6kg/mm^2)	Electrical conductivity (%IACS)	0.2% Offset yield strength (10^3kg/mm^2)	Stress relaxation
Brass (C26000)	11.2	28	40/60	Poor
Phosphor Bronze (C51000)	11.2	20	50/70	Good
Beryllium-Copper (C17200)	13.3	20/26	55/95	Excellent

Separable interfaces For separable interfaces, the main function of a contact spring is to provide the required contact normal force, the force between the two mating surfaces. The material properties which are important in this context are the elastic or Young's Modulus and yield strength as these properties strongly influence the deflection characteristics of the spring and the amount of deflection which can be supported while remaining elastic. Stress relaxation resistance is also important because it can reduce the contact normal force over time. A brief treatment of these variables and their effect on contact normal force is in order.

A simple cantilever beam can be used to indicate how variations in materials properties affect the contact normal force. For a cantilever beam, Figure 8.10, the force versus deflection equation takes the form:

$$F = \frac{D}{4} E W \left(\frac{T}{L}\right)^3 \tag{8.4}$$

where F is the force resulting from a beam deflection D, E is the elastic modulus of the spring material, and W, L, and T represent the width, length, and thickness of the beam, respectively. Many contact spring configurations approximate this cantilever geometry.

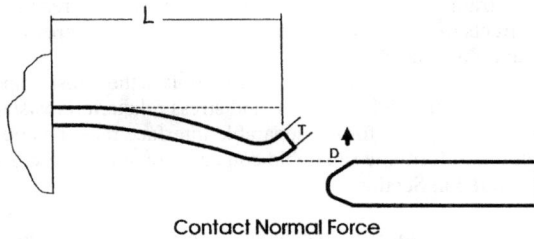

Contact Normal Force

$$F = \left(\frac{D}{4}\right) E \; W \left(\frac{T}{L}\right)^3$$

$$F = \left(\frac{\sigma}{6}\right) W \; T/L^2$$

FIGURE 8.10 Schematic illustration of a cantilever beam. Deflection of the cantilever beam provides the contact force which creates and maintains the contact interface. (*Courtesy AMP Incorporated.*)

The only materials property in Equation 8.4 is the elastic modulus. The contact force for a given deflection of the beam increases as E increases. For this reason, a high value of E is generally desirable because it provides a higher normal force for a given deflection. Once again, however, for the copper alloys most often used in connectors, E falls in a relatively narrow range of 11 to 14 gigapascals, as indicated in Table 8.2.

A second equation which is useful in illustrating the effects of material properties relates the contact force to the stress in the beam:

$$F = \frac{1}{6} \sigma T^2 \left(\frac{W}{L}\right) \tag{8.5}$$

This equation shows that the contact force, F, depends on the stress, σ, in the beam. This relationship is particularly important because the maximum normal force, which can be realized by a contact spring while remaining elastic, is determined by the maximum stress the spring can support which is the yield strength of the spring material.

Equations 8.4 and 8.5 show that the material properties having the greatest impact on contact force performance are the elastic modulus, which, in combination with the contact dimensions, determines the spring rate, and the yield strength, which determines the maximum normal force which can be achieved under elastic loading of the beam. The variation in yield strength capability among the copper alloys commonly used in connectors is significant as indicated in Table 8.2. It is also important to note that the formability of copper alloys in the higher strength tempers used in connectors also varies significantly among the alloys. Equations 8.4 and 8.5 also indicate the importance of the contact geometry as all three dimensions of the spring enter into the equations for calculating contact force. Loewenthal et al.[27] provide a detailed analysis of a contact spring design in terms of materials properties. The range of contact spring designs in connectors, for both separable and permanent connections is very large. A few examples are included in Section 8.6.

Another important materials property relevant to contact force is stress relaxation. The effect of stress relaxation is apparent in Equation 8.5. Stress relaxation causes a reduction of the stress in the beam under load as a function of time and temperature because of the conversion of elastic strain to plastic strain. The reduction in stress results in a reduction in normal force. The effects of stress relaxation are materials and temperature dependent, as discussed by Horn and Zarlingo.[28]

Thus, from a contact spring/normal force viewpoint, the most important materials properties are elastic modulus, yield strength, and stress relaxation resistance.

The spring properties required for the separable interface must be counterbalanced by the formability required to form the separable contact geometry. These geometries can be quite complex, as covered in Section 8.6.

Permanent connections Creation of mechanical permanent connections involves the same materials properties, but in a different context. In some cases, such as crimped connections, good formability is required to support the extensive plastic deformations which occur in the crimping process. For others, for example IDC, spring characteristics are important to create the high contact forces needed to ensure connection resistance stability. The various types of permanent connections are covered in Section 8.7.

8.4.3 Contact Spring Material Selection

Selection of a material for a contact spring requires consideration of both ends of the contact, the permanent connection and the separable interface. The requirements on each are related, but different. A listing of materials properties of three basic families of contact spring materials is provided in Table 8.2. In each case, a number of alloys in the family can be used, the data given are for the most commonly used composition. Copper alloy C26000, cartridge brass, is used for low-cost and commercial applications. Alloy C51000, a typical phosphor bronze, is a general purpose alloy used in a large variety of contact springs. C172000, beryllium copper, is used when stress relaxation resistance becomes very important. Bersett[29] and Spiegelberg[30] provide additional discussions and properties of contact spring materials.

In Table 8.2, the tensile strength ranges listed are those typically used in connectors and represent a trade-off between spring performance and formability in manufacturing. A few brief comments on each alloy are in order.

Brass Cartridge brass (70% copper-30% zinc) is a low-cost material accounting for its wide usage in consumer applications. Brass has adequate spring properties for room temperature applications and more robust contact designs. Although the electrical conductivity of brass is quite good, it has poor stress relaxation resistance, which limits its application temperature to the order of 75°C in most cases.

Phosphor bronze A number of phosphor bronze alloys are used in connectors with C51000 (4% tin balance copper) being the most common. Phosphor bronzes are used in a wide range of contact spring designs because of a very good combination of formability, electrical conductivity, and spring performance, along with reasonably good stress relaxation resistance at a moderate cost.

Beryllium copper The use of beryllium copper is limited by its cost, although lower-cost alloys have recently reached the market. Where high strength (miniaturized contacts) and high stress relaxation resistance (elevated temperature applications) are necessary, however, beryllium copper is often the material of choice.

Other alloys There are, of course, many other alloys used in connectors. One connector application that might drive selection of alloys, other than those already covered, is power contacts/connectors. In such applications, the electrical conductivity becomes an important consideration and high-conductivity alloys, such as C19400 (copper-2.4% iron) and C19500 (copper-1.5% iron-0.6% tin), are more suitable. These alloys have conductivities of 65 and 50 percent of the conductivity of copper respectively and result in lower bulk resistance and Joule heating in high-current applications. C70250 (copper-3.0 nickel%-0.75% silicon) is another stress relaxation resistant alloy seeing increased use in higher temperature applications such as automotive connectors.

8.4.4 Summary

Selection of an appropriate contact spring material for a given connector application requires particular attention to the operating temperature and power requirements, which the connector must satisfy. Ensuring the generation and stability of an adequate contact force and sufficient formability to form the separable and permanent contact interfaces are the primary trade-offs in material selection.

8.5 CONNECTOR HOUSINGS

A connector housing has obvious electrical and mechanical functions, but it also can provide environmental benefits. The electrical and mechanical performance of a connector housing depends on the polymer from which the housing is molded, in most cases thermoplastic polymers. The environmental benefit depends primarily on the design of the housing. Consider each separately.

8.5.1 Electrical

The electrical function of a connector housing is to insulate the individual contact springs from one another. The surface and volume resistivities of the polymer, as well as its dielectric breakdown voltage, are the major properties of interest for this function. Most engineering thermoplastics have electrical properties that readily meet typical connector requirements. Polymer materials do differ, however, in the stability of these properties with exposure to temperature, humidity, and chemical exposures. Electrical stability is one selection parameter for polymer materials.

8.5.2 Mechanical

There are two types of mechanical requirements that a connector housing must satisfy. The first relates to support of the contact spring within the housing and the second to dimensional control of the housing.

Mechanical support   In most cases, the contact spring is latched into the housing which must therefore support the spring deflection to varying degrees depending on the housing and spring designs. Polymers differ significantly in their mechanical properties and, in particular, in the variation of those properties with temperature. Although stress relaxation is an important consideration in contact springs, the corresponding degradation mechanism in polymers is creep. Creep also results from the conversion of elastic to plastic strain. During creep the material flows away from the point of stress as a function of time and temperature. Creep can result in a loss in contact force or a displacement of the contact from its intended position in the housing.

Another aspect of mechanical stability relates to the handling of the connector during assembly and application. In many cases, the connector housing contains guiding features, to reduce the potential for abusive mating, and latches, to ensure proper engagement of the two halves of the connector and eliminate undesired connector separation in high-vibration environments.

Dimensional control   Dimensional control is important in maintaining the proper spacing of the contact springs at both the separable interface and permanent connection. Variations in dimensions at the separable interface can have a significant impact on the connector mating force. Variations at the permanent connection end can result in an inability to make the required permanent connection, in particular in soldered connections to PWB or IDC connections to small centerline ribbon cable. Dimensional stability depends on the molding characteristics of the polymer, which affect shrinkage and bow/warp of the housing, and the thermal expansion and moisture absorption characteristics of the polymer. The assembly process which the housing must support also affects dimensional stability. Surface-mount (SM) soldering requirements are particularly demanding with respect to temperature stability.

8.5.3 Environmental shielding

Although materials properties are critical to ensuring the required electrical and mechanical performance, housing design determines the effectiveness of environmental shielding. Figure 8.11 demonstrates the shielding provided by the housing against a corrosive environment. The coupon shown is silver plated and was exposed while mated to the card-edge connector shown, to an operating environment containing sulfur, which is highly reactive with silver. The corrosion of the silver is apparent on the exposed section of the coupon as is the shielding provided where the coupon is within the housing. Figure 8.12 demonstrates the effect of shielding on connector resistance of connectors exposed to a mixed flowing gas environment intended to simulate an industrial environment. The test envi-

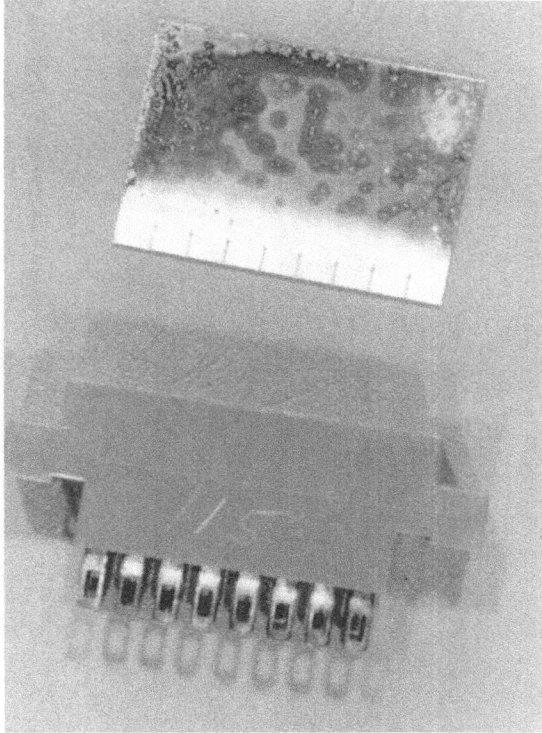

FIGURE 8.11 Photograph of a silver-plated coupon exposed to an environment containing sulfur and chlorine while mated to the card-edge connector shown. (*Courtesy AMP Incorporated.*)

ronment contains parts per billion (ppb) levels of chlorine, hydrogen sulfide and nitrogen oxide. The connector was a two-piece post-receptacle design and was exposed to the test environment in both mated and unmated conditions. In addition durability cycling of the connectors was performed before exposure. Data for both gold and tin finished connectors are shown. Note the significant increases in contact resistance for the gold finished connectors exposed unmated and the stability of the resistance of those exposed in the mated condition. The effectiveness of the housing in providing environmental shielding is apparent. Note also the stability of tin finished connectors in both mated and unmated conditions. Tin is non-reactive in this environment because of the protective nature of the tin oxide. It must be noted, however, that this exposure does not stimulate the dominant tin degradation mechanism, fretting corrosion.

8.5.4 Application requirements

Two types of application requirements must be considered: those during assembly of the product and the functional requirements during use. Assembly processes, in particular soldering, might dictate the selection of the polymer. This is especially true for connectors

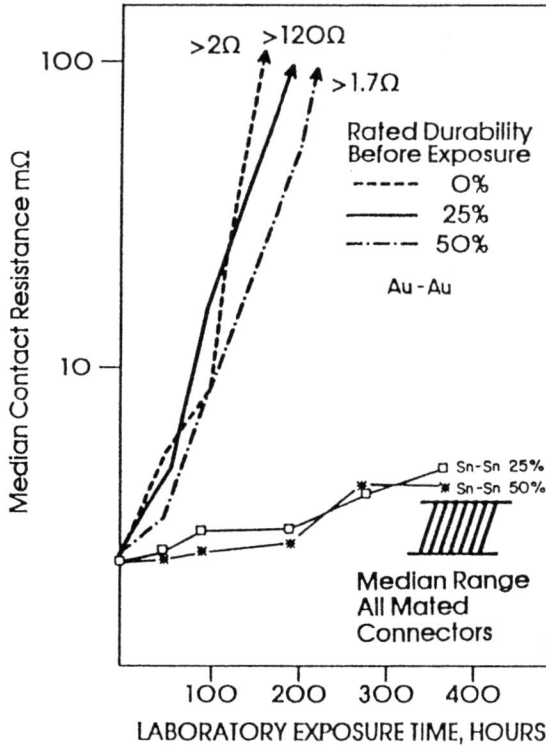

FIGURE 8.12 Connector resistance versus exposure time for tin and gold finished 25 square post connector system exposed mated and unmated to a test environment which simulates an industrial environment. The effect of housing shielding in attenuating the environment is apparent. (*Courtesy AMP Incorporated.*)

that must be surface mounted. Temperature stability against both warping and degradation is required. In use requirements must also be considered but, in most cases, are less demanding than assembly requirements. The application temperature might, however, limit the materials options because of polymer creep and its associated effects on mechanical support and dimensional changes. Moisture absorption and sensitivity to humidity and solvents also can be important.

Additional mechanical requirements that could arise involve latches/locks, hinges, and snap-fit features. These might be necessary in assembly of the connector or attaching it to its intended circuitry.

8.5.5 Material Selection

Selection of an appropriate connector housing material, as for a contact spring material, depends on a number of property/performance trade-offs. Before specific materials and trade-offs are considered, a brief discussion of polymer structure as it affects connector performance is in order. Alvino[31] provides a general discussion of plastics for electronic applications.

Polymer structure The word *polymer* means "many mers," where a "mer" is the basic structural unit(s) from which the polymer is synthesized, and is descriptive of polymer structure. In general, polymers consist of long chains of carbon-carbon bonds with a variety of side groups distributed along the carbon-carbon backbone. The polymer type determines the side groups and the length of the carbon backbone chains, the molecular weight, determines the processing and mechanical properties of the polymer.

There are two classes of polymer materials: thermosets and thermoplastics. As the names imply, thermosetting polymers take a permanent set during molding (because of cross-linking or bonding between the chains) and, therefore, cannot be reprocessed. Thermoplastic polymers, which exhibit little if any cross-linking, become plastic (can flow under pressure) at increasing temperatures, which allows them to be readily injection molded and reprocessed. It should be noted, however, that the material for reprocessing (the regrind, sprues, and runners from the injection molding, for example) has lower properties, because of a reduction in molecular weight, than the virgin material. This effect might limit the amount of regrind that can be used, depending on any reprocessing of the regrind and the application requirements. Thermoplastic polymers dominate in connector applications because of their processing advantages and cost effectiveness.

There are two basic classes of thermoplastic polymers: amorphous and crystalline. Once again, the names are descriptive. Amorphous polymers show little internal ordering of the carbon chains; in crystalline polymers the long chain molecules are aligned to a significant degree. These structural differences give rise to chemical, mechanical, and thermal differences between the two types. In general, amorphous polymers are isotropic, which provides advantages in molding and dimensional stability. These positive traits are counterbalanced by greater thermal sensitivity, lower strength and, generally, lower chemical stability than crystalline polymers.

The alignment of the polymer chains provides enhanced mechanical strength, thermal stability, and solvent resistance to crystalline polymers. The alignment also leads to anisotropy, which is reflected in molding stresses that might result in dimensional instability.

The performance of both amorphous and crystalline polymers is generally enhanced by a variety of additives. Such additives include flame retardants, reinforcing agents (primarily glass fibers) for mechanical strength, lubricants for molding ease, and other additives to enhance thermal stability.

When all the trade-offs are taken into account, crystalline thermoplastics dominate the materials used in connector housings. With this limited background, a few of the more commonly used connector housing materials are covered briefly.

Table 8.3 provides a brief compilation of properties of interest in connectors for a small sampling of polymers commonly used in connector housings. A brief discussion of these materials providing general comments on their suitability for connector applications follows. Temperature capabilities, both application temperature and processing, in particular surface mounting technology (SMT) capability, are particularly important in material selection.

Polyamides (Nylon) Although there are many types of nylon, at one time 6/6 nylon was a dominant connector housing material. 6/6 nylon possesses a good balance of mechanical, electrical, and thermal characteristics, good chemical resistance, and good processibility. It has, however, one severe limitation, a tendency to absorb moisture. Moisture absorption has a negative effect on the mechanical properties of nylon and a significantly negative effect on its dimensional and electrical stability. This limitation, in combination with the development of a wide variety of alternative materials, including other nylons such as 4/6, has diminished the range of applications for 6/6 nylon as a connector mater-

TABLE 8.3 Selected Properties of Typical Connector Housing Materials

	Flexural modulus (10^6kg/mm^2)	Heat deflection temperature @264 psi (°F)	UL temperature index (°C)	Dielectric strength (v/m)
Polyamide (6/6)	0.7	466	130	17.4
Polybutylene Terephthalate	0.8	400	130/140	24.4
Polyethylene Terephthalate	1.0	435	150	26.0
Polycyclohexane Terephalate	0.9	480	130	25.4
Polyphenylene Sulfide	1.2	500	200/230	18.0
Liquid Crystal Polymers	1.5	650	220/240	38.0

ial. 4/6 nylon has a lower sensitivity to moisture and higher temperature capability—a heat deflection temperature (HDT) of 543°F—which makes it SMT compatible.

Polybutylene terepthalate (PBT) PBT is arguably the most commonly used connector housing material because of its combination of processibility and functional/performance characteristics. PBT is a crystalline material and possesses good electrical and mechanical properties, dimensional and chemical stability, good solvent resistance, and generally acceptable temperature capability. Temperature capability, however, does limit the suitability of PBT for SMT applications (an HDT of 400°F). PBT is generally used in glass reinforced grades, (20 to 40%) for improved mechanical properties.

Polyethylene terepthalate (PET) PET shares the structural and performance characteristics of PBT with an improvement in temperature capability. On the negative side, the creep and warpage of PET are greater than PBT. Creep under load is a concern in connectors in which the contact spring is preloaded against the housing. Once again, glass reinforced grades are used to improve the mechanical properties. PET resins are generally reinforced with 30 to 50 percent glass. The HDT of PET (435°F) is marginal for SMT.

Polycyclohexylenedimethylene terepthalate (PCT) PCT shares the generally good electrical and mechanical characteristics of PBT and PET and has higher temperature capability (HDT 480°F), which makes it suitable for SMT applications.

Polyphenylene sulfide (PPS) PPS is also crystalline. It has high-temperature capabilities (HDT 500°F and a continuous-use temperature of 210°C). The mechanical properties of PPS are also very good, particularly with respect to stiffness. The negative side of this strength, however, is a tendency toward brittleness.

Liquid crystal polymers (LCPs) LCPs are relatively new materials, available in a number of compositions, which are characterized by a highly aligned rod-like structure which is maintained even in the melt state (the "liquid" for a polymer). The highly crystalline structure results in high stiffness, mechanical strength, and electrical properties that are maintained at elevated temperatures, with UL temperature index values in excess of 220°C. LCPs are SMT compatible with HDT values above 600°F.

8.5.6 Summary

The process of selection of a polymer for a connector housing is complicated by the variety of materials available and the rapid rate of introduction of new materials. The electrical characteristics of engineering polymers are far in excess of typical connector requirements in most cases. Mechanical strength and dimensional stability as well as application factors, including assembly processes such as soldering, are the considerations most likely to influence material selection. In general, more than one material will be acceptable, and selection is often influenced by experience and familiarity as well as technical advantages in a particular process, such as SMT soldering. Walczak, McNamara, and Podesta,[32] Gupta,[33] and Hawley[34] provide overviews of polymer materials from different connector application perspectives.

8.6 SEPARABLE CONNECTIONS

This section provides a brief overview of separable contact interfaces in terms of spring design parameters, such as contact force and contact geometry. The interaction of these variables on important performance characteristics, such as mating force and durability, are reviewed. Permanent connections are covered in Section 8.7, but is referenced in this section as well because all contacts have both separable and permanent connection requirements.

8.6.1 Contact Designs

A wide variety of separable interface contact designs are used in connectors. The separable connection always has two sides. In general, one side (usually the receptacle) is a spring member, and the other (the plug) is a solid contact, a PWB, post, or pin. Posts and pins differ primarily in geometry, posts being square or rectangular and pins being round. Generically, these systems are referred to as card-edge, post/receptacle or pin/socket connectors.

PWBs, posts, and pins Examples of typical plug contacts are provided in Figure 8.13. A PWB, Figure 8.13a, is part of a connector system in many level-3 and some level-4 applications. In most cases, the contact area for separable connections to the PWB is a gold/nickel plated pad or land. It is important that the platings on the PWB and the receptacle contact be compatible in performance characteristics.

The 25 square post (0.025 in on a side), Figure 8.13b, is currently the most common post geometry, although applications using smaller posts, 1 mm and 15 square, are increasing. 25 square technology is used in connectors from levels 3 through 5 and in noble metal and tin finished versions. Noble metal finishes are used in high-performance applications, and tin is used in both electronic and commercial products. Many post contacts are duplex plated with a tin or tin-lead finish for the permanent connection, in particular for soldered applications, and a noble finish at the separable interface.

In levels 3 and 4, the posts can be inserted directly into the board or included in headers, shrouded or unshrouded, depending on application requirements. The posts are connected to the board by soldering or press-in connections.

Although pins (Figures 8.13c and 8.13d) have limited use in level 4, their prime areas of application are in levels 5 and 6, often as wire-to-wire connections. The majority of pin contacts are designed for crimped permanent connections. They are available in a number of sizes (diameters) depending on the application. Two general types of pins are used: screw machined (Figure 8.13d) and stamped and formed, (Figure 8.13c). Screw-machined contacts are generally gold plated and used in military and high-performance systems; the stamped

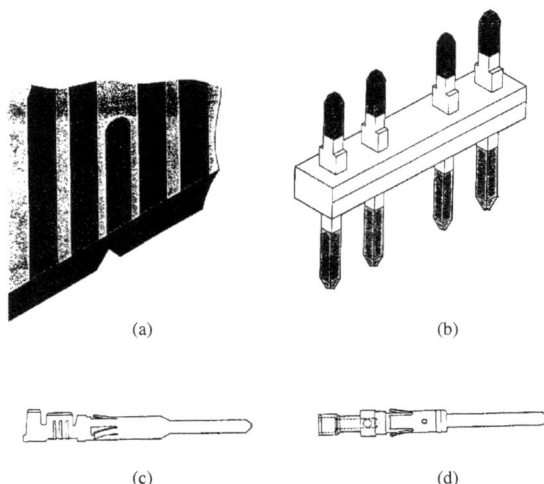

(a)   (b)

(c)   (d)

FIGURE 8.13   Schematic illustration of plug contacts, including (a) a printed wiring board (PWB), (b) a post, (c) stamped and formed, and (d) screw-machined pin contacts. (*Courtesy AMP Incorporated.*)

and formed pins are gold or tin plated and used in electronic and commercial applications. The major differences between screw machined and stamped and formed pins are the presence of a seam in the stamped and formed pin and differences in dimensional tolerances.

Receptacle contacts   The majority of receptacle contact designs are cantilever beam geometries, although very complex compound beams are also seen. A few of the more common geometries are shown in Figures 8.14 through 8.16. The simplest receptacle contact design is a cantilever beam, as exemplified by a card-edge contact, Figure 8.14a, although some card-edge contacts take on more complex geometries such as the bellows design shown in Figure 8.14b.

For post/receptacle systems, dual-contact beams are generally used, in open or box geometries, as shown in Figures 8.15a through 8.15d. In these examples, the dual-cantilever beams make contact with opposing sides of the post. Open twin-beam contacts predominate over box contacts in commercial applications because of lower cost. Four styles of twin-beam contacts are common, but with many variations. The flat stamped contact in

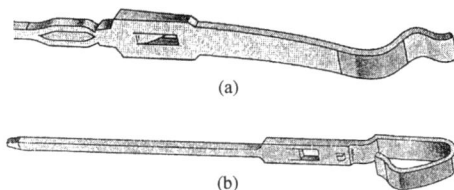

(a)

(b)

FIGURE 8.14   Schematic illustration of card-edge contacts: (a) simple and (b) compound cantilever contacts are shown. (*Courtesy AMP Incorporated.*)

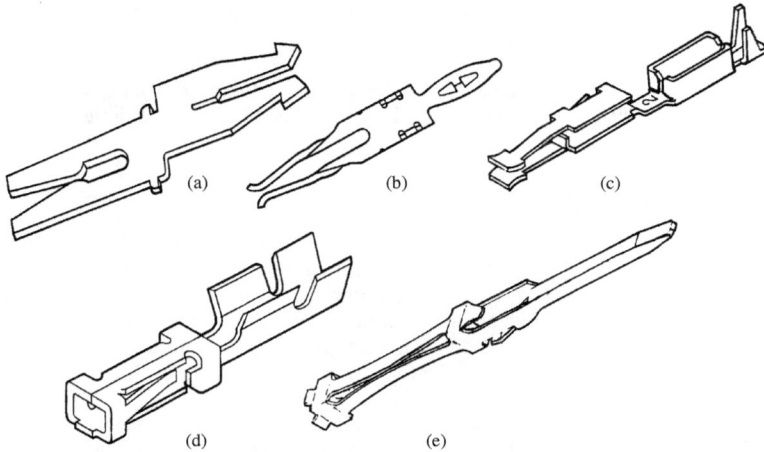

FIGURE 8.15   Schematic illustrations of a variety of receptacle contact geometries that mate to 25 square posts. (*Courtesy AMP Incorporated.*)

Figure 8.15a is commonly referred to as tuning fork. The contact in Figure 8.15b is commonly used in Eurocard connectors. In Figure 8.15c, an open twin cantilever-beam contact is illustrated. Figures 8.15d and 8.15e illustrate box type designs, a twin-beam in Figure 8.15d and a four-beam version in Figure 8.15e.

The contacts differ in manufacturing practice (and therefore in cost) and cantilever beam design. The first four designs provide dual redundant contacts and are used with both noble and tin finishes. The tuning fork contact mating surface is a sheared surface. For tin-finished contacts, this is acceptable but, for gold finishes, attention to the surface roughness might be necessary to ensure adequate mating durability. All other contacts mate to the mill rolled surface of the contact beams. The dual-beam box contact provides additional advantages in lead in protection to minimize the effects of misalignment on mating and, in the design shown, anti-overstress protection for the contact spring. All of these designs are available in tin and gold finishes, including duplex finishes, to satisfy a variety of commercial and electronic applications. The four-beam box contact is primarily used in high-performance and military applications and is generally gold plated. The full range of permanent connection technologies is available for receptacles, although not all designs lend themselves to each of the technologies. For example tuning fork contacts are generally intended for IDC applications.

Socket contacts are also available in screw machined and stamped and formed versions for the same markets as for the pins. Examples of screw machined and stamped and formed sockets are shown in Figure 8.16a and 8.16b, respectively. Screw-machined contacts are generally gold plated, whereas stamped and

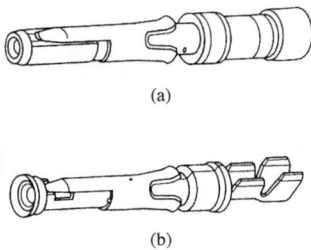

FIGURE 8.16   Schematic illustration of (a) screw machined, and (b) stamped and formed socket contacts. (*Courtesy AMP Incorporated.*)

formed versions are available with noble and tin finishes. Socket contacts, as mentioned, are generally intended for wire-to-wire applications, and crimped connections are the most common permanent connection technology.

There are many other receptacle contact designs in use intended to address specific application requirements. These are not included herein because of space limitations.

8.6.2 Application Issues

Selection of an appropriate connector for a given application requires consideration of a number of design performance interactions and trade-offs. Among those important from the separable connection perspective are the connector mating force and the contact durability. These performance characteristics, in turn, depend on design parameters such as the contact finish, the normal force, and the contact geometry. Criteria for contact finish selection were included in Section 8.3. The following section addresses the other issues.

Connector mating mechanics Mating mechanics, in particular the peak mating force, has a significant impact on two important performance factors, the capability of mating high pin count connectors, and contact durability.

Normal force, contact geometry, and engagement length, and their role in the connector mating process, is briefly described using a post/receptacle contact system as an example. A schematic illustration of the connector mating process is provided in Figure 8.17. The position of the post in the receptacle and the mating force versus insertion depth relationship are shown simultaneously. There are two phases to the mating process, insertion of the post, phase 1, and sliding, to the final contact location, after the receptacle beams are fully deflected, phase 2.

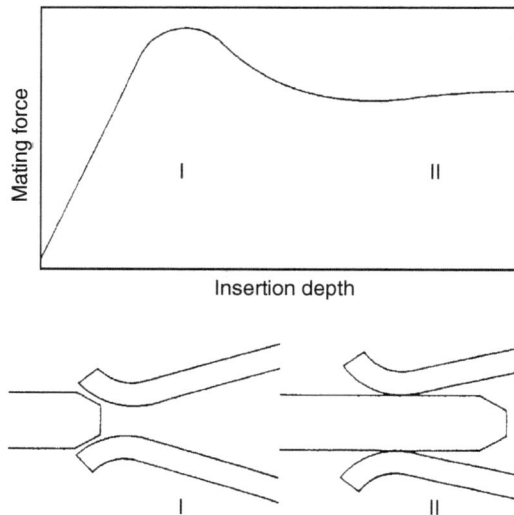

FIGURE 8.17   Schematic illustration of the mating mechanics of a separable interface. Two phases of mating, insertion, and sliding are indicated. (*Courtesy AMP Incorporated.*)

Phase I: The post enters the receptacle and begins deflection of the contact beams. The insertion force increases with a slope dependent on the coefficient of friction, which depends on the contact finish, the deflection characteristics, or spring rate, of the beam system and the contact interface geometry. During this phase normal force and contact geometry are the key factors which influence the dynamic frictional forces (the state of lubrication of the interface is also important because it has a significant effect on the coefficient of friction). For a given state of lubrication, the force/geometry relationship determines the maximum insertion force of the contact system. The peak force depends on the normal force, the angle between the mating surfaces and the coefficient of friction.

This section covers the peak contact mating force. The mating force for the connector is the sum of the contact mating forces plus any additional force necessary to overcome misalignment of the mating halves of the connector and dimensional variances in the housing. Such effects can add considerably and might even dominate connector mating force.

Also note that, under certain conditions of coefficient of friction and geometry, stubbing of the contacts can occur, making mating impossible without damage to the contacts. This "friction lock" condition is addressed by control of the mating geometry.

Phase II: The contact spring has been fully deflected, and the insertion force is that produced by sliding friction, which depends on normal force and the dynamic coefficient of friction. The withdrawal, or retention, force is also determined by these same parameters with the peak force being dependent on the static coefficient of friction, which is higher than the dynamic value. Contact geometry, however, can also influence the magnitude of the retention force.

Connector mating mechanics and contact durability   The relationship between mating forces and contact durability is different in the two phases. In phase I, surface roughness and improper initial mating geometry can lead to catastrophic wear, including skiving of the finish. In phase II, high friction forces can result in adhesive wear and brittle fracture because of high shear forces at the interface. More typically, in particular for lubricated contacts, the sliding wear process, which is dependent on normal force, coefficient of friction and contact geometry results in gradual burnishing wear of the contact surface. Under these conditions, the durability also depends on the length of engagement of the connector, a distance which is typically of the order of two millimeters.

Mating of the connector involves relative motion of the contact surfaces against one another. This motion is referred to as wipe and has an important impact on film and contaminant displacement. The effectiveness of wipe on film disruption and contaminant displacement are discussed by Williamson[3] and Brockman, Sieber, and Mroczkowski,[35,36] respectively. Here, it is sufficient to note that wiping action is necessary for reliable connector performance. In most applications, the engagement length of the connector provides wipe far in excess of what is necessary for film and contaminant displacement. In some designs, zero insertion force (ZIF) connectors in particular, wipe distances can be much smaller, and a zero wipe condition must be avoided. The minimum wipe distance depends on normal force and geometry, as discussed in Mroczkowski.[37] For typical connector normal forces and geometries, a wipe distance on the order of 10 mils is adequate in most applications.

Contact normal force   In most connectors, the contact normal force is generated by deflection of the receptacle contact beams by the plug. Contact normal force is arguably the most important connector design parameter. A "conventional wisdom" value for minimum normal force is 100 g.[38] As connectors miniaturize and pin counts increase, new limits on normal force come into play so that it becomes necessary to reconsider the context in which the 100 g minimum "requirement" was developed. The case for this statement is argued by Whitley and Mroczkowski[39] who also provide an overview of normal force requirements for precious metal plated connectors. This section briefly summarizes that reference.

The normal force performs two key functions. First, it provides the force that establishes the contact interface as the connector is mated. Second, it maintains the stability of the interface against mechanical disturbances in the application environment. The normal force required for each of these functions is different. To establish a contact interface with an "acceptable" value of contact resistance (a few milliohms) requires only a few grams, as indicated in Figure 8.18. Maintaining the mechanical stability of the interface is where the remainder of the normal force requirement arises. The value required depends on the contact finish, the design of the connector and the application environment. Fifty grams provides a benchmark for precious metal contacts if the connector design addresses all the relevant application considerations. One hundred grams is a lower limit for tin connectors[40] to provide the mechanical stability necessary to avoid the important degradation mechanism of fretting corrosion.

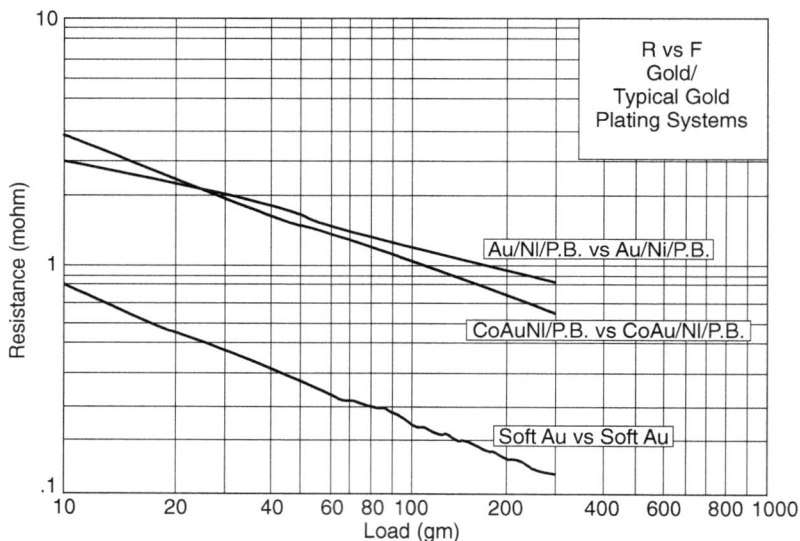

FIGURE 8.18   Contact resistance, R_c, versus contact force, F_n, for a gold-plated contact interface. (*Courtesy AMP Incorporated.*)

Contact geometry   Another important design parameter that interacts strongly with normal force is contact geometry. Wiping effectiveness, mating force, durability, and contact resistance all vary with the normal force-geometry combination of the separable interface.[41] It has been proposed by Kantner and Hobgood[42] that a parameter identified as "Hertz stress," which includes normal force, contact geometry, and finish and elastic modulus in a simple model, provides a good design guidelines for predicting connector performance. This proposal has been covered by Fluss[43] and Mroczkowski.[41] Eammons et al.[44] provide another useful study relating connector design to performance. In the authors opinion, these references indicate that a Hertz stress requirement does not provide unambiguous indications of the performance capability of contact interfaces.

8.6.3 Summary of Separable Interfaces

The basic requirements on the separable interface are described in Section 8.1. The connector must provide a low and stable value of resistance while meeting mating requirements, in particular mating, force and the number of mating cycles it must withstand without affecting performance. Separable interfaces include a wide variety of contact geometries and forces to satisfy these requirements under an increasing diversity of application conditions.

8.7 PERMANENT CONNECTIONS

Separable interface to permanent connections occur between the connector and the electronic component or system to which it provides a connection. The two basic classes of permanent connections are mechanical and metallurgical. Mechanical connections include crimped, insulation displacement, press-in, and wrapped connections. Soldered connections are the predominant metallurgical connections, but brazed and welded connections also are used.

Permanent connections are sometimes described as an extension of the conductor to which they are connected. In that context, the resistance introduced by the connection is expected to be of the same order as a section of the terminated conductor of the same length as the connection. As mentioned previously, permanent connection resistances are of the order of tens to hundreds of microohms, compared to the milliohm or so of separable connections. The magnitude of permanent connection resistance depends on the resistance of the conductor being terminated.

In this section, permanent mechanical connections are considered in two classes, connections to wire/cable and to a PWB. Prior to discussion of the permanent connection technologies, a brief discussion of wire/cable and PWB construction is in order.

8.7.1 Overview of Wire and Cable

The terminology "wire" is used in this section, but these remarks also apply to cables. In this context, a cable is viewed as a protected wire (jacketed cable) or a number of wires arranged in a "controlled" geometry (ribbon cable). Coaxial cable are also considered.

A wire consists of a conductor, the current-carrying component, and its insulation, if any. Characteristics of a wire include:

- *Conductor material*   In the majority of electrical and electronic applications, the conductors are annealed copper.
- *Conductor finish*   In many applications, the conductors are bare copper. Common finishes include tin, silver, and nickel. Tin finishes enhance solderability and crimped and IDC permanent connections. Silver is used for high-frequency applications, and nickel for high-temperature applications.
- *Conductor geometry*   Round conductors predominate, and flat and foil conductors are used in specialized applications. This description is limited to round conductors.
- *Conductor construction*   The two basic round conductor constructions are solid and stranded. Wires are usually characterized by their cross-sectional area, in millimeters squared or American Wire Gauge (AWG). Stranded wires are made up of multiple strands of conductors to achieve the total cross section desired. For example, a 26AWG

wire can be constructed in several ways as illustrated in Figure 8.19. The solid wire consists of a single 26AWG conductor having a nominal diameter of 0.0159 in. The other two 26AWG wires shown are 7/34 and 19/38 constructions. The 7/34 construction consists of seven 34AWG conductors having a diameter of 0.008 in to make up the nominal 26AWG cross section. Similarly, the 19/38 construction includes nineteen 38 AWG conductors having a diameter of 0.004 in. The geometric arrangement of the conductors can take several forms in terms of the relative wire positions and the twist or lay of the conductors. The constructions illustrated are concentric constructions in which the geometric relationship of the strands to one another are constrained. Stranded conductors predominate in most applications because of their higher flexibility compared to solid conductors, though they have a slightly lower current-carrying capacity than the corresponding solid wire. High conductor count wires are frequently bunch stranded, a construction in with the individual conductors are randomly distributed in the bundle. Bunch stranding provides greater wire flexibility and lower cost than concentric constructions.

- *Wire insulation*   Insulations are characterized by their material and thickness. Common insulating materials include polyvinyl chloride (PVC), polyethylene (PE), polypropylene (PP), and polytetrafluoroethylene (PTFE). The choice of insulation material and thickness depends on the function the insulation is intended to perform. The two most common functions are electrical insulation and mechanical protection. In some applications the dielectric characteristics of the insulation also become important, for example in coaxial cables intended for high-speed applications. In such cases, foamed insulations are seeing increasing usage to decrease the effective dielectric constant and thereby maximize the propagation velocity of the cable.

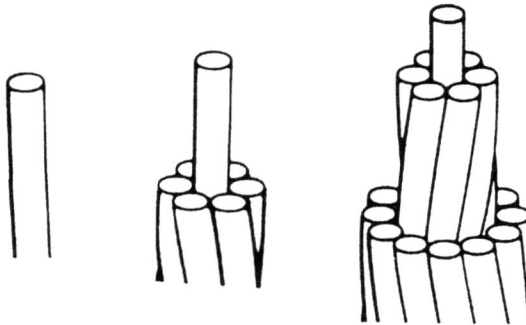

FIGURE 8.19   Schematic illustration of three-strand configurations used in stranded conductor wire constructions. (*Courtesy AMP Incorporated.*)

Schematic illustrations of some common cable constructions are provided in Figures 8.20 and 8.21.

Jacketed cable, Figure 8.20a, consists of a collection of discrete wires contained within a single protective jacket. Such cables provide mechanical protection of the wires and reduced wire handling during manufacturing and assembly processes. The wires within the jacket might be of a variety of constructions including discrete, shielded, coaxial, and twisted pair.

Ribbon cable refers to a planar arrangement of multiple wires as shown in Figure 8.20b. Both round and flat-wire ribbon cable are used, but round wire dominates. In addi-

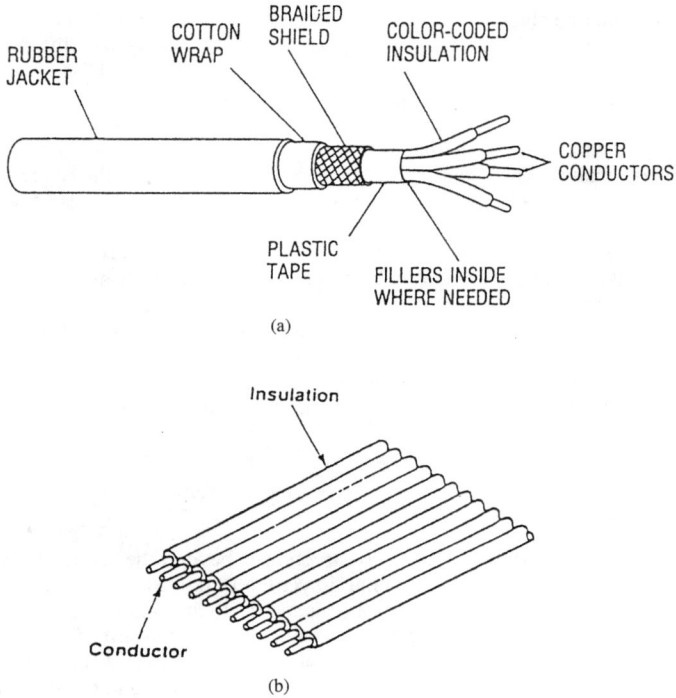

FIGURE 8.20 Schematic illustration of (a) jacketed and (b) ribbon cable constructions. (*Courtesy AMP Incorporated.*)

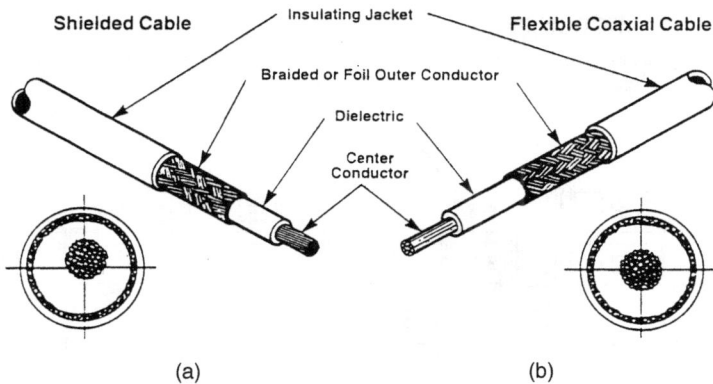

FIGURE 8.21 Schematic illustration of (a) shielded and (b) coaxial cable constructions. (*Courtesy AMP Incorporated.*)

tion, the wires can be shielded, coaxial, twisted pair, or transmission-line constructions. Ribbon cable also simplifies wire handling and, in addition, reduces the potential for wiring errors by fixing the relative positions of the individual wires. Discrete wire ribbon cable lends itself to mass termination, usually using insulation displacement connection (IDC) technology.

Twisted-pair wires are exactly that: two wires twisted around one another for improved electrical performance. The details of the improvement are beyond the scope of this discussion, but suffice it to say that the signal is split between the two wires so as to reduce noise and interference effects on signal integrity.

Transmission line cables consist of alternating signal and ground conductors, which are generally of different sizes, to improve high-speed digital data transmission performance and provide controlled impedance.

Shielded and coaxial cable are identical in construction, as indicated in Figure 8.21, with the essential difference that the concentricity and dimensions of the coaxial cable are much more tightly specified. These geometrical parameters, along with the dielectric properties of the insulation material, determine the impedance of the cable. Coaxial connectors are covered in Section 8.9.

8.7.2 Overview of PWB Construction

Detailed information concerning PWB construction is beyond the scope of this chapter. Figure 8.22 contains a cross section of a multilayer PWB to illustrate the features of PWB construction relevant to making permanent connections to the board. Pads or lands on the board surface allow the making of separable connections to the board edges, and soldered connections to lands distributed over the board surface, surface mount connections (SMT). Plated through holes allow for through hole soldered connections and mechanical press-in connections. Also shown is a press-in connection with the tail extending through the board. Such exposed tails provide sites for both connectors or wrapped connections.

With this context, attention now turns to the structure of permanent connections. Although the focus is on mechanical permanent connections, the subject of soldered connections is briefly covered.

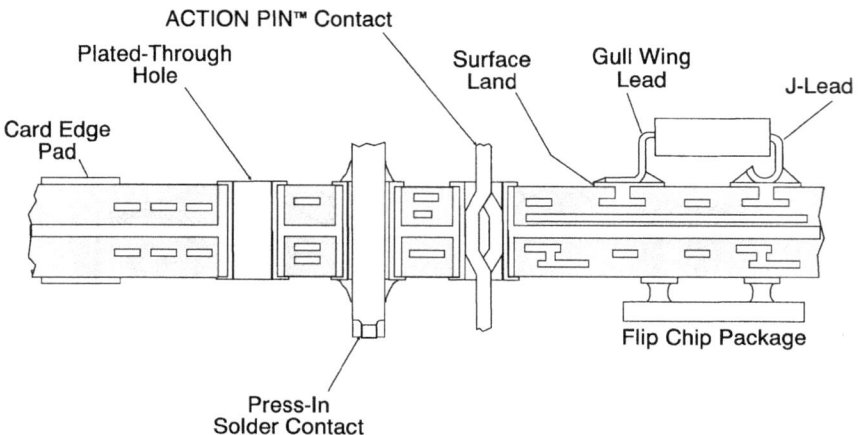

FIGURE 8.22 Schematic cross section of a multilayer printed wiring board. (*Courtesy AMP Incorporated.*)

8.7.3 Mechanical Permanent Connections to Wire/Cable

The two major wire/cable permanent connections are crimped and insulation displacement connections.

Crimped connections The principles underlying crimped connections are quite simple. Crimp terminal design and the crimping process, however, require careful attention to the overall crimping system. This system approach cannot be overemphasized, because the controlled deformation of wire and terminal that is responsible for consistent and reliable crimped connection performance derives from ensuring the proper system. The crimping system consists of the proper combination of wire, crimp terminal, and crimp tooling.

Although crimped connections can be made to solid wires, stranded wire applications dominate. Copper conductors represent the majority of applications in both bare and tin-plated versions. Annealed copper wire is the most common case, but applications with alloy wire also exist. There are two basic requirements on the wire. First, it must be the correct size; that is, it must fall in the range of wire sizes the crimp terminal is designed to terminate. Second, the insulation must be stripped from the wire to the proper length without damaging the conductors or the stranding excessively. Damage of the conductors will reduce the mechanical strength of the crimped connection and might affect the crimp resistance. Excessive disturbance of the stranding could compromise the insertion of the stripped wire into the terminal, particularly in automatic machinery.

There are two basic styles of crimp terminals: open and closed-crimp barrels, as illustrated in Figure 8.23. Open-barrel terminals are used in high-volume machine-applied applications because the open-barrel facilitates automatic insertion of the stripped wire into the open barrel which provides a large target area. For this reason, open-barrel crimp terminals are used in the majority of crimped connections. Closed-barrel terminals are often semiautomatically or hand-applied in smaller volume or repair applications. The stripped wire must be inserted axially into the smaller target area of the closed barrel. The crimped connection is made by controlled deformation of the conductors in the crimp barrel.

Figure 8.23 also illustrates a second important element of crimp terminals, the insulation support, or insulation grip. In the insulation support barrel the insulation is gripped by

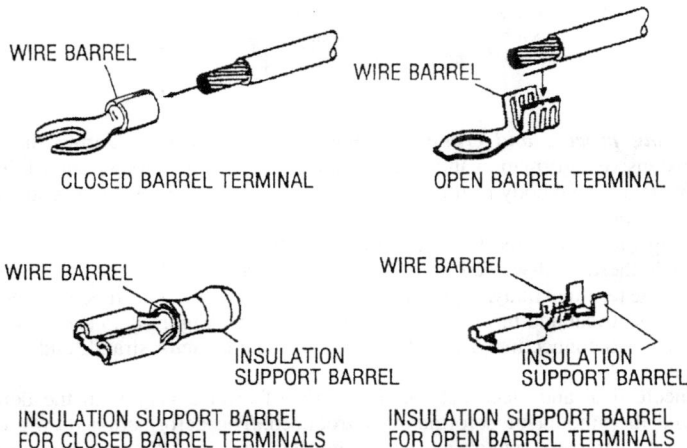

FIGURE 8.23 Schematic illustration of open- and closed-crimp barrel constructions. (*Courtesy AMP Incorporated.*)

deformation of the tabs to provide a strain relief for the crimped connection. In addition to the strain relief, the insulation support provides protection against the effects of vibration on the crimped connection. The majority of crimp terminals take advantage of the benefits of an insulation support barrel.

The function of the crimp tooling is to control the deformation of the conductor and crimp terminal because controlled deformation is the source of crimped connection performance. Figure 8.24 schematically illustrates the crimping process in an open barrel terminal. The conductor/crimp terminal cross sectional area is fixed by selection of the appropriate wire/crimp terminal combination. The crimp tooling then controls the deformation by controlling the crimped connection geometry. Two controls on geometry are used. Clearly the shape of the crimp tooling, and the terminal geometry, fix the overall geometry of the crimped connection. Control of the crimp height, illustrated in the inset of Figure 8.24, is the final control and is accomplished by controlling the closing action of the crimp tooling.

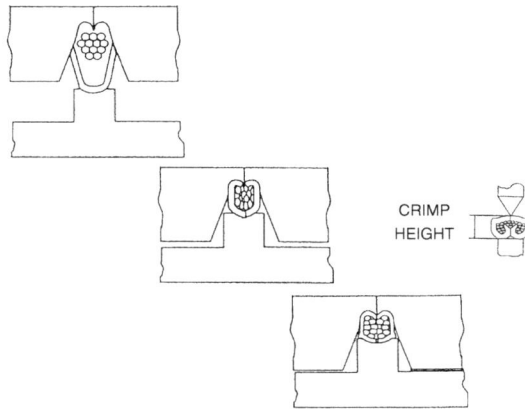

FIGURE 8.24 Schematic illustration of the crimping process for an open-barrel crimp termination. (*Courtesy AMP Incorporated.*)

The crimping process and crimped connection performance During the crimping process, extensive deformation and relative motion of the conductor(s) against themselves and the crimp terminal body results in a disruption of surface films on the conductors and terminal body producing new film free surfaces. Under the high forces of crimping, contact between these film free surfaces results in the formation of numerous micro cold welds.[45] It is these cold-welded joints, between the conductor strands and between the strands and the terminal body, supported by an appropriate residual stress distribution, that provide the electrical and mechanical integrity of crimps. Figure 8.25 shows a cross section of a crimped connection. The deformation of the conductor strands and the terminal are evident.

The mechanical and electrical characteristics of crimps vary with the deformation process and the amount of deformation introduced. The deformation during crimping, therefore, is a parameter which must be controlled to ensure that the properties of the crimp remain in the "optimum" range.

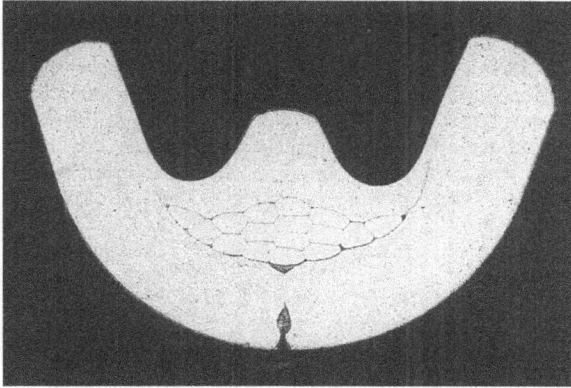

FIGURE 8.25 Photomicrograph of a crimped-connection cross section. Note the deformation of each of the individual conductor strands. (*Courtesy AMP Incorporated.*)

A number of crimp geometries are widely used, examples of which are provided in Figure 8.26. The F, for folded, crimp (Figure 8.26a) is a commonly used open barrel crimp geometry. W crimps (Figure 8.26b) are used for large wire size closed barrel crimped connections. Military applications often use indent crimping processes as illustrated in Figure 8.26c. The crimp geometry shown is a 4-8 indent crimp.

Summary If the appropriate wire/terminal/tooling combination is used, crimp deformation will be controlled and a repeatable and reliable crimped connection will result.[46] Properly crimped connections provide mechanical strengths approaching that of the wire

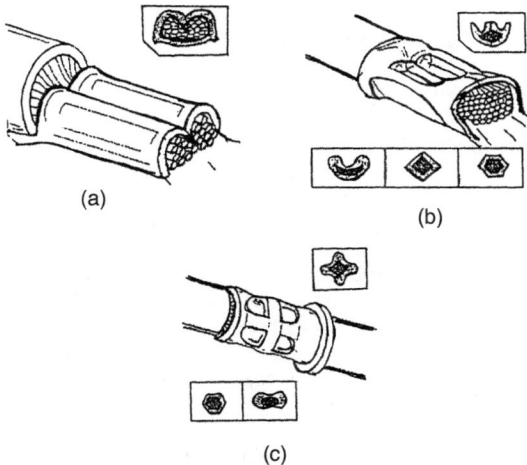

(a)

(b)

(c)

FIGURE 8.26 Schematic illustrations of selected crimped connection cross sections: (a) F crimp, (b) W crimp, and (c) 4-8 indent crimp. (*Courtesy AMP Incorporated.*)

being terminated and electrical resistances comparable to that of an equivalent length of the crimped wire.

Inspection and process control for crimped connections is simple. Visual inspection of the crimped connection should verify proper stripping of the insulation and insertion of the conductors into the crimp barrel as indicated in the sample inspection document in Figure 8.27. In addition to the visual examination, the crimp height also serves as a process verification, on crimped connections, and, on some automatic crimping equipment, as a process monitor during the crimping process. A periodic verification of the tensile strength can also be performed.

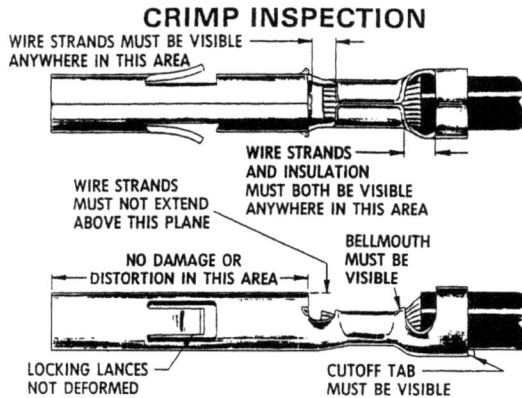

CRIMP INSPECTION

WIRE STRANDS MUST BE VISIBLE ANYWHERE IN THIS AREA

WIRE STRANDS MUST NOT EXTEND ABOVE THIS PLANE

WIRE STRANDS AND INSULATION MUST BOTH BE VISIBLE ANYWHERE IN THIS AREA

NO DAMAGE OR DISTORTION IN THIS AREA

BELLMOUTH MUST BE VISIBLE

LOCKING LANCES NOT DEFORMED

CUTOFF TAB MUST BE VISIBLE

FIGURE 8.27   Example of visual crimp inspection criteria. (*Courtesy AMP Incorporated.*)

8.7.4 Insulation Displacement Connections

The second mechanical wire/cable permanent connection is insulation displacement connection (IDC) technology. The principles of IDC technology are schematically illustrated in Figure 8.28. A wire is simply inserted into a controlled geometry slot in the terminal. As the wire slides down the slot, the wire insulation is displaced and the conductor cross section deformed to create the desired metallic permanent connection interface.

Although IDC connections are used on discrete wire, the dominant application is to cable, in particular ribbonized cable (Figure 8.21b). The reason for this dominance becomes apparent when the termination process for a ribbon cable is considered. First, there is no requirement to strip the insulation which reduces wire handling. Second, it is possible to make all the connections at one time using housings in which the contacts are preloaded. In addition to increasing throughput, wiring errors are eliminated compared to the crimp and insert process necessary for crimped connection approaches. Figure 8.29 shows a variety of IDC cable connector applications.

IDC terminals   There are a wide variety of insulation displacement terminal designs, some of which are shown in Figure 8.30. The designs differ primarily in the compliance of the beams of the IDC slot which impacts on the residual force capability and range of wire sizes on which the terminal can be used.

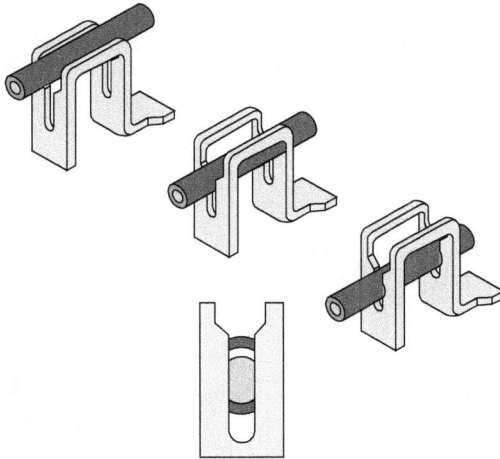

FIGURE 8.28 Schematic illustration of insulation displacement termination technology. (*Courtesy AMP Incorporated.*)

FIGURE 8.29 Photograph of a variety of ribbon-cable IDC termination types. (*Courtesy AMP Incorporated.*)

The simplest IDC terminal design, which is used to illustrate the principles of insulation displacement connection mechanics, is the slotted beam, as shown in Figure 8.30a. Two separate functions are critical to successful ID terminations:

1. the displacement of the insulation.

2. the formation and maintenance of the contact area.

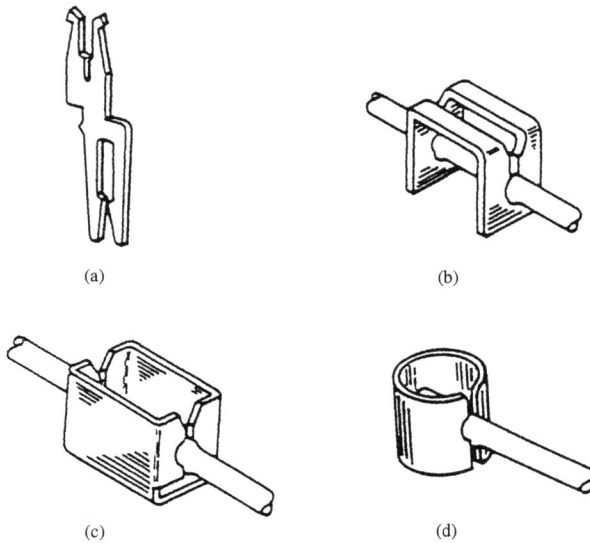

(a)   (b)

(c)   (d)

FIGURE 8.30   Schematic illustration of a variety of IDC terminal designs. (*Courtesy AMP Incorporated.*)

Figure 8.31 schematically depicts these two functions. The insulation of the wire musts be displaced at or near the top of the slot to prevent overstressing of the contact beams, Figure 8.31a. The effectiveness of insulation displacement depends on the geometry of the lead-in section of the termination and the thickness and properties of the insulating material on the wire. The terminal lead-in must penetrate the insulation and strip it away from the conductor surface. Lead-in geometry is important, but the hardness of the insulation and the adherence of the insulation to the conductor are equally important. Wire insulations vary significantly in the repeatability with which they can be displaced. PVC insulations dominate in current IDC applications.

After displacement of the insulation, the conductor continues downward in the slot to its final position, Figure 8.31b. It is during the downward motion that conductor deformation takes place producing a wiping action on the beam sides and the conductor(s). The effectiveness of surface film removal during wiping depends on the design of the IDC slot and the surface finish. Deformation of the conductors provides intrinsic cleaning and these surfaces make contact with the sides of the termination beams.

At the final conductor location, the displacement of the beams determines the residual normal force which maintains the integrity of the contact interface. The allowable deformation of the conductor is limited by two factors. The lower limit is dependent on the requirement to establish a contact area, A in Figure 8.31b, which results in an acceptable

DURING INSERTION

(a)

AFTER INSERTION

**INSULATION
DISPLACED**

(b)

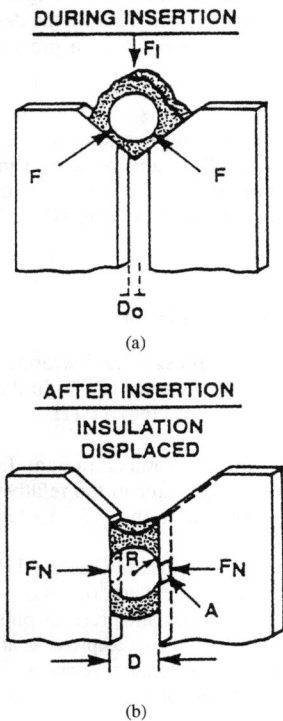

FIGURE 8.31 Schematic illustration of the IDC termination process. Two stages are shown: (a) insulation displacement and (b) conductor termination. (*Courtesy AMP Incorporated.*)

value of connection resistance. The upper limit of conductor deformation is because of the weakening of the wire and its effect on handling of the connectors during assembly and in their intended application.

Strain relief The force required to insert a conductor into an IDC slot is relatively low. It follows that a force applied parallel to the slot might disturb, or even dislodge, the wire. For this reason, most IDC connectors include some mechanism for protecting the wire from such forces. The protection can be built into the terminal, housing or separate covers, as covered in Mroczkowski.[47]

Solid and stranded wire IDC technology is suitable for both stranded and solid wire, but more attention to termination design is necessary when stranded wire is used because control of the conductor deformation is more difficult with stranded conductors because the strands can move relative to one another during the insertion process. This effect is illustrated in Figure 8.32, where examples of solid and stranded wire IDC connections are provided. The rearrangement of the conductor strands can have a detrimental effect on both the magnitude and stability of IDC resistance. The effect on resistance stability is arguably more significant because the strand rearrangement reduces the beam deflection and, therefore, the contact normal force which provides the desired mechanical stability. These concerns can be addressed, however, as indicated by the fact that the majority of IDC application are on stranded wire.

FIGURE 8.32 Photomicrograph illustrating insulation displacement connections to solid and stranded wire. (*Courtesy AMP Incorporated.*)

Conductor finish IDC connections are readily made to both bare and tin-coated conductors. The wiping action on insertion serves to clean the conductor surfaces, enabling the desired metallic interface to be created. Tin-finished conductors, however, show a greater consistency in performance.

Summary IDC technology is primarily used in cable connector applications to take advantage of enhanced productivity through reduced wire handling and reduction in wiring errors. A variety of IDC terminal designs exists to address differing application requirements. For additional discussion of IDC connection technology, see Mitra,[48] Mroczkowski,[47] Wandemacher,[49] and Key.[50]

8.7.5 Mechanical Permanent Connections to Printed Wiring Boards

Two mechanical permanent-connection technologies are covered, press-in and wrapped connections. Press-in connections are made directly to plated through holes (PTH) in the PWB, and wrapped connections are made to pins pressed-in or soldered into a PTH.

Press-in connections The two types of press-in connections are: rigid and compliant. In both cases, residual forces maintain the contact normal force. They differ in the relative amounts of elastic and plastic deformation that occurs in the pin, and in the deformation produced in the PTH in the PWB.[51]

Rigid press-in pins undergo little elastic deformation when they are inserted in the PTH and cause plastic deformation and damage to the PTH. Little elastic recovery, or residual force, capability is retained in such connections, which is why compliant press-in pins were developed. A major driving force for compliant pin technologies was to provide an alternative to soldering for through-hole connections. A mechanical permanent connection technology eliminates the potential degradation/contamination effects of the high temperatures and fluxes used in soldering.

All compliant press-in geometries are intended, in one way or another, to produce elastic deformation of the pin, and thereby a residual force to maintain the integrity of the interface between the press-in section of the compliant pin and the PTH. Many designs of compliant press-in pins are in use. Three of the major variants are shown in Figure 8.33.

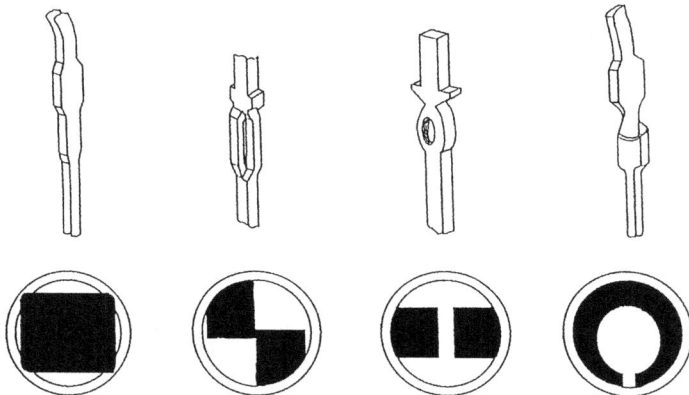

FIGURE 8.33 Schematic illustration of press-in contact geometries. (*Courtesy AMP Incorporated.*)

There are four basic requirements on the press-in section:

- Establishing and maintaining a normal force
- A limit on insertion force
- A minimum retention force
- No unacceptable damage to the PTH

The normal force requirement is, of course, intended to ensure a low and stable connection resistance. The insertion force limit is necessary to permit the insertion of multiple pin count connectors as well as to limit PTH damage. The minimum retention force requirement depends on the function of the tail of the pin, which extends from the PWB. The requirement on hole damage is to ensure that damage to the PTH and lands of the PWB, including buried layers, does not occur to the point where the performance of the PWB is compromised. Needless to say, some of these requirements are in direct opposition, and a number of compliant press-in section designs have evolved to address these issues and meet these requirements.

Transverse and longitudinal cross sections of the AMP ACTION PIN compliant pin geometry in a PTH are shown in Figure 8.34. The normal force is created by the deflection of the compliant section diagonal of the pin. The original diagonal dimension is shown as the dotted line. The transverse contact occurs at two points. The longitudinal contact extends along the axis of the compliant section.

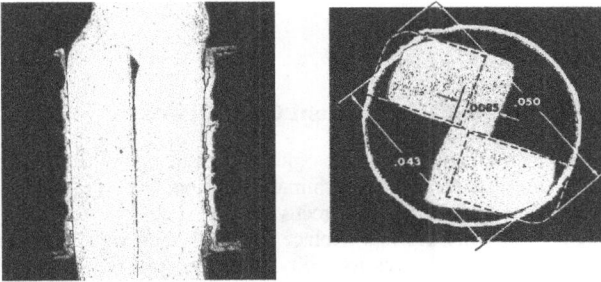

FIGURE 8.34 Photomicrograph of a compliant in termination to a plated-through hole. (*Courtesy AMP Incorporated.*)

The 25 square technology is currently dominant, but smaller sizes are seeing increasing use as circuit miniaturization and I/O density requirements continue to increase. Additional discussion on the design and function of compliant press-in pins is provided by Scaminaci,[52] Goel,[51] and Dance.[53]

Wrapped connections Wrapped connections are made to the tails of press-in connections to PTH which extend from the board surface. Figure 8.35 schematically illustrates a wrapped connection. The example shown is a modified wrap in that the first contact of the wire to the board occurs to the insulation. The electrical connections are formed to the stripped wires as the wrapping process continues. The force responsible for creating and maintaining the integrity of the electrical connection is the residual tensile stress in the conductors as they come in contact with the corners of the wrap post. These corners have a controlled radius requirement to facilitate the formation of a metallic contact interface. The tension on the wire is maintained as the conductor is wrapped around the post corners and, therefore, is retained in the conductor sections between the post corners. Wrapped connections exhibit a high redundancy of contact interfaces as readily seen in Figure 8.35.

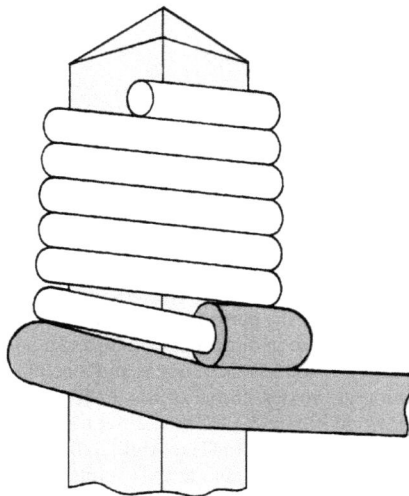

FIGURE 8.35   Schematic illustration of a modified wrapped connection. (*Courtesy AMP Incorporated.*)

8.7.6 Summary of Mechanical Permanent Connections

Four mechanical permanent connection technologies have been mentioned. Crimped and insulation displacement connections are wire termination technologies. Press-in and wrapped connections are connections to PWB, either directly or indirectly. Each technology is widely used in the electrical and electronics marketplace, although they vary in presence across these markets. Each technology derives its utility from the creation and preservation of metallic contact interfaces through controlled deformation of the conductor, wire or PTH, and the terminal. Control of the deformation, through tooling and process controls, is necessary to ensure repeatable permanent connections regardless of the technology used.

8.7.7 Soldered Connections

Numerous texts are available on soldering technology (e.g., Manko).[54] In this section, only some of the aspects of soldering relevant to connector technology are reviewed. There are two major types of connectors for solder connections. They differ in how the solder terminations of the connector interface to the board. The older technology, through-hole technology (THT) relies on soldering the lead into a PTH. Surface mount technology (SMT), as the name implies, relies on soldered connections made to a pad or land on the surface of the PWB. The two technologies differ in several respects.

Through-hole technology   The THT soldering process is typically accomplished through wave soldering. The solder supply and the soldering heat are provided by the solder wave. This is a significant difference from SMT soldering. The solder joint in THT is also much larger than that in SMT, extending through the PWB and into a fillet above and below the surface of the board. In addition to this solder joint, however, the soldered post is also mechanically supported by the PWB. These two differences, the solder source and the mechanical support, have a major effect on the design and materials requirements on THT as compared to SMT.

Surface-mount technology For SMT soldering, the solder leg of the connector is pressed down into a solder paste which is then reflow soldered, by a variety of techniques, to create the solder joint. The solder source, the amount of solder used, the source of the soldering heat and the structure of the solder joint all differ significantly from that of THT soldering. The solder source is typically a solder paste which is screened or printed onto the PWB in a secondary operation. Control of the amount of solder is critical to SMT performance. Too little solder could affect mechanical strength and too much could lead to solder bridging and shorting of contact pads. The solder heat is generally provided externally to reflow the solder paste, though some wave soldering is used. Reflow technologies which have been used include vapor phase, infrared, hot gas, and convection processes in addition to wave soldering. The solder joint is made only to the surface pad or land with no mechanical support from the PWB itself. The solder joint must support all subsequent mechanical stresses. This fact leads to concerns about solder creep and relaxation during thermal cycling of the PWB during application.

SMT versions of many electronic components, such as chip forms of resistors and capacitors and special IC chips, are available. These components differ significantly from connectors in two respects—size and the stresses they must support. Connectors are much larger than the other components mentioned, which leads to issues in retaining the connectors on the boards prior to reflow and consistent soldering of a larger number of leads than in other components. Connectors are also subjected to larger mechanical loads, particularly because of the mating requirements. These differences create many new and challenging materials, design, and processing requirements. Among the design/material and process issues that must be addressed to ensure the repeatability of SMT soldering processes are:

- Registration of the connector leads to the pads
- Coplanarity of the leads to ensure contact to the pads
- Adequate connector hold down prior to soldering
- Solderability of the leads
- The solder supply, sufficiency, and control of bridging
- Thermal stability of the housing to withstand the heat of the process
- The ability of the solder joint to withstand mating stresses
- Inspection of the solder joints

The economic and performance advantages that can be derived from SMT connectors will lead to continued development of both connectors and processes. Economic advantages include smaller and more cost-effective boards resulting from the elimination of PTH. Performance advantages are also realized by the smaller boards, resulting in reduced signal path lengths and the potential for improved conductor routing on the board to provide enhanced signal transmission quality and speed.

Summary Soldering technologies remain widely used in assembly processes with connectors. THT is being replaced, when economics and performance dictate, by SMT, but both will coexist into the foreseeable future.

8.8 CONNECTOR APPLICATIONS

In this section, connector signal and power applications are briefly reviewed. In signal applications, the focus is on maintaining the integrity of the signal—in particular, the signal rise time and amplitude. For power applications, minimizing the effects of the connector on the distribution of power, primarily current, within the system is the major concern.

8.8.1 Signal Applications

Signal applications can be either analog or digital. In this discussion, the focus is on digital applications because connector requirements for digital circuits are generally more stringent than those for analog.

The explosive development of microprocessor technology has created two major driving forces for connector technology: miniaturization and "high speed." The effects of miniaturization span virtually all aspects of connector design. With respect to digital applications miniaturization, and the associated reduced spacings between conductors, on boards and in cables and connectors, results in increased crosstalk. Crosstalk, in turn, affects signal integrity. In addition, control of the magnitude and variation of the connector impedance, because it depends on dimensional tolerances, has become a significant connector design consideration.

There are two aspects to "high speed," the signal pulse frequency, or clock frequency, and the rise time of the pulse. In general, the pulse rise time is the dominant consideration in defining a high-speed application. This is so because the frequency content of a fast-rising pulse includes frequencies higher than the clock frequency. The maximum frequency, f_m, in a pulse of rise time t_r can be approximated by:

$$f(\text{max}) = \frac{0.35}{t(r)} \tag{8.6}$$

For a 1-ns pulse rise time, a maximum frequency of 350 MHz is calculated. This frequency exceeds typical clock frequencies, of the order of 100 MHz in 1995 vintage personal computers.

An application must be considered high-speed when transmission line design rules become necessary for any of the components in the system. In general, the transmission line properties of an interconnection become important when the length of the component becomes comparable to the electrical length of the pulse. There are many approaches to selecting an appropriate electrical length. The approach used here relates the component length to the electrical rise length of the pulse, which is given by:

$$L(r) = v(prop)\, t(r) \tag{8.7}$$

where the signal propagation velocity, v_{prop} is given by:

$$v(prop) = c(\varepsilon_{eff})^{1/2} \tag{8.8}$$

where C is the speed of light and ε_{eff} is the effective dielectric constant of the propagation medium. The effective dielectric constant takes into account material composition variances, such as the amount of air in a foamed dielectric or the geometry of a polymer/air cavity in a connector housing. An appropriate average of the individual dielectric constant is necessary in such cases.

When the component length is larger than $0.3L_r$, the component should be considered as a transmission line. For connectors and printed circuit boards, an effective dielectric constant of about 4 is appropriate. For 1- and 10-ns rise-time pulses, the critical component length becomes 2 and 20 cm, respectively.

From these lengths, it is clear that printed circuit boards and cables, which are typically tens of centimeters in length, should be considered as transmission lines for pulses faster than 10 ns. This is in fact the case, and transmission line design rules have been applied to PWB and cable design for several years. Microstrip geometries on the PWB and controlled-impedance cables are commonly used. Connectors are reaching the point where they must be considered as transmission lines as signal rise times break into the subnanosecond regime. In such applications, transmission line considerations such as characteristic impedance and crosstalk replace contact resistance as "key" performance parameters.

For a general discussion of high-speed applications, refer to Katyl and Simed,[55] and Chang.[56]

Crosstalk and controlled impedance   Crosstalk and characteristic impedance both depend on the materials and geometries of the components so that design changes affect both characteristics, but not necessarily to the same degree or in the same direction. This allows designs to be optimized, depending on whether crosstalk or characteristic impedance is the more important parameter. In cables, crosstalk is arguably the more important because of the length of typical cables; in connectors, impedance control is dominant because of the inevitable changes in geometry and materials along the length of a connector.

Crosstalk   Crosstalk results from the electromagnetic coupling of the field surrounding an active conductor into an adjacent conductor, called the *victim*. The strength of the coupling depends on the proximity and geometry of the active and victim conductors. Crosstalk can be minimized by increasing the separation between conductors and keeping conductor lengths as short as possible. It is this second feature that limits the crosstalk introduced by connectors, because connectors tend to be short in the direction of the signal.

An additional method for reducing crosstalk is to introduce grounds between the signal conductors. Grounds, in effect, attenuate the electric fields and reduce the coupling between signal lines. This approach can be, and is, used in connectors, but is even more important in cables because of their greater length. The use of ground pins in connectors is, in fact, intended more as a means of controlling the impedance of a connector.

Controlled impedance   Impedance control is an important requirement for at least two reasons. Constancy of impedance is necessary to ensure the quality of signal transmission over distance, and matched impedance at interfaces, such as connectors, is critical to minimizing signal reflections. Control of signal integrity is particularly important in digital applications where false triggering or missed triggering of devices can occur because of reflections, positive or negative, caused by impedance mismatches.

The impedance of a component depends on the geometry and materials of the component. Constant impedance requires constant geometries and consistent materials. Impedance control in cables is relatively straightforward in that cables tend to be consistent in both materials and geometry. As cables become miniaturized, however, dimensional tolerances and materials consistency become more difficult to maintain. For connectors, it is a very different situation. Impedance control in a connector is complicated with respect to both dielectrics, a varying mix of air and the polymer of the connector housing, and changes in conductor geometries, both in cross section and curvature. Under these conditions a "constant impedance" connector is intrinsically impossible.

This difficulty leads to two approaches to controlled impedance in connectors which rely on adding grounds to moderate the variations in characteristic impedance. The two approaches are the addition of dedicated ground pins and introducing ground planes within the connector. Examples of these two cases are provided in Figure 8.36.

Open Pin Field Connectors   Varying the number of pins dedicated to grounds in an open pin field connector was the first approach to controlling the impedance and reducing the

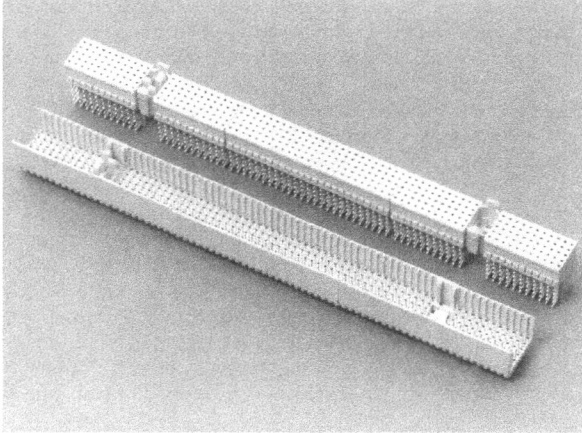

(a)

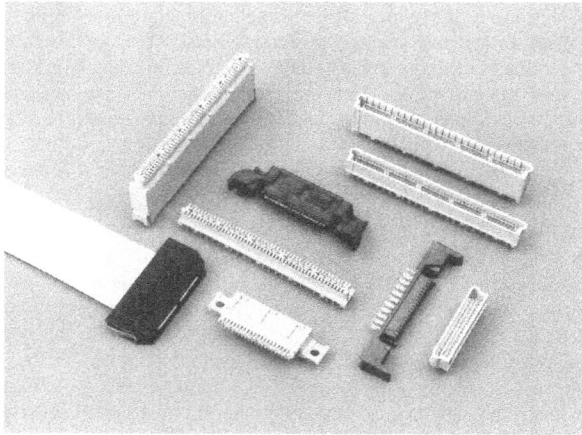

(b)

FIGURE 8.36   Examples of (a) open pin field and (b) ground-plane connector constructions. (*Courtesy AMP Incorporated.*)

crosstalk of board-to-board connectors. Many factors, primarily circuit and application related, influence the number of grounds necessary to ensure the required signal integrity. The number of grounds needed is often expressed in terms of the signal/ground ratio. As the signal/ground ratio decreases the electrical performance increases, but the number of pins in the connector which are available to satisfy the I/O requirements decreases. The arrangement of the grounds with respect to the signal lines is also important.

Ground-plane connectors The introduction of ground planes into the connector is one method for reducing the impact of grounding requirements on the number of available signal pins. Stripline and microstrip geometries have been used in PWB design for a number of years and are now available in connectors. Such connectors are usually designed for 50-Ω impedance and can be used in applications with rise times below a nanosecond. In addition to adding the ground planes, other important design considerations include keeping the conductor lengths as short as possible and minimizing changes in geometry through the connector. Figure 8.37 provides examples of connectors with stripline and microstrip geometries.

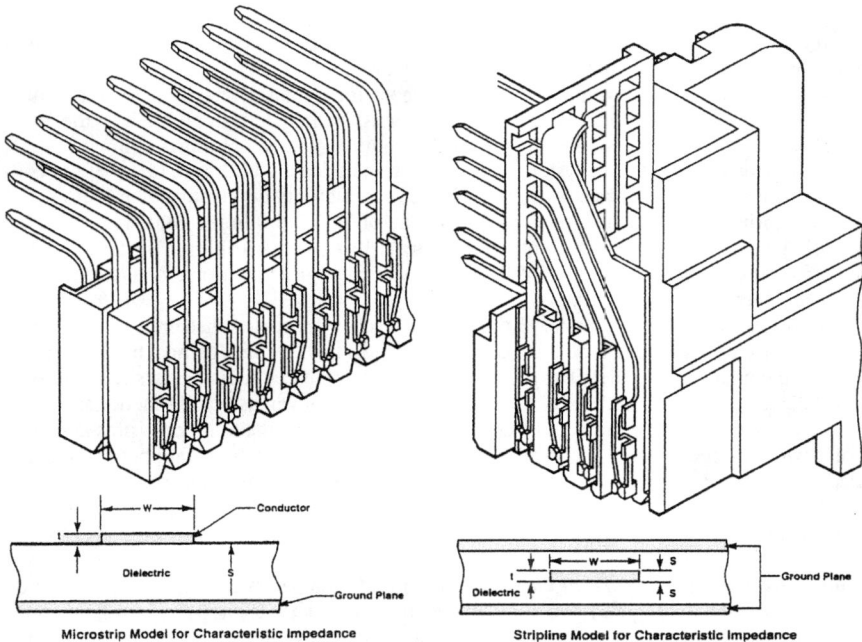

Microstrip Model for Characteristic Impedance   Stripline Model for Characteristic Impedance

FIGURE 8.37 Schematic illustration of (a) microstrip and (b) stripline ground plane connector constructions. (*Courtesy AMP Incorporated.*)

Ground-plane connectors offer advantages in both performance and density. Performance advantages result from the impedance control and reduced crosstalk because of the better shielding provided by a plane as compared to a pin. Density is improved by the fact that all the pins are available for signal applications. For a 100-position connector, an open pin filed connector with a 1:1 signal/ground ratio has only 50 pins for signals. A ground plane connector would use all 100 pins as signals and can often be built to the same, or only slightly larger, footprint as the open pin field connector. Another potential advantage is that the ground plane can also be used for power distribution providing additional effective density increases.

Summary Signal applications, in particular nanosecond and subnanosecond rise time digital pulse applications, require that connectors be considered as transmission lines.

Grounding becomes critical and various grounding schemes must be evaluated with application requirements in mind. In addition, shielding and electromagnetic interference must also be considered. These topics are not considered here, but were covered by Southard,[57] Sucheski and Glover,[58] and Aujla and Lord.[59]

8.8.2 Power Applications

Power distribution places considerably different requirements on contacts and connectors than those for signal distribution. Power distribution includes both voltage and current considerations. This section, however, focuses on current because the Joule heating, which accompanies current flow, introduces many application related concerns. Power connector design requirements to address voltage related issues involve mainly material selection and dimensions to withstand voltage breakdown effects.

Power connector requirements differ from those for signals primarily in the magnitude and stability of the connector resistance which must be realized. In addition to the Joule heating, I^2R, mentioned previously, power distribution connectors must minimize the IR drop they introduce into the system. The IR drop includes the bulk resistance of the connector as well as the interface resistance. Higher-conductivity copper alloys are often chosen as spring materials for power contacts in order to minimize bulk resistance as mentioned in Section 8.4. Millivolt drop budgets in power distribution systems are becoming increasingly stringent as PWB systems take on higher functionality and require increasing currents to the board.

There are two fundamentally different approaches to power distribution. They are, first, the use of one or a few dedicated power contacts and, second, using a higher number of signal contacts in parallel. The development of high pin count connectors, especially at level 3, has led to increased use of the second option. However, as board functionality continues to increase both current and I/O counts increase so that the number of pins available for power distribution becomes limited. Hybrid connectors which contain both signal and dedicated power contacts are seeing increasing use to address this concern.

The section begins with some fundamental principles of design for power connectors following Corman and Mroczkowksi.[60]

Connector design and current capacity   Important parameters include current (both dc and ac—and the duty cycle), the allowable resistance (both interface and bulk), and temperature (both contact and ambient). The discussion begins with the fundamentals of current flow across a contact interface.

For the purposes of this discussion, the most important aspect of the asperity structure of the contact interface is the size of the asperities because it dictates very high current densities. Joule heating at the asperity contacts can lead to a very high localized temperature referred to as the *supertemperature, T_s*. The supertemperature must be distinguished from the bulk temperature of the contact, T_b, and the temperature of the application, T_a. Three comments on supertemperature are important.

- The supertemperature depends on the voltage drop across the interface, therefore, all asperity contacts are at the same temperature.

- The response time of the asperity contacts is of the order of microseconds so supertemperature is determined by the peak current for any currently used power distribution frequency.

- A critical voltage exists at which the asperity contacts will attain a supertemperature sufficient to cause melting of the asperity interface. This voltage limits the peak current the interface can support. Because the voltage is given by IRc, the peak current capability can be increased by lowering the resistance of the interface. This is generally accomplished by increasing the normal force which also increases the contact resistance stability. The critical melting voltages for gold, silver, and tin are 430, 570, and 130 mV, respectively.

The contact bulk temperature will be determined by the balance of the Joule heating of the contact, which depends on the current and the bulk resistance of the spring, and the heat dissipation, which depends on the heatsinking provided by the permanent connection to the circuit component. This bulk heating effect is generally referred to as the contact T-rise as measured against the ambient or application temperature.

The application temperature is the temperature at which the equipment is operating at the connector location. It might be the ambient temperature or it might be a higher temperature, caused by heat generation by the equipment itself.

Supertemperature and T-rise are influenced by two different resistances, supertemperature by the contact interface resistance, and T-rise by the bulk resistance of the contact spring. Interface resistance limits the peak current the contact can carry, as mentioned, but bulk resistance limits the current rating. The T-rise that is generally, and arbitrarily, used to establish the current rating of a contact is 30°C.

For dc, the peak and "average" current are the same. For ac, the "average" current, with respect to T-rise, depends on the shape of the current waveform and the duty cycle. The duty cycle comes into play when it is realized that heat generation depends on power and time, so continuous ac and pulsed ac of the same "average" current results in different values for T-rise, depending on the duty cycle. Wise[61] discusses these issues.

Current rating Current rating of a contact is relatively straightforward. The same is not necessarily true of a connector. Current rating of multiple signal contacts used in parallel to distribute current can be a complex matter because it depends on interaction of the current rating of the individual contacts, the number of contacts carrying current and the distribution of the contacts in the connector housing. This subject is covered by Corman and Mroczkowski,[60] and is only briefly reviewed here.

The current rating of a contact is generally based on T-rise considerations which depend on the balance of heat generation, generally the Joule heat, and heat dissipation into the medium to which the contact is connected—the PWB or wire. A 30°C T-rise is commonly, and arbitrarily, used as a criterion and is derived from UL practice. The current rating is determined experimentally and, therefore, is highly dependent on the test conditions as discussed by Wise.[62] The current rating also depends on the application temperature because the combination of T-rise and ambient must not exceed the temperature rating of the connector, usually 105°C.

Because connectors contain multiple contacts, current distribution and the thermal effects between the contacts must be considered. The key factors include:

1. The current distribution among the contacts. A true parallel circuit, in which all contacts carry the same current, is preferred. Unequal distribution of current can lead to failures of an overloaded contact which can cascade through the connector.

2. The current-carrying contacts should be distributed throughout the connector, if possible, to minimize thermal coupling between the contacts. The loading of the connector and the percentage of the contacts carrying current affect the current rating for the same reason.

Power contact/connector summary Power applications of connectors share all the usual concerns of connector design with a few extra. Special attention must be paid to the magnitude and stability of contact resistance to ensure that voltage drops and temperature increases are minimized. The magnitude of the contact resistance determines the peak current the connector can support, and the overall resistance determines the T-rise. In parallel-contact power distribution, attention must also be paid to circuit design to ensure current distributions are uniform, or nonuniformities must be accounted for by allocating additional contacts.

8.9 CONNECTOR TYPES

The discussion of connector types in this section follows a format similar to that of levels of interconnection. Connector types are reviewed in terms of the permanent connections between the circuit elements being connected. This leads to three different categories: board to board, wire to board, and wire to wire.

8.9.1 Board-to-Board Connectors

The class of board-to-board connectors is often referred to as printed wiring board (PWB) connectors. Level-3 connections are board to board, by definition, but many level-4 connections are also board to board, either directly or via cable assemblies. Level-3 connectors must meet both high-performance and high-density requirements—high performance in the sense that they must often meet transmission line requirements, and high density in that the increasing functionality of PWB units has resulted in a significant increase in I/O requirements, and miniaturization has maintained or reduced connector size. The end result of these trends in an increase in connector density. Level-3 board-to-board connectors are available in pin counts over 1000 and at contact pitches of 2.54 and 1.27 mm, with 1-mm pitches in development. Level-4 board-to-board connectors see similar high-performance requirements but are generally under 200 positions and most commonly on 1.27- and 1.0-mm pitches.

The typical level-3 packaging structure is a daughter card to mother board connection, with the daughter cards plugged in perpendicular to the mother board and parallel to one another.

In level-3 applications, the mother board is generally a bussing system. In level-4 applications, the mother board might contain functional elements in addition to being a bussing system between the daughter cards. The daughter cards might be different subassemblies of the overall system (level 4).

Such systems might also be configured with parallel boards by using stacking board-to-board connectors, or baby boards plugged in parallel to the daughter boards as illustrated in Figure 8.38.

Because of their high pin counts, level-3 board-to-board connectors generally include alignment guides, polarizing, and keying features. In addition, mating forces become a major consideration. Level-3 board-to-board connectors are usually used by skilled personnel.

Level-4 connectors, as mentioned, have lower pin counts which reduces mating force concerns.

The two major classes of board-to-board connectors are card-edge and two-piece (Figure 8.39). The advantages and limitations of each style can be simply summarized.

The major advantage of card-edge connectors is cost, because of the fact that only a receptacle is needed because the PWB provides the plug half of the connector system. The major limitation arises from the tolerances, both thickness and warpage, on printed wiring

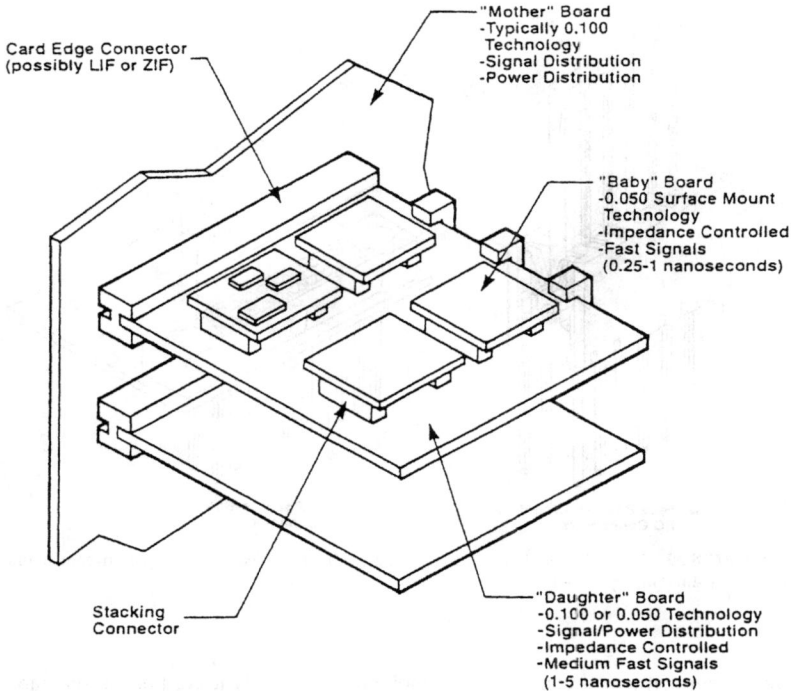

FIGURE 8.38 Schematic illustration of a variety of level-4 board-to-board connection possibilities. (*Courtesy AMP Incorporated.*)

boards. Variances in board thickness and warpage require that the receptacle contact be able to accommodate a very large deflection range. This, in turn, places severe limitations on the contact spring design to ensure that minimum contact forces and wipe distances are realized. Another packaging related limitation is the fact that such connectors can mate only to the edges of the board.

For two-piece connectors the situation is exactly opposite. Cost, relative to card-edge, is the major disadvantage of two-piece connectors. Among the major advantages is tolerance control. Two-piece connectors use stamped/formed posts and receptacles which can meet much tighter tolerances than a PWB. Two-piece connectors are seeing increasing usage for three reasons.

1. Shrinking connector centerlines
2. Surface mounting requirements
3. Increasing performance requirements including controlled impedance and lower signal/ground ratios

Each type of connector is covered in more detail.

Card-edge connectors Through-hole mounting is the most common method of providing the permanent connection of card-edge connectors to mother boards. Both soldering and

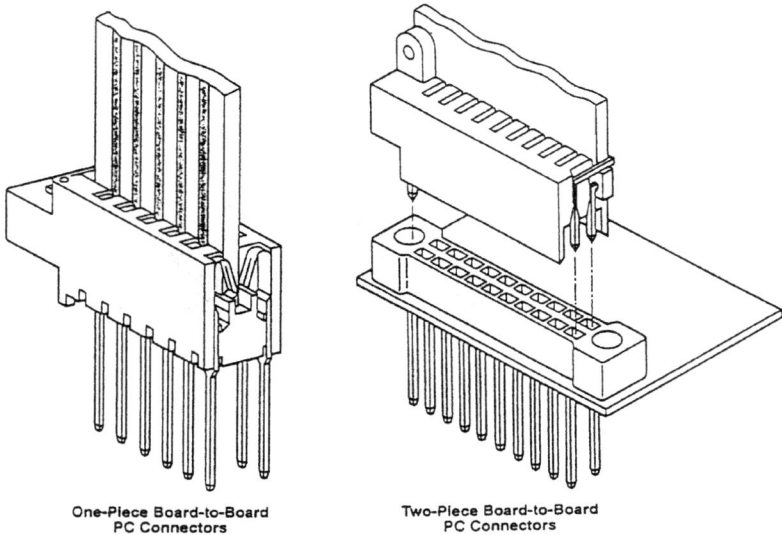

One-Piece Board-to-Board
PC Connectors

Two-Piece Board-to-Board
PC Connectors

FIGURE 8.39   Examples of (left) card-edge and (right) two-piece board-to-board connectors. (*Courtesy AMP Incorporated.*)

compliant pin permanent connection technologies are widely used. For the separable interface, card-edge connectors generally consist of two rows of spring contacts that mate with pads on the two sides of the printed circuit board. To meet increasing I/O requirements on the PWB, dual-level contacts are also used. An example of a dual-level connector is shown in Figure 8.40.

Although card-edge connectors are still in wide use, and innovative designs have been employed to enhance their performance, including low insertion force (LIF) and zero insertion force (ZIF) designs, there is an increasing trend toward two-piece connectors as I/O requirements continue to increase.

Two-piece connectors   Two-piece connectors offer advantages with respect to increasing pin count requirements. As mentioned, card-edge connectors are generally two-row connectors, which limits the density of contact pins. Two-piece connectors are not subject to this limitation and are available in up to eight rows and in configurations including ground planes. In addition, two-piece connectors can be placed anywhere on the board, which facilitates increasing packaging density through stacking connectors.

Two-piece connector development is advancing on two fronts. First, surface-mountable connectors are now available, and development efforts are continuing. Second, high-performance or high-speed connectors are available which provide improved electrical performance through the use of integral ground planes and controlled impedance as well as improved shielding. In addition, connectors with integrated filters that reduce noise and protect signal integrity are also available.

Two types of two-piece board-to-board connectors are predominant in the marketplace. These are Eurocard and high-density connectors.

Eurocard connectors derive their name from their origin and widespread popularity in Europe. True Eurocard connectors are available in 32, 64, and 96 positions based on one,

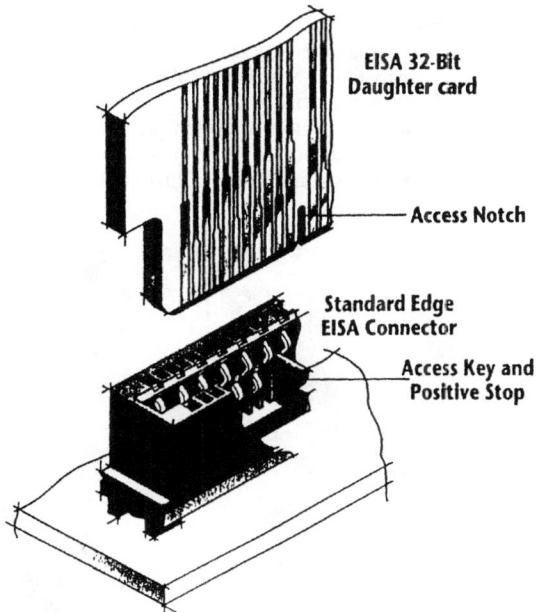

FIGURE 8.40 Schematic illustration of a dual-level card-edge PWB connector construction. (*Courtesy AMP Incorporated.*)

two, or three rows of the standard 32-contact array on 2.54-mm centerlines. Examples of Eurocard connectors are shown in Figure 8.41.

The Eurocard contact system consists of a 25 square post and a twin beam receptacle. The standard Eurocard uses a vertical receptacle and a right angle pin connector although

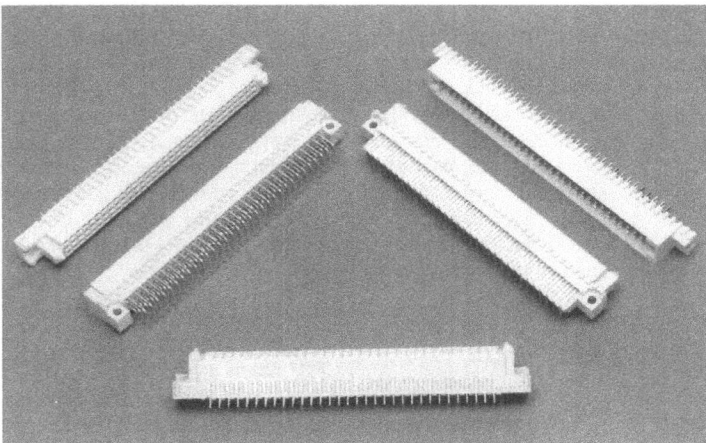

FIGURE 8.41 Examples of Eurocard connectors. (*Courtesy AMP Incorporated.*)

an inverse Eurocard is also used. Three performance levels are defined based on both mating cycles, contact resistance and the qualification test programs which are required.

High-density connectors are available with over 1000 positions and in lengths of over 50 cm. Many styles are available using from two to eight rows of two or four beam receptacle contacts. The high pin count connectors are often "sectionalized;" that is, groups of contacts are separated by spaces to incorporate guide pins to assist in connector alignment on mating. Examples of high-density connectors are illustrated in Figure 8.42.

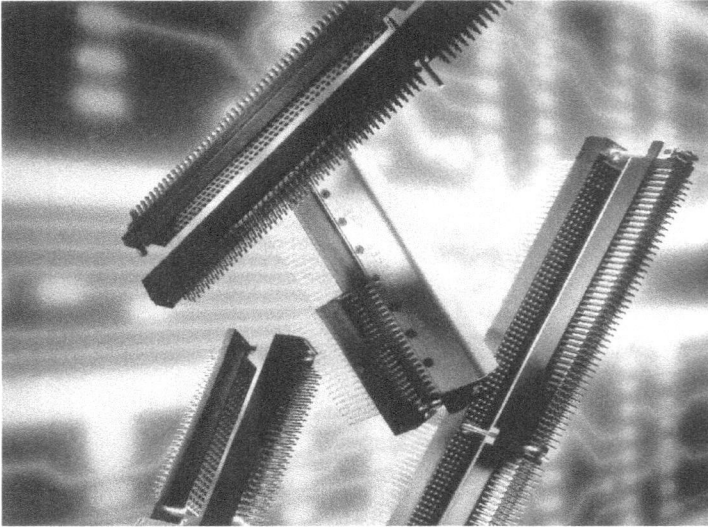

FIGURE 8.42   Examples of high-density connectors. (*Courtesy AMP Incorporated.*)

High-density connectors are also available in hybrids that contain special modules for coaxial, high-current, and fiber optic contacts, as schematically illustrated in Figure 8.43.

Sequential mating of ground, power, and signal can be accommodated by varying the lengths of the mating pin contacts. Along with density improvements, high-density connectors have also been developed to provide improved electrical performance in two ways. First, the high pin count capability increases the ability to decrease signal-to-ground ratios in open-field connectors. Second, ground planes have been incorporated in a number of configurations to provide microstrip and stripline geometry connectors, included in Section 8.8.

Two-piece connectors are seeing increasing usage because of higher I/O and performance capabilities. I/O capacity is realized by multiple rows and performance by the addition of ground planes to the connector.

8.9.2 Wire-to-Board and Wire-to-Wire Connectors

Wire-to-board and wire-to-wire connectors are combined in this discussion because the focus is on the permanent or wire connection. They differ at the separable interface as defined by the mating element, a board (or board mounted connector) or another wire or cable connector. A dominant technology in wire-to-board connectors is the 25 square

Straight pin connector
with cable-mount
power or coax contacts

Ports accept either
power or coax contacts

Power contact

Coaxial contact

Right-angle receptacle
connector with board-mount
power or coax contacts

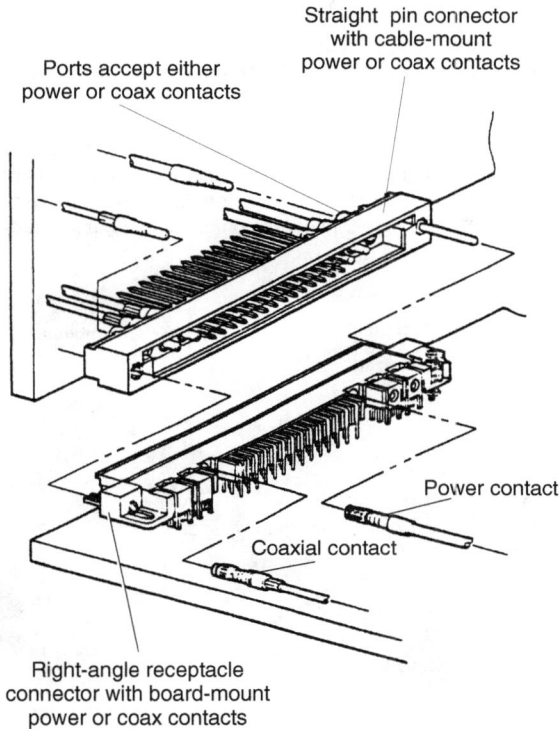

FIGURE 8.43 Schematic illustration of a hybrid connector construction. Signal, coax, and power contacts are shown. (*Courtesy AMP Incorporated.*)

technology. Wire-to-wire connectors can also be 25 square, but pin/socket connectors are also common.

The terminology, "wire," is used in this section, but these remarks also apply to cables. In this context, a cable is viewed as a protected wire (jacketed cable) or a number of wires arranged in a "controlled" geometry (ribbon cable). Wire-to-board connections are used in level 4, and wire-to-wire connectors are predominately levels 5 and 6.

The termination technologies used in wire-to-wire and the wire half of wire-to-board connectors are primarily crimping and IDC, covered in Section 8.7. Soldering is used to a limited extent, but is not considered here. For wire-to-board connectors, the connection technologies to the board (THT and SMT) have been described in Section 8.7. The information concerning wire-to-wire and wire-to-board types of connectors is combined because the basic structure of the connectors is similar in both applications. And, in fact, some types of connectors are used in both applications.

Wire/cable connector types are listed in order of their predominant level of application, from level 4 through level 6.

25 Square technology (levels 4 and 5) Receptacle connectors mating to a 25 square post represent a significant portion of electronic connectors in both high-performance and commercial applications. At level 4, connectors for both discrete wire and ribbon cable are common. The majority of these applications are based on 2.54-mm centerlines. As the need for miniaturization grows, similar 1.27- and 1.0-mm centerline connector systems based on 15 square posts are emerging.

The variety of level-4 applications ranges from interconnection of high-speed digital electronic subassemblies to connection of processors to peripherals and includes high-current power distribution. 25 square technology is used in board-to-board, wire-to-board, and wire-to-wire applications. This range of functionality also contributes to the variety of connector types. Examples of some 25 square connectors and applications are illustrated in Figure 8.44. Usage in level 5 is limited.

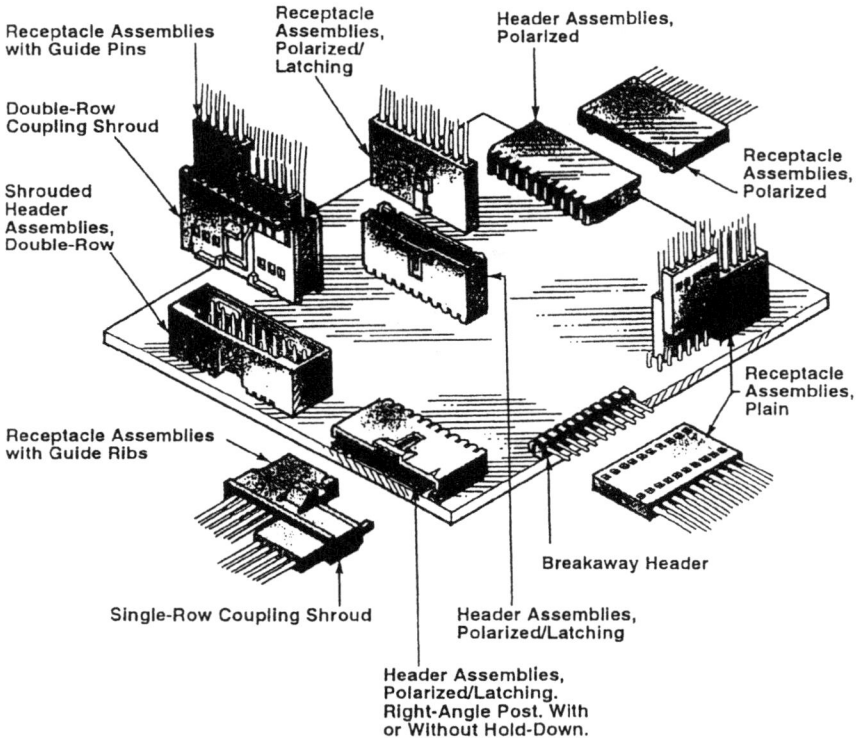

FIGURE 8.44 Schematic illustration of examples of 25 square connector applications. (*Courtesy AMP Incorporated.*)

The popularity of 25 square technology derives from its versatility.

- Modularization of connectors is readily accomplished.
- Common posts provide intermatability across a variety of connector types.
- Connectors are available in both precious metal and tin finishes and with a variety of spring and housing materials.

- Cost effectiveness is realized by the availability of a wide range of automatic assembly equipment.
- Performance capabilities are adequate for a broad range of applications.

Discrete wire connectors use both crimp-snap and IDC contacts. In crimp-snap applications, individual wires, discrete or bundled, are stripped and crimped into loose contacts, generally by automatic equipment, and the contacts are then snapped into the receptacle housing, again automatically. IDC contacts for discrete or bundled wire are generally preloaded into the housing and mass terminated.

Receptacle contact geometries include a variety of designs, some of which were illustrated in Figure 8.16. Receptacle contacts are generally made from phosphor bronze or beryllium copper alloy.

Both precious metal (gold or palladium alloy based) and tin finishes are used. Duplex plating, gold at the separable interface and tin at the permanent connection, is common on these connectors. Housings are made from a variety of polymers dependent on the application, with PBT as a major housing material.

Board-mounted post contacts can be discrete, but most commonly are mounted in headers, shrouded or unshrouded. Brass and phosphor bronze are the most common post materials. Precious metal and tin finishes are both used, including duplex plated versions.

Ribbon cable connectors are characterized by mass termination of multiple conductor cable into preloaded housings via IDC technology. In addition to the economics of mass termination, ribbon cable reduces wiring errors because the individual wires are not handled, and wire locations are fixed by the cable construction. Ribbon cable connectors allow easy daisy chaining because the connectors can be applied anywhere on a cable, which permits bussing from a single master cable. The majority of ribbon cable used today uses stranded copper conductors, 26 through 30 AWG, on 1.27 mm centerlines. Coaxial and shielded ribbon cable are also used. Cable preparation is more complicated for such applications. Examples of ribbon cable connectors are provided in Figure 8.45.

Receptacle contact geometries include tuning fork, twisted tulip, and open twin beam contacts. The dominant ribbon cable IDC connection technology is a slotted beam. The contacts are generally manufactured from phosphor bronze or beryllium copper alloys. Precious metal and tin finishes are used.

The same comments on posts apply as for discrete wire connectors. Some ribbon cable connectors are card-edge connectors and mate to PCBs directly.

Pin and socket connectors (levels 4, 5, and 6)   Pin and socket connectors are differentiated from post/receptacle designs by geometry. Pins, and the majority of sockets, are round rather than square. In contrast to 25 square posts, pins are seldom directly mounted to a PWB. Instead, connectors containing the pins are board mounted. Pin/socket contacts, in various sizes and types are inserted into a wide variety of housing sizes and shapes to meet the diverse needs of level-4 and level-5 application requirements. Among the most common of these connectors are rectangular, circular, soft shell, and subminiature D (also called *D sub*) connectors.

These connector families have many common components, and a number of hybrid versions with mixed power and signal contacts are used. Coax and fiber optic contacts are less common. The majority of signal contacts are pin/socket designs, both stamped and formed and screw machined. Stamped and formed contacts dominate commercial applications, and screw-machined contacts are preferred in military and aerospace applications. The contact materials include phosphor bronze and beryllium copper alloys for the receptacle contacts and phosphor bronze and brass for the pins. Precious metal and tin finishes are used in military/electronic and commercial applications, respectively. Screw machined

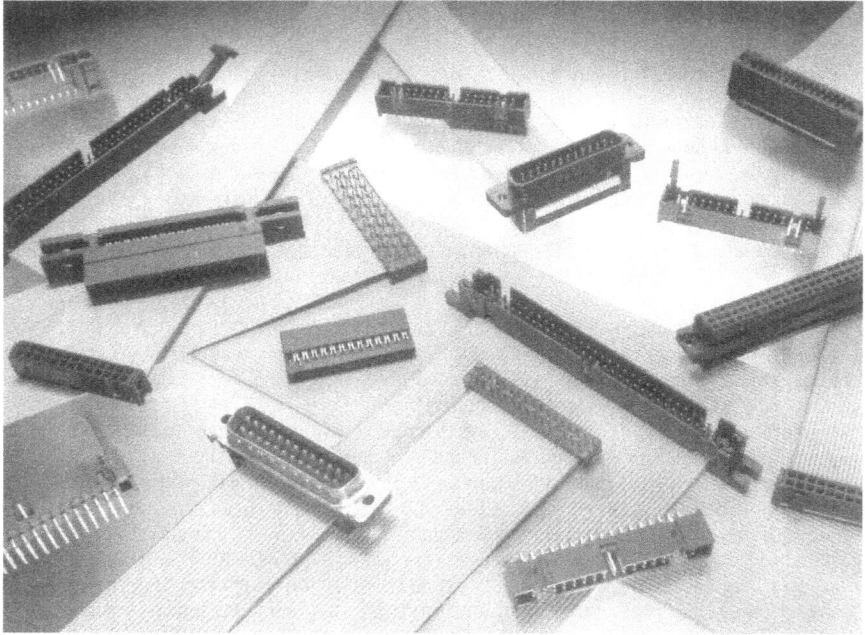

FIGURE 8.45   Examples of ribbon cable connectors. (*Courtesy AMP Incorporated.*)

versions are generally gold plated. Housings are more complex for these families in that many versions contain inserts. Housing materials include PBT and PPS as well as metal shells with inserts varying from polypropylene to thermosets, depending on whether the application is commercial or military.

Rectangular connectors are common in levels 4 through 6. They are characterized by their modularity and diversity with a wide variety of positions and contact combinations, both pin/socket and other geometries available, including signal, power, coax, and fiber optic contacts. These connectors also come in shielded and metal shell versions, primarily for levels 5 and 6. Rectangular connectors are generally discrete wire/cable connectors, regardless of whether they are wire-to-wire or wire-to-board, although ribbon cable use is increasing. Rectangular connectors are used in military, electronic, and commercial applications. Examples of rectangular connectors are provided in Figure 8.46.

With the exception of the shape of the housing, the same general comments apply to circular connectors. Most circular connectors are used in levels 5 and 6, and include many hybrid connectors to combine power distribution and signal functions. Examples of circular connectors are provided in Figure 8.47.

Subminiature D is the predominant level-5 connector, seeing wide use as an I/O connector for printers, disk drives, video, and even including low-cost video games. It also sees applications in level 6. The popularity of the line is based on a number of features.

• The line dates back to military connectors of the 1950s, providing a history of reliable performance.

- They are rugged and easily used, with the D-shaped shell providing polarization and the metal shell shielding benefits.
- They are covered in many standards.

In a typical level-5/6 D sub-application, one-half of the connector will be bulkhead mounted and the other end connected to an internal PWB by a cable. The level-6 half, which mates to the bulkhead connector, will also be a cable and connects to the desired peripheral through another connector that might be a different style than D sub, depending on the peripheral connected.

Subminiature D connectors come in a wide variety of styles, but the standards are based on two pin sizes, 20 (0.04 inch diameter) and 22 (0.03 inch diameter) with five different shell sizes to hold a number of contacts ranging from 9 to 50 in size 20 and 15 to 72 in size 22 contacts, with 25 being the most common number.

Both screw machined and stamped and formed pins and sockets are used. Brass pins and phosphor bronze sockets are most common in D sub connectors. D sub permanent connections include solder, crimped, and IDC, with crimping the most common wire connection. Board mounted versions include through-hole and SMT soldered connections with some compliant pin applications.

D sub connectors use an insert within the shell to hold the contacts. Inserts are usually nylon for commercial connectors and diallyl phthallate (DAP) or polyester for military connectors. Gold finishes predominate. Steel shells plated with tin or zinc are most common, but some plastic shells are used in low-cost versions. Examples of D sub connectors are provided in Figure 8.48.

FIGURE 8.46   Examples of rectangular connectors. (*Courtesy AMP Incorporated.*)

FIGURE 8.47 Examples of circular connectors. *(*Courtesy AMP Incorporated.*)*

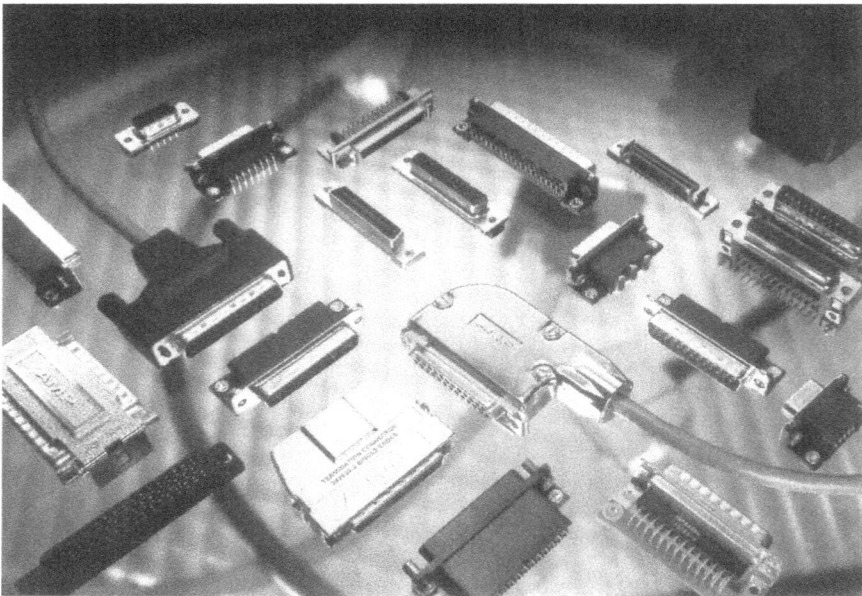

FIGURE 8.48 Examples of D subminiature connectors. *(Courtesy AMP Incorporated.)*

Miniature ribbon connectors are used in telecommunications, computer peripheral, test equipment, and medical industries. There are numerous standards and some de facto standards covering these connectors.

The miniature ribbon connector family includes board mount and cable connectors in both shielded and unshielded versions. There are also a number of specialty connectors in general use such as gender menders and back-to-back connectors that address particular application requirements. The contact spacing is on 0.085 in centerlines in standard pin counts of 14, 24, 36, 50, and 64 positions. The standard contact finish is gold over nickel, with a 200 mating cycle durability requirement. The mating interface is keystone shaped for polarization.

Cable assemblies might consist of a pair of miniature ribbon connectors or a miniature ribbon and D sub connector, depending on the application. Examples of miniature ribbon connectors are shown in Figure 8.49. Although miniature ribbon connectors are generally used in relatively low-speed applications, some shielded versions are suitable for pulse rise times down to 3 ns with proper attention to electrical design.

Soft-shell connectors were developed to fill a need for a small, lightweight, low-cost connectors with current capacities up to 15 A per contact for level-4 and level-5 applications in white goods, computers, medical equipment, and automotive markets, among others. To satisfy this diversity of applications, a wide variety of designs and configurations are available, some of which are illustrated in Figure 8.50.

Soft-shell connectors are generally low pin count, up to about 36 positions, and subject to multiple matings by manufacturers and consumers. The connectors are designed to meet safety and abuse conditions more severe than those of the connectors covered up to this point. Toward this end, they incorporate a variety of assembly/application aids including polarization systems, lead in chamfers, locking latches, strain reliefs, and contact protection features.

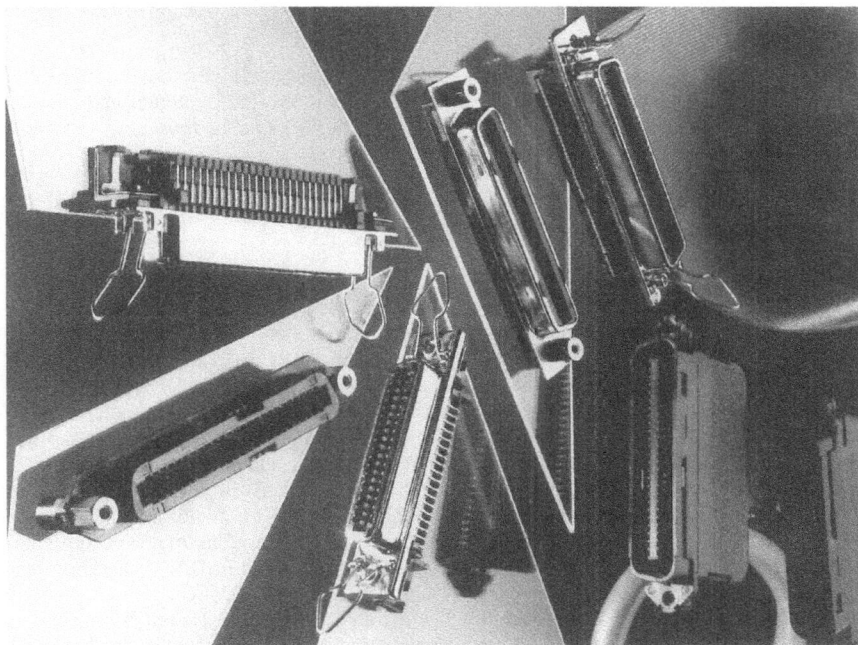

FIGURE 8.49   Examples of miniature ribbon cable connectors. (*Courtesy AMP Incorporated.*)

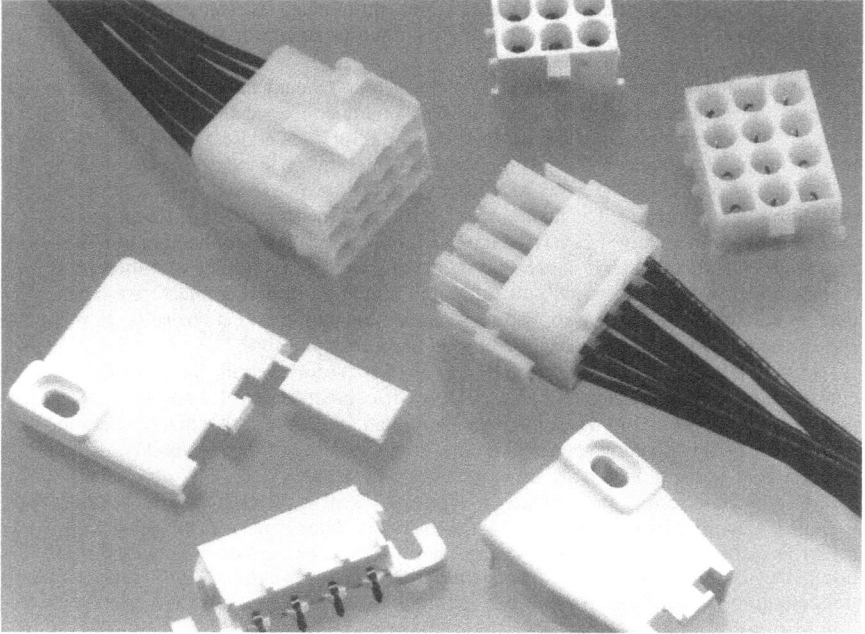

FIGURE 8.50 Examples of soft-shell connectors. (*Courtesy AMP Incorporated.*)

The contacts are generally pin/socket configurations and made from brass or phosphor bronze. Tin finishes predominate in commercial applications, but computer and medical connectors often use selective gold over nickel finishes. Nylon is the most common housing material, with UL 94V-O a common requirement.

8.9.3 Coaxial Connectors

Coaxial connectors differ from the connectors thus far in the chapter in that they are specifically intended for high-frequency applications. This difference is reflected in their design and materials of manufacture. A brief discussion of coaxial cable is also included.

Coaxial cable Coaxial cable is the dominant transmission medium for high-frequency level-5 and level-6 applications for two reasons. First, it is a controlled-impedance transmission medium that offers high performance at high frequencies, in the gigahertz range. Second, the coaxial construction provides inherent shielding and electromagnetic interference (EMI) and radio frequency interference (RFI) control. Even at lower frequencies, when the transmission characteristics are not required, EMI/RFI shielding benefits make coaxial cable the medium of choice in demanding applications. The major design challenge for coaxial connectors is to maintain the transmission and EMI/RFI shielding performance of the cable through the connection process and in application.

The basic structure of coaxial cable was illustrated in Figure 8.21, but is repeated as Figure 8.51. The four basic parts of a coaxial cable are the center conductor, the dielectric, the outer conductor, and the protective jacket. Using this basic construction, over 1000 dif-

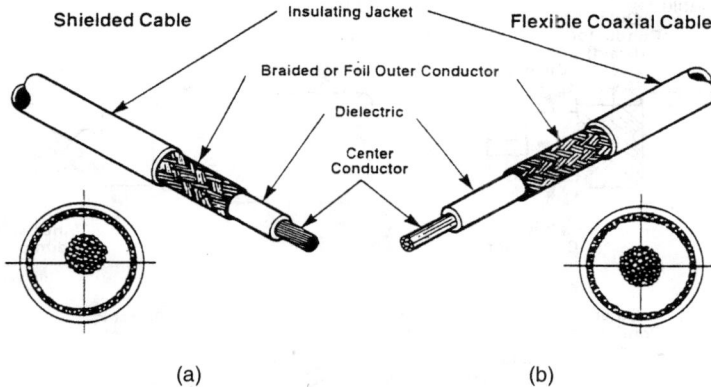

FIGURE 8.51   Schematic illustration of (a) shielded and (b) coaxial cable constructions. (*Courtesy AMP Incorporated.*)

ferent coaxial cables have been developed to meet a broad range of application requirements. Military applications dominated early developments, and many coaxial cables are described and specified in MIL-C-17. In addition, there are numerous commercial specifications, some highly specialized and some minor variants, to reduce costs of the military specification. Examples of common coaxial cable constructions, which extend beyond the basic geometry to meet increasingly demanding applications, are presented in Figure 8.52.

The most common coaxial cable, flexible cable, is illustrated in Figure 8.52a. It uses a braided outer conductor of small gauge wires. Although the braided construction provides flexibility, it does not afford complete shielding, because of gaps in the braid. To improve shielding, several braid layers can be used. In some cases, foil shields are also included. A shielding effectiveness of 85 percent is often considered the minimum acceptable. Some flexible cables use only a foil shield, which is lower cost but not as mechanically rugged as the braid construction.

It should also be noted that several technologies have been developed to incorporate air into the dielectrics to reduce the effective dielectric constant and thereby increase the velocity of propagation of coaxial cables in all constructions.

Twinax cable, shown in Figure 8.52b, consists of two inner conductors with a common outer conductor. The inner conductors can be twisted-pair or parallel-line geometries. The principal application of twinax is balanced-mode multiplex transmission in large computer systems. The signal is split between the two conductors and inverted to provide improved noise immunity. Similar to twinax, twin coaxial cable (Figure 8.52c) consists of two coaxial cables surrounded by a common outer conductor.

Coaxial ribbon cable (Figure 8.52d) consists of individual coaxial cables in a ribbon cable construction. It differs from ribbon or transmission cable in the presence of an outer conductor, which is generally a foil and drain wire combination.

A few words on signal transmission in coaxial cable are in order. It is important to note that the electromagnetic field is transmitted through the dielectric while current flows on the outside of the inner conductor and on the inside of the outer conductor. Performance characteristics therefore depend on the conductor and dielectric materials and the geometric relationships of the cable construction.

Coaxial cable performance parameters of general interest include characteristic impedance, capacitance, voltage standing wave ratio (VSWR), velocity of propagation (V_p), and attenuation. These parameters are not included here.

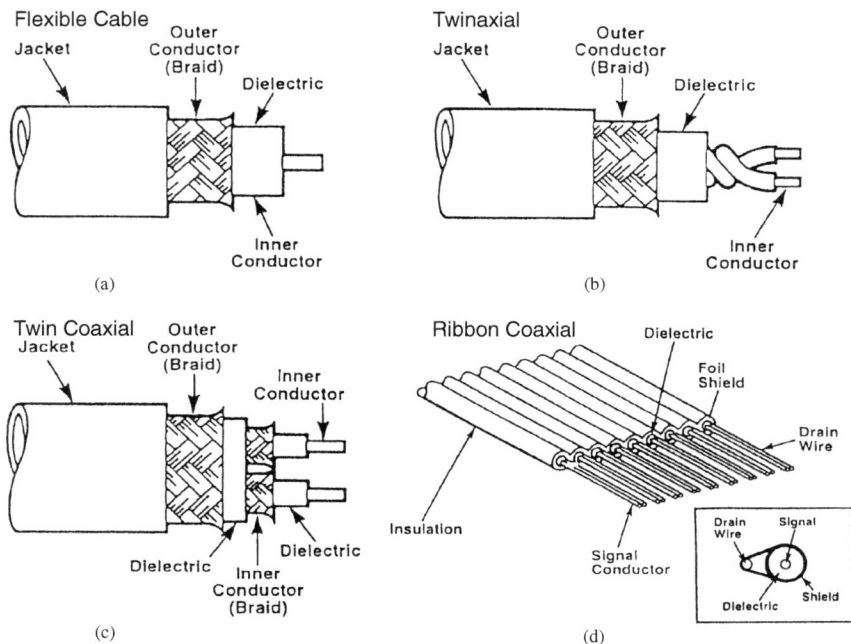

FIGURE 8.52 Schematic illustration of cross sections of selected coaxial cable constructions. (*Courtesy AMP Incorporated.*)

Coaxial connectors A coaxial connector ideally should appear as an extension of the cable. In other words, the connection process should not have an unacceptable impact on the performance parameters just listed. In practice, however, the connector will always introduce some nonuniformity in impedance—even for controlled-impedance connectors. Sources of nonuniformity include variations in the cross section of the contacts, variations in the dielectric medium, and abrupt changes in direction of the contacts. The effect of these nonuniformities will be frequency dependent with the magnitude of the variation increasing with frequency. VSWR is a measure of the connector nonuniformity, and a value of 1.3 is often specified as a limit.

As for cables, a MIL specification, MIL-C-39012, covers many types of coaxial connectors although numerous commercial specifications and proprietary designs are also in common use. Coaxial connector selection is generally determined by the frequency range, cable size, and the coupling method.

The maximum frequency of the application, signal rise time dependent frequency, is a primary consideration. Connector characteristics that are strongly impacted by the application frequency include the coupling mechanism and the connection method.

The three major coupling mechanisms are: bayonet, snap-on, and threaded. Bayonet and snap-on coupling, although simple, carry a penalty in overall shielding effectiveness and in noise during vibration. Threaded couplings offer a higher degree of resistance to shock and vibration as well as increased shielding performance because of the continuous shielding provided by this coupling mechanism.

Connection methods are arguably even more important in coaxial connectors than in conventional connectors because of cable preparation requirements and the VSWR penalties that accompany improper connections.

The three main methods of making connections to coaxial cable are:

• Solder/clamp
• Solder/crimp
• Crimp/crimp

The first term describes the connection technique for the center conductor, and the second the connection to the braid. The center contacts are usually pin/socket configurations, as these are the most spatially uniform contact geometries.

In solder/clamp connections the center conductor is soldered to the contact, and the braid is clamped to the connector shell. This method provides easy field repairability but is operator sensitive and does not lend itself well to high-volume production.

Solder/crimp connections are the most common with the center conductor being soldered to the contact and the braid crimped, generally with a ferrule, to the connector shell. This method lends itself to volume production but requires tooling for field repair.

Crimp/crimp connections are especially suited to high-volume manufacturing with dedicated tooling. In addition, crimp/crimp connections provide the most consistent VSWR performance and reliability.

The connector size is determined by two factors. The obvious factor is the cable size. Coaxial connectors are described as subminiature, miniature, small, and large. Cable size, in turn, is affected by both physical constraints and attenuation considerations. Signal attenuation, of course, increases as cable size decreases.

Material selection for coaxial connectors affects both cost and performance. The base materials for the contacts are similar to other connectors with brass, phosphor bronze, and beryllium copper being the usual contact spring materials. Connector shells and ferrules include tin- and nickel-plated brass and stainless steel.

Gold is the most common center contact finish, although silver and tin are also used in lower-cost applications.

The dielectric material is usually either PTFE or polypropylene. PTFE is the superior material because of a wider temperature range and a lower, less lossy, and more stable dielectric constant. Polypropylene is generally used only in commercial applications.

Coaxial connector types are defined in terms of the coupling mechanism, the connection mechanism, and the cable style to which they are connected. Examples of the connector types to be described are presented in Figure 8.53.

BNC connectors are bayonet coupling connectors, most of which are 50-Ω impedance and suitable for applications up to 4 GHz. Connection styles depend on the application (military or commercial) and include single and dual crimp, single and dual hex crimp, and other proprietary connections. BNC connectors are used primarily as I/O connectors in networks, instrumentation, and computer peripheral applications, level 5.

TNC connectors differ from BNC in that the coupling is threaded, which provides improved performance to 11 GHz and enhanced vibration and shock resistance. TNC connectors are available in essentially the same variety of styles as BNC.

SMA connectors were developed for subminiature coaxial cable and are widely used in avionics, radar, and high-performance test instrumentation applications. They are 50 Ω threaded coupling connectors and meet MIL-C-39012 up to 12.4 GHz. Special-crimped SMA connectors are suitable up to 26 GHz. They are available in a similar range of styles to BNC.

SMB connectors are a snap-on coupling version of SMA and, as such, have a lower performance capability, to 4 GHz.

SMC connectors are also 50-Ω threaded coupling connectors, but because of their smaller size they have a lower performance rating than SMA. The reduction in performance comes from the proportionally larger tolerance variation as connector size is reduced. These variations lead to an increased VSWR.

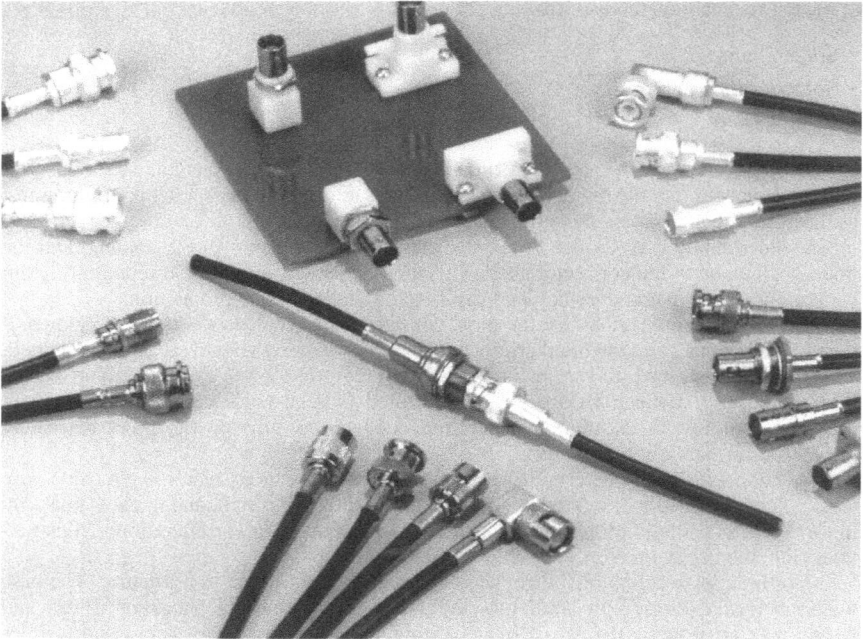

FIGURE 8.53   Examples of coaxial connectors. (*Courtesy AMP Incorporated.*)

Coaxial connectors for twinax and other coaxial cable variants are also available. Space precludes discussion of this variety of connector types, many of which are proprietary.

The number and variety of high-performance, high-frequency applications is continually increasing. Coaxial connector types and performance requirements are changing along with this trend. A wide variety of connectors is available for applications including satellite and microwave communications, electronic counter-measures, computer networking, and increasingly sophisticated test equipment.

8.9.4 Summary of Connector Types

This section has covered only some basic types of connectors and variants. The variety of connector types exists and keeps increasing to satisfy the ever expanding and increasingly sophisticated functional requirements of electronics technology.

8.10 CONNECTOR TESTING

Connector testing is performed for several reasons, primary among which are development or performance verification and product qualification. In addition, reliability testing is receiving increased attention.

8.10.1 Connector Testing

In general, a test consists of a combination of an exposure or conditioning process, intended to simulate an application condition or environment, followed by a measurement, generally related to some performance characteristic. For example, a heat-age test might consist of an exposure to 1000 hours at 125°C to assess the effect of stress relaxation on contact normal force, followed by a contact resistance measurement—a measure of performance.

The conditioning provided by a given exposure can be mechanical (vibration, shock), thermal (steady-state heat age or temperature cycling), or environmental (mixed flowing gas or humidity). Measurements include electrical (contact resistance, dielectric withstanding voltage), mechanical (pull strength, mating force), and visual (dimensions, workmanship).

Connector testing might consist of individual or sequential tests, depending on the purpose. This section is limited to an overview, but with an emphasis on qualification testing. Additional issues of concern for reliability assessment are also included.

Qualification testing is intended to verify that a given product, produced under defined manufacturing processes, is capable of meeting the performance and dimensional requirements of the manufacturers product specification or of an industry derived specification. In general, the specification requirements will have been developed to ensure that the connector will meet the application requirements of the particular market/application for which it was developed.

Types of tests In this section, the type of test is defined by the exposure condition (e.g., vibration followed by contact resistance is a mechanical test), and heat age followed by mating force is an environmental test. A qualification test program consists of at least the following:

1. Sample validation
2. Mechanical tests
3. Electrical tests
4. Environmental tests

Sample validation includes verification of dimensions, color, polarization mechanisms, and other general features of the product family as appropriate, including quality of workmanship. In addition, the materials of manufacture and design parameters, such as contact finish thickness and normal force, can be verified. The purpose of these evaluations is verify that the samples are representative of the product that is to be qualified.

Mechanical exposures include:

- Durability cycling
- Vibration, sinusoidal or random
- Shock
- Acceleration
- Cable bending

  Basic mechanical measurements include:

- Contact engaging and separating force
- Connector mating and unmating force
- Contact retention force in the housing
- Connection tensile strength
- Latching mechanism strength

Other mechanical tests are used and might be standard or proprietary tests to meet specific product or industry requirements.

Electrical exposures include:

- Overload current
- Current cycling
- Dielectric breakdown/withstanding

Electrical measurements cover a broad range, including

- Contact resistance
- Overall resistance
- Termination resistance
- Insulation resistance
- T-rise versus current
- Capacitance
- VSWR

The same comments apply as for mechanical testing. Contact resistance measures the interface resistance, termination resistance measures the resistance of the permanent connection, and overall resistance measures the resistance across the total connector and insulation resistance the performance of the housing material. T-rise, capacitance, and VSWR are application-dependent tests.

Environmental tests are intended to the verify stability of connector performance under the anticipated conditions of the application environment. The measurements just described might be performed, but only after conditioning, such as humidity or temperature exposure. Examples of environmental conditioning include:

- Humidity (steady-state and/or cyclic)
- Temperature/humidity cycling
- Mixed flowing gas
- Salt spray

Test sequences   A qualification testing program generally consists of a number of test sequences which are performed serially or in parallel. The test sequences and test groups are usually specific to an exposure condition, i.e., mechanical, electrical, or environmental. Different sets of samples are used for different sequences because different measurements and fixturing are required (Table 8.4)

Summary   The relevance of qualification testing is dependent on ensuring that the test exposures and measurement methods and procedures are appropriate for the product and the intended application. Testing in addition to product qualification can be directed toward verifying the reliability of the connector, in which case application considerations become even more important.

8.10.2 Estimating Connector Reliability

A definition of reliability appropriate to this discussion is:

Reliability is the probability of a product performing a function, for which designed, under specified conditions for a specific period of time.

In evaluating connector reliability, the "probability" would be the required reliability, or failure rate, that the connector must provide. "Function" relates to performance criteria, examples of which include the allowable change in contact resistance, T-rise, or the specified number of mating cycles. The "specified conditions" include the operating environment and application requirements. Operating environment conditions of importance include the application temperature, chemical, and mechanical considerations. Application requirements include the required number of mating cycles and stability of connector resistance. The time factor could be the intended lifetime of the product in which the connector is used or an acceptable maintenance interval.

The three major methods for assessing the reliability of a connector are: comparative, prediction based on design and materials of manufacture, and a physics of failure approach (which includes an appropriate test program, followed by statistical analysis of the relevant data).

Comparative reliability Reliability assessment via comparison has many limitations. If a particular reference connector, connector A, has a "known" field history, connector B, if it has common materials of manufacture and design parameters to connector A, might be assumed to have a similar reliability. This approach is often tacitly used by comparing connector product specifications, which are based on qualification test programs, against one another. The validity of the assumption of equivalent reliability has become more questionable as connector performance requirements and connector designs have become more sophisticated and demanding.

To address this increased performance sophistication, a comparative approach can be made more relevant by the use of comparative testing. A test program that exercises the reference connector and the alternative connectors in a manner relevant to the intended operating environment can be used to assess relative performance. Such comparisons, al-

TABLE 8.4 Typical Connector Qualification Test Sequence

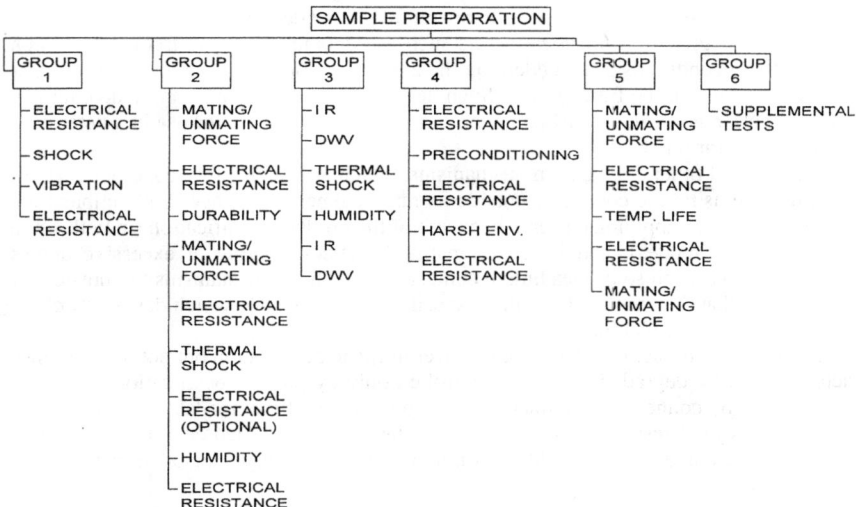

though they might provide a hierarchy of performance, cannot provide a quantitative measure of reliability. Even if a "reliability" based on field experience exists for connector A, the validity of extrapolation of this number to the comparative testing data is dependent on the correspondence of the test program to the field conditions.

Reliability prediction In military and aerospace applications, component and connector reliability is often predicted based on generic failure rate data bases and modeling, usually according to MIL-HDBK-217, *Reliability Prediction of Electronic Equipment*. Although this approach might be appropriate for some components, its application to connectors is questionable for at least two reasons. First, the data base for failure rate values is, in general, not well documented either with respect to failure criteria or operating environment. Second, the data do not generally distinguish among connector materials and design features intended to enhance connector performance. For these reasons, this approach is not considered further. For additional information, see Mroczkowski and Maynard.[63]

Physics of failure The physics of failure approach to connector reliability prediction requires consideration of the following:

1. The active degradation mechanisms must be identified and categorized with respect to their importance in the application of interest.
2. Appropriate tests, acceleration factors, and exposures must be known, defined, or determined for these degradation mechanisms.
3. Failure criteria appropriate to the application of interest must be established.
4. The statistical approach to determining, or calculating, reliability values must be agreed upon.

Degradation or failure of a contact or connector can occur in many ways. It is convenient to divide the mechanisms into two categories: intrinsic and extrinsic. Intrinsic degradation mechanisms refer to those mechanisms, which are related to the design and materials of construction of the contact or connector. Extrinsic mechanisms are those which are related to the application of the contact or connector.

Examples of intrinsic degradation are corrosion, loss of normal force through stress relaxation, and Joule heating leading to temperature related degradation.

Extrinsic degradation includes factors such as contamination and fretting corrosion. Each of these conditions is dependent on the application of the connector, both in manufacturing and use in the final system. Such degradation mechanisms can be qualitatively assessed but, in general, are difficult if not impossible to quantify for use in a determination of connector reliability.

Examples of other degradation mechanisms, which are outside the scope of our definition, include using the connector outside its rated temperature range (both ambient and enclosure related), applying currents in excess of the product specification (in both single and distributed modes), and improper mating practices (mating at excessive angles, pulling on cables, and so on) leading to contact abuse. These mechanisms fall outside our definition in that they violate the "under specified conditions for which designed" clause of our definition of reliability.

Connector manufacturers have control over intrinsic degradation but not over extrinsic factors. Extrinsic degradation can be controlled only by proper specification of product performance by connector manufacturers and proper use of the available information by the user. This joint responsibility ensures that the "under specified conditions for which designed" section of the reliability definition is met. Extrinsic degradation cannot be straightforwardly analyzed, so coverage of the topic is limited to a few aspects of intrinsic degradation to provide a suggested approach to connector reliability evaluation.

Only one failure mode, increases in contact resistance, is covered, although many others exist, both mechanical (broken latches, bent pins, and so on) and electrical (crosstalk, leakage between contacts, and so forth).

Consider three degradation mechanisms: corrosion, stress relaxation, and wear. Each of these mechanisms can lead to connector failure, directly or indirectly, through increased contact resistance. Corrosion of the contact interface directly results in increases in contact resistance. Stress relaxation results in loss of contact normal force which, in turn, can lead to contact resistance increases, either directly or through increased susceptibility to mechanical or corrosive degradation. Plating wear can lead to contact resistance degradation if wear through occurs to the contact spring material which might be susceptible to corrosion.

These degradation mechanisms are well established, and the failure cause associated with them—increased contact resistance—is an obvious factor in contact reliability. Thus, information is available relevant to our first concern, degradation mechanisms. Experience with the product, or similar products or applications, will allow us to categorize and rank the degradation mechanisms.

Determination of an acceleration factor for a test exposure requires that a relationship between test methods/data and field experience be developed. Simply put, the objective is to be able to state that X days of test A are equivalent to Y years in application B. In other words, an acceleration factor between the test exposure and field performance can be defined. Such a capability is necessary because one of the components of the definition of reliability is performance "for a specified period of time." Unfortunately, there are few tests for which such a statement can be made. One possible example is the work done at Battelle Columbus Laboratories,[21] which provides an approach and data base from which an acceleration factor for mixed flowing gas (MFG) corrosion testing of noble metal finished connectors can be developed.

Given an acceleration factor, a statistical method to use the data obtained to quantify reliability is needed. There are also a number of additional statistical analysis issues which must be addressed to assess connector reliability. Mroczkowski and Maynard[62] discuss this topic in detail. The type of data, variables or attribute and the acceptance criteria are of particular interest.

In a corrosion test, such as MFG, contact resistance can be measured, and variables data and the distribution of the data can be determined. Alternatively, the contact resistance can be compared to some acceptance criterion and recorded as pass/fail attribute data. The statistical treatment of the data and the sample size needed for a given statistical confidence level varies significantly for these two options.

One parameter which enters into both methods of data handling is the "acceptance" or "failure" criterion applied to the data. For attribute data, it appears directly. In the case of variables data, it will appear in the K factor, a statistical parameter, which reflects the relationship between sample size and the reliability, and confidence level, which can be obtained from a test program. In either case, the choice of "acceptance" or "failure" criterion directly influences the calculated reliability.

As an example of a failure criterion, consider the allowed change in contact resistance after exposure to the test program. What is the appropriate value of contact resistance to use in calculating contact reliability? Consider two possibilities, the product specification and an application related value.

The product specification contact resistance is not the proper choice. This value has a "reliability" aspect to it in the sense that the manufacturer has tested the product design with respect to ensuring that the specified contact resistance will be maintained in its intended range of applications. In effect, the product specification value includes a "safety factor."

In a particular application, a user will have established a value of contact resistance at which the system of interest will cease to function. The value might be 100 mΩ or more in a signal application, or 0.5 mΩ or less for a power contact. The "failure" criterion for contact resistance should be based on this application specific value rather than the product specification contact resistance. The desired requirements on confidence limit and reliability should then be applied with this resistance value as the upper limit of acceptability.

From contact to connector reliability The discussion up to now has been concerned with contact reliability and contact resistance as a degradation mode. The ultimate concern is with connector reliability, which would generally include a number of contacts and degradation modes. A methodology for translating contact reliability, as determined for the desired range of degradation mechanisms, into connector reliability is required. The issue is complicated by two interacting factors. The resistance data is obtained for individual contacts, but the effect of the test on contact resistance is influenced by the fact that the contacts are, generally, in a connector and are part of a system. In other words, the effects of connector housing design and the particular connector system on contact resistance are included in the data. Under these conditions, the reliability being measured is the contact reliability in the specific connector. The contact reliability might not be the same in a different housing geometry or connector system. An important part of developing a reliability testing program is to account for this fact. Aside from this consideration, which is far from trivial, there is another factor that must be considered before contact reliability is used to determine connector reliability, and that is the issue of multiple degradation mechanisms and multiposition connectors.

One approach to this issue is to treat all the degradation mechanisms and individual contact reliabilities as independent. Under this assumption, the contact reliability would be the product of the reliabilities for the individual degradation mechanisms. The connector reliability, in turn, would be the contact reliability to the power of the number of positions in the connector, assuming a failure at any position in the connector results in a system failure.

For example, the reliability for a 16-position connector, which is to be qualified for five degradation mechanisms, would be given by:

$$R = \{R(t_1)R(t_2)R(t_3)R(t_4)R(t_5)\}^{16}$$

where $R(t_n)$ is the contact reliability as determined for test n, which is intended to simulate degradation mechanism n. If it is assumed that the contact reliability happened to be the same for each test, the connector reliability will be given by the contact reliability raised to the fifth power for the five tests, and to the 16th power for the number of positions. For the connector then, the contact reliability will be raised to the 80th power!

This 80th-power dependence will significantly impact at least two aspects of qualification programs, cost and contact reliability requirements. For a connector reliability of the order of 0.9999, the contact reliability requirement will be of the order of 0.999999. Achieving a reliability of this magnitude is one problem (a significant one, indeed), but another is verifying that it has been achieved. The sample size for such a verification could be very large and affect program cost.

Summary of reliability assessment The purpose of this discussion was to present some of the issues and considerations which are pertinent to determining and calculating connector reliability on a statistical basis. In such an approach, a reliability qualification program consists of the following steps:

1. Determine an application specific contact resistance acceptance criterion. A criterion will also be required for any other failure mode that is to be included in the qualification program.
2. Develop a test program to address the expected degradation mechanisms operative in the application. The ranking of failure modes can be considered in this process.
3. Derive acceleration factors, when possible, for the tests to be specified. When this cannot be done, no reliability prediction can be made. In such cases, only comparative performance capabilities can be provided.
4. Decide on the statistical treatment appropriate to the data generated in the qualification program.
5. Calculate the component reliability. It must be emphasized that both the connector manufacturer and the user should agree on the content, approaches, and values to be specified in these steps individually, and in the qualification program in general. In particular, mutually agreed engineering judgements must be made to select appropriate "acceptance/failure" criteria and acceleration factors for the program.

8.10.3 Summary of Connector Testing

The purpose of this section has been to provide an overview of connector testing and reliability evaluation. General practices and procedures have been reviewed, although it must be noted that specific product and application requirements could dictate different approaches. In general, however, the major purpose of connector testing is to verify product qualification so that selection of the product for a given application based on product specification values will result in satisfactory performance in the field.

8.11 REFERENCES

1. Granitz, R.F., "Levels of Packaging," *Inst. Control Syst.*, Aug., 1992.
2. Mroczkowski, R.S., "Materials Considerations in Connector Design," in *Proc. 1st Elec. Mat. Conf. of the American Soc. for Materials*, Chicago, IL, 1988.
3. Williamson, J.P.B., "The Microworld of the Contact Spot," in *Proc. 27th Ann. Hole Conf. on Elec. Contacts*, Chicago, IL, 1981.
4. Bowden, F.P. and Tabor, D., *The Friction and Lubrication of Solids*, Clarendon Press, Oxford, U.K., 1986.
5. Holm, R., *Electric Contacts*, Springer-Verlag, New York, NY., 1967.
6. Antler, M., "Wear of Contact Finishes: Mechanisms, Modeling and Recent Studies of the Importance of Topography, Underplate and Lubricants," in *Proc. 11th Ann. Conn. and Interconn. Tech Symposium*, Cherry Hill, NJ, 1978.
7. Durney, L.J., ed. Electroplating Engineering Handbook, Van Nostrand Reinhold, New York, 1981.
8. Reid, F.H. and Goldie, W., eds., *Gold Plating Technology*, Electrochemical Publications Ltd., 1974.
9. Harlan, C., "Overview: Inlay Clad Metal and Other Electrical Contact Materials," *Metals Prog.*, Nov., 1979.
10. Mroczkowski, R.S., "Connector Contacts: Critical Surfaces." *Adv. Mat. and Processes*, Vol. 134, Dec. 1988.

11. Zimmerman, R.H., "Engineering Considerations of Gold Electrodeposits in Connector Applications," presented at *10th Technical Convention on Electronic Components*, Milano, Italy, 1973.

12. Antler, M. and Sproles, E.S., "Effect of Fretting on the Contact Resistance of Palladium," *Trans. IEEE CHMT*, 5, 1982.

13. Whitley, J.H., Wei, I.Y., and Krumbein, S.J., "A Cost Effective High Performance Alternative to Conventional Gold Plating on Connector Contacts," in *Proc. 33rd Elec. Comp. Conf.*, Orlando, FL, 1983.

14. Whitelaw, K.S., "Gold Flashed Palladium Nickel for Electronic Contacts," *Trans. Inst. Met. Finish.*, 64, 1986.

15. Antler, M., Drozdowicz, M.H., and Haque, C.A., "Connector Contact Materials: Effect of Environment on Clad Palladium, Palladium-Silver Alloys and Gold Electrodeposits," *Trans. IEEE CHMT*, 4, 1981.

16. Bock, E.M. and Whitley, J.H., "Fretting Corrosion in Electrical Contacts," *Proc. 20th Ann. Holm Conf. on Elec. Contacts*, Chicago, IL, 1974.

17. Antler, M., "The Tribology of Contact Finishes for Electronic Connectors: Mechanisms of Friction and Wear," *Plat. and Surf. Fin.*, 75, Oct., 1988.

18. Antler, M., "The Tribology of Contact Finishes for Electronic Connectors: The Effect of Underplate, Topography and Lubrication." *Plat. and Surf. Fin.*, 75, Nov., 1988.

19. Bader, F.E., Sharma, S.P. and Feder, M., "Atmospheric Corrosion Testing of Connectors—A New Accelerated Test Concept," in *Proc. 9th International Conf. on Elec. Contact Phenomena*, 1978.

20. Abbott, W.H., "Field Versus Laboratory Experience in the Evaluation of Electronic Components and Materials," in *NACE Corrosion/83*, Anaheim, CA, Paper 234, 1983.

21. Abbott, W.H., "The Development and Performance Characteristics of Mixed Flowing Gas Test Environments," in *Proc. 33rd IEEE Holm Conf. on Elec. Contacts*, 1987.

22. Mroczkowski, R.S., "Corrosion and Electrical Contact Interfaces," in *NACE Corrosion/85*, Paper 235, 1985.

23. Geckle, R.J. and Mroczkowski, R.S., "Corrosion of Precious Metal Plated Copper Alloys Due to Mixed Flowing Gas Exposure," *Trans. IEEE CHMT*, 14(3), 1991.

24. Wagar, H.N., "Principles of Electronic Contacts," in *Physical Design of Electronic Systems*, Vol. III, Chap. 8. Prentice-Hall, Englewood Cliffs, NJ, 1971.

25. Abbott, W.H., and Schreiber, K.L., "Dynamic Contact Resistance of Gold, Tin and Palladium Connector Interfaced During Low Amplitude Motion," in *Proc. 27th Ann. Holm Conf. on Elec. Contacts*, Chicago, IL, 1981.

26. Antler, M., "Contact Materials for Electronic Connectors: A Survey of Current Practices and Technology Trends in the U.S.A.," *Plat. and Surf. Fin.*, 78, June, 1991.

27. Lowenthal, W.S., Harkness, J.C., and Cribb, W.R., "Performance Comparison in Low Deflection Contacts," *Proc. Internepcon/UK84*, 1984.

28. Horn, K.W. and Zarlingo, S.P., "Understanding Stress Relaxation in Copper Alloys," *Proc. 15th Ann. Conn. and Interconn. Tech. Symposium*, 1983.

29. Bersett, T.E., "Back to Basics: Properties of Copper Alloy Strip for Contacts and Terminals," *Proc. 14th Ann. Conn. and Interconn. Technology Symposium*, Philadelphia, PA, 1981.

30. Spiegelberg, W.D., "Elastic Resilience and Related Properties in Electronic Connector Alloy Selection," in *Proc. ASM Int'l 3rd Electrical Materials Conf.*, San Francisco, CA.

31. Alvino, W.M., *Plastics for Electronics: Materials, Properties and Design Applications*, McGraw-Hill, New York, NY, 1995.

32. Walczak, R.S., McNamara, P.F., and Podesta, G.P., "High Performance Polymers, Addressing Electronic Needs for the '90's." *Proc. 21st Int. Conn. and Interconn. Tech. Symposium*, Dallas, TX, 1988.

33. Gupta, G.W., "High Performance Resins for VPS/IR Reflow Applications," *Proc. 22nd Ann. Conn. and Interconn. Tech. Symposium*, Philadelphia, PA, 1989.
34. Hawley, J., "Design, Tooling and Material Considerations for Connector Housings Made of Glass-filled Crystalline Materials" *Proc. 25th Ann. Conn. and Interconn. Tech. Symposium*, San Jose, CA, 1992.
35. Brockman, I.H., Sieber, C.S., and Mroczkowski, R.S., "A Limited Study of the Effects of Contact Normal Force, Contact Geometry, and Wipe Distance on the Contact Resistance of Gold Plated Contacts," *Trans. IEEE CHMT*, 11(12), 1988.
36. Brockman, I.H., Sieber, C.S., and Mroczkowski, R.S., "The Effects of the Interaction of Normal Force and Wipe Distance on Contact Resistance in Precious Metal Plated Contacts," in *Proc. 34th IEEE Holm Conf. on Elec. Contacts*, Chicago, IL, 1988.
37. Mroczkowski, R.S., "Contact Wiping Effectiveness: Interactions of Normal Force, Geometry and Wiping Mode," *Proc. Int'l. Conf. on Elec. Contacts and Electromechanical Components*, Beijing, China, 1989.
38. Van Horn, R.H., "The Design of Separable Connectors," *Proc. 20th Elec. Comp. Conf.*, Las Vegas, NV, 1970.
39. Whitley, J.H. and Mroczkowski, R.S., "Concerning Normal Force Requirements for Precious Metal Plated Connectors," *Proc. 20th Ann. Conn. and Interconn. Tech. Symp.*, Philadelphia, PA, 1987.
40. Whitley, J.H., "The Tin Commandments," *Plating and Surface Finishing*, Oct., 1981.
41. Mroczkowski, R.S., "Concerning Hertz Stress as a Connector Design Parameter," *Proc. 24th Ann. Conn. and Interconn. Tech. Symposium*, San Diego, CA, 1991.
42. Kantner, E.A. and Hobgood, L.D., "Hertz Stress as an Indicator of Connector Reliability," *Connection Technology*, Mar., 1989.
43. Fluss, H.S., "Hertzian Stress as a Predictor of Connector Reliability," *Connection Technology*, Dec., 1990.
44. Eammons, W., Chang, J., Stankos, J., Abbott, W., Sharrar, R., Wutka, T., and Stackhouse, A., "Connector Stability Test for Small System Connectors," *Proc. 15th Int'l. Conf. on Elec. Contacts and 36th IEEE Holm Conf. on Elec. Contacts*, Montreal, Canada, 1990.
45. Mroczkowski, R.S. and Geckle, R.J., "Concerning Cold Welding in Crimped Connections," *Proc. 41st IEEE Holm Conf. on Elec. Contacts*, Montreal, Canada, 1995.
46. Whitley, J.H., "The Mechanics of Pressure Connections," presented at EDN Regional Engineers meeting, New York, Dec., 1963.
47. Mroczkowski, R.S., "Conventional versus IDC Crimps," *Connection Technology*, July, 1986.
48. Mitra, N.K., "An Evaluation of the Insulation Displacement Contact," *Proc. 25th Holm Conf. on Elec. Contacts*, 1979.
49. Wandemacher, K.S., "Insulation Displacement Connector Reliability," *Proc. 19th Ann. Conn. and Interconn. Tech. Symposium*, 1986.
50. Key, E., "A New Concept in Wiring Device Termination—Using Zero-Gap Insulation Displacement," *Proc. 21st Ann. Conn. and Interconn. Tech. Symposium*, Dallas, TX, 1988.
51. Goel, R., "Analysis of Press Fit Technology," *Proc. 31st Elec. Comp. Conf.*, Atlanta, GA, 1981.
52. Scaminaci, J., "Solderless Press-Fit Connections—A Mechanical Study of Solid and Compliant Contacts," *Proc. Ninth Annual Connector Symposium*, Cherry Hill, NJ, 1976.
53. Dance, F.J., "The Electrical and Reliability Characteristics of Compliant Press-Fit Pin Connectors," *IPC Tech. Rev.*, Dec., 1987.

54. Manko, H.H., *Soldering Handbook for Printed Circuits and Surface Mounting*, Van Nostrand Reinhold, New York, NY, 1986.

55. Katyl, R.H. and Simed, J.C., "Electrical Design Concepts in Electronic Packaging." Chapter 3 *in Principles of Electronic Packaging*, D. Seraphim, R. Lasky and C. Li, eds. McGraw Hill, New York, 1989.

56. Chang, C.S., "Electrical Design Methodologies," in *ASM Electronic Materials Handbook*, Vol. 1, Packaging, ASM, Materials Park, 1989.

57. Southard, R.K., "High Speed Signal Pathways from Board to Board," *Electron. Eng.*, Sept., 1981.

58. Sucheski, M.M. and Glover, D.W., "A High Speed, High Density Board to Board Stripline Connector," *Proc. 40th Elec. Comp. and Tech. Conf.*, Las Vegas, NV, 1990.

59. Aujla, S. and Lord, R., "Application of High Density Backplane Connector for High Signal Speeds," *Proc. 26th Ann. Conn. and Interconn. Symposium and Trade Show*, San Diego, 1993.

60. Corman, N.E. and Mroczkowski, R.S., "Fundamentals of Power Contacts and Connectors," *Proc. 23d Ann. Conn. and Interconn. Tech. Symposium*, Toronto, Canada, 1990.

61. Wise, J.H., "A Method for End of Life Contact Current Rating," *Proc. 21st Ann. Conn. and Interconn. Tech. Symposium*, Dallas, TX, 1988.

62. Wise, J.H., "End of Life Contact Current Rating Method for Non-sinusoidal Wave Forms," *Proc. 23d Ann. Conn. and Interconn. Tech. Symposium*, Toronto, Canada, 1990.

63. Mroczkowski, R.S., and Maynard, J.M., "Estimating the Reliability of Electrical Connectors," *IEEE Trans. on Reliability*, 40 (5), 1991.

8.12 RECOMMENDED READING

1. Stearns, T.H., *Flexible Printed Circuitry*, McGraw-Hill, New York, 1995.

CHAPTER 9
ELECTRONIC DEVICE COOLING

Richard F. Porter and Ohbin Kwon
Northrop Grumman Corporation, Baltimore, MD

9.1 INTRODUCTION

Electronics of all types are becoming more a part of everyday life for everyone. Household electronics of some type represent a large portion of family disposable income and compose approximately a third of the cost of most aircraft. The buyers and users need and expect long-lasting quality in their products. Temperature control is one of the major contributors to the reliability we all seek.

This chapter presents tools, techniques, and some devices that can be used to define, control, and reduce temperatures in commercial and military electronics. The difference between the two is not addressed directly because the material can be applied to either. The difference is in the environments in which the electronics are expected to operate that define the boundary conditions for the thermal engineer.

9.2 HEAT TRANSFER THEORY

The basic theory of heat transfer is found in many references; however, to make this chapter a complete reference, pertinent theory and information is presented in a practical way to ensure the volume can be used without or at least with minimal referral to external sources.

All forms of heat transfer are used in the packaging of electronics and a basic understanding of them is necessary to produce the best cooling system for a given application. To this end conduction, convection, radiation, and ebullient heat transfer theory is presented in the following sections.

9.2.1 Conduction

Conductive heat transfer in its most general form is defined by the energy balance.

$$k_x\frac{\partial^2 T}{\partial x^2} + k_y\frac{\partial^2 T}{\partial y^2} + k_z\frac{\partial^2 T}{\partial z^2} + q = \rho c_p \frac{\partial T}{\partial t} \tag{9.1}$$

If the material conductivity, k, is uniform in all directions, this equation reduces to:

$$\frac{\partial^2 T}{\partial x^2} + \frac{\partial^2 T}{\partial y^2} + \frac{\partial^2 T}{\partial z^2} + q/k = \frac{\rho c_p}{k}\frac{\partial T}{\partial t} \tag{9.2}$$

and if the system is time invariate (steady state) it becomes:

$$\frac{\partial^2 T}{\partial x^2} + \frac{\partial^2 T}{\partial y^2} + \frac{\partial^2 T}{\partial z^2} + q/k = 0 \tag{9.3}$$

However, this equation is generally reduced to the Fourier equations, which express the heat transfer in each direction independently. Thus, a set of equations, essentially one di-mensional in nature, can be used to represent a three-dimensional system using a network of conductances between points. This is often called the *lumped parameter method* in which each node of the network is approximated as a single temperature; heat is transferred be-tween nodes at different temperatures, just as current is transferred between nodes of dif-ferent voltages in an electrical circuit and is, in practice, used in a finite difference format:

$$q_z = -kA\frac{\partial T}{\partial z} \tag{9.4}$$

$$q_y = -kA\frac{\partial T}{\partial y} \tag{9.5}$$

$$q_x = -kA\frac{\Delta T}{\Delta x} \tag{9.6}$$

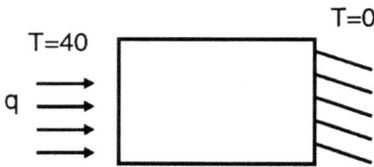

FIGURE 9.1 Plate conduction problem.

The use of the Fourier equation can be demonstrated by a simple plate problem as shown in Figure 9.1. In this illustration, one side of the plate is at 0°C and the opposite sur-face is at 40°C.

To calculate the amount of heat transferred through the plate to maintain this temperature at steady state the following calculation is per-formed.

$$q = \frac{kA}{x}(40 - 0) \tag{9.7}$$

If the area of the plate, the conductivity of the plate and the thickness of the plate, x, are substituted into the equation, the heat flow (in watts) required to maintain this temperature difference can be computed.

The conduction equations presented so far have been in Cartesian coordinates. It is of-ten necessary to analyze parts using a cylindrical coordinate system, such as represented in Figure 9.2. Here, the Fourier equation becomes:

$$q = -kA\frac{dT}{dr} \tag{9.8}$$

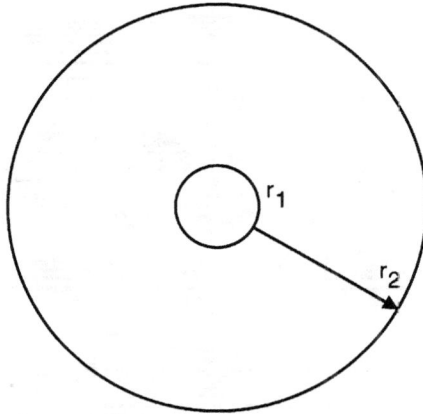

FIGURE 9.2   Radial conduction through a cylinder or annulus.

but now $A = 2\Pi rt$, where t is the thickness of the material. Substituting, the equation becomes:

$$q = -2\Pi k\, tr\, \frac{dT}{dr} \tag{9.9}$$

and integrating from r_1 to r_2 yields:

$$q = \frac{2\Pi k\, t(T_1 - T_2)}{\ln\dfrac{r_2}{r_1}} \tag{9.10}$$

In some occasions, conduction occurs in only a portion of a cylinder and can be tangential as well as radial. The radial conductance is easily found by integrating from θ_1 to θ_2, instead of over 2Π, as in Eq. 9.10. The tangential conductance is found by defining the area, $A = tdr$ and defining x as $rd\theta$. The resulting integration yields:

$$q = kt\, \frac{\ln\dfrac{r_2}{r_1}}{\theta_2 - \theta_1}(T_1 - T_2) \tag{9.11}$$

One element of conduction that is treated very differently is contact conductance. *Contact conductance* is defined as the conductance across an interface of two surfaces in contact. The value of this conductance is a function of the pressure applied, flatness, surface finish (smoothness), hardness, and malleability of the two surfaces. Although contact conductance can be predicted, so much information is needed about the surfaces in contact that it is not practical to use. As a result, contact conductance is determined empirically. Because many practical equations used in heat transfer are based on empirical data, this isn't a detriment. The contact conductance, G, is expressed in units of watts/°C-M^2 and the temperature drop across the interface is then:

$$\Delta T = \frac{Q}{GA} \tag{9.12}$$

Data exists for many combinations of materials and is usually expressed in terms of conductance versus pressure as in Figure 9.3.

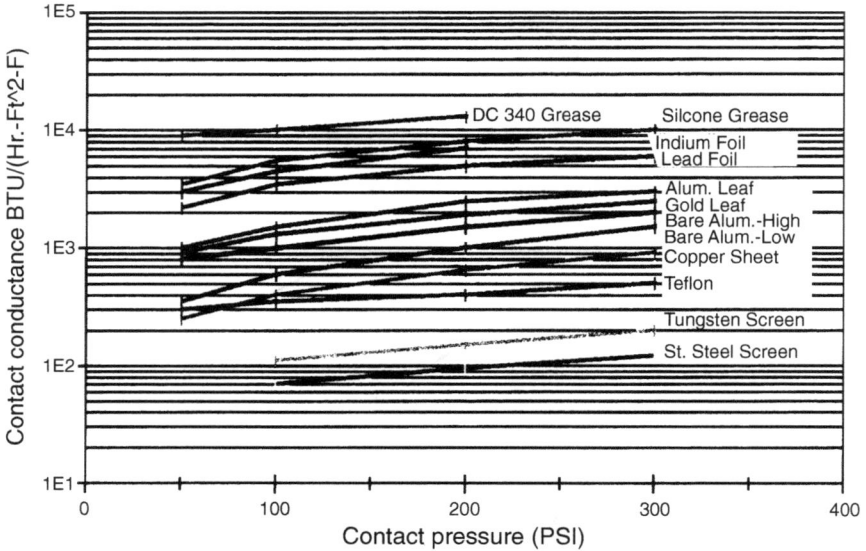

FIGURE 9.3   Contact conductance versus pressure for several interstitial materials.

The actual pressure over the area in contact is often difficult to determine. The pressure applied can be determined by computing the tensile force on the bolt or screw:

$$P = \frac{T}{KD} \qquad (9.13)$$

Where
P is the tensile force or tension (Lb$_f$)
T is the torque applied to the bolt (ft.-Lb$_f$)
D is the diameter of the bolt (ft.)
K is a friction coefficient = 0.15

However, the pressure reduces with distance from the bolt or screw. A good general rule is that the pressure can be considered to apply over an area defined by a two bolt-head-diameter radius.

Many materials have been developed to enhance and worsen the contact conductance. Materials that improve contact conductance are composed of different types of soft materials that conform to the irregularities of the surface, which is never perfectly smooth and to fill gaps caused by imperfect flatness. Of course, the smoother and flatter the two surface are, the better the conductance will be. Some interstitial materials can be used to reduce the conductance in cases where increased resistance is desired; usually, a hard porous material, such as steel screens, is used.

Some new materials change phase (i.e., melt and are trapped in the interface, which is reportedly as good or better than thermal grease, which has been the best when applied correctly). These phase-change fillers wet the surfaces of the interface materials to provide high transfer areas over very small distances. They can also have (although not yet proven) the advantage of being less sensitive to pressure, which eliminates the need to control this parameter in thermal designs.

9.2.2 Convection

Convection is extremely important in cooling electronics because it represents a very efficient method of removing heat that can be easily designed into the system. The challenge
in convection heat-transfer analysis is the determination of the convective heat-transfer
coefficient, h. Convection cooling in electronics can occur through either natural convection or forced convection. Natural convection involves the movement of a fluid (usually
air) over a warm body or surface induced by the buoyancy effect of the air being heated. It
is, therefore, free, whereas forced convection requires the inclusion in the design of a forcing mechanism, such as a fan or pump.

Forced convection   Forced convection can occur externally (e.g., over the outside of an
electronic box) or internally (e.g., flow through a tube). The flow can be either laminar or
turbulent. This is important because of the difference in the boundary layer (where the heat
transport from the surface actually takes place and changes the characteristics of heat
transfer) which affects the coefficient, h, which expresses this mechanism.

Several dimensionless parameters are important in convection:

Reynolds Number, $Re = \rho V d / \mu$

Prandtl Number, $Pr = C_p \mu / k$

Nusselt Number, $Nu = hd/k$

Stanton Number, $St = Nu/(RePr)$

The transition from laminar to turbulent flow takes place at Re values of 5×10^5 for external flow and 2100 for internal flow. The velocity profile of the fluid in the boundary layer
is different for laminar and turbulent flows. Figure 9.4 illustrates the laminar boundary layer
velocity profile and Figure 9.5 shows the turbulent profile. Although the turbulent boundary
layer has a laminar sublayer, it is much thinner than pure laminar boundary layers; therefore,
heat transfer is improved. The transition from laminar to turbulent is shown in Figure 9.6.

Although the turbulent boundary layer is thicker than the laminar boundary layer, the
laminar sublayer is thinner, which improves the transport of heat into the fluid; heat-transfer coefficients are thus higher.

The laminar heat-transfer coefficient for external flow over a flat plate is the most often used and useful expression for computing external flow heat transfer. For a plate
heated over its entire surface, the Nusselt number is:

$$Nu_x = 0.332 Pr^{1/3} Re_x^{1/2} \tag{9.14}$$

Where x is measured along the flow direction on the plate. The average heat-transfer
coefficient for the plate is obtained by integrating over the length of the plate to obtain:

$$\overline{Nu} = \frac{\overline{h}L}{k} = .664 Pr^{1/3} Re_L^{1/2} \tag{9.15}$$

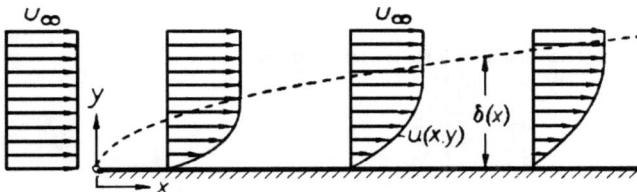

FIGURE 9.4   Fluid velocity profile in the laminar boundary layer.

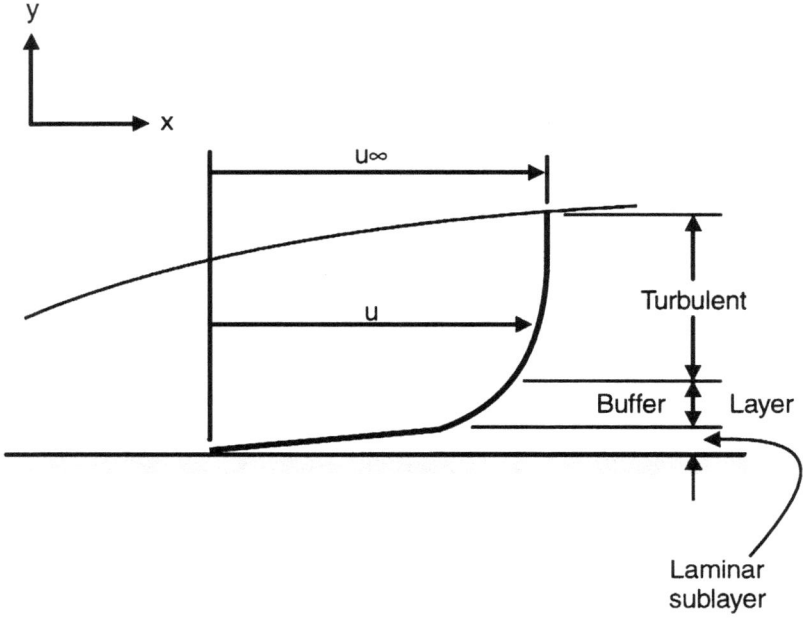

FIGURE 9.5 Fluid velocity profile in the turbulent boundary layer.

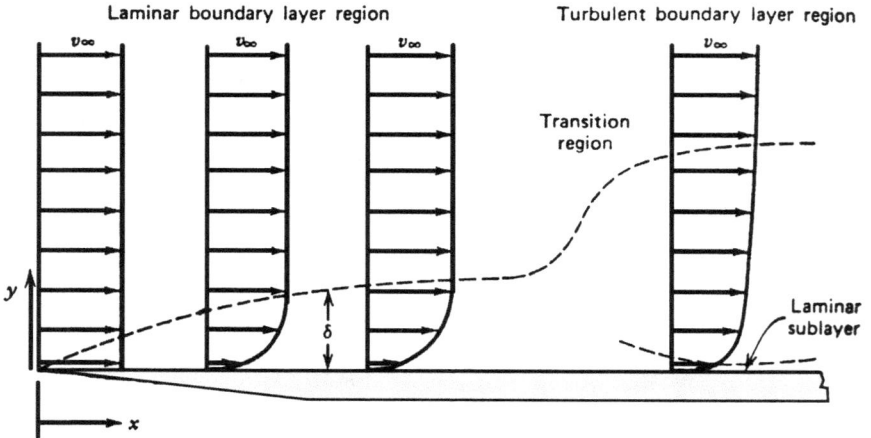

FIGURE 9.6 Fluid flow transition from laminar to turbulent.

For turbulent flow, the local Nusselt number is:

$$Nu_x = 0.0288 Re_x^{4/5} Pr^{1/3} \tag{9.16}$$

and the average is:

$$\overline{Nu} = 0.036 Re_L^{4/5} Pr^{1/3} \tag{9.17}$$

Heat-transfer coefficients and how to compute them are detailed in *Engineering Heat Transfer* by James Welty and *Heat Transfer* by J.P. Holman. These include cylinders and spheres, as well as banks of tubes, which can be useful in the design of heat exchangers.

Internal heat transfer is most often defined by flow through a tube. Seider and Tate[1] have correlated experimental data to the expression for laminar flow:

$$Nu = 1.86 \left(RePr \frac{D}{L} \right)^{1/3} \left(\frac{\mu_m}{\mu_o} \right)^{0.14} \tag{9.18}$$

Where μ_m is evaluated at the mean fluid temperature and μ_o is evaluated at the wall temperature. Then for turbulent flow in a tube, the most general expression is given by McAdams[21] as:

$$St = 0.023 Re^{-0.2} Pr^{-2/3} \left(\frac{\mu_m}{\mu_o} \right)^{0.14} \tag{9.19}$$

where:

1. All fluid properties are evaluated at the average fluid temperature, except μ_o, which is evaluated at T_o, the wall temperature.
2. $Re > 10^4$
3. $0.7 < Pr < 17,000$
4. $L/D > 60$

These tube expressions are accurate for circular tubes (Figure 9.7), but can be used for non-circular tubes with reasonable accuracy. Information on corrections for non-circular tubes can be found in *The Handbook of Heat Transfer*[1] by Rohsenow and Hartnet.

Many abbreviated expressions have been defined for convection heat-transfer coefficients when the cooling fluid is air at or near sea level conditions. Some of these are presented here. All are in units of BTU/hr.-°F-ft.[2] In these equations, the following units must be used:

ρ: Density in lb./ft.^{3}
V: Gas velocity in ft./hr.
L or D: Length or diameter in feet
C_p: Specific heat in BTU/lb.-°F

1. Laminar flow over a flat plate with velocities less than 15 ft./second:

$$h = 0.8 + \frac{V}{322} \tag{9.20}$$

2. Turbulent flow over a flat plate with velocities greater than 15 ft./second:

$$h = \frac{V^{0.75}}{830 \, L^{0.25}} \tag{9.21}$$

3. Turbulent flow in a tube:

$$h = 0.0144 \frac{C_P (V\rho)^{0.8}}{D^{0.2}} \tag{9.22}$$

(a)

(b)

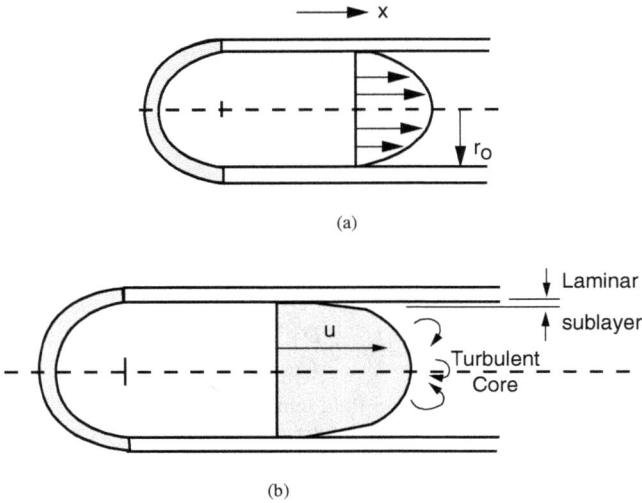

FIGURE 9.7 A comparison of laminar-flow and turbulent-flow velocity profiles in a tube.

Natural convection Natural convection occurs because the movement of fluid is caused by buoyancy effects as a result of the heating of the fluid. Obviously, buoyancy depends on the external action of body forces, such as gravity or some sort of acceleration, so that no natural convection can occur in the microgravity environment of space or in orbiting spacecraft. Whereas forced convection is characterized by the Reynolds number, in natural convection, the Grashof Number is the defining parameter, which is involved in defining the laminar vs. turbulent boundary flow. The critical Grashof number for a vertical flat plate is approximately 4×10^8 and the Grashof number is defined as:

$$Gr_x = \frac{g\beta(T_w - T_\infty)x^3\rho^2}{\mu^2} \tag{9.23}$$

Where
  g is the acceleration of gravity
  β is the coefficient of volumetric expansion, which is $1/T$ for a perfect gas
  T is the absolute temperature of the gas
  T_w is the wall temperature
  T_∞ is the fluid bulk temperature
  x is the distance from the start of flow or heating
  ρ is the fluid density
  μ is the fluid viscosity

The average free convection heat transfer coefficient can be expressed as:

$$\overline{Nu} = C(GrPr)^m \tag{9.24}$$

where all the fluid properties are evaluated at the film temperature, $T_f = (T_\infty + T_w)/2$, T_∞ is the free stream or ambient fluid temperature and T_w is the wall temperature. Values for C and m are given in Table 9.1 for isothermal surfaces. For vertical plates, the characteristic dimension for calculating the Grashof number is the height or length of the plate; for hor-

TABLE 9.1 Coefficient for Above Equation for Isothermal Surfaces from Holman

Geometry	GrPr	C	m
Vertical planes and cylinders	$10^{-1} - 10^{-4}$	Use Fig. 9.8	Use Fig. 19.8
	$10^4 - 10^9$	0.59	¼
	$10^9 - 10^{13}$	0.021	⅖
	$10^9 - 10^{13}$	0.10	⅓
Horizontal cylinders	$0 - 10^{-5}$	0.4	0
	$10^{-5} - 10^4$	Use Fig. 9.9	Use Fig. 9.9
	$10^4 - 10^9$	0.53	¼
	$10^9 - 10^{12}$	0.13	⅓
Upper surface of heated	$2 \times 10^4 - 8 \times 10^6$	0.54	¼
plates or lower surface of cooled plates	$8 \times 10^6 - 10^{11}$	0.15	⅓
Lower surface of heated plates or upper surface of cooled plates	$10^5 - 10^{11}$	0.58	⅕

izontal cylinders, the diameter should be used. Often, the vertical plate coefficient is used for vertical cylinders if the boundary-layer thickness is small compared to the cylinder diameter. The criterion for determining whether the plate equation can be used is:

$$\frac{D}{L} \geq \frac{35}{Gr_L^{1/4}} \tag{9.25}$$

For vertical surfaces with constant heat flux instead of isothermal surfaces, the Nusselt number can be expressed in terms of a modified Grashof number:

$$Gr^{*x} = Gr_x N_x = \frac{g\beta\rho^2 q_w x^4}{k\mu^2} \tag{9.26}$$

Where q_w is the wall heat flux in energy per unit area.

Vertical Plates and Cylinders

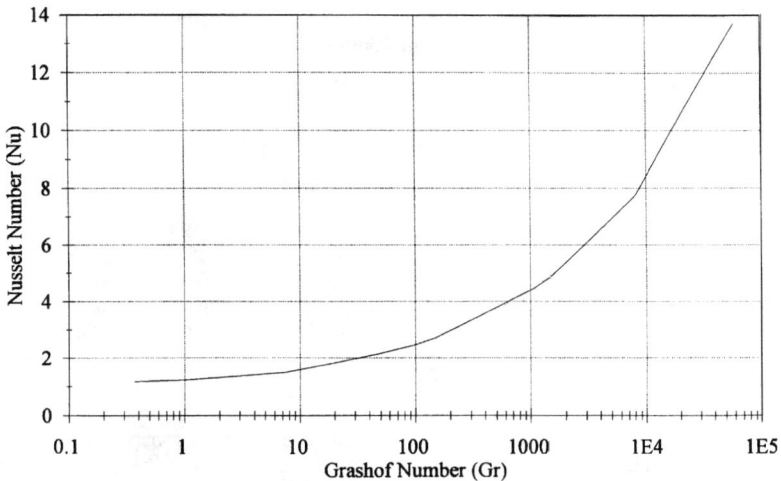

FIGURE 9.8 Nusselt Number for vertical plates and cylinders (Data from Ref. 3).

For laminar flow, the Nusselt number expressed in terms of the modified Grashof number is:

$$N_{xf} = \frac{hx}{k_f} = 0.60(Gr_x^*Pr_f)^{1/5} \tag{9.27}$$

$$10^5 < Gr_x^* < 10^{11}$$
$$q_w = \text{constant}$$

Horizontal Cylinders

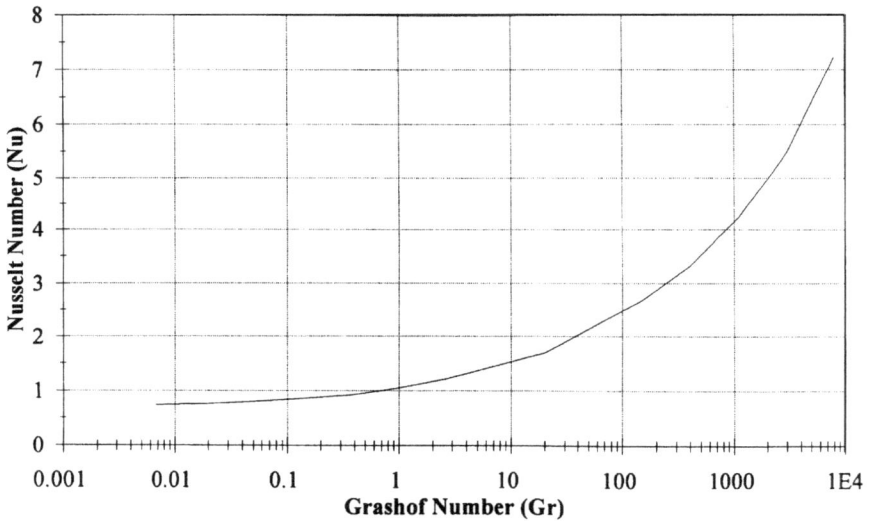

FIGURE 9.9   Nusselt Number for horizontal cylinders (Data from Ref. 3).

For turbulent flow, the Nusselt number is:

$$Nu_x = 0.17(Gr_x^*Pr)^{1/4} \tag{9.28}$$

$$2 \times 10^{13} < Gr_x^* < 10^{16}$$

$$q_w = \text{constant}$$

Again, in most cases the designer is interested in the average heat transfer coefficient of the surface and this can be obtained for laminar flow by integrating Eq. 9.28 to get:

$$\bar{h} = \frac{5}{4}h_{x=L} \tag{9.29}$$

and for turbulent flow, J.P. Holman[2] shows that the coefficient is constant with x.

For use with air at one atmosphere, some simplified expressions have been developed and are shown in Table 9.2.

TABLE 9.2 Simplified Equations for Natural Convection in Air at One Atmosphere[3].

Surface	Laminar $10^4 < Gr_fPr_f < 10^9$	Turbulent $Gr_fPr_f > 10^9$
Vertical plate or cylinder	$h = 1.42(\Delta T/L)^{1/4}$	$h = 0.95(\Delta T)^{1/3}$
Horizontal cylinder	$h = 1.32(\Delta T/d)^{1/4}$	$h = 1.24(\Delta T)^{1/3}$
Horizontal plate:		
Heated plate facing upward or cooled plate facing downward	$h = 1.32(\Delta T/L)^{1/4}$	$h = 1.43(\Delta T)^{1/3}$
Heated plate facing downward or cooled plate facing upward	$h = 0.61(\Delta T/L^2)^{1/5}$	$h = 0.61(\Delta T/L^2)^{1/5}$

h = heat transfer coefficient, W/M² - °C
ΔT = $T_w - T\infty$, °C
L = vertical or horizontal dimension, M
d = diameter, M

Time constant   The time constant of a system is defined as the length of time required for the system to reach 63.2% of its final, steady-state temperature from some initial condition. In very complex systems, many time constants are involved and different parts of the system are reaching steady state at different times. However, it is very often useful to treat a system as a single degree of freedom lumped mass. The time constant of a single degree of freedom system is defined as:

$$\tau = WC_pR \qquad (9.30)$$

Where
  W is the weight
  C_p is the weight specific heat
  R is the thermal resistance of the lumped mass to the sink

The time constant equation for a system at an initial temperature exposed to a sudden sink temperature change is:

$$\frac{T - T_i}{T_s - T_i} = (1 - e^{-t/\tau}) \qquad (9.31)$$

Where
  T_i is the initial temperature
  T_s is the sink temperature
  t is time

and if $t = \tau$, then:

$$\frac{\Delta T}{\Delta T_{final}} = (1 - e^{-1}) = .63212 \qquad (9.32)$$

However, in the design of electronics cooling systems, the question is most often related to how fast the equipment will heat up as a result of its own internal power dissipation. In this situation, the following equation is more helpful.

$$WC_p\frac{dT}{dt} + G(T - T_s) = Q \qquad (9.33)$$

This can be rewritten as:

$$\frac{dT}{dt} + \frac{(T - T_s)}{WC_pR} = \frac{Q}{WC_pR} \tag{9.34}$$

Which can be solved by separation of variables to yield:

$$T = [T_i - T_s - RQ]e^{-t/\tau} + T_s + RQ \tag{9.35}$$

Radiation   Radiation heat transfer predicts energy exchange between two surfaces. The exchange is determined by the surface properties of the two surfaces, the temperatures of the surfaces and how well each can see each other, called the *view factor*. Theoretical approaches usually present black-body theory, followed by gray-body theory, and some information about real surfaces. Radiation heat transfer, in practice, uses gray-body theory, and only in extreme cases makes correction for real surfaces. In fact, most surfaces approximate gray bodies reasonably well. The ability to radiate energy from a surface is determined by its emissivity, ε, which is the ratio that the surface can radiate compared to a theoretical black body which is 1. Thus, all real emissivities are less than one. Absorption of energy is defined by α, the absorptivity. The fraction of energy reflected from the surface is defined by ρ, the reflectivity. If the surfaces in question are transparent, energy can also be transmitted through the surface. In most cases of electronic cooling situations, transmittance is not a consideration and is not addressed in any detail here. Detailed examinations can be found in textbooks, such as *Heat Transfer* by J.P. Holman and *Engineering Heat Transfer* by James R. Welty.

The total energy that can be radiated by a black body, $E_b = \sigma T^4$, where σ is the Stefan-Boltzmann constant defined as:

$$\sigma = 5.669 \times 10^{-8} W/m^2 \, {}^\circ K^4 (1.714 \times 10^{-8} BTU/hr \, {}^\circ R^4) \tag{9.36}$$

Grey bodies radiate energy per unit area, according to $E = \sigma\varepsilon T^4$ (i.e., $\varepsilon\%$ of a black body). A comparison of emissive power, E of black, gray, and real surfaces is shown in Figure 9.10.

The difference between black body, gray body, and real surfaces has a more fundamental difference than the emissive power. Both black body and grey body theory presume that the surface radiates using Lambert's cosine principle (i.e., the radiative intensity varies with the cosine to the normal to the surface). This gives a smooth hemispherical intensity relationship, which can be contrasted with real surfaces (Figure 9.11). Notice that organic or nonconductive surfaces reasonably approximate the cosine distribution, except at large θ, whereas metallic surface differ greatly.

The spectral range of radiant energy from gray bodies is represented in Figure 9.12. An object begins to glow red hot at approximately 800°C. Where electronics are involved, radiation occurs below 200°C. To illustrate the radiative properties of real materials, Figures 9.13 and 9.14 (taken from *The Handbook of Infrared Radiation Measurement* by Barnes Engineering) show the transmissibilities of several organic materials as a function of wavelength. These materials have low emissivities where their transmissibility is high and, because their reflectance is very low, have high emissivities ($\alpha = \varepsilon$) in the high-absorption bands. One technique to keep a surface cool in sunlight is to have a low α_s/ε_t, where α_s is the absorptivity to the solar spectrum and ε_t is the emissivity to the thermal spectrum. Using this technique, one of the best surfaces for reflecting solar radiation while having a high emissivity is a second surface mirror, using quartz instead of glass. Figure 9.15, also from the Barnes Engineering handbook, shows the transmissibility of quartz and glass. The transmis-

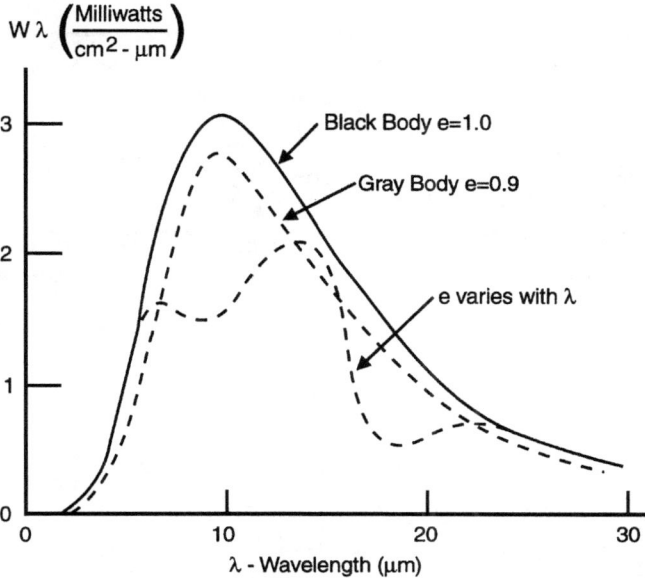

FIGURE 9.10 Comparison of emissive power for black, gray, and real surfaces.

sibility of quartz, especially in the 2.75- to 5-micron range allows a high percentage of solar radiation to pass through and reflect off the silvered back surface. However, at thermal wavelengths above 5 microns, the quartz has a high emissivity to effectively radiate energy.

Where two surfaces are involved, all the energy leaving one surface might not strike the other (Figure 9.16). This geometric relationship between the two surfaces is defined by the view factor, F. The view factor between two surfaces, F_{1-2}, is defined as the fraction of the total energy leaving surface 1 which strikes surface 2. Mathematically, the view factor for an area dA_1 to a surface dA_2 is presented as:

$$dA_1 dF_{dA_1-dA_2} = \frac{\cos\theta_1 dA_1 \cos\theta_2 dA_2}{\Pi r^2} \tag{9.37}$$

the view factor from dA_1 to A_2 is thus:

$$dA_1 F_{1-2} = dA_1 \int_{A2} \frac{\cos\theta_1 \cos\theta_2 dA_2}{\Pi r^2} \tag{9.38}$$

then integrating over A_1 gives the total view factor from surface 1 to surface 2.

$$A_1 F_{1-2} = \int_{A1}\int_{A2} \frac{\cos\theta_1 \cos\theta_2 dA_1 dA_2}{\Pi r^2} \tag{9.39}$$

In practice, numerical integration is easily accomplished by replacing the integration in Eq. 9.23 with a summation of finite subdivisions or nodes over surface two. This done for each subdivision of surface one and summed. With the reciprocity theorem:

$$A_1 F_{1-2} = A_2 F_{2-1} \tag{9.40}$$

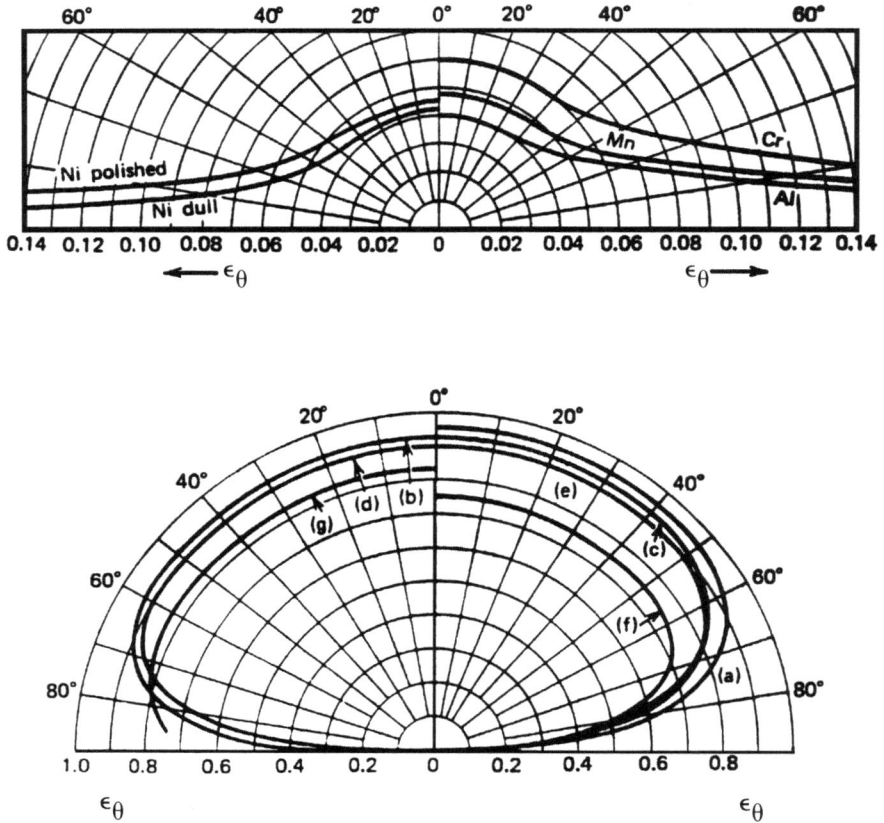

FIGURE 9.11 Radiative intensity distributions for real surfaces (a) ice; (b) wood; (c) glass; (d) paper; (e) clay; (f) copper oxide; (9) aluminum oxide

the view factor from either surface is known. The reciprocity theorem is a consequence of both surfaces radiating with the same directional intensity distribution (i.e., gray or black body theory). If real surfaces are involved, it is still true if the distribution is the same; however, if one surface is bare metallic and the other painted, the effect of non-Lambertian radiation should be considered, but is beyond the scope of this chapter.

Here, it is convenient to consider the total energy exchange per unit time, Q, instead of the energy exchange per unit time per unit area, E. The computation of the energy exchange

$$Q = \sigma A_1 F_{1-2}(T_1^4 - T_2^4) \tag{9.41}$$

between two surfaces therefore depends on the surface properties, ε, α, ρ, τ whose sum must equal one, the temperature of the two surfaces and the view factor of the two surfaces. If the surfaces are black bodies, this interchange can be expressed as, and F_{1-2} is called the *black body view factor*.

However, when nonblack surfaces are involved, the emissivity and other properties must be considered and is usually done by altering the black body view factor to a "script F," $\mathscr{F}$.

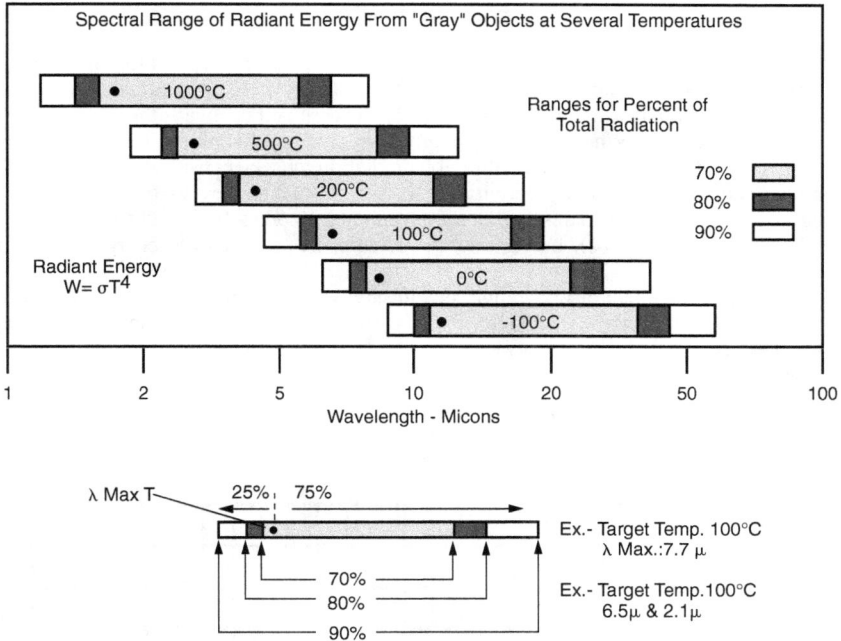

FIGURE 9.12   Gray-body radiation bandwidths.

For example, consider two infinite surfaces; the energy leaving surface one can strike surface two, where some is absorbed and some is reflected back to surface 1. Surface 1 then absorbs part of this reflected energy back to surface two, which reflects it again. This result is an infinite series, which represents an "effective emissivity" between the two surfaces expressed as

$$\mathfrak{F}_{1-2} = \frac{1}{\dfrac{1}{\varepsilon_1} + \dfrac{1}{\varepsilon_2} - 1} \tag{9.42}$$

and the energy exchange between the surfaces is:

$$Q = \sigma A_1 \mathfrak{F}_{1-2}(T_1^4 - T_2^4) \tag{9.43}$$

Table 9.3 lists $\mathfrak{F}$ expressions for some commonly encountered surface geometries. Complicated geometries will require actual integration and infinite series evaluation. Computer programs available to do this are TRASYS and NEVADA.

Other geometries have been evaluated in NACA Technical Note 2836. J.P. Holman presents the equation: $F_{A1-A2} = D^2/(4R^2 + D^2)$ for a small area, A_1 to a larger disk, A_2, corresponding to Figure 9.17.

Figures 9.18, 9.19, and 9.20 from Holman's Heat Transfer define black body view factors for parallel rectangles and disks and for perpendicular rectangular surfaces. A particularly useful view factor is given in the NACA Technical Note 2836 (Figure 9.21) reproduced from that document for a small surface, dA_1, relative to any infinite plane, A_2 where the plane of A_2 and dA_1 intersect at an angle. For this situation, the view factor is:

$$F_{1-2} = \frac{1}{2}(1 + \cos\theta).$$

Space-radiation considerations   The radiation between surfaces of different temperature, as presented to this point, presumes that the radiation is of approximately the same wavelength, in which case $\varepsilon = \alpha$. This is usually the case and is always the case inside of electronic enclosures. When considering problems involving sunlight, the effect of the high spectral intensity of visual and ultraviolet radiation must be considered because many materials can have different absorptivities to this spectrum than they do to infrared or thermal radiation (i.e., $\varepsilon \neq \alpha$). In this case, it is convenient to express the interchange of energy, not in terms of $(T_1^4 - T_2^4)$, but as energy radiated, εT_1^4, minus the energy absorbed, $H\alpha\cos\theta$, as in Figure 9.22.

To illustrate the effect of wavelength on emissivity, values for a few materials are presented in Table 9.4. At any wavelength, $\varepsilon = \alpha$, so the values in this table can be interpreted as the thermal emissivity at the surface temperature and the solar radiation absorptivity at the solar wavelength.

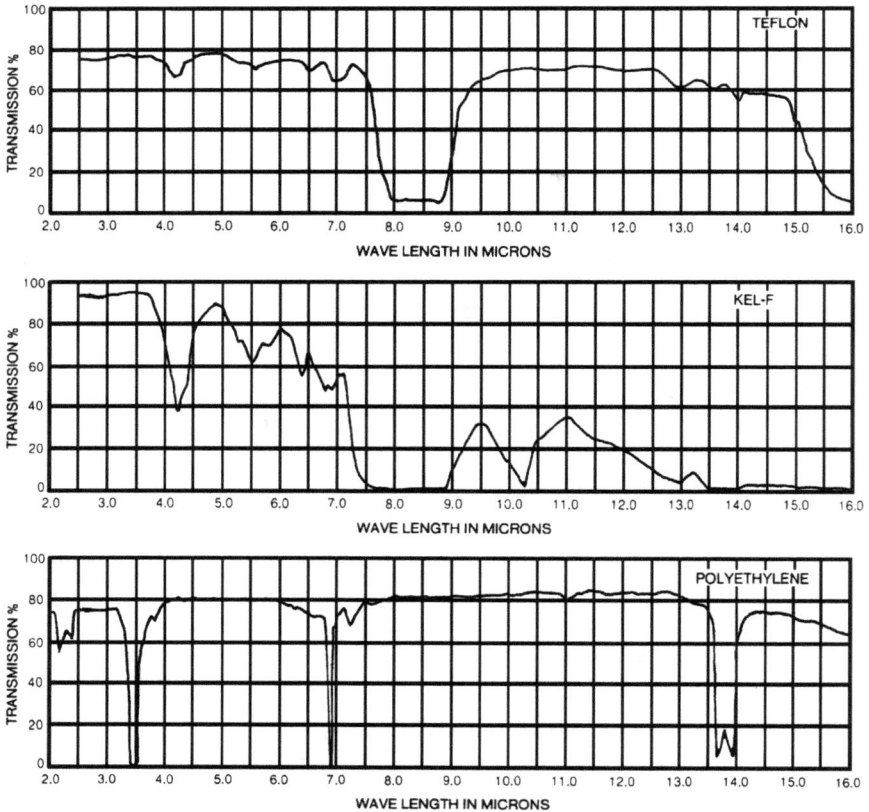

FIGURE 9.13   Transmissibility variation with frequency.

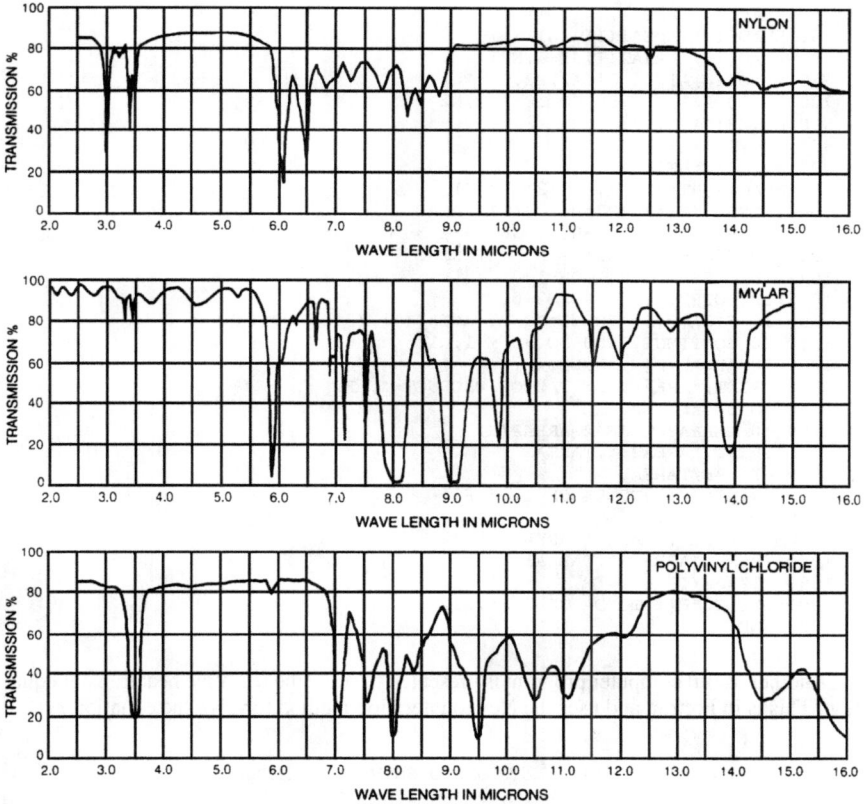

FIGURE 9.14 Transmissibility variation with frequency.

The following computer program computes the view factor for two plane surfaces oriented at any angle to each other. Data input is the center of each sub-area of the two surfaces, the areas of each sub-surface and coefficients defining the planes of the surfaces to describe an equation in the form:

$$Ax + By + Cz = 0$$

```
50 DIMENSION X1(225),Y1(225),Z1(225),X2(225),Y2(225),Z2(225)
100 DIMENSION AREA1(225),AREA2(225),F(225
110 DATA TAREA1/0./,TAREA2/0./,AF/0./
120 DATA F/225*0./
150 CALL OPENF(1,"COORD")
C N1 and N2 are the number of sub-areas on each surface
200 READ (1,)N1,N2
250 READ(1,)(X1(I),I=1,N1)
300 READ(1,)(Y1(I),I=1,N1)
350 READ(1,)(Z1(I),I=1,N1)
400 READ(1,)(X2(I),I=1,N1)
450 READ(1,)(Y2(I),I=1,N1)
```

```
500 READ(1,)(Z2(I),I=1,N1)
550 READ(1,)(AREA1(I),I=1,N1)
600 READ(1,)(AREA2(I),I=1,N2)
650 READ(1,)A1,B1,C1,A2,B2,C2
700 DO 20 J=1,N1 750 DO 10 I=1,N2
800 R1=X2(I)-X1(J)
850 R2=Y2(I)-Y1(J)
900 R3=Z2(I)-Z1(J)
950 RMAG=SQRT(R1**2+R2**2+R3**2)
1000 COSB1=(R1*A1+R2*B1+R3*C1)/RMAG
1010 R1=-R1 1020 R2=-R2 1030 R3=-R3
1050 COSB2=(R1*A2+R2*B2+R3*C2)/RMAG
1052 IF(COSB1.LT.0.)COSB1=0.
1055 IF(COSB2.LT.0.)COSB2=0.
1100 10 F(J)=F(J)+COSB1*COSB2*AREA2(I)/(3.14159*(RMAG**2))
1150 20 CONTINUE 1200 DO 30 M=1,N1
1250 30 AF=AF+F(M)*AREA1(M)
1300 PRINT,"AREA-VIEW FACTOR PRODUCT=",AF
1350 DO 50 K=1,N2
1400 50 TAREA2=TAREA2+AREA2(K)
1450 PRINT,"AREA2=",TAREA2
1500 F21=AF/TAREA2
1600 PRINT,"F(2 TO DA OF 1)=",F21
1700 DO 60 K=1,N1
1750 60 TAREA1=TAREA1+AREA1(K)
1800 PRINT,"AREA1=",TAREA1
1850 F12=AF/TAREA1
1900 PRINT,"F(1 TO DA OF 2)=",F12
9000 END
```

Another useful computer program is presented to solve the radiative heat transfer equation. This is in Fortran and uses the Newton iteration method to solve the equation.

```
C     NEWTON'S METHOD ITERATION TO SOLVE NON-LINEAR EQUATIONS
C     Q+H*Area*alpha=sigma*Area*epsilon*T**4+Con*(T-T0)
C     AREA IS IN INCHES, A IS ALPHA, EP IS EPSILON, CON IS CONDUCTANCE
      DATA EP/.03/,A/.1/,AREA1/28.1/
      DATA SUN/430./,T2/733./,POW/0./
      DATA AREA/134.55/,CON/.0105/
      F(E)=1.713E-9*EP*AREA*E**4/144.+CON*(E-T2)-SUN*A*AREA1/144.-POW
      FPRIME(E)=4.*1.713E-9*EP*AREA*E**3/144.+CON
      FPRIM2(E)=12.*1.713E-9*EP*AREA*E**2/144.
      WRITE(*,*)'Enter an initial temperature as a starting point'
5     READ(*,*)X
      IF(ABS(F(X)*FPRIM2(X))/FPRIME(X)**2-1.)10,15,15
10    WRITE(*,50)
      I=0
      Y=F(X)
      WRITE(*,20)I,X,Y
      DO 11 I=1,30
      X1=X-(Y/FPRIME(X))
      Y=F(X1)
      WRITE(*,20)I,X1,Y
      IF(ABS((X1-X)/X1).LT.1.E-5)GOTO 5
11    X=X1
      WRITE(*,40)
      GO TO 5
15    WRITE(*,30)X
      GOTO 5
```

```
20      FORMAT(7X,I2,8X,E14.8,2XE14.8)
30      FORMAT(3H X=,F12.4,27H WILL NOT CAUSE CONVERGENCE)
40      FORMAT(/36H FAILED TO CONVERGE IN 30 ITERATIONS)
50      FORMAT(//41H ITERATION    X1      F(X1))
        END
```

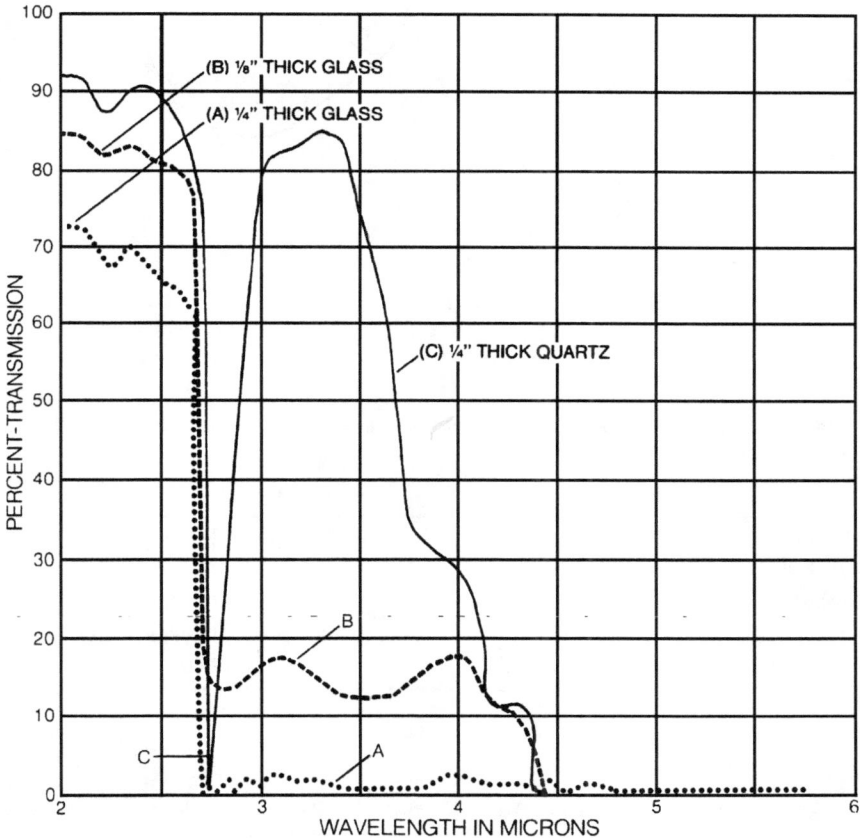

FIGURE 9.15 Transmissibility of glass and quartz.

9.3 PRESSURE DROP ANALYSIS

The two terms to consider in predicting the pressure drop of a system are the friction and the losses caused by expansions, contractions, and turns. Liquids and gasses are handled identically, but friction losses are much higher in liquids and the effect of viscosity is much more important. In either fluid, there is a source of motion (such as a blower or pump) and a source of impedance, which is discussed in this section. In designing a system, these two forces must be matched to obtain the desired flow rates.

The pressure losses covered in this section refer to system energy losses that are not recoverable. The total pressure is composed of static and dynamic pressure and the values of

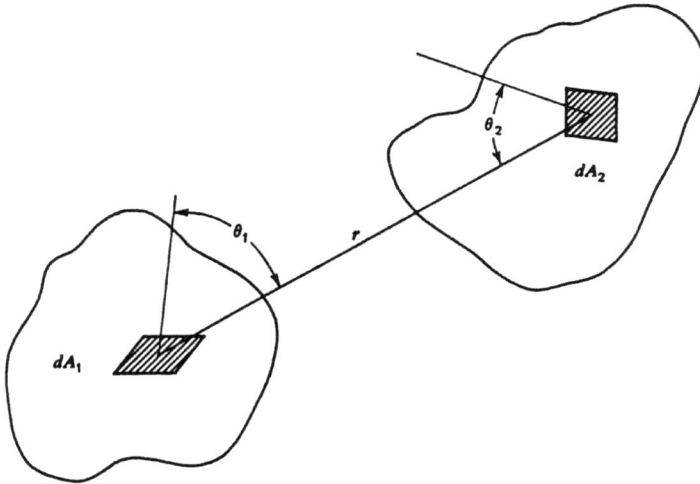

FIGURE 9.16 Black-body view factor between two surfaces.

Table 9.3 Values for Commonly Encountered Script F Geometries Data from *Cooling Electronic Equipment*, Alan Kraus

Surfaces	Area, A	f
Infinite parallel plates	Either area	$1/(1/\varepsilon_1 + 1/\varepsilon_2 - 1)$
Small body, A_1, enclosed by a large body, A_2	A_1	ε_1
Large body, A_1, completely enclosed by a larger body, A_2	A_1	$1/(1/\varepsilon_1 + /\varepsilon_2 - 1)$
Concentric spheres, A_1 inside A_2	A_1	$1/(1/\varepsilon_1 + A_1/A_2(1/\varepsilon_2 - 1))$
Infinite concentric cylinders, A_1 inside A_2	A_1	$1/(1/\varepsilon_1 + A_1/A_2(1/\varepsilon_2 - 1))$

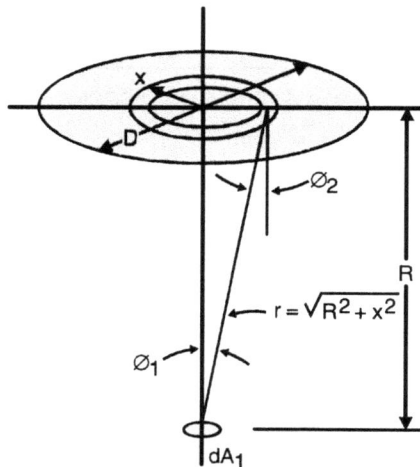

FIGURE 9.17 Radiation from a small area to a disk.

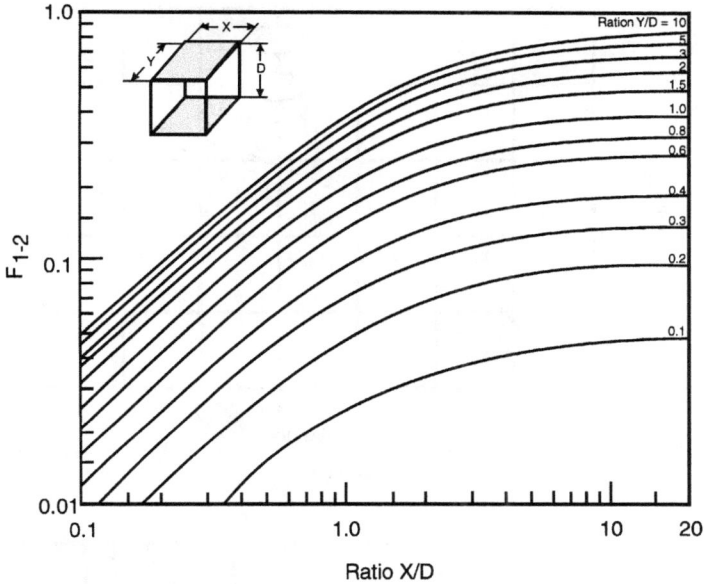

FIGURE 9.18   View factors for parallel rectangles.

these can change through a system of ducts by converting static pressure to dynamic pressure by contracting the flow and the reverse by expanding the flow. Pressure loss for electronic equipment, even though not specified in some cases, is meant to be loss in total pressure.

Friction loss   The two commonly used equations for friction loss are identical, except for the definition of the friction factor.

Using the Fanning friction factor, the equation is:

$$\Delta P = \frac{fL\rho V^2}{2gd} \tag{9.44}$$

Using the Darcy friction factor, the equation is:

$$\Delta P = \frac{4fL\rho V^2}{2gd} \tag{9.45}$$

Where, in both equations:

L = Length in inches
ρ = Density of the fluid in lb/inch^3
V = Velocity of the fluid in inches/sec.
g = Acceleration of gravity (384 inch/sec^2)
d = Hydraulic diameter (inch)

It is obvious that the Fanning friction factor is four times the Darcy friction factor. This is mentioned as a caution for the designer to use the correct friction factor in the equation. The Fanning friction factor is plotted in Figure 9.23. It is useful in most cases to define the friction factor for laminar flow (i.e., $R_e < 2100$ as $64/R_e$ for Eq. 9.44 and $16/R_e$ for Eq. 9.45.

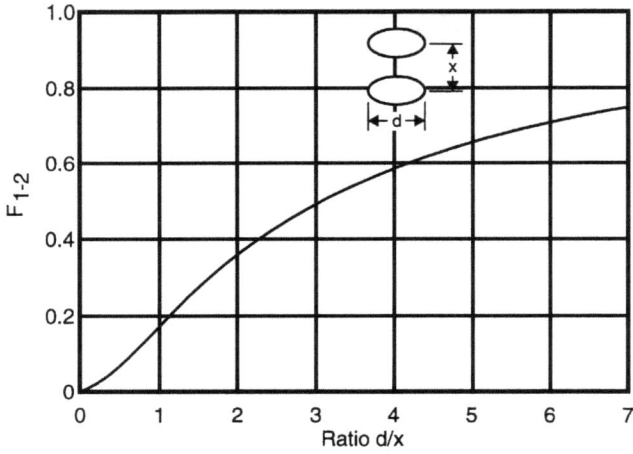

FIGURE 9.19 View factors for parallel disks.

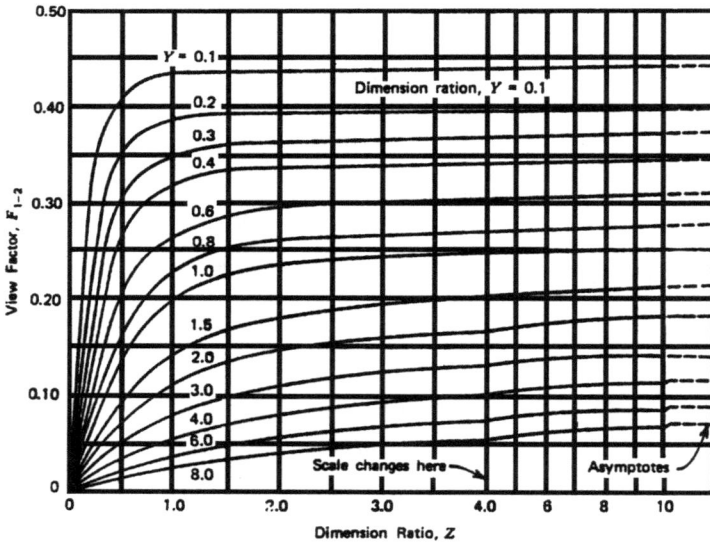

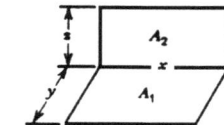

FIGURE 9.20 View factors for perpendicular rectangular surfaces.

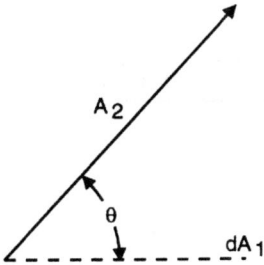

FIGURE 9.21 The view factor from a small plane surface to an infinite plane whose places intersect at an angle θ.

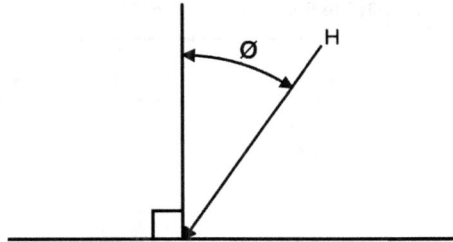

FIGURE 9.22 Solar radiation striking a flat plate.

Expansions, contractions, and turns   When flowing fluid is forced to change direction by turning or, to a lesser extent, by contractions and expansions, additional energy is lost in the form of pressure and is expressed as

$$\Delta P = \frac{K\rho V^2}{2g} \tag{9.46}$$

Where K is a loss coefficient, defined empirically and dependent on the severity of the turn. Some useful definitions are given in Figures 9.24 through 9.26 and Table 9.5. The total pressure loss in a duct is thus defined as:

$$\Delta P = \frac{\rho V^2}{2g} \left(\frac{fL}{d} + K\right) \tag{9.47}$$

An examination of the values of K in Figures 9.24 through 9.26 leads to the conclusion that the turns should be as gradual as possible and changes in area should be gradual. In fact, many tests and references talk about diffusers and recommend that transitions between two duct areas be made at 12- to 15-degree angles. Although true, there is usually not room for this gradual transition inside of electronic units.

9.4 CIRCUIT CARD ASSEMBLY (CCA) COOLING

The CCA is the basic building block of most electronic systems. The exceptions are usually high-power transmitters (built around traveling-wave tubes) or high-voltage power supplies (built around isolated floating decks, where arc lengths and voltage isolation usually require emersion in a dielectric liquid or gas). The thermal designer is therefore most often concerned with removing heat from heat-dissipating components mounted to a CCA. Radiation is of little use and is used only as a safety factor except in some low power space applications. Heat is removed from the components either by conduction or convection or a combination of the two. Where heat is removed by conduction to the circuit card or a heatsink on the circuit card, methods of removing the heat from the card must be utilized.

Typically in the analysis of CCAs, the components are characterized as a one-dimensional heat source and resistance connected to the card (Figure 9.27). The resistance is called θ_{jc}, which is a misnomer because θ refers to temperature difference in most heat-transfer texts. However, the term originated in the electronics industry and now is tradition.

Table 9.4 Emissivities of Materials at Different Wavelengths

Material	9.3 μ 100° F	5.4 μ 500° F	3.6 μ 1000° F	1.8 μ 2500° F	0.6 μ Solar α_s
Asbestos paper	.93	.93			
Red brick	.93				
White paper	.95				.28
Water	.96				
Carbon, T-carbon	.82	.8	.79		
Polished Aluminum	.04	.05	.08	.19	.3
Oxidized	.11	.12	.18		
Anodized	.86	.42	.6	.34	
Polished chromium	.08	.17	.26	.4	.49
Polished copper	.04	.05	.18.	.17	
Oxidized	.87	.83	.77		
Polished gold	.02	.018	.035		.14
Polished Iron	.06	.08	.13	.25	.45
New galvanized	.23			.42	.66
Dirty galvanized	.28			.9	.89
Oxide	.96		.85		.74
Rusted		.69			
Lead, Grey anodized	.28				
Magnesium	.07	.13	.18	.24	.3
Molybdenum			.09	.15	.2
Polished silver	.01	.02	.03		
Polished stainless steel	.15	.18	.22		
Weathered	.85	.85	.85		
Bright in		.043			
Paints					
Aluminum Lacquer	.65	.65			
Cream	.95	.88	.7	.42	.35
Flat Black (lacquer)	.96	.97	.97	.97	.97
Red	.96				.74
White (ZnO)	.95				.18
White	.8–.95				.19–.33
White Epoxy	.9–.95				.3–.35
Grey	.84–.91				.5–.57
Glass	.9–.95				

This one-dimension characterization is usually adequate, but it must be recognized that it is not in all cases and detailed analysis of the chip and its mounting and conduction to the component case are often necessary (covered in Section 9.5). However, as system-level thermal designers, there is often little knowledge of how the component is constructed; we have to use either the manufacturer's stated θ_{jc} or a MIL-SPEC value. The MIL-SPEC values usually give conservative (too high) temperature estimates because they are set at a level to cover all manufacturer's values; thus, requiring a design for the worst-case manufacturer. In an increasingly more commercial environment, it is becoming

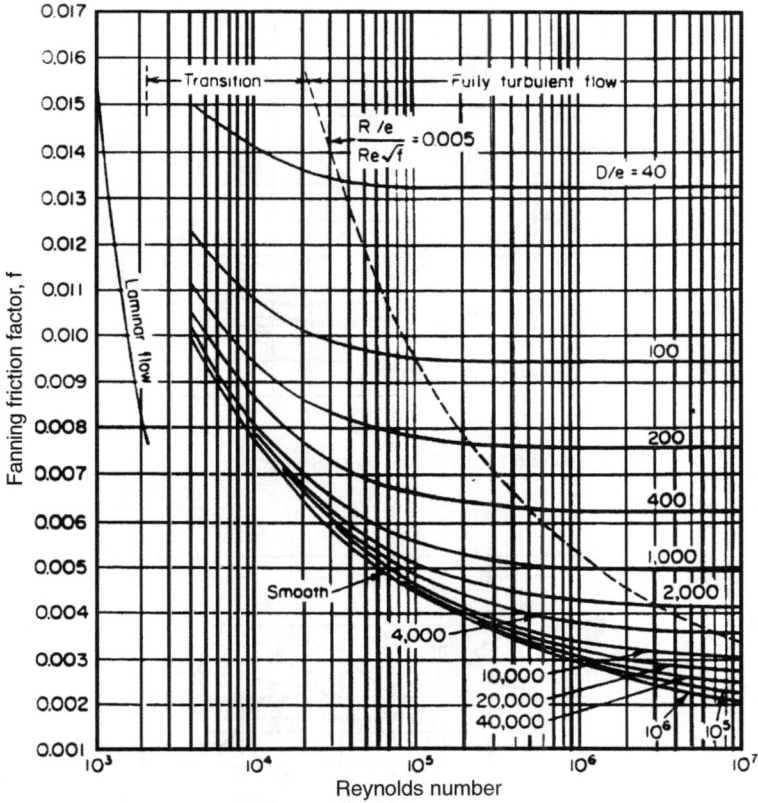

Fanning friction factor plot for smooth and rough tubes.

FIGURE 9.23 Fanning friction factor chart (Moody diagram).

more acceptable to use the manufacturer's values and limit the use of certain parts to acceptable suppliers when necessary.

9.4.1 Conduction-Cooled CCAs

Conduction is used both as the primary cooling method for CCAs and to augment convection cooling. CCAs designed to be cooled by conduction generally are of three different types:

1. Heatsinks under the board
2. Heatsinks on top of the board
3. No heatsink: the board is used to conduct heat to a sink

Heatsinks under the board are used with surface-mount components, such as flat packs, quad packs, and multichip modules (Figure 9.28). Either a single CCA can be bonded to the

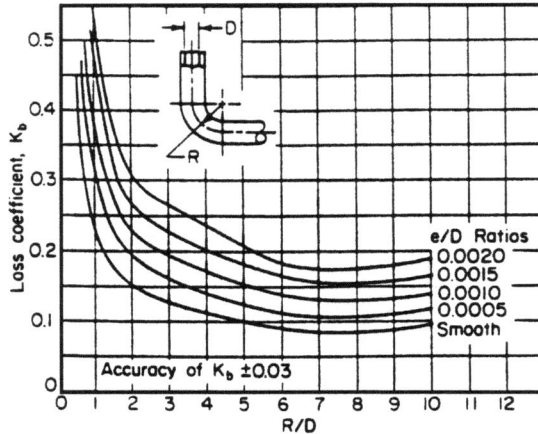

FIGURE 9.24 Loss coefficients for smooth 90° bends.

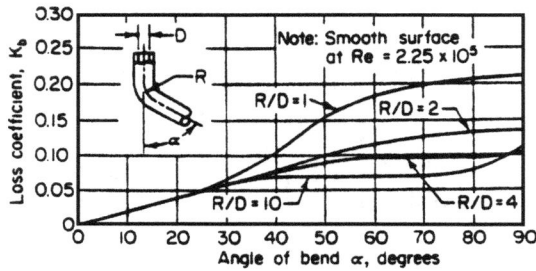

FIGURE 9.25 Loss coefficients for smooth bends in smooth tubes versus bend angle.

sink or a CCA on each side of the heatsink can be used. This type of electronic module packaging has been defined in specific sizes called *Standard Electronic Modules (SEMs)*, as illustrated in Figure 9.29. The heatsink is a solid sheet of material, which extends beyond the edges of the circuit card to be clamped into the chassis to provide a good conductive path to the chassis. A common clamping mechanism made by Wedge-Lock is shown in Figure 9.30. Resistance through this interface depends on the number of sections of Wedge-Lock used in the mechanism. Values for 3- and 5-piece mechanisms are defined by Table 9.6.

Providing a uniform heat distribution is assumed, the conduction heatsink provides a parabolic temperature distribution according to Eq. 9.48 and is illustrated in Figure 9.31:

$$\Delta T = \frac{q'x^2}{2kA} \tag{9.48}$$

where q' is in watts per inch

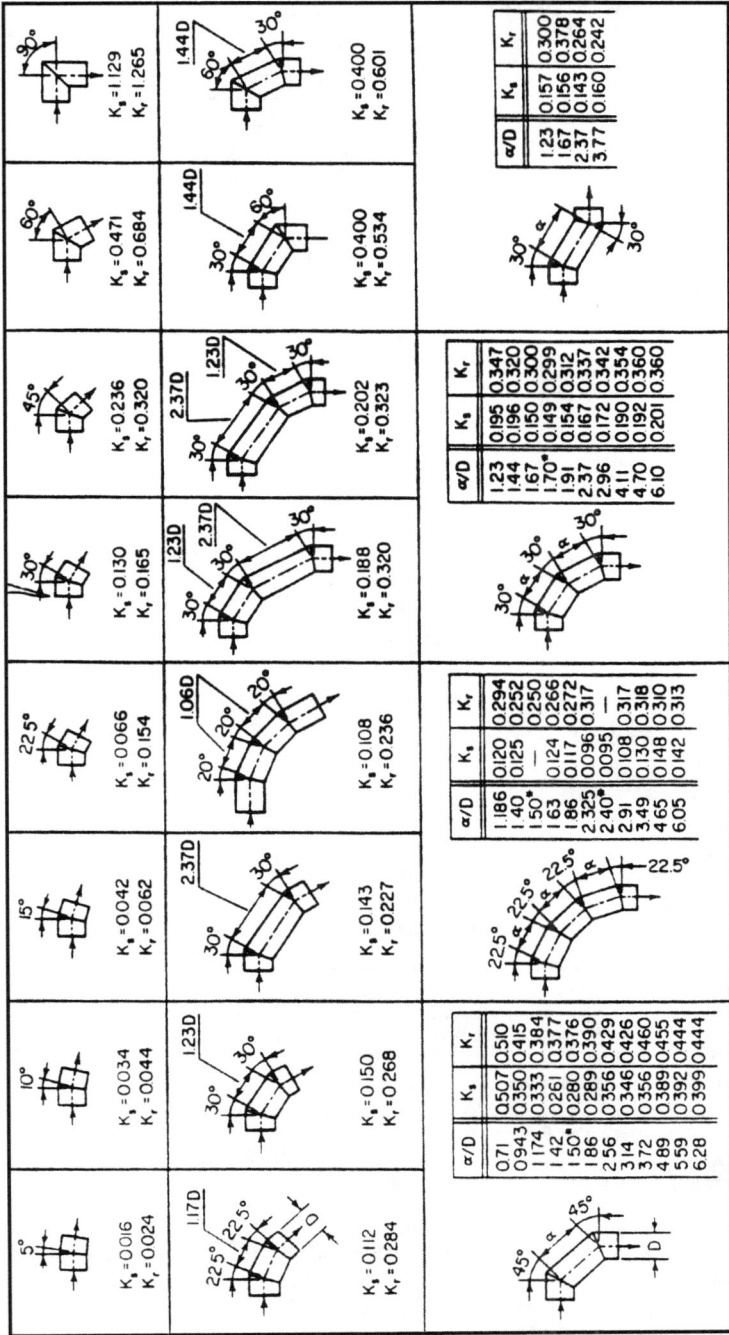

FIGURE 9.26 Loss coefficients for miter bends of various angles and straight sections.

TABLE 9.5 Pressure Drop for Turbulent Flow through Screwed Valves and Fittings

	Pressure drop*	
Type of valve or fitting	Equivalent L/D	Equivalent velocity head (K_v)
Couplings	2	0.04
Unions	2	0.04
Gate valves†		
Open	7	0.1
¾ open	40	0.8
½ open	200	4
¼ open	800	15
Globe valves,† bevel seat		
Open	350	6
½ open	550	10
Composition disk		
Open	330	6
½ open		
Plug disk	500	9
Open	500	9
¾ open	700	13
½ open	2000	10
¼ open	6000	110
Angle valves†		
Open	170	3
Y or blow-off valves†		
Open	170	3
Check valves‡		
Swing	110	2
Disk	500	10
Ball	3500	65
Water meters		
Disk	400	8

*Pressure drop is expressed as a straight pipe of same nominal size as the fitting. Where the pressure drop is expressed as L/D, L is the equivalent length (feet) of straight pipe and D is the inside diameter (feet). Where the pressure drop is expressed as velocity heads, the velocity in the pipe is based on the nominal diameter of the fitting.

†Flow direction through valves has negligible effect on pressure drop.

‡Values apply only when check valve is fully open, which generally is attained at pipe velocities over 3 ft/sec for water.

SOURCE: Reprinted from "Velocity Head Simplified Flow Computation," by C.E. Lapple, by permission from *Chem. Engr.*, 56, May 1949.

The average temperature is ⅔ of the maximum T and is only ½ of the temperature computed when all the heat is applied at the center. Thus, the maximum temperature can be modeled by putting all of the heat at ¼, the length of the plate from each edge.

Obviously, the temperature of the components on the CCA depends on the conductive path through the circuit card to the heatsink and the conduction in the heatsink to the chas-

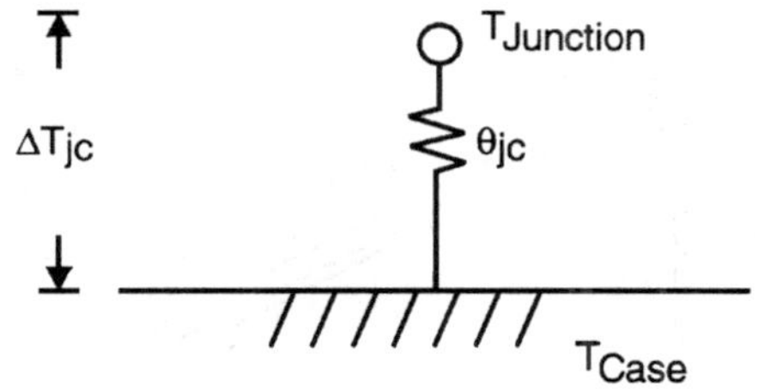

FIGURE 9.27 One-dimensional characterization of an electrical component.

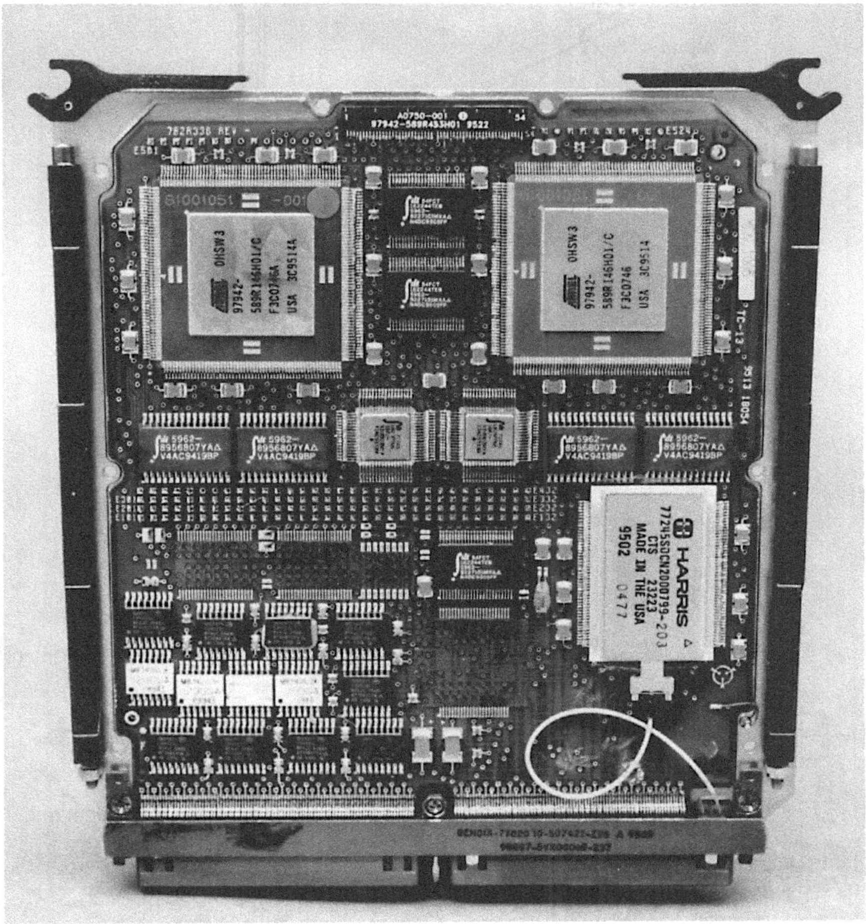

FIGURE 9.28 A conduction-cooled surface-mount-type CCA.

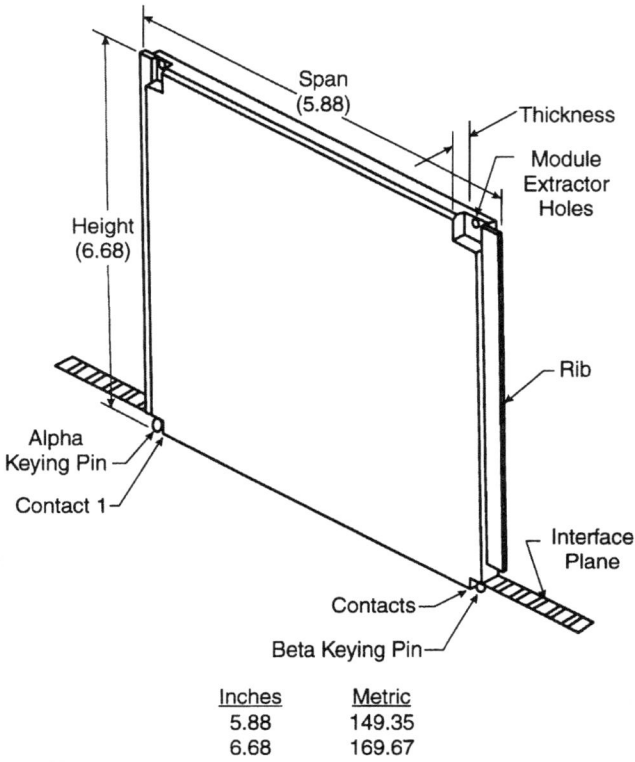

Inches	Metric
5.88	149.35
6.68	169.67

Notes:
1. Dimensions are in inches.
2. Metric equivalents are given for general information only.

FIGURE 9.29   A standard electronic module: "E" size.

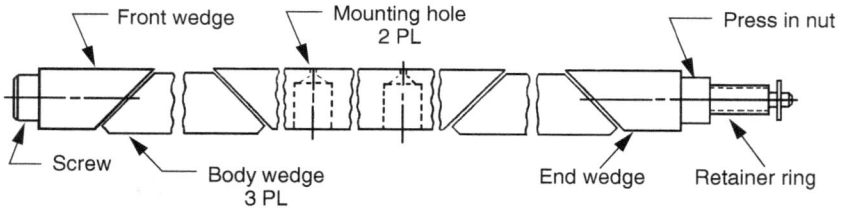

FIGURE 9.30   A CCA-clamping device made by Wedge-Lock to provide conduction from the CCA to the electronic chassis.

TABLE 9.6 Contact Resistance Provided by the Wedge-Lock Clamp at the Edge
of the Module[14]

Wedgelock type	°C-M^2/watt	°C-in^2/watt
3 Segment, chemical film finish per MIL-C-5541, Class III	0.0003459	0.5
3 Segment, Type II, Class 2 anodize finish per MIL-A-8625	0.0003079	0.45
5 Segment, chemical film finish per MIL-C-5541, Class III	0.0002864	0.417

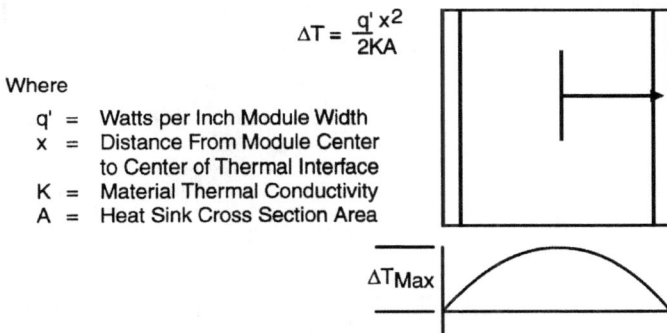

$$\Delta T = \frac{q' x^2}{2KA}$$

Where

q' = Watts per Inch Module Width
x = Distance From Module Center
    to Center of Thermal Interface
K = Material Thermal Conductivity
A = Heat Sink Cross Section Area

ΔT_{Max}

FIGURE 9.31 Conduction-cooled CCAs provide a parabolic temperature distribution.

sis. These values can be affected favorably by good design or negatively by poor design. The conduction through the board can be enhanced by the addition of vias. In some extreme cases, the board is cut away so that the part can sit directly on the heatsink. The material of the heatsink is also important. Aluminum and copper are commonly used, but new composite materials are starting to be used. Composite materials have the definite advantages of being lighter, having higher conductivities, and having adjustable thermal expansion coefficients, which stress the CCA less. Their disadvantage is cost, but this is coming down. Composite Optics, Inc. compiled figures to show the cost of building composite heatsinks, compared to the 1993 cost of raw composite (carbon) fiber (Figure 9.32). Though still more expensive than aluminum or copper, where weight or particularly good thermal performance is needed, the cost is low enough to merit serious consideration (Figure 9.33). Properties of some materials are shown in Table 9.7.

The effective conductivity of laminated composite heatsinks can be adjusted through fiber orientation. Figure 9.34 shows this effect for P-100/Epoxy made for a SEM-E CCA heatsink. Also note that a similar graph could be plotted for thermal expansion.

Heatsinks on top of the board are referred to as *rail* or *ladder cooled* because the configuration of the heatsink has parallel rails or rungs, like a ladder. A CCA of this type is shown in Figure 9.35. This type of heatsink requires through-hole mounting components, such as dual-inline packages, and is generally used in military equipment, where direct con-

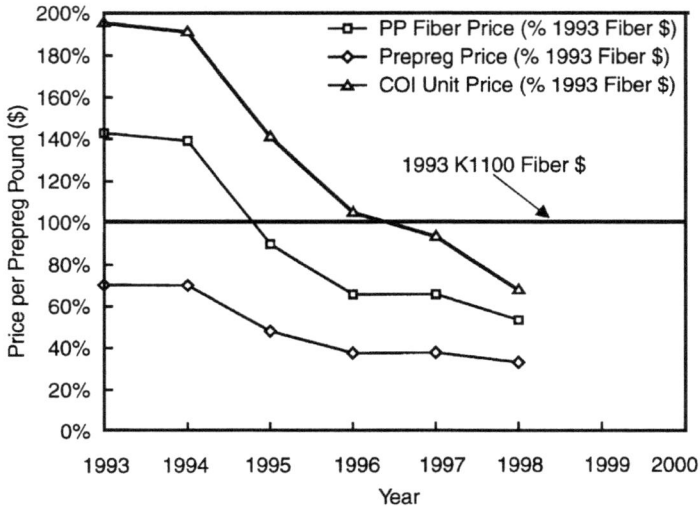

FIGURE 9.32 Reducing cost trends of fabricated advanced thermal composite hardware relative to 1993 cost of raw material *(Courtesy of Composite Optics Inc.)*.

- Design & Fabrication by COI
- Size: 233mm x 160mm
- Configuration: 600 Watt VME Format
    Airborne Power Supply
- Materials: K1100/CE
    XN80/CE
- Weight: a.) 1.7 lbs Copper Baseline
    0.10" thick
    b.) 0.38 lbs K1100/CE
    0.020" thick
- Tested by E-Systems: 400 Watt Power
    Input
- Performance: 18°C Greater for
    K1100/CE than Cu at
    22% of Cu's Weight

FIGURE 9.33 A composite-material VME-size conduction module core *(Courtesy of Composite Optics Inc.)*.

tact of air on the CCAs isn't desirable because of contaminants and moisture in the air or where natural convection isn't adequate and conduction to the electronic chassis is needed.

Like the SEM-type CCAs, the board is actually smaller than the heatsink so that the edges of the sink can be clamped into the chassis to provide a conductive path to the chassis. The temperature profile on each rail can be approximated as a uniformly distributed heat load to again give a parabolic temperature distribution. The SEM conductive plate has the advantage of conducting in two dimensions, but the rail has the advantage of having

TABLE 9.7 Comparison of Advanced Composite Heatsink Materials to Aluminum

Material	Type	Density (g/cm³)	Thermal Conductivity		
			One direction (W/M–°K)	Two directions (W/M–°K)	Third direction (W/M–°K)
Aluminum	2024, isotropic metal	2.7	170	170	170
K13710	CFRP, quasi-isotropic	1.75	40	40	2.1
P120	CFRP, quasi-isotropic	1.75	190	190	2.1
K1100	CFRP, quasi-isotropic	1.75	300	300	2.1
Thermal graph 8000	Unidirectional carbon-carbon	1.76	700	20	20
Thermal graph EWC 600	High k fabric	1.65	160–200	160–200	~1.5
TPG	Pyrolytic graphite, Grade "T"	1.9	1300	1300	<20
TC 1050	TPG encapsulated in aluminum	2.5	1050	1050	<20
TPG/resin composite	TPG bonded into conductive CFRP	1.85	~1100	~1100	~2

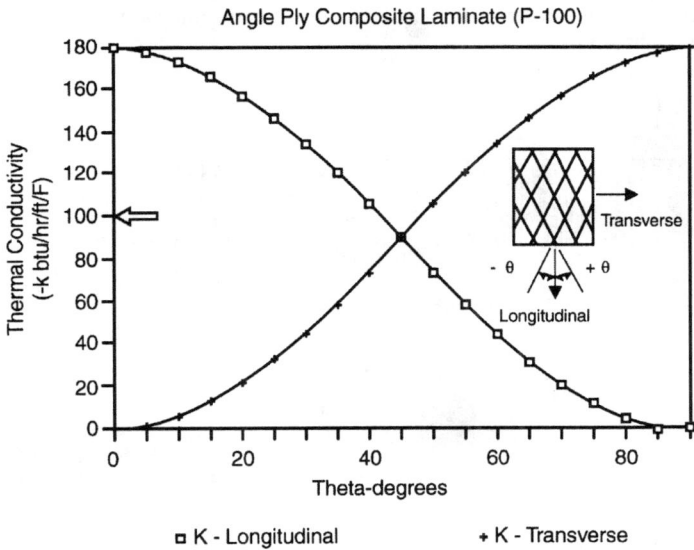

FIGURE 9.34 Effective thermal conductivity of a P-100 epoxy-composite CCA heatsink.

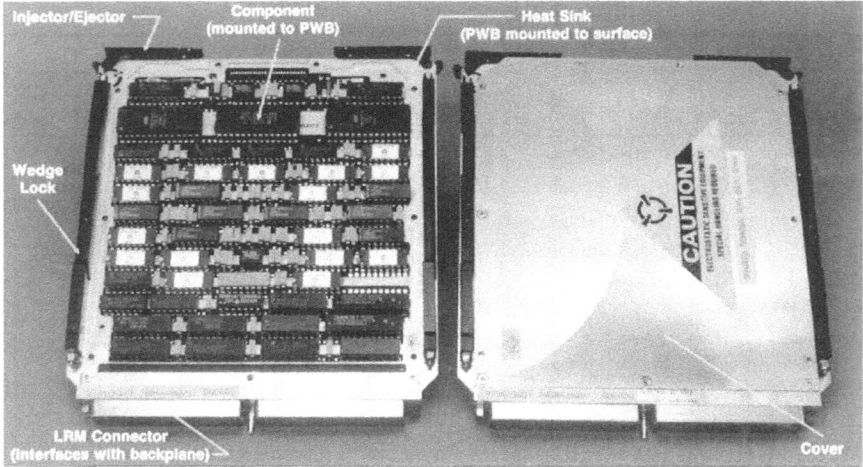

FIGURE 9.35   Conduction-cooled through-hole-type CCA.

the component mounted directly onto it so that the problem of conducting through the circuit card is eliminated.

Conduction in the heatsink wall of a chassis can also be improved by using advance thermal composites. Figure 9.36 is a chassis wall developed by Composite Optics Inc., which produced temperatures 20°C cooler than the aluminum equivalent with 400 watts heat sunk to the wall.

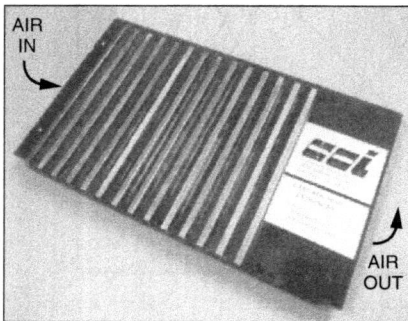

- Design & Fab by COI
- Size: 165mm x 297mm
- Configuration: Single Pass Air Cooled
  Heat Exchanger, 12 Card Capacity
- Weight: 890 g (Aluminum)
          527 g (COI Composite)
          41% Weight Reduction
- Materials: P75S/954-3 Gr/CE Tape
           K1100/ERL1999 Tape
           6061-T6 Aluminum Rails
- Tested by NAWC: 400 Watt Total Power
          Input at All 14 Aluminum Rails
- Performance: 20 deg. Lower Operating
           Temperature than Aluminum

FIGURE 9.36   Advanced composite chassis wall designed to replace aluminum walls *(Courtesy of Composite Optics Inc.)*.

9.4.2 Forced Convection-Cooled CCAs

Direct impingement of air on CCA components   Most applications of forced convection cooling to CCAs involves blowing air over the components on the circuit card. Although very effective, accurate prediction of the heat transfer is difficult because of the irregular surface presented to the air flow. Flow can be quite turbulent, even when Reynolds Numbers are low. Tall components can shield downstream components. Typically, you can use undeveloped flow over each component and account for temperature rise as the air or cooling fluid passes down the length of the CCA. A correlation by Wills (May 1983 *Electronic Production*) has been shown by experiment to be accurate in the absence of component shielding.

$$h_x = 0.22 Re_x \frac{Pr^{1/3}k}{x} \tag{9.49}$$

Where
$\quad$ x is the location in the flow direction along the CCA in inches
$\quad$ k is the conductivity of air evaluated at the local air temperature in BTU hr-ft.-°F

Numerical thermal analysis is today's standard and many finite element and finite difference programs are available. Only through numerical analysis can a high level of accuracy be achieved. A computer program using the method of Direct Numerical Simulation (DNS) was used to analyze a forced convection CCA. The DNS method is explained in detail in Section 9.5. An example of the use of DNS using Eq. 9.49 to define the heat-transfer coefficient is shown in Figures 9.37 through 9.41. Figure 9.37 shows the CCA with many components mounted on its surface. The following four figures show the junction temperatures based on the manufacturers θ_{jc} values, each component case temperature, the board temperature under the components, and finally the air temperature as it passes across the CCA in the x direction.

Flow-through-board pair cooling   Forcing coolant to flow between a pair of CCAs is a method used extensively in military electronics. This method is effective for several general reasons; it keeps the coolant from depositing contaminants on the board and components, it gets the coolant very close to the heat source and the electronic part, and liquid or gas can be used as a coolant. In field conditions, air can contain salt, fuel vapors, water vapor, dirt, and other contaminants. Keeping this contamination off the electronic circuitry can add many hours to the reliability of complex equipment. The construction consists of coolant fins sandwiched between two thin faceplates to which two CCAs are bonded so that the components are facing outward. A thin layer of insulation, such as polyimide or mylar, is used to insulate the CCA circuitry from the metal faceplates. The cooling method and construction is cold plate, which is described in Section 9.6. As such, this type of cooling lends itself to more accurate temperature prediction than direct air impingement, which, in turn, allows less-conservative designs, thus saving weight and power.

$\qquad$ The DNS method was again used to analyze a board using the equations defined in Section 9.7. This analysis is shown in Figures 9.42 through 9.45. Figure 9.42 shows one CCA on one side of the cold plate. The following three figures show the junction temperatures, the coolant temperature as it passes in the y direction, and a temperature cross-section plot showing temperatures at different levels under the components.

JASDF-DDP Parallel FFT Stage Board (3D62180)

Size =   6.500 in x   7.300 in

Total Power =   7.6662 Watts

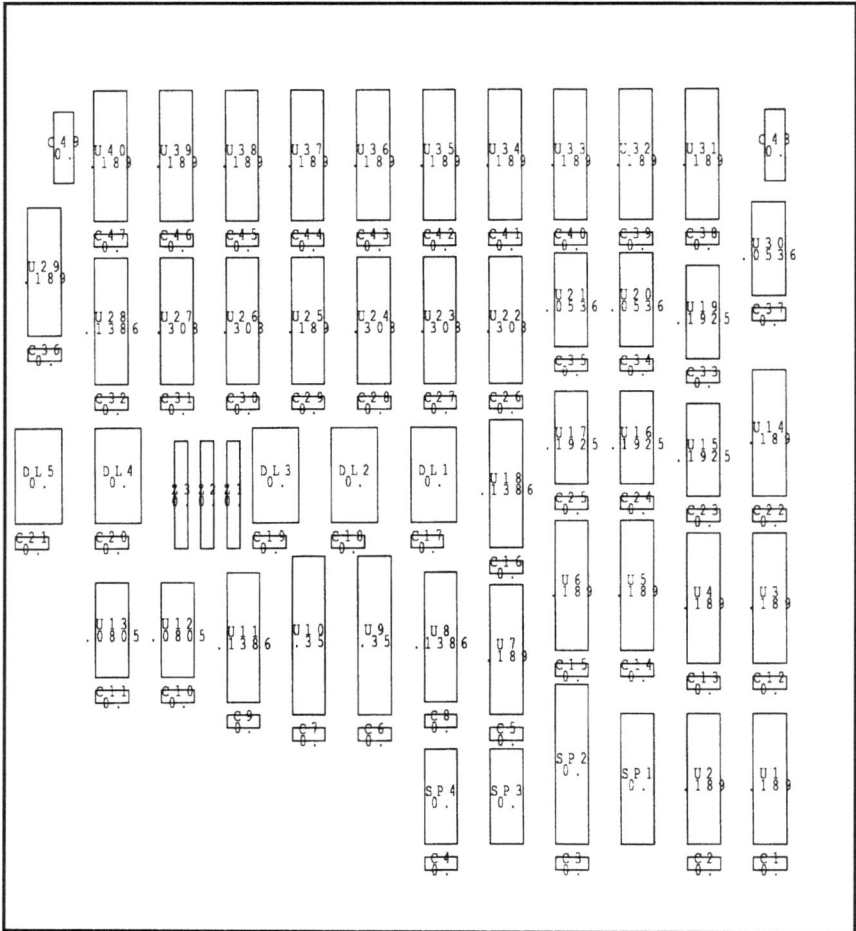

FIGURE 9.37 A CCA used for DNS analysis of direct-impingement cooling of parts on the board.

Microchannel and "macrochannel" cooling. Microchannel cooling defines a technology dealing with very small fins. The absolute dimensions that define microchannel versus simply small fins are not well defined. For general purposes, when the space between fins is 0.005 inches (0.127 mm) or less, the fins can be called *microchannels*. It is always

JASDF-DDP Parallel FFT Stage Board (3D62180)
Junction & Top Layer
Power = 7.666 Watt

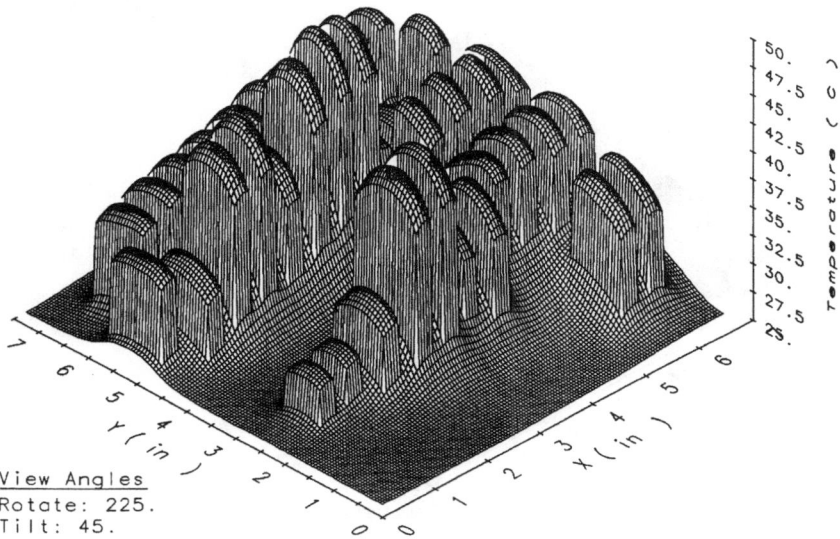

View Angles
Rotate: 225.
Tilt: 45.

FIGURE 9.38 Computed part junction or hot-spot temperatures.

JASDF-DDP Parallel FFT Stage Board (3D62180)
Case & Top Layer
Power = 7.666 Watt

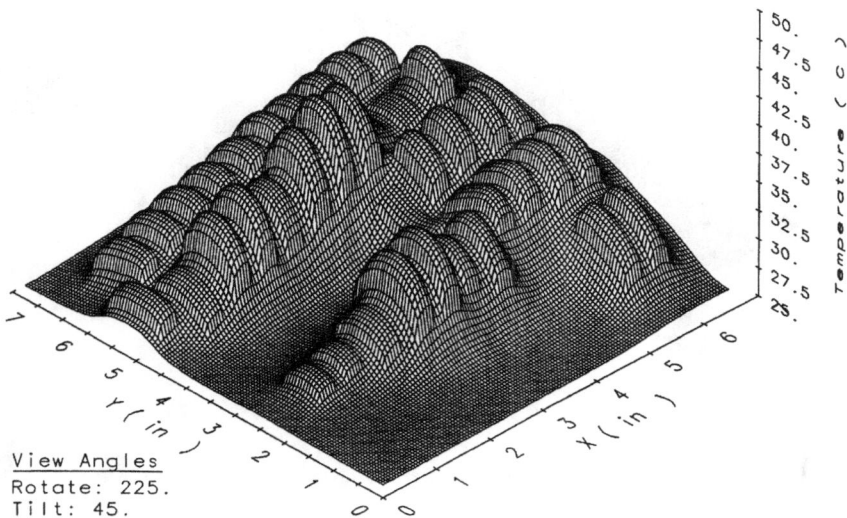

View Angles
Rotate: 225.
Tilt: 45.

FIGURE 9.39 Temperature of part cases and the CCA board surface.

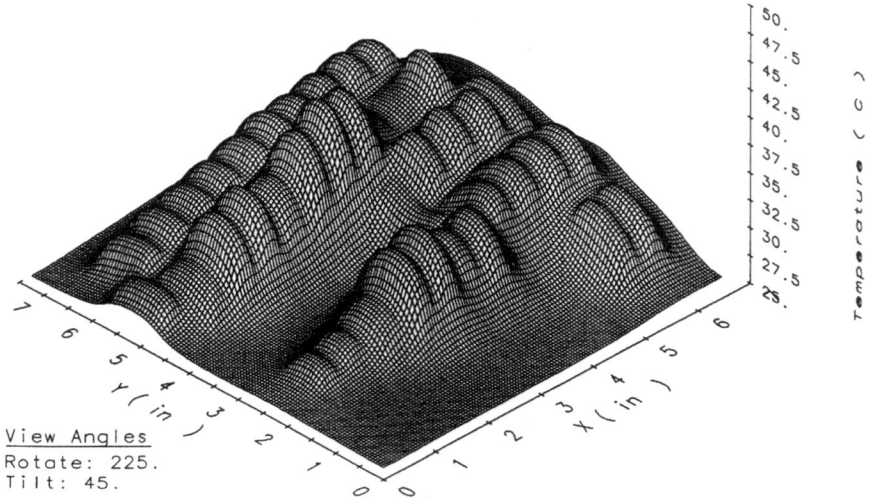

FIGURE 9.40   Temperature of the circuit card surface under the electronic parts.

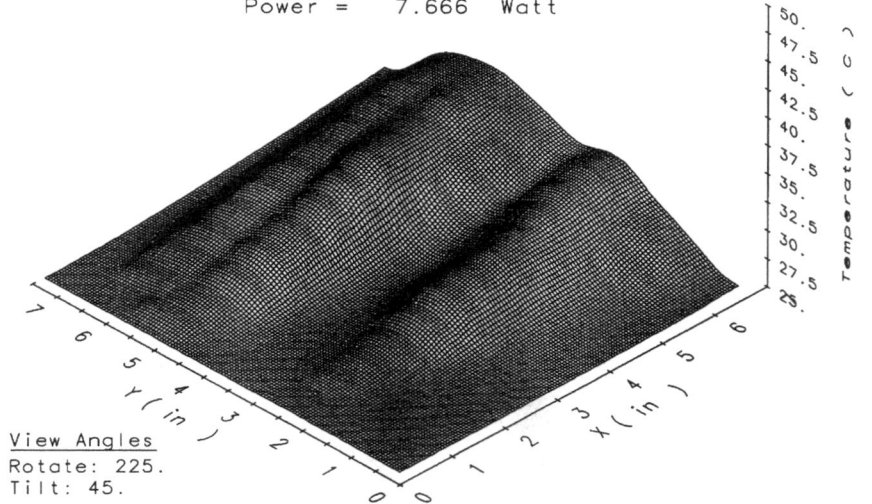

FIGURE 9.41   Air-temperature increase as it cools the electronics passing in the x direction.

RCVR/SDA (1.0 C/W interface, Base line)

Size =   5.000 in x   5.500 in

Total Power =   48.0000 Watts

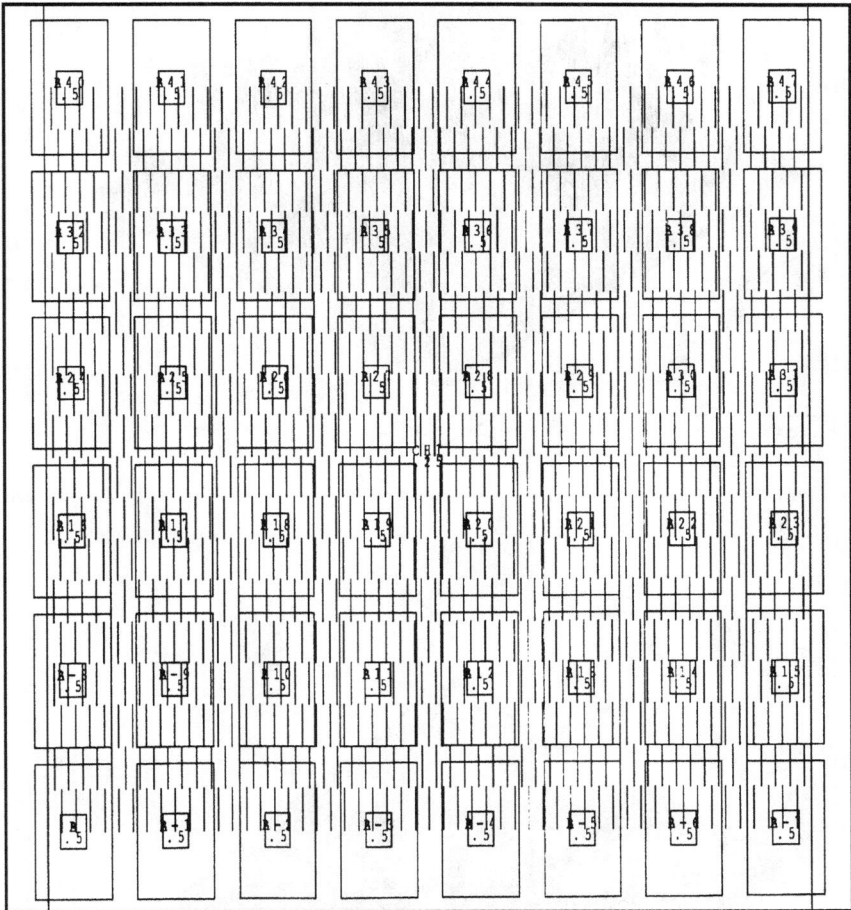

FIGURE 9.42   One side of a CCA pair used to analyze a flow-through-cooled module using DNS.

the case that cooler heat source temperatures can be achieved by getting the coolant closer to the heat source. In electronic packaging, the heat source is a chip, resistor, or capacitor, and getting the coolant directly to the part has historically been difficult to impossible. There are several reasons for the difficulty (such as contamination of the chips and circuitry, possibility of shorts where liquids are concerned, corrosion, space, and lack of ac-

RCVR/SDA (1.0 C/W interface, Base line)
Junction & Top Layer
Power = 48.000 Watt

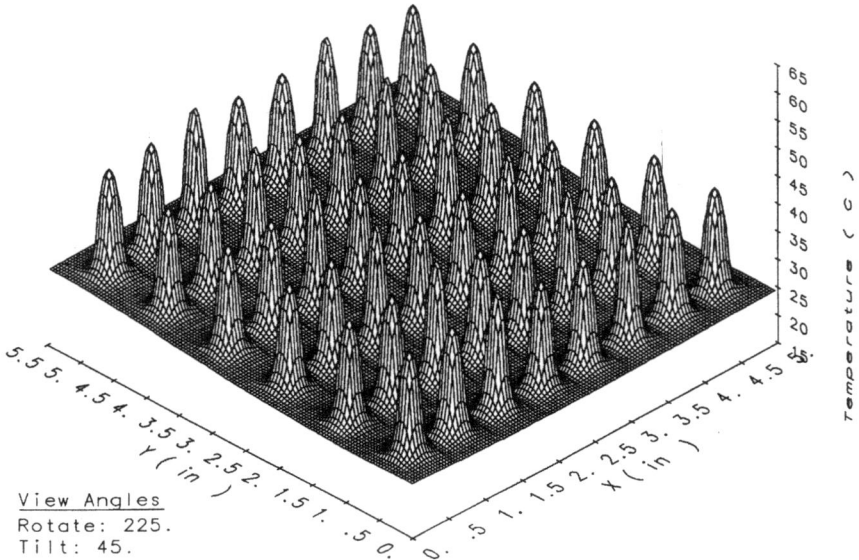

FIGURE 9.43   Junction/hot-spot temperatures of electronic parts on a surface-mount type, flow-through-cooled CCA pair.

RCVR/SDA (1.0 C/W interface, Base line)
Cooling Fluid @Aluminum-1
Power = 48.000 Watt

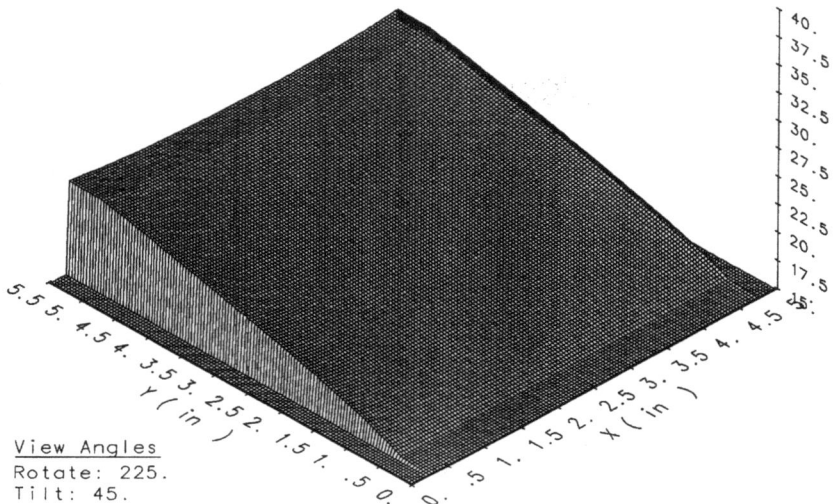

FIGURE 9.44   Temperature increase of air as it passed through the flow-through module.

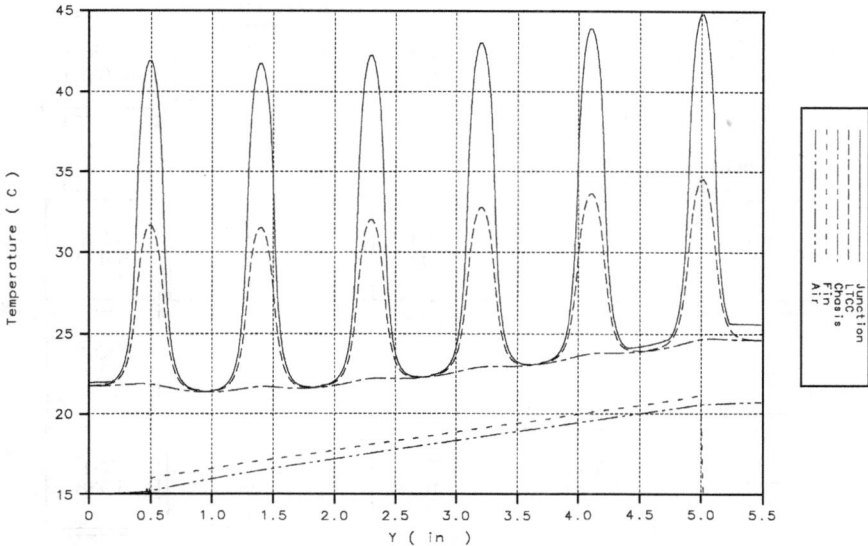

FIGURE 9.45   Temperature profiles at different levels in the CCA.

cess to modify chip designs by the higher-level packaging engineers). Thus, many chips have to be used as purchased from the foundry. The development of microchannel fin structures[12,13] has shown the possibility of placing fine-fin structures on the chip or on the substrate on which the chip is mounted.

The large number of fins inherent in microchannel cooling enables a very large heat transfer area at or very near the heat source. It is still not practical to place the fins actually at the heat source because the fins are usually placed on the bottom of the chip while heat is dissipated on or very near, within 2 or 3 microns, of the top of the chip; still, it is a great improvement.

Design problems must be overcome to successfully use microchannel cooling. The distribution of the liquid to the chip locations in multi-chip assemblies, such as CCAs and MCPs, can make the assemblies very expensive. The possibility of leaks inside electronic assemblies might not be desirable and can affect yields. Pressure drops can be high. Fluids must be kept very clean to avoid clogging or fouling the small passages and reducing the effectiveness of the fins.

One class of fins is larger than microchannel (characterized by over 100 fins per inch) and is characterized by more than 25 standard fins per inch. This class is called "macrochannel" for convenience. These fins are too large to be on a chip or an individual device, but are more practical for the package level designer to implement. An example of this type of structure is presented in Figures 9.46 through 9.49. These figures show the layout of the parts on a power-supply heatsink, which is dissipating 606 watts, and the flow path and fin structure inside the heatsink. Figures 9.48 and 9.49 show the temperatures of the parts and the refrigerated liquid coolant as it passes through the heatsink and absorbs heat. These results were again the result of a DNS analysis.

PALS New Power Supply – micro channel
Size = 4.500 in x 4.000 in
Total power = 606.0040 Watts

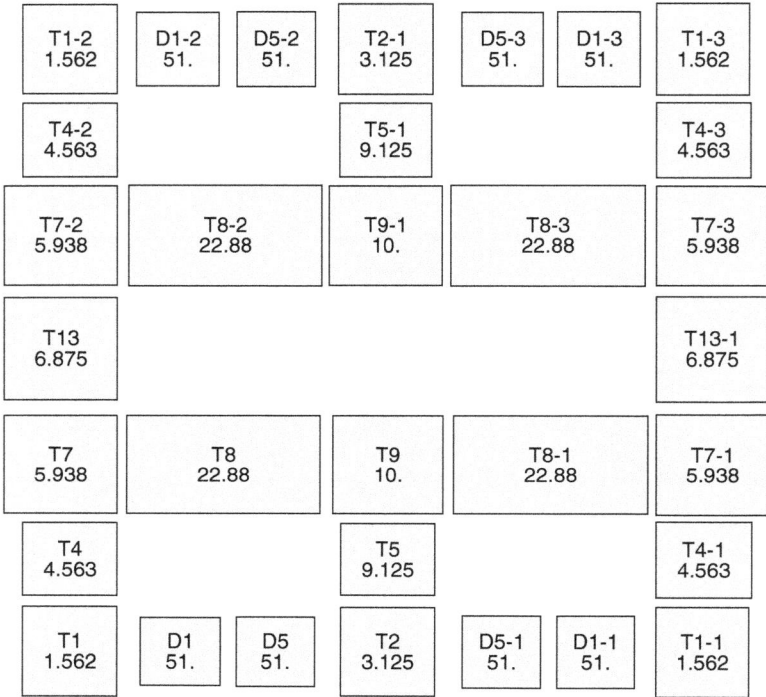

T1-2 1.562	D1-2 51.	D5-2 51.	T2-1 3.125	D5-3 51.	D1-3 51.	T1-3 1.562

T4-2 4.563			T5-1 9.125			T4-3 4.563

T7-2 5.938	T8-2 22.88	T9-1 10.	T8-3 22.88	T7-3 5.938

T13 6.875				T13-1 6.875

T7 5.938	T8 22.88	T9 10.	T8-1 22.88	T7-1 5.938

T4 4.563			T5 9.125			T4-1 4.563

T1 1.562	D1 51.	D5 51.	T2 3.125	D5-1 51.	D1-1 51.	T1-1 1.562

FIGURE 9.46 Parts layout of a microchannel-cooled power supply.

A promising, but as yet not demonstrated, technology to implement in microchannel cooling is the use of micromachines, such as micropumps (to place the entire coolant loop, including the pump, inside the multichip package or on a circuit card). One concept is shown in Figure 9.50, where a micropump the size of a dual-inline package, provides the work to circulate coolant through a high-power device and the heat exchanger required to cool the liquid after it absorbs the heat from the device.

9.4.3 Natural Convection-Cooled CCAs

Circuit-card assemblies are often mounted vertically to induce and enhance natural convection flow of air over the assembly. The natural convection equations presented can be used to define the heat-transfer coefficient at any location on the assembly, but presume a knowledge of the air bulk temperature. Most electronic boxes contain multiple circuit

PALS New Power Supply–micro channel
Size = 4.500 in x 4.000 in
Aluminum-1

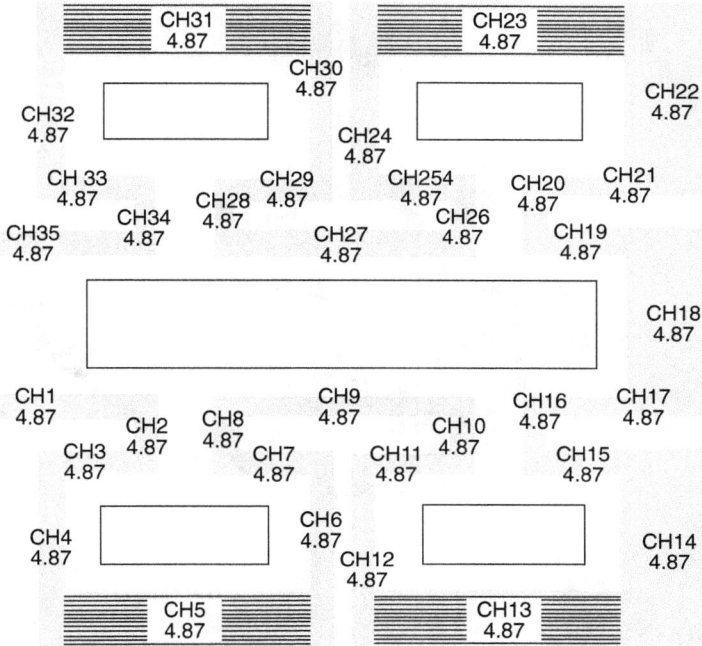

FIGURE 9.47 Fluid channels and microchannel-fin structure in the power-supply heatsink.

cards and air trapped between them increases in temperature as it rises, according to the expression:

$$\Delta T_x = \frac{Q_x}{\dot{m}C_p} \tag{9.50}$$

Where:

Q_x = The total heat on the assembly upstream of position x
T_x = The air temperature at location x
$\dot{m}$ = The mass flow rate of air
C_p = The specific heat of air

The temperature rise depends on the mass flow rate, which is driven by the temperature difference between the air between the circuit cards and the outside surrounding air. The air flowing over a single circuit card can be estimated by the expression:

$$\dot{V} = 4.06\mu^{0.5}T_f^{0.5}(T_\infty + 460)^{-0.25}Pr^{-0.5}(0.952 + Pr)^{-0.25}\Delta T_w^{0.25}L^{0.75} \tag{9.51}$$

FIGURE 9.48 DNS-method computed part temperatures.

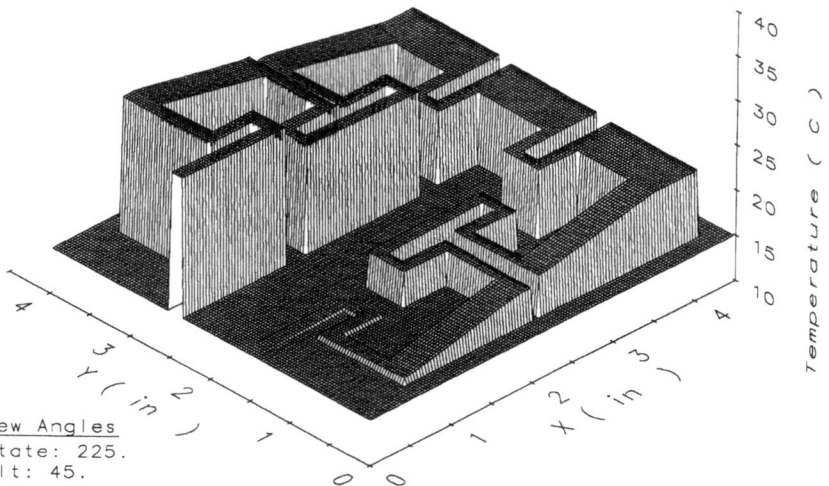

FIGURE 9.49 Liquid coolant (PAO) temperature as it passes through the cooling channels and microchannel fins.

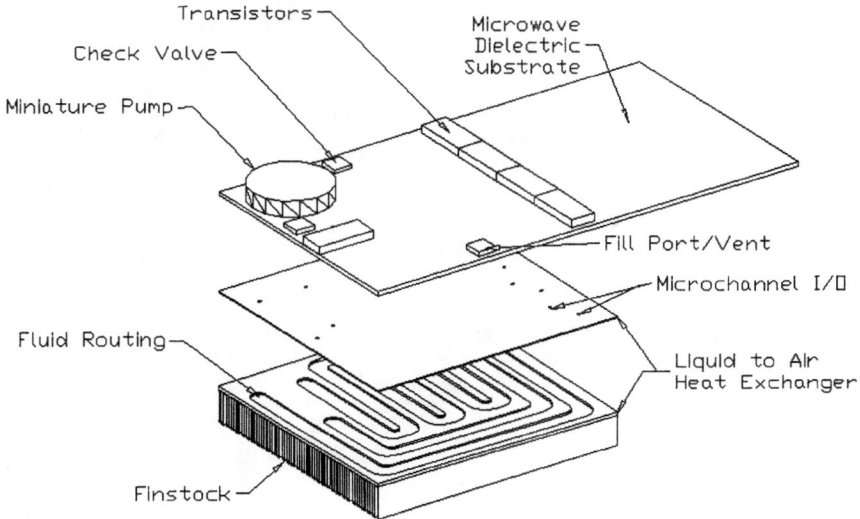

FIGURE 9.50　Micromachine pumping using microchannel cooling.

Where:
$\dot{V}$ is the volume flow rate in ft.^{3}/min.
T_f is the average film temperature in °F
T_w is the temperature rise of the surface above the ambient
T_∞ is the ambient air temperature in °F

However, when air is channelled between boards, the flow rate is determined by a balance between the buoyancy forces and the friction forces. Thus, the actual flow rate is determined by the pressure developed by buoyancy in Eq. 9.3 balanced against the resistance to flow:

$$\Delta P = 2.67 \rho L \left[1 - \frac{(T_\infty + 460)}{\Delta T} \, ln\left(1 + \frac{\Delta T}{T\infty + 460} \right) \right] \qquad (9.52)$$

Where:
P is the pressure difference in inches of water
L is the height of the flow channel in inches
T is the rise in air temperature through the channel in °F
T_∞ is the ambient air outside the channel or at the entrance of the channel in °F

A simple computer program to define the temperature rise between two circuit cards is presented. Note that the expressions presented and used in the computer program are for flat plates, which are smooth, compared to circuit cards, covered with components of varying heights and sizes. It is therefore an approximation. Some studies have been done for the specific case of component-covered circuit-card assemblies and a perusal of *Advances in Thermal Modeling of Electronic Components and Systems* by Bar-Cohen and Kraus will give additional information. Also be aware that the computer program is not self-limiting:

If the circuit cards are moved too far apart, the program will provide flow rates that are too high. The maximum flow rate should be limited to Eq. 9.51.

```
Program is to calculate flow rate and temp. rise of air in natural
'convection between closely space PC boards, which are uniformly
heated.
INPUT "Input ambient temp.(C)"; T
INPUT "Input heat dissipation of board (watt)";
Q Cp = 1 / .1316
Pr = .72
'Estimated film temperature
Tf = T * 1.8 + 32 + 15
INPUT "What is the flow height (inch)"; H
INPUT "What is the air flow space between boards (inch)"; b
INPUT "What is the width of the boards (inch)"; L
A = b * L
Dh = 4 * A / (2 * (L + b))
DT = 10
'k is in watts/inC
1 k = (.0132 + .000024 * Tf) * 1.8 * 12 / 3.413
'Visc is in #/ft.hr.
visc = (3.378E-07 + 6E-10 * Tf - 2.4E-13 * Tf ^ 2) * 115812
'converting visc to #/min.inch
visc = visc / 12 / 60
'rho is in #/in^3
rho = (.086 * 460 / (Tf + 460)) / 1728!
mdot = Q / Cp / DT
Tf = (T + DT / 2) * 1.8 + 32
Re = mdot * Dh / A / visc
f = 16 / Re
'Buoyancy pressure available
DP1 = 27.67 * rho * H * (1 - (Tf + 460) / (DT * 1.8) * LOG(1 + (DT *
1.8) / (Tf + 460)))
'Friction flow resistance
DP2 = 27.67 * 2 * f * H * mdot ^ 2 / Dh / 386.4 / 3600 / rho / A ^ 2
IF ABS(DP1 - DP2) < .00001 THEN GOSUB pout
IF DP1 > DP2 THEN DT = DT - .05 ELSE DT = DT + .06
GOTO 1
END
pout:
PRINT "Pressure drops =", DP1, DP2
form1$ = " ####.## ###.## #.### ###.##"
PRINT " Height        Width Spacing Tambient"
PRINT USING form1$; H; L; b; T
form2$ = " ##.###      ##.#######      ####.## "
PRINT " mdot    DT              Tf"
PRINT USING form2$; mdot; DT; Tf
STOP
END
SUB pout
END SUB
```

Enhancing natural convection. There are cases where natural convection is not adequate to cool the circuit-card assembly and a cooling enhancement is required without resorting to the installation of a fan or completely changing the design to liquid. One method has been devised and marketed (AAVID) to provide localized enhancement by sandwiching a liquid-filled bag (Figure 9.51) between the circuit card and the chassis

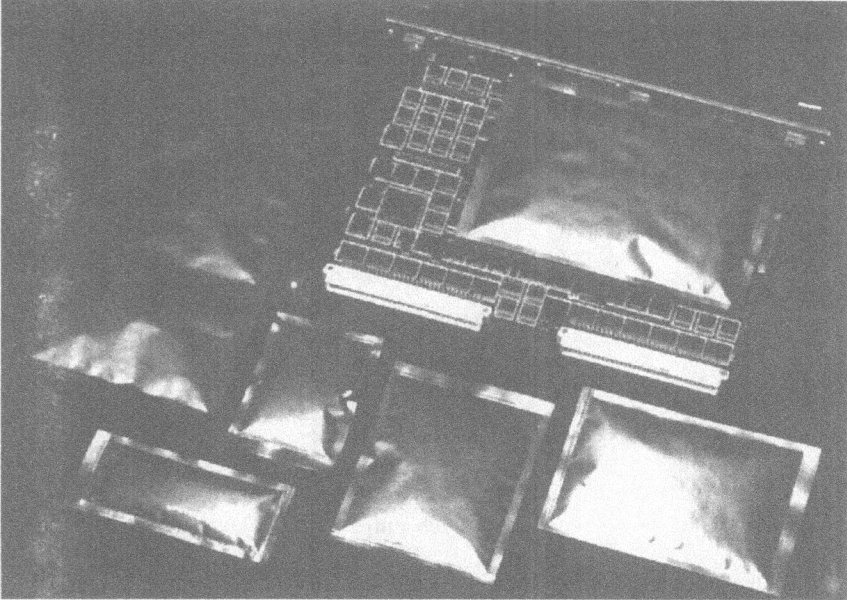

FIGURE 9.51   Liquid-filled heatsink bag used between a circuit card assembly and a chassis wall to provide heat transfer when natural convection is not adequate *(Courtesy of AAVID, Inc.)*.

wall or cover or an adjacent circuit card, which has a means of dissipating the heat to a sink. Because natural convection is buoyancy driven, the higher density of liquid provides improved circulation. In addition, the thermal conductivity can be much higher in liquids, thus providing a higher heat-transfer coefficient. These bags are very flexible, so good contact is made with each component on the board. The convective heat transfer in the enclosed space of these liquid-filled bags can be computed, but depends on whether they are oriented vertically or horizontally. Also heat is being transferred to the liquid and then to the other, cooler surface, so the "effective" heat-transfer coefficient is different from the coefficient from a single surface to a fluid only. Also, if the bag is oriented horizontally and the upper surface is warmer, no convection currents can be developed and energy transfer must be through conduction only.

For the case of the vertically oriented liquid-filled bags, the Grashof number is calculated as:

$$Gr_\delta = \frac{g\beta(T_1 - T_2)\delta^3\rho^2}{\mu^2} \tag{9.53}$$

Where δ is the internal space between the two surfaces of the bag (assuming a uniform distance).

At low Grashof numbers, the circulation is weak and the temperature difference between the two surfaces is nearly linear through the fluid and is largely a conduction phenomenon. At any Grashof number between 3×10^4 and 1×10^6, a high-velocity circulation is established. For a given geometry, the Grashof number is a function of ΔT; for a desired temperature difference, the Grashof number can be increased by increasing δ or ρ or decreasing μ. The Nusselt number is defined from empirical correlations[18] as:

$$N_\delta = 0.42(Gr_\delta Pr)^{1/4}Pr^{0.012}(L/\delta)^{-0.03} \tag{9.54}$$

for constant heat-flux conditions and where $10^4 < Gr_\delta Pr < 10^7$

$$1 < Pr < 20000$$

$$1 < L/\delta < 40$$

$$N_\delta = 0.046(Gr_\delta Pr)^{1/3} \tag{9.55}$$

Where:

$$10^6 < Gr\delta Pr < 10^9$$

$$1 < Pr < 20$$

$$1 < L/\delta < 40$$

Then, the heat transfer between the two walls is calculated as:

$$Q = ANu_\delta \frac{k}{\delta}(T_1 - T_2) \tag{9.56}$$

Where A is the area of the heat-dissipating surface.

When the orientation of the two surfaces is horizontal the following equation can be applied.

$$N_\delta = 0.012(Gr_\delta Pr)^{0.6} \tag{9.57}$$

When $1700 < Gr_\delta Pr < 6000$ and $1 < Pr < 5000$

$$N_\delta = 0.375(Gr_\delta Pr)^{0.2} \tag{9.58}$$

When $6000 < Gr_\delta Pr < 37,000$ and $1 < Pr < 5000$

$$N_\delta = 0.057(Gr_\delta Pr)^{0.333} \tag{9.59}$$

When $37,000 < Gr_\delta Pr < 10^8$ and $1 < Pr < 20$

$$N_\delta = 0.13(Gr_\delta Pr)^{0.3} \tag{9.60}$$

When $Gr_\delta Pr > 10^8$ and $1 < Pr < 20$

The liquid-filled bags can also be used in a two-phased cooling mode (i.e., boiling and condensing usually referred to as *ebullient cooling*). This is covered in the section on ebullient cooling.

Another product designed to aid the transfer of heat to a cover or other structural plate is a sheet of flexible foam-type material (Figure 9.52). The standard material made by the Bergquist Company of Minneapolis comes in standard widths of 0.02 to 0.125 inches in thickness and is designed to be sandwiched between a CCA and chassis structure that can reject heat to the outside ambient. It has a thermal conductivity of 1.75 W/M-°C (at 10 psi) and conforms to the component shape to provide intimate contact to conduct heat to the cover.

9.4.4 Ebullient Cooling of Electronic Devices

Although less often used, ebullient or boiling of a coolant is used in high-power situations to provide very efficient heat transfer. This type of heat transfer is very appealing because of the high heat-transfer coefficients it provides, but its implementation requires some special considerations. The simplest method is to totally immerse the electronics in the liquid. This requires that the liquid be a dielectric; in fact, high-voltage applications often use this method because the immersion in a dielectric for arc suppression is already a requirement. The correct liquid must be one with the right dielectric and heat-transfer properties.

FIGURE 9.52 The Bergquist Gap Pad material designed to conduct heat from CCA components to a cover or structure *(Courtesy of The Bergquist Company.).*

Usually, the heat transfer via boiling is so efficient, compared to other methods, that the liquid is not critical; however, it can be.

The pressure that the container or electronic box must withstand must be considered. The box must be evacuated before filling, so it must withstand atmospheric pressure on the outside. Once filled, the box must withstand the internal vapor pressure of the boiling liquid. Because air has been removed from the box, the pressure is the vapor pressure of the liquid. By definition, a liquid boils when its vapor pressure equals that of its environment. Because its environment is its own vapor pressure, it is always at the boiling point. The maximum pressure the box must withstand is the liquid vapor pressure at the maximum temperature expected. Some vapor pressure curves of dielectric fluids are presented in Figure 9.53.

To maintain the boiling process, condensation must occur to return the vapor to the fluid state and keep the liquid level covering the electronics. Basic equations for boiling and condensation are now presented, along with some information on surface and fluid parameters required to predict the heat-transfer coefficients.

The equation for nucleate pool boiling was developed by Rohsenow.[22]

$$\Delta T = \frac{C_{sf} h_{fg} Pr_1^{1.7}}{Cp_1} \left[\frac{q/A}{\mu_1 h_{fg}} \sqrt{\frac{\sigma}{g(\rho_1 - \rho_v)}} \right]^{0.33} \tag{9.61}$$

Where:

C_{sf} is constant determined from experiment for combinations of liquid and surface material

Cp_l is the specific heat of the liquid

h_{fg} is the heat or enthalpy of vaporization of the liquid

Pr is the Prandtl number of the saturated liquid

q/A is the heat per unit area

μ_1 is the viscosity

σ is the surface tension of the liquid-vapor interface

ρ_1 is the density of the saturated liquid

ρ_v is the density of the saturated vapor

g is the gravitational constant

Values for C_{sf} for water and various surfaces can be found in Ref. 3. A simplified formulation is also given in Ref. 9 with coefficients for Freons and fluorocarbons.

A critical heat flux occurs with all boiling liquids at which the surface becomes coated with vapor and any increase in flux will cause dramatic increases in temperature; essentially a thermal runaway situation. This critical flux should be avoided and is defined by Zuber[23] and repeated in many texts as:

$$q/A = \frac{\Pi}{24} h_{fg} \rho_v \left[\frac{\sigma g^2 (\rho_1 - \rho_v)}{\rho_v^2} \right]^{.25} \left(1 + \frac{\rho_v}{\rho_1} \right)^{.5} \tag{9.62}$$

Other types of two-phase cooling, which utilize solid-to-liquid and solid-to-gas phase change, are used in electronics. These are not explored here.

Boiling heat transfer can be used in situations where the vapor is expendable, but the usual situation is one where condensation replenishes the liquid so that a close loop system is used to transfer the heat (via boiling and condensing) to a prime coolant, such as air.

Vapor Pressure Curves

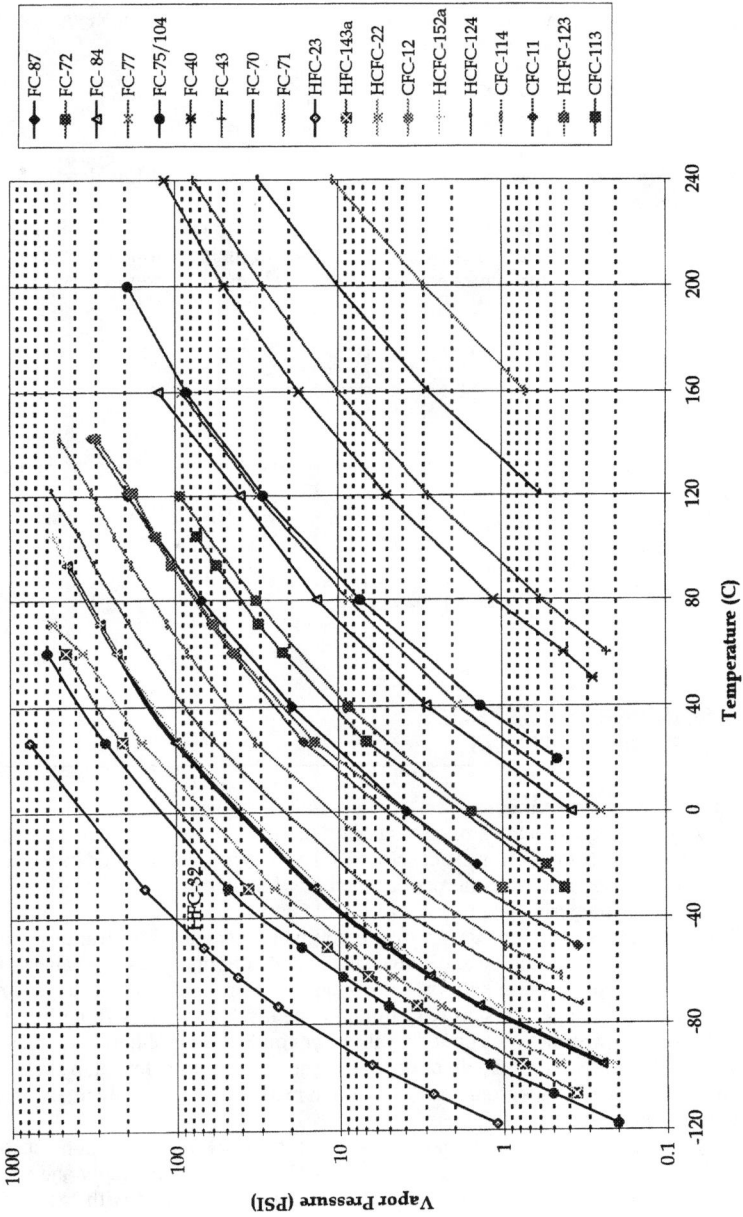

FIGURE 9.53 Vapor pressure curves of selected thermal fluids (*Data provided by 3M and DuPont.*).

Legend:
FC-87
FC-72
FC-84
FC-77
FC-75/104
FC-40
FC-43
FC-70
FC-71
HFC-23
HFC-143a
HCFC-22
CFC-12
HCFC-152a
HCFC-124
CFC-114
CFC-11
HCFC-123
CFC-113

In condensation flow, the transition to turbulent flow occurs at a Reynolds number of 1800. The Reynolds number is calculated as:

$$Re_f = \frac{4\dot{m}}{x\mu_f} \qquad (9.63)$$

Where x is either the perimeter (if the flow is in a tube) or the flow depth (if the flow is on a vertical wall).

For laminar flow on a vertical wall or tube, the following equation is suggested by McAdams[21], which includes a 20-percent increase in the heat-transfer coefficient to account for common disturbances in the flow, which make it depart from true laminar flow.

$$h = 1.13 \left[\frac{\rho_1(\rho_1 - \rho_v)gsin\phi h_{fg}k^3}{L\mu(T_g - T_w)} \right]^{.25} \qquad (9.64)$$

Where L is the length of the vertical wall or fin and ϕ is the angle between the wall and vertical, if it is inclined.

If the flow is turbulent, the following equation by Kirkbride[23] is suggested.

$$h = .0077 \, Re_f^{0.4} \left[\frac{\rho_1(\rho_1 - \rho_v)gk^3}{\mu^2} \right]^{1/3} \qquad (9.65)$$

9.5 NUMERICAL METHODS IN THERMAL ANALYSIS

With the advent of digital computers, it is possible to solve many heat-transfer problems by numerical procedures. Such methods permit thermal analysis to be carried out in situations that are too complex for exact, closed-form solution methods or hand calculation. The present section describes numerical methods for thermal analysis of the electronic packages. Many numerical approaches are available.

Many computer programs are available commercially. In general, a physical continuum is replaced by information at discrete spatial and temporal locations called *nodes*. The nodes can be arranged along patterned grid lines, as with finite-difference schemes, or in almost-arbitrary arrangements, as with finite-element schemes. Both finite-difference and finite-element approaches are commonly used for conduction problems, whereas the finite difference method is more widely used for convective heat transfer. Many references are available that provide background information in mathematical fundamentals, conduction, convection, and fluid mechanics.[1,3,4,6,10,21,22]

For those planning to use numerical methods, take several precautions at the outset. First, various means should be used, where possible, to check the quality and validity of a numerical solution. These include mesh refinements, comparison with experimental results, comparison with analytical solutions, and hand checking of results. Accurate prediction of chip-junction temperature is critical in the design process to optimize for cost, weight and electrical performance, while meeting reliability and integrity requirements. Direct Numerical Simulation (DNS) software provides multi-chip module thermal analyses with automatic mesh generation, high accuracy, high feature resolution, reduced analyst and cycle times, large node capability (6 million nodes for VAX 2100), and numerous options for the visualization of results. In this section, Direct Numeral Simulation is introduced, along with a detailed thermal trade study of material stack-ups for a high-power GaAs FET used in a multi-chip module design.

Material choices for substrates, solders, epoxies, and thermal spreaders or tabs are considered in each multi-chip module design and an accurate thermal-analysis tool is required to predict the impact of each material selected. For example, it would be less expensive to bond a substrate to a thermal spreader with conductive epoxy than soldering with a gold-tin preform, but the epoxy can increase junction temperatures significantly. The use of thermal vias in low-temperature or high-temperature ceramic substrates will improve thermal performance, but will increase cost, weight, and possibly lower the yields. A study by T. Poulin, et al.[1] empirically determined the conductance of Green Tape via arrays for a specific material stack-up. However, it is shown that the conductance of the entire material stack-up with via arrays depends highly on the interface materials used.

Thermal analysts typically utilize numerical modeling, which is tedious and error prone, when trying to resolve to length scales of chip junctions less than 1.0 micrometer. The methods of modeling complex systems are usually inadequate, are subject to assumptions based on user experience, and can only be validated by comparison with experiment. In contrast, DNS requires no assumptions; thus, the results are more accurate than measurements because of the inaccuracies associated with controlling test parameters, instrumentation, and equipment. Temperature profiles of electronic devices are typically measured with infrared-scanning or liquid-crystal techniques,[2] where infrared scanning is simple, but spatial resolution is limited and liquid-crystal phase-transition visualization is somewhat subjective. DNS software has been compared to problems that have exact mathematical solutions and has been proven to be accurate and mathematically valid.

Solution method   The wide spread of the powerful computer has made DNS extremely valuable for solving problems that are not susceptible to the analytic methods. DNS basically solves the three-dimensional unsteady heat equation, where the equation in tensor notation is:

$$\rho Cp \frac{DT}{Dt} = \frac{\partial}{\partial x_j}\left(k \frac{\partial T}{\partial x_j}\right) + Q \tag{9.66}$$

Or where repeated index means summation, Q is the heat generation per unit volume, ρ, Cp and k are the density, specific heat and thermal conductivity, respectively. If the equation

$$\rho Cp\left(\frac{\partial T}{\partial t} + U_j \frac{\partial T}{\partial x_j}\right) = \frac{\partial}{\partial x_j}\left(k \frac{\partial T}{\partial x_j}\right) + Q \tag{9.67}$$

is rewritten in Cartesian coordinates:

$$\rho Cp\left(\frac{\partial T}{\partial t} + U\frac{\partial T}{\partial x} + V\frac{\partial T}{\partial y} + W\frac{\partial T}{\partial z}\right) = \frac{\partial}{\partial x}\left(k \frac{\partial T}{\partial x}\right) + \frac{\partial}{\partial y}\left(k \frac{\partial T}{\partial y}\right) + \frac{\partial}{\partial z}\left(k \frac{\partial T}{\partial z}\right) + Q \tag{9.68}$$

Where U, V, and W are velocities in x, y, and z direction, respectively. Most analyses are performed for steady-state operation. The result is the following equation, after integrate with respect to volume:

$$\iiint_V \nabla \cdot k\nabla T d\overline{V} + Q \cdot Vol = 0 \tag{9.69}$$

Where ∇ represents the Laplacian operator. Using Green's divergence theorem equation (Eq. 9.69) can be simplified to:

$$\iint_S k\nabla T \cdot \vec{n}\,dS + Q \cdot Vol = 0 \qquad (9.70)$$

where $\vec{n}$ is a normal vector of the surface. The integral on the left represents the net flow of heat into any arbitrary fixed region. Generally speaking, a closed-form solution for this heat equation can be obtained via the separation of variable technique for the simple boundary condition. The final form of the solution, however, is expressed as double infinite series of the eigen functions, which can only be evaluated numerically. Therefore, instead of seeking an expansive exact solution, we only wish to establish the temperatures at any of the nodal points within the body. Applying Fourier's law of conduction to evaluate the inflow of heat through each of the four boundaries of the control volume about point (i,j,k) gives:

$$-k_x\Delta y\Delta z\frac{\partial T}{\partial x}\rangle_{i-1/2,j,k} + k_x\Delta y\Delta z\frac{\partial T}{\partial x}\rangle_{i+1/2,j,k} - k_y\Delta x\Delta z\frac{\partial T}{\partial y}\rangle_{i,j-1/2,k} + k_y\Delta x\Delta z\frac{\partial T}{\partial y}\rangle_{i,j+1/2,k}$$

$$(9.71)$$

$$- k_z\Delta x\Delta y\frac{\partial T}{\partial z}\rangle_{i,j,k-1/2} + k_z\Delta x\Delta y\frac{\partial T}{\partial z}\rangle_{i,j,k+1/2} + Q_{i,j,k}\Delta x\Delta y\Delta z = 0$$

The ½ in the subscripts refers to evaluation at the boundaries of the control volume which are halfway between mesh points. Approximating the derivatives by central difference:

$$-k_x\Delta y\Delta z\frac{(t_{i-1,j,k} - T_{i,j,k})}{\Delta x} + k_x\Delta y\Delta z\frac{(T_{i+1,j,k} - T_{i,j,k})}{\Delta x} + k_y\Delta x\Delta z\frac{(T_{i,j-1,k} - T_{i,j,k})}{\Delta y}$$

$$+ k_y\Delta x\Delta z\frac{(T_{i,j+1,ik} - T_{i,j,k})}{\Delta y} + k_z\Delta x\Delta y\frac{(T_{i,j,k-1} - T_{i,j,k})}{\Delta z} + k_z\Delta x\Delta y\frac{(T_{i,j,k+1} - T_{i,j,k})}{\Delta z} + \quad (9.72)$$

$$Q_{i,j,k}\Delta x\Delta y\Delta z = 0$$

Dividing through by the finite volume,

$$\Delta x\Delta y\Delta z$$

a descretized equation can be obtained for the heat-conduction equation:

$$k_x\frac{(T_{i-1,j,k} - 2T_{i,j,k} + T_{i+1,j,k}}{(\Delta x)^2} + k_y\frac{(T_{i,j-1,k} - 2T_{i,j,k} + T_{i,j+1,k}}{(\Delta y)^2}$$

$$+ k_z\frac{(T_{i,j,k-1} - 2T_{i,j,k} + T_{i,j,k+1}}{(\Delta z)^2} + Q_{i,j,k} = 0 \qquad (9.73)$$

This descretized equation can be solved numerically. A large number of text books have been written about the numerical method. For a typical module analysis, i and j can run up to 528 and k can go up to 24 to resolve the minimum-length scale (order of micrometer) involved in the module design.

Electronic circuit geometry and material information is entered into a Direct Numerical Simulation data file that is user friendly and simple to modify. Computer-generated layouts from electrical design or drafting can be transferred directly to DNS data file format to significantly reduce analysis cycle time. Trade studies are generated by simply changing one line in the data file.

9.5.1 Component and Bare Chip Assemblies

In the previous section on the analysis of CCAs, the internal temperatures of the electronic components was characterized by θ_{jc}. Often, the θ_{jc} characterization is overly simplistic. This is especially true when dealing with GaAs MMICs, which respond very quickly to power pulses. In these, the transient nature of the internal temperature rise of the chip must be considered. Also, the characterization of the chip temperature rise by using θ_{jc} implies that the chip is isothermal. Although we know it is not, there is often no information to define it better. This lack of information leads to the treating of the chip as isothermal, which maximizes the heat-transfer area used for the chip to CCA or the case to CCA. This results in optimistic temperature predictions.

The need for chips to do more per square inch has resulted in annual increases in the power density that the thermal designer is faced with having to cool. The increased use of GaAs in microwave applications has pushed the watt density (i.e., heat flux) to extreme levels because of the reduced size of the chip circuitry (often sub-micron dimensions) and increasing absolute-power dissipations. The actual point of maximum power dissipation in these chips is a few microns below the surface of the chip and often reaches a flux greater than 6000 watts/inch.[2] The dimensions are so small that measurement by physical means is nearly impossible. Liquid-crystal measurements have been able to define temperatures in the area around the source/drain sites, but at best give an average. The best measurement is the calibration and measurement of the base-to-emitter voltage, V_{be}, which is sensitive to temperature. These measurements have proven reasonably accurate, although some averaging does occur. Analysis using the DNS method, has given excellent results when node sizes are small enough to resolve the minute geometry of the chip circuitry.

In contrast to the high-power GaAs chips, new high-temperature superconducting circuits challenge the thermal designer to keep chips with small power dissipations at or near 75 K (−198°C).

9.5.2 Chip Analysis and Characterization

Accurate representation of bare chips requires that a chip thermal analysis, which requires detailed knowledge of the chip construction and layout.

Many new chip technologies are emerging using new chip materials. This is especially true for very high-power and very high-frequency devices. Some of the new materials being used in chip development are listed here, along with the device technologies of current interest or showing the most promise. Some of these materials have very high thermal conductivities, which looks nice for the thermal designer, except that they are being made with the expectation of being able to dissipate many times more power than present devices. This puts the cooling problem right back where it started, with the additional problem of having to absorb much more power once it gets off the chip.

- *GaAs*   MESFET, HBT, PHEMT, CHBT, IMPATT
- *InP*   HEMT, HBT, PIN
- *SiC*   SIT, MESFET

- *InSb* HEMT
- *SiGe* HBT
- *GaN* IGFET
- *Diamond* MESFET, MEMS, VME
- *YBCO* High-temperature, superconducting Josephson junction

The thermal analysis of a chip of any sort is complicated. Silicon chips, in relation to GaAs MMIC chips, have been easier because of their relatively high thermal conductivity and large heat-dissipating areas. GaAs chips can have extremely small heat-generating volumes with high heat densities. This makes their time constants very short and the evaluation of the response to peak powers during power pulses must be examined. In contrast, silicon chips have time constants long enough that a steady-state analysis using average power dissipation is usually adequate.

The complexity of chip structures necessitates analysis using either symmetry or some substructure that is far enough from other structures that no interaction takes place. This type of chip characterization is shown by the analysis of two cells of a GaAs MMIC having the structure shown in Figure 9.54. This shows the structure like an X-ray picture through the 0.004" thickness of the chip around the two cells that each have two sources. The circular structures are vias that extend through the chip, just like plated-through holes in a CCA. The entire structure shown here is 200×230 µm. Figures 9.55 through 9.66 show alternately deeper layers into the chip and the temperature profiles of each layer. The actual heat dissipation is several layers down inside the chip and this can be seen in the temperature cross-section view of Figure 9.67, which shows the hottest location well inside the chip. Notice that this DNS analysis is steady state, but a transient analysis must also be done to define the temperature rise of this hot spot during a peak power pulse. Once the rise is known, a θ_{jc} can be defined as:

$$\theta_{jc} = \frac{\Delta T_{peak}}{\overline{P}} \tag{9.74}$$

This resistance can be used for the next higher-level analysis (i.e., chip package, CCA, or multi-chip package).

9.5.2 Multi-Chip Packages/Chip-On-Board CCAs

This section applies to assemblies referred to by a variety of names (such as MCP, MCM, COB, or IMA), which are all multiple chips on a substrate. The substrate can be Duroid, high-temperature ceramic (HTC and HTCC), low-temperature cofired ceramic (LTCC), or various types of circuit materials used for different applications.

The MCM shown in Figure 9.68 is constructed using low-temperature cofired ceramic (LTCC) as the substrate and has a variety of passive and active devices, both digital and microwave. There are also both silicon and GaAs active devices. Figure 9.69 shows how this chip was modeled using DNS. This view shows how vias under high power-dissipating devices are placed to enhance heat transfer through the relatively low conductivity substrate. The effect of the low conductivity of the substrate combined with the comparative short distance for heat to travel downward is illustrated graphically in Figures 9.70 to 9.72. These show very little interaction between chips, except those that are very close together.

GaAs MMIC (Andy Ezis)

Size = 200.000 um x 230.000 um

Total Power = 0.1500 Watts

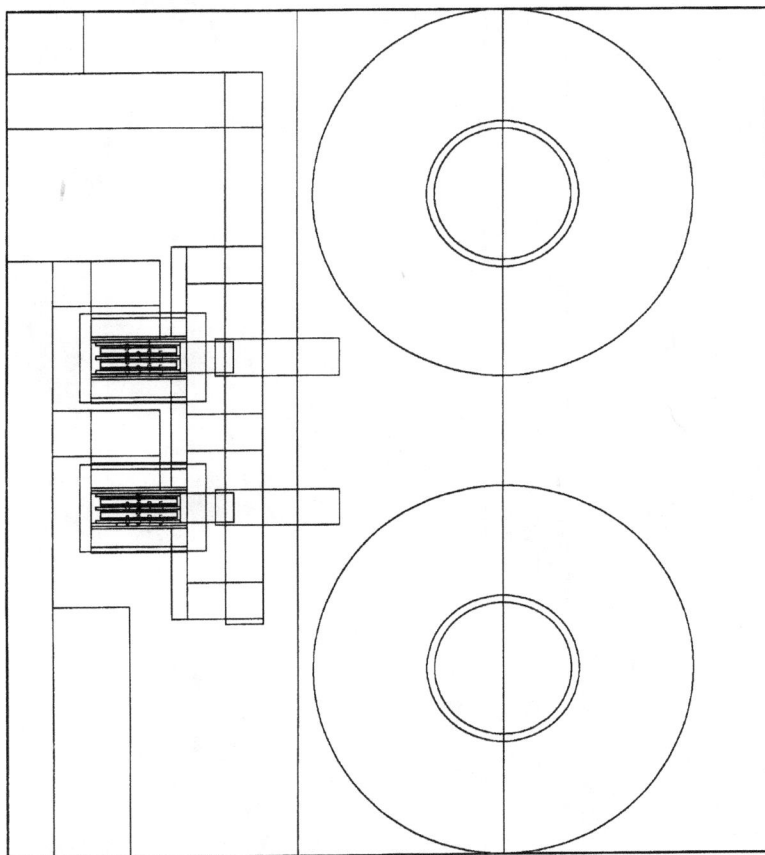

FIGURE 9.54 GaAs chip structure.

Finally, Figure 9.73 shows a comparison between the physical thickness of each layer used in the construction and the "thermal thickness" or temperature rise caused by thicknesses of the materials under component U6 (refer to Figure 9.69 for the location of U6).

9.6 AIR HANDLING

Air motion provides greatly improved heat transfer, compared to natural convection. Used in electronic packaging, the impetus for air motion is provided by either the vehicle or facility in which the electronics is installed or, most usually by fans or blowers.

GaAs MMIC (Andy Ezis)
Size = 200.000 μm × 230.000 μm
Gold

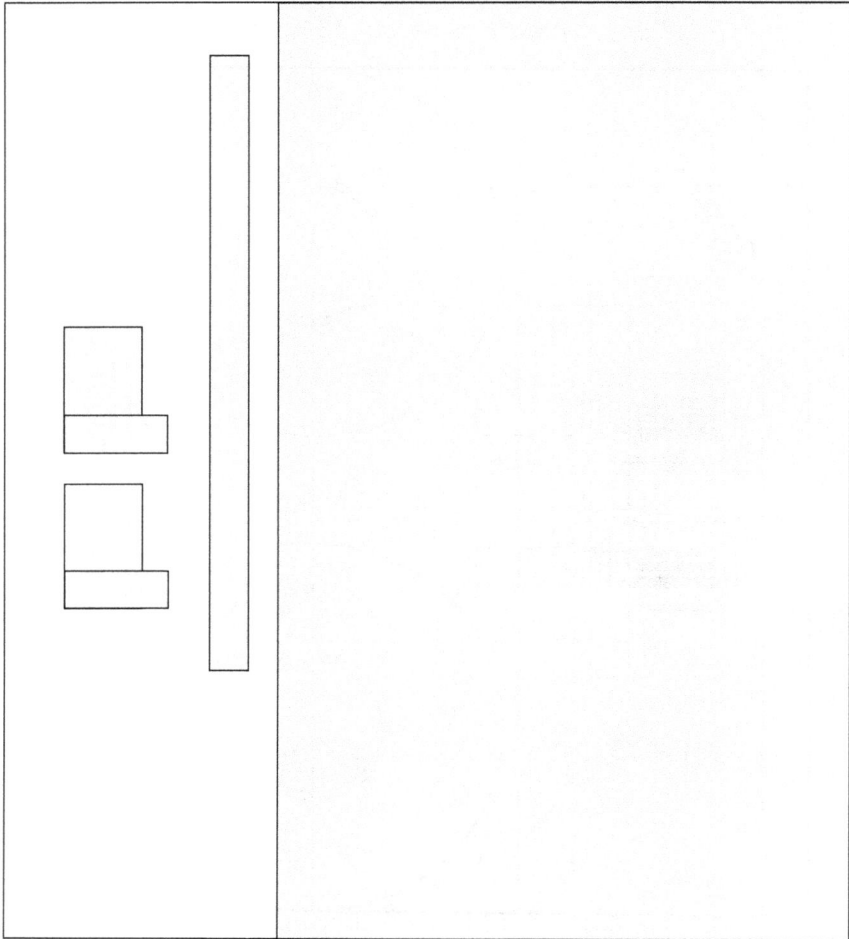

FIGURE 9.55 Top layer of the chip.

9.6.1 Fans and Blowers

Fans and blowers are governed by a simple set of laws, known as the fan laws, which characterize the performance of a blower, with respect to speed, flow, pressure, density, and power. These are:

$$W \quad = K_0 \rho N \tag{9.75}$$

$$CFM = K_1 N \tag{9.76}$$

$$\Delta P \quad = K_2 \rho N^2 \tag{9.77}$$

$$HP \quad = K_3 \rho N^3 \tag{9.78}$$

Where:

W is the mass flow rate in lb_m/minute

ρ is the air density in lbm/ft.^{3}

N is the fan speed in RPM

CFM is the volumetric flow rate in cubic feet per minute

ΔP is the pressure developed by the fan in inches of H_2O

HP is horsepower required by the fan motor

K_0 to K_3 are proportionality constants that depend on the fan size and design

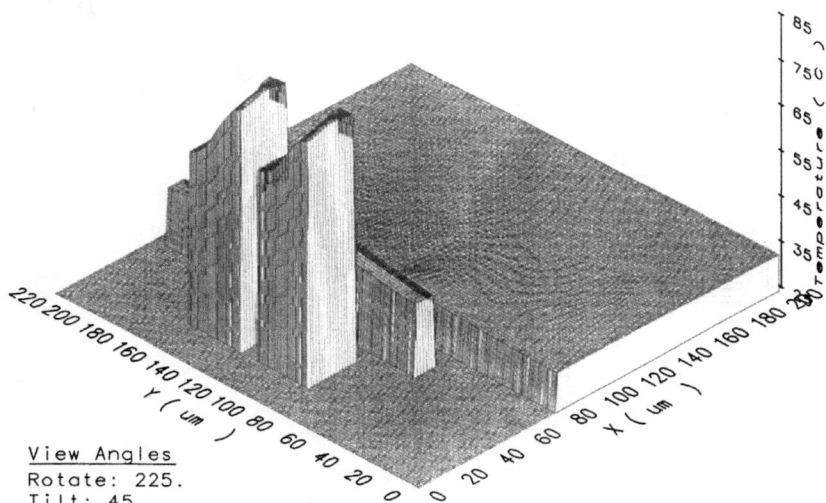

FIGURE 9.56 Temperature of the top of the chip.

The fan laws can be used to find the properties of a desired fan based on known fan properties when the fan type is unchanged. For example:

$$CFM_2 = CFM_1 \left(\frac{Size_2}{Size_1}\right)^3 \left(\frac{RPM_2}{RPM_1}\right) \tag{9.79}$$

$$\Delta P_2 = \Delta P_1 \left(\frac{Size_2}{Size_1}\right)^2 \left(\frac{RPM_2}{RPM_1}\right)^2 \frac{\rho_2}{\rho_1} \tag{9.80}$$

$$HP_2 = HP_1 \left(\frac{Size_2}{Size_1}\right)^5 \left(\frac{RPM_2}{RPM_1}\right)^3 \frac{\rho_2}{\rho_1} \tag{9.81}$$

Equations courtesy of EG&G ROTRON, Lancaster, PA.

From Eqs. 9.79 through 9.81, it is easy to see that obtaining high flow rate has a high cost. It is less costly to increase the speed than fan size. Also notice that the CFM is independent of density. When dealing with fan performance at altitude, it is important to remember that the fan is a constant-CFM machine.

GaAs MMIC (Andy Ezis)
Size = 200.000 μm × 230.000 μm
Polyimide

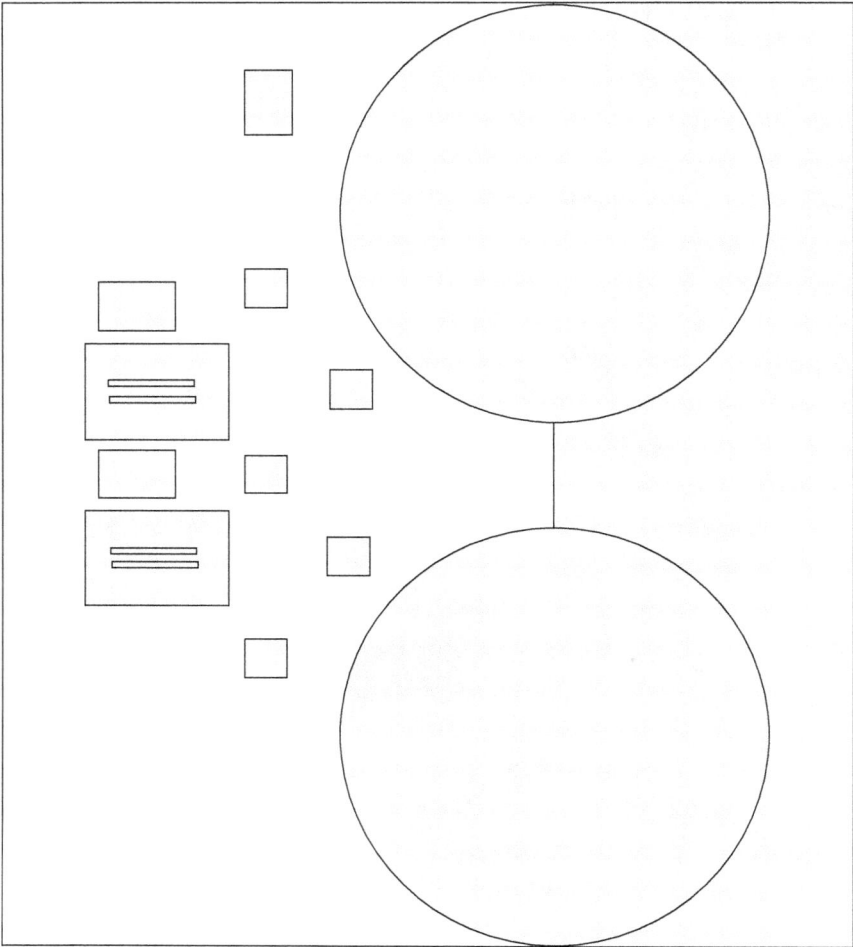

FIGURE 9.57   Chip structure at the polyimide layer.

When choosing a fan, the flow rate and pressure capability of the fan must be matched to the pressure drop characteristics of the electronic unit or system being cooled. This is illustrated in Figure 9.74, which shows the pressure of the fan curve dropping with increasing flow and the pressure resistance of the electronic unit increasing with increasing flow rate. Where the two lines cross is the operating point of the system. Knowing the desired flow rate, a fan must be chosen to cross the system operating curve as near the desired operating point as possible. One method of characterizing fans, which is an aid in choosing the most efficient fan for an application, is specific speed, N_s. This is defined as:

$$N_s = \frac{N(CFM)^{1/2}}{P^{3/4}} \qquad (9.82)$$

Where

N is the RPM

CFM is the volume flow rate in ft.^{3}/min.

P is the operating point static pressure in in. H_2O

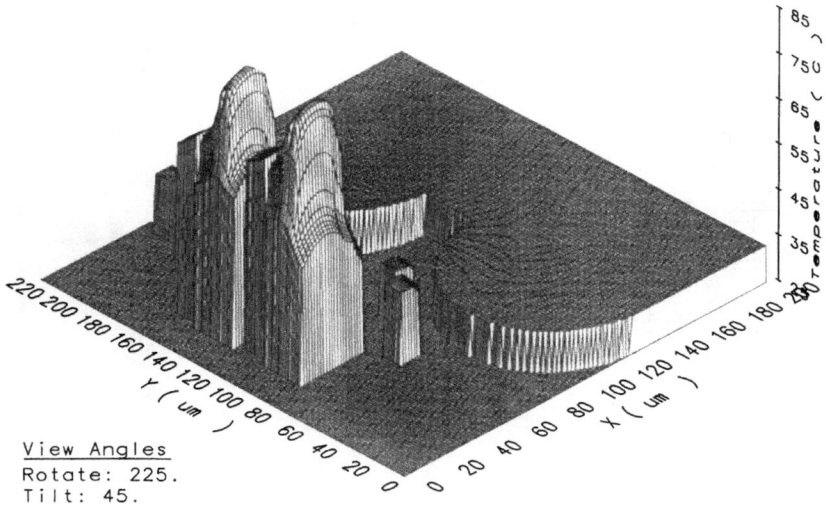

FIGURE 9.58   Temperatures at the polyimide layer of the chip.

The maximum $N = 0.9 \times 120 \times f/P$, where f is the line frequency and P is the number of poles (always even).

The desired pressure and CFM allows the computation of a specific speed, which can be matched to fan characteristics (Figure 9.75).

The fan curve in Figure 9.74 shows a smooth fan curve with monotonically decreasing pressure capability as the flow rate increases. This is true for many types of fans and is largely a function of the blade design. Figure 9.76 illustrates the relative shape of the performance curve for three types of impeller designs. Notice that the more forward curved the impeller design, the higher is the free delivery-pressure capability. When a fan has a dip in the performance curve, as is characteristic of forward-curved centrifugal and vaneaxial blowers, it is important to be sure that the system pressure curve does not fall in that area. A lower CFM and or higher pressure blower must be chosen to move the system pressure curve to the right (i.e., into the negative slope area of the blower performance curve).

9.6.2 Altitude Considerations

In the previous section, it was stated that the fan or blower is a constant volume device. Blower motors turn at a constant speed that is determined by the number of commutators

GaAs MMIC (Andy Ezis)
Size = 200.000 μm × 230.000 μm
Polyimid-1

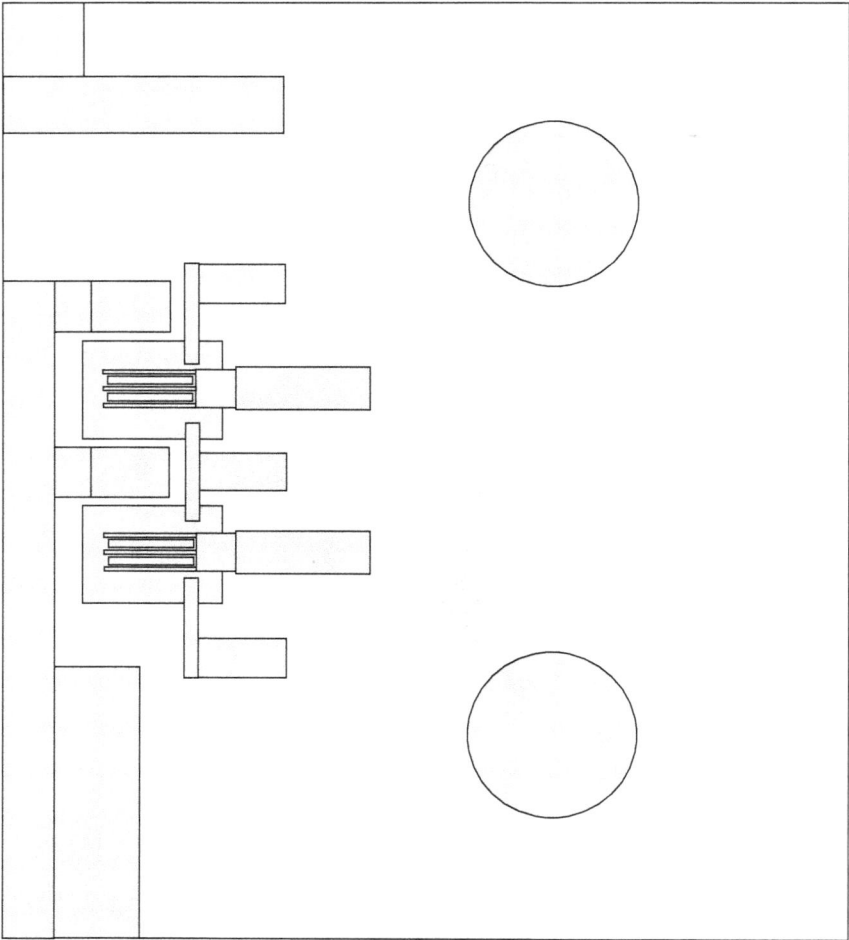

FIGURE 9.59   Chip structure at the top of the GaAs.

and the power line frequency. This means that they always deliver a specified volume flow rate. Because cooling is determined by mass flow rate, the amount of cooling thus varies with density; at altitude, the mass flow rate is, therefore, greatly reduced. The system pressure resistance must be calculated at the lower density and the blower chosen to provide the required CFM, defined by that condition. However, the blower will deliver the same volume flow at sea level (usually, much more than is needed) and the motor, blower size, and power requirements can be very large. As an example, consider the following situation.

Suppose that a 500-watt unit must be cooled at 30,000 feet. It is desired that the air temperature rise be limited to 30°C and a fan chosen to accomplish this.

$$\dot{m} = \frac{.1316 * 500}{30} = 2.193 \ lb/min.$$

At 30,000 feet, the air density is 0.0261 lb./ft.^{3} and the unit pressure drop is calculated to be 2 in. of H_2O at this density and flow rate.

At sea level, the density is 0.0706 lb./ft.^{3} Because both the unit-pressure drop and the fan-pressure capability are directly proportional to density, the fan must supply 84 ft.^{3}/min $\times$ 0.0706 lb./ft.^{3} = 5.93 lb./min. at sea level and will require 2 in. of $H_2O \times 0.0706/0.0261$ = 5.4 in. of H_2O. The unit still dissipates 500 watts and the air-temperature rise is only 11°C (i.e., overcooled). Ignoring the motor and impeller efficiency, the power required is:

$$HP = \frac{84 * 5.4 * 144}{27.67} \ ft\text{-}lb./min. * 3.03 \times 10^{-5} = .072 HP$$

whereas only 0.0266 HP is actually needed to cool the unit (from Equation 9.81).

To use a fan with the size and power requirements better matched to the cooling needs of the system, two schemes are usually used. The most straightforward solution is to use a two-speed blower. The blower can run at low speed, delivering less CFM at sea level, then, at higher altitudes, can increase its speed to provide the needed cooling. From the fan equations, power requirements increase as RPM cubed. Therefore, the fan should be designed or chosen for the proper size, speed, and altitude to provide low-altitude cooling.

Another method is to use a "limited-slip" blower. This type of construction allows the impeller shaft to turn slower than the motor when there is a high load as a result of high-density air on the impeller. This allows a smaller, lower-power motor because it does not

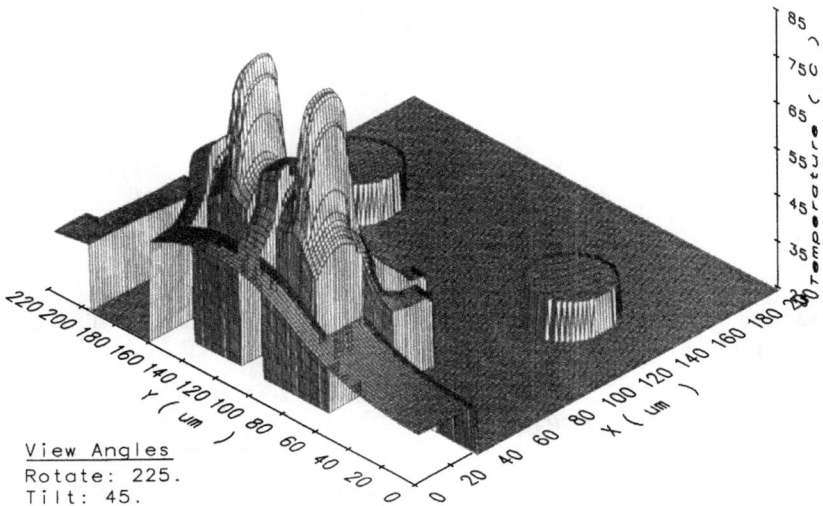

FIGURE 9.60 Temperatures at the top of the GaAs chip.

GaAs MMIC (Andy Ezis)
Size = 200.000 μm × 230.000 μm
Total Power = 0.1500 Watts

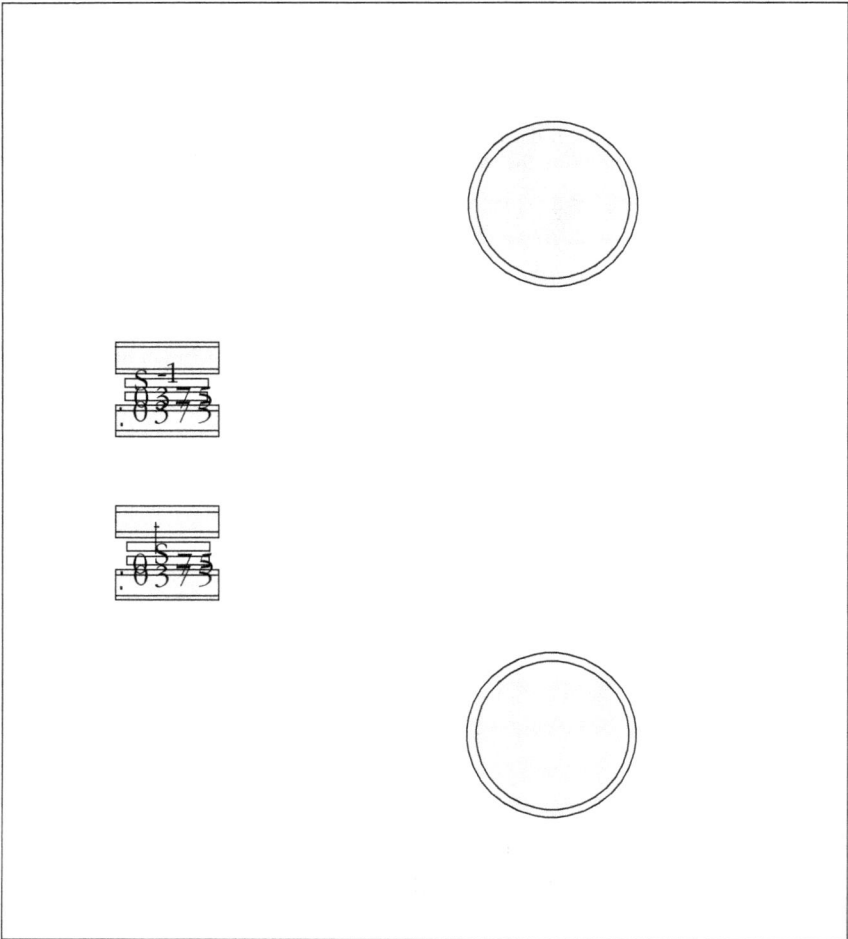

FIGURE 9.61   Structure 0.4 microns inside the GaAs where power is dissipated.

have to do as much work as a non-slip blower at sea level. The amount of slip is determined by the manufacturer.

9.7 COLD-PLATE DESIGN

The cold plate is and has been an important device in cooling electronics. Although its use has been mostly in military electronics, it is now being used more in high-powered areas of commercial electronics, such as power supplies and transmitters. The cold plate is a

GaAs MMIC (Andy Ezis)
GaAs*
Power =   0.150   Watt

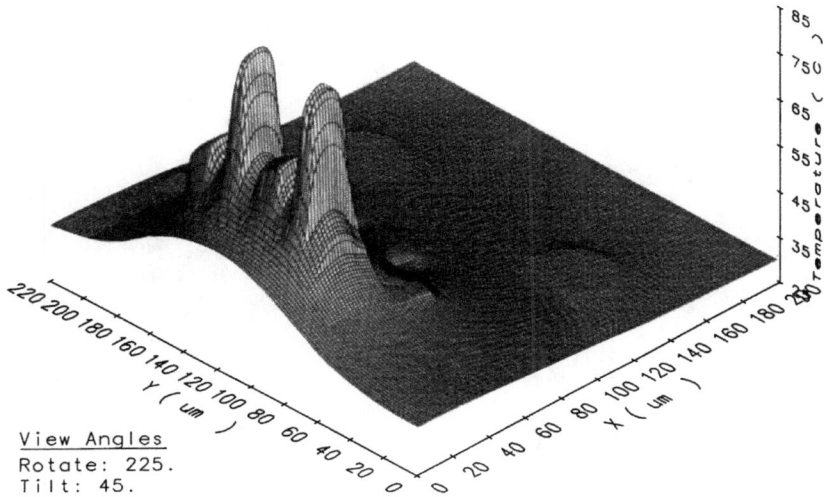

FIGURE 9.62   Temperature at 0.4 microns inside the chip.

forced convection-cooled plate that absorbs heat from one or both surfaces (Figure 9-77). These two surfaces are used to mount electronics. It can be considered to be a simplified heat exchanger and the same methods of design are used.

In practice, the heat sources are Circuit Card Assemblies (CCAs) or subassemblies of electronics, such as occur in power supplies (i.e., transformers, diode assemblies, and power transistors). In any configuration, the heat can be applied fairly uniformly or highly nonuniformly, and can be applied on one side or both sides. In addition, there are two standard methods of the design of the cold plate; the most common is called the *NTU method*, which stands for the *number of thermal units* (also called the *effectiveness method*) and the other is commonly called the *rating method*. In either case, two equations generally define the final configuration of the cold plate; the rate equation:

$$Q = hS\Delta T_m \tag{9.83}$$

Where

h = The heat-transfer coefficient in watts/ft.^{2}-°C)
S = The heat-transfer surface area (wetted area) in ft.^{2}
T_m = The log mean temperature difference °C

and the energy equation:

$$Q = \dot{m} C_p(T_{out} - T_{in}) \tag{9.84}$$

Where:

m = Mass flow of fluid in lb$_m$/min.
C_p = Specific heat of the fluid in watt-min./lb$_m$-°C)

GaAs MMIC (Andy Ezis)
Size = 200.000 μm × 230.000 μm
GaAs*-1

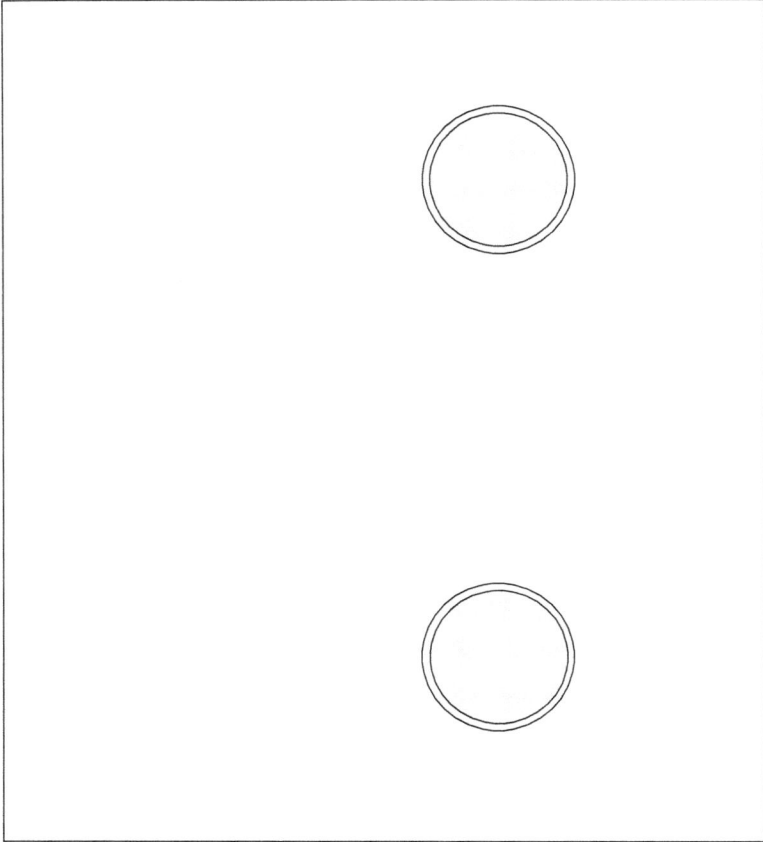

FIGURE 9.63 Chip structure at 1.1 microns.

T_{out} = The outlet temperature of the fluid in °C
T_{in} = The inlet temperature of the fluid in °C

Obviously, these two equations must balance when the log mean temperature is defined. The definition depends on the heat distribution or the temperature distribution, as applied to the cold-plate surface. The most useful and straight-forward definition is based on an isothermal surface (Figure 9.78). For this configuration, the log mean temperature difference is defined as:

$$\Delta T_m = \frac{(T_2 - T_1)}{\ln\dfrac{T_0 - T_1}{T_0 - T_2}}$$

(9.85)

Where the temperatures are defined in Figure 9.78.

By combining Eqs. 9.83, 9.84, and 9.85:

$$Q = \frac{hS(T_2 - T_1)}{\ln\dfrac{T_0 - T_1}{T_0 - T_2}} = \dot{m}C_p(T_2 - T_1) \tag{9.86}$$

from which the definition of NTU is derived:

$$NTU = \frac{hS}{\dot{m}C_p} = \ln\frac{T_0 - T_1}{T_0 - T_2} \tag{9.87}$$

GaAs MMIC (Andy Ezis)
GaAs*-1
Power = 0.150 Watt

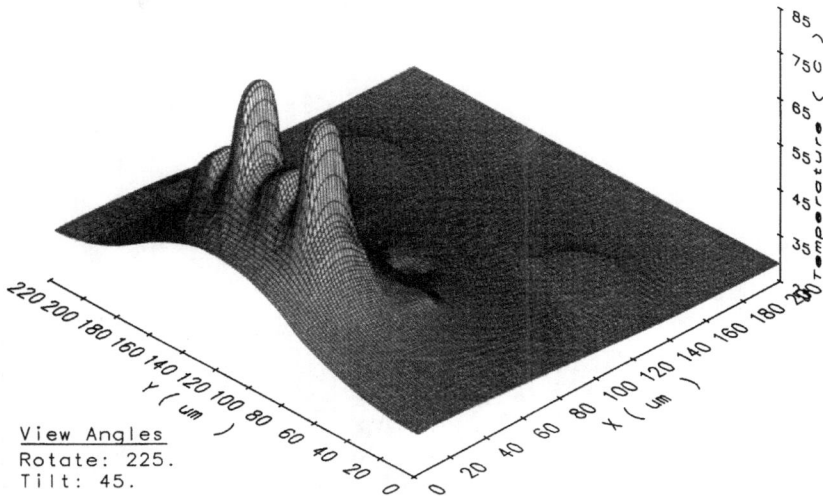

View Angles
Rotate: 225.
Tilt: 45.

FIGURE 9.64   Chip temperature at 1.1 micron.

Although this equation expresses all the properties of the cold plate, the actual design is a trial-and-error process, defined by the available space, fluid temperatures, and flow rates available; amount of pressure available; and, of course, the desired temperature of the cold plate, T_0. Often, a trial configuration is defined, based on the space available to define an approximate hS, the flow rate is defined by the pressure drop (as covered in Section 9.2.3) and T_1 and T_2 (as defined by the energy equation, 9.84). This allows solution for T_0. Equation 9.87 is often used in a different form by taking the antilog to get:

$$e^{NTU} = \frac{T_0 - T_1}{T_0 - T_2} \tag{9.88}$$

A method used to get an initial trial sizing is to start with the definition of effectiveness for a cold plate, ε:

$$\varepsilon = (1 - e^{-NTU}) \tag{9.89}$$

GaAs MMIC (Andy Ezis)
Size = 200.000 μm × 230.000 μm
GaAs*-2

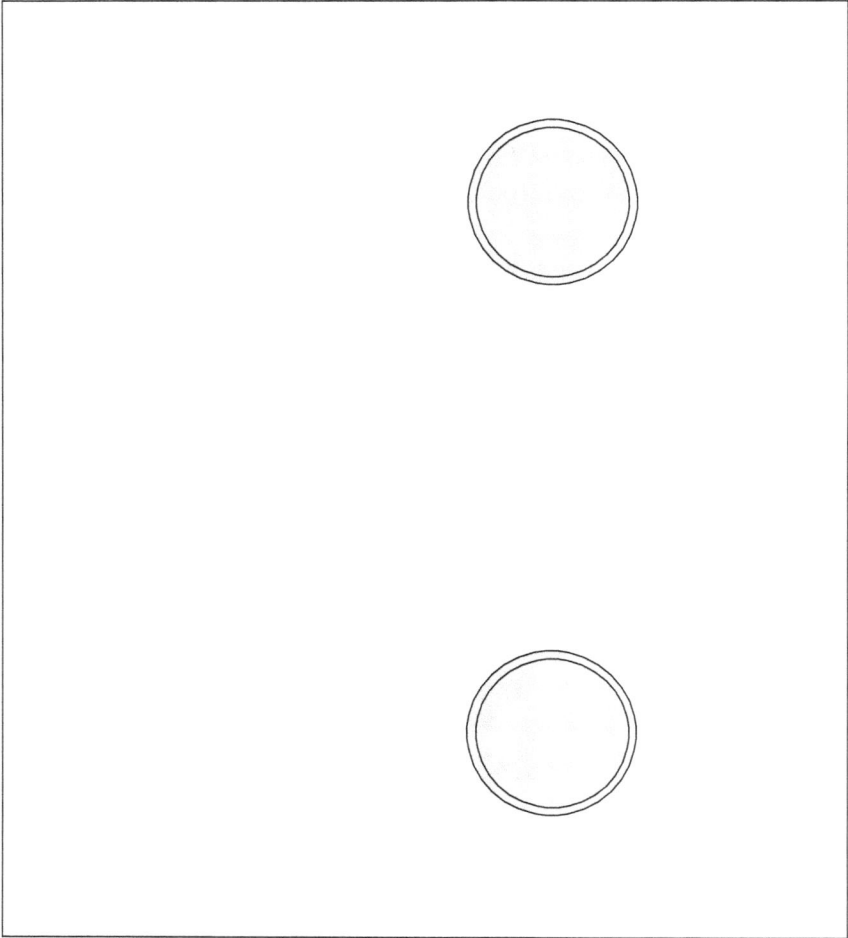

FIGURE 9.65 Chip structure at 2.1 microns.

and choose a reasonable effectiveness (between 0.5 and 0.85) to solve for NTU and use Eq. 9.86 or 9.87 to solve for other parameters. If the design is started with a calculation of the required effectiveness, it is called the *NTU* or *effectiveness method.* If the design is started from an assumed configuration and all heat-transfer parameters are used, then refined to converge on a final configuration, the method is called the *rating method* (because it is based on the rate equation).

The previous text is based on an isothermal surface and is a good approximation of actual surface temperatures in most cases. The only way to obtain a truly isothermal cold

GaAs MMIC (Andy Ezis)
GaAs*-2
Power =   0.150   Watt

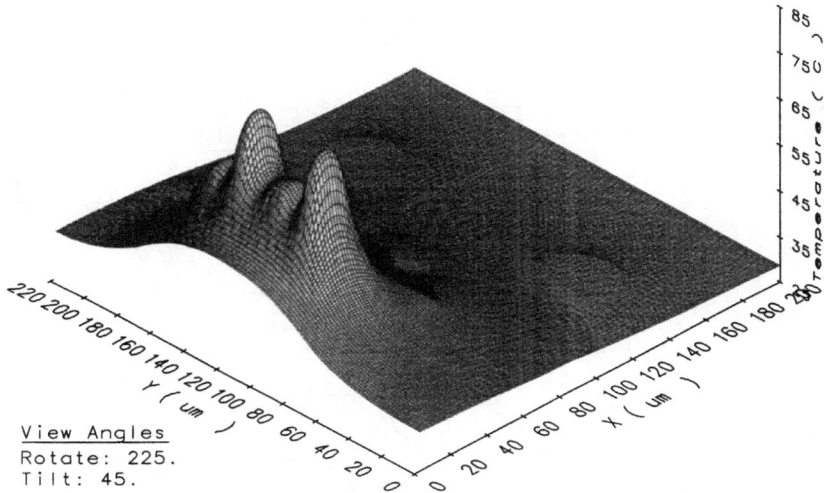

FIGURE 9.66 Chip temperature at 2.1 microns.

plate is to have a very high thermal-conductivity material or to simulate it with a heat pipe. A condenser heat exchanger also provides an isothermal surface. Actual surface temperatures can vary and depend on the uniformity of the heat dissipation and the thickness and conductivity of the cold-plate surface material (faceplate thickness). For thin faceplates, the temperature profile might look more like that shown in Figure 9.79, which reduces the actual log mean temperature difference and requires a slightly larger value of NTU to yield the same cold-plate surface temperature. In this case, the more general definition of the log mean temperature difference used in heat-exchanger design should be used.

$$\Delta T_m = \frac{(T_{01} - T_1) - (T_{02} - T_2)}{\ln\dfrac{T_{01} - T_1}{T_{02} - T_2}} \tag{9.90}$$

The choice of the fins for the core of the cold plate is the heart of the design effort. The core material and the flow rate determine the hS and ΔP from which all the temperatures are determined. The fins can be pin, straight, wavy, strip (lanced-and-offset), or some combination. A comprehensive treatment of test data for many types of fins is presented in *Compact Heat Exchangers* by Kays and London.

Data for a strip-fin core is presented in Figure 9.80 from Kays and London. This presents the fanning friction factor on the upper curve and the heat transfer expressed as the product of the Stanton number and the Prandtl number to the 2/3 power (i.e., the Colburn Modulus). A lot of analysis of test data has been performed to define correlation equations usable for design, based on extensive test data available for air. The most common type of cold-plate core

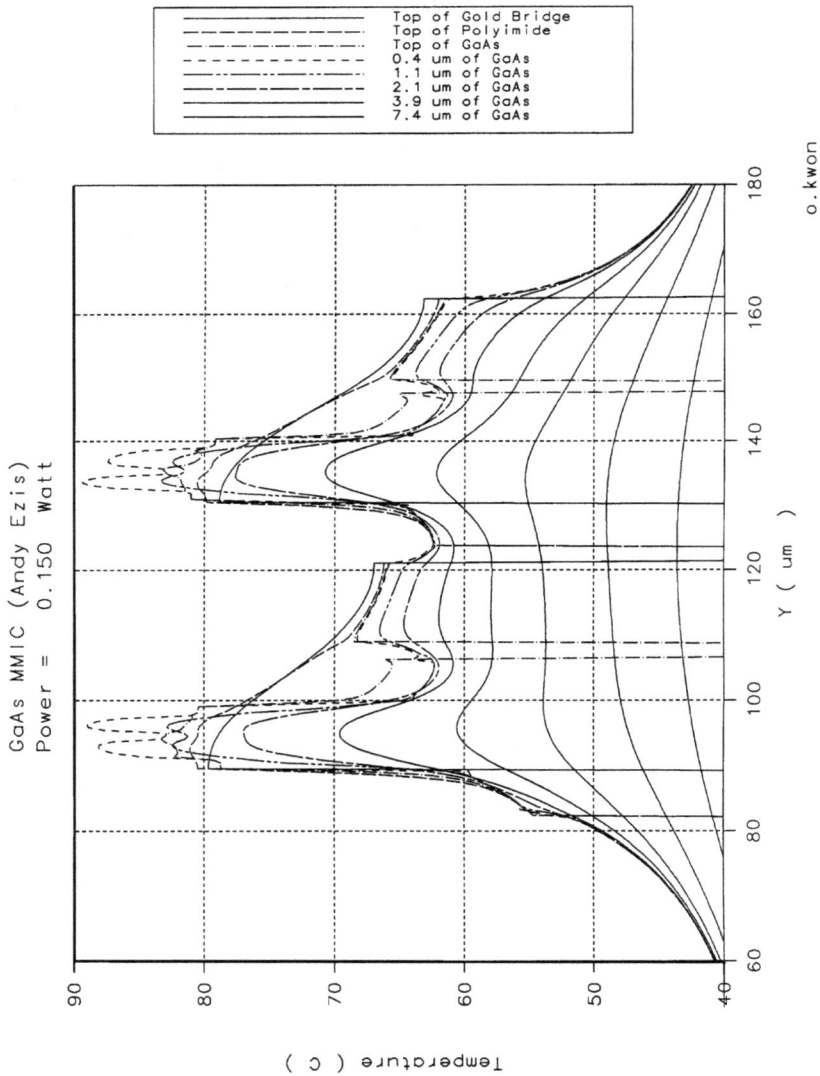

FIGURE 9.67 Chip-temperature profiles at all analysis levels showing the hot spots to be well inside the chip surface.

FIGURE 9.68 MultiChip Module (MCM) using a Low-Temperature Cofired Ceramic (LTCC) substrate.

is either the strip fin or the straight fin. For straight fins having an aspect ratio, α (defined as w/h in Figure 9.81), of 3.0 to 4.0, F. Altoz[9] gives the following expressions:

$Re \leq 2000$ and $L/D_h = 100$

$$f = \frac{18}{Re} \tag{9.91}$$

$$j = \frac{5.0}{Re} \tag{9.92}$$

For $Re > 8000$:

$$f = \frac{0.048}{Re^{0.2}} \tag{9.93}$$

$$j = \frac{0.025}{Re^{0.22}} \tag{9.94}$$

For lanced, and offset or strip fins, Weiting[17] gives the following relationships:

For $Re \leq 1000$:

$$f = 7.661 \left(\frac{x}{D}\right)^{-0.384} \alpha^{-0.092} Re^{-0.712} \tag{9.95}$$

$$j = 0.483 \left(\frac{x}{D}\right)^{-0.162} \alpha^{-0.184} Re^{-0.536} \tag{9.96}$$

and for $Re \geq 2000$:

$$f = 1.136 \left(\frac{x}{D}\right)^{-0.781} \left(\frac{t}{D}\right)^{0.534} Re^{-0.198} \tag{9.97}$$

$$j = 0.242 \left(\frac{x}{D}\right)^{-0.322} \left(\frac{t}{D}\right)^{0.089} Re^{-0.368} \tag{9.98}$$

All of the parameters of these equations are defined in Figure 9.81, except α, which was defined as w/h, and D, which is the hydraulic diameter of a single opening between fins.

The pressure loss through the cold plate is the sum of the pressure losses caused by entrance and exit contractions and expansions, and to the viscous losses in the core. Entrance- and exit-loss coefficients, K_c and K_e, for a multiple square tube and triangular tube cores are shown in Figures 9.82 and 9.83.

Many references discuss the entrance- and exit-pressure drops expressed as:

$$\Delta P_c = \frac{\rho}{2g}(\overline{V}^2(1 - \sigma^2) + K_c \overline{V}^2) \tag{9.99}$$

$$\Delta P_e = \frac{\rho}{2g}(\overline{V}^2(1 - \sigma^2) + K_e \overline{V}^2) \tag{9.100}$$

Where σ is the area ratio of open areas at the inlet and exit of the fins.

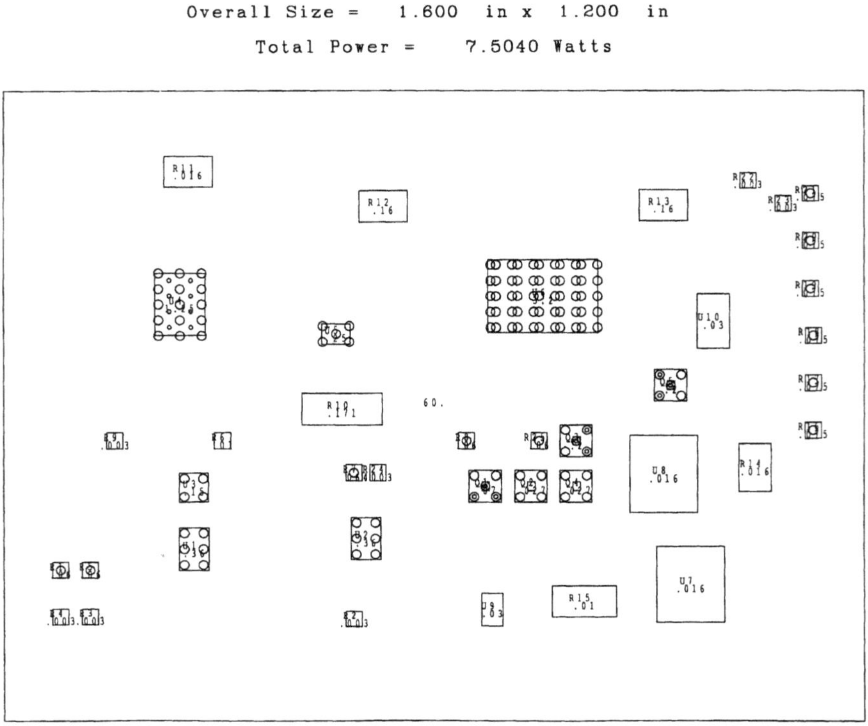

```
Overall Size =    1.600  in x  1.200  in
Total Power =    7.5040 Watts
```

FIGURE 9.69   Thermal model of the MCM shown in Fig. 9.68 used for DNS thermal analysis.

Junction & Top Layer
Power = 7.860 Watt

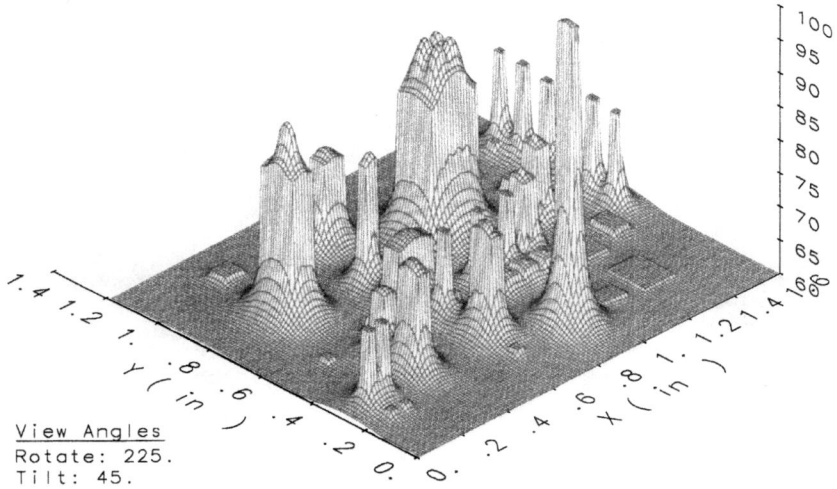

FIGURE 9.70 Junction and hot-spot temperatures of all the components mounted on the LTCC substrate.

Top of LTCC
Power = 7.860 Watt

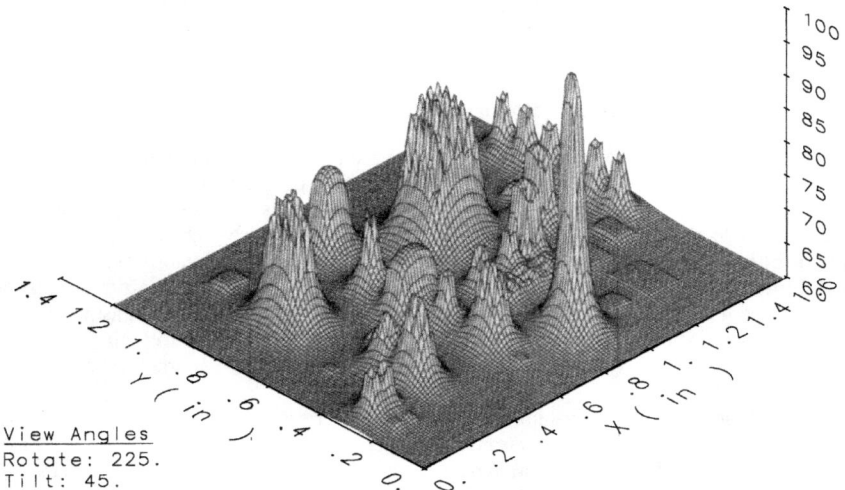

FIGURE 9.71 Top of the LTCC showing the effect of vias on depressing temperatures under chips.

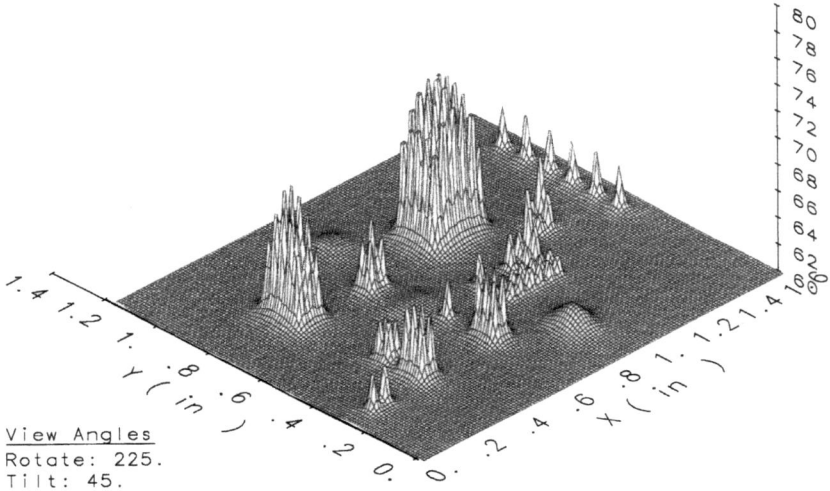

FIGURE 9.72 Bottom of the LTCC showing the increased temperature at via locations because of localized heat transfer supplied by the vias.

At the exit, pressure is recovered so that Eq. 9.100 represents an increase in pressure and must be subtracted from Eq. 9.99. However, the sign convention allows the recoverable terms to cancel and the irreversible losses to add. The friction loss of the core must be added to Eq. 9.99 and 9.100 to define the total pressure loss of the cold plate.

9.7.2 Cabinet and Large-Enclosure Cooling

Large cabinet and electronic enclosures, such as are used in many ground-based systems, are usually cooled by natural or forced convection. The choice is determined by the heat density in the cabinet and the heat flux of the individual assemblies inside the cabinet. Natural convection has the advantage of being quiet, not requiring power, and never failing. Forced convection can handle much higher heat densities and can provide a greater flexibility in the internal arrangement and packaging of the cabinet. The air inlet must be at the bottom of the cabinet and the exit at the top. Both should be as large as possible to allow maximum air flow. A maximum value for natural convection cooling given by George Ross[15] is 5.5 W/M^2-°C. A more-general value is given by Frank Altoz[9] as 100 watts/ft.^{3}

As with any convection system, the air temperature rises as it picks up heat. For proper cooling, it is desirable for each electronic assembly to get cool air and not air that is exiting from another assembly. This can often be accomplished by proper arrangement and including baffles to direct the air, such as in Figure 9.77. Whether air flow is forced by a fan or buoyancy, the flow through the cabinet is a balance of the pressure loss through the system

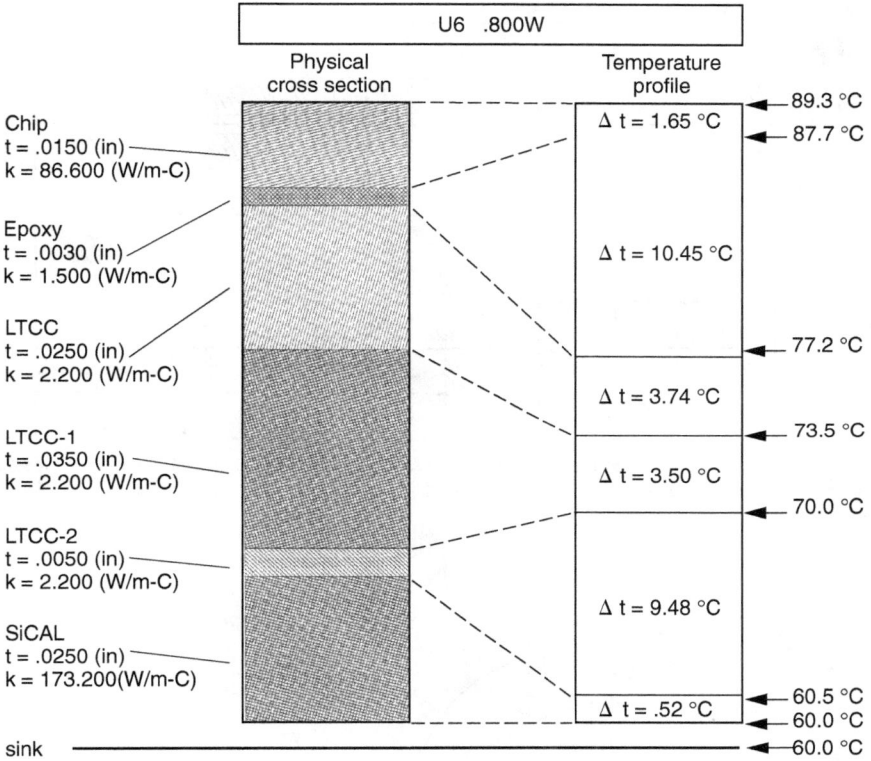

FIGURE 9.73 Physical and thermal stack-up diagram illustrates and helps visualize the sources of high resistance.

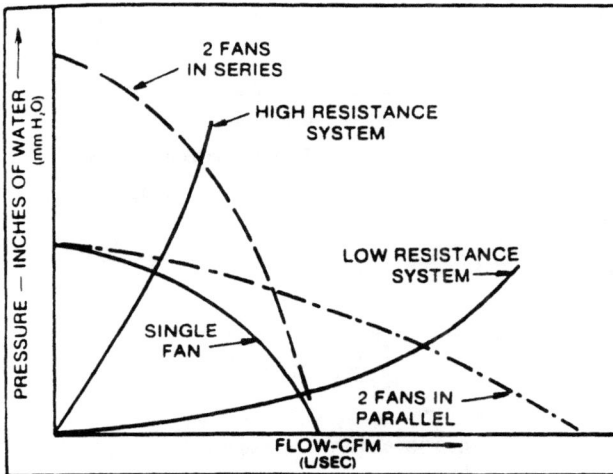

FIGURE 9.74 Fan pressure and system resistance must be matched to determine actual air flow *(Courtesy of E.G.& G. Rotron.).*

SPECIFIC SPEED

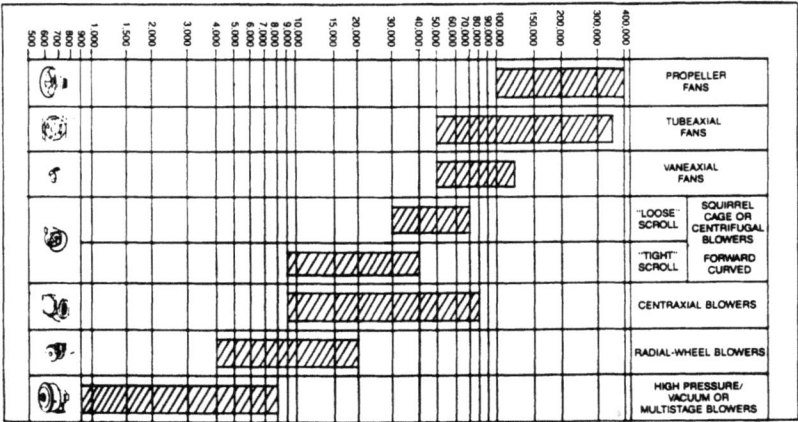

							PROPELLER FANS
							TUBEAXIAL FANS
							VANEAXIAL FANS
				"LOOSE" SCROLL			SQUIRREL CAGE OR CENTRIFUGAL BLOWERS
				"TIGHT" SCROLL			FORWARD CURVED
							CENTRAXIAL BLOWERS
							RADIAL-WHEEL BLOWERS
							HIGH PRESSURE/ VACUUM OR MULTISTAGE BLOWERS

FIGURE 9.75 Specific speed characteristics of the fan or blower is an initial indication of the type of blower needed for the system *(Courtesy of E.G.& G. Rotron.)*.

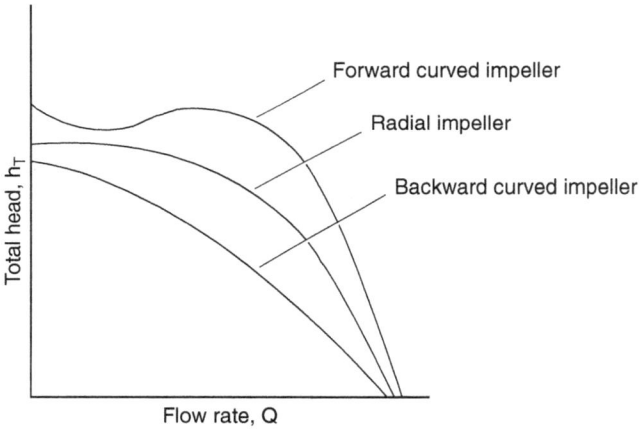

Forward curved impeller

Radial impeller

Backward curved impeller

Total head, h_T

Flow rate, Q

FIGURE 9.76 Characteristic shapes of fan curves relative to blade design.

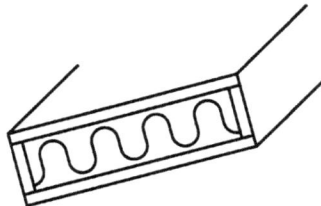

FIGURE 9.77 Typical cold-plate construction.

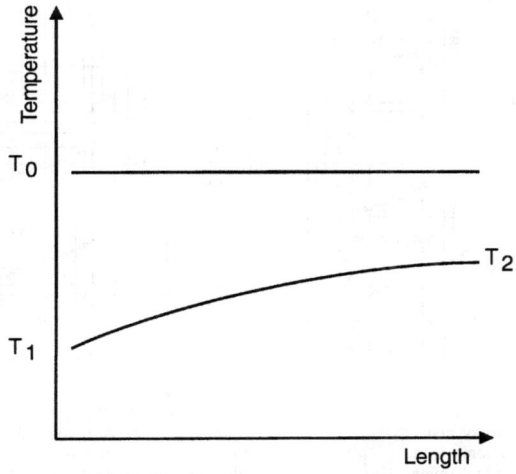

FIGURE 9.78 Typical fluid and cold-plate temperature profile.

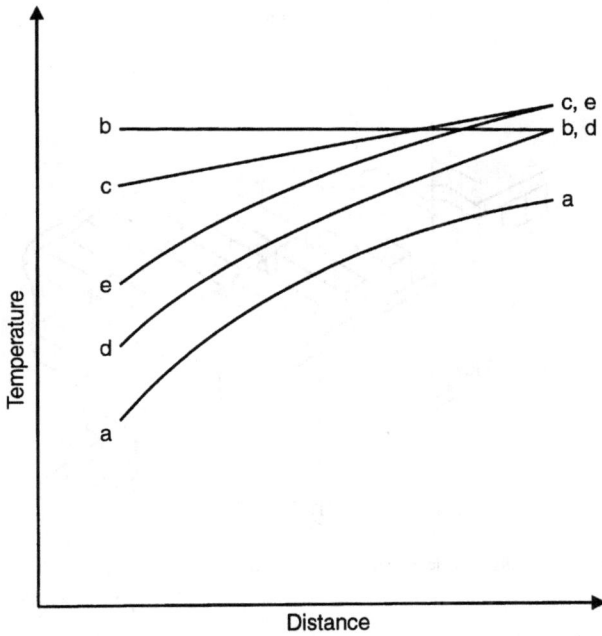

FIGURE 9.79 Some temperature vs. location profiles for cold plates.

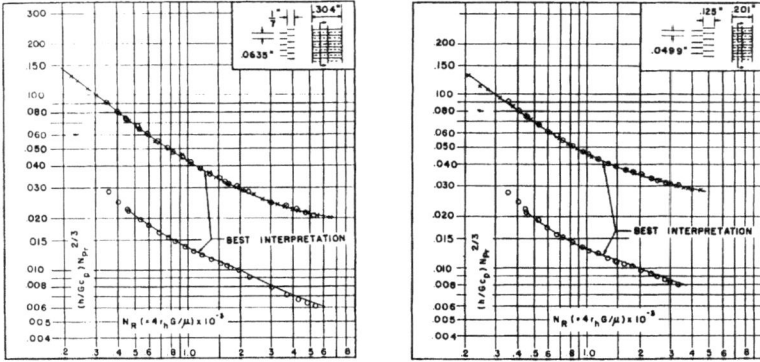

Fin pitch = 15.75 per in.
Plate spacing, b = 0.304 in.
Splitter symmetrically located
Fin length flow direction = $1/7$ in.
Flow passage hydraulic diameter,
   $4r_h$ = 0.006790 ft
Fin metal thickness = 0.004 in., aluminum
Splitter thickness = 0.006 in.
Total heat transfer area/volume between plates,
   β = 526 ft^2/ft^3
Fin area (including splitter)/total area = 0.859

Fin pitch = 20.06 per in.
Plate spacing, b = 0.201 in.
Splitter symmetrically located
Fin length flow direction = 0.125 in.
Flow passage hydraulic diameter,
   $4r_h$ = 0.004892 ft .0587 in.
Fin metal thickness = 0.004 in., aluminum
Splitter metal thickness = 0.006 in.
Total heat transfer area/volume between plates,
   β = 698 ft^2/ft^3
Fin area (including splitter)/total area=0.843

FIGURE 9.80 Strip-fin friction factors and Colburn Moduli. (From Kays & London, "Compact Heat Exchanger")

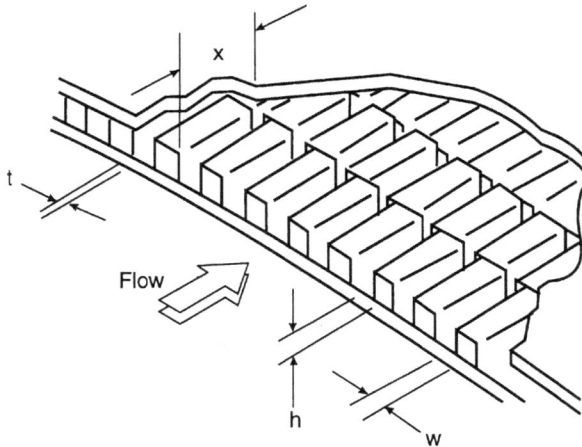

FIGURE 9.81 Typical rectangular strip fin.

and the forces driving the air. The draft equation[9] give the forcing pressure in inches of water.

$$\Delta P = 0.192 \, H\rho\ln\left(\frac{T_2}{T_1}\right) \tag{9.101}$$

Where:

ρ is the average density of air in the cabinet in lb/ft.[3]

T_2 is the exit absolute temperature of the air temperature in °R or K

T_1 is the entrance or ambient absolute air temperature in °R or K

H is the height of the column of air from the centroid of heat to the top of the cabinet in feet.

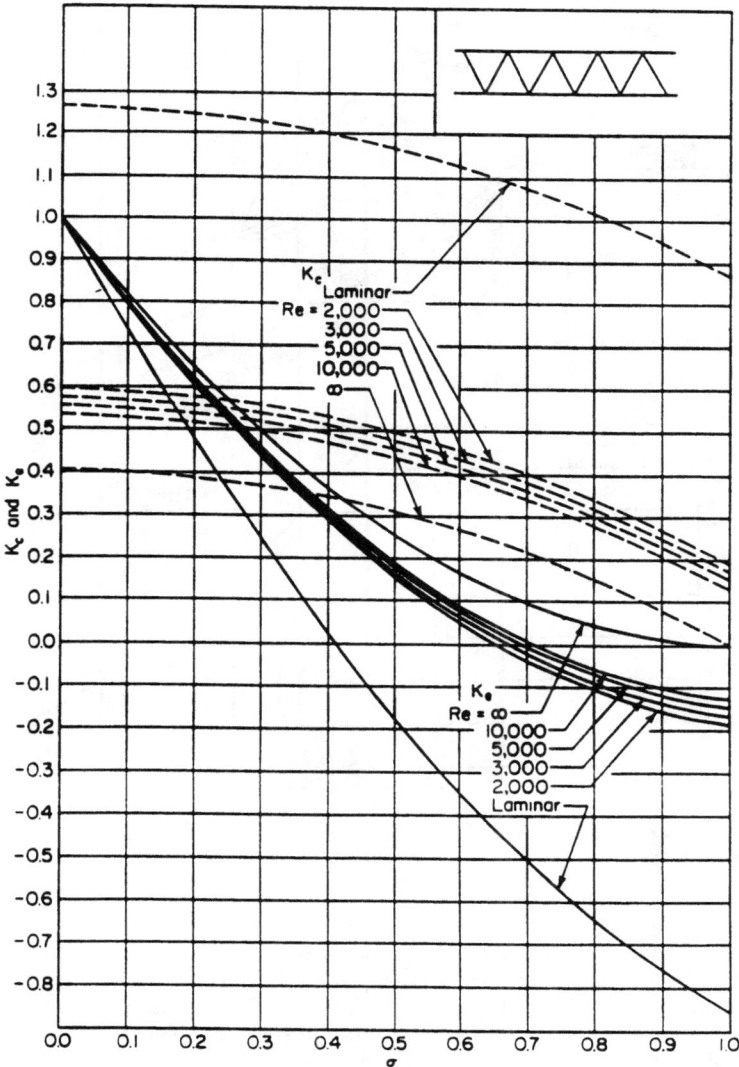

FIGURE 9.82   Entrance and exit pressure loss coefficient for a triangular tube with abrupt contraction entrance and abrupt expansion exit.

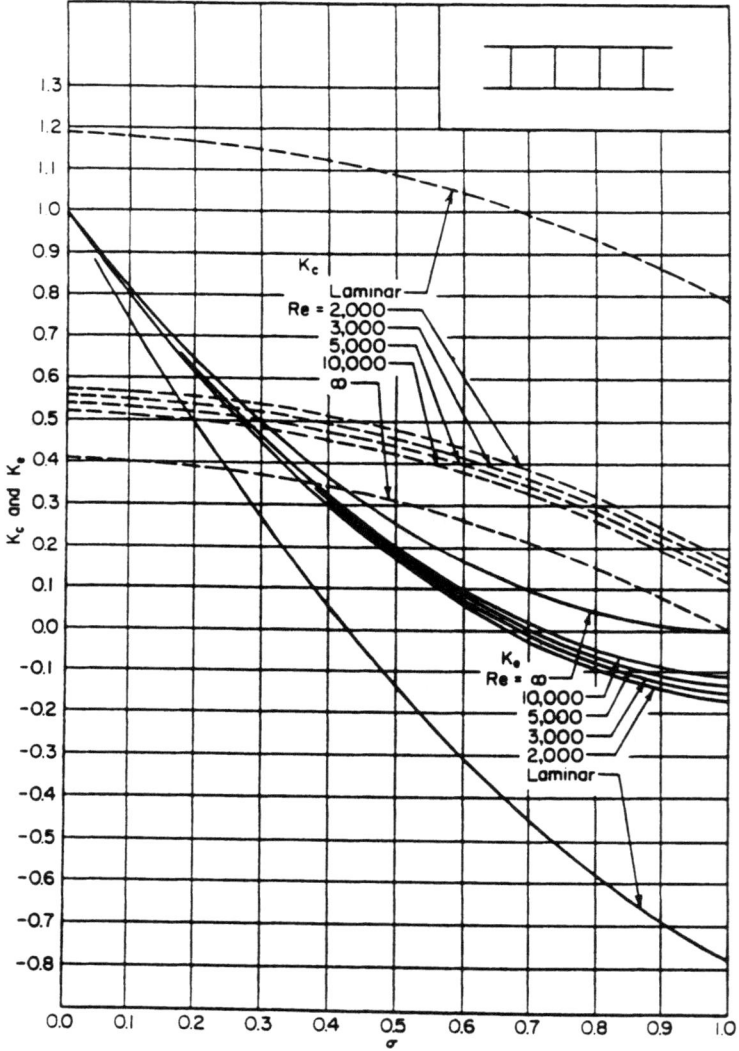

FIGURE 9.83 Entrance and exit pressure loss coefficient for a rectangular tube with abrupt contraction entrance and abrupt expansion exit.

Notice that putting the heat low in the cabinet generates a higher draft pressure. This could be a consideration in planning the layout of the cabinet. A computer program to generate both the draft pressure and pressure loss curves vs. flow rate is available.[9] Also, be aware that this draft equation, 9.83, presumes that the entire air volume is heated in a series of heat absorptions. When the heat sources are arranged, as in Figure 9.77, the equation should be applied for each rack not the entire cabinet (Figure 9.84).

Warm Air Exit at Top

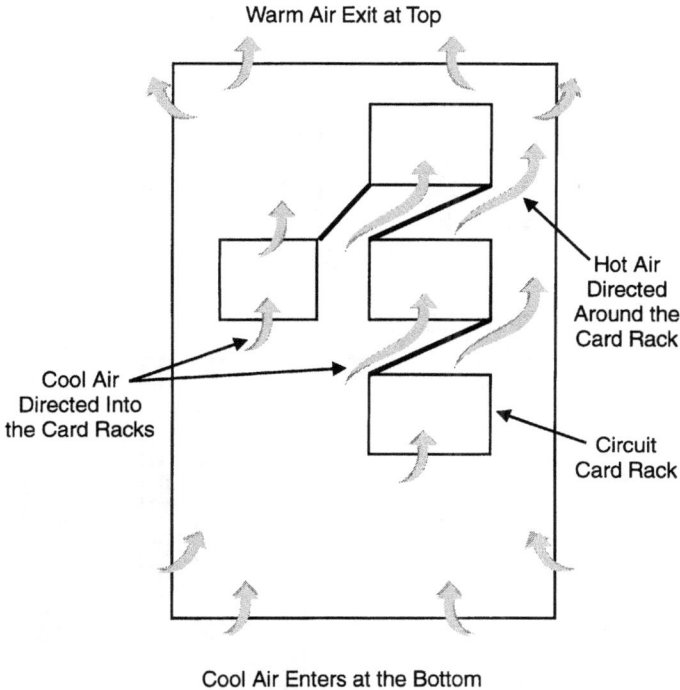

Cool Air Enters at the Bottom

FIGURE 9.84 Circuit card racks arranged to induct cool air into each level.

Improper placement of large components within the cabinet can generate dead-air areas or local circulations of air that are warmer than desired. A detailed examination of internal flows can be performed using computational fluid-dynamics computer programs, such as Flowtherm[17] or Icepack.[18] These programs combine computational fluid dynamics with heat transfer to predict fluid velocity, pressures, and temperatures. Figure 9.85 is an analysis done using IcePack to define air flow and provide a visual representation so that dead or blocked areas can be easily located.

When fans are used to cool large enclosures, there is a choice of forcing air through the cabinet or drawing air through the cabinet. Each has advantages. A draw-through system allows entry points to be placed where wanted to bring cool air into the cabinet at desired locations. However, air entering the cabinet must be filtered; thus, multiple filters are required. Also leaks in the cabinet might bring unwanted air and dust into the enclosure, which can reduce the air flow in other areas [because the fan(s) will only handle a certain amount]. Pressurizing the cabinet is a more usual scheme and has the advantage of not ingesting dust from leaks and not being as sensitive to leakage, but filters must still be used to keep the air entering the cabinet dust free. Fans at the entrance of a pressurized cabinet are handling cooler air, so they are more reliable and smaller fans are needed because the air is denser. The fans heat the air as it passes through the fans, so the air in the cabinet is slightly warmer. As explained in Section 9.6, it is important to match the fan to the system pressure drop at the desired flow rate. Entrance and exit losses at the inlet and outlet are often a large part of the pressure loss in large cabinets. Be sure that they are both large enough that the desired flow rates are not overly restricted.

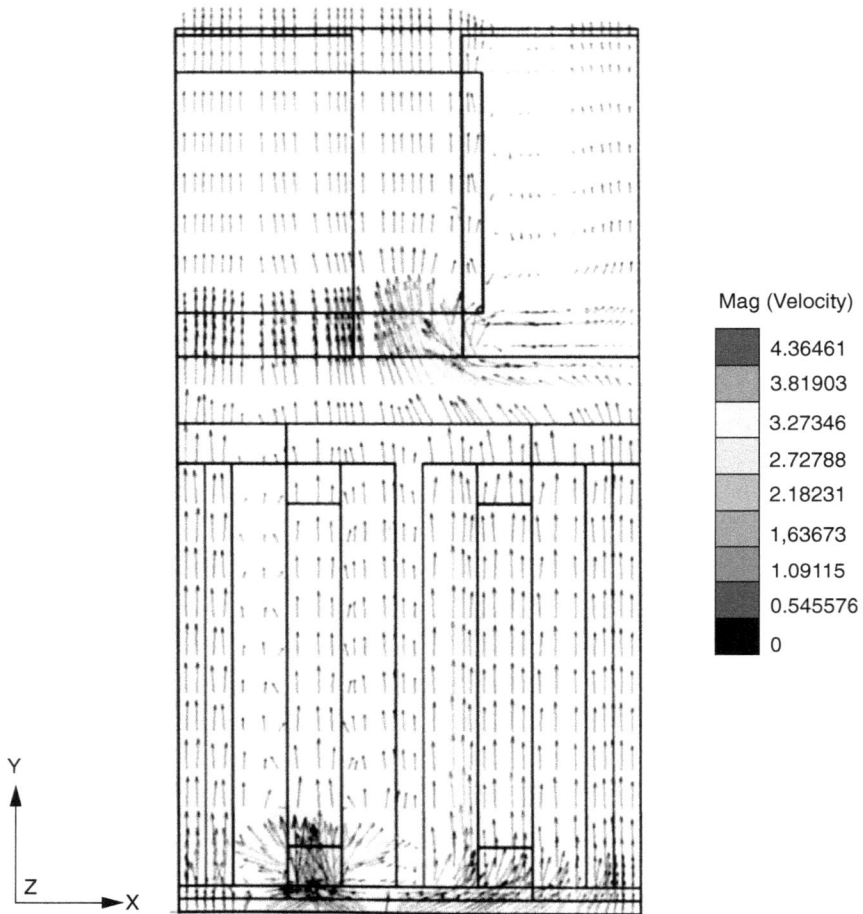

FIGURE 9.85 Computational Fluid Dynamics (CFD) used to define flow fields and temperature *(Courtesy of Fluid Dynamics, Inc.).*

Fans for enclosures can be located at the bottom of the cabinet and supply air for the entire cabinet or individual trays of fans can be used for each card rack. Most of these are designed to slide out for easy removal and replacement of both fans and filters (Figure 9.86).

It is not necessary that an enclosure be either natural or force convection cooled. It can be both and this is often the optimum solution—especially where local hot spots require higher flow rates. This is the case with most desktop personal computers, which are generally natural convection cooled, but have a small fan to cool the power supply and an even smaller fan to cool the processor chip. These fans don't make the air in the enclosure any cooler because they don't increase the flow through the enclosure. But they do reduce the temperature of individual pieces of electronics inside the enclosure.

FIGURE 9.86   Fan tray for large cabinets.

9.8 SPECIAL THERMAL-MANAGEMENT DEVICES

9.8.1 Thermoelectric Devices

Thermoelectric devices operate on the Peltier principle, which states that a current forced through the junction of two dissimilar materials will produce a cooling or heating effect at the junction, depending on the direction of the current flow. This is expressed mathematically as

$$Q = \alpha TI \tag{9.102}$$

Where:
  Q = The heat generated or removed
  α = The Seebeck coefficient in volts/K
  T = The absolute temperature at the junction in K
  I = The current in amperes

The device is the opposite of a thermocouple for which an EMF is induced by temperature differences at the junctions. The operation of the thermoelectric device is a balance of opposing heating and cooling terms; the Peltier effect, the Joule effect (I^{2R} heating in the conductors), and the Thompson effect:

$$Q = \tau I \Delta T \tag{9.103}$$

Where:
  τ is the Thompson coefficient in volts/K
  I is the current
  T is the temperature difference between the hot and cold junctions

The sign of the Thompson coefficient changes with current direction so that at one junction, it cools and heats at the other.

Modern devices are made of semiconductor material, usually some form of Teluride (such as Bismuth, Antimony, or Lead) oriented so that the current flows through N-P junctions on one side and P-N junctions on the other. This forms the "cold" side and the "hot" side junctions and looks much as shown in Figure 9.87.

The amount of heat that can be absorbed on the cold side, Q_c, is expressed as:

$$Q_c = n[I(\alpha_c T_c + 1/2\tau\Delta T) - 1/2I^2 R_m - G_m\Delta T] \tag{9.104}$$

Where:
  R_m is the average electrical resistance of the couple
  α_c is the Seebeck coefficient at the cold-junction temperature

τ is the Thompson coefficient at the average temperature of the couples $(T_c + T_h)/2$
G_m is the thermal conductance (kA/x) of the device couples and k is evaluated at the average temperature
ΔT is the temperature difference between the hot and cold sides
n is the number of couples

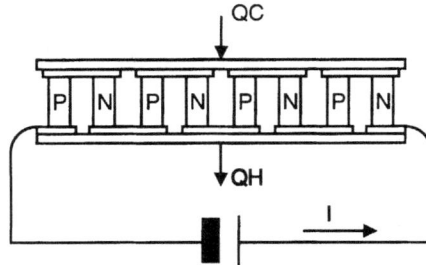

FIGURE 9.87 Construction and polarity of a refrigerating ThermoElectric Device (TED).

Obviously, the Peltier and Thompson effects are providing the cooling while the Joule effect and the first law of thermodynamics are opposing energy sources.

In practice, Eq. 9.104 can be simplified because the Thompson coefficient can be expressed in terms of the Seebeck coefficient. This allows the average value of the Seebeck coefficient, α_m, to be used to eliminate a term in the equation and yield:

$$Q_c = n[I\alpha_m T_c - 1/2I^2R_m - G_m\Delta T) \qquad (9.105)$$

The Thompson and Peltier effects are reversible so that at the hot side junction, the heat flow is:

$$Q_h = n[I\alpha_m T_h + 1/2I^2R_m - G_m\Delta T) \qquad (9.106)$$

where the Seebeck coefficient is defined at the average couple temperature. For practical purposes, the heat flow at the hot junction is more simply defined as the sum of Eq. 9.104 plus the input power, $V \times I$ and the voltage is determined from:

$$V = n(\alpha_m\Delta T + IR_m) \qquad (9.107)$$

Where:
n is the number of couples connected in series

Caution must be used when using these devices. The most important consideration is the ability of the heatsink on the hot side of the device to absorb heat without changing temperature. The heat rejected at the hot side is the sum of the heat absorbed and the power input to the device. If this heat causes the heatsink temperature to rise, any gains in the temperature suppression of the cold side will be offset and possibly negated by the sink rise. Once this hurdle is passed, a secondary, but equally important consideration, is the interfaces. These should be very flat and smooth with a very thinly applied thermal-grease interstitial filler.

The operational efficiency is expressed as with most refrigeration systems as the coefficient of performance in refrigeration (COPR). Because the device can be used as a heater or heat pipe by reversing the current, a coefficient of performance in heating (COPH) is also used. These are simply the ratio of the amount of cooling or heating obtained to the input power.

$$COPR = \frac{Q_c}{nI(\alpha_m\Delta T + IR_m)} \tag{9.108}$$

$$COPH = \frac{Q_h}{nI(\alpha_m\Delta T + IR_m)} \tag{9.109}$$

Devices can be cascaded or stacked in smaller and smaller sizes to increase the ΔT capability. Each device must be able to absorb the heat from the device above it and still maintain a desired temperature depression so that the amount of heat that can be absorbed by the final stage is much smaller.

When designing with thermoelectric devices, it can be useful to know certain behavioral characteristics. Because the material properties of the couples vary with temperature, the performance of thermoelectric devices also vary with temperature (Figure 9.88). Notice that the performance of the device improves with temperature up to approximately 100°C, then drops off. This is useful in the electronics cooling industry because electronics are often operating in the 70°C to 110°C range.

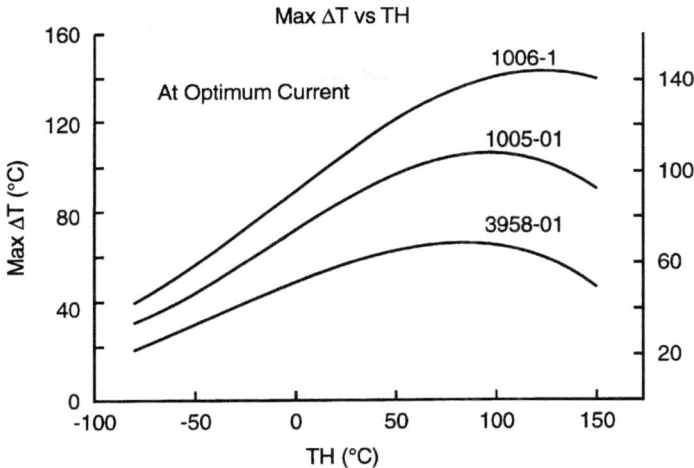

FIGURE 9.88 Performance characteristics vary with temperature.

Another useful characteristic is illustrated in Figure 9.89, which shows the variation of *COPR* and Q_C with ΔT. The *COPR* drops below 1 when $\Delta T > 20$°C. It is useful to know that a 20°C temperature reduction can be obtained for the same power input as the power being removed.

The coefficients depend on the material, and precise values are somewhat guarded by suppliers. However, the following values are useful for design calculations. Precise performance tables and curves can be supplied by manufacturers.[26]

$$\alpha_m = 1 \times 10^{-9}(22224 + 930.6T - 0.9905T^2) \tag{9.110}$$

$$R_m = 1 \times 10^{-8}(5112 + 163.4T - .6297T^2) \tag{9.111}$$

$$G_m = 1 \times 10^{-6}(62605 - 227.7T + 0.4131T^2) \tag{9.112}$$

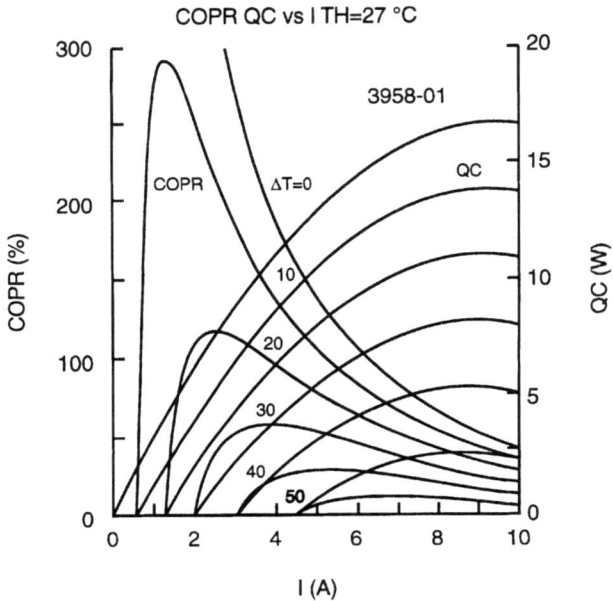

FIGURE 9.89   Large T_s or large Q_c require high currents and produce low efficiencies.

The following computer program can be used to make design estimates. It uses the coefficient values presented in the previous paragraph.

```
5 REM ON ERROR GOTO 670
10 CLS:COLOR 20,0:PRINT" Use Ctrl-PrtSc if you want a hard copy of output"
20 COLOR 7,0:PRINT:KEY OFF
30 READ S0,S1,S2,P0,P1,P2,K0,K1,K2
40 DATA 22224.,930.6,-0.9905,5112.,163.4,.6279,62605.,-277.7,.4131
41 KEY 15,CHR$(&H4)+CHR$(114)
42 ON KEY(15) GOSUB 180
43 KEY(15) ON
50 PRINT" This program contains five analysis options.":PRINT
60 PRINT"1. Generate performance properties similar to the curves in
the catalog."
70 PRINT"2. For a given voltage, determine refrigeration capacity and
current for a range of delta T's."
80 PRINT"3. For a given voltage and required refrigeration capacity,
determine the current and delta T."
85 PRINT"4. Determine a near optimum configuration for your voltage and
refrigeration requirements."
86 PRINT"5. Determine the proper device to use as a generator.":PRINT
```

```
90 PRINT" YOUR CHOICE":PRINT
100 LOCATE 15,15:COLOR 1,7:PRINT" 1 ":COLOR 7,1:LOCATE 15,19:PRINT"
for option 1"
105 LOCATE 15,40:COLOR 1,7:PRINT" 5 ":COLOR 7,1:LOCATE 15,44:PRINT"
for option 5"
110 LOCATE 17,15:COLOR 1,7:PRINT" 2 ":COLOR 7,1:LOCATE 17,19:PRINT"
for option 2
120 LOCATE 19,15:COLOR 1,7:PRINT" 3 ":COLOR 7,1:LOCATE 19,19:PRINT"
for option 3
125 LOCATE 21,15:COLOR 1,7:PRINT" 4 ":COLOR 7,1:LOCATE 21,19:PRINT"
for option 4"
127 LOCATE 23,15:COLOR 1,7:PRINT" E ":COLOR 7,1:LOCATE 23,19:PRINT" to
end"
130 A$=INPUT$(1):COLOR 7,0:IF A$="1" THEN CLS:GOTO 200
140 IF A$="2" THEN CLS:GOTO 460
150 IF A$="3" THEN CLS:GOTO 710
153 IF A$="4" THEN CLS:GOTO 2000
154 IF A$="5" THEN CLS:GOTO 1000
155 IF A$="E" OR A$="e" THEN CLS:END
160 PRINT" That option not available. Enter 1,2 or 3.":GOTO 130
180 LPRINT CHR$(27)"N"CHR$(6)
190 RETURN
200 PRINT"Program option 1-Generate performance curves"
210 INPUT"ENTER NUMBER OF COUPLES";N:NE=2*N
220 INPUT "ENTER LENGTH/AREA RATIO OF ELEMENT (1/cm.).
Default=[13.015]";LAMBDA
225 IF LAMBDA=0 THEN LAMBDA=13.015
230 INPUT "ENTER Th (Kelvin)";TH
240 INPUT"ENTER MAXIMUM CURRENT OF DEVICE-(AMPS)-";CURMAX
250 PRINT:PRINT"TABLES WILL BE PRINTED FOR CURRENT STEPS SPECIFIED UP
TO THE MAXIMUM CURRENT. YOU MUST ENTER THE CURRENT STEP YOU DESIRE."
260 INPUT "ENTER CURRENT STEP-";ST1
270 INPUT"ENTER STARTING CURRENT-";STC
280 FOR J=STC TO CURMAX STEP ST1:I=J:PRINT:PRINT "——FOR
CURRENT=";J:PRINT
290 FOR M=0 TO 4:TC=TH-10*M:DT=TH-TC
300 T=(TC+TH)/2:PRINT "DELTA T (KELVIN)—";DT
310 SM=(S0+S1*T+S2*T^2)*1E-09
320 RHOM=(P0+P1*T+P2*T^2)*1E-08
330 KM=(K0+K1*T+K2*T^2)*.000001
340 Z=SM^2/(RHOM*KM)
350 S=SM*NE:DT=TH-TC
360 R=RHOM*LAMBDA*NE:K=KM*NE/LAMBDA
370 'note that voltage equation has a fudge factor to match curves
380 V=(I*LAMBDA*RHOM+SM*DT)*NE*1.1
390 QC=S*I*TC-I^2*R/2-K*DT
400 PRINT "QC=";QC,"VOLTAGE=";V
410 NEXT M:NEXT J
411 PRINT:COLOR 2,0:PRINT"Do you want to repeat option 1? 'Y' or 'N'"
412 Y$=INPUT$(1):IF Y$="Y" OR Y$="y" THEN COLOR 7,0:CLS:GOTO 200
413 PRINT"Do you want terminate this program?"
415 Y$=INPUT$(1):IF Y$="y" OR Y$="Y" THEN END ELSE COLOR 7,0:CLS:GOTO
50
460 PRINT"Program option 2-Performance at fixed voltage"
470 INPUT"ENTER NUMBER OF COUPLES";N:NE=2*N
480 INPUT "ENTER LENGTH/AREA RATIO OF ELEMENT (1/cm.).
Default=13.015";LAMBDA
485 IF LAMBDA=0 THEN LAMBDA=13.015
490 INPUT "ENTER Th (Kelvin)";TH
500 INPUT"ENTER OPERATING VOLTAGE OF DEVICE—";V
```

```
510 PRINT:PRINT"TABLE WILL BE PRINTED FOR CURRENT, HEAT ABSORBED AND
DELTA T." 520 PRINT:PRINT" FOR TED VOLTAGE OF ";V;" VOLTS"
530 PRINT" QC  CURRENT DELTA T (KELVIN)"
540 FOR M=0 TO 40:TC=TH-M:DT=TH-TC
550 T=(TC+TH)/2
560 SM=(S0+S1*T+S2*T^2)*1E-09
570 RHOM=(P0+P1*T+P2*T^2)*1E-08
580 KM=(K0+K1*T+K2*T^2)*.000001
590 Z=SM^2/(RHOM*KM)
600 S=SM*NE:DT=TH-TC
610 R=RHOM*LAMBDA*NE:K=KM*NE/LAMBDA
620 'note that voltage equation has a fudge factor to match curves
630 I=(V/NE/1.1-SM*DT)/LAMBDA/RHOM
640 QC=S*I*TC-I^2*R/2-K*DT
650 PRINT QC,I,DT
655 NEXT M:PRINT:COLOR 2,0:PRINT"Do you want to repeat option 2? 'Y'
or 'N'"
660 Y$=INPUT$(1):IF Y$="Y" OR Y$="y" THEN COLOR 7,0:CLS:GOTO 460
665 PRINT"Do you want terminate this program?"
666 Y$=INPUT$(1):IF Y$="y" OR Y$="Y" THEN END ELSE COLOR 7,0:CLS:GOTO
50
670 A$="PLEASE TURN PRINTER ON"
680 IF ERR=24 OR ERR=25 OR ERR=27 THEN PRINT A$ ELSE 700
685 PRINT"YOU MAY HAVE TO RE-ENTER 'Ctrl-PrtSc' BECAUSE PRINTER WAS
OFF."
690 INPUT"PRESS <ENTER> WHEN PRINTER IS READY";Y$:RESUME
700 PRINT" ERROR NO.-";ERR:STOP
710 PRINT"Program option 3-Specific operating point"
720 INPUT"ENTER NUMBER OF COUPLES";N:NE=2*N
730   INPUT   "ENTER   LENGTH/AREA   RATIO   OF   ELEMENT   (1/cm.).
Default=13.015";LAMBDA
735 IF LAMBDA=0 THEN LAMBDA=13.015
740 INPUT "ENTER Th (Kelvin)";TH
750 INPUT"ENTER OPERATING VOLTAGE OF DEVICE-";V
760 INPUT"ENTER HEAT TO BE ABSORBED (QC)-";QC
770 PRINT" V   Qc     CURRENT Th Tc DELTA T "
780 PRINT:Z$="##.# ##.#### ##.#### ### ### ###"
790 T=TH:DT=0:TC=TH
800 SM=(S0+S1*T+S2*T^2)*1E-09
810 RHOM=(P0+P1*T+P2*T^2)*1E-08
820 KM=(K0+K1*T+K2*T^2)*.000001
830 Z=SM^2/(RHOM*KM)
840 S=SM*NE
850 R=RHOM*LAMBDA*NE:K=KM*NE/LAMBDA
860 'note that voltage equation has a fudge factor to match curves
870 I=(V/NE/1.1-SM*DT)/LAMBDA/RHOM
880 QC2=S*I*TC-I^2*R/2-K*DT
885 REM PRINT QC2,I,DT
890 IF QC2>QC+.2 THEN T=T-.2:TC=2*T-TH:DT=TH-TC:GOTO 800
910 PRINT USING Z$;V,QC2,I,TH,TC,DT
920 PRINT" INPUT POWER=";I*V
930 PRINT:COLOR 2,0:PRINT"Do you want try another voltage and Qc?-'Y'
or 'N'"
940 Y$=INPUT$(1):IF Y$="Y" OR Y$="y" THEN COLOR 7,0:CLS:GOTO 750
941 PRINT"Do you want to repeat option 3 (change the TED parameters)?"
942 Y$=INPUT$(1):IF Y$="y" OR Y$="Y" THEN COLOR 7,0:CLS:GOTO 710
950 PRINT"Do you want terminate this program?"
960 Y$=INPUT$(1):IF Y$="y" OR Y$="Y" THEN END ELSE COLOR 7,0:CLS:GOTO
50
1000 PRINT"Program option 5-Power generation"
```

```
1005 INPUT"ENTER REQUIRED CURRENT";I
1006 INPUT"ENTER REQUIRED OPERATING VOLTAGE-";V
1010 INPUT "ENTER Th (Kelvin)";TH
1015 INPUT"ENTER Tc (Kelvin)";TC
1020 PRINT" V CURRENT Th Tc DELTA T "
1025 Z$="##.# ##.#### ### ### ###"
1030 T=(TH+TC)/2:DT=TH-TC
1035 SM=(S0+S1*T+S2*T^2)*1E-09
1040 RHOM=(P0+P1*T+P2*T^2)*1E-08
1045 KM=(K0+K1*T+K2*T^2)*.000001
1050 Z=SM^2/(RHOM*KM)
1055 S=SM*NE
1060 PRINT USING Z$;V,I,TH,TC,DT:PRINT
1065 EOC=2*V
1070 R=V/I
1075 NE=EOC/SM/DT
1080 LAMBDA=R/RHOM/NE
1085 NC=NE/2
1090 PRINT"LOAD RESISTANCE=DEVICE RESISTANCE=";R
1100 PRINT"NUMBER OF COUPLES REQUIRED=";NC
1105 PRINT"LAMBDA REQUIRED=";LAMBDA
1110 QC=NE*KM*DT/LAMBDA:EFF=V*I*100/QC
1115 PRINT"HEAT TRANSPORTED THROUGH THE GENERATOR (WATTS) =";QC
1120 PRINT"GENERATOR EFFICIENCY=";EFF
1130 PRINT:COLOR 2,0:PRINT"Do you want try another voltage and cur-
rent?-'Y' or 'N'"
1140 Y$=INPUT$(1):IF Y$="Y" OR Y$="y" THEN COLOR 7,0:CLS:GOTO 1000
1150 PRINT"Do you want terminate this program?"
1160 Y$=INPUT$(1):IF Y$="y" OR Y$="Y" THEN END ELSE COLOR 7,0:CLS:GOTO
50
2000 PRINT"Program option 4 to design optimum configuration"
2010 N=71:NE=2*N:LAMBDA=13.015 'These values are meaningless
2040 INPUT"ENTER DESIRED DELTA T-";DT:NE=2*N
2050 INPUT "ENTER Th (Kelvin)";TH
2060 TC=TH-DT:T=(TC+TH)/2
2070 INPUT"ENTER SUPPLY VOLTAGE AVAILABLE-";V
2080 INPUT"ENTER TOTAL HEAT TO BE ABSORBED (QC)-";QC
2090 GOSUB 2200 'Series optimization
2100 GOSUB 3000 'Parallel optimization
2110 GOTO 3900
2200 'Subroutine for series optimization
2210 GOSUB 4000
2212 ILAM=KM*DT*(1+SQR(1+Z*T))/SM/T
2214 NE=V/(ILAM*RHOM+SM*DT):N=NE/2
2215 LAMBDA=.5:GOSUB 4000
2220 GOSUB 4050
2230 IF QC2<QC+.25 AND QC2>QC-.25 THEN 2420 ELSE LAMBDA=LAMBDA+.1:GOTO
2220
2420 PRINT" V Qc     CURRENT Th Tc DELTA T COUPLES LAMBDA"
2430 PRINT:Z$="##.# ##.#### ##.#### ### ### ###       ### ##.###"
2440 PRINT USING Z$;V,QC2,I,TH,TC,DT,N,LAMBDA
2460 CNTR=0:PRINT" INPUT POWER=";PWR
2470 RETURN
3000 'Subroutine for parallel optimization
3020 PRINT:PRINT" For two Parallel strings:"
3030 QC1=QC:QC=QC/2
3060 GOSUB 2220
3100 'subroutine for three parallel strings
3120 PRINT:PRINT" For three parallel strings:"
3130 QC=QC1/3
```

```
3160 GOSUB 2220
3170 RETURN
3900 CNTR=0:PRINT:COLOR 2,0:PRINT"Do you want to try another voltage
and Qc?"
3910 Y$=INPUT$(1):IF Y$="y" OR Y$="Y" THEN QC2=0:QC3=0:COLOR
7,0:CLS:GOTO 2070
3920 PRINT"Do you want to repeat option 4 for another design point?"
3930 Y$=INPUT$(1):IF Y$="Y" OR Y$="y" THEN COLOR 7,0:CLS:GOTO 2010
3940 PRINT"Do you want to terminate the program?":Y$=INPUT$(1)
3950 IF Y$="Y" OR Y$="y" THEN END ELSE COLOR 7,0:CLS:GOTO 50
4000 SM=(S0+S1*T+S2*T^2)*1E-09
4010 RHOM=(P0+P1*T+P2*T^2)*1E-08
4020 KM=(K0+K1*T+K2*T^2)*.000001
4030 Z=SM^2/(KM*RHOM)
4040 S=SM*NE
4050 R=RHOM*LAMBDA*NE:K=KM*NE/LAMBDA
4060 'note that voltage equation has a fudge factor to match curves
4070 I=(V/NE/1.1-SM*DT)/LAMBDA/RHOM
4080 QC2=S*I*TC-I^2*R/2-K*DT:PWR=V*I
4085 REM PRINT "QC2=";QC2,"I=";I
4086 CNTR=CNTR+1:IF CNTR>500 THEN PRINT:PRINT"**QC too high or V too
low**"
4087 IF CNTR>500 THEN CNTR=0:GOTO 3900
4090 RETURN
```

9.9 HEAT PIPES

The heat pipe is an extremely efficient method of transferring heat from one area to an-other, such as from an electronic component to the edge of a CCA. Heat pipes have been considered and designed for electronics for many years, but haven't yet been commonly used, except in space applications, although they are used in extreme problem areas, where cost permits and volume imposes restrictions on cheaper methods. The principle of opera-tion of the heat pipe is based on boiling and condensing of a working fluid inside the pipe (see Section 9.4.4). This process is illustrated in Figure 9.90, wherein heat enters the evap-orator end of the pipe via conduction through the pipe material and conduction into a liq-uid-filled wick structure, which causes the liquid to change into a vapor phase. A small pressure differential exists between the evaporator area and the condenser area, caused by a small temperature difference. This pressure difference causes the vapor to flow to the condenser area, where it condenses within the wick structure and releases heat to be con-

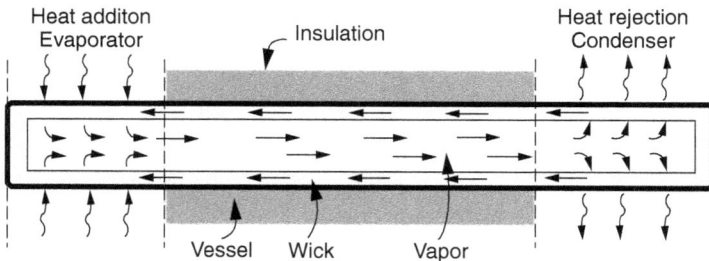

FIGURE 9.90 Typical heat-pipe operation.

ducted out through the pipe walls to the ultimate sink. The liquid in the condenser wick travels back to the evaporator area via capillary action.

The wick can be made of many structures, such as wire screen, felt, sintered material, or just grooves in the pipe wall small enough to promote capillary pumping (Figure 9.91). Of course, the cheapest heat-pipe fabrication is to eliminate the wick by placing the condenser at the top and letting gravity return the liquid.

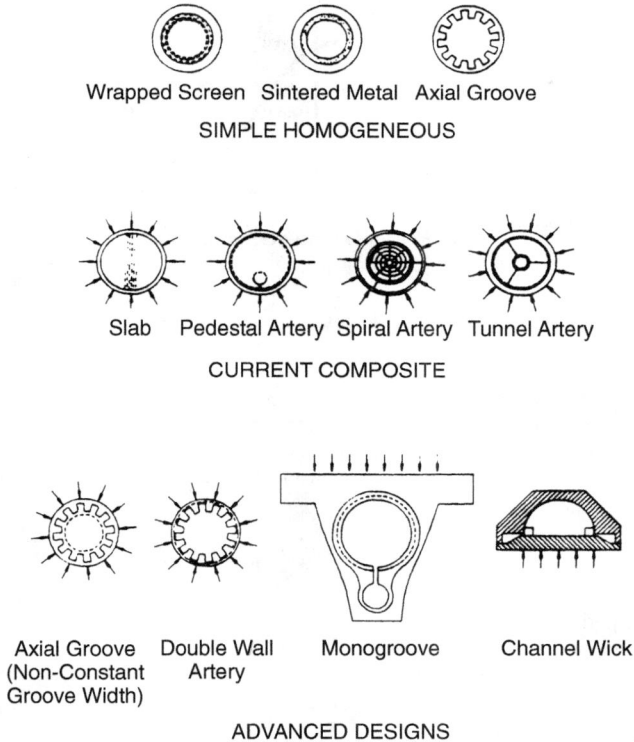

Wrapped Screen Sintered Metal Axial Groove

SIMPLE HOMOGENEOUS

Slab Pedestal Artery Spiral Artery Tunnel Artery

CURRENT COMPOSITE

Axial Groove   Double Wall   Monogroove   Channel Wick
(Non-Constant   Artery
Groove Width)

ADVANCED DESIGNS

FIGURE 9.91 Heat-pipe wick structures. (From Reference 10.)

Several processes occur inside the heat pipe, which can limit its operation. The most prevalent of these is the limit on the capillary pumping action acting against gravity. Where the heat pipe is horizontal or used in space, this might not apply. However, too much heat can require more fluid to be returned to the evaporator area than a given wick structure can supply. This results in a drying out of the evaporator wick and total loss of heat transfer in the heat pipe; a catastrophic failure. The forces involved include: the capillary pressure, axial hydrostatic pressure, radial hydrostatic pressure, liquid pressure drop, vapor pressure drop, and choked flow of the vapor. The capillary pressure must be greater than the fluid pressure drop and the axial hydrostatic pressure, in order for liquid to flow from the condenser area to the evaporator area. The axial hydrostatic pressure can help or impede flow, depending on the orientation of the evaporator and condenser to the gravity vector. If the condenser is higher than the evaporator, gravity assists operation by adding

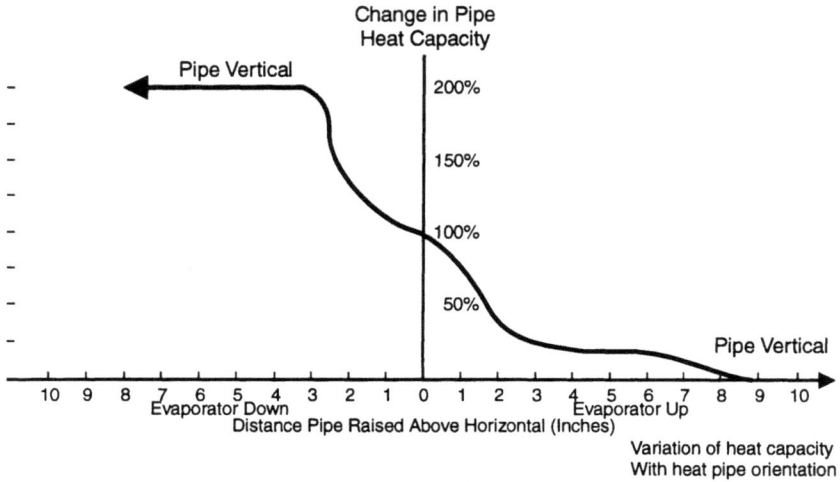

FIGURE 9.92   Pumping against gravity reduces the capacity of the heat pipe.

to the capillary pressure. Figure 9.92 shows the impact of gravity on a typical heat pipe operation, with respect to orientation to a 1-g gravity field.

The capillary pressure, as expressed by Bar-Cohen and Kraus[10], is a function of the meniscus radii in evaporator and condenser wick structures (as shown in Figure 9.93 and as expressed in Eq. 9.113).

$$\Delta P_c = 2\sigma \left(\frac{1}{r_e} - \frac{1}{r_c} \right) \tag{9.113}$$

Where:

   σ is the liquid surface tension
   r_e is the evaporator meniscus radius
   r_c is the condenser meniscus radius

In a properly operating heat pipe, the condenser is being flooded, so r_c is considered infinite and Eq. 9.113 reduces to:

$$\Delta P_c = \frac{2\sigma}{r_e} \tag{9.114}$$

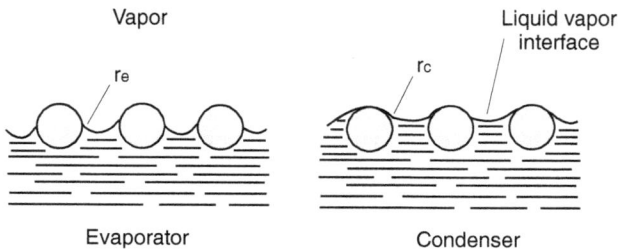

FIGURE 9.93   Radius of curvature of the meniscus in the evaporator and condenser vapor/liquid interface.

and effective r_e for a few-wick structure is given in Table 9.8.

The axial hydrostatic pressure is expressed by Bar-Cohen and Kraus as:

$$\Delta P_{ah} = \rho g L sin\theta \tag{9.115}$$

Where:

e is the liquid density
g is acceleration of gravity
L is the length of the heat pipe
θ is the angle between the heat pipe axis and horizontal

TABLE 9.8 Expressions for Effective Capillary Radius (from Chi, 1976, and Bar-Cohen and Kraus)

Structure	r_e	Data
Circular cylinder Artery or tunnel wick	r	r = Radius of liquid passage
Rectangular grooves	w	w = Groove width
Triangular grooves	$w/cos\,\beta$	w = Groove width
		β = Half included angle
Parallel wires	w	w = Wire spacing
Wire screens	$(w + d_w)/2 = \frac{1}{2}N$	d = Wire diameter
		N = Screen mesh number
		w = Wire spacing
Packed spheres	$0.41 r_s$	r_s = Sphere radius

The pressure drop of the liquid flowing through the wick is expressed by Bar-Cohen and Kraus as:

$$\Delta P_f = \left(\frac{\mu}{K A_w h_{fg} \rho} \right) L_{eff} q \tag{9.116}$$

Where:

μ is the liquid viscosity
ρ is the liquid density
h_{fg} is the latent heat of vaporization
$L_{eff} = 0.5 L_e + L_a + 0.5 L_c$ and L_e is the evaporator length
L_a is the length between the evaporator and condenser areas
L_c is the condenser length
q is the heat flow rate
K is the wick permeability defined below in Table 9.9

The heat-flow rate determines the mass flow rate and, in steady-state operation, the mass flow rate of the liquid and vapor must be equal.

Expressions for the effective evaporator radius for some wick structures are given in Table 9.8 and values of K in Eq. 9.116 are given in Table 9.9 (both by permission of Chi, 1976).

TABLE 9.9 Wick Permeability (from Chi, 1976 and Bar-Cohen and Kraus).

Structure	K	Data
Circular cylinder (artery or tunnel wick)	$r^2/8$	r = Radius of liquid flow passage
Open rectangular grooves	w/s	w = Groove width s = Groove pitch
Circular annular wick	$2r_{\mathrm{h}}^2/fR_{\mathrm{e}}{}^*$	r_{h} $r_1 - r^2$
Wrapped screen wick	$0.0082d^2\varepsilon^3/(1-\varepsilon)^2$	d = Wire diameter $\varepsilon = 1 - (1.05\ \Pi Nd/4)$ N = Mesh number
Packed sphere	$0.0267r^2\varepsilon^3/(1-\varepsilon)^2$	r = Sphere radius ε = Porosity (dependent on packing mode)

$^*fR_{\mathrm{e}} = 4r_{\mathrm{h}}\tau(\varepsilon A\rho\lambda)^2/\mu q^2$
where τ is the friction shear stress at the liquid-solid interface
  ε is the wick porosity
  A is the wick cross section area
  ρ is the liquid density
  h_{fg} is the latent heat of vaporization
  μ is the liquid viscosity

Other forces and phenomena taking place in the heat pipe are not considered here, but a much more detailed treatise is given in Refs. 17 and 18. Variations of the basic heat-pipe structure have been successfully used to provide temperature control and one-directional (diode) heat transfer. Figure 9.94 shows some of these applications in which either excess gas or excess liquid is used to limit heat transfer. In Figure 9.94a, a noncondensible gas is added to the working fluid. Because the vapor pressure curve is much steeper than the perfect gas law, the working fluid can easily compress the noncondensible gas into the reservoir and use the entire evaporator area. When the temperature of the evaporator starts to drop, the gas expands into the condenser area to reduce the heat transfer and maintain some temperature control. Where this passive scheme is not accurate enough, the method illustrated in Figure 9.94b adds a heater to the reservoir to enhance the expansion of the gas. Excess liquid is used in 9.94d to make a diode heat pipe, whereby if the condenser end ever becomes hotter than the evaporator end, condensed liquid is collected in the reservoir and depletes the working liquid, which renders the heat pipe inoperative.

A major concern in the construction of heat pipes is material compatibility. Working fluids can be chosen for the temperature range and heat transport characteristics and the materials used in the wick and pipe structure must be compatible to avoid corrosion. Table 9.10 lists a few compatible material combinations.

The best heat-transfer liquid, in most cases, is water and, when used with copper, provides excellent conduction properties to get heat into and out of the heat pipe. Figure 9.95 shows the operation of a copper-water heat pipe with a screen wick at different orientations, with respect to gravity.

Heat pipes are used on CCAs as individual heat pipes on rail-cooled boards (see conduction-cooled CCAs, Section 9.3) or are embedded in a frame structure (Figures 9.96 and 9.97). The effectiveness of heat pipes in reducing the temperature to which components are subjected is illustrated in Figure 9.97. In this figure, a heat-pipe rail is constructed (as in Figure 9.98) is compared to aluminum and copper rails of equal size. The main temperature rise for the heat pipe rail occurs at the ends, where the heat flux is high, conducting out of the pipe and into the CCA heatsink.

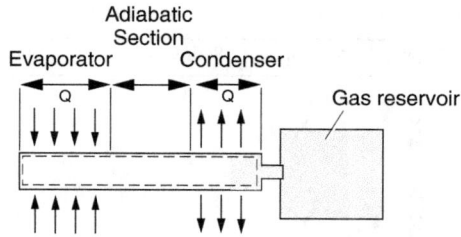

a. Passively controlled gas-loaded heat pipe

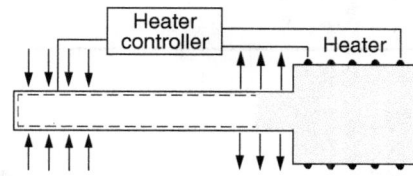

b. Actively controlled gas-loaded heat pipe

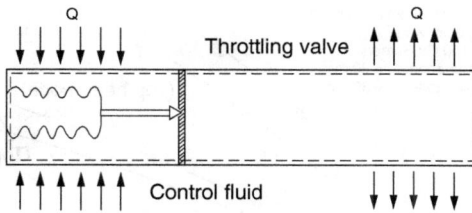

c. Vapor-modulated variable conductance heat pipe

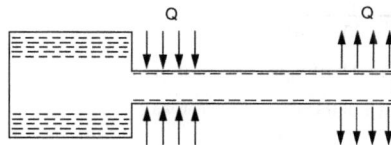

d. Liquid-modulated heat pipe

FIGURE 9.94 Variable conductance heat pipes.

TABLE 9.10 Acceptable Combinations of Heat Pipe Material and Fluids

Fluid	Copper	Aluminum	Stainless	Steel	Nickel	Refrasil
Water	X		X			X
Acetone	X	X				X
Ammonia		X	X		X	X
Methonol	X				X	X
Dow-A	X	X	X		X	X
Dow-E	X	X	X		X	
Freon-11	X	X	X			X
Freon-113	X	X	X			

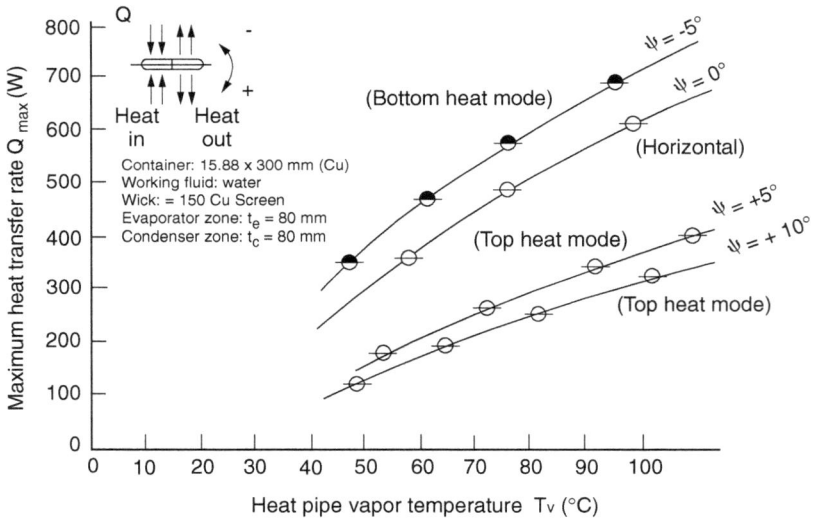

FIGURE 9.95 Maximum heat transfer of a copper-water heat pipe with a screen wick.

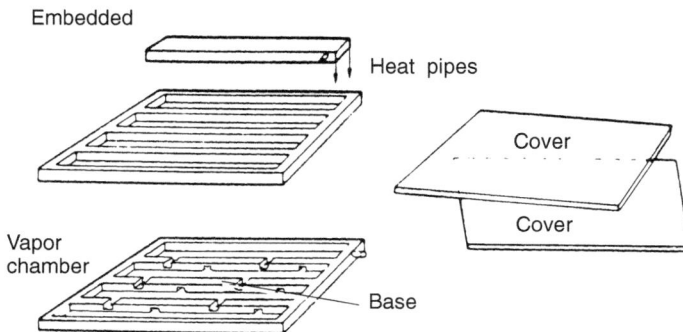

FIGURE 9.96 Basic embedded and vapor chamber concepts for cooling CCAs.

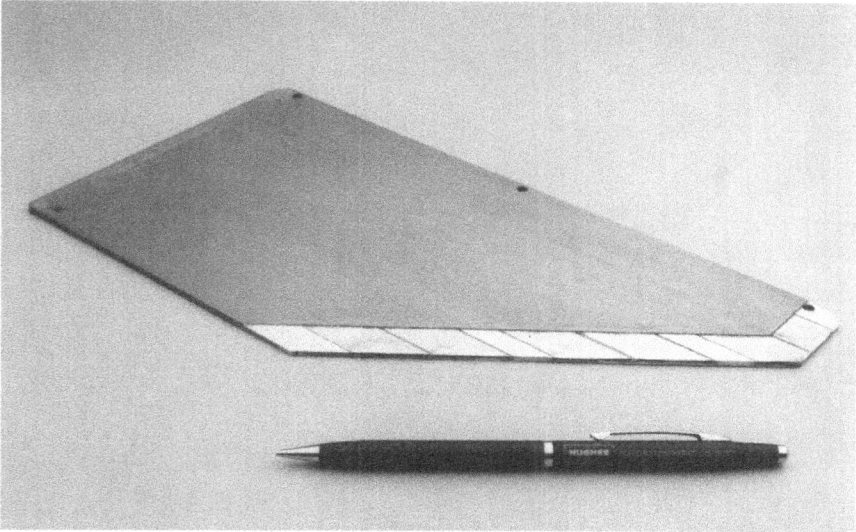

FIGURE 9.97   Embedded heat pipe in a CCA-pair heatsink.

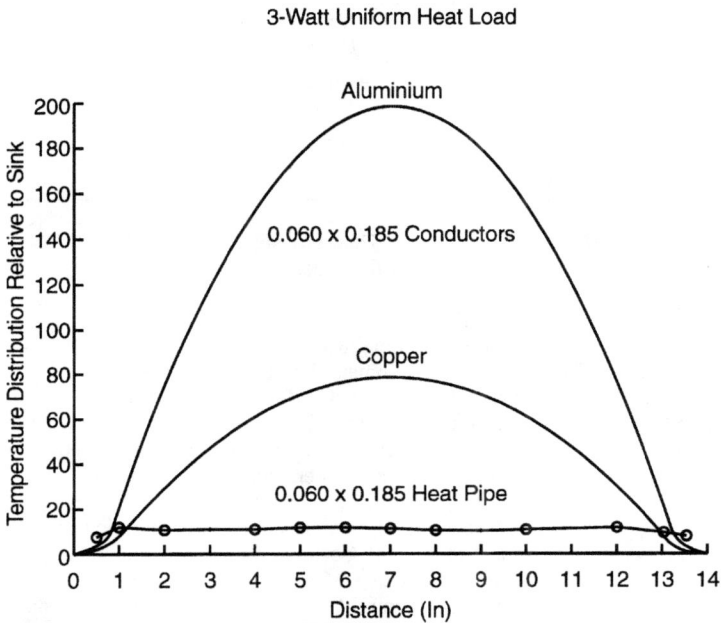

FIGURE 9.98   Comparison between the CCA heat pipe and metal conductor rails with a 3-watt uniform load.

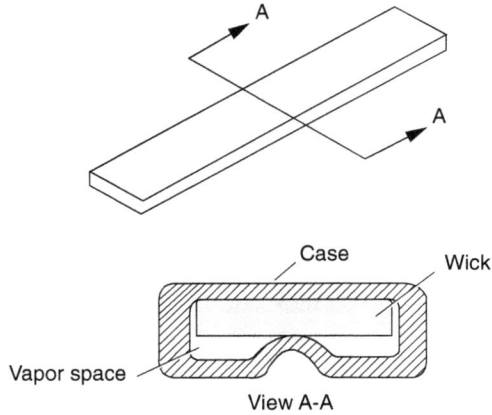

FIGURE 9.99   Flat heat-pipe rail design.

PHYSICAL DATA

- Working fluid - methanol
- Case material - type 304 stainless steel
- Wick material - sintered type 304 stainless steel
  wire rovings
- Length - 6 in. to 14 in.
- Cross-sectional dimensions
    — 0.076 in. x 0.220 in.
    — 0.060 in. x 0.185 in.
    — 0.040 in. x 0.160 in.

9.10 REFERENCES

1. *Rohsenow and Hartnett,* Handbook of Heat Transfer, McGraw-Hill, 1973.
2. *Steinberg, D.S.,* Cooling Techniques for Electronic Equipment, Wiley Interscience, 1980.
3. *Holman, J.P.,* Heat Transfer, Fourth edition, McGraw-Hill.
4. *Schlichting, H.,* Boundary-Layer Theory, McGraw-Hill, 1968.
5. *Kraus, A.D.,* Cooling Electronic Equipment, Prentice Hall, 1968.
6. *Conte, S.D.,* Elementary Numerical Analysis, McGraw-Hill, 1965.
7. *Barnes Engineering Co.,* Handbook of Infrared Radiation Measurement, 1983.
8. *Kays and London,* Compact Heat Exchangers, 2nd Edition, McGraw-Hill, 1964.
9. *Harper, C.A.,* Electronic Packaging and Interconnection Handbook, McGraw-Hill, 1991.
10. *Bar-Cohen and Kraus,* Advances in Thermal Modeling of Electronic Components and Systems, Vol. 1, Hemisphere Publishing Corporation, New York, 1988.
11. *Chi, S.W.,* Heat Pipe Theory and Practice, McGraw-Hill, New York, 1976.
12. *Philips, R.J.,* Forced-Convection, Liquid-Cooled, Microchannel Heatsinks, Masters Thesis, MIT, Cambridge, Massachusetts, 1987.

13. *Tuckerman, D.B.,* Heat-Transfer Microstructures for Integrated Circuits, Ph.D. Thesis, Stanford University, Stanford, CA, 1984.
14. *Welty, J.R.,* Engineering Heat Transfer, Wiley and Sons, 1974.
15. *Token, K.H.,* A New Avionics Thermal Control Concept, ASME Intersociety Conference on Environmental Systems, San Francisco, CA July 11 – 14, 1977.
16. *Brennan, P.J. and Kroliczek, E.J.,* Heat Pipe Design Handbook, Prepared by B & K Engineering, Inc. for NASA, Goddard Space Flight Center, June 1979.
17. *Weiting, A. R., "Empirical Correlations for Heat Transfer and Flow Friction Characteristics of Rectangular Offset-Fin Plate-Fin Heat Exchangers,"* ASME Journal of Heat Transfer, pp. 488 – 490, Aug. 1975.
18. *Basilius, A. and Minning, C.P., "Improved Reliability of Electronic Components Through the Use of Heat Pipes,"* 37th National Aerospace and Electronics Conference, Dayton, Ohio, May 20 – 24, 1985.
19. *Basilius, A., Tanzer, H., and McCabe, S., "Heat Pipes for Cooling of High Density Printed Wiring Boards,"* Proc. 6th International Heat Pipe Conf., Grenoble, France, pp. 531 – 536, 1987.
20. *Token, K., "Trends in Aircraft Thermal Management,"* Proc. Printed Wiring Board Heat Pipe Workshop, Hughes Aircraft Co., Electron Dynamics Division, Torrance, CA, 1986.
21. *McAdams, W.H.,* Heat Transmission, 3rd edition, McGraw-Hill, New York, 1954.
22. *Rohsenow, W.M., "A Method of Correlating Heat Transfer Data for Surface Boiling Liquids,"* Trans. ASME, Vol. 74, p. 969, 1952.
23. *Zuber, N., "On the Stability of Boiling Heat Transfer,"* Trans. ASME, Vol. 80, p. 711, 1958.
24. *Kirkbride, C.G., "Heat Transfer by Condensing Vapors on Vertical Tubes,"* Trans. AIChE, Vol. 30, p. 170, 1934.
25. *NACA Technical Note 2836, "Radiation Interchange Configuration Factors,"* D.C. Hamilton and W.R. Morgan.
26. Thermoelectric Handbook, Cambridge Thermionic Corporation, Cambridge, MA, Second Edition, 1972
27. *H.C. Hottel, "Radiant Heat Transmission,"* Mechanical Engineering, 52 (1930).
28. *Murose, T., Yoshida, K., Fujikake, J., Koizumi, T., and Ishida, N., "Heat Pipe Heat Sink HEAT KICKER for cooling of semi-conductors,"* Furukawa Review 2, 24–33, 1982.

CHAPTER 10
COMPONENT HANDLING WITH ESD CONTROL

Lucian H. Harris
Consultant; ESD Control
Oviedo, Fl.
Phone: (407) 366-3199
FAX: (407) 365-1597

10.1 INTRODUCTION

Engineering involvement with electronic components does not end with the design of the component. Components must be built, tested, and assembled into electronic hardware with an educated awareness that electrostatic discharge (ESD) can destroy the functional integrity of most devices. An ESD control program should be in place to manage the people, the equipment, and the materials involved in the handling of electronic components during the manufacturing cycle—from component build to black box assembly.

Even a high level of automation in handling components during subassembly does not mitigate the need for ESD controls. In fact, automated equipment can be related to the severe problem of the charged device discharging to ground and a charged machine discharging to the device.

10.1.1 The Effect of Circuit Scale-Down

The constant requirements for faster and smaller chip sets translate into thinner gate oxide, shorter channel length, and reductions in junction depth for all types of components. It is unrealistic to assume that protective circuitry, by itself, will be able to provide ESD protection in a manufacturing/assembly environment that is not using the procedures of a well thought-out ESD control program.

Junction thermal breakdown   It is estimated[1] that the limits for the scale down of silicon circuits used in very large scale integration (VLSI) devices is 0.1 µm for channel length, 0.05 µm for junction depth and 40 A for oxide thickness. To go from a 1-µm baseline for channel length to the 0.1-µm limit is scaling down by a factor of 10. Thermal modeling shows that scaling down by a factor of 10 at the planar junction increases ESD sensitivity for thermal breakdown by a factor of 5.5, when using a human body model (HBM) ESD pulser. HBM ESD pulse with 150-ns decay time.

At the cylindrical junction, ESD sensitivity is increased by a factor of 5 for the same scale down. The 1-μm baseline has been validated by an AT&T CMOS technology with an NMOS protection circuit. The 0.1-μm technology is a hypothetical one that is reduced from the baseline technology by a factor of 10.

This is useful theoretical data, to impress on the manufacturing engineering community the pragmatic side of ESD controls that are often challenged as an unnecessary cost burden to the manufacturing process. Thermal breakdown proceeds as follows: as an ESD voltage increases from zero to a large value, the voltage across the depletion region in the pn junction will also rise. Until the breakdown voltage of the pn junction is reached, little current will flow. However, the 1500 Ω of the HBM circuit will develop a large IR drop across the pn junction when the breakdown voltage is reached. The sum of this voltage drop and the breakdown voltage across the pn junction equals the ESD voltage. The silicon temperature starts to rise significantly. Over time, the temperature rise can reach 1412°C and thermal breakdown occurs.

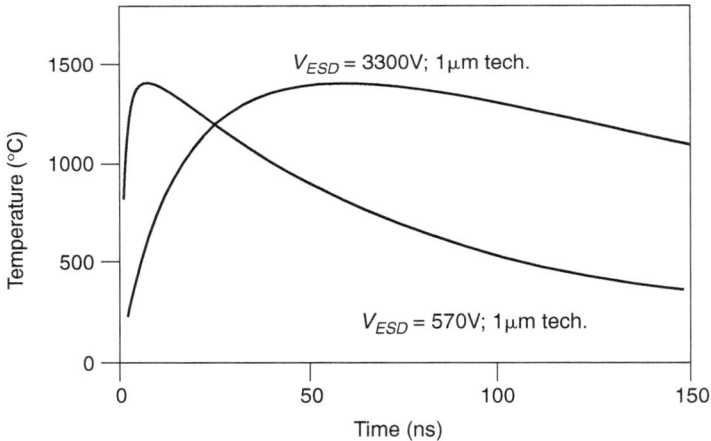

FIGURE 10.1 Temperature profile at planar pn junction.

See Figures 10.1 and 10.2 for the temperature profiles at the planar pn junction and the cylindrical pn junction, respectively, for the two junction depths. Figures 10.3 and 10.4 show how the maximum temperature at the hottest silicon spot changes as ESD voltage increases, for the planar and cylindrical junctions respectively.

Gate oxide breakdown In MOSFET technology, the gate electrode and the silicon semiconductor are separated by the silicon oxide used as the gate dielectric. The electric field generated by the gate dielectric controls the channel behavior. Silicon dioxide is obviously the most important insulator for the electronics industry, today.

Lin[1] developed a dielectric breakdown model for thick oxides that shows how the dielectric breakdown field changes with the oxide thickness. According to the Lin model, the calculated gate-oxide dielectric-breakdown sensitivity increases by a factor of 10 by scaling the thickness from 400 A to 40 A, if the area is reduced proportionally.

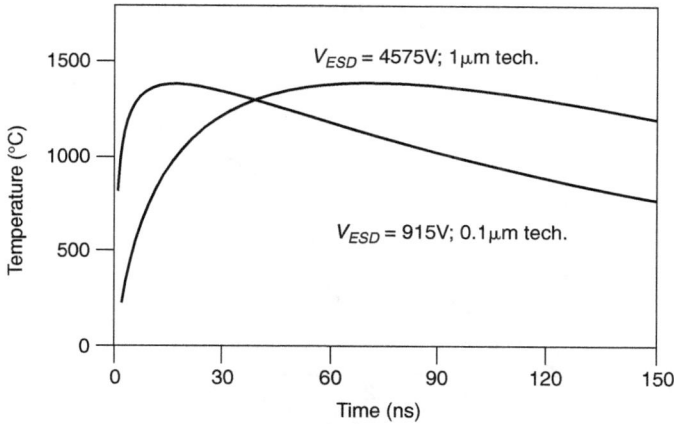

FIGURE 10.2   Temperature profile at cylindrical pn junction.

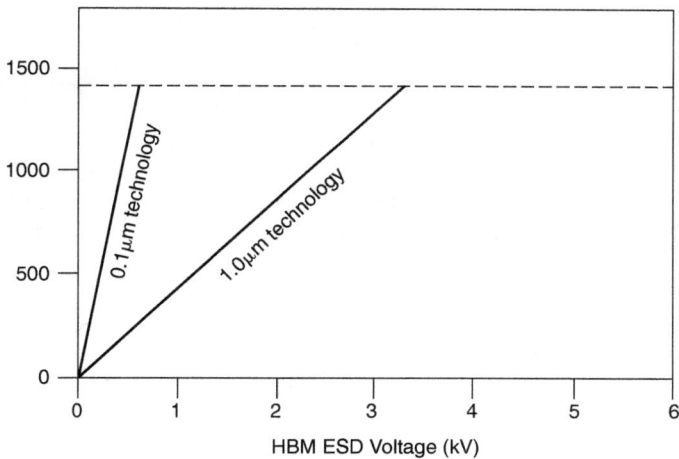

FIGURE 10.3   Maximum temperature at planar pn junction.

When plotting voltage breakdown versus oxide thickness (A), Figure 10.5, the result is a linear decrease in the voltage required for dielectric breakdown versus oxide thickness that confirms the Lin calculation. This plot was obtained from the data in Figure 10.6[2], derived from the measurements of Sodan, Wand, and Osburn. The dielectric strength (*MV/CM*) was multiplied by oxide thickness to obtain the breakdown voltages.

10.2   BASIC ESD THEORY

10.2.1 Static Charge Generation

The atom   The smallest particle of any substance that has the properties of that substance is a molecule. Molecules are composed of atoms, which are the smallest particles of the

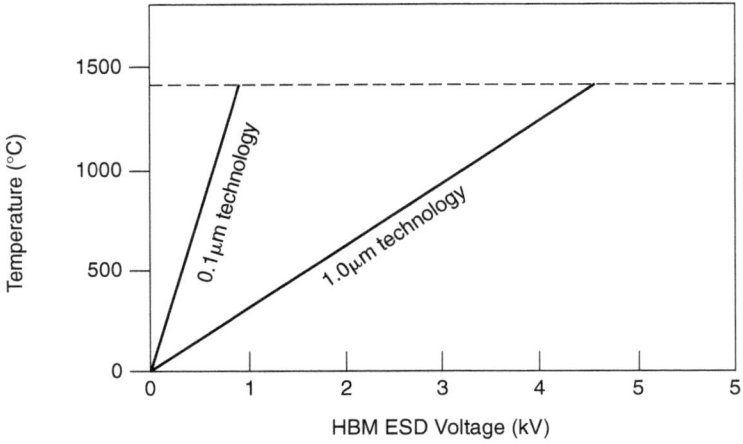

FIGURE 10.4 Maximum temperature at cylindrical pn junction.

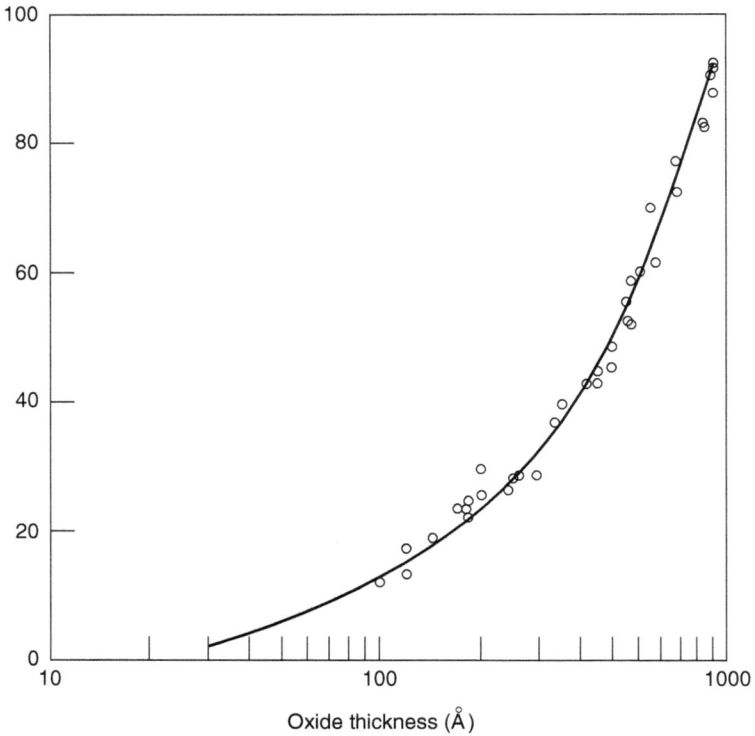

FIGURE 10.5 Dielectric breakdown voltage.

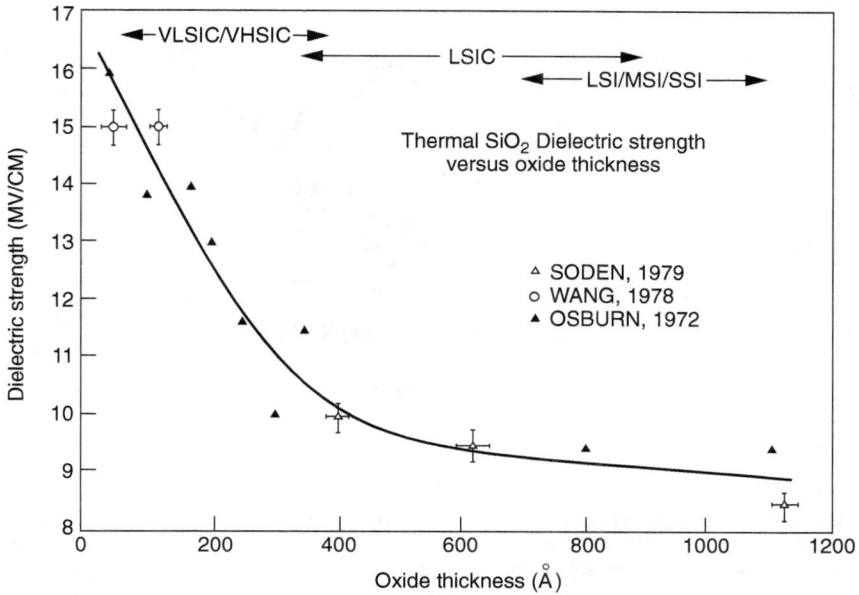

FIGURE 10.6 Gate dielectric breakdown.

chemical elements that can enter into combination with other elements. Atoms have a dense, central nucleus, which contains protons and neutrons, and which is surrounded by a cloud of electrons arranged in shells or energy levels.

Each electron has a definite amount of energy, which determines the orbital in which it moves. An orbital is not an orbit in the sense that the planets orbit around the Sun. An orbital is a probable pattern of movement characteristic of the energy of the electron.

Groups of orbitals in an atom are usually called *shells* and they are frequently designated by letters of the alphabet, beginning with K and running sequentially to Q in the most complex elements. They can also be given numbers from 1 to 7. The number of orbitals in a shell is the square of the shell number (1st. shell, 1 orbital; 2nd. shell, 4 orbitals; 3rd. shell, 9 orbitals; etc.). The maximum number of electrons that can occupy an orbital is 2; thus, the maximum number of electrons that can occupy a shell is two times the number of orbitals in the shell.

This model of the atom is useful in explaining the physics involved in triboelectric charging. Negative charges, represented by the electrons orbiting around the nucleus, are exactly balanced by the positive charges of the protons of the nucleus. The net electric charge of the atom and the molecular structure of the material is zero (Figure 10.7)[3]. If subjected to heating or friction, a material will either give up electrons from its outer shell and develop a positive charge; or the material will acquire electrons and develop a negative charge.

The charge e and $(-e)$ is called a (positive/negative) *elementary charge*. It is the smallest existing amount of charge and can be expressed in coulombs, C, as: $e = 1.602 \times 10^{-19}$. To put that in perspective, if you walked across a carpet and received a shock when you touched the doorknob, your charge would be at least 10^{-7}. More than 10^{12} electrons have been exchanged in the charging.

Trioelectric charging If you rub two materials together, such as a Teflon rod and a wool shirt, the Teflon will take on a negative charge and the wool will become positively

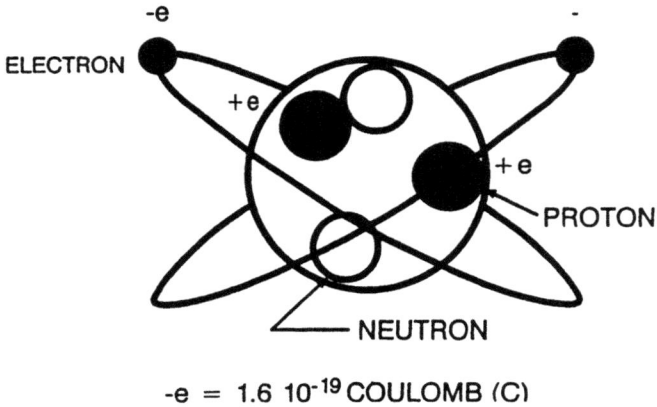

FIGURE 10.7 The atom.

charged. These charges are called *electrostatic charges* or *charges at rest*, from the Greek: *Statikos*, "to make stand." The two materials are nonconductors; therefore, the charges are not free to redistribute themselves over the material. The charges remain at the points of contact. Another term used to describe these charges is also from the Greek: Tri-bein, meaning to rub. Today, we substitute a combining form: Tribo, meaning friction, to get the term *triboelectricity*. If you rubbed the wool shirt over a steel bar, instead of the Teflon, the wool shirt would still become positively charged and the steel bar (if ungrounded) would take on a negative charge. Unlike Teflon, the steel bar being a conductor would redistribute its charge evenly over itself until grounded.

The ability of two materials to be electrified in this manner, when at least one of the materials is a nonconductor, is generally represented as a triboelectric series, such as the one shown in Table 10.1.[4] The materials will take on a positive charge every time they come into contact with a material lower on the scale, and are suddenly separated. Human hands and clothing material are near the top of the list, and widely used packaging materials are near the bottom. Large charge transfers occur when these materials are in contact; these electrostatic charges would represent a danger to components in an uncontrolled manufacturing environment.

This scale is not inclusive of all such materials and should be verified with laboratory testing for specific situations.

Induction Charging by induction is the process whereby an uncharged conductor becomes charged when grounded in the presence of an electrostatic field. The basic law of electrostatics, sometimes referred to as the *first law of electrostatics* states that like charges repel and unlike charges attract. The field diminishes or increases as the square of the distance between the charges.

If a charged object, such as an insulator, is brought in proximity to a neutral ungrounded conductor, such as an IC, the conductor will become polarized by the electrostatic field. That is, the charge on the object will repel the similar charges on the conductor causing (inducing) them to move to the opposite set of leads until all of the conductor is at the same potential (Figure 10.8).[4] If the conductor is temporarily grounded and the charged object is not moved, current will flow (i.e., ESD event). The IC will be left with a net charge of opposite polarity to the static field. After the charged object is removed, the charge will redistribute evenly over the conductor or IC.

TABLE 10.1 Triboelectric Series

Materials	Polarity (+ or −)
Asbestos	Acquires a more positive charge
Acetate	
Glass	
Human hair	
Nylon	
Wool	
Fur	
Lead	
Silk	
Aluminum	
Paper	
Polyurethane	
Cotton	
Wood	
Steel	
Sealing wax	
Hard rubber	
Acetate fiber	
MYLAR*	
Epoxy glass	
Nickel, copper, silver	
UV resist	
Brass, stainless steel	
Synthetic rubber	
Acrylic	
Polystyrene foam	
Polyurethane foam	
SARAN†	
Polyester	
Polyethylene	
Polypropylene	
PVC (vinyl)	
Teflon*	Acquires a more negative charge
Silicone rubber	

* Trademark of E.I. du Pont de Nemours
† Trademark of Dow Chemical U.S.A.

It will be left with the net residual charge until it dissipates into the air, or is grounded during processing. If the device is again grounded the result is another ESD event, opposite in polarity, and equal in charge to the previous one.

The energy transmitted in these ESD events depends on the field strength and the conductor's (device's) coupling to the field.

10.3 COMMON MODES FOR ESD EVENTS

Most available test data used for classifying ESD susceptibility of components is based on the Human Body Model (HBM). The Reliability Analysis Center (RAC) at Rome, NY

A) Device in static field

B) First event-grounding in field

C) Device left with net charge

D) Second event-grounding in later step

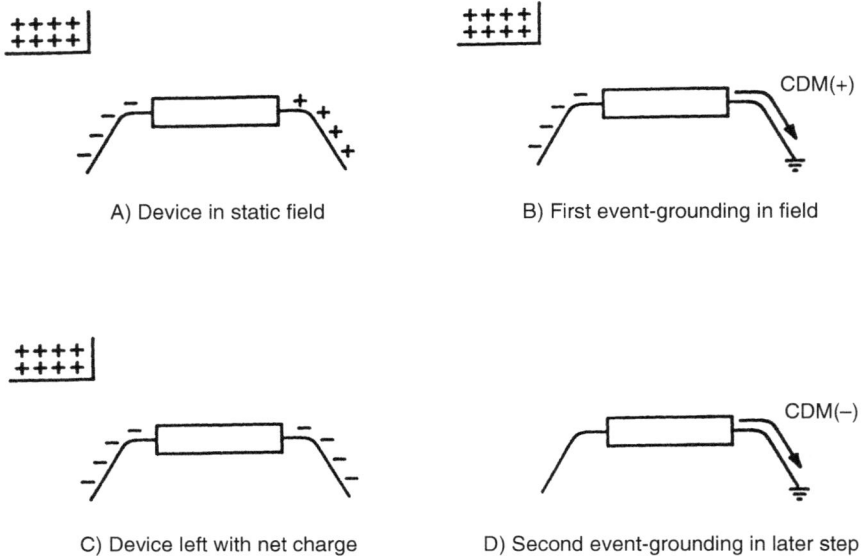

FIGURE 10.8   ESD by induction.

periodically publishes data compiled from industry tests and their own testing that shows breakdown voltages for a wide range of components based on the HBM. No such data base presently exists for a Charged Device Model (CDM) and a Machine Model (MM). This has been because of a general lack of awareness of charged devices (components) and charged machines (equipment) as a serious ESD threat. Today, the electronic industry is well aware of this threat to production yields and product reliability. Therefore, the question arises: what is the correlation between the RAC data base and the voltage levels at which a component would fail under a charged device or a charged machine scenario?

A round-robin ESD study[5] was recently completed that attempted to answer this question for CMOS devices, as well as the question of distinct failure signatures for each ESD model. See Table 10.2 for the CMOS devices used in the experiment. It was concluded, that there was no correlation between the failure signatures of the three ESD models, although failure analysis revealed the presence of common ESD failure signatures (Table 10.3).

Based on the "first failure" failure threshold information developed in Table 10.4, correlation analysis revealed a 93% correlation coefficient between the HBM and MM ESD models. Regression analysis was performed that showed that the HBM component-failure threshold is roughly 12× higher than the MM failure threshold for these devices (Table 10.5). This correlated closely with previous results obtained using "best estimate" failure thresholds, prior to obtaining test data. This information should cause concern for ESD programs that are geared for 100-V ESD sensitivity levels, based on HBM data.

10.3.1 The Human Body

The human body was the most common source for ESD events until the wide-spread introduction of the grounded wrist strap. Today, in the field, away from the protection of in-plant ESD controls, the human body is still a big contributor to unwanted ESD events.

TABLE 10.2 Devices Used in Round Robin Experiment

Device	Package type	Technology	Description
X1	44 PLCC	0.9 μm 3 volt/5 volt CMOS	Echo cancellor
X2	100 EIAJ	0.9 μm .5 volt CMOS	ASIC for disk drive system
X3	40 PDIP	1.5 μm 5 volt CMOS	Audio applications
X4	28 PLCC	1.5 μm 5 volt CMOS	Bus interface for communication and data
X5	28 PDIP	1.2 μm 5 volt CMOS	Controls module communications for vehicle
X6	24 PDIP	1.5 μm 5 volt CMOS	2K × 8 bit static RAM

Unless tight controls on wrist straps are implemented and maintained in the plant, human-body failure modes can still contribute to serious product yield problems.

Human body model The standard human-body model is a low-loss capacitor charged to 100 pF and discharged through a series 1500-Ω resistor via a nonbounce relay (Figure 10.9).[6] Most specifications specify a maximum allowed current rise time of 10 ns (Figure 10.10).[7]

All of the standard specifications associated with ESD and/or components use the HBM as their reference for component sensitivity. The EOS-ESD Association Standard S5.1, *ESD Sensitivity Testing, Human Body Model (HBM)—Component Level*[8] classifies component sensitivity ranges (Table 10.6) as an aid to determine the level of ESD protection that will be required, including design hardening, during manufacturing and electronic assembly procedures. This standard contains a detailed procedure for testing components to the HBM. MIL-STD-1686 currently uses the EOS-ESD Association component classification ranges.

Figure 10.11[9] shows some of the ways that the human body can build up static charges through triboelectric charging. This is the most obvious and direct way that components can be exposed to ESD during the manufacturing and assembly process.

Test data compiles only the relative ESD sensitivity between different components and will not cover every variable present in real-world ESD events associated with the human body (i.e., the human body can vary from the HBM standard).

The human body itself can have a capacitance up to several thousand pF, but is typically 50 to 200 pF. A study[10] performed on human capacitance indicated that approximately 80 percent of the population tested had a capacitance of 100 pF or less. Variations in human capacitance are because of factors, such as variations in the amount and the type of clothing and shoes worn by personnel. The type of floor material also plays a role.

Human resistance is normally between 1000 and 5000 Ω. This variation is caused by conditions, such as the amount of moisture, salt, and oils at the skin surface, skin contact area, and contact pressure. A properly fitted and grounded wrist strap, with at least daily monitoring, as part of a comprehensive ESD control program is the most widely used method to protect ESD-sensitive components from human body charges, regardless of any real life deviations from the standard HBM.

TABLE 10.3 Device Failure Mode Results

Test sample	ESD model & stressing company	Failing pin (Input, output, power)	Failing circuitry (protection internal)	Failing structure (PMOS, NMOS, diode)	Failure signature
	HBM/Delco	Output	Protection	Diode	FMA inconclusive
	HBM/Ford	Output	Protection	Diode	FMA inconclusive
X1	MM/Delco	Ouptut	Protection	Diode	Junction damage
	MM/Ford	Output	Protection	Diode	Junction damage
	FCDM/AT&T	VDD	Protection	PMOS	Gate oxide
	HBM/Delco	I/O	Protection	NMOS	Gate oxide
	HBM/Ford	I/O	Protection	NMOS	Gate oxide
X2	MM/Delco	I/O	Protection	Diode	Junction damage
	MM/Ford	I/O	Protection	Diode	Junction damage
	FCDM/AT&T	No fail	No fail	No fail	No fail
	HBM/Delco	Output	Protection	Resistor	Contact spiking
	HBM/Ford	GND	Protection	GND runner	Metal burn-out
X3	MM/Delco	Input, output, & GND	Protection	GND runner, resistor, & NMOS	Metal burn-out, contact spiking, & gate oxide
	MM/Ford	GND	Protection	GND runner	Metal burn-out
	FCDM/AT&T	I/O	Internal	NMOS	Gate oxide
	HBM/Delco	I/O & VDD	Internal	NPN & PMOS	Contact spiking & gate oxide
	HBM/Ford	Input	Internal	NMOS	Gate oxide
X4	MM/Delco	VDD	Internal	PMOS	Gate oxide
	MM/Ford	I/O & VDD	Internal	PMOS & NMOS	Gate oxide & contact spike
	FCDM/AT&T	Input	Internal	NMOS	Gate oxide

X5	HBM/Delco	I/O	Protection	NMOS	Poly-extrusion
	HBM/Ford	VDD	Internal	NMOS	Gate oxide
	MM/Delco	VDD	Protection	NMOS	Metal melt
	MM/Ford	VDD	Protection	NMOS	Metal melt
	FCDM/AT&T	VDD	Internal	PMOS	Gate oxide & poly-filament
	HBM/Delco	I/O	Protection	NMOS	Gate oxide
	HBM/Ford	I/O	Protection	NMOS	Gate oxide
X6	MM/Delco	No fail	No fail	No fail	No fail
	MM/Ford	I/O	Protection	NMOS & PMOS	Metal melt & contact spiking
	FCDM/AT&T	I/O	Protection	PMOS	Gate oxide & poly-filament

TABLE 10.4 "First Failure" Failure Thresholds

Device	HBM ESD (V)	MM ESD (V)	FCDM ESD (V)
X1	500	50	1000
X2	1500	100	2500
X3	4000	400	1000
X4	1500	100	1000
X5	2500	250	1000
X6	4500	300	1000

TABLE 10.5 ESD Model Correlation Coefficients

ESD model to ESD model	Correlation coefficient	Regression analysis
*HBM to CDM	0.28	$V_{HBM} = (1.63) \times V_{CDM}$
*HBM to MM	0.92	$V_{HBM} = (11.73) \times V_{MM}$
*CDM to MM	0.42	$V_{CDM} = (3.37) \times V_{MM}$
**HBM to MM	0.93	$V_{HBM} = (11.64) \times V_{MM}$

*Best estimate failure threshold
**From Table 10.4 1st failure

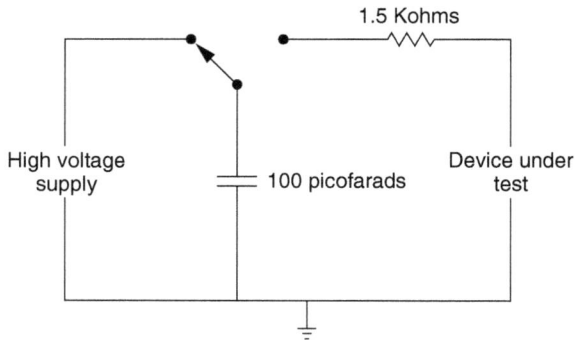

FIGURE 10.9 Typical human-body model circuit.

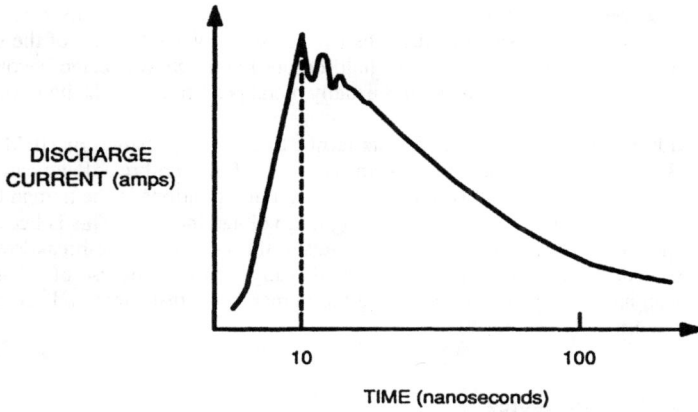

FIGURE 10.10   Idealized current waveform HBM ESD pulses.

TABLE 10.6   ESDS Component
Sensitivity Classification—Human
Body Model

Class	Voltage range
Class 0	0 volts to <= 249 volts
Class 1A	250 volts to <= 499 volts
Class 1B	500 volts to <= 999 volts
Class 1C	1000 volts to <= 1,999 volts
Class 2	2000 volts to <= 3,999 volts
Class 3A	4000 volts to <= 7,999 volts
Class 3B	>= 8000 volts

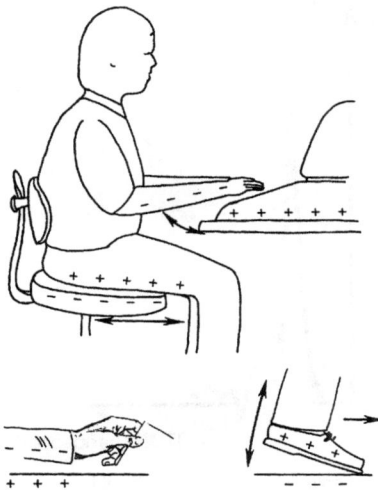

FIGURE 10.11   Some of the classic ways a
human can generate ESD.

Component failure mode   In well-designed circuits,[11] failures are generally semiconductor junction related. Well-designed structures will resist many ESD pulses of the same or higher voltage after the first damage threshold before the junction leakage becomes excessive. The ultimate failure mechanism is usually metal penetration of the junction, eventually resulting in a short.

Gate oxide breakdown in MOS circuits is not likely, but possible, in HBM testing (Figure 10.3) because of the slow waveform rise time. However, breakdown can readily occur with the typical real-world HBM ESD with certain variations in the human RC network; such as a lower skin resistance, resulting in a voltage increase. This is because the rise time is generally faster with high-current pulses allowing the oxide-breakdown voltage to be exceeded before any protection circuits fully turn on. The use of a hand-held metal tool also has the effect of intensifying the human body discharge and producing a very rapid rise-time.

10.3.2 The Charged Device

In recent years, the importance of failure from a charged device discharging onto a conductor has been recognized. This has come about with the widespread use of auto-insertion equipment and other auto assembly equipment. The charged device waveform is much steeper and of a shorter duration than the HBM (Figure 10.12).[7]

The CDM is not complex, but is very difficult to reproduce in practice. There are many opportunities during the manufacturing and assembly cycles for a device to acquire a charge on its surfaces—either through induction or by tribocharging. We are mainly concerned with the charge on conductive parts, such as the lead frame. This charge is mobile. If one or more of the device pins contacts a grounded conductive surface a high current, fast rise-time discharge occurs that lasts less than 2 nanoseconds with rise times in the hundreds of picoseconds.

Because most ESD-sensitive devices incorporate protection schemes, this is not a serious ESD threat. Instead, the serious ESD threat is the secondary discharge following charge transfer (covered in the Induction section).

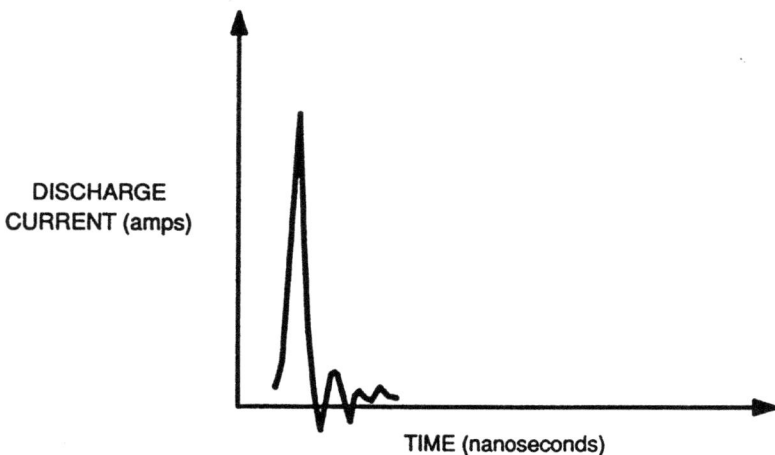

FIGURE 10.12   Idealized current waveform CDM ESD pulses.

The failure threshold is not just chip dependent.[11] The package size plays a major role. A chip packaged in small outline plastic (SOP) and "quad" packages has a lower failure threshold than the same chip in a standard dual-in-line plastic package of up to 2:1. The SOP and "quad" packages have lower lead inductance, thinner dielectric and higher effective capacitance. Even a larger package, such as a pin grid array (PGA), also tends to be more susceptible because of the higher package capacitance.

Charged device model   In EOS/ESD-DS5.3 1993,[12] the EOS/ESD association has a draft standard for *Electrostatic Discharge (ESD) Sensitivity Testing Charged Device Model (CDM) Component Testing*, which includes detailed test procedures for testing and classifying components for ESD failure thresholds. See Table 10.7 for the range of classifications. These classifications are also included in MIL-STD-1686.

TABLE 10.7   ESDS Component
Sensitivity Classification—Charged
Device Model

Class	Voltage range
Class C0	>0 volts to <125 volts
Class C1	125 volts to <250 volts
Class C2	250 volts to <500 volts
Class C3	500 volts to <1,000 volts
Class C4	1,000 volts to <2,000 volts
Class C5	>2,000 volts

The charged device waveform is much steeper and of a shorter duration than the HBM. The test procedures use three test modes for evaluating components: contact mode socketed (cs), contact mode, nonsocketed (cn), and noncontact discharge (nd). The socketed and nonsocketed modes have waveforms that are influenced by parasitics.

See Figure 10.13 for the noncontact CDM ESD waveform for the standard test modules. This waveform represents a discharge that is initiated by an arc between a probe tip and a component pin. It more nearly replicates the conditions of a CDM ESD event under electronic assembly conditions. However, it requires a controlled environment for close-tolerance reproduction. Provision is made for the use of the "cs," "cn," and "nd" suffix to identify how a component was tested for ESD withstand voltage.

Component failure mode   The CDM discharge[6] is so fast and the current so high that the major failure mechanism encountered with CDM is oxide rupture. In MOS circuits, the discharge is fast enough that the normal ESD plasma-related protection structures do not turn on fully to a low-resistance state. Also, the current is high enough that voltage drops on internal metal lines become a significant part of the stress. The CDM discharge often causes damage in the oxide gate area (Figure 10.14).[13]

10.3.3 The Charged Machine

A variation of the HBM is the 200-pF, 0-Ω machine model that originated in Japan. This model represents the discharge occurring from the charged cables of a device or board tester, the discharge occurring from the human body through a hand-held metal object, or the accumulated static charge from furniture.

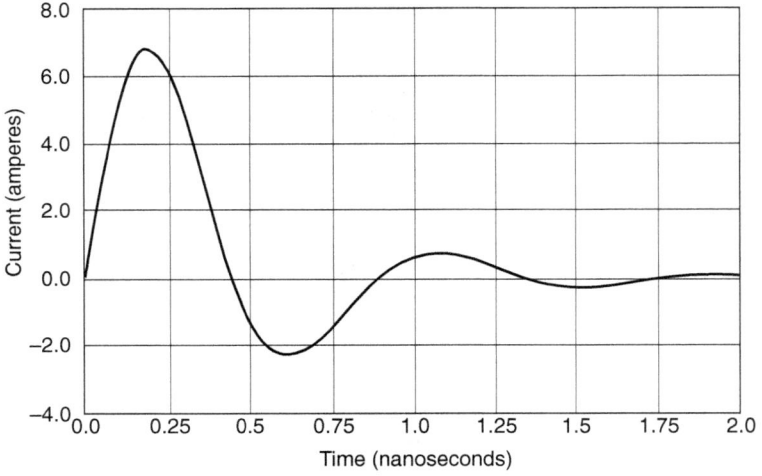

FIGURE 10.13   Noncontact CDM ESD waveform.

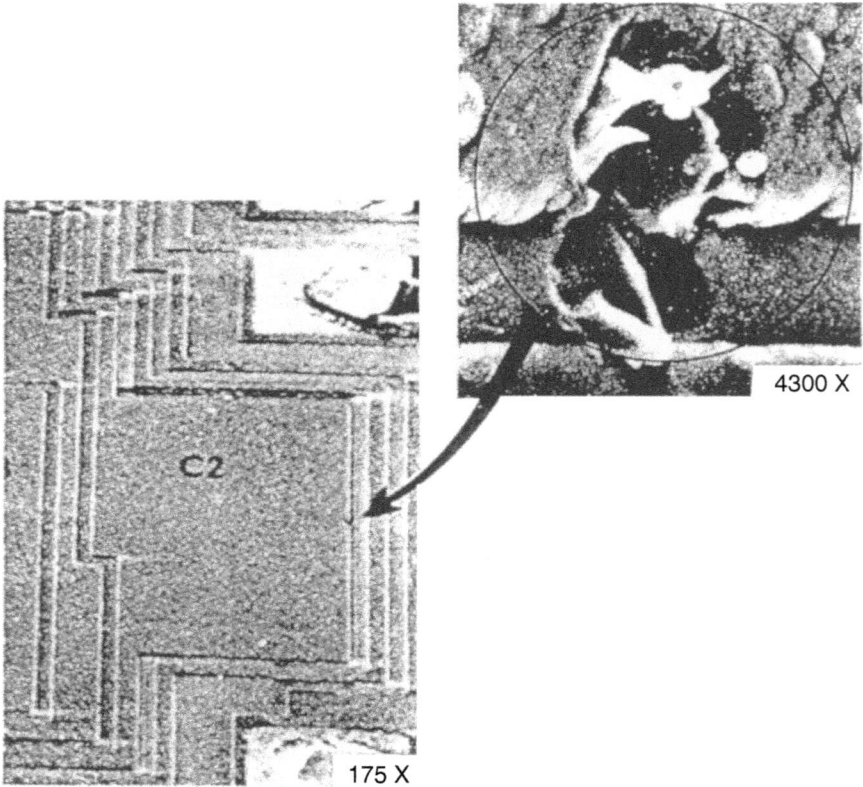

FIGURE 10.14   Damage caused by ESD.

Again, the machine model (as with the HBM) cannot fully represent the real-world variations in resistance and capacitance found in manufacturing/assembly areas. Ultimately, ESD-sensitive devices are protected by the implementation and continued support of appropriate ESD controls.

Machine model ANSI ESD-S5.2-1994 *Electrostatic Discharge Sensitivity Testing— Machine Model,*[14] establishes the testing procedure for evaluating and classifying the ESD sensitivity to the MM, as defined in the standard (Table 10.8).

TABLE 10.8 ESDS Component Sensitivity Classification—Machine Model

Class	Voltage range
Class M0	>0 volts to < 25 volts
Class M1	25 volts to < 100 volts
Class M2	100 volts to < 200 volts
Class M3	200 volts to < 400 volts
Class M4	400 volts to < 800 volts
Class M5	>800 volts

The machine model used by the ESD Association consists of a RC series network using a charged 200-pF capacitor and (nominally) 0 Ω of resistance. Series resistance and inductance are variations specified in terms of the current waveform through a shorting wire. See Figure 10.15[6] for a typical machine-model circuit and Figure 10.16[14] for current waveform through a short at 400 V.

Component-failure mode The failure mode for the MM is basically the same as the failure mode for the HBM. The MM damage thresholds are approximately an order of magnitude lower than for the HBM. See Sections 10.3 and 10.4.

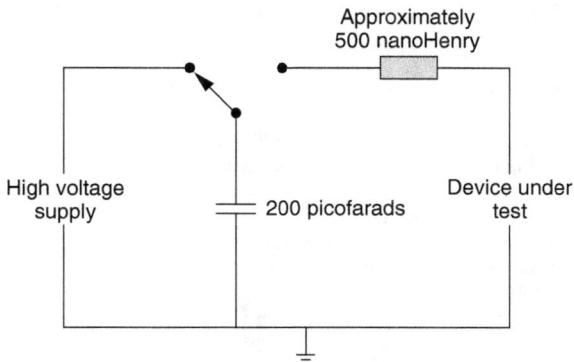

FIGURE 10.15 Typical machine model circuit.

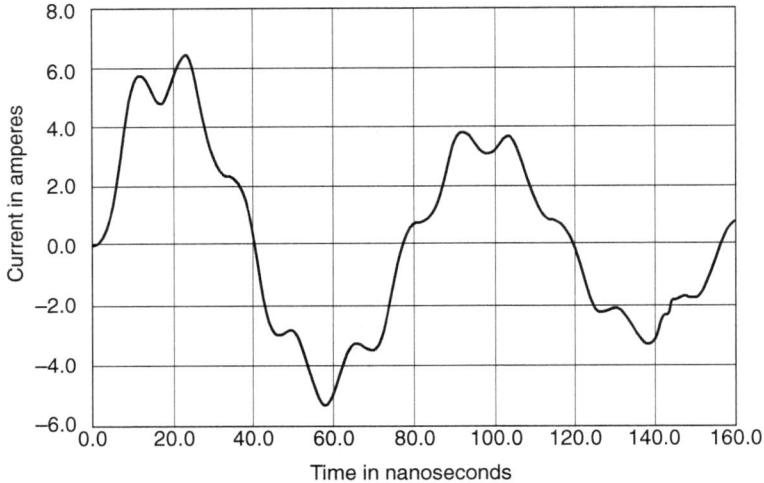

FIGURE 10.16   MM current waveform through a short circuit.

10.4 CONTROL OF ESD IN MANUFACTURING, TESTING, AND ENGINEERING

The principles underlying ESD control of components and assemblies in both the engineering development, manufacturing test laboratories, and the manufacturing/assembly areas are the same. Often, the need for tight ESD controls in the engineering labs and the manufacturing test areas is not understood. The laws of physics are not suspended when components are outside of the manufacturing/assembly areas. The basic ESD controls must be in place whenever components are being manufactured, put into assemblies, or tested.

Engineering labs should have access to the manufacturing stock rooms for ESD control supplies, including bags, totes, mats, wrist straps, etc. The primary reason that labs often do not meet manufacturing criteria for ESD control is their failure to budget for ESD-control supplies and their failure to use the same control items as manufacturing uses, when they do have a budget. Cross charges should be set up into the manufacturing stockrooms, if necessary.

10.4.1 The Effect of Relative Humidity

In general, high humidity tends to reduce triboelectric generation and increase surface conductivity, allowing the charge to bleed off to ground more easily. When the relative humidity is high, a conductive layer of water will be absorbed on the surface of hydrophillic materials (i.e., many plastic formulations).

The myth is that no static electricity problems occur when the humidity is high. This is probably because the voltages encountered in high-humidity environments are generally less than the threshold of perception of the average person, around 3 to 4 kV.

Practically all the voltages generated at high humidity are below the threshold of human perception (air discharge), but are well within the range of ESD damage to common components. The question is, what is meant by "high" humidity. Certainly, not so high as to cause water to condense on hardware and packaging materials.

TABLE 10.9 Typical Electrostatic Voltages

	Electrostatic voltages	
Means of static generation	10 to 20 percent relative humidity	65 to 90 percent relative humidity
Walking across carpet	35,000	1,500
Walking over vinyl floor	12,000	250
Worker at bench	6,000	100
Vinyl envelopes for work instructions	7,000	600
Common poly bag picked up from bench	20,000	1,200
Work chair padded with polyurethane foam	18,000	1,500

Table 10.9[10] indicates that 65 to 90% RH would be required to get static voltage, of most likely encountered ESD events in the electronic environment, down below the human perception range of 3 to 4 kV. But data compiled by Guisti[15] indicates that 40 to 60% RH would be just as effective, and much more realistic for most manufacturing environments. Indeed, MIL-HDBK-263B[10] does recommend using the 40 to 60% RH range.

Guisti examined the VZAP-1 *Electrostatic Discharge (ESD) Susceptibility of Electronic Devices* data published by the Reliability Analysis Center (RAFB) in the Spring of 1983. This included histograms of voltage failures using the HBM for various integrated circuit technologies.

Data from the literature was used to compile the human body voltage for walking across a carpeted floor and for standing up at a work station, at various percentages of relative humidity (Figures 10.17 and 10.18, and Table 10.10).

Modeling the VZAP-1 data statistically, Guisti arrived at an expected failure voltage for CMOS, NMOS, and "all other" technologies (Table 10.11). From this, the probability of ESD failure for various technologies and conditions was calculated (Tables 10.12 and 10.13). The data indicates that relative humidity can drastically reduce voltages generated by triboelectric charging (i.e., walking across a carpet, in this instance). At 40% RH, the static voltage generated from walking across a carpet is 1.5 kV, a 79% reduction from voltage

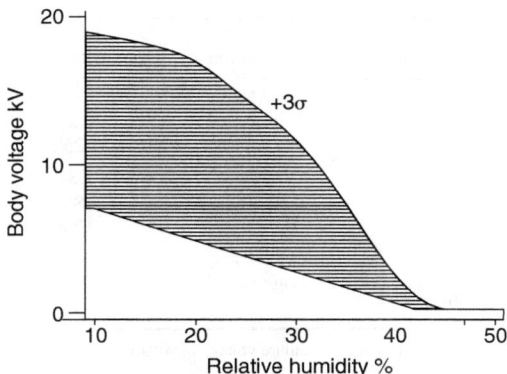

FIGURE 10.17 Effect of RH on HB voltage distribution; walking across carpet.

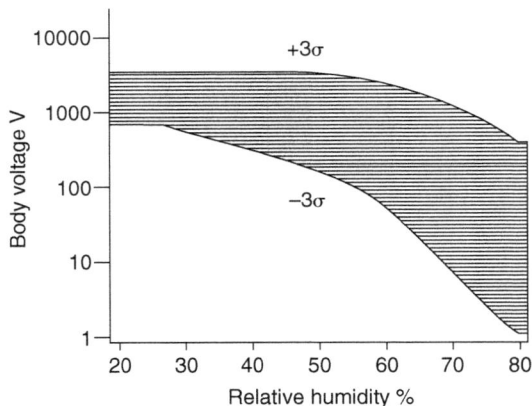

FIGURE 10.18 Effect of RH on HB voltage distribution; standing up at W.S.

TABLE 10.10 Results of Human Body Voltage Analysis

		Relative humidity (%)				
Activity		20	30	40	50	60
Walking across	μ	10.9K	7250	1500	–	–
carpet	σ	2033	1517	333	–	–
Standing up at work	μ	–	1970	1850	1695	1175
station	σ	–	478	517	515	375

(All values are in volts.)

TABLE 10.11 Summary of the ESD Failure Voltage Analysis

Technology	Best fit distribution	Distribution parameters
CMOS	Log-normal	$\mu = 7.4$ $\sigma = 0.8$ $\overline{V} = 2250$
NMOS	Weibull	$\eta = 1.34$ $\theta = 1850$ $\overline{V} = 1700$
All microcircuits	Log-normal	$\mu = 7.54$ $\sigma = 0.98$ $\overline{V} = 3040$

($\overline{V}$ is the expected failure voltage in volts.)

TABLE 10.12 Walking on Carpet

IC type	%RH	Probability of failure		
		20	30	40
CMOS		.977	.961	.449
NMOS		.957	.936	.803
All microcircuits		.937	.900	.402

Failure probabilities calculated for various IC technologies and humidities while walking across a carpeted area.

TABLE 10.13 Standing Up at Work Bench

IC type	%RH	Probability of failure			
		30	40	50	60
CMOS		.57	.54	.50	.34
NMOS		.87	.86	.84	.76
All microcircuits		.50	.48	.44	.31

Failure probabilities calculated for various IC technologies and humidities while standing up at a work station.

generated at 30% RH. The voltage generated from standing up at the work station was reduced from 1970 V at 30% RH to 1175 V at 60% RH, a 40% reduction. Not as dramatic a reduction because of the reduction in effective body capacitance when the operator stood up at the work station. This indicates that a reasonable relative humidity range for minimizing triboelectrification is 40% to 60% RH. Controlled humidity will not preclude damage to components if standard ESD-control procedures are not in place.

10.4.2 Packaging

Bags In the control of ESD, the resistance of a material plays an important role. For thin materials conducting along the surface over a relatively insulative base material, such as ESD bags, this is expressed as surface resistivity, measured in ohms per square. It is the ratio of direct current (dc) voltage to the current that passes along the surface of the material.

The resistance of a square-sized sample, whether it is 1 cm/2 or 1 ft/2, yields the same results. It is independent of the size of the square. Surface-resistivity ranges are used to describe materials classified as insulative (static prone), dissipative (static free), or conductive (Table 10.14). MIL-HDBK-263 recommends using the ASTM D 257 method of using the unitless correction factor (L/W) for the measured resistance in ohms:

$$(P_s) \text{ ohms per square}) = R \text{ (ohms)} \times \frac{L}{W}$$

where:

P_s = Surface resistivity in ohms per square
R = Total resistance measured in ohms
W = Distance between the probes
L = Probe length

TABLE 10.14 Material Classification for ESD Control

Material classification	Volume resistivity Ohm-Cm (ASTM 991)	Surface resistance Ohms (ANSI ESD-S11.11)	Surface resistivity Ohms sq. (ASTM 267)	Tendency to discharge to ground
Insulative	$<10^{11}$	$>10^{11}$	$>10^{12}$	Poor
Dissipative	$\geq 10^4$ to $\leq 10^{11}$	$>10^4$ to $\leq 10^{11}$	$\geq 10^5$ to $\leq 10^{12}$	Good to fair
Conductive	$<10^4$	$\leq 10^4$	$<10^5$	Excellent
Shielding	$\leq 10^3$	$\leq 10^3$	$\leq 10^4$	Excellent

The military handbook recognizes that materials of complex construction (i.e., not clearly bulk or surface conductive) can pose measurement problems. Testing facilities should document all test parameters for surface and volume resistivity so that the tests can accurately be repeated at that facility.

The ESD Association recommends the test procedure for surface resistance as detailed in ANSI ESD S11.11-1993 *Surface Resistance Measurement of Static Dissipative Planar Materials*.[16] This is a direct-current test method for measuring electrical resistance, expressed in ohm-cm, designed specifically for static-dissipative planar materials (such as bags used to package ESDS parts, including components) (Table 10.14).

Static dissipative   The bag that is used to carry or store non-ESDS items does not have to give static-charge shielding protection to the items carried. It does have to have static-dissipative properties that minimize charge build-up on the bag itself from handling and allow quick charge dissipation when put on a grounded workstation; or is handled by a grounded person.

The most common type of construction for this type of bag is a polymer, such as polyethylene, filled with approximately 35% of a hydrophillic surfactant, which migrates to the surface of the bag and adsorbs moisture from the air to provide a conductive layer (Figure 10.19).[13]

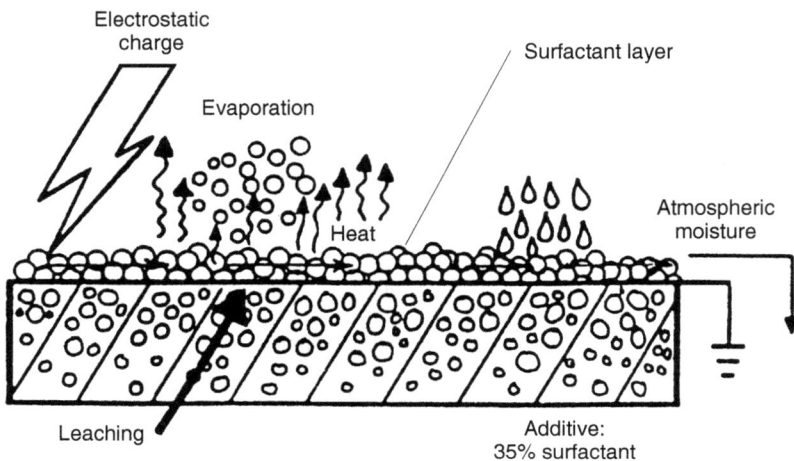

FIGURE 10.19 Surfactant-filled material.

This material is not recommended for lengthy storage of electronic items, as some additives in the polymer can be corrosive to platings, affect precision bearings, and outgass to deposit on optical surfaces. Seam strength and punch-through strength are important mechanical strength considerations in the selection of this type of bag.

Static shielding In 1843, Michael Farraday, an English physicist, conducted his famous "ice-pail" experiment and concluded that the charge on a conductor rests entirely on its outer surface.

This Farraday shield concept is utilized in the modern static-shielding bag, in which a conductive layer isolated within or on the packaging material, provides a shield to attenuate external fields. Any charge accumulation will be present on the outside of the conductive surface, only; thereby, protecting ESDS items contained within the sealed package from ESD.

The modern three- or four-layer metallized bag was not only designed to give Farraday-cage protection to sensitive components, but also to provide dissipative surfaces on the outer and inner layers of the bag, to minimize triboelectric charging from handling (Figure 10.20).[13] Seam strength and punch-through strength are equally as important for the static-shielding bag as for the dissipative bag.

Totes Totes must be considered an integral part of the packaging of ESDS parts and components. Carelessness in the use of totes can result in damage to parts that are considered safely packaged. The volume resistivity of the tote material expressed as ohm-cm determines the classification of the material as either insulative, dissipative, conductive, or (importantly) static shielding (Table 10.14). The current standard test methods are ASTM D257 for dissipative materials and ASTM D991 for conductive materials.

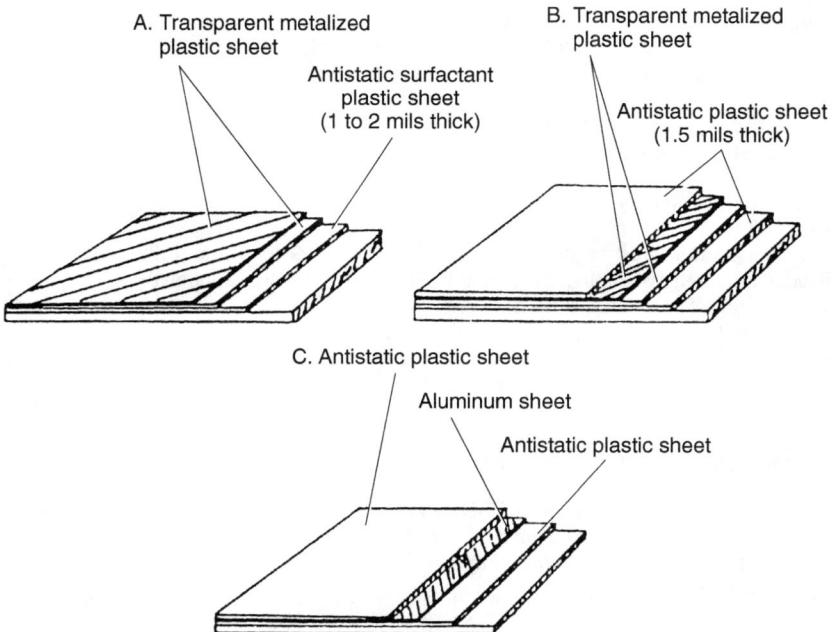

A. Transparent metalized plastic sheet

Antistatic surfactant plastic sheet (1 to 2 mils thick)

B. Transparent metalized plastic sheet

Antistatic plastic sheet (1.5 mils thick)

C. Antistatic plastic sheet

Aluminum sheet

Antistatic plastic sheet

FIGURE 10.20 Laminated sheet and bag material.

Insulative totes An insulative tote presents a real ESD danger to sensitive components that it might contact. A plant using insulative totes must package all ESDS items in shielded bags or other shielded containers prior to placing the ESDS items in the totes. During transport, the tote can easily pick up a charge from personnel handling. Subsequent grounding of personnel at the ESD-protective workstation will not drain off charges from the insulative surface. Therefore, all ESDS items must be removed from the workstation prior to placing the tote on the mat or adjacent to the mat. The tote must be unloaded and removed from the station prior to unpackaging the ESDS items it contained. If logistics will not permit this procedure, then forced-air ionization is necessary.

Dissipative totes The dissipative tote shares one thing in common with the insulative tote. It does not provide static shielding for ESDS parts. Therefore, it can only be allowed to carry ESDS items that are packaged with shielding material. Beyond that one item, the dissipative tote presents fewer problems for material handling and ESDS items at the workstation.

The dissipative tote will quickly dissipate charges accumulated during handling, if touched by a grounded person or put down on a grounded mat. Care must be taken, however, that the tote not be put down next to exposed ESDS items on the mat prior to being touched by a grounded person.

Conductive totes The conductive tote with a volume resistivity less than $1.0E + 3$ ohm-cm does one thing that neither of the other two types of totes can do. It provides a farraday shield for otherwise unpackaged ESDS items. This allows the transport of populated printed boards, and other large ESDS items that are inconvenient to put into shielded bags. The drawback is that a conductive latched lid must be in place at all times during transport and storage.

Other drawbacks exist if the conductive tote itself, is used as the shielding package. The lid can only be opened if the person opening it is grounded, unless complex labeling techniques are used to identify the packaging level of the contents. The conductive tote can pick up a charge as it is handled by ungrounded personnel. The handling restraints that apply to the dissipative tote also apply to the conductive tote.

10.4.3 Workstations

For the most commonly recommended grounding set-up see Figure 10.21, per MIL-HDBK-263.[10]

The workstation must be designed to dissipate charges to ground (Table 10.15) at a controlled rate (i.e., fast enough to prevent ESD damage, yet slow enough to prevent shock to the operator). The area must be large enough to present an uncluttered space to assemble ESDS parts. Ancillary items essential to the manufacturing process (such as metal tools, plastic squeeze bottles, masking tape, etc.) that can be triboelectrically charged or charged by induction must be handled in an ESD-safe manner. Workstations must not be used as tote staging areas.

Electrical parameters The test methods for the following electrical tests are included in:

• ESD Association Draft Standard ESD DS4.1-1995 *ESD Protective Work Surfaces*[17] The test methods are derived from ASTM F150-72, *Standard Test Method for Electrical Resistance of Conductive Resilient Flooring.*

• MIL-W-87893 *Workstation, Electrostatic Discharge (ESD) Control*[18] The test methods are derived from ASTM F150-72, *Standard Test Method for Electrical Resistance of Conductive Resilient Flooring.*

- *Resistance to ground* The industry standard for a current-limiting resistance between the mat grounding post and the factory hard-ground connection is 1 MΩ.
- *Mat surface resistance, point to point* The static-dissipative worksurface must possess a resistance equal to or greater than $1.0E + 6$ Ω, per ESD DS4.1; or $1.0E + 6$ Ω to $1.0E + 9$ Ω, per MIL-W-87093.
- *Total resistance to the mat ground point from the mat surface* $1.0E + 6$ Ω to $1.0E + 9$ Ω.

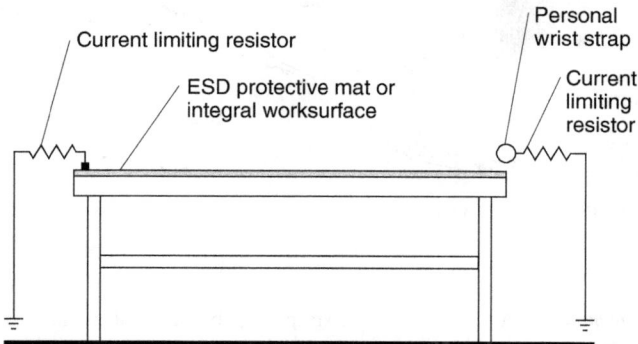

Note: ESD protective mat current limiting resistor optional

FIGURE 10.21   ESD protective work bench.

TABLE 10.15   Effects of Electrical Current on the Human Body

Current value in milliamps		Effects on human body
AC (60 Hz)	DC	
0–1	0–4	Perception
1–4	4–15	Surprise
4–21	15–80	Reflex action
21–40	80–160	Muscular inhibition
40–100	160–300	Respiratory block
Over 100	Over 300	Usually fatal

10.4.4 Wrist Straps

A properly fitted and properly grounded wriststrap, worn by the operator, is the single most effective deterrent to ESD damage (Figure 10.22).[13] It is well worth the effort for companies to replace wrist straps at the first sign of wear; to require daily checks of wrist strap integrity; to routinely sweep through the plant to verify that the wrist strap is being worn properly; and to check that it is, indeed, grounded. Grounding jacks for wrist straps should be available at every tote staging area, oven, storeroom, QA station, Rec./Insp., etc. Constant wrist strap monitors are a recommended technology to reduce random loss of grounding continuity between the operator and the wristband.

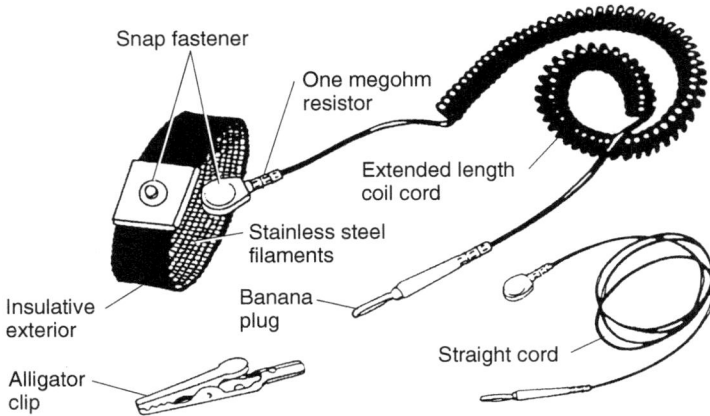

FIGURE 10.22   ESD protective wriststrap.

Electrical parameters   An effective wrist strap must be made of metallized conductive fabric or a flex metal band, which intimately contacts the wearers skin around the wrist/forearm area. It is not advisable to wear a wrist strap with exposed metal when working around live circuitry, such as in a test area. Contact could result in shorting.

A current-limiting resistor should be in the ground wire to limit current flow to 5 mA maximum and thus protect the operator if defective live electrical equipment is contacted (Table 10.15). The industry-standard resistor value is 1 MΩ. The amount of resistance needed to protect personnel is calculated, as:

$$\frac{\text{Highest voltage near operator}}{5 \text{ milliamps } (.005)} = \text{resistance for safe ground}$$

10.4.5 Grounded Floors and Mats

Grounded floors are very useful in a busy storeroom or other areas where ESDS items are constantly being handled, the floor area is large and a wrist strap would require a long grounding cord. This flooring material is available in the form of conductive or dissipative carpeting, and vinyl materials in either sheet or tile form. Vinyl flooring should be applied with conductive adhesive.

Grounded floor mats are very practical around wave soldering machines or automatic-insertion equipment, where operator movement is confined to the immediate equipment area. Again, this type of grounding precludes the use of long entangling wrist strap cords, which can present a safety hazard. Special footwear is required to ground the operator to the conductive floor or mat.

The use of grounded floors in a manufacturing area seems to be an unnecessary expense, unless the aisles between workstations are so narrow that visitors might contact exposed ESDS parts. Also, in a manufacturing area, operators and quality-control personnel must still wear a grounded wrist strap when seated at an ESD-controlled workstation, regardless of whether grounded floors or mats are used. Grounded floor mats are the best bet for special situations when wrist straps are too restrictive on operator movement.

Electrical parameters
- *Resistance to ground*   The same requirements apply as apply for workstation mats.

- *Surface resistance, point to point*   The same requirements apply as apply for workstation mats. The resistance between footwear and the mat/floor must be taken into consideration when establishing electrical parameters.

10.4.6 Air Ionization

Balanced air ionization is essential at an active ESD-controlled workstation when non-conductors, such as tapes and insulated totes must be brought in. When tapes are being pulled from a roll prior to use, intense triboelectric charging results and must be neutralized. An air ionizer is not a substitute for a grounded wrist strap.

Electrical parameters   Positive and negative ions are formed by the ionizer to dissipate electrostatic charges. Charge neutralization results when the positively charged ions are attracted to negatively charged items and negatively charged ions are attracted to positively charged items. Forced-air (i.e., fan-aided) ionizers are preferred over convection-type air ionizers.

The ionization source should be balanced as close to 0 V as possible, so as not to leave a residual charge on parts in the ion path. ANSI EOS/ESD S3.1-1991, *Ionization*,[19] is the existing industry standard for evaluating the performance of air ionizers.

10.4.7 Personnel Training

All personnel entering or working in engineering, testing, or manufacturing areas should have received training in basic ESD control. The level of training could vary according to the level of contact that an individual is expected to make with ESDS parts. Recertification training should be on a yearly basis, as a minimum. The requirements for training of all personnel in ESD-control procedures should be documented as a company policy.

10.4.8 In-House Documentation

Once a company decides to implement an ESD control program, the details of the program must be documented on all pertinent levels; either before or doing implementation of the program. At a minimum, there should be procedures for manufacturing, quality/receiving inspecting, and production control. These procedures must be kept current with changes in the ESD-control program.

10.5 COST BENEFIT OF ESD CONTROLS

10.5.1 Probability of ESD Events

ESD is a statistically variable phenomenon. You cannot accurately predict the voltage at which measurable damage will occur or the exact nature of the damage and the location of the discharge path. In addition, the damage can be hard to identify as ESD damage, per se. How then, do you gain the conviction that installing an ESD-control program in your plant will give you a return on your investment, without an expensive throughput analysis to correlate device failures to ESD? Moss[20] reasoned that ESD events are just like any other random event, such as the probability of a stone hitting your car windshield while out driving, and that the way to deal with this question was to use statistical methods. Using the Poisson distribution, where $P(r)$ is the probability of exactly r events, and y is the average number of occurrences of that event, the equation is:

$$P(r) = \frac{e^{-y} \times y^r}{r!}$$

Applying this formula to an ESD-control program and trying to calculate the probability of zero ESD hits ($r = 0$), the formula then becomes:

$$P(0) = e^{-y}$$

To make this useful, first suppose that y is the average number of electrostatic discharges in your facility or process without ESD control. This defines how "bad" your environment is. Second, define factor C, the percentage of ESD hits the program prevents, which is a measure of its effectiveness. Finally, for the measure of the savings a program generates, compute factor S, the percentage change in the probability of ESD damage:

$$S(\%\text{savings}) = 100 \left\{ \frac{1 - 1 - e^{-y(1 - c/100)}}{1 - e^{-y}} \right\}$$

Figure 10.23 is a plot of this equation for various values of y – the average number of ESD "hits" in the process. Even in an ESD-free environment (where $y < 0.1$), an ESD-control program with $C = 90\%$ effectiveness is necessary to achieve a savings of $S = 90\%$ of ESD losses. Where you have a process with a high number of hits on the average ($y = 20$), a 99.5%-effective program is needed to achieve 90% savings.

Even if you are not able to calculate the exact value of y, Moss points out that it is clear a half-hearted ESD control program will not do—especially if your plant has low humidity, extensive use of plastic packaging and state-of-the-art electronic components that are sensitive to ESD. See Table 10.16 for real-world examples of improvements before and after.

10.5.2 Potential Cost Savings

In 1990, an industry survey indicated that the average respondent saved just under $250,000 as a result of installing static controls. The range of savings ran from as little as $20,000 to as much as $2,000,000, with expenditures in the 2 to 3% range of revenues (Table 10.17).[21]

$$S(\%\text{savings}) = 100 \left[1 - \frac{1 - e^{-y(1 - c/100)}}{1 - e^{-y}} \right]$$

Consulting data from the same time frame of 13 defense contractors indicates that returns on investment (ROI) exceeding 20:1 and 30:1 are common[22]. ROIs of 40:1 can be achieved by implementing a technically well-rounded ESD-control program in the right sequence over a short time frame.

10.6 ESD SPECIFICATIONS

10.6.1 Military/Government

The use of military specifications and standards are being phased out, in accordance with the memo by Secretary of Defense William J. Perry, dated June 29, 1994. Commercial specifications and standards are to be used where they meet the user's needs. The use of

TABLE 10.16 The Payoff for Tighter ESD Controls

Area	Device affected	Before ESD controls	After ESD controls	Changes
IC manufact.	High-speed bipolar LSI	— Assembly/test yields —		Conductive containers Wriststraps.
		22%	100%	
	NMOS-LSI circuits.	—	Losses cut 50%	
	MOS-IC	—	Losses cut 50%	
Electronics mfg. (Stores/ incoming)	Bipolar transistor arrays	13%	100%	Packaging
	Low power Schottky-TTL ICs	0.35%	0.11%	Various
	Linear ICs (op amps)	2.6%	0.26%	Various plus vendor ESD
(Assembly & test)	CMOS logic	—	Losses cut 70%	Various
	Beam-lead diodes/ hybrid circuitry	—	Losses cut 79%	Various
	Bipolar digital logic	0.18%	0.11%	Varous
Electronic hardware (field reliability)	CMOS. Low-power Schottky-TTL ICs	—	Losses cut 50%	Protection circuit

TABLE 10.17 Savings on Average Static Control Practices

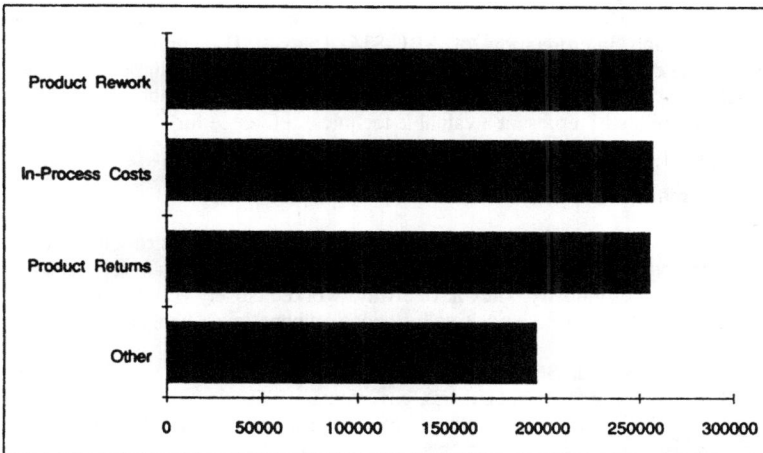

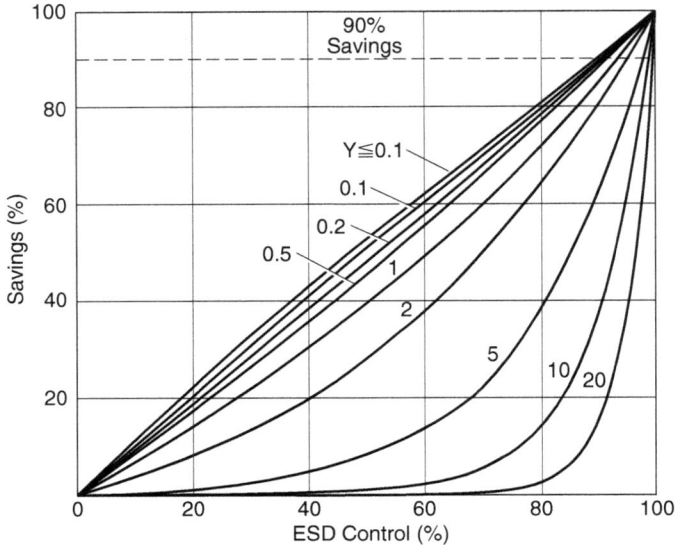

FIGURE 10.23   Plot of ESD control versus cost savings.

military specifications and standards will require waivers approved by the Milestone Decision Authority, as defined in Part 2 of DoD Instruction 5000.2. Currently, most government contracts negotiated prior to 1994 require the use of military specifications and standards for ESD control. Some contractors might elect to continue using the military documents with which they are familiar.

Since 1980, MIL-STD-1686 (then DOD-STD-1686), *Electronic Discharge Control Program* . . . has been the controlling document for ESD control for all military contracts. It was designed to be comprehensive enough in its scope to allow contractors to tailor its requirements to fit their particular manufacturing environment. MIL-HDBK-263 is complementary to MIL-STD-1686 and contains useful back-up technical data, as well as a paragraph-by-paragraph interpretation of MIL-STD-1686 ESD-control requirements. From the applicable documents list in these two specifications will flow all pertinent military specifications involved in ESD controls and tests. When these documents are no longer in general use, they will still represent a valuable technical reference for ESD control.

10.6.2 Commercial

The ESD Association is dedicated to presenting a forum on ESD issues through its yearly ESD Symposium, and to creating ESD test and control standards that are accepted throughout the electronics industry. They have always worked closely with the military on ESD-control issues and have been a positive influence on beneficial changes made to key government specifications.

Today, the ESD Association Standards Committee is working to create additional specifications to either augment or supplant existing government ESD-related documents. These documents represent the best thinking on this subject in the electronics industry and government, today. Test methods and control procedures are being challenged to represent improvements in measurement methods and the way that ESD is controlled to protect sensitive components and assemblies. See Tables 10.18 through 10.22 for representative cross references of current ESD Association standards, and for other standards (including military, EIA, and ASTM), and ESD-control properties that are related to component protective packaging.

TABLE 10.18 Antistatic Packaging Materials

Property	Test method	General result/attribute
Triboelectrification	EOS/ESD-11.20 series (in development)	
Contact corrosion	MIL-B-81705 FTMS 101C method 3005	Subjective observation of test coupons
Outgassing	ASTM E 595	Condensable volatile mat'l. total mass loss
Static decay	FTMS 101C, method 4046 EIA - 541 EOS/ESD-11.13 (in development)	Charge/discharge time
Plastic compatibility	EIA-564	Affect on polycarbonate
Seam seal strength	Mil-B-81705	Package integrity
Puncture resistance	FTMS 101C method 2065	Package integrity

TABLE 10.19 Dissipative Packaging Materials

Property	Test method	General result/attribute
Surface resistance	EOS/ESD-S11.11	Ohms
Volume resistivity	ASTM D991 EOS/ESD-DS11.12 (in development)	Ohms-cm Ohms
Triboelectrification	EOS/ESD 11.20 series (in development)	
Contact corrosion	MIL-B-81705 FTMS 101C method 3005	Subjective observation of test coupons
Outgassing	ASTM E 595	Condensable volatile mat'l. total mass loss
Static decay	FTMS 101C, method 4046 EIA - 541 EOS/ESD-11.13 (in development)	Charge/discharge time
Plastic compatibility	EIA-564	Affect on polycarbonate
Seam seal strength	Mil-B-81705	Package integrity
Puncture resistance	FTMS 101C method 2065	Package integrity

TABLE 10.20 Static Shielding Materials

Property	Test method	General result/attribute
Exterior resistance	EOS/ESD-S11.11 ASTM D257	Ohms Ohms/square (higher range)
Interior resistance	EOS/ESD-S11.11 ASTM D 257	Ohms Ohms/square (higher range)
Shielding	MIL-B-81705(C) EOS/ESD-DS11.31 (in development)	Volts Volts/joules

Transparency	Optical densitometer
	   described in ANSI PH2.18 and
	   ANSI PH2.19
		Percent

Triboelectrification	EOS/ESD-11.20 series (in development)	
Contact corrosion	MIL-B-81705 FTMS 101C method 3005	Subjective observation of test coupons
Outgassing	ASTM E 595	Condensable volatile mat'l. total mass loss
Static decay	FTMS 101C, method 4046 EIA - 541 EOS/ESD-DS11.13 (in development)	Charge/discharge time
Plastic compatibility	EIA-564	Affect on polycarbonate
Seam seal strength	Mil-B-81705	Package integrity
Puncture resistance	FTMS 101C method 2065	Package integrity

TABLE 10.21 Cushioning Materials

Property	Test method	General result/attribute
Surface resistance	EOS/ESD DS 11.11 ASTM D 257 (some forms)	Ohms Ohms/square
Volume resistivity	ASTM D 991	Ohm-cm
Resistance	EOS/ESD DS 11.12 (in development)	Ohms
Tensile strength	ASTM D 3575 (test E pascals)	Lbs./in^2
Elongation	ASTM D 3575 (test E)	Percent
Compression set	ASTM D 3575 (test A)	Return to original thickness
Compression strength	ASTM D 3575 (test B)	Force to 25% compression
Density	ASTM D 3575 (test C)	Kg/m^3
Corrosion	FTMS 101, method 3005	Rating scale
Physical properties as above		

TABLE 10.22 Rigid Materials

Property	Test method	General result/attribute
Surface resistance	EOS/ESD DS 11.11 ASTM D 257 (Some forms)	Ohms Ohms/square
Volume resistivity	ASTM D 991	Ohm-cm
Triboelectrification	EOS/ESD 11.20 series (in development)	
Appropriate physical properties		
Abrasion resistance	ASTM D 4060 or equivalent	Material removed

10.7 REFERENCES

1. D.L. Lin, *ESD Sensitivity and VLSI Technology Trends: Thermal Breakdown and Dielectric Breakdown*, EOS/ESD Symposium Proceedings, pp. 73–81, 1993.
2. R.K. Pancholy, *The Effects of VLSI Scaling on EOS/ESD Failure Threshold*, EOS/ESD Symposium Proceedings, p. 85, 1981.
3. B. Unger, "ESD Basics," *Electrostatic Discharge Tutorial*, ESD Association, Inc., pp. A2–A72, 1990.
4. AT&T, *Discharge Control Handbook*, 1989.
5. M. Kelly, T. Diep, S. Twerefour, G. Servais, D. Lin, and G. Shah, *A Comparison of Electrostatic Discharge Models and Failure Signatures for CMOS Integrated Circuit Devices*, EOS/ESD Symposium Proceedings, pp. 175–185, 1995.
6. *Electrostatic Discharge Control Handbook*, ESD ADV-2.0-1994. ESD Association, Rome, NY.
7. T. Dangelmeyer, *ESD Program Management*, AT&T, Van Nostrand Reinhold, 1990.
8. ANSI EOS/ESD-S5.1-1993, *Electrostatic Discharge Sensitivity Testing Human Body Model*, ESD Association, Rome, NY.
9. M. Mardiguian, *Electrostatic Discharge*, Interference Control Technologies, Inc., 1986.
10. MIL-HDBK-263, *Electrostatic Discharge Control Handbook For Protection of Electrical and Electronic Parts, Assemblies and Equipment (Excluding Electrically Initiated Explosive Devices) (Metric)*.
11. L.R. Avery, *Device ESD Testing*, Electrostatic Discharge Tutorial, ESD Association, Inc., pp. I1–I27, 1990.
12. EOS/ESD-DS5.3-1993, *Electrostatic Discharge Sensitivity Testing. Charged Device Model*, ESD Association, Rome, NY.
13. B. Matisoff, *Handbook of Electrostatic Discharge Controls*, Van Nostrand Reinhold, 1986.
14. ANSI ESD-S5.2-1994, *Electrostatic Discharge Sensitivity Testing. Machine Model*, ESD Association, Rome, NY.
15. J.H. Guisti, *The Probability of an ESD Event in an Unprotected Environment*, EOS/ESD Symposium Proceedings, p. 62, 1986.
16. ANSI ESD S11.11-1993, *Surface Resistance Measurement of Static Dissipative Planar Materials*, ESD Association, Rome, NY.
17. ESD DS4.1-1995 Revised, *Worksurfaces. Resistive Characterization*, ESD Association, Rome, NY.
18. MIL-W-87893, *Workstation, Electrostatic Discharge (ESD) Control*.
19. ANSI EOS/ESD S3.1-1991, *Ionization*, ESD Association, Rome, NY.
20. R. Moss, *Cost Benefit Analysis Part 1; How Much is Enough*, EOS/ESD Technology, August/September, 1989.
21. J.Brinton, *Frost and Sullivan Report Scrutinizes ESD Practice*, EOS/ESD Technology, April, 1991.
22. R. Peirce, *Static Control Pays*, EOS/ESD Technology, April/May, 1990.

INDEX